NATURE-INSPIRED MOBILE ROBOTICS

NATURE-INSPIRED MOBILE ROBOTICS

editors

Kenneth J Waldron

Stanford University, USA & University of Technology Sydney, Australia

Mohammad O Tokhi

University of Sheffield, UK

Gurvinder S Virk

University of Gävle, Sweden

World Scientific

NEW JERSEY • LONDON • SINGAPORE • BEIJING • SHANGHAI • HONG KONG • TAIPEI • CHENNAI

Published by

World Scientific Publishing Co. Pte. Ltd.
5 Toh Tuck Link, Singapore 596224
USA office: 27 Warren Street, Suite 401-402, Hackensack, NJ 07601
UK office: 57 Shelton Street, Covent Garden, London WC2H 9HE

British Library Cataloguing-in-Publication Data
A catalogue record for this book is available from the British Library.

The front cover image has been provided by: Arne Rönnau, FZI Research Center for Information Technology, Karlsruhe, Germany.

NATURE-INSPIRED MOBILE ROBOTICS
Proceedings of the 16th International Conference on Climbing and Walking Robots and the Support Technologies for Mobile Machines

ISBN 978-981-4525-52-7

In-house Editor: Rhaimie Wahap

Printed in Singapore by B & Jo Enterprise Pte Ltd

PREFACE

It is a pleasure to be in and to welcome everybody to beautiful Sydney, Australia for the 2013 CLAWAR (Climbing and Walking Robots) conference, the 16th in the series. It is the first time the conference has been held in Oceania. In fact, it is the very first time it has been held in the Southern Hemisphere.

As with many technical fields, robotics in Australia has its own distinctive character, to match the unique flora and fauna of the country. As befits an economy built on mining and agriculture, field robotics is very strong here, with a consequent emphasis on mobile systems. That has led to developments that have had impacts throughout the world, notably SLAM. It has also led to spectacular successes in infrastructure like the complete automation of the Port of Brisbane, and a series of infrastructure maintenance projects, papers on some of which are featured in the conference proceedings.

The CLAWAR 2013 proceedings present state of the art research findings around the theme of biologically-inspired robotics in 104 technical articles by authors from 29 countries from throughout the five continents and nearly all of robotics R&D is based on nature in some way. This is particularly true of systems that walk, run, climb, or crawl. As is usual at a CLAWAR conference there is a strong stream of papers on legged locomotion with numbers of legs from two on up. There is also a strong collection of papers on systems that climb walls, poles, or more complex structures continuing the distinctive CLAWAR themes.

Another particularly strong theme of this year's conference is robot-human interaction. This includes "natural" methods for programming humanoid robots. It also includes a very strong stream of papers on human assist devices, notably exoskeletal and prosthetic devices, but also personal care robots and mobility assistance devices designed to meet the growing challenges due to the global ageing society. There are also several papers on motion capture of humans, or animals, and on musculo-skeletal biomechanics.

Finally, but very importantly, there are papers on various aspects of the societal impact of robotics. These are issues of creating standards to create and strengthen emerging markets, and ethical considerations. It is time to move beyond Asimov's Laws of Robotics. Armed UAV's are in use every day and armed autonomous machines for use on land, in the air, and under the sea are just around the corner. This all creates enormous ethical issues and we, who understand the technology must step forward and lead the way in creating not only the new robot technologies but have a strong say in its market developments and the urgently needed new societal rules to govern their capabilities.

The editors would like to thank members of the International Program Committee, International Advisory Committee and National Organising Committee for their efforts in reviewing the submitted articles, and the authors in addressing the comments and suggestions of the reviewers in their final submissions. It is intended that the CLAWAR 2013 proceedings will be a valuable source of reference for research and development in rapidly growing area of mobile service robotics.

K. J. Waldron, M. O. Tokhi, G. S. Virk

CONFERENCE ORGANISERS

University of Technology Sydney
Sydney, Australia

CLAWAR Association

A charitable non-profit making membership based professional organization serving the robotics community
www.clawar.org

CONFERENCE COMMITTEES AND CHAIRS

Conference Chairs

of Sixteenth International Conference on Climbing and Walking Robots and the Support Technologies for Mobile Machines

K. J. Waldron (General Chair)	– Stanford University, USA, & University of Technology Sydney, Australia
M. O. Tokhi (Program Chair)	– University of Sheffield, UK
G. S. Virk (IAC Chair)	– University of Gävle, Sweden

International Advisory Committee (IAC)

of Sixteenth International Conference on Climbing and Walking Robots and the Support Technologies for Mobile Machines

H. Levent Akin	– Turkey
Manuel Armada	– Spain
A. K. M. Azad	– USA
Karsten Berns	– Germany
Philippe Bidaud	– France
Bryan Bridge	– UK
N. J. Cowan	– USA
H. Fujimoto	– Japan
Krzysztof Kozlowski	– Poland
Giovanni Muscato	– Italy
Ming Xie	– Singapore

International Programme Committee (IPC)

of Sixteenth International Conference on Climbing and Walking Robots and the Support Technologies for Mobile Machines

Abdallah M.	– USA
Ahmadabadi M. N.	– Iran
Alam M. S.	– Bangladesh
Aldbrez F. M.	– UK
Arkin R.	– USA
Baras J. S.	– USA
Benmar F.	– France
Bhattacharya B.	– India
Billingsley J.	– Australia
Bonsignorio F.	– Italy
Bostelman R. V.	– USA
Bouazza-Marouf K.	– UK
Cameron S.	– UK
Cheng H.	– China
Cheng X	– China
Chevallerau C.	– France
Chugo D.	– Japan
Cruse H.	– Germany
Dehghani A.	– UK
de Santos P. G.	– Spain
Dillmann R.	– Germany
Dodd T. J.	– UK
Dogramadzi S.	– UK
Dubowsky S.	– USA
Dutra M. S.	– Brazil
Dutta A.	– India
Fortuna L.	– Italy
Fu Y.	– China
Garcia E.	– Spain
Gatsoulis Y.	– UK
Goher K. M.	– Oman
Goswami A.	– USA
Gradetsky V.	– Russia
Gross R.	– UK
Guan Y.	– China
Gupta S. K.	– USA
Halme A.	– Finland
Steinicke L.	– Belgium
Tan Y.	– China
Heyes N.	– UK
Hossain M. A.	– UK
Howard D.	– UK
Ion I.	– Romania
Jailani R.	– Malaysia
Jarvis R.	– Australia
Kadar E. E.	– UK
Kiriazov P.	– Bulgaria
Kumar V.	– USA
Lefeber D.	– Belgium
Lewis M. A.	– USA
Li C.	– USA
Lopez-Coronado J.	– Spain
Ma H.	– China
Madhavan R.	– USA
Marques L.	– Portugal
Massoud R.	– Syria
Mochiyama H.	– Japan
Mohamed Z.	– Malaysia
Moon S.	– Korea
Nakamura T.	– Japan
Oetomo D.	– Australia
Pratihar D. K.	– India
Qian J.	– China
Rachkov M.	– Russia
Ribeiro M.	– Portugal
Rodriguez H. L.	– Columbia
Santos V.	– Portugal
Sattar T. P.	– UK
Schilling K.	– Germany
Schmiedeler J.	– USA
Shaheed M. H.	– UK
Shamsudin H. M. A.	– Malaysia
Sharma S. K.	– UK
Siddique N. H.	– UK
Silva F.	– Portugal
Skibniewski M.	– USA
Xie S.	– New Zealand
Xiong R.	– China

CONFERENCE SPONSORS AND CO-SPONSORS

IFToMM

International Federation for the Promotion of Mechanism and Machine Science
http://www.iftomm.org/

CONTENTS

Preface ... v
Conference organisers ... vii
Conference committees and chairs ... viii
Conference sponsors and co-sponsors ... xi
Table of contents ... xiii

Section–1: Plenary presentations

TIRAMISU: FP7-Project for an integrated toolbox in humanitarian demining, focus on UGV, UAV, technical survey and close-in-detection ... 3
Y. Baudoin and TIRAMISU Consortium

Anthropomorphic biological equipment ... 6
Y. Nakamura

Muscle coordination of human locomotion ... 7
M. Pandy

Exoskeleton systems for medical and civilian applications ... 8
H. Kazerooni

Section–2: Assistive robotics

Optimization-based gait planning for wearable power-assist locomotor by specifying via-points ... 11
C. Sung, T. Kagawa and Y. Uno

The designing of the exoskeleton leg with pneumatic drives ... 19
A. N. Sukhanov, N. N. Bolotnik, V. G. Gradetsky, I. L. Ermolov, M. M. Knyazkov and E. A. Semyonov

Motion control algorithms for the exoskeleton equipped with pneumatic drives ... 27
A. A. Kryukova, V. G. Gradetsky, I. L. Ermolov, M. M. Knyazkov, E. A. Semyonov and A. N. Sukhanov

Development of a lower extremity exoskeleton system for walking assistance while load carrying 35
W. S. Kim, H. D. Lee, D. H. Lim, C. S. Han and J. S. Han

Dynamic coupling characteristics of a semi-active knee prosthesis 43
M. Awad, A. Dehghani, D. Moser and S. Zahedi

Covered area detection based on brightness change of inner camera images for crawler robot 51
K. Tokuda, T. Hirayama, T. Kinugasa, T. Haji and H. Amano

Actuator with adjustable-rigidity and embedded sensor for an active orthosis knee joint 59
M. Cestari, D. Sanz-Merodio, J. C. Arevalo, X. Carrillo and E. Garcia

A personal robot integrating a physically-based human motion tracking and analysis 68
C. Granata, Ph. Bidaud, R. Ady and J. Salini

Humanoid robot programming through face expressions 77
A. Uribe-Quevedo, H. Leon and B. Perez-Gutierrez

Dynamics of human lower limbs using CGA data and BSIP predictive equations 85
A. B. Filho, V. H. Fernandes, A. H. Maia, E. Garcia, A. Frizera and T. Bastos

Variable impedance control of a parallel robot for ankle rehabilitation 93
S. Q. Xie and Y. H. Tsoi

A mobile 3D vision-based embedded system for robust estimation and analysis of human locomotion 101
C. Zong, Ph. Bidaud and X. Clady

A simple load estimation of a patient during a standing assistance motion 109
T. Yamada, D. Chugo, S. Yokota and H. Hashimoto

Exoskeleton for assisting human walking 117
I. N. Indrawibawa and G. S. Virk

Hopping machine: A tool for human running simulation and for legged robots and exoskeleton design 125
A. H. Maia, S. Laquini Jr, A. B. Filho, V. H. B. Fernandes and E. Garcia

Design of a reconfigurable wheelchair with a sit-to-stand facility for a disabled kid 133
K. M. Goher, M. Shafiq and A. Al-Yahmadi

Integrated design, modelling and analysis of two-wheeled wheelchair for disabled 141
A. Aula, T. M. Altalmas, S. Ahmad, R. Akmeliawati, S. N. Sidek and M. O. Tokhi

A comparison of a passive and variable-damping controlled leg prosthesis in a simulated environment 153
J. Zhao, K. Berns, R. D. S. Baptista and A. P. L. Bó

Section–3: Autonomous robots

DODEKAPOD as universal intelligent structure for adaptive parallel spatial self-moving modular robots 163
S. Sayapin, A. Karpenko and D. X. Hiep

Aquatic multi-robot system for lake cleaning 171
P. Agrawal and B. Bhattacharya

Hybrid spiral dynamic bacterial chemotaxis optimisation for hybrid fuzzy logic control of a novel two wheeled robotic vehicle 179
A. M. Almeshal, K. M. Goher, A. N. K. Nasir, M. O. Tokhi and S. A. Agouri

BFA optimisation of control parameters of a new structure two wheeled robot on inclined surface 189
S. A. Agouri, M. O. Tokhi, A. M. Almeshal and K. M. Goher

Section–4: Biologically-inspired systems and solutions

Compliant backbone in mobile robots 199
N. I. Kern, R. J. Bachmann, R. J. Michols, R. J. Triolo and R. D. Quinn

Development of two types of maintenance robots for a jacket inside a mixing tank 207
T. Tanaka, T. Nakamura, D. Sannohe, Y. Morisita and M. Tanaka

Adaptive control of leg position for hexapod robot based on somatosensory feedback and organic computing principles 215
A. Al-Homsy, J. Hartmann and E. Maehle

Implementation of an adjustable compliant knee joint in a lower-limb exoskeleton 223
D. Sanz-Merodio, M. Cestari, J. C. Arevalo and E. Garcia

New propulsion system with pneumatic artificial muscles 232
I. Veneva

Development of a wall climbing robot using the mobile mechanism of traveling waves propagation 241
Y. Mizota, K. Takahashi, Y. Goto and T. Nakamura

Bio-inspired elbow impedance modulation using a compliant technical joint drive 249
S. Annunziata, J. Paskarbeit and A. Schneider

A resilient robotic actuator based on an integrated sensorized elastomer coupling 257
J. Paskarbeit, S. Annunziata and A. Schneider

Derivation of mathematical models of the peristaltic crawling robot for maintenance of a mixing tank 265
Y. Morishita, D. Sannohe, T. Osawa, T. Tanaka and T. Nakamura

Improvement of energy consumption by movement of center of rotation of joint 273
K. Tanaka, S. Nishikawa and Y. Kuniyoshi

A simplified variable admittance controller based on a virtual agonist-antagonist mechanism for robot joint control 281
X. Xiong, F. Wörgötter and P. Manoonpong

Section–5: HMI, inspection and learning

Determination of trajectories using non-invasive BCI techniques in 3D environments 291
T. Garcia-Egea, C. A. Díaz-Hernández, J. Lopez-Coronado and J. L. Contreras-Vidal

Development of a peristaltic crawling inspection robot with pneumatic artificial muscles for a 25A elbow pipe 301
T. Kishi, T. Nakamura and M. Ikeuchi

Examination of surface feature analysis and terrain traversability for a wall-climbing robot 309
D. Schmidt, M. Jung and K. Berns

Development of humanoid robot teaching system based on a RGB-D sensor 317
C. Leng, Q. Cao, B. Fang, Y. Yang and Z. Huang

Section–6: Innovative design of CLAWAR

Development of worm-rack driven cylindrical crawler unit 327
J.-Y. Nagase, K. Suzumori and N. Saga

Unconventional five-legged robot for agile locomotion 335
M. Wasik, M. Wasielica and P. Skrzypczynski

Climbing robot equipped with a postural adjustment mechanism for conical poles 343
Y. Ishigure, H. Kawasaki, T. Kato, K. Hirai, N. Iinuma and S. Ueki

Caminante: A platform for sensitive walking 351
V. Chernyak, E. Claretti, S. S. Nestinger, and E. Torres-Jara

Layered body for flexible mono-tread mobile track 359
T. Haji, S. Araki, T. Kinugasa, T. Ito, K. Yoshida, H. Amano, R. Hayashi, K. Tokuda and M. Iribe

Surface adaptation robot for defect detection by performing continuously an ultrasound wheel probe 367
H. Leon-Rodriguez

Optimal design of a magnetic adhesion system for climbing robots 375
P. Ward, D. Liu, K. Waldron and M. Hassan

Compliant foot system design for bipedal robot walking over uneven terrain 383
N. Wu, B.-H. Tan, C.-M. Chew, A.-N. Poo

New design of peristaltic crawling robot with an earthworm muscular structure 392
H. Nozaki, S. Tesen, N. Saga, H. Dobashi and J.-Y. Nagase

HEX-PIDERIX: A six-legged walking climbing robot to perform inspection tasks on vertical surfaces 399
X. Y. Sandoval-Castro, M. A. Gracia-Murillo, J. P. Zavala-De Paz and E. Castillo-Castaneda

Section–7: Locomotion

Reinforcement learning of bipedal lateral behaviour and stability control with ankle-roll activation 411
B. Hengst

Obstacle/gap detection and terrain classification of walking robots based on a 2D laser range finder 419
P. Kesper, E. Grinke, F. Hesse, F. Wörgötter and P. Manoonpong

Motion control for the 6-legged robot in extreme conditions 427
Y. F. Golubev, V. V. Korianov, V. E. Pavlovsky and A. V. Panchenko

Development of hexapod robot supported mechanically using pneumatic rubber artificial muscles 435
H. Tomori, Y. Hirata, T. Nakamura and H. Osumi

Local reflex generation for obstacle negotiation in quadrupedal locomotion 443
M. Focchi, V. Barasuol, I. Havoutis, J. Buchli, C. Semini and D. G. Caldwell

Event driven ground-impedance identification for legged robots 451
J. C. Arevalo, M. Cestari, D. Sanz-Merodio and E. Garcia

The Turtle, a legged submerged inspection vehicle 459
J. Billingsley

An actuated continuous spring loaded inverted pendulum (SLIP) model for the analysis of bouncing gaits 463
D. A. Jacobs, L. J. Park and K. J. Waldron

Design and prototyping of active-suspension-based 4-legged running robot with anteroposterior asymmetry in body 471
K. Nishimura, K. Tsutsumi, M. Mizoguchi, S. Sakio and K. Shibuya

Design of leaping behavior in a planar model with three compliant and rolling legs 479
Y.-C. Chou, K.-J. Huang, W.-S. Yu, and P.-C. Lin

Walking despite the passive compliance: Techniques for using conventional pattern generators to control intrinsically compliant humanoid robots 487
P. Kryczka, K. Hashimoto, A. Takanishi, H. O. Lim, P. Kormushev, N. G. Tsagarakis and D. G. Caldwell

A behavior-based library for locomotion control of kinematically complex robots 495
M. Langosz, L. Quack, A. Dettmann, S. Bartsch and F. Kirchner

Effects of neck swing motion on the body posture of a four-legged robot....... 503
K. Shibuya and K. Matsuhira

Actuator sizing for highly-dynamic quadruped robots based on squat jumps and running trots 511
H. Khan, C. Semini, D. G. Caldwell and V. Barasuol

Previewed impedance adaptation to coordinate upper-limb trajectory tracking and postural balance in disturbed conditions 519
A. Ibanez, P. Bidaud and V. Padois

A hip joint structure for biped robot with reduced DoF's of motion 529
S. Ito, A. Nakazawa, T. Onozawa, S. Nishio and M. Sasaki

Exploring the Lombard paradoxon in a bipedal musculoskeletal robot 537
K. Radkhah and O. von Stryk

An automated system for systematic testing of locomotion on heterogeneous granular media 547
F. Qian, K. Daffon, T. Zhang and D. I. Goldman

Inverse dynamics for a quadruped robot locomoting along slippery surfaces 555
S. Zapolsky, E. Drumwright, I. Havoutis, J. Buchli and C. Semini

LAURON V: Optimized leg configuration for the design of a bio-inspired walking robot 563
A. Roennau, G. Heppner, L. Pfotzer and R. Dillmann

Shuffle turn of humanoid robot simulation based on EMG measurement 571
M. Koeda, T. Serizawa and Y. Matsui

Studies of total adhesive force of multiple magnetic wheels for a climbing robot 578
A. Boonyaprapasorn, K. Thung-Od, R. Silapunt and T. Maneewarn

Quantitative kinematic performance comparison of reconfigurable leg-wheeled vehicles 585
A. Alamdari, R. Hérin and V. N. Krovi

Enhancing personal electric vehicles using reconfigurable design........ 593
R. S. Hérin, Y.-C. Chen, A. Alamdari and V. N. Krovi

Hand-manoeuvred wheelchair using wheels fitted with feet for enhanced mobility 599
T. Okada, H. Wada, N. Mimura, T. Shimizu and K. Nagata

Design and locomotion modes of a small wheel-legged robot........ 609
I. Doroftei and I. Ion

Compliance contact control of Mecanum wheeled mobile robot for improving safety........ 617
N. Nishigami, N. Takesue, R. Makino, K. Kikuchi, K. Fujiwara and H. Fujimoto

Neural control of a three-legged reconfigurable robot with omnidirectional wheels........ 625
P. Manoonpong, F. Wörgötter and P. Laksanacharoen

The passive dynamics of walking and brachiating robots: Results on the topology and stability of passive gaits........ 633
N. Rosa Jr. and K. M. Lynch

Section–8: Manipulation and gripping

Modeling and analysis of robotic grasping using soft fingertips........ 643
A. Khurshid, A. Ghafoor, M. A. Malik and Y. Ayaz

Force sensing and control during movement and object manipulation in MERO walking robots 651
I. Ion, A. Curaj, A. Vasile, D. Iulia and S. Grigore

Object grasping with dual robot manipulator using genetic algorithms........ 659
M. A. H. Hassan and M. O. Tokhi

A novel hybrid spiral-dynamics random-chemotaxis optimization algorithm with application to modelling of a flexible robot manipulator........ 667
A. N. K. Nasir and M. O. Tokhi

Section–9: Modelling and simulation of CLAWAR

Modelling of movement of the three-link robot with operated friction forces on the horizontal surface 677
S. F. Jatsun, L. Y. Volkova, G. S. Naumov and A. S. Yatsun

Transition analysis of a biped pole-climbing robot – Climbot 685
H. Zhu, Y. Guan, W. Wu, X. Zhou and H. Zhang

Design of SMA based actuators used in robotics 693
M. Öztürk, B. Koçer and R. Uygun

Finite element modelling of the adhesion system of a vortex based climbing robot 701
F. Bonaccorso, S. D'Urso, D. Longo and G. Muscato

Development of a quadruped robot model in Simmechanics™ 709
M. F. Silva, R. S. Barbosa and T. S. Castro

Section–10: Planning and control

Model-based elastic tendon control for electrically actuated musculoskeletal bipeds 719
K. Radkhah and O. von Stryk

Velocity control of serial elastic actuator based on EKF estimator and neural network 729
Y. Mao, R. Xiong, Q. Zhu and J. Chu

Low-dimensional user control of autonomously planned whole-body humanoid locomotion motion towards brain-computer interface applications 740
K. Bouyarmane, J. Vaillant and J. Morimoto

State to state motion planning for underactuated systems using a modified rapidly exploring random tree algorithm 749
R. Shvartsman, Y. Tan and D. Oetomo

Biped walking over rough terrain by adaptive ground reference map 761
N. Wu, C.-M. Chew and A. N. Poo

Section–11: Positioning, localization and perception

Method for estimating location and yaw angle of a six-legged robot for omni-directional walking control 773
H. Uchida, K. Takahashi and M. Suzuki

Cooperative multi-robot localization with radio-based sensors 781
L. Pfotzer, S. Bohn, J. Oberlaender, G. Heppner, A. Roennau and R. Dillmann

Efficiently using RGB-D data to self-localize a small walking robot in man-made environments 789
P. Skrzypczyński

Recognition of 3D objects for walking robot equipped with Multisense-SL sensor head 797
J. Będkowski, K. Majek, A. Masłowski and P. Kaczmarek

Section–12: Sensing and sensor fusion

Human motion localization and interaction based on wearable inertial and contact sensors 807
Q. Yuan and I.-M. Chen

Real-time detection of the activity of a dog 815
G. Lemasson, P. Lucidarme and D. Duhaut

Estimation of the trunk attitude of a humanoid by data fusion of inertial sensors and joint encoders 822
S. Khandelwal and C. Chevallereau

Section–13: Service robot standards and standardization

Challenges of the changing robot markets 833
G. S. Virk, C. Herman, R. Bostelman and T. Haidegger

OFEX 2.0: Database for performance evaluation of object feature extraction algorithms 841
K.-Y. Sung and S. Moon

Ethical assessment of robots 849
M. O. Tokhi

Author index 857

SECTION–1

PLENARY PRESENTATIONS

TIRAMISU: FP7-PROJECT FOR AN INTEGRATED TOOLBOX IN HUMANITARIAN DEMINING, FOCUS ON UGV, UAV, TECHNICAL SURVEY AND CLOSE-IN-DETECTION

YVAN BAUDOIN and TIRAMISU CONSORTIUM
www.fp7-tiramisu.eu

The TIRAMISU project aims at providing the foundation for a global toolbox that will cover the main mine action activities, from the survey of large areas to the actual disposal of explosive hazards, including mine risk education and training tools. After a short description of some tools, particular emphasis will be given to the possible use of UAV (or UGV/UAV) in Technical survey and/or Close-in-Detection

1. UGV

The idea of using robotics for humanitarian demining and/or advanced light-weight modular low-cost robots for technical survey and/or close-in-detection has been proposed by numerous authors [1]. Even though none of the current robotics platform seem to have reached production on a larger scale (except for military EOD/IEDD tasks), many lessons may be learned from several projects and experiences. Robotics solutions properly sized with suitable modularized structure and well adapted to local conditions of dangerous unstructured areas can greatly improve the safety of personnel as well as the work efficiency, productivity and flexibility. In this sense, mobile systems equipped with manipulators for detecting and locating antipersonnel landmines are considered of most importance towards autonomous/semi-autonomous mine location in a proficient, reliable, safer and effective way. Conventional vehicle-mounted mine detector systems employ an array of sensor devices to achieve a detection swath typically 2~4m wide. Some systems employ more than one type of sensor technology. These systems, while being very useful are often expensive, unsafe, complex and inflexible. Nevertheless, several IARP (International Advanced Robotics Programme – http://iarp.isir.upmc.fr) workshops [2 – 5] have on the contrary shown that the use of Robotics Systems could improve the safety and the clearance efficiency and that they may be considered as promising tools.

One goal of the TIRAMISU project is to develop an advanced lightweight detecting system for humanitarian demining purposes consisting of an integration of three different sensors, a MD, a GPR and an explosives vapor detector (EVD). The final prototype (tool) would consist of a manned or unmanned vehicle-based system capable of detecting explosive devices (ED) with a higher clearance rate than that of the current level of technology.

2. UAV

The Micro-mini UAV (MAV) and Rotary Wings seem to be good candidates for Technical Survey and, possibly, Close-in-detection in Humanitarian Demining while the MALE (Medium Altitude Large Endurance) and SR (Short Range) or MR (Medium Range) could be envisaged for the Non-Technical survey.

The small size of MAVs has several implications on their performance capabilities. First and foremost, they have limited payload capacity (size, weight, and power) and are therefore unable to carry significant computational resources or sensors of the highest accuracy and capability. Second, their small size and relatively low flight speeds make them susceptible to degraded performance caused by high winds and wind gusts. These challenges, imposed by the small size of MAVs, must be overcome for MAVs to be utilized successfully. Future advancements in miniaturization and performance of sensors and computers will enable increased success. However, efforts must also be made to utilize existing sensor and computer capabilities in novel and innovative ways to enhance the utility of MAV systems in the immediate future [6].

The ability to land a MAV accurately on a designated landing point is useful for several applications including the safe recovery of the MAV in risky conditions (a.o. minefields). Several challenges make precision landing/strike difficult for small UAVs. The disturbance imposed by wind is always an issue with MAVs, and for the present, sensor errors such as those common in the measurement of attitude and altitude cause difficulties. For precision landing, the most critical information is to know the location of the MAV relative to the desired landing location. Developing reliable methods for measuring this information is central to the success of a precision landing approach. The value of target (Explosive Device) tracking and localization from a small MAV platform is obvious and significant. This capability is attractive, but presents challenges.

The most significant challenges associated with accurate target localization stem from the lack of precise attitude estimates for the MAV platform. Pitch and roll are difficult to estimate with a high degree of accuracy with the Micro-electromechanical systems (MEMS) sensors typically used for MAV control. Measurements of heading are not available – heading is often approximated by the course of the MAV which is estimated from successive GPS measurements. In high-wind conditions (relative to the desired airspeed), the course is often a poor approximation of the heading.

The ability to operate in constrained environments is of great importance for MAV that must be able to fly amidst natural and man-made obstacle. Air-to-ground detection systems could provide an interesting contribution to the technical survey as well as to the close-in-detection aims for so far technical and operational improvements are precisely defined. Both the aerial platform (a sensor's carrier) and the Remote Pilot Station have to be considered, both in the context of evolving legal (national, European and/or international) regulations.

References

1. E. Colon,Y. Baudoin, Humanitarian demining and robotics: A difficult challenge, CLAWAR 1999, Portsmouth, UK (1999).
2. M. A. Armada, J. Cobano, E. Garcia and P. Gonzalez de Santos Configuration of a legged robot for humanitarian de-mining activities. *Proc. of the IARP International Workshop on Robotics and Mechanical Assistance in Humanitarian Demining* (HUDEM2005), pp. 131-135 (2005).
3. E. Fukushima, et al. Teleoperated buggy vehicle and weight balanced arm for mechanization of mine detection and clearance tasks, *IARP WS HUDEM'2005*, Tokyo, pp. 58-63 (2005).
4. T. Fukuda, et al. Environment-adaptive antipersonnel mine detection system – Advanced mine sweeper', *IARP WS HUDEM'2005*, Tokyo, pp. 33-38 (2005).
5. R. Fernández, C. Salinas, H. Montes, M. Armada. Omnidirectional stereo tracking system for humanitarian demining training. *International Symposium Humanitarian Demining*, Šibenik, Croatia, pp. 113-116 (2011).
6. Gerard ten Buuren, 2011: Small UAS for Law Enforcement (Netherlands Police Agency KLPD), UAS Global Perspective 2011/2012 – www.uvs-info.com

ANTHROPOMORPHIC BIOLOGICAL EQUIPMENT

YOSHIHIKO NAKAMURA
University of Tokyo, Japan

Evolution of the human brain emerged by the pressure for communication. It is hard to believe that the pressure of evolution exempted the features of motion patterns from using them for communication. The communication skill is therefore anthropomorphic.

The author would like to call the natural function of the brain to perceive, recognize, understand and respond to the human-like motion patterns, the Anthropomorphic Biological Equipment. The natural and general human-machine interface can be established by developing the similar function on the machine side, which we may call the Anthropomorphic Artificial Equipment. Our research started from the mathematical model of mirror neurons and continues to acquisition of semiology of human behaviors based on technologies such as unsupervised segmentation, iterative clustering, and construction of state transition network. The current research interests target to connect the behavioral semiology to the semiology of a natural language to develop a statistical system that evokes mutual association. This talk introduces the scope of our research and overviews the direction of research.

MUSCLE COORDINATION OF HUMAN LOCOMOTION

MARCUS PANDY

Department of Mechanical Engineering, University of Melbourne, Australia

Gait-analysis techniques have been used for more than a century to provide information on the kinematics and kinetics of human locomotion, yet the ability of this technique to quantify function is limited because it cannot be used to discern the actions of individual muscles. Computational modelling is the only means available to determine musculoskeletal loading non-invasively. Rapid increases in computing power combined with recent advances in imaging and more efficient algorithms for modelling the musculoskeletal system have enabled more detailed analyses of lower-limb muscle function. This presentation will illustrate how computational modelling has been used to study muscle coordination of human locomotion across a wide spectrum of walking and running speeds. The results have important applications in medicine (orthopaedics), sports (training and injury prevention) and robotics.

EXOSKELETON SYSTEMS FOR MEDICAL AND CIVILIAN APPLICATIONS

HOMAYOON KAZEROONI
University of California, Berkeley, USA

For widespread use, exoskeletons must be accessible. The medical wearable robotic exoskeletons allow people with paraplegia or other mobility disorders to be upright and mobile, preventing secondary diseases and enhancing their quality of life. These systems will be used for in-home care and everyday use, as well as within hospitals and rehabilitation centers. The industrial wearable robotic systems minimize spinal compression forces of workers who repeat various maneuvers on the job. These devices will be used in auto assembly plants, factories, manufacturing facilities, distribution centers, warehouses, and delivery services. These systems decrease the severity and number of work-related injuries, while enhancing worker safety. The quest to develop accessible exoskeleton orthotic systems suggests less hardware while placing more emphasis on the intelligence and cleverness during both the design stage and the device operation. This talk will describe new engineering developments to realize accessible exoskeleton systems.

SECTION–2

ASSISTIVE ROBOTICS

OPTIMIZATION-BASED GAIT PLANNING FOR WEARABLE POWER-ASSIST LOCOMOTOR BY SPECIFYING VIA-POINTS*

CHANGHYUN SUNG, TAKAHIRO KAGAWA, YOJI UNO

Department of Mechanical Science and Engineering, Nagoya University, Furo-cho, Chikusa-ku, Nagoya, Japan

We propose a planning method of gait for a wearable robot. In gait planning of a wearable robot, it is important to reflect desired stride length and foot clearance according to motor function of a user. Furthermore, there are a number of constraints such as fall prevention, collision avoidance and so on. We specify some via-points to deal with these conditions in gait planning. Our method is applied to generate the walking motion of Wearable Power-Assist Locomotor (WPAL). We confirmed that the adequate gait patterns were planned by the proposed method.

1. Introduction

In recent years, rehabilitation systems of paraplegics with spinal cord injuries have been developed for daily life support. Life in a wheelchair often results in secondary complications. Reconstruction of walking is desirable to improve not only their locomotor functions but also the physiological problems. We are developing Wearable Power-Assist Locomotor (WPAL) for paraplegic patients [1]. In control of WPAL, the design of an appropriate gait pattern is important to maintain smooth walking and to reduce the risks of falling. In this study, we propose a gait planning method by specifying via-points. In gait planning, redundancy problem of joints have to be solved subject to various constraints such as maintaining balance and range of motion.

In the previous method, the locomotion of WPAL is planned using inverse kinematics [2]. Then, joint trajectories were preliminary planned and confirmed that the singular postures could be avoided. On the other hand, the singular posture problem does not have to be taken into account for the planning in the joint space in the proposed method. In addition, the via-point representation of joint angle trajectories plays crucial roles to deal with the constraints during movements [3]. Various conditions which have to be satisfied over whole motion duration are transformed into the conditions of via-points. We confirmed

* This study was supported by a Grant-in-Aid for Scientific Research (C) No. 23560526 and for Young Scientists (B) 25820085.

that the desired gait patterns satisfied with given stride length and foot clearance could be planned under stability conditions. Furthermore, we examined planned gait patterns by applying to WPAL.

2. Motion Planning by Specifying Via-point

2.1. *Representation of Joint Angle Trajectory*

Minimum jerk criterion is typically used to generate smooth motion of the robot manipulators [4]. A simple minimum jerk trajectory can be represented by a fifth order polynomial function (t), which has six parameters

$$\theta(t) = a_5 t^5 + a_4 t^4 + a_3 t^3 + a_2 t^2 + a_1 t + a_0 \tag{1}$$

Generally, the minimum jerk trajectory is determined by the initial and final states (joint angle, angular velocity and angular acceleration). Therefore, the trajectory satisfying the constraints cannot be represented by this simple function.

In this research, various motions are expressed by a minimum jerk trajectory with a via-point. We make use of the via-point to consider the various constraints. Among the constraints, the target information is the most important condition. Let t_{via} be the time that joint angle trajectory passes the via-point. The minimum jerk trajectory with via-point constraints of joint angle $\theta(t_{via})$ and angular velocity $\dot{\theta}(t_{via})$ can be formulated using Lagrange multipliers Π_1 and Π_2 [5].

$$\theta(t_{via}) = \begin{cases} a_5 t^5 + a_4 t^4 + a_3 t^3 + a_2 t^2 + a_1 t + a_0 \\ \qquad (0 \le t \le t_{via}) \\ a_5 t^5 + a_4 t^4 + a_3 t^3 + a_2 t^2 + a_1 t + a_0 \\ + \frac{1}{24}(t - t_{via})^4 \Pi_1 + \frac{1}{120}(t - t_{via})^5 \Pi_2 \\ \qquad (t_{via} \le t \le t_f) \end{cases} \tag{2}$$

where t_f is the end time of the movement.

Equation (2) has eight parameters, which are determined by the six conditions of each joint angle, angular velocity $(=0)$ and angular acceleration $(=0)$ at the initial and final states, and two conditions of joint angle $\theta(t_{via}) = \theta_{via}$ and angular velocity $\dot{\theta}(t_{via}) = \dot{\theta}_{via}$ at the via-point. Briefly, we can express

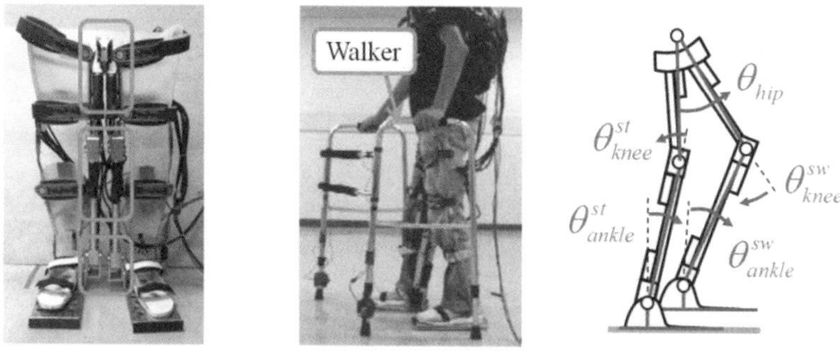

(a) Wearable robot WPAL (b) Walking with walker (c) Joints angles

Figure 1. Wearable Power-Assist Locomotor (WPAL) and sensing system.

various trajectories by specifying joint angle θ_{via} and angular velocity $\dot{\theta}_{via}$ at the via-point.

2.2. *Optimization Based Motion Planning*

The motion planning problem is usually solved by representing the set of joint angle trajectories. We can determine the joint angle trajectories by determining joint angle parameters of start, end and via-points, and joint angular velocity parameters of via-point. For reducing the computation time, we divide the calculation of motion planning parameters in two steps. In this research, the joint angle parameters are calculated using inverse kinematics equations [6]. Additionally, joint angular velocity parameters are optimized to satisfy various constraints. Specifically, this problem is solved as Semi Infinite Programming (SIP) which is one of the constrained optimization problems [7]. We can obtain safe motions that are satisfied with constraints by considering motion planning as the optimization problem. For more details, we address in gait planning for WPAL as follows.

3. Gait Planning for WPAL

3.1. *Wearable Robot WPAL*

Figure 1 shows the wearable robot WPAL. Figure 1(a) shows the robotic system with 6 actuators assisting extension and flexion movements of the hip and knee joints, and plantarflexion and dorsiflexion movements of the ankle joints. Figure 1 (b) shows a walking movement with WPAL assisting leg movements. To maintain the upright posture during walking, the arm supports using a walker are required. Figure 1(c) shows the joint degrees of freedom. We define five joint angles of WPAL for gait planning. The ranges of joint motion are -50 ~ 50 deg (hip extension and flexion), 0 ~ 120 deg (knee flexion), and -35 ~ 35 deg (ankle dorsiflexion and plantarflexion).

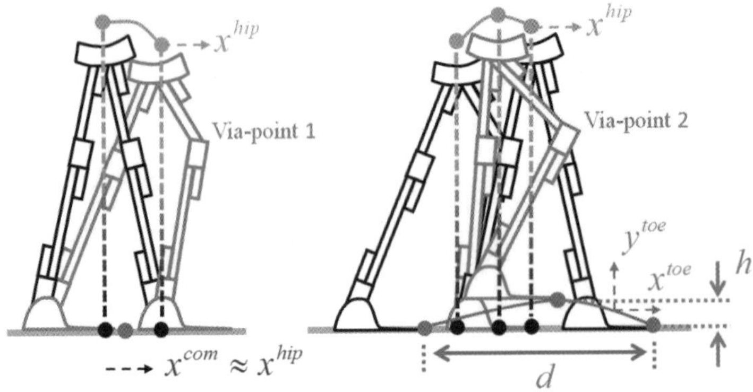

(a) Loading Phase (b) Swing Phase

Figure 2. Gait planning by specifying two via-points (d: stride length, h: foot clearance).

3.2. *Gait Phases*

A user handles a walker and moves one's arm to displace the walker. While the arm movement with the walker is performed, the leg joints of WPAL are not moved to maintain the balance. After the arm movement, the joints of WPAL perform the walking motion. In our research, the walking motion is divided into two phases that are defined as 'loading phase' and 'swing phase' as shown Figure 2. After the arm movement, movements of the leg joints should be ready to toe and heel off from the ground. We call this period as 'loading phase' as shown Figure 2(a). In this phase, the horizontal hip position moves forward while keeping a double support. The COM moves forward to satisfy the condition of forward progression [8]. Finally, the COM locates within the area of single support foot as Figure 2(a). Next, WPAL gets into a movement phase from the toe and heel off of the swing leg. This phase is defined as 'swing phase' as shown in Figure 2(b). The swing leg moves forward and the ankle of the stance leg continues to the dorsiflexion movement so that the movement distance of the foot is equal to the given stride length and foot clearance. In this phase, it is also necessary to satisfy the condition for forward progression over whole motion duration [8].

3.3. *Gait Planning*

In the gait planning of WPAL, the conditions of stride length and foot clearance should be considered to reduce some risks. In addition, a redundancy problem has to be solved in motion planning of WPAL because there are infinite trajectories which satisfy the given stride length and foot clearance. In the previous works, joint angle trajectories are determined by inverse kinematics equations with a preliminarily determined position trajectory of toe, heel and hip [2]. Therefore, the only examined trajectories could be used because of the singular posture problem in the inverse kinematics equations.

In our method, each joint angle trajectory is directly determined by the minimum jerk trajectory with via-point constraints that represent important points of the walking motion. Figure 2 shows our strategy for gait planning by specifying via-points. We select two via-points; the first and second via-points are specified at the end time of loading phase and the middle time of the swing phase, respectively. Especially, the condition of foot clearance is satisfied at the time of second via-point. Indeed, the trajectories of "loading phase" and "swing phase" are planned individually. For safety, we allow to stop the WPAL at the end of "loading phase". Specifically, the trajectories of each phase are represented by Equations (1) and (2), respectively.

Each gait pattern of WPAL is generated by calculating potions of initial and final states, and via-point parameters that are joint angle and angular velocity. Via-point parameters consist of the kinematic quantities related to position and velocity of the robot. First we calculate the joint angle parameters of start, end and via-points using inverse kinematics equations. The conditions of stride length and foot clearance are satisfied at this process. Next, we optimize joint angular velocity parameters based on the results of calculated joint angle parameters. The joint angular velocity parameters of each via-point are optimized to be satisfied with various constraints including the fall prevention, collision avoidance with the ground and so on. This calculation problem can be transformed into Semi Infinite Programming by setting some cost function. In this paper, we select the cost function for minimizing energy consumption of each joint angle motor.

An advantage of our method is that we can deal with specific conditions based on certain kinematic quantities without changing all parameters of motion planning. Furthermore, the computation time would be reduced by separating the calculations of parameters related to position and velocity. Consequently, we obtain suitable gait patterns with low computation cost (about two seconds).

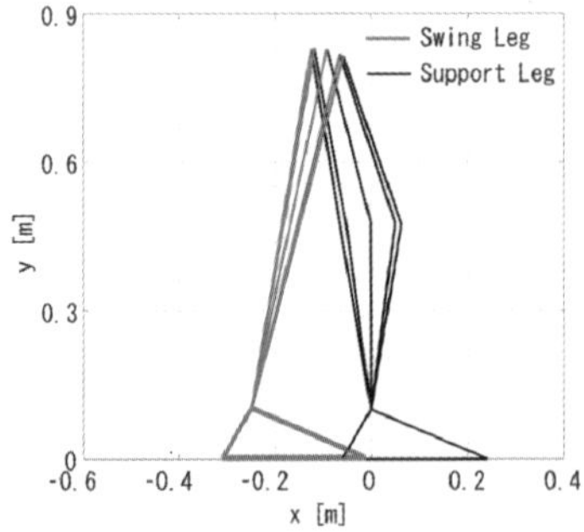

(a) Stick picture of loading phase

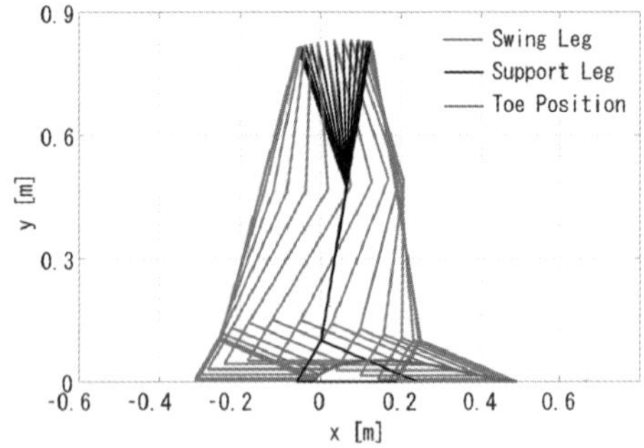

(b) Stick picture of swing phase

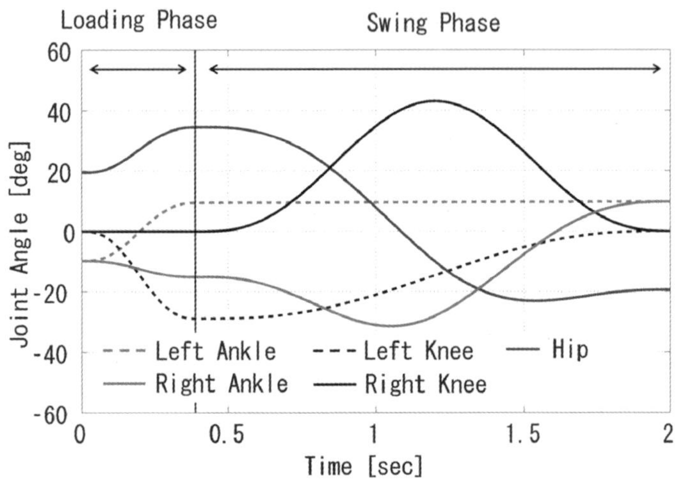

(c) Planned joint angle trajectory

Figure 3. Planned gait pattern with stride length 50 [cm] and foot clearance 5 [cm].

4. Experiments with Planned Gait Patterns

Figure 3 shows a planned gait pattern with stride length 50 [cm] and foot clearance 5 [cm] by the proposed method. The stick pictures of "loading phase" and "swing phase" are illustrated in Figure 3(a) and (b), respectively. The stick pictures show that the desired gait patterns satisfy the given stride length and foot clearance (Figure 3b). Furthermore, the constraints of fall prevention and collision avoidance with the ground are satisfied during the movement duration (Figure 3(a) and (b)). Finally, the planned joint angle trajectories were smooth as shown in Figure 3(c). We also confirmed that the walking motions satisfied with

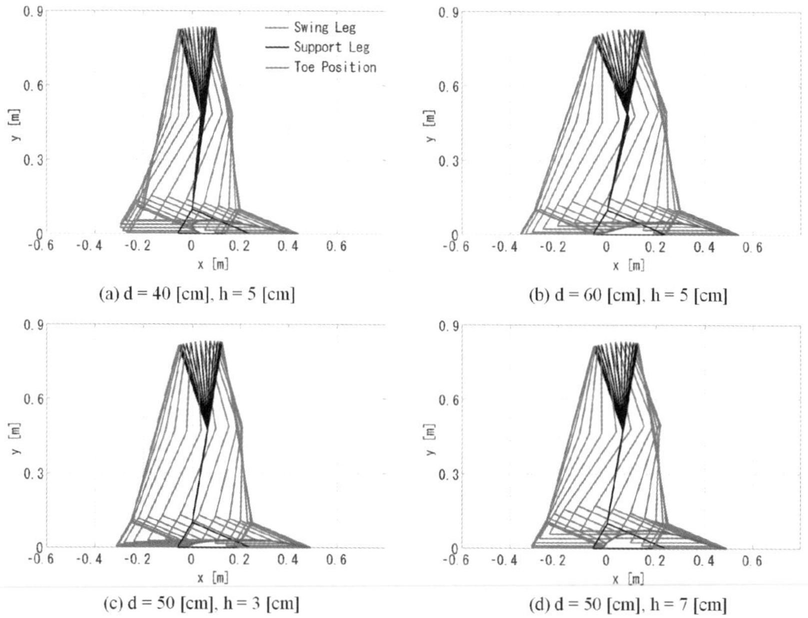

(a) d = 40 [cm], h = 5 [cm]

(b) d = 60 [cm], h = 5 [cm]

(c) d = 50 [cm], h = 3 [cm]

(d) d = 50 [cm], h = 7 [cm]

Figure 4. Planned gait patterns with various conditions of stride length and foot clearance (d: stride length, h: foot clearance).

given values of stride length and foot clearance, and constraints such as maintaining balance, were successfully planned (Figure 4).

We performed the experiments of walking on level ground to evaluate the feasibility of the planned trajectory. Figure 5 shows the pictures of walking motion under the conditions of stride length 50 [cm] and foot clearance 5 [cm]. We confirmed that a user could walk with WPAL in the conditions of given parameters by planning proposed method.

5. Conclusion

We proposed a planning method for walking motion with a wearable assistant robot. The motions satisfying various constraints were planned by specifying via-point constraints. The via-point constraints that consist of each joint angle and angular velocity of the robot were determined by an optimization process for appropriate locomotions. We demonstrated the feasibility of our planning method corresponding to the desired stride length and foot clearance of WPAL.

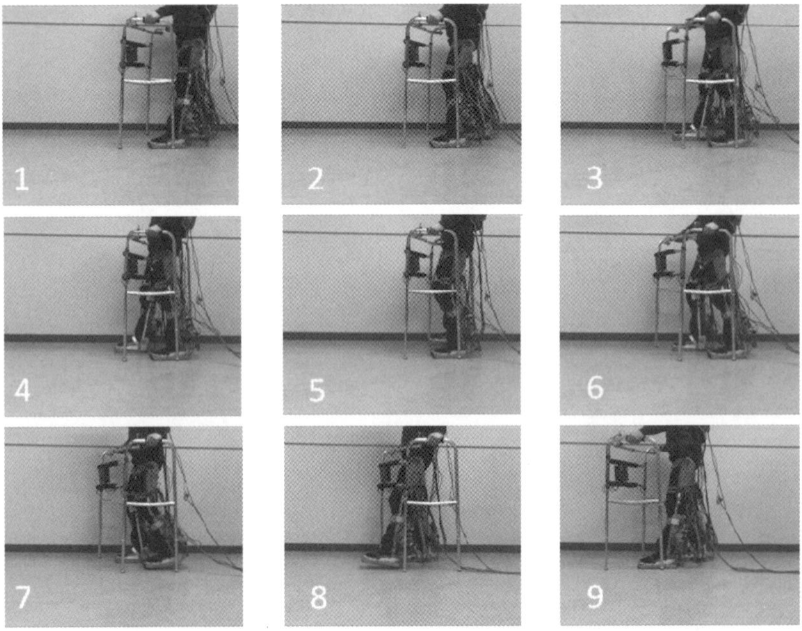

Figure 5. Experimental scenes of walking motion with conditions of Figure 3.

The computation cost of gait planning was still expensive about two seconds. For practical implementation, the motions should be planned in the shorter time. It is future works to develop an on-line control system with our planning method combining with a sensing system of WPAL.

References

1. Y. Muraoka et al. *The 4th World Congress for International Society of Physical and Rehabilitation Medicine, (2007).*
2. H. Kitamura et al. *Emerging Trends in Mobile Robotics,* 1047 (2010).
3. CH. Sung et al, *The 8th International Conference on Ubiquitous Robots and Ambient Intelligence,* (2011).
4. A. Piazzi and A. Visoli, IEEE Transactions on Industrial Electronics, **1-47** (2000)
5. T. Flash and N. Hogan, *Emerging Journal of Neuroscience,* **5-7**, 1688 (1985).
6. M. Stilman, IEEE Transactions on Robotics, **3-26** (2010)
7. R. Hettich and K.O. Kortanek, SIAM Review, **3-53** (1993)
8. T. Kagawa and Y. Uno, *Human Movement Science*, **29-6**, 964 (2010).

THE DESIGNING OF THE EXOSKELETON LEG WITH PNEUMATIC DRIVES*

ARTYOM N. SUKHANOV
Institute for Problems in Mechanics
of the Russian Academy of Sciences
101-1, Prospect Vernadskogo, 119526 Moscow, Russia
gradet@ipmnet.ru

NIKOLAI N. BOLOTNIK, VALERY G. GRADETSKY, IVAN L. ERMOLOV, MAXIM M. KNYAZKOV, EUGENIY A. SEMYONOV
Institute for Problems in Mechanics
of the Russian Academy of Sciences
101-1, Prospect Vernadskogo, 119526 Moscow, Russia
gradet@ipmnet.ru

The subject matter of this paper is the definition of the design structure for an exoskeleton for personal use and the investigation of the behavior of its leg in the loading-unloading operation mode. The design includes a mechanical leg system, a recuperation system, a control system, and an upper exoskeleton system equipped with fluidic muscles.

A mathematical model for the exoskeleton's legs is constructed. This model is used to determine the design parameters of the legs and control algorithms. The motion of the exoskeleton with pneumatic drives and pneumatic muscles, treated as a multibody system, is analyzed. The results of the mathematical simulation of the operation of the mechanical system of the exoskeleton and the data of the experimental study of its physical prototype are presented to demonstrate the properties of the system as a force-torque amplifier. Comparison of various mechanical systems enables obtaining a reasonable solution.

1. Introduction

The study of locomotion of humans and animals draws attention of many researchers. The importance of this problem increases in connection with application of exoskeletons in medicine as an orthopedic apparatus designed for motion-impaired patients. Mathematical modeling and simulation of the motion of the legs of medical exoskeletons is helpful for improving gaits of the patients.

---

*This study is partially supported by Program "Scientific Fundamentals of Robotics and Mechatronics" of the Department of Power Engineering, Machine Building, Mechanics and Control Processes of the Russian Academy of Sciences

Of particular interest is studying the dynamics of pneumatically driven exoskeleton legs treated as multibody systems.

The experience of many physicians in the field of therapy for patients with foot arthritis and paralysis of higher or lower extremities justifies the need for rehabilitation devices such as exoskeletons that are highly effective in aiding patients to overcome the ill effects of paralysis [1, 2]. Meanwhile, exoskeletons could be very useful in loading operations or rescue services.

The first prototypes of exoskeletons [3-6] were constructed to implement forced limb motion of paralyzed upper or lower extremities of patients governed by feedforward (no feedback) global control, with minimum information about operating parameters acquired from sensors. The task for a pre-planned motion was introduced by an operator before the motion had started; end the execution of the motion was checked after it stopped.

The previous exoskeleton systems were redundant in terms of degrees of freedom at the joints of the mechanical system and task execution [6]. It was, therefore, necessary to develop the first prototypes of such systems to study the complex behavior of the exoskeleton-man system. To satisfy a patient's needs, it is sometimes unnecessary to build a redundant system and only a limited number of degrees of freedom are required, which can be provided by using the modular method.

This method was discussed for the modular-design evolution and mechatronic approach to the application of the exoskeleton system concentrating on important practical requirements when a full complex exoskeleton is not necessary, and only a number of controlled exoskeleton modules are required for the patient.

The proposed approach takes into account the requirements of every patient and allows for designing several versions of the exoskeleton from modules at lower cost.

In this paper, a modular design of an exoskeleton multilink system with pneumatic drives is analyzed.

The main results of the study are outlined below.

- Based on the fundamental research results [1-4], a multi-link leg structure was developed to enable the exoskeleton to perform desired motions.

- A mathematical model of the multi-link mechanical system of the leg with pneumatic drives was created.

- Physical prototypes of the system were constructed and experimental studies with these prototypes were performed.

2. The mechanical model of the exoskeleton

The physical prototype of the system (Fig.1) includes several modules of pneumatic drives, the mechanical system of the legs and sensors for estimating the angles of the construction. The mechanical model of the exoskeleton's legs is analyzed. It is used for designing the parameters of the control algorithms.

Figure 1. Physical prototype of the exoskeleton

Each exoskeleton's leg can reach various positions (Fig.2). These positions were taken into consideration when developing the mathematical model of the system.

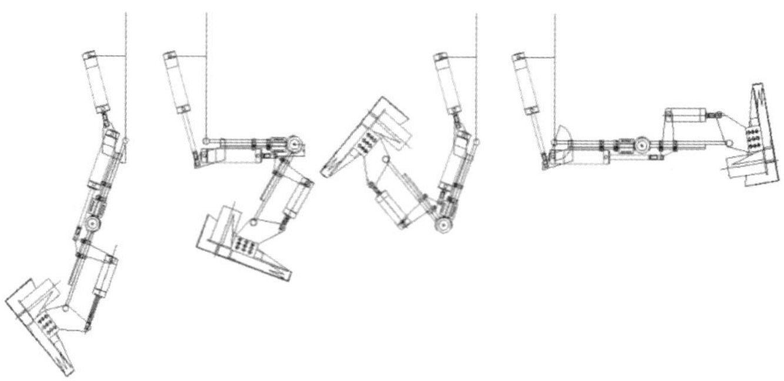

Figure 2. Exoskeleton leg with pneumatic drives

3. The mathematical model and simulation of the exoskeleton legs

A mathematical model for a kinematic structure under consideration (Fig. 3) with pneumatic cylinders was developed and used for simulating the behavior of the system.

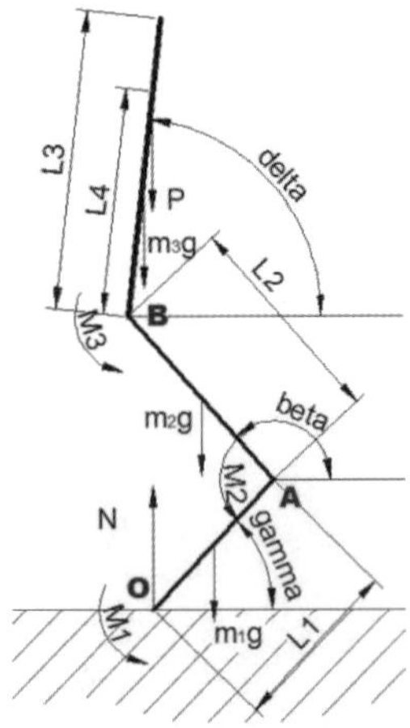

Figure 3. The multilink kinematic system with 3 DOF in the sagittal plane

The system is in equilibrium in the sagittal plane, if

$$\begin{cases} M_1 = -\dfrac{m_1 L_1 g}{2}\cos(\gamma) - m_2 g\left[L_1\cos(\gamma) + \dfrac{L_2}{2}\cos(\beta)\right] - \\ \quad -m_3 g\left[L_1\cos(\gamma) + L_2\cos(\beta) + \dfrac{L_3}{2}\cos(\delta)\right] - \\ \quad -P[L_1\cos(\gamma) + L_2\cos(\beta) + L_4\cos(\delta)] \\ M_2 = -m_2 g\dfrac{L_2}{2}\cos(\beta) - m_3 g\left[L_2\cos(\beta) + \dfrac{L_3}{2}\cos(\delta)\right] - \\ \quad -P[L_2\cos(\beta) + L_4\cos(\delta)] \\ M_3 = -m_3 g\dfrac{L_3}{2}\cos(\delta) - PL_4\cos(\delta) \end{cases} \tag{1}$$

where M_1, M_2 and M_3 are the torques at the joints of the multilink system produced by the pneumatic actuators, P is the external force loading the system, m_1, m_2 and m_3 are the masses of the links, g is the acceleration due to gravity, γ, β and δ are the angles between the links and the horizon, L_1, L_2, L_3 and L_4 are the geometrical design parameters of the links.

The forces due to the pneumatic actuators are expressed as follows:

$$\begin{cases} F_1 = \dfrac{M_1}{L_{arm1}} \\ F_2 = \dfrac{M_2}{L_{arm2}} \\ F_3 = \dfrac{M_3}{L_{arm3}} \end{cases} \tag{2}$$

where L_{arm1}, L_{arm2} and L_{arm3} are the arms of the forces of the pneumatic actuators.

Figure 4 shows the torques and forces, respectively, as functions of the angle between the upper link, which models the patient's back, and the horizon:

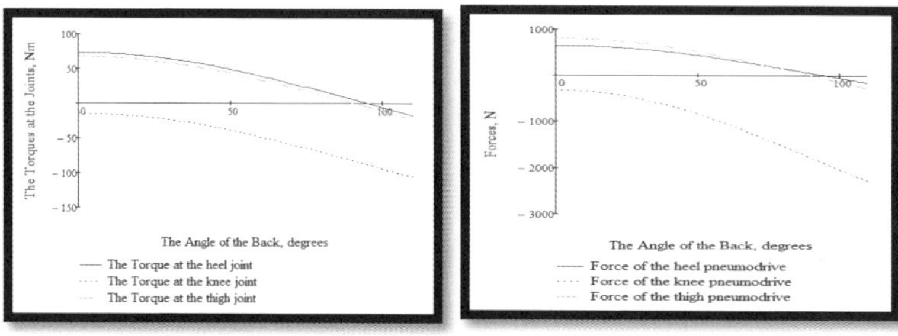

Figure 4. The joint torques and the forces of the pneumatic actuators

Due to the design of the exoskeleton, the arm of the pneumatic drive force at the heel joint is constant, while the variations of the arms of the forces at the knee and thigh joints are presented in Fig. 5:

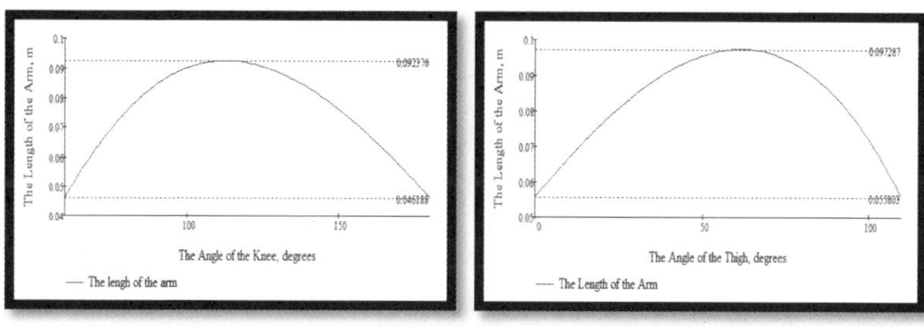

Figure 5. The arm of the knee pneumatic drive force and the arm of the force at the thigh joint

For the case when the angle between the back link and the horizon is constant (e.g., 90 degrees) in the "Stand Up"-operation, the joint torques and the pneumatic drive forces are shown in Fig. 6.

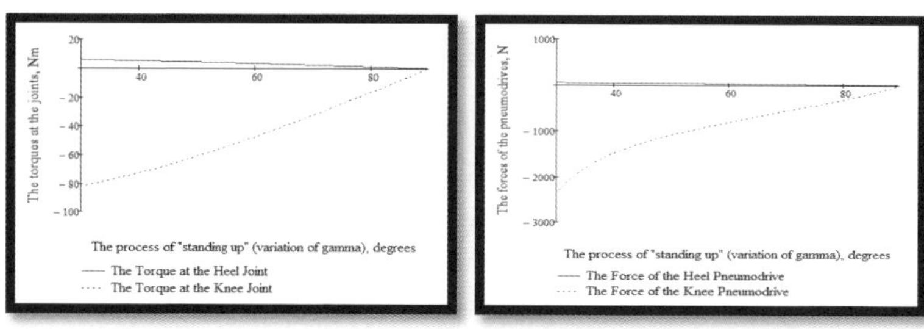

Figure 6. Variation of the joint torques and the pneumatic drive forces

By varying the angles we can obtain 3D-diagrams of the simulation. The results of the simulation are presented in Fig. 7.

The heel joint

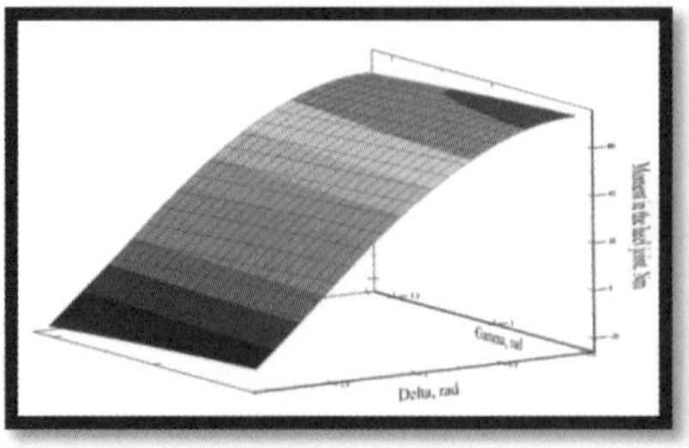

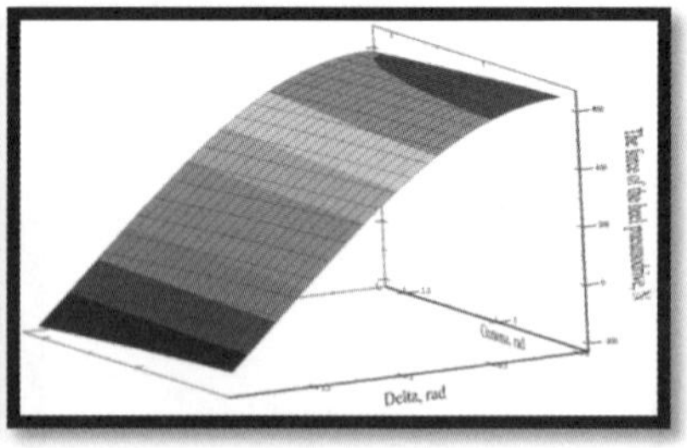

1 2

The knee joint

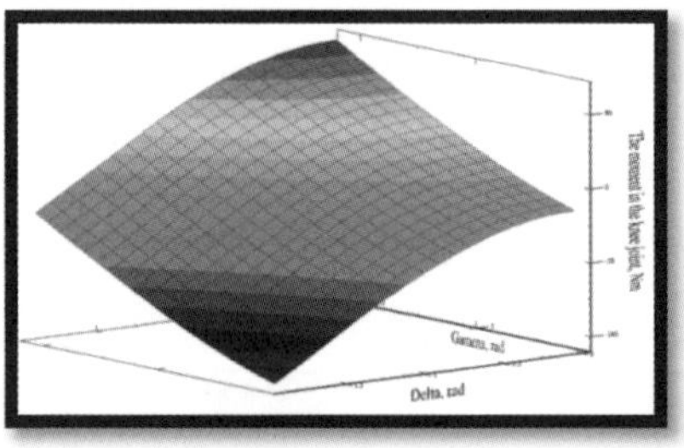

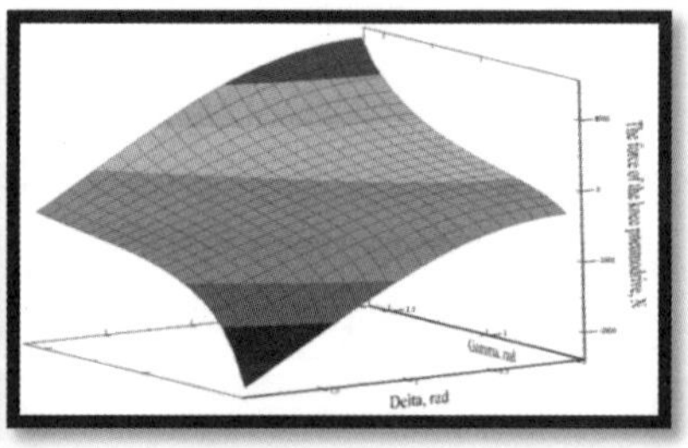

3 4

The thigh joint

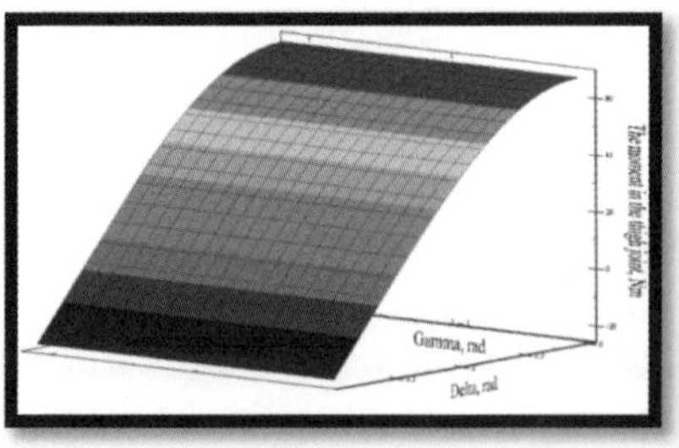

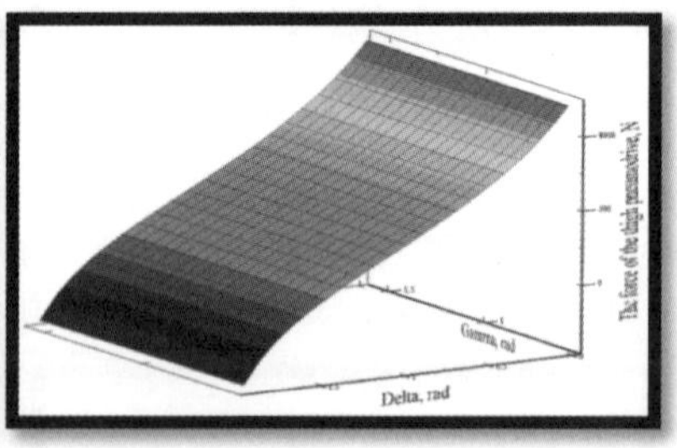

5 6

Figure 7. Variations of the joint torques (1,3,5) and forces (2,4,6)

The physical prototype of the system has been created at the Robotics and Mechatronics Laboratory of the Institute for Problems in Mechanics of the Russian Academy of Sciences. The results of the first experiments showed reasonability of the use of servo pneumodrives in the design. The results of the last experiment are:

The mass of the exoskeleton – 48 kg

The payload is up to 62 kg

Air consumption is 15.68 l/min

Maximum walking speed – 16 m/min

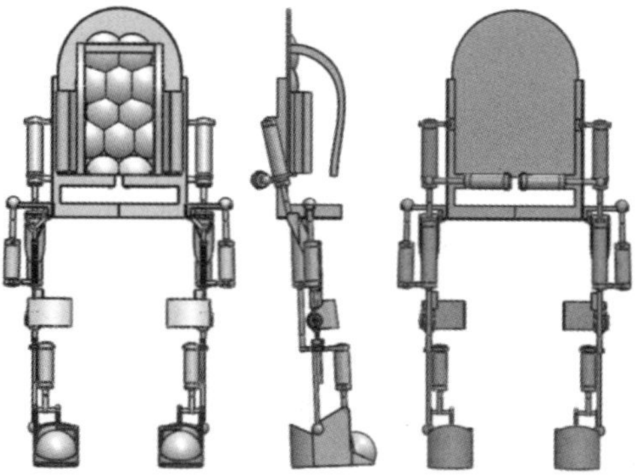

Figure 8. SolidWorks 3D modeling of the exoskeleton system

4. Conclusion

A mathematical model for an exoskeleton's legs is constructed. This model is used to determine the design parameters of the legs and control algorithms. The motion of the exoskeleton with pneumatic drives and pneumatic muscles, treated as a multibody system, is analyzed. The results of the mathematical simulation of the operation of the mechanical system of the exoskeleton and the data of the experimental study of its physical prototype are presented to demonstrate the properties of the system as a force-torque amplifier. Comparison of various mechanical systems enables obtaining a reasonable solution.

References

1. Vukobratovič M., Locomotion Robots and Anthropomorphic Mechanisms (in Russian), "Mir", Moscow, 1976, p. 541.

2. Vukobratovič M., Stokič D., Scientific Fundamentals of Robotics Control of Manipulation Robots: Theory and Application, Springer-Verlag, 1982, Vol. 2.

3. Vukobratovič M., Borovač B., Surla D., Stokič D., Scientific Fundamentals of Robotics, Biped Locomotion: Dynamics, Stability, Control and Application, Springer-Verlag, 1989, Vol. 7.

4. Gradetsky V., Kalinichenko S., Kravchuk L., Lopashov V. "Modular Design and Mechatronic Approaches to the Exoskeleton System", Lecture Notes of the ICB Seminars. Biomechanics of the Musculoskeletal System. Medical Robotics, Polska Akademia Nauk, Warsaw 2000, pp. 260-269.

5. Jeffrey Word, Thomas Sugar, John Standeven, Jack Engsberg "Stroke Survivor Gait Adaptation and Performance After Training on a Powered Ankle Foot Orthosis", Proc. 2010 IEEE International Conference on Robotics and Automation, May 3-8, 2010, Anchorage, Alaska, USA, pp. 211-216.

6. H. Sihmidt, F. Piorko, R. Bernhardt, J. Kruger, S. Hesse "Synthesis of Perturbations for Gait Rehabilitation Robots", in Proceedings of IEEE International Conference on Rehabilitation Robotics, 2005, pp.74-77.

7. G.L.Belforte, L. Gastaldi, S.Pastorelli, M.Sordy "Active gait orthosis with pneumatic proportional control", Proceed. Of the 4th Conference of Climbing and Walking Robots, CLAWAR 2001 24-26 September 2001, Germany, pp. 47-54.

8. G.L.Belforte, L. Gastaldi, S.Pastorelli, M.Sordy "A novel actuation for the pneumatic active gait orthosis", Proceed. Of the 5th Conference of Climbing and Walking Robots, CLAWAR 2002 23-27 September 2002, France, pp. 691-698.

9. S.Charooni, M.Tokhi "Dynamics of springs leg and control of swinging leg in paraplegic walking", Proceed. Of the 5th Conference of Climbing and Walking Robots, CLAWAR 2002 23-27 September 2002, France, pp. 681-690.

MOTION CONTROL ALGORITHMS FOR AN EXOSKELETON EQUIPPED WITH PNEUMATIC DRIVES*

ANASTASIYA A. KRYUKOVA, VALERY G. GRADETSKY, IVAN L. ERMOLOV, MAXIM M. KNYAZKOV, EUGENIY A. SEMYONOV and ARTYOM N. SUKHANOV

Institute for Problems in Mechanics
of the Russian Academy of Sciences
101-1, Prospect Vernadskogo, 119526 Moscow, Russia
gradet@ipmnet.ru

The particular paper is devoted to motion control algorithms for an assist walker robot. Some problems of the motion stability are considered to satisfy the exoskeleton leg dynamics by means of a multilevel control system. The interaction of the user, the exoskeleton, and the environment is considered to improve the maneuverability and velocity of the robot's links. An appropriate motion control algorithm is proposed for an exoskeleton with pneumatic drives. The results of the experimental study of the exoskeleton's physical prototype are presented.

1. Introduction

The study of locomotion of anthropomorphous robots draws attention of many researchers. Of particular interest is studying the dynamics of pneumatically driven exoskeleton legs treated as multibody systems. With the development of the structure kinematics and the architecture of microcomputers, the tasks for the interaction of the user and the exoskeleton became more complex. The increase in the number of DOFs allows creating more maneuverable and flexible platforms, which is necessary for solving many practical tasks. Designing anthropomorphous robots requires studying the dynamics of their behavior in different situations, where maneuverability of these robots is needed. To increase the maneuverability it is necessary to reduce the stability of such robots. These structures should have rather small supporting areas and lean only on one foot when walking. The efficiency of such robots should be provided with the dynamical stability with high speeds of links of the exoskeleton and a multilevel real time control system.

*This study is partially supported by Program "Scientific Fundamentals of Robotics and Mechatronics" of the Department of Power Engineering, Machine Building, Mechanics and Control Processes of the Russian Academy of Sciences

The R&D of different anthropomorphous robot's locomotion by means of theory and mathematical modeling was carried out in many countries. The fundamental work on research and optimization of the motion energy and developing control algorithms of biped robots moving in various dynamic modes of operation: walking, running, jumping, are performed [1-2]. The experimental model of the robot-exoskeleton has been created (Fig.1) and the algorithms of control considered in this paper are applied [3].

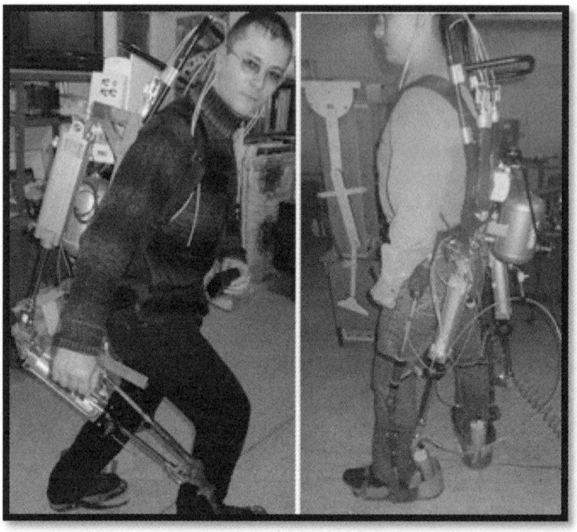

Figure1. The working prototype of the exoskeleton with pneumatic drives

2. The structure of the user-exoskeleton system

The first prototypes of exoskeletons [4-6] were constructed to implement forced limb motion of paralyzed upper or lower extremities of patients governed by feedforward (no feedback) global control, with minimum information about operating parameters acquired from sensors. The task for a pre-planned motion was introduced by an operator before the motion had started; and the execution of the motion was checked after it stopped.

The previous exoskeleton systems were redundant in terms of degrees of freedom at the joints of a mechanical system and task execution [6]. It was, therefore, necessary to develop the first prototypes of such systems to study the complex behavior of the exoskeleton-man system. To satisfy a user's needs, it is sometimes unnecessary to build a redundant system and only a limited number of degrees of freedom are required, which can be provided by using the modular method.

This method was discussed for the modular-design evolution and mechatronic approach to the application of the exoskeleton system concentrating on important practical requirements when full complex exoskeleton is not necessary, and only a number of controlled exoskeleton modules are required for the user. At the same time, it is possible to make use of the advantages of a modern control system, namely, to apply a distributed supervision control system using, for example, fuzzy logic processor and a biological feedback to determine the positions of the exoskeleton mechanical modular system in different phases as a result of exchange of information between the operator and the control system in supervision operating mode.

The general structure of the "user-exoskeleton" system is presented in figure2.

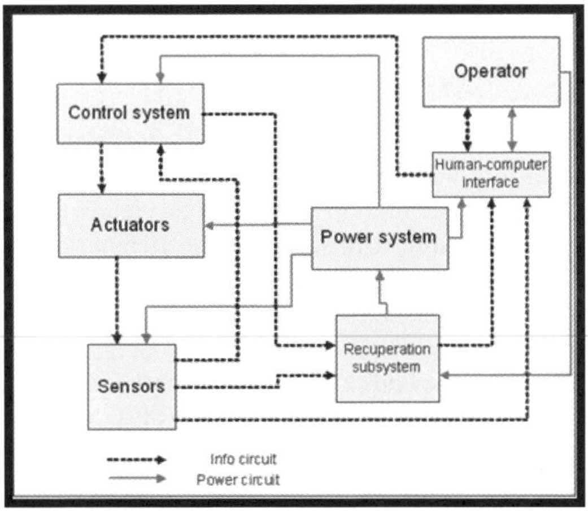

Figure2. The general structure of the "user-exoskeleton" interaction

Here the operator controls exoskeleton limbs via the control system by means of the human-computer interface. The human-computer interface may include program of master-slave control and force feedback. The control system could control actuators attached to the limbs of the device. The information from sensors may go straight to the human-computer interface and to the control system to organize feedback. The power system of the exoskeleton-robot may be autonomous or dependent. To increase the life cycle and the efficiency of the device the integration of the recuperation subsystem in the power circuit is needed. The rod of each pneumatic drive is equipped with a linear sensor to define current positions of the rod. Each foot of the exoskeleton has several piezosensors, which are able to measure the pressure of the operator's foot.

3. Motion control algorithms

Based on the results of Dai Owaki and Hiroki Fukuda work [10] it is decided to equip each foot of the exoskeleton with 5 buttons: the heel button, 2 buttons on different sides of the foot, the toe button, placed under the user's foot, and another toe button, placed above the user's toe (Fig.3).

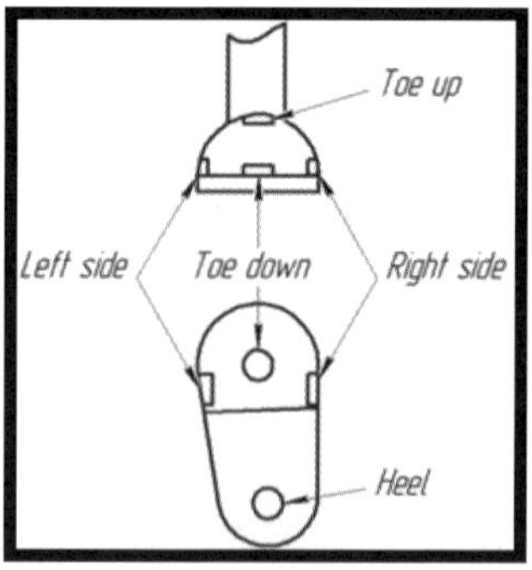

Figure3. The piezosensors' location

The kinematic structure of the designed exoskeleton has 2 legs and a torso. Each leg has a foot with 2 DOFs, a knee joint and a ball-and-socket joint in the torso. To control the motion of the exoskeleton limbs properly it is necessary to equip the structure with sensors of sufficient sensitivity to trace the behavior of the operator.

The virtual model of the assist walker robot with 12 DOFs is created. Its mechanical structure is rigid and can move without an operator under remote control or under algorithms specified for different tasks. Such a structure can be considered as a multilink manipulator with 12 DOFs [9].

Thus, the task of the route planning will be reduced to the search for the trajectory of the operating point of the manipulator. In this case one foot of the exoskeleton is proved to be the base of the manipulator and the toe of another foot turns into the operating point of the manipulator. After executing the program of movement of one foot the algorithm is repeated, but with new conditions: now the "Base" becomes "the last link of the manipulator" and the former "last link" becomes the new "Base". It is necessary to estimate the force reactions due to the ground properties. The mathematical model of the ground is described by 2 forces: F_i is the force defining elastic and non-elastic interaction of the foot and the ground, and R_i is the force defining a viscous friction [7].

$$F_i = \begin{cases} c_z z_i U_i, & \text{if } \dot{z}_i \le 0 \\ k_{cz} c_z z_i U_i, & \text{if } \dot{z}_i > 0 \end{cases} \tag{1}$$

$$R_i = \begin{cases} \mu_z \dot{z}_i U_i, & \text{if } \dot{z}_i \le 0 \\ 0, & \text{if } \dot{z}_i > 0 \end{cases} \tag{2}$$

Here c_z is the normal rigidity of the ground; $k_{Cz}=z_{imax}/(z_{imax}-h)$ is the factor characterizing the increase in rigidity of the ground due to its plastic deformation; z_{imax} is the maximum deformation of the ground; h is the depth of the footprint; μ_z is the factor of viscous resistance; U_i is the unit function that characterizes the condition of the leg; Ui=1 when the ith leg is a supporting leg, and Ui=0 if the ith leg is transferring.

It is considered that $U_i = 1$ when a foot leans on the ground.

$$U_i = \begin{cases} 0, & \text{if } N_i \le 0 \\ 1, & \text{if } N_i > 0 \end{cases} \tag{3}$$

Here the normal reaction of the ground under the foot is the sum of the elastic and viscous components.

$$N_i = F_i + R_i \tag{4}$$

While planning of trajectories, one of the following two approaches is usually applied [7]:

1. The exact set of limits on the position, speed and acceleration of the generalized co-ordinates of a leg of the exoskeleton at some trajectory points is defined. The trajectory scheduler chooses from some set of functions (as a rule, among the multinomial which degree does not exceed some prescribed number n) a function that passes through the nodal points and satisfies the constraints at these points. The constraints and the trajectory to be planned are defined in terms the joint coordinates.

2. The desirable trajectory of movement of a foot is defined in the form of analytically described function, as, for example, a rectilinear trajectory in Cartesian coordinates. The scheduler makes an approximation of the desired trajectory in terms of the joint or Cartesian coordinates.

The planning in terms of the joint variables has several advantages:

- The behavior of the variables is set at once;
- Trajectory planning can be carried out in real time;
- Trajectories in the joint variables are easier for planning [8].

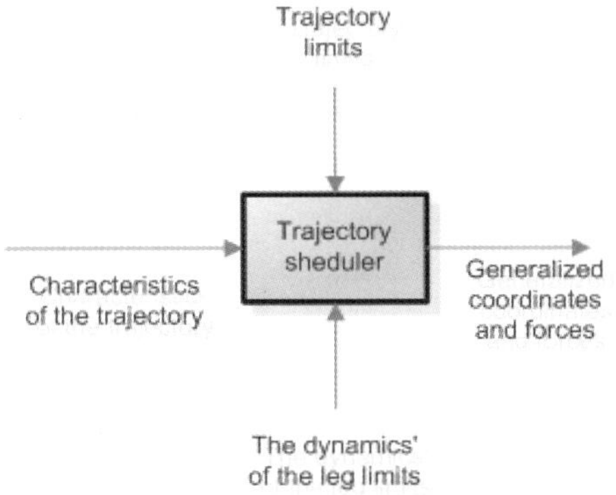

Figure4. The scheme of the trajectory scheduler

The disadvantage of the method is complexity of definition of the positions of the links. It is necessary to prevent collisions while moving.

In general, the basic algorithm of formation of the nodal points of a trajectory in terms of the joint variables is rather simple:

- $t = t_0$;
- loop: wait for the next correction event;
- $t = t_0+\Delta t$;
- $h(t)$ – the position of the leg in the space of joint variables at time t
- if $t = t_f$, leave the loop;
- loop.

Here Δt is the time interval between two consecutive moments of parameter correction of movement of the robot-exoskeleton legs.

All calculations are made for definition of the trajectory function $h(t)$ which should be updated at each point of link movement parameter correction.

In adaptive control it is supposed that movement along the desired trajectory in the Cartesian system of coordinates is a function of time.

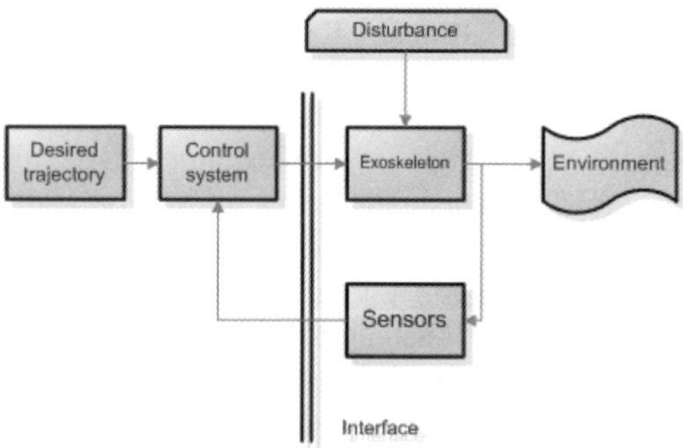

Figure5. The general control scheme of exoskeleton.

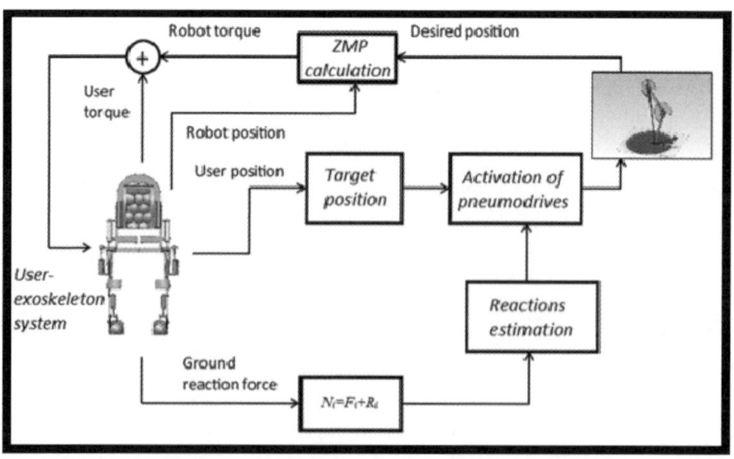

Figure6. The interaction model of user, exoskeleton and the environment

The next algorithm describes the rules of motion planning under the user control.

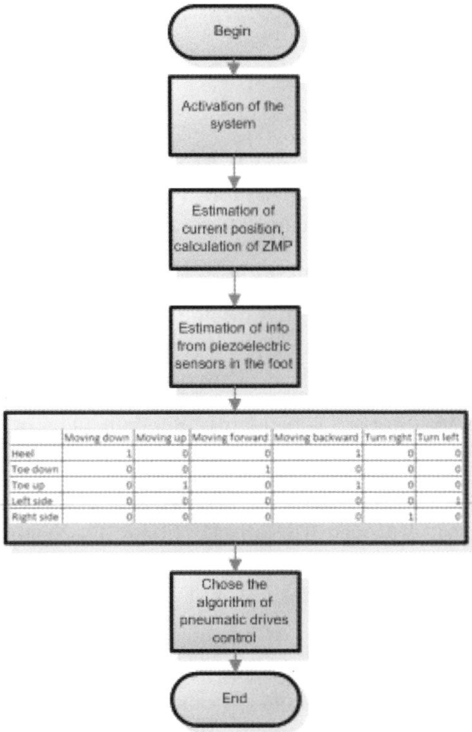

Figure7. Simplified control algorithm

This algorithm is based on the data received from buttons integrated in the feet of the exoskeleton. The experiments showed the following results:

Table 1. The results of the experiments

Mass of the system	48 kg
Payload	32-62 kg (4atm-8atm)
Air consumption	15,68 l/min
Time of the full step	3 sec
Walking speed	16 m/min

The future work involves the use of strain gauges instead of buttons to measure the pressure of the foot to control the velocity of the exoskeleton's legs.

4. Conclusion

Some problems of the motion stability are analyzed to satisfy the exoskeleton leg dynamics by means of multilevel control system. The interaction of the user, the exoskeleton and the environment is analyzed to improve the maneuverability and the velocity of the robot's links. An appropriate motion control algorithm is developed for the exoskeleton with pneumatic drives. The results of the experimental study of the exoskeleton's physical prototype are presented.

References

1. E. S. Briskin, A. V. Maloletov, A. M. Kolesov, I. P. Vershinina "About control of the biped robot motion by means of a reaction wheel" Volgogorad State Technical University, Izvestia 2010, pp 9-13.
2. A.K. Kovalchuk "Control algorithms and stabilization analysis of biped robots in dynamics", MSTU, Moscow, 2008. p150.
3. Vukobratovič M., Locomotion Robots and Anthropomorphic Mechanisms (in Russian), "Mir" Edition Press, Moscow, 1976, p. 541.
4. Vukobratovič M., Stokič D., Scientific Fundamentals of Robotics, Control of Manipulation Robots: Theory and Application, Springer-Verlag, 1982, Vol. 2.
5. Vukobratovič M., Borovač B., Surla D., Stokič D., Scientific Fundamentals of Robotics, Biped Locomotion: Dynamics, Stability, Control and Application, Springer-Verlag, 1989, Vol. 7.
6. Gradetsky V., Kalinichenko S., Kravchuk L., Lopashov V. "Modular Design and Mechatronic Approaches to the Exoskeleton System", Lecture Notes of the ICB Seminars Biomechanics, Biomechanics of the Musculoskeletal System Medical Robotics, Polska Akademia Nauk, Warsaw 2000, pp. 260-269.
7. S. Davis and D. G. Drawin, "The design of an anthropomorphic dexterous humanoid foot," in Proc. of IROS 2010, pp. 2200–2205, 2010.
8. Gradetsky V., Knyazkov M, Semyonov E, Ermolov I, Sukhanov A. "The designing of the exoskeleton leg with pneumatic drives", Proc. CLAWAR – 2013 Conference.
9. Gradetsky V., Knyazkov M, Semyonov E, Ermolov I, Sukhanov A. "The use of unloading elements in the structure of the exoskeleton", MAU №11, 2012.
10. Dai Owaki, Hiroki Fukuda, Akio Ishiguro. Soft Deformable Feet Yield Sensory-motor Coordination for Adaptive Bipedal Walking, in Proc. of IROS 2010, pp. 4257-4263.

DEVELOPMENT OF A LOWER EXTREMITY EXOSKELETON SYSTEM FOR WALKING ASSISTANCE WHILE LOAD CARRYING

W. S. KIM, H. D. LEE, D. H. LIM and C. S. HAN*

Mechanical Engineering, Hanyang University,
Ansan-si, Gyeonggi-do, South Korea
**E-mail: cshan@hanyang.ac.kr*
www.hanyang.ac.kr

J. S. HAN

Mechanical System Engineering, Hansung University,
Seoul, South Korea
E-mail: jshanjs61@gmail.com

This study developed an underactuated lower extremity exoskeleton system to carry a heavy load. To synchronize that system with a user, a feasible modular-type wearable system and its corresponding sensor systems are proposed. To operate the system with a user, human walking analysis and intention signal acquisition methods for actuating the proposed system are developed. In particular, a sensing data estimation strategy is applied to correctly synchronize the exoskeleton system with a user. Finally, several experiments were performed to evaluate the performance of the proposed exoskeleton system by measuring the electromyography signal of the wearers muscles.

Keywords: Exoskeleton; Muscle Volume Sensor(MVS); Quasi-Passive Mechanism; Muscle Activation; Human-robot interface

1. Introduction

The exoskeleton robot system is a mechanical structure that is attached to the exterior of a human body to improve the muscular power of the wearer. Lower extremity exoskeleton robots can be categorized by the purpose of muscle strength support: power assistance and power augmentation systems. Power assistance systems are exoskeleton robots that directly assist power exerted by the human body, thereby giving the wearer greater strength. EKSO and HAL are notable exoskeleton robots currently under development.[1,2] These exoskeleton robots are mainly developed for persons who are aged, feeble, or disabled to assist them in their daily lives.

Meanwhile, power augmentation systems to amplify the power of wearers, enabling them to perform tasks that they otherwise cannot easily perform by themselves. The University of California at Berkeley has also been developing an exoskeleton system referred to as the Berkeley Lower Extremity Exoskeleton (BLEEX).[3] Rockheed martin has developed the Human Universal Load Carrier (HULC) that based on the BLEEX system. For the same purpose, an exoskeleton called HEXAR(Hanyang EXoskeletal Assistive Robot) was developed in South Korea.[4] For the purpose of industrial usage, however, the operational convenience and compactness of the system must be strongly considered. This means that the system has to be designed so that it is easy to handle and manage and so that it can be easily synchronized in order to correspond to human movement.[5]
This study proposes an underactuated lower exoskeleton system used to carry a heavy load. A sensor system was developed to determine the measurements of muscle activation in order to synchronize the exoskeleton with the wearers movements. To this end, this study developed the mechanical structure and the control algorithm using the developed sensor, called the muscle volume sensor (MVS). Moreover, this study verified the efficiency of the developed system based on several experimental results for measuring muscle fatigue.

2. Mechanical System for Exoskeleton

2.1. *Design Concept*

The proposed exoskeleton has a lower exoskeleton and a lower ground support structure. The total degree of freedom is 3 for each leg in the sagittal plane; furthermore, one active joint and two quasi-passive joints are integrated. This research focused on the load carrier that performs the anti-weight function of the load while walking on a level surface or walking up steps. According to this scope, when the exoskeleton is operated on level ground, the leg in the stance phase is only required to support the load bearing along the z-axis direction, which is normal to the ground.[6,7] During the segue from the heel strike to the stance phase, an energy conservative ankle joint can simultaneously perform load-bearing functions against the weight of the total system and the gait propulsion function.

2.2. *Structure Design*

Figure 1 shows the exoskeleton assisting the muscular strength of the lower extremities by arranging the joint mechanism and joint functions. Devel-

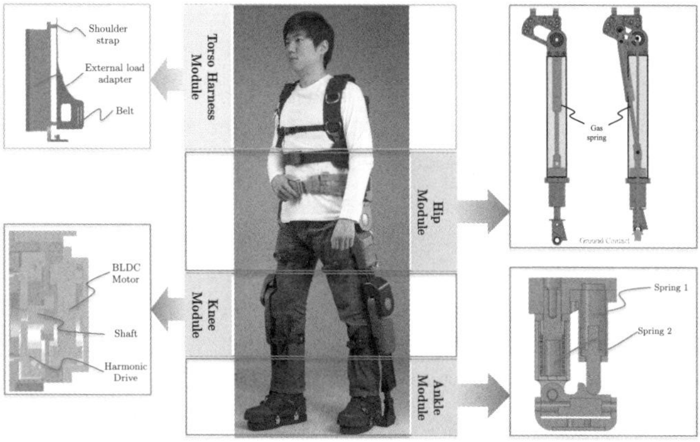

Fig. 1. Exoskeleton System Configuration According the Mechanism Type.

oped exoskeleton system consists of the torso harness, the hip joint, the knee joint, and the ankle joint module that combine the quasi-passive and active mechanism. Active mechanism is only provided to the knee joint for extension/flexion movement, and thus a hip and ankle joint module are consisted by a quasi-passive mechanism. The torso harness module serves to link the exoskeleton to the wearer through the shoulder strap and belt. The torso hares module is connected to the hip joint module through the backbone, and the hip joint module bears the total weight of the upper mechanism through the constant force mechanism. The hip joint module consists of the quasi-passive mechanism that uses the gas spring and the sliding part of the exoskeleton. Then the hip joint module is combined with actuated knee joint module. The knee joint module was composed of an active mechanism that uses an electrical BLDC motor, which can produce extension/flexion motions in the sagittal plane, and a harmonic drive combination at a gear ratio of 100: 1. The ankle joint module was designed to minimize the torque generated by the loads by designing it so that the ankle joint of the exoskeleton is close to the ground. In addition, a mechanism using springs was developed in order to compensate the torque generated in the stance phase and to generate propulsion force in the toe-off phase.

2.3. *Design of the Passive Mechanism*

Proposed passive mechanism should be designed to transmit the weight of the robot and the external loads to the ground when the stance phase.

Therefore, knee and hip module was designed to support the external load and own mass by proposed a quasi-passive mechanism that consist of spring. Figure 2 shows the kinematic model of hip and ankle joint module. Through the hip mechanism, the moment arm of the gas spring becomes 0 in the swing phase, whereas l_{ma} is generated when the end of the gas spring is slid along the roller to position P_{sl} in the stance phase. The moment arm, l_{ma} is generated and force τ_s is generated by the spring that becomes identical to τ_l during the external loads. The ankle joint module was designed to be as closed to the ground. As a result, the range of motion of the robot and human were different. Therefore, the range of motion was a calculate through the exoskeleton of the ankle joint module of kinematic analysis. Through the kinematic analysis of the range of motion for the exoskeleton ankle joint was -30 to 20°. And also to compensate for the torque that occurs in the stance phase to toe-off of the propulsion force generated using the spring mechanism was developed.

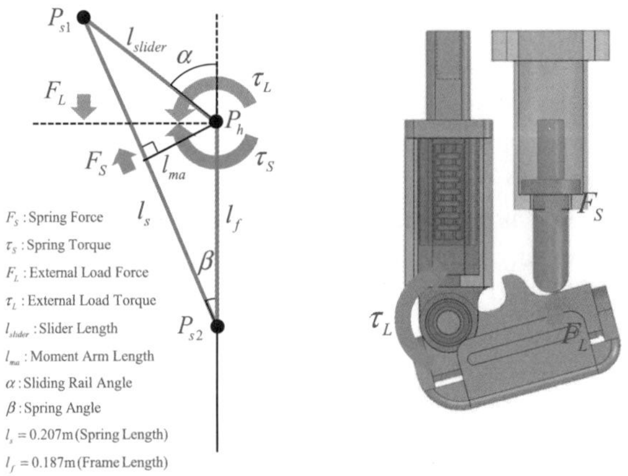

Fig. 2. Kinematic Analysis for the Quasi-Passive Mechanism; Hip (left), Ankle (right)

3. Control Algorithm using HRI

3.1. *HRI Sensor*

In this study, a muscle volume sensor (MVS) to acquire the signal for the degree of expansion of the muscle and insole sensor are developed. The measured signal is utilized as the wearers intent signal to operate the proposed exoskeleton system.(Fig. 3) Changes in muscle volume that occur through muscle activation may be expressed as physiological cross-sectional area (PCAS) according to muscle length and muscle strength, F may be examined through PCAS.[8] So developed the Muscle Volume Sensor (MVS) that may measure Δx, changes in muscle circumference that occurs due to muscle activation. MVS developed to measure muscle activation is divided into a measurement part that may measure changes in length and a fixed part to deliver changes in muscles and maintain wearing form. To this end, the measurement part was designed so that changes in the circumference of the cross sectional area was converted into rotational changes. For the purpose of gait control for knee active joint through the MVS, selected the quadriceps muscle group while walking. And insole sensor developed to detect the gait phase is consist of the force sensor. For the purpose of the detect the gait phase through the insole sensor, arrange the sensor position and measure the COP (center of pressure) of the foot.

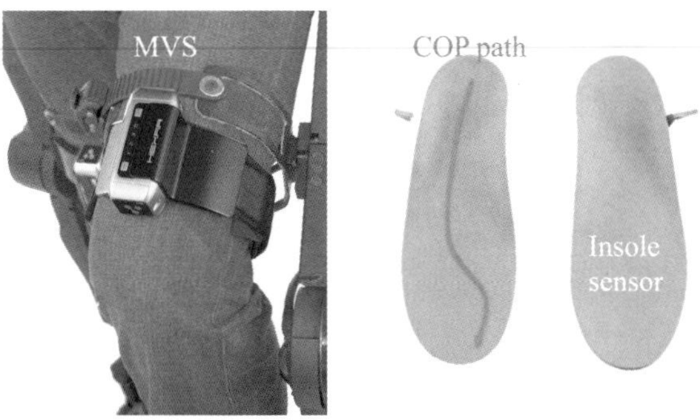

Fig. 3. Muscle Volume Sensor (left) and Insole Sensor (right)

3.2. *Control Strategy using the HRI Sensors*

The sensor was attached to the quadriceps muscle group in order to measure the intent signal for generate the commands of knee joint control. However, since different signal characteristics appear between the stance/swing phases, controls for different gait phases were configured using insole sensors. For controls using an MVS, a one-time calibration is necessary at the beginning of the process. Calibration should be conducted for the stance phase and the swing phase separately and the polynomial coefficients, a_n should be noted and the polynomial fittings should be determined by MVS the amount of changes in the volume of the muscles ,x_n and the human knee joint angles, θ_n. The command signals for the exoskeleton robot are generated by calculating knee joint angles, θ_d as shown in expression Eq. (1); where, C is the gain for calibration and k is the gain for reception which can be adjusted by the users. The θ_d generated in this method is used as inputs for the PID controller of the exoskeleton robot (Fig. 4).

$$\theta_d = a_1(x_n + k\dot{x}_n + \theta_n^C) + a_2(x_n + k\dot{x}_n + \theta_n^C) + a_3 \tag{1}$$

$$where,\ 0 < C \leq 1$$

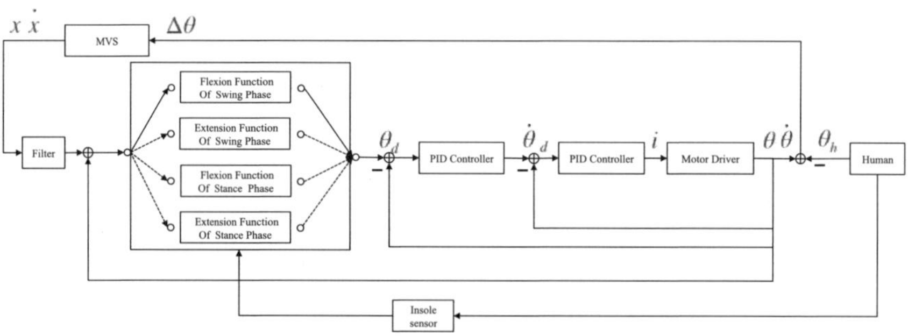

Fig. 4. Control Algorithm for Operating the Knee Joint Module using the HRI Sensors.

4. Experiment and Results

Figure 5 shows the calibration process for the generate the command signal using the MVS. This result show that the performance of the MVS is similar to the conventional EMG sensor, such as the hysteresis when extension/flexion. Then, polynomial curve for the apply the calibration gain, C and reception gain, k to normalized sensor signal was shown in Figure 5. The EMG electrodes are attached to the quadriceps and gastrocnemius muscle groups. The EMG signal is gathered during locomotion and compared to the results of the two conditions of not wearing the exoskeleton and wearing the exoskeleton, when changes occur in the external load condition. Then, the percent maximum voluntary isometric contraction (%MVIC) is measured and calculated using a standardized, objective, and sensitive tool designed to measure muscle strength. The average decrement of %MVIC as

40.5% while walking on a level ground and 12.5% when walking up stairs. values decreased as much as 32% to 49% when the user was not wearing the exoskeleton while walking on a level surface. When the wearer was walking up stairs, the %MVIC of the muscle power was reduced by 11% to 24%. Fig. 6 The degree of the wearers muscle power assistance affects the level walking and stairway conditions, and this Therefore, this study found that the degree of the wearers muscle power assistance is more effective in the level walking condition than it is in the stairway condition.

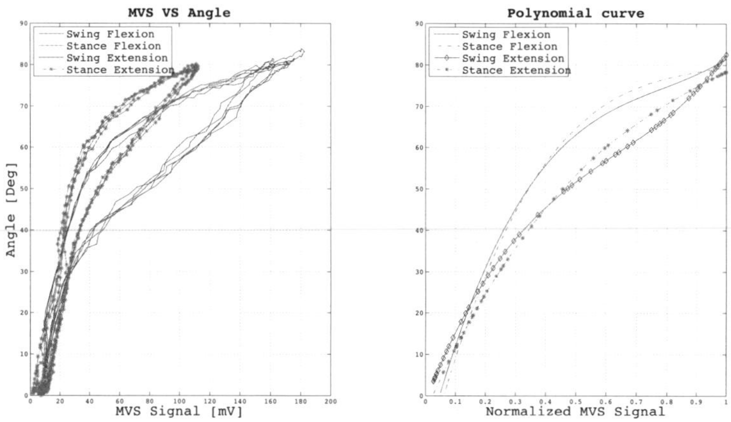

Fig. 5. MVS Signal and Angle Data (left), Polynomial Curve for the Stance and Swing Phase (right)

5. Conclusion

We reported on a robotic system designed using a minimized actuator and a newly designed quasi-passive mechanism. Moreover we developed an MVS that can be used with an HRI sensor. However, the degree of the wearers effort is extremely reduced considering the carried weight of the total system. Moreover, although the proposed exoskeleton system is an underactuated robotic system, the results of this study show that it delivers an acceptable performance.

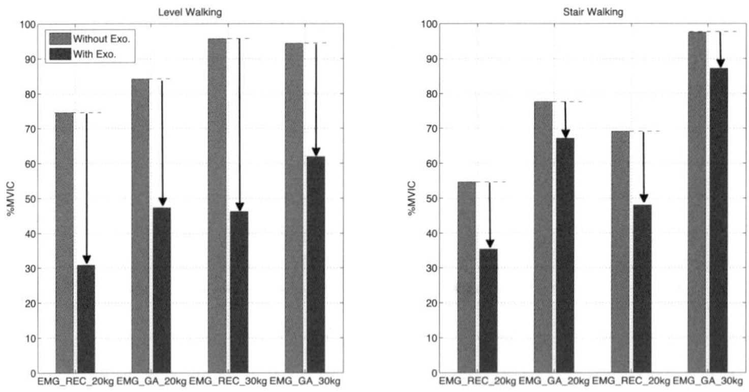

Fig. 6. Percent Maximum Voluntary Isometric Contraction (%MVIC) Value while Walking Experiments (REC.:Rectus Femoris Muscle, GA.: Gastronemius muscle).

Acknowledgements

This research was supported by Basic Science Research Program Through the National Research Foundation of Korea(NRF) funded by the Ministry of Education, Science and Technology (2012008672) and Korea under the Advanced Robot Industrial Promotion Agency (NIPA) (NIPA-2012-H1502-12-1002).

References

1. S. Lee and Y. Sankai, *Advanced Robotics* **19**, 773(August 2005).
2. K. a. Strausser and H. Kazerooni, The development and testing of a Human Machine Interface for a mobile medical exoskeleton, in *IEEE/RSJ International Conference on Intelligent Robots and Systems (IROS)*, (Ieee, September 2011).
3. H. Kazerooni and R. Steger, *Journal of Dynamic Systems, Measurement, and Control* **128**, p. 14 (2006).
4. H. Lee, W. Kim, J. Han and C. Han, *International Journal of Precision Engineering and Manufacturing* **13**, 1491(July 2012).
5. a. R. Seo, H. Y. Jang, W. S. Kim, C. S. Han and J. S. Han, *International Journal of Precision Engineering and Manufacturing* **13**, 899(May 2012).
6. S. N. Yu, H. D. Lee, S. H. Lee, W. S. Kim, J. S. Han and C. S. Han, *Advanced Robotics* **26**, 561(January 2012).
7. C. Walsh, K. Endo and H. Herr, *International Journal of Humanoid Robotics* **4**, 487 (2007).
8. R. Lieber, *Skeletal muscle structure, function, and plasticity: the physiological basis of rehabilitation* (Lippincott Williams & Wilkins, 2002).

DYNAMIC COUPLING CHARACTERISTICS OF A SEMI-ACTIVE KNEE PROSTHESIS

MOHAMMED AWAD*, ABBAS DEHGHANI
Institute of Engineering Systems and Design, Schoolof Mechanical Engineering, University of Leeds, Leeds, LS2 9JT, United Kingdom.

DAVID MOSER, SAEED ZAHEDI
Chas A Blatchford & Sons Ltd, Lister Road, Basingstoke, Hants, RG22 4AH, United Kingdom.

This paper characterises the dynamic coupling interaction between the thigh and an electrical semi-active above knee prosthesis. This prosthetic knee works as a passive knee in part of the gait cycle phases and as an active knee prosthesis during other parts. The knee prosthesis is designed to be back-drivable in passive mode. The results of the experimental work carried out to evaluate the dynamic coupling effect in both passive and active mode show that the prosthesis weight has an important effect on the performance.

1. Introduction

Around the entire world, several thousand go through lower limb amputation operations every year due to: complications of diabetes, vascular disease e.g. poor blood circulation in which lower limb parts do not receive adequate amount of oxygen and nutrition, accidents which result from trauma and cancer in limb segments [1]. The loss in mobility results in a degradation of the quality of life of the amputees because it affects many aspects of their personal and professional lives. Hence, the transfemoral (TF) prosthesis should be fitted to the amputee to improve his/her mobility and to restore the lost functions. The best transfemoral prosthesis is the one that can mimic the functions of the biological leg.

1.1. *Background*

In the past, walkers, crutches, open socket peg legs and wheelchairs were used, but they were limited in the degree of mobility they could restore. Hence,

---

* Assistant Professor, Mechanical Engineering, Faculty of Engineering, Ain Shams University, Cairo, Egypt.

there was a real need to develop an efficient prosthetic to restore the lost mobility functions and to assist the amputees to participate actively in their daily lives.

Over the last few decades, a technological revolution in the prosthetic industry took place due to state-of-the-art advancements in materials, electronics, sensing, and actuators. Accordingly, the currently available lower limb prostheses can be divided into three main groups: purely passive, actively controlled and actively driven or powered prostheses. Purely passive prostheses depend only on mechanical systems, such as polycentric knee joint, four bar linkages, locking mechanisms and passive hydraulic cylinders. These types of prostheses require a significant voluntary control effort from amputees.

Actively controlled (energetically passive) knee prostheses were presented for the first time in 1997 by Otto Bock with the release of the C-Leg. This C-leg controls the damping effect by using hydraulic cylinder. Unlike C-Leg, other knee prostheses use pneumatic swing control units, such as smart IP, IP+ microprocessor knee and smart adaptive [2]. Moreover, Biedrmann et al. [3] developed a knee prosthesis that uses smart Magnetorheological (MR) fluid for stance and swing control. This smart fluid is ideal for successful integration into advanced prosthetics devices for real time gait control to provide adaptive stance and swing control. More advanced intelligent actively controlled prostheses were presented recently, such as the Orion microprocessor knee and Genium microprocessor knee. These advanced prostheses use a wide variety of sensors to measure the load transfer and the knee angle in order to determine the knee security requirements.

On the other hand, actively driven prostheses are fully actuated, such as the Victhom knee [4-6] which is commercially known as Power Knee. These actively driven prostheses are actuated using either brushless dc motors [7, 8], or by pneumatic actuators [9].

1.2. *Motivation*

In the case of actively controlled prostheses, these prostheses cannot provide the positive power required during some tasks or phases. Therefore, above-knee amputees often compensate for the loss of function in both the knee and the ankle by applying extra input energy via the hip joint. On the other hand, the actively driven prostheses are able to supply positive power, but they consume more power than the human muscle, as they are continuously powered. In contrast, the dynamics of human lower limbs and muscles shows that muscles of the lower limb are not continuously activated during normal walking [10, 11].

Therefore, there is a real need for an advanced knee prosthesis that can deliver positive energy, and regenerates the negative energy when it is required. Hence, one of the primary challenges is developing a semi-active prosthetic knee that can back-drive easily in the passive mode to mimic the human muscle. Moreover, the dynamic coupling interaction between the lower limb segments and the prosthesis can be used to provide more efficient and comfortable walking for above knee amputee with a little power applied to the prosthetic knee at the right time. This research aims to assess dynamic coupling interaction between a mechanical thigh and the semi-active prosthetic knee. Moreover, the prosthetic weight effect on the dynamic coupling performance is studied.

2. Methodologies

2.1. *A Semi-Active Prosthetic Knee*

The semi-active knee prosthesis, shown in Figure 1, is designed based on the considerations suggested by the authors [12] to provide an efficient knee prosthesis. This prosthetic knee is driven by a 42V DC motor in the active mode which drives the ball screw through a 1:1 timing belt. The detailed design was presented by the authors in [13].

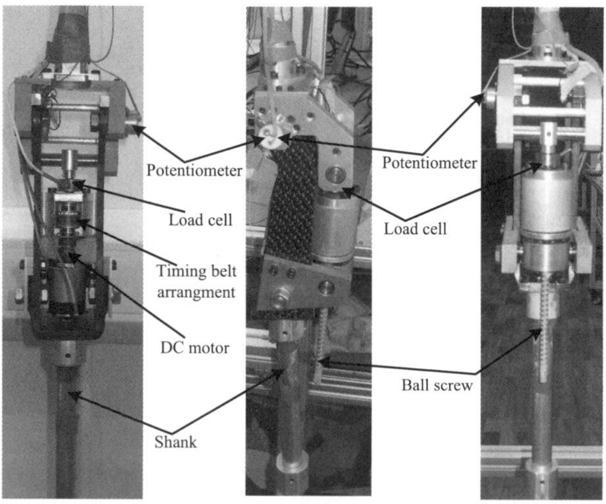

(a) Back (b) Side (c) Front

Figure 1: The back-drivable semi-active knee prosthesis.

This prosthetic knee is equipped with a potentiometer to sense the knee angle, and a load cell to measure the driving and the resistance torque. The motor supplies assistive power in the active mode to assist the dynamic coupling energy that is produced by both the amputee's hip and the sound leg. Two limit switches are connected in series with the motor terminals and attached to the extreme positions of the prosthetic knee to act as a failsafe in the event of a failure.

2.2. *Experimental Setup*

To test the dynamic coupling behavior in the swing phase, the prosthetic knee is connected to a custom built mechanical thigh and hip test rig as shown in Figure 2. A rotary hip mechanism was designed to generate the kinematics parameters for the thigh. This mechanism is equipped with a low noise potentiometer to measure the kinematic data.

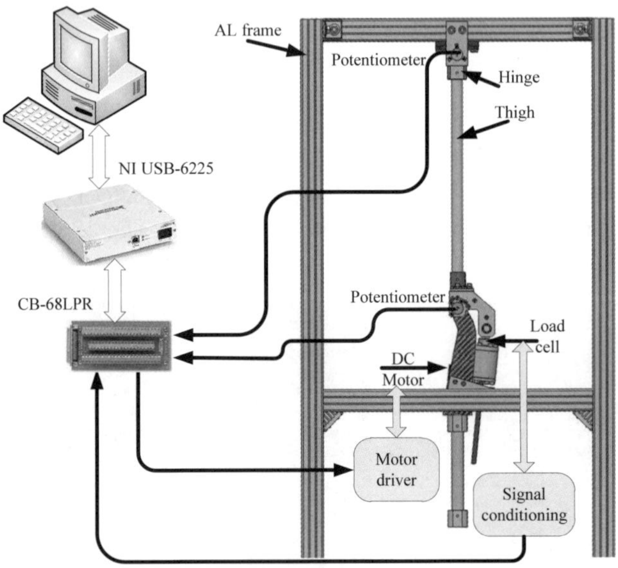

Figure 2: Experimental setup for dynamic coupling effect.

2.3. *Experiment Protocol*

The developed test rig is used to evaluate the prosthetic knee performance when it is subjected to both the dynamic coupling interaction and its gravity components. The dynamic coupling interaction is created by manually swinging of the mechanical thigh at different acceleration and deceleration rates. This experiment is carried out for the passive mode when the motor terminals are not connected to the motor driver (open circuit (O.C.)), and when the motor terminals are connected to each other (short circuit (S.C.)). In order to study the effect of prosthetic knee's weight on the energy transferred from the mechanical thigh to the prosthetic knee and on the dynamic coupling elements, experiments were carried out for three different weights (prosthetic knee weight only, 0.564kg added weight, 1.079kg added weight). The use of the prosthetic knee weight plus an additional weight is a representative for taking into account the foot weight.

A global dynamic coupling index, shown in equation (1), is developed to indicate and measure the amount of the natural dynamic coupling between the prosthetic knee and mechanical hip at all moving path in the joint space. This index can be also used to quantify how much acceleration is possibly required at the mechanical hip ($\ddot{\boldsymbol{\theta}}_h$) to drive and produce a desired acceleration at the prosthetic knee ($\ddot{\boldsymbol{\theta}}_k$). The greater the dynamic coupling index, the easier for the hip (amputee's hip) to drive the prosthetic knee. In this research, the norm of the acceleration vector is used to measure the overall global coupling index as follows:

$$\boldsymbol{G_c} = \frac{\|\ddot{\boldsymbol{\theta}}_k\|}{\|\ddot{\boldsymbol{\theta}}_h\|} \tag{1}$$

3. Results

The results of the dynamic coupling experiments indicated that the energy transfers from the mechanical thigh to the shank and generates kinematics motion parameters on the knee. The prosthetic knee mechanism impedance is small in the passive mode because the motor terminals are open circuit (O.C.). Figure 3 shows that the knee resists the transferred energy at small angles rather than the large angles as the friction is higher. It was observed that the gravity component could work with or against the coupling between the thigh and the shank to support the knee rotation or resist it. However, when extra weights are added to the prosthesis shank, it was found that less energy is required from the thigh to rotate the knee.

In the active mode, the prosthetic knee did not respond to low input voltages applied to the knee's DC motor. However, the dynamic coupling can help the knee to overcome this impedance and rotate. This shows that a little energy applied to the prosthetic knee's motor in addition to the dynamic coupling could help the prosthetic to move with less energy consumption. Moreover, the prosthetic weight plays an important role in the active mode as well as in the passive mode. More added weight means that more kinetic energy is stored in the prosthetic knee. However, extra weight will generate a resistive gravity weight force which will make the amputee's movement difficult. This means that either too small or too large weights are not the best choices for an efficient knee prosthesis.

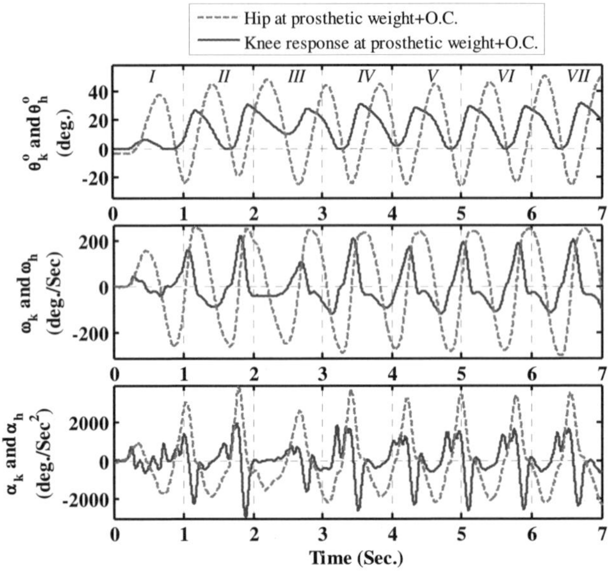

Figure 3: Prosthetic knee kinematics performance in passive mode.

The global dynamic coupling index (G_c), which was developed to quantify the amount of coupling in the knee mechanism based on the norm of the overall acceleration transmission ratios, was calculated for both passive and active modes according to the operating modes and the prosthetic knee weight as shown in Table 1.

Table 1: Global dynamic coupling performance index.

		Performance index			
		Applied voltage	**Prosthetic weight**	**0.564kg added weight**	**1.079kg added weight**
Testing Mode	passive	**(O.C.)**	0.6013	0.6782	1.5578
		(S.C.)	0.2079	0.2361	0.4317
	active	**0V**	0.2393	0.2969	0.5104
		-0.5V	0.2962	0.4285	0.5466
		-1V	0.3623	0.4483	0.6410

4. Discussions

The results of the conducted experiments showed that the regenerative braking concept that is used in hybrid cars could be useful with knee prostheses.

This will lead to a new generation of lower limb prostheses that can generate power and as a result, its battery could sustain longer. In addition, this shows that the semi-active back-drivable lower limb prosthesis is more efficient than stiff powered knee. The results also showed that the hip kinematics could potentially be used for an efficient control of the knee prosthesis by using the dynamic coupling concept.

According to Table 1, the open circuit (O.C.) in the passive mode with 1.079kg added weight has the largest global coupling index. This index is more than one which means that the acceleration produced on the prosthetic knee ($\ddot{\boldsymbol{\theta}}_{\boldsymbol{k}}$) will be bigger than hip acceleration ($\ddot{\boldsymbol{\theta}}_{\boldsymbol{h}}$). As the hip movement is the main source of energy, this means that less energy is required from the amputee hip at 1.079kg added weight in comparison to less weight. The prosthetic weight effect on the prosthesis performance plays an important role. Therefore, an extensive optimization study is required to identify the optimal weight for the prosthetic knee. Moreover, the movement of knee prosthesis can be regulated by controlling the acceleration of the amputee's stump.

In the active mode, the dynamic coupling effect could be used to reduce the power consumption from the motor if the knee accelerates due to the dynamic coupling in the same direction of the motor due to the applied voltage. In this case, the knee motor is back-driven by the dynamic coupling forces which will reduce the total energy consumption.

5. Conclusions

This paper presented an investigation into assessment and characterisation of an efficient back-drivable semi-active prosthetic system. The semi-active prosthetic knee is based on previous research provide key design parameters of the knee which affect the prosthesis performance. The experimental works were carried out to test the dynamic coupling interaction between the thigh and the prosthetic knee. From the experimental results it was noticed that the prosthesis weight has an important effect on the passive dynamic walking performance. It was also observed that the dynamic coupling helps to reduce the power consumption in the prosthetic system. The outcome of this research can lead to a new generation of lower limb prostheses that can generate power and as a result more energy efficient prosthetic systems could be developed. In such systems it is also envisaged that the hip kinematics could potentially be used for an efficient control of the prosthesis.

References

[1] A. Cristian, "*Lower Limb Amputation: A Guide to Living a Quality Life*," Demos Health, 2005.

[2] S. Zahedi, A. Sykes, S. Lang, and I. Cullington, "Adaptive prosthesis- a new concept in prosthetic knee control," *Robotica*, vol. 23, pp. 337-344, 2005.

[3] L. Biedermann, W. Matthis, and C. chulz, "Leg prosthesis with an artificial knee joint and method for controlling a leg prosthesis," United States Patent, 2004.

[4] S. Bedard, "Control system and method for controlling an actuated prosthesis," United States Patent, 2004.

[5] S. Bedard, "Control device and system for controlling an actuated prosthesis," United States Patent, 2006.

[6] S. Bédard and P.-o. Roy, "Actuated leg prosthesis for above-knee amputees," United States Patent, 2008.

[7] K. Fite, J. Mitchell, F. Sup, and M. Goldfarb, "Design and Control of an Electrically Powered Knee Prosthesis," *Rehabilitation Robotics, 2007. ICORR 2007. IEEE 10th International Conference on*, pp. 902-905, 2007.

[8] F. Sup, H. A. Varol, J. Mitchell, T. Withrow, and M. Goldfarb, "Design and control of an active electrical knee and ankle prosthesis," *Biomedical Robotics and Biomechatronics, 2008. BioRob 2008. 2nd IEEE RAS & EMBS International Conference on*, pp. 523-528, 2008.

[9] F. C. Sup and M. Goldfarb, "Design of a Pneumatically Actuated Transfemoral Prosthesis," *ASME Conference Proceedings*, vol. 2006, pp. 1419-1428, 2006.

[10] B. R. Durward, G. D. Baer, and P. J. Rowe, "*Functional human movement : measurement and analysis* ", Butterworth-Heinemann, Oxford, 1999.

[11] J. Perry, "*Gait analysis : normal and pathological function*," SLACK inc., Thorofare, N.J., 1992.

[12] M. Awad, K. Tee, A. Dehghani, D. Moser, and S. Zahedi, "Design of An Efficient Back-Drivable Semi-Active Above Knee Prosthesis," *The 14th International Conference on Climbing and Walking Robots (CLAWAR2011)*, Paris, France, pp. 35-42, 2011.

[13] Mohammed Awad, Kian Sek Tee, Abbas A. Dehghani-Sanij, David Moser, and S. Zahedi, "Analysis and Performance of A Semi-Active Prosthetic Knee," *Proceedings of the International Conference on Mechanical Engineering and Mechatronics (ICMEM 2012)*, Ottawa, Ontario, Canada, 16-18 August 2012, 2012.

COVERED AREA DETECTION BASED ON BRIGHTNESS CHANGE OF INNER CAMERA IMAGES FOR CRAWLER ROBOT

Kenichi TOKUDA* and Tatsuya HIRAYAMA

Faculty of Systems Engineering, Wakayama University ,
930 Sakaedani, Wakayama, 640-8510, Japan
**E-mail: tokuda@sys.wakayama-u.ac.jp*

Tetsuya KINUGASA

Okayama Univrsity of Science,
1-1, Ridai-cho, Okayama, 700-0005, Japan
E-mail: kinugasa@mech.ous.ac.jp

Takafumi HAJI

Matsue college of Technology,
14-4, Nishiikuma-cho, Matsue, 690-0865, Japan

Hisanori AMANO

National Research Institute of Fire and Disaster, Japan,
4-35-3 Jindaiji-higashimachi, Chofu, Tokyo, 182-0012, Japan
E-mail: amano@fri.go.jp

A rescue robot with crawler has few camera spaces. Our purpose is development how to create the space inside the crawler belt. However it makes robot vision's image fragmentation to attach inside the cameras such the space. Our previous report showed that a patching method can generate complete image by using two cameras. This paper describes additional method based on brightness changes to solve a low-light problem. Experimented result shows that the method is effective.

Keywords: crawler robot, tele-operation, image processing

1. Introduction

FMT, Flexible mono-tread mobile track,[1] is a rescue robot system, which is covered by crawler track except the both side surfaces(Fig.1). It is inappropriate to attach a camera on the side of the robot because the camera system structure is easily caught by rubbles. The purpose of this study is to put the camera system onto the front of the robot. Running track is moving across the face of the front camera system. The images from the camera often catch the crawler track region in the view. The botheration disrupts remote control of the robot system with the nuisance by crawler track region.

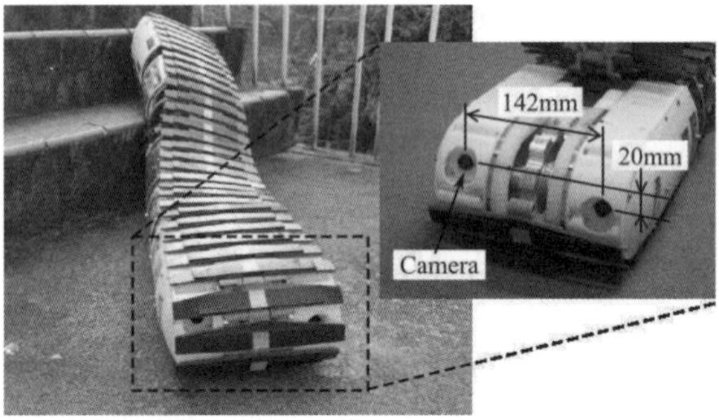

Fig. 1. FMT overview

Our previous report[2] proposed a method to complement an image's patch by using two cameras mutually in real time(Fig.2). However, there leaves a problem that an environmental brightness changes decreases the complement precision. Our algorithm is intended to be improved by positively using a change of the environmental brightness.

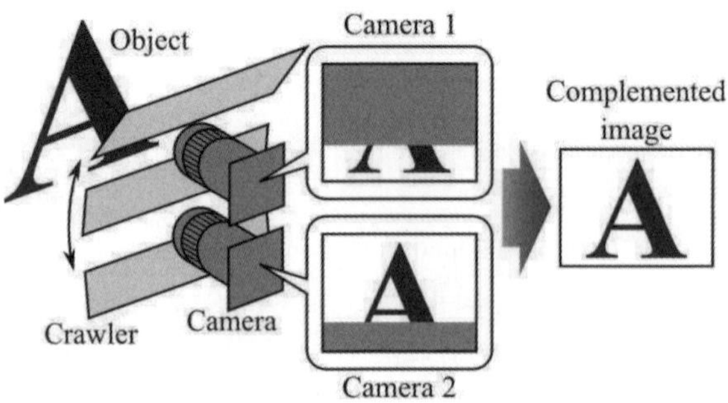

Fig. 2. Concept of image complement

2. Covered area detection

2.1. *Technique by using brightness change*

We propose a cover detection technique using by the brightness change in images. As stated above, a brightness change between adjacent pixels is small in the covered domain. In contrast, the brightness change between pixels to be adjacent by objects such as the debris coming out in the region that is not covered grows big. This property enable the covered area detection. The area detection method is advantageous in that the camera images are not influenced.

2.2. *Extraction of brightness changes*

We find the brightness conjugation of images by a differential calculus filter and multiply a value every line of the image sequence. The multiplication level of this brightness change has a property to grow big small in the line without a cover in the line with the cover. It is that the condition found in a differential calculus filter here is strong for the noise in the difference of the multiplication level being clear in a cover, a non-covered area, the dark place. About the latter, this is because a brightness change becomes big in a noise part by the rise of sensitivity (a gain) in the dark place.

We assumed an input example image (320 × 240pixel) and made a differential calculus filter. In addition, the input image worked to filter it after having been gray-scaled(Fig.3).

Fig. 3. Example image

In a covered area and the non-covered area, we can find that a difference occurs in the multiplication level of the brightness change. It may be said that separableness is good so that **A** in a figure is big then. **A** is differences between maximum (Max) of the cover domain (Row=0 - 84) and minimum (Min) of the non-cover domain (Row=94 - 239).

0	0	0
-1	0	1
0	0	0

First derivative filter

0	0	0
1	-2	1
0	0	0

Second derivative filter

0	1	0
1	-4	1
0	1	0

Laplacian(4) filter

1	1	1
1	-8	1
1	1	1

Laplacian(8) filter

-1	0	1
-1	0	1
-1	0	1

Prewitt filter

-1	0	1
-2	0	2
-1	0	1

Sobel filter

Fig. 4. Derivative Filters

Table 1 Comparison of filters

Filter	Max	Min	A	A/Max
First derivative	312	548	236	0.76
Second derivative	336	414	78	0.23
Laplacian(4)	734	966	232	0.32
Laplacian(8)	1688	2456	768	0.45
Prewitt	766	1398	632	0.83
Sobel	1030	1908	878	0.85

3. Experiment and evaluation

We made a experiment for effectiveness determination by using real FMT robot system. The environment is set in indoors and we decided an evaluation value E as following,

$$(\text{Mean value})\,\bar{x} = \frac{1}{n}\sum_{i=1}^{n} x_i \tag{1}$$

$$(\text{Standard deviation})\,s = \sqrt{\frac{1}{n-1}\sum_{i=1}^{n}(x_i - \bar{x})^2} \tag{2}$$

$$(\text{Evaluation value})\,E = \bar{x} - s \tag{3}$$

where i is frame number, n is total frame, x is a visible dots rate in a frame, $\bar{x}$ is a average of visible dots, s is a standard deviation.

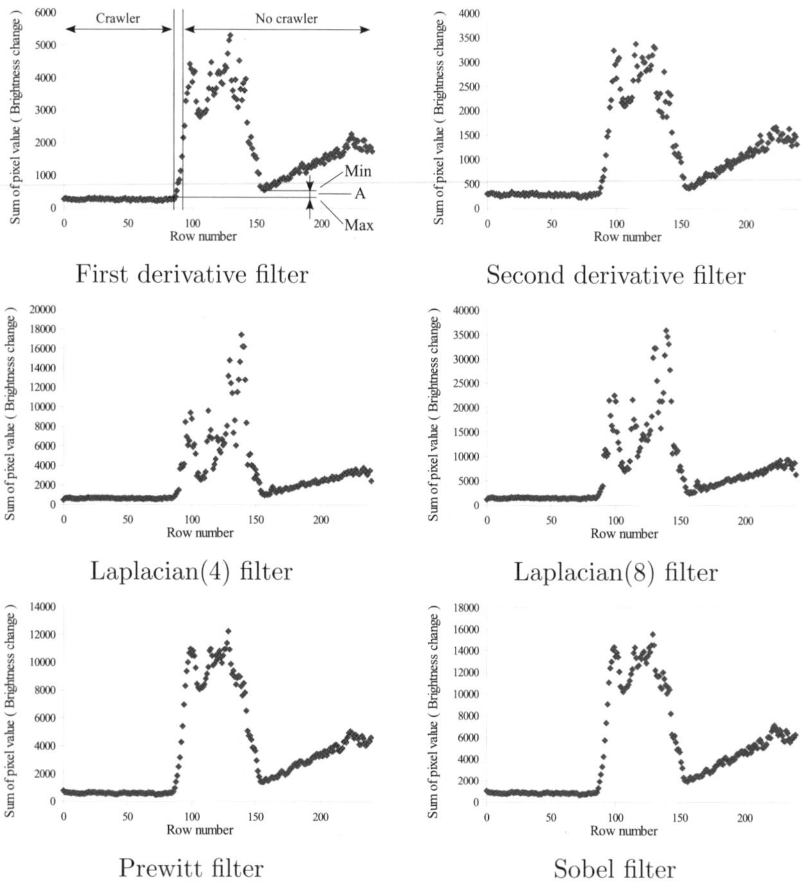

Fig. 5. Effects by using derivative Filters

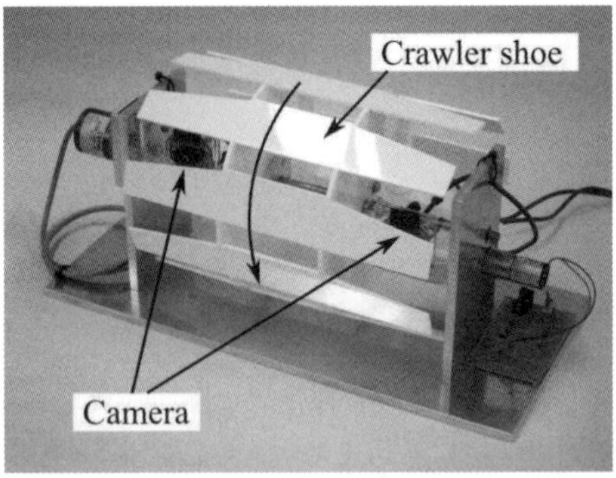

Fig. 6. Experiment device

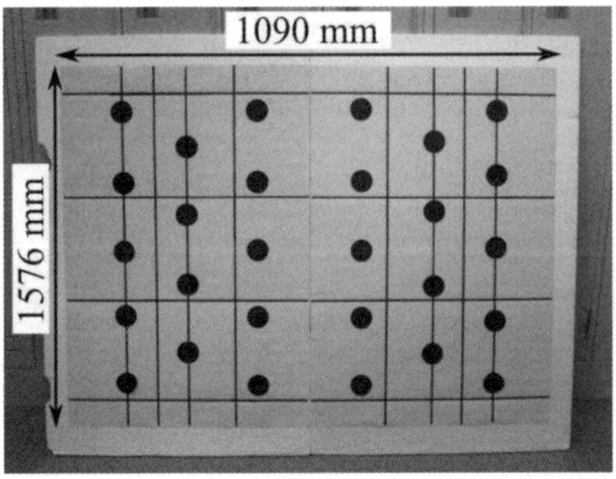

Fig. 7. Target for ecaluation

Table 3 Evaluation of image processing result

Method	Min	Max	$\bar{x}$	s	E
Nothing	0.0	1.0	0.48	0.44	0.05
Previous	0.0	1.0	0.91	0.20	0.71
Proposed	0.0	1.0	0.95	0.13	0.82

Experimental results(Table 3 and Fig.9-11) showed that our proposal technique is effective.

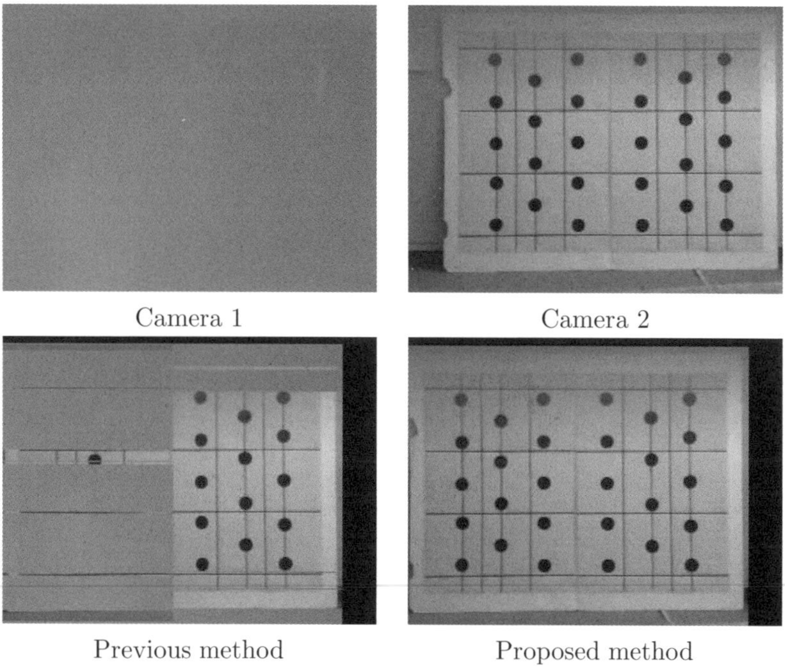

Fig. 8. Experiment result by using evaluation target

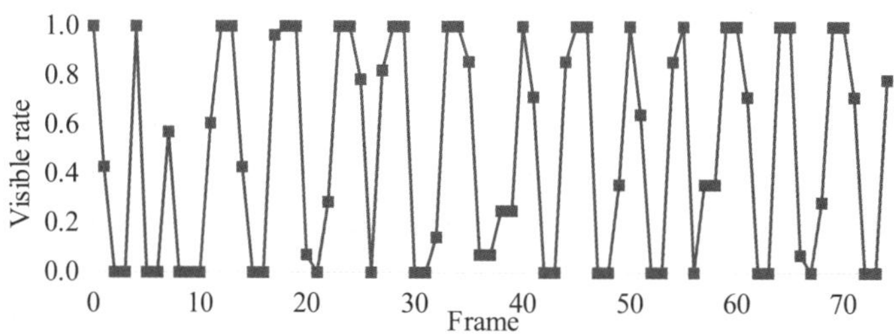

Fig. 9. Original data

4. Conclusion

In this paper, we proposed a detection technique to use only the brightness change in the image. This method is effective with an image photographed in dusky environment. Our technique evaluated by the experiment showed that performance of the obstructed area detection is good.

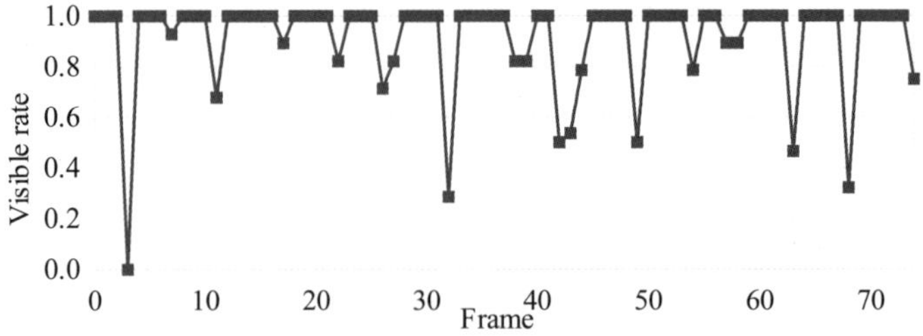

Fig. 10. Data by using previous method

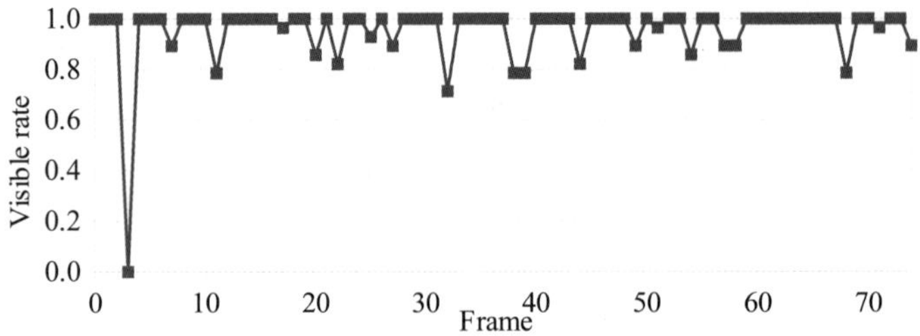

Fig. 11. Data by using proposed method

Acknowledgment

This work was supported by JSPS KAKENHI Grant Number 23560289.

References

1. T. Kinugasa, Y. Otani, T. Haji, K. Yoshida, K. Osuka and H. Amano: "A Proposal of Flexible Mono-tread Mobile Track", Proc. of IEEE International Conference on Intelligent Robots 08, 1642 /1647 (2008)
2. Kenichi TOKUDA, Tatsuya HIRAYAMA, Tetsuya KINUGASA, Takafumi HAJI and Hisanori AMANO"Remote control system for rescue robot FMT: Removal algorithm of obstacles in the images", Proc. of CLAWAR2011, 363 / 370 (2011)

ACTUATOR WITH ADJUSTABLE-RIGIDITY AND EMBEDDED SENSOR FOR AN ACTIVE ORTHOSIS KNEE JOINT

M. CESTARI[1], D. SANZ-MERODIO, J.C. AREVALO,
X. CARRILLO AND E. GARCIA
Centre for Automation and Robotics (CAR UPM-CSIC)
Madrid, Spain
[1]*m.cestari@csic.es*

In the growing field of powered wearable devices for gait assistance and gait rehabilitation, adjustable compliant actuators are being designed and implemented because of their ability to minimize large forces due to shocks, to safely interact with the user, and their ability to store and release energy in passive elastic elements. Many simulation researches have been performed evaluating the benefits of incorporating compliant joints in active orthosis and exoskeletons. In this paper we present an Actuator with Adjustable Rigidity and Embedded Sensor (ARES). This actuator integrates in its mechanism a force sensor that allows force control at the joint, taking advantage of the elastic elements that simultaneously regulate the joint stiffness. Mass and inertia of the actuator are minimized by the compact design and the utilization of the different components for more than one utility. A main advantage of ARES is that by tuning the stiffness, the precision of the force sensor is also adjusted; therefore no bandwidth limitation is expected.

Keywords: Embedded sensor, compliant actuator, torque control.

1. Introduction

Powered wearable devices such as active orthoses are portable robotic devices worn by an operator [1] that operate in parallel with human legs assisting the human motion. These powered devices have energy requirements for long periods that current batteries cannot cope with and this has become a major challenge in the development of these portable devices. Animals are capable of autonomously producing a wide range of stable movements in environments with unpredictable disturbances [2]. They use muscles as energy-efficient, a dynamic control of joint stiffness during the different phases of the gait, allowing for the exchange of kinetic and potential energy [3]. By modulation of joint stiffness a significant reduction of energy expenditure and transmitted forces can be achieved [4]. The superior power-to-weight ratio, force-to-weight ratio, compliance, and control of the muscle, when compared with traditional robotic actuators, are the main barriers for the development of machines that can match motion, safety, and energy efficiency of humans [5]. In the growing field of active orthoses and exoskeletons for gait assistance and gait rehabilitation, adjustable compliant actuators are being designed and implemented because of their ability to minimize large forces due to shocks, to safely interact with the user, and their ability to store and release energy in passive elastic elements [5]. Many simulation have been performed evaluating the benefits of incorporating compliant joints in active orthoses and exoskeletons. In [3] a study of the gait phases is performed and a proposal of joint control implementing compliance and elastic elements is evaluated resulting in an energy consumption reduction in the power required during locomotion. A study of two different configurations of compliant actuators is performed in [6] where the required power peaks at the joints are shown to be lower when compliant actuators are used. In this paper we present an Actuator with Adjustable Rigidity and Embedded Sensor (ARES), to be implemented in the knee joint of the ATLAS active orthosis [7]. This actuator takes advantage of the elastic elements in its mechanism for reducing energy requirements during gait and incorporates an embedded sensor system for force-control. Mass and inertia of the actuator are minimized by the compact design and the utilization of the different components for more than one utility. A main advantage of ARES is that, by tuning the stiffness, the precision of the force sensor is also adjusted; therefore no bandwidth limitation is expected as it is known to occur in compliant actuators as the traditional SEA [8] and in the devices where this kind of actuator is implemented as LOPES [9]. The actuator is focused on achieving a good force tracking, by implementing an appropriate force control, taking advantage of leg dynamics and achieving a good adaptation to the terrain. The paper is organized as follows. A description of the joint mechanics and working principle is given in Section 2. Tracking force simulations were performed with software of mechanism dynamics and presented in Section 3. The prototype preliminary test setup and experimental results are presented and discussed in Section 4 and conclusions and future work in Section 5.

2. Joint Mechanics

2.1. *Design*

A compliant joint prototype based on an actuator with adjustable-rigidity and embedded sensor (ARES) has been conceived as a knee joint of the ATLAS active orthosis. This first prototype of ARES is intended to be force-controlled. To achieve an effective impedance control that allows for a desired behavior of the exoskeleton knee joint, i.e. guarantying better adaptability to different terrains, and the implementation of energy saving control strategies, the actuator should be a high-precision force source [9]. The components of the actuator present an arrangement along the structure to reduce lateral volume.

The actuator can be divided into two main components: the stiff actuator set and the compliant mechanism, see Figure 1. A flat Maxon 90W brushless DC motor in combination with a Harmonic Drive unit with 1:100 reduction ratio connected to the joint conform the stiff set. A slotted bar couples the stiff mechanism with the compliant mechanism. The compliance of the joint is achieved by elastic elements in contact with the sliding pivot coupled with the slotted bar. An encoder is placed over the pivot, measuring the displacement associated with the joint deflection. Taking advantage of the compliance of the system a torque measure at the joint can be obtained.

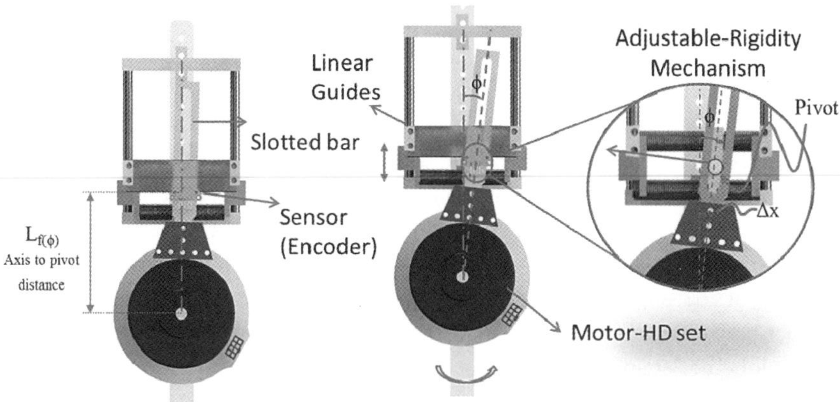

Figure 1. ARES main components

2.2. *Principle of operation*

ARES stiffness regulation system is based on the pivot displacement principle. The main difference with previous designs lies on the arrangement along the structure instead of a compact design that would result in a high lateral volume besides, the elastic elements that give the compliance to the joint are simultaneously used to obtain a force/torque measure at the joint. This force/torque measure is achieved by decoupling the stiff set of the joint and the compliant mechanism, see Figure 2. When the motor produce a joint torque or an external force acts at joint, the slotted bar transmit the reaction force to the cylindrical protuberance in the sliding pivot, and the elastic elements in contact with them.

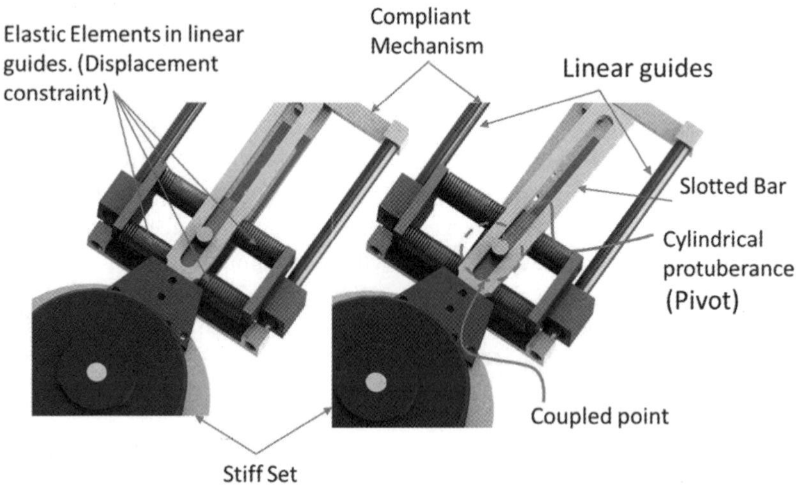

Figure 2. ARES coupled point

The cylindrical protuberance can be displaced along the slotted bar to modify the stiffness of the compliant mechanism, the adjustment of the pivot position change the distance to the joint axis increasing or decreasing the arm length. The displacement of the sliding pivot is constrained by some linear guides to avoid bending of the elastic elements. As the elastic elements do not bend and the contact between the slotted bar and the cylindrical protuberance is not blocked during joint deflection, the torque measure is directly related to the displacement of the elastic element by the relationship below.

$$\tau = \frac{2 \cdot \Delta X \cdot K_{equiv} \cdot L_{f(\phi)}}{\cos(\phi)} \quad (1)$$

Where:

ΔX, correspond to the elastic elements compression, sensed by the encoder (See Figure 1).

K_{equiv}, is the equivalent rigidity of the elastic elements, depending of the arrangement.

ϕ is the deflection angle between the actuator and the joint, and it is a function of the displacement of the sliding pivot.

The arm length, $L_{f(\phi)}$, would slightly vary with the increase of the deflection as a consequence of the mechanism to avoid bending of the elastic elements.

As the joint deflection increases, the angle ϕ would influence on the torque measure and the required force to adjust the joint rigidity would increase, because a larger component of the joint force has to be overcome to vary the arm length. The pivot displacement principle and the characteristics of the contact in the couple point where the cylindrical protuberance may slide through the bar slot; keep reduced the power necessary for the adjustment.

3. Simulation Evaluation

The actuator prototype was designed with a set of constrains that intended to achieve a good force measurement by taking advantage of the elastic elements needed to introduce compliance at the joint. To evaluate the proposed working principle a series of simulations were perform with the help of Mechanism-CAD Software. The dynamic behavior of the actuator was simulated when a commanded torque was applied at the joint and the compression of elastic elements was measured for further calculations as the data expected to be collected with the encoder in the sliding pivot. In Figure 3 and 4, the torque measure obtained by using Eq. (1) and the elastic elements compression data is presented and contrasted with the reference torque, the tracking torque in open loop was evaluated for different inputs and good response can be observed.

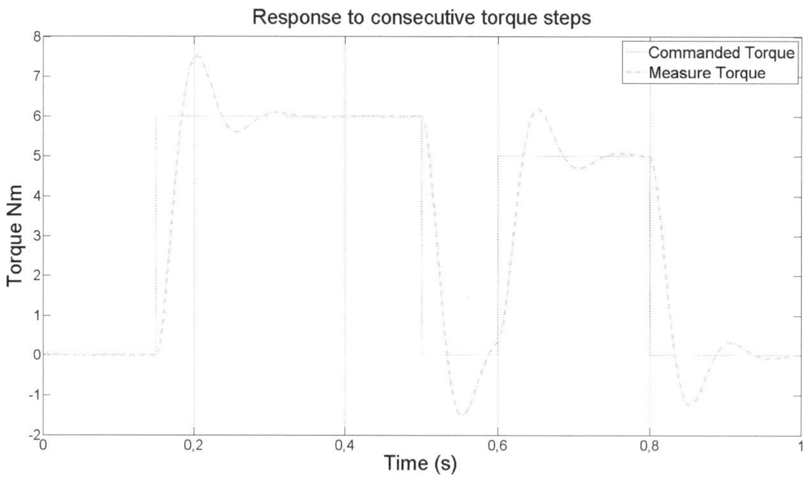

Figure 3 Torque Tracking in open loop at 2 consecutive steps.

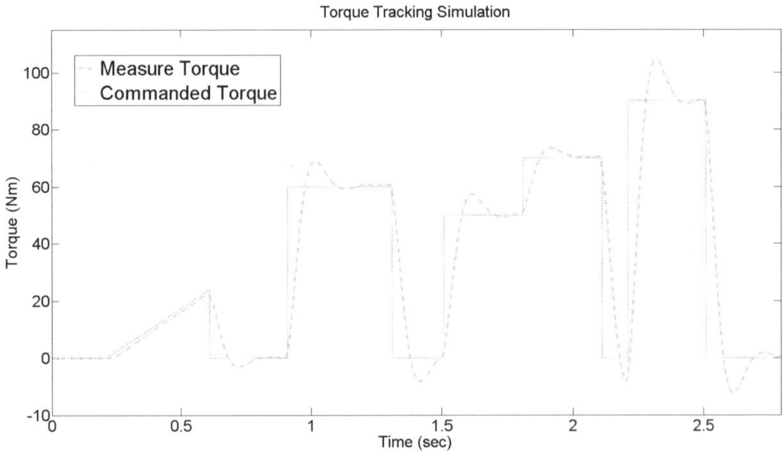

Figure 4 Torque Tracking in open loop at different and combine inputs

4. ARES Prototype Evaluation

The actuator prototype was evaluated in a knee joint, see Figure 5; different rigidity levels were manually adjusted to evaluate de behavior of the force sensor embedded in the design.

Figure 5. ARES Prototype

The torque reference was made with the help of an external spring, see Figure 6. An extension spring; attach to the shank at a known distance, L_{ext}, from the axis joint, allows us to determine the real torque at the joint and then contrasted with the torque measured by the actuator. In Figure 7 the calculated torque with the embedded sensor, as a function of the compression of the elastic elements when the compliant joint is deflecting is shown.

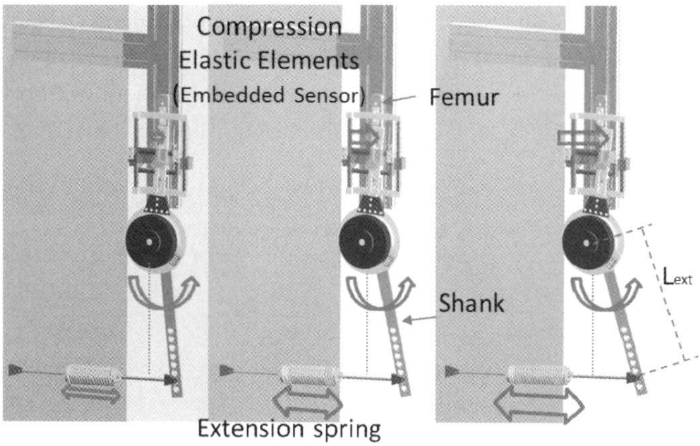

Figure 6. Experimental Setup-Torque reference measure

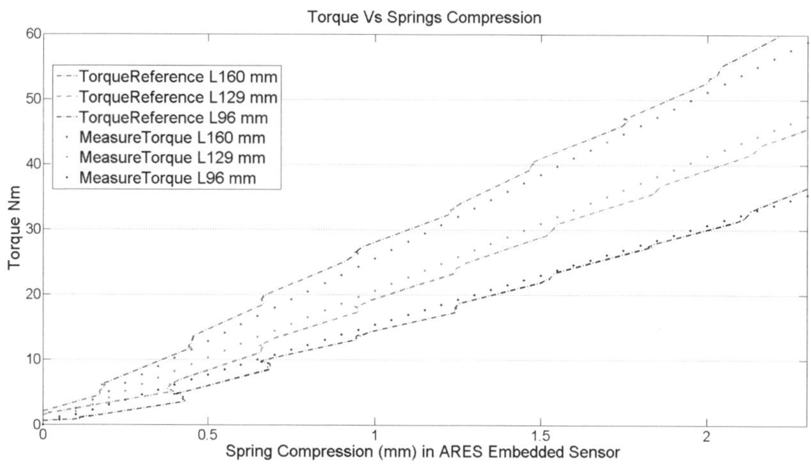

Figure 7. ARES Prototype-Torque tracking for different arm length

Different arm length were manually fixed during trials to achieve a variation in the ARES equivalent stiffness, the Figure 7 compares the torque measured with the embedded sensor as a function of the elastic elements compression for three different arm lengths. During our experiments for low torques, below 10Nm; a fuzzy area is observed, where the reference data is affected by the bandwidth of the extension spring used as reference. The striped data representing the torque reference gets more precise over the 10 Nm. The torques measured with the actuator present a good precision when the arm length and the joint torque are low as the linearity of the data for L96mm shows, and over the 10 Nm fits to the reference data. The data of the embedded sensor obtained during trials shows a good differentiation for different arm length and by adjusting the distance of the sliding pivot a variation of the torque measure sensibility can be achieved. Some of the ARES general specifications are presented in Table 1.

Table 1. ARESA General Specifications

Peak Torque	Up to 76 Nm
Max Deflection	$\pm 8^{\circ}$
Weight	900 gr
Length	220 mm
Width	50 mm
Power	90 W

5. Conclusions & Future Works

A compliant actuator designed to be force-controlled was presented. This actuator takes advantage of the different elements included in it to achieve different tasks, such as the compliant behavior due to the elastic elements and their utilization for achieving a good torque measure. The arrangement of the elements allow a reduced lateral size, this is an important element to take into consideration for the implementation in wearable powered devices like ATLAS.

The adjustable rigidity is tightly related to the torque measurement scale. By a proper adjustment the actuator can be controlled for different functions, as rehabilitation or walk at different speeds. The obtained results during trials suggest that adjusting the rigidity during operation would allow us to implement different control strategies for a better exploitation of the gait dynamics. Future works are oriented to improve the mechanism for acquiring more smooth measures, implementation of the actuator in the ATLAS active orthosis and the incorporation of an extra actuator in the joint for autonomous stiffness-adjustment during a gait cycle.

Acknowledgments

This work has been partially funded by the Spanish National Plan for Research, Development and Innovation through grant DPI2010-18702.

Mr. Cestari would like to thank the Spanish Ministry of Economy and Competitiveness for funding his PhD research.
Mr. Arevalo would like to thank the Spanish National Research Council for funding his PhD research.

References

[1]A. M. Dollar and H. Herr, "Lower extremity exoskeletons and active orthoses: Challenges and state-of-the-art," *IEEE Transactions on Robotics*, vol. 24, no. 1, pp. 144–158, FEB 2008.
[2]S. Migliore, E. Brown, and S. DeWeerth, "Biologically inspired joint stiffness control," in *IEEE International Conference on Robotics and Automation (ICRA), Vols 1-4*. IEEE, 2005, pp. 4508–4513.
[3]J. C. A. Daniel Sanz-Merodio, Manuel Cestari and E. Garcia, "Control motion approach of a lower limb orthosis to reduce energy consumption," *International Journal of Advanced Robotic Systems*, vol. 9, 2012.
[4]C. English and D. Russell, "Mechanics and stiffness limitations of a variable stiffness actuator for use in prosthetic limbs," *Mechanism and Machine Theory*, vol. 34, no. 1, pp. 7–25, JAN 1999.
[5]R. Ham, T. Sugar, B. Vanderborght, K. Hollander, and D. Lefeber, "Compliant actuator designs," *Robotics Automation Magazine, IEEE*, vol. 16, no. 3, pp. 81 –94, september 2009.
[6]S. Wang, W. van Dijk, and H. van der Kooij, "Spring uses in exoskeleton actuation design," in *Rehabilitation Robotics (ICORR), 2011 IEEE International Conference on*, 29 2011-july 1 2011, pp. 1 –6.
[7]E. Garcia, D. Sanz-Merodio, F. Sanchez, J. Arevalo, and P. G. de Santos, *Development of the ATLAS lower-limb active orthosis*. [Online]. Available: http://www.worldscientific.com/doi/abs/10.1142/9789814374286_0116
[8]G. Pratt and M. Williamson, "Series elastic actuators," in *IEEE/RSJ International Conference on Intelligent Robots and Systems*, Ikeuchi, K and Khosla, P, Ed. IEEE, Ind Electr Soc; IEEE, Robot & Automat Soc; Robot Soc Japan; Soc Instrument & Control Engineers; New Technol Fdn, 1995, pp. 399–406, 1995 IEEE/RSJ International Conference on Intelligent Robots and Systems - Human Robot Interaction and Cooperative Robots, PITTSBURGH, PA, AUG 05-09, 1995.
[9]H. Vallery, J. Veneman, E. van Asseldonk, R. Ekkelenkamp, M. Buss, and H. van der Kooij, "Compliant actuation of rehabilitation robots," *Robotics Automation Magazine, IEEE*, vol. 15, no. 3, pp. 60 –69, september 2008.

A PERSONAL ROBOT INTEGRATING A PHYSICALLY-BASED HUMAN MOTION TRACKING AND ANALYSIS

C. GRANATA and Ph. BIDAUD and R. ADY and J. SALINI

Institut des Systèmes Intelligent et de Robotique
Université Pierre et Marie Curie (CNRS - UMR7222), Paris
E-mails: granata, bidaud, ady, salini@isir.upmc.fr

The capture and analysis of human motor activity in daily life situations can provide a diagnosis tool or simply allow the evaluation of a person's psychological and physiological state (cognitive and emotional). This paper proposes a method based on the use of a physical model where embedded Kinect sensors on a mobile robot are employed for the detection and tracking of people. We describe a method for determining the individual's anatomical parameters and the algorithms developed for the digital animation model based on the measurement of a number of characteristic points obtained by the Kinect. The implementation of this method is validated experimentally from locomotion exercises where the positions of the Center of Pressure, measured and calculated using a force platform, are compared.

Keywords: Human activity analysis; motion capture; social personal robots; dynamic human body modeling.

1. Introduction

Monitoring elderlies' activity in daily life tasks or during cognitive/physical exercises, can be an efficient way for a prompt diagnosis in order to detect motion anomalies and prevent falls. Recording and analyzing the person's activity while walking and standing can provide temporal-spatial parameters useful to identify gait deviations, make diagnoses, determine appropriate therapy and assess patient progress.[1] In the last few years many technical solutions have been developed in monitoring systems for human activity analysis. In healthcare applications, the most commonly used methods are wearable device-based. The wearable sensors are worn on a person's body (i.e. heart rate sensors and wrist watches[2]) or mounted inside clothes to detect displacements and to recognize certain activities.[3] Other studies focused on the extraction of different parameters (e.g. the Center of Mass

(CoM) and the Center of Pressure (CoP)) related to posture and locomotion analysis. In these cases, trademills, instrumented surfaces such as force platforms or particular shoes were exploited.[4,5] Unfortunately, the use of this kind of technology requires the data to be gathered in controlled laboratory conditions and is not suitable at all for daily monitoring in real life environments. Other approaches were based on the capture and estimation of 3D human motion by using single or multi cameras. When only a single view is used,[6] self-occlusions and depth ambiguities can occur. These limitations can be alleviated by using multiple cameras,[7,8] but in these cases, an important calibration process is required. The most accurate methods for 3D motion capture exploit marked optical technology, such as Phasespace,[9] MoCap from Vicon[10] and CodaMotion,[11] to turn multiple camera observations of a moving subject into 3D pose information of the body. These active motion analysis systems are very accurate, but they are very expensive and have a long set up time (they require the installation of several cameras and a long marker placement process). Thus again, these approaches are not suitable for daily tasks monitoring. The attempt to create a non-invasive, portable and inexpensive system to analyze the human activity was made by Gonzáles[12] who proposed the use of 2 portable sensors, a Wii balance board (as a force platform substitute) and a Kinect (for 3D pose reconstruction) to compute the CoM, but the Wii board is restricted to the analysis of only on spot movements. In our knowledge, there is not system that can be used for an in-depth analysis of human activity (especially of posture balance and stability) in non controlled environmental conditions, without any environmental equipment and in a non-invasive fashion (marked-free method). In order to meet this need, we have developed a system exploiting a Kinect embedded in a mobile personal robot. Indeed, personal robots can provide a support able to perceive and track humans while moving and to collect information about their activity. This paper proposes a method based on the use of a physical model where embedded Kinect sensors on a mobile robot are employed for the detection and tracking of people. The implementation of this method is validated experimentally on different exercises during which the positions of the Center of Pressure, measured and calculated using a force platform, are compared.

2. Interactive human following

The robot used for implementation is a commercial robot, Kompaï (Fig. 2 right) to which we have added a Microsoft Kinect and an extra laptop for sharing computation in order to increase its potential. A multimodal human detector and a system for interactive human following, developed in previous works,[13,14] are employed to make the robot move while taking into account a person's displacement. The use of a mobile platform that tracks the human while analyzing his activity at the same time, allows the overcome all limitations due to wearable sensors, instrumented environment and occlusions. Indeed the human following system integrates a human state predictive model that solves occlusions and lack of detection problems. During person following, the Kinect embedded in the robot is able to detect the skeleton (by employing the Microsoft Kinect SDK). The collected data is then processed as described in Section 3.

3. Motion capture and filtering method

If the data provided by the employed skeleton detector can be used to assess general trends in movement, it will not be accurate enough for a reliable quantitative estimation. One of the main drawbacks to this detector is in a non-anthropometric kinematic model with variable limb lengths (the lengths of the detected skeleton limbs don't remain constant but change at each frame). Indeed, the variability of the algorithm in pose estimation is about 0.1 m.[15] Thus, since serious errors exist in the estimated skeleton, an improved skeletonization solution to a space-time constraint problem is required. In previous works, the pose correction was approached in different ways: according to physical-model,[16] by using Kalman-like filters,[17] by adopting different kinds of regressors (random forest regressor,[18] nearest neighborhood regressor,[19] Gaussian process regression[20] etc.). We propose a method for pose correction that takes into account both the coherence of the anthropomorphic model of the skeleton and the spatial-temporal motion consistency. This method is composed of 3 steps (Fig. 1):

Step 1: Initialization. When the person is in the Kinect field of view, his skeleton is detected and the 3D positions of all his joints are provided (Fig. 2 left). During the initialization phase, we use a first set of Kinect measurements to define a reference model of the involved body. In particular, the system computes the anthropometric kinematic model of the body according to the mean distances between the consecutive joints.

Step 2: Physical model calibration. This step is required to keep the human physical skeleton consistent with the reference model (in anthropomorphic point of view). Our approach relies on a standard mathematical problem, constrained optimization. It consists of minimizing the scalar value quadratic objective function $f(X)$ while satisfying a certain number of constraints $g(X)$ (Equation 1).

$$\begin{aligned} minimize\ f(X) \\ subject\ to\ g(X) = 0 \end{aligned} \tag{1}$$

In the objective function $f(X)$ (Equation 2), X is a vector residing in $(3 * nJ)$-dimensional space (3D coordinates for nJ joints), a_i, b_i, c_i are the weights associated with the coordinates of the joint i (in our case we consider equal weights on the 3 dimensions).

$$f(X) = \sum_{i=1}^{nJ} a_i(\bar{x}_i - x_i)^2 + b_i(\bar{y}_i - y_i)^2 + c_i(\bar{z}_i - z_i)^2 \tag{2}$$

$(\bar{x}_i, \bar{y}_i, \bar{z}_i)$ and (x_i, y_i, z_i) are respectively the coordinates of the joint i provided by the skeleton detector algorithm and the coordinates that we are searching for in order to minimize the objective function while respecting the constraints in Equation 3.

$$\begin{aligned} g_j(X) = d^2(J_j; J_{j+1}) - d^2(M_j; M_{j+1}) = \\ = (x_j - x_{j+1})^2 + (y_j - y_{j+1})^2 + (z_j - z_{j+1})^2 - d^2_{j,j+1} = 0 \end{aligned} \tag{3}$$

with $j = 1, ...nJ - 1$, $(J_j; J_{j+1})$ consecutive detected joints, $(M_j; M_{j+1})$ consecutive joints in the model, and $d_{j,j+1}$ the distance between M_j and M_{j+1}.

All the constraints are equality constraints and are fixed according to the reference model (Initialization step).
The optimization of the objective function is made by using the quadratic interior point method (QIPM), which is based on the improvement of initial conditions (measurements) for solving quadratic programming problems.

Step 3: Model Simulation and Parameters Analysis. Finally, the optimized data is used to animate a physical model on a dynamic simulator, Arboris-python. Arboris-python is a rigid multibody dynamics and contacts simulator written in the programming language Python. It includes a generic and easily extensible set of joints (singularity-free multi-DoF joints, non-honolomic joints, etc.) for modeling arborescent rigid body mechanisms with a minimal set of state variables. It gives access to modeling, controllers and constraints. The equations of motion of these systems are obtained from the Boltzmann-Hamel equations[21] integrated in Arboris-python software,[22] which computes the first-order approximation of the model. The integration is done using a time-stepping method and a semi-implicit Euler integration scheme. In this way, it is possible to solve the additional constraints, i.e. the kinematic loops, which can be unilateral like contacts or bilateral like a revolute joint, with a Gauss-Seidel algorithm.[23] The use of the dynamic simulator allows to respect the spatial-temporal consistency of the sensed motion. This means that the simulated mannequin naturally filters the skeleton detection aberrations by limiting the articular velocities of the joints in a range of physically possible values. Moreover, the dynamic model integrated in the mannequin can take into account the real mass of all the body parts while reproducing the observed activity. The mass distribution is computed according to the reference model built during the Initialization step and takes into account the person's weight. The use of the reference model specifically built upon the observed body is quite new with respect to previous works. In fact, the estimation of a body's parameters was widely based on body segment data computed according to antropometric tables, such as those proposed by Leva.[24] But using these kind of methods may cause important errors in the estimation of body posture due to significant mismatches between the physical body and the antropometric data.[25] By considering a consistent mass distribution, the dynamic simulation provides some parameters such as the CoM and CoP trajectories that represent important indexes for posture balance and stability.[5,12,26] An additional advantage of using a virtual mannequin is the possibility to having a visual feedback of the person's activity, which is very helpful for caregivers and therapists.

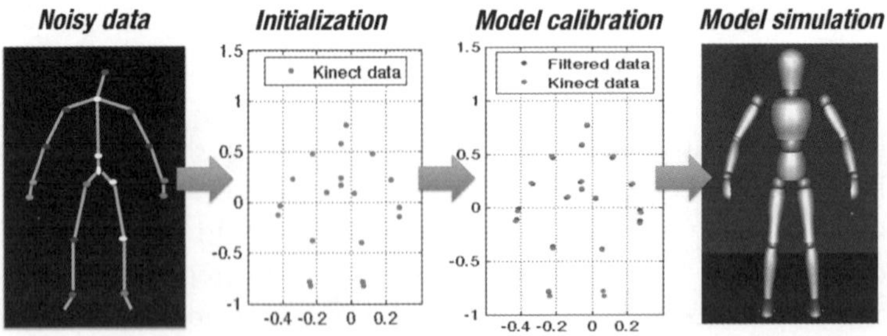

Fig. 1. The steps followed for pose correction.

4. Experiments

4.1. *Procedure*

In order to assess the reliability of the system, a preliminary set of experiments were made in a laboratory setting (Fig. 2 right). Five healthy subjects were asked to execute 3 different movements (arm movement, rocking movement and side steps) on a force platform while wearing 13 CodaMotion markers, to validate the consistency of our system (see Fig. 2 center for marker placement). The Kinect embedded in the robot was simultaneously used for skeleton detection. The goal of this set of experiments was to assess the CoP trajectory provided by Arboris by comparing it with the CoP trajectory measured with a force platform (ground truth). The CoP was chosen because it's the only parameter that can be directly measured by a force platform. Both CodaMotion and Kinect data are collected and replayed by dynamic simulation and the resulted CoP trajectory was recorded and compared with the force platform measurements. All data was filtered using a second order lowpass Butterworth filter with cutoff frequency of $10Hz$. For Kinect data, the pose correction method described in Section 3 was applied.

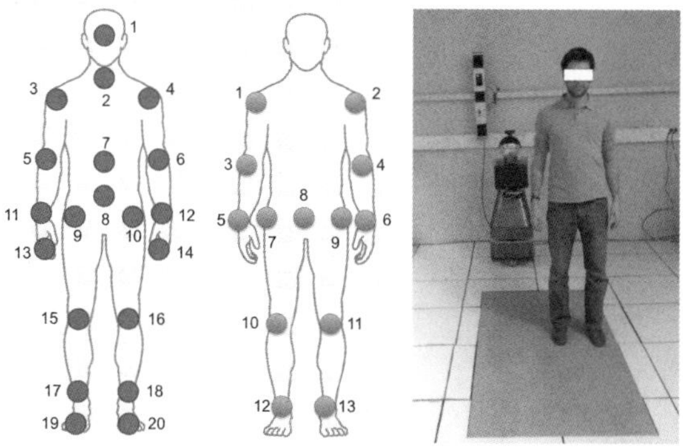

Fig. 2. The anatomical joints detected by the Microsoft Kinect SDK algorithm (left), the CodaMotion markers placement on the subjects' bodies (center) and the Kompaï robot during the experiments (right).

4.2. *Results*

To evaluate the consistency of our system with respect to the ground truth, the CoP trajectories computed by Arboris replaying both the Kinect and the CodaMotion data, and the CoP trajectories measured by the force platform were compared. An example of the 3 trajectories corresponding to the rocking movement of a tested subject is reported in Fig. 3. The mean errors between the 3 trajectories were computed for each singularly assessed movement (see Fig. 4 and 5). The differences in the trajectories were provided by Arboris by fitting on the mannequin first, the Kinect and then the CodaMotion data (max mean error $= 0.07m$), were also due to the approximative correspondence between the skeleton joints and the CodaMotion markers (for imprecise marker placement). The max mean error between the trajectories provided by Arboris by using only the Kinect data (with the correction pose algorithm) and the ground truth was about $0.08m$ with a standard deviation of $\pm 0.03m$. These results support the idea that it is possible to estimate some important parameters related to human activity by using our system with satisfactory accuracy. Indeed, not only the overall motion was well preserved during the simulation, but also the posture assessment showed good results.

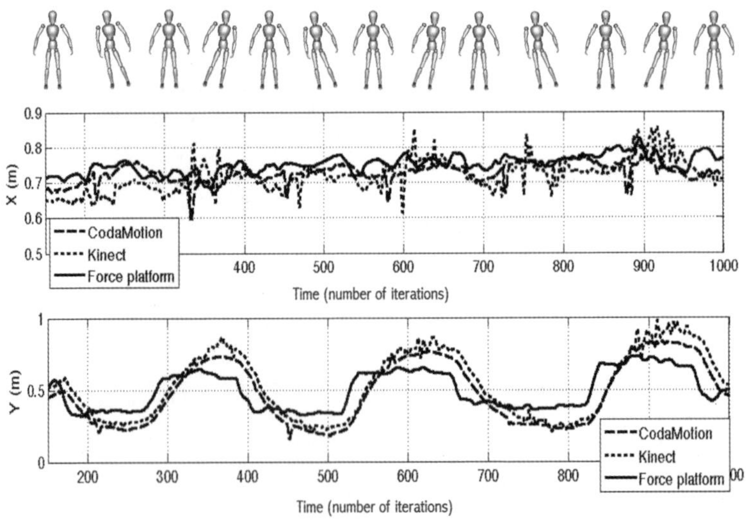

Fig. 3. CoP trajectory results of rocking movement for a subject. The trajectories issued from Kinect and CodaMotion data is calculated by Arboris.

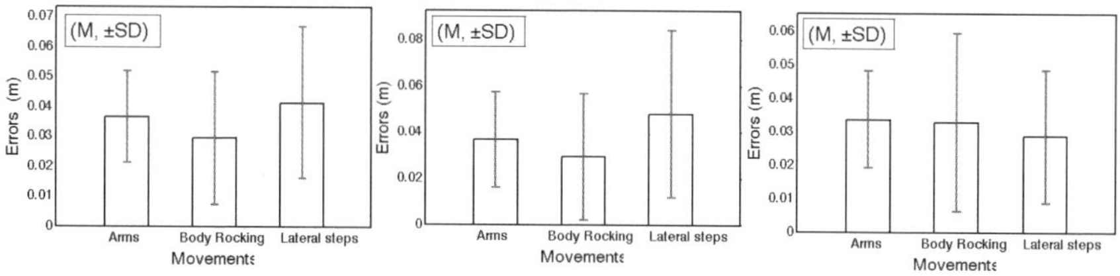

Fig. 4. For all tested movements: CoP error on X between the trajectories provided by Arboris replaying Kinect data and CodaMotion (left); between the force platform measurements and the Arboris results with Kinect data as input (center); between the force platform measurements and the Arboris results with CodaMotion data as input (right).

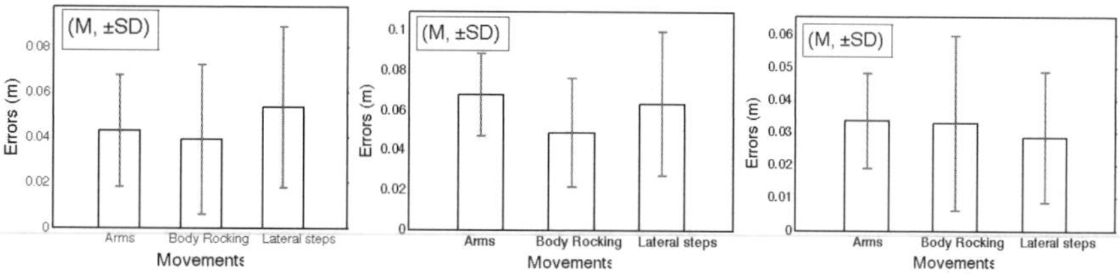

Fig. 5. For all tested movements: CoP error on Y between the trajectories provided by Arboris replaying Kinect data and CodaMotion (left); between the force platform measurements and the Arboris results with Kinect data as input (center); between the force platform measurements and the Arboris results with CodaMotion data as input (right).

References

1. C. K. Balasubramanian, M. G. Bowden, R. R. Neptune, S. A. Kautz and other, *Archives of physical medicine and rehabilitation* **88**, 43 (2007).
2. A. Sarela, I. Korhonen, J. Lotjonen, M. Sola and M. Myllymaki, Ist vivago®-an intelligent social and remote wellness monitoring system for the elderly, in *Information Technology Applications in Biomedicine, 2003. 4th International IEEE EMBS Special Topic Conference on*, 2003.
3. S. Amini and P. Narasimhan, *CyLab* , p. 26 (2009).
4. Y. Brenière *et al.*, *Journal of motor behavior* **28**, p. 291 (1996).
5. H. M. Schepers, E. van Asseldonk, J. H. Buurke and P. H. Veltink, *Biomedical Engineering, IEEE Transactions on* **56**, 1189 (2009).
6. C. Sminchisescu and B. Triggs, *The International Journal of Robotics Research* **22**, 371 (2003).
7. L. Kakadiaris and D. Metaxas, *Pattern Analysis and Machine Intelligence, IEEE Transactions on* **22**, 1453 (2000).
8. L. Ren, G. Shakhnarovich, J. K. Hodgins, H. Pfister and P. Viola, *ACM Transactions on Graphics (ToG)* **24**, 1303 (2005).

9. A. Aristidou, J. Cameron and J. Lasenby, Real-time estimation of missing markers in human motion capture, in *Bioinformatics and Biomedical Engineering, 2008. ICBBE 2008. The 2nd International Conference on*, 2008.
10. S. Nakao, K. Komatsu, W. Sakai, M. Kashihara, Y. Masuda, K. Nishikawa, T. Okahisa, S. Kondo, T. Osawa, R. Kaji *et al.*, *The journal of medical investigation: JMI* **58**, 264 (2011).
11. J. Hidler, W. Wisman and N. Neckel, *Clinical Biomechanics* **23**, 1251 (2008).
12. A. González, M. Hayashibe and P. Fraisse, Estimation of the center of mass with kinect and wii balance board, in *In Proc IEEE/RSJ Int Conf on Intelligent Robots and Systems (IROS)*, 2012.
13. C. Granata and P. Bidaud, A framework for the design of person following behaviors for social mobile robots, in *Intelligent Robots and Systems (IROS), 2012 IEEE/RSJ International Conference on*, 2012.
14. C. Granata, P. Bidaud, M. Chetouani and N. Melchior, Multimodal human detection and fuzzy decisional engine for interactive behaviors of a mobile robot, in *Cognitive Infocommunications (CogInfoCom), 2012 IEEE 3rd International Conference on*, 2012.
15. Š. Obdrzálek, G. Kurillo, F. Ofli, R. Bajcsy, E. Seto, H. Jimison and M. Pavel.
16. S. Tak and H. Ko, Physically based motion retargeting filter(December 23 2005), US Patent App. 11/317,844.
17. J. Lee and S. Shin, Motion fairing, in *Computer Animation'96. Proceedings*, 1996.
18. W. Shen, K. Deng, X. Bai, T. Leyvand, B. Guo and Z. Tu, Exemplar-based human action pose correction and tagging, in *Computer Vision and Pattern Recognition (CVPR), 2012 IEEE Conference on*, 2012.
19. R. Duda, P. Hart and D. Stork, Pattern classification and scene analysis 2nd ed.1995.
20. B. Schölkopf, A. Smola, R. Williamson and P. Bartlett, *Neural computation* **12**, 1207 (2000).
21. V. Duindam and S. Stramigioli, Lagrangian dynamics of open multibody systems with generalized holonomic and nonholonomic joints, in *IEEE/RSJ International Conference on Intelligent Robots and Systems*, nov. 2007.
22. S. Barthélemy, J. Salini and A. Micaelli, Arboris-python `https://github.com/salini/arboris-pyhton`.
23. T. Liu and M. Wang, *IEEE Transactions on Automation Science and Engineering* (nov. 2005).
24. P. De Leva, *Journal of biomechanics* **29**, 1223 (1996).
25. M. Jaffrey, Estimating centre of mass trajectory and subject-specific body segment parameters using optimization approaches, PhD thesis, Victoria University, 2008.
26. D. Lenzi, A. Cappello and L. Chiari, *Journal of biomechanics* **36**, 1335 (2003).

HUMANOID ROBOT PROGRAMMING THROUGH FACE EXPRESSIONS

ALVARO URIBE-QUEVEDO, HERNANDO LEON
Industrial Engineering, Military Nueva Granada University, Cra 11#101-80 Bogota, Colombia, Zip: 110111.

BYRON PEREZ-GUTIERREZ
Mechatronics Engineering, Military Nueva Granada University, Cra 11#101-80 Bogota, Colombia, Zip: 110111.

Accessibility plays a fundamental role for interacting with several current technologic devices, advances in this area are focused in features for natural interfaces that benefits regular and handicapped users. Robot motion and programming requires configuring a set of actions to be executed by each joint, in humanoid didactic kits these tasks are achieved from preloaded commands, or sliding bars for positioning each joint. Current user interfaces based on image processing allows complementing robot motion and programming through gestures without requiring mouse or keyboard thus, making more appealing the experience for students by offering a more natural form of interaction through face or hand gestures. This project implemented a low-cost framework for programming a humanoid robot with 14 degrees of freedom through facial and hand gestures as means to facilitate accessibility to robot motion and programming without the keyboard or mouse. Results proved that moving and programming sequences with gestures is attractive no new users, however, actions based on eyebrow movement such as click or double click, along error in the expression detection affected user interactions, however, interactions through hand motion was more comfortable and was only affected by the sensors precision.

1. Introduction

Didactic humanoid robotics is characterized for its simply motion and programming which has been a success for attracting non-specialists to robotics [1]. Many available kits on the market offer various types of robots whose features vary from Degrees of Freedom (DOF), modularity and ease of programming [2] [3] [4]. Interaction with these robots is commonly done through keyboards, mouse or remote controls with preprogrammed functions for moving each joint, however, newer form of user interfaces are already being used as alternatives to traditional means of input, image processing, neuroheadsets or even electromechanical sensors are ways to overcome some problems associated with occupational health and accessibility [5].

Current technological advancements in image processing and computing power have yielded to noninvasive gesture interactions using depth of field cameras (Primesense and Kinect) for tracking users [6] [7]. Ease of access has expanded through computers, tablets and gaming benefiting users with some sort of disability [8], in rehabilitation [9], in computer interaction [5], in navigation [10] and in educational aids [11]. Particularly in robot programming, the keyboard and mouse are important elements; input data often requires alphanumerical entries along with navigation through the user graphic interface (GUI).

In robotics several systems have been using the Kinect, Yanik et al. used a growing neural gas algorithm for improving robot response from recognized gestures [12]. Intuitive robot operation has also become a field of interest, Marinho et al. proposed a control system based on the dual quaternion framework, for operating a robotic arm performing pick and place tasks based on hand tracking [13]. This trend in using Kinect in robotics shows how the gap for a natural interaction o even human motion mapping to several types of robots are currently taking place [14] [15].

This work presents the implementation of a face and hand gesture-based tool for programming a humanoid robot as an alternative to keyboard, keypad or even body gestures. The goal of this project is to offer a programming alternative to traditional inputs for students learning robotics. The paper is organized in sections as follows, in Section 2 a brief review of face expression characteristics and how the Kinect tracks them; in Section 3 the humanoid robot characteristics is presented; in Section 4 the system architecture, development and integration is covered; in Section 5 the results and adjustments are presented; and finally in Section 6 the conclusions are discussed along future works.

2. Gesture tracking

Gesture tracking allows natural forms of interactions as it takes advantages of our ergonomics, face and hand gestures can be programmed as alternative inputs. The human face can express numerous emotions that are possible thanks to the muscles, ligaments and tendons allowing changes in the forehead, eyebrows, nose, eyelids, libs and chin [16]. Facial expression can be involuntary or controlled across different scenarios, for our implementation, the focus is centered on voluntary muscle motion given that, a particular set of muscle motion will determine what command is executed.

The Kinect sensor allows tracking faces and skeletons through several APIs [17] [18]. The tracking is done in real time; the sensor algorithms calculate the head and body position and facial expressions. The origin of the tracking coordinate system is centered in the sensor; in the case of face tracking 83 facial points are detected and mapped them over a grid as presented in Figure 1; in the case of skeletal tracking only arms are of interest, the sensor tracks the shoulders, elbows and neck as presented in Figure 1.

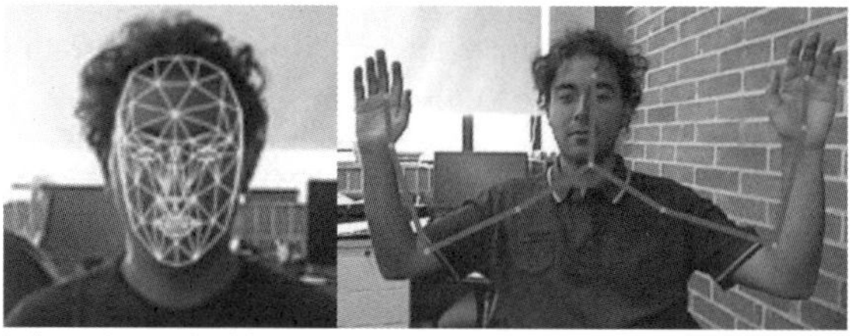

Figure 1: 3D mesh from the tracked face expressionless and seated skeleton tracking.

3. Humanoid robot

The robot has 14 servo motors for 2 DOF in the arms and 2 DOF in the legs, thus mimicking the human form. Figure2 presents the configuration of the humanoid which allows it for performing tasks such as raising the arms, walk, sit and stand up. The structure was constructed using aluminium for reducing the weight and maintain rigidity, due to the fact that the motions to be executed are slow, dynamic effects are not considered and the structure's goal is to support servos and allow the robot to move as previously stated. The distribution of elements is symmetric so the weight is uniformly resulting in a gait suitable device that does not require advanced control for walking.

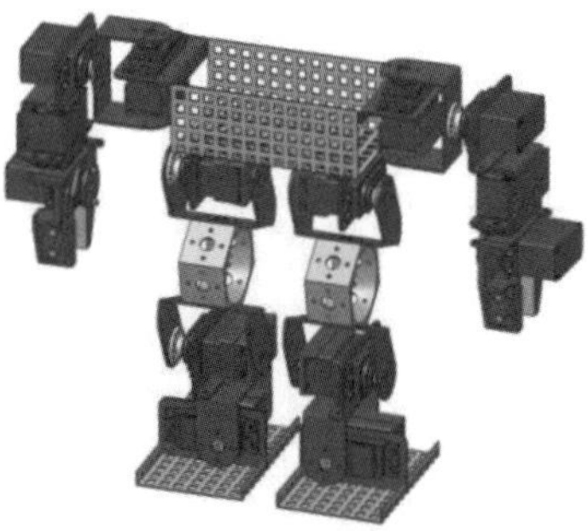

Figure 2: Designed humanoid robot

The main control system is connected to the controllers of the servos through a RS485 bus. The programming is achieved through its proprietary software communicating to the computer through the RS232 serial port. The motion sequences and code is generated from moving each servo to the desired position thus, creating a list sequential order for executing. The data is managed and configured using an Atmega128 microcontroller with a boot-loader which allows user to change the code and access directly to the servo controller parameters. The microcontroller also allows retrieving the output signals from the Humanoid Robot servos; these provide information regarding the actual servos angular position, angular velocity, DC current and voltage. The robot is powered by high-current Lithium-polymer rechargeable batteries, which are located in its pelvis, the cells last for about 30 minutes of operation offering and acceptable autonomy for a learning environment.

The robot can run also autonomously, this type of operation deals with all kinds of movements, storage functions, online posture adjustment and feedbacks. The robot has a remote control which manages several numbers of programs sequentially. In addition, to the required aesthetics of the robot, the method of actuation of joints also plays a large part in determining mechanical design.

4. System development and integration

The proposed system architecture consists of the user, the Kinect, a computer with the servo programming tools and the humanoid robot. Mouse interactions are mapped to *X Y* motions related to pitch and yaw head rotations and hand motions, click actions are configured to be accomplished by pulling up the eyebrows or pushing closer the hand to the camera. These events were selected as sufficient after analyzing the graphic user interface for programming the humanoid robot which offers the selection of configurable commands through buttons, sliders for configuring the position, direction and servo identification, and check buttons for gain, performance and time response aspects. Even though the Kinect has an operational field of view covering 2 m from the sensor, after tracking several persons, 0.5 m was chosen to be an appropriate distance as closer positions to the camera or incorrect posture generate tracking error information given the sensor limitations as pitch optimal tracking is less than 10°, roll less than 45° and yaw less than 30° [19], when using the seated tracking mode the detection presented problems when the user distance was below 0.5 m.

Tracking information was analyzed for defining which face feature would allow mapping click actions without causing discomfort or fatigue. The SDK allows through its shape units (SU) and animation units (UI) to identify expressions based on the *Candide 3* model [20], however, minor movements of the eyebrows, lips and mouth can be falsely detected just by inclining the head over any axis, which would lead to an unsatisfactory interaction. After detecting these non-present expressions, the next step was to determine which of the facial features would provide comfortable motion for mapping the click action. Within a social environment, talking is very common, so considering lip and mouth movements would be counterproductive, even during work, several involuntary mouth motions is present, smiling and yawing are some examples. By disregarding the lips and the mouth, the only significant feature left are the eyebrows, which only by extreme surprise are fully extended upward on the forehead. Eyebrow tracking is significantly better recognized and was chosen for act as a mouse button with a configured range of action proportional to tracked face; it is detected by the AU4 equal to -1 which recognizes the eyebrows completely rose. The system configuration is presented in Figure 3 where the head and hand motion is mapped for horizontal and vertical mouse scrolling.

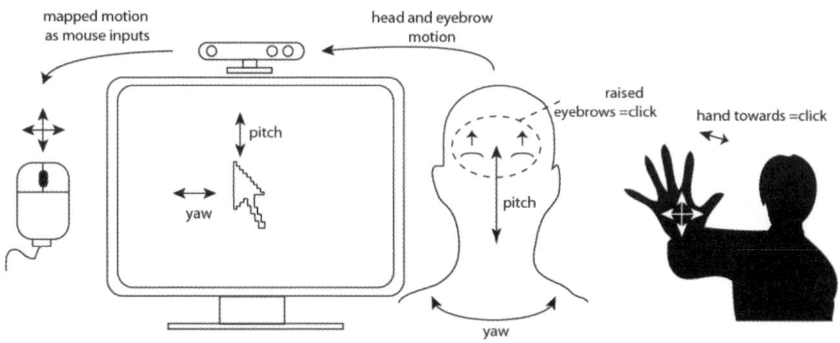

Figure 3: Motion mapping interaction diagram

5. Results

The implemented gesture interaction was tested for validating its suitability, comfort and potential as an alternative for moving and programming a humanoid robot through its interface, the set up goal is to move and program arm motion for rising them up, at this stage, neither efficacy nor efficiency were analyzed. Mouse scrolling across the windows was successfully achieved; however, when tracking the face a scaling factor was required for reaching all screen corners without having to perform full head rotations that were not recognized by the sensors, this limitation was not encountered when using the hand. Given the simplicity of the GUI for programming the robot, the left mouse button action mapped to eyebrow movement was sufficient for the selecting ratio buttons, checkboxes and slider controls. Mouse cursor control based on face tracking resulted satisfactory; users moved the cursor around the screen comfortably, although continuous clicking resulted in eyebrow fatigue given the number of DOF of the humanoid robot; however this difficulty was overcome when using the hand tracking. Tracked positions, mouse control over robot GUI programming, and robot upper member motion are presented in Figure 4 and Figure 5, were each servo's motion is configured and the code generated.

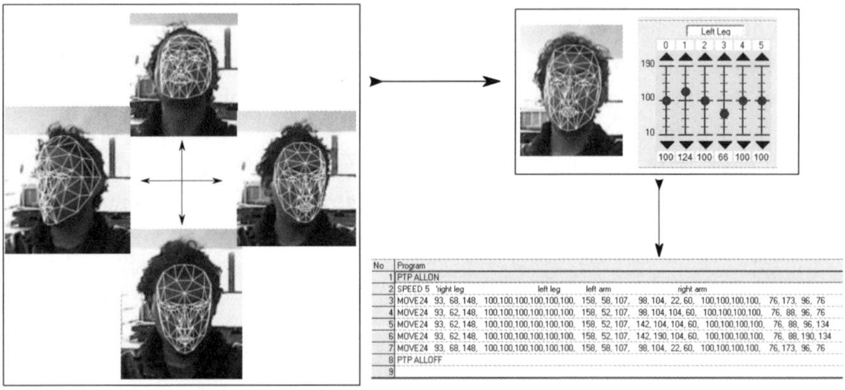

Figure 4: Programming sequences selected using head motion.

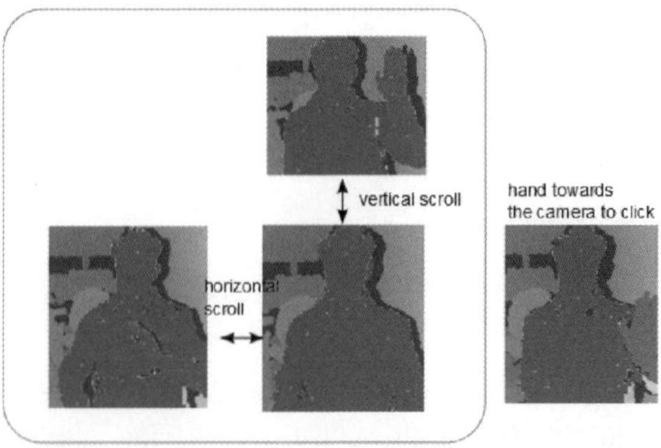

Figure 5: Programming sequences selected using hand motion

The resulted motion sequence in both scenarios was successful and its representation is presented in Figure 6.

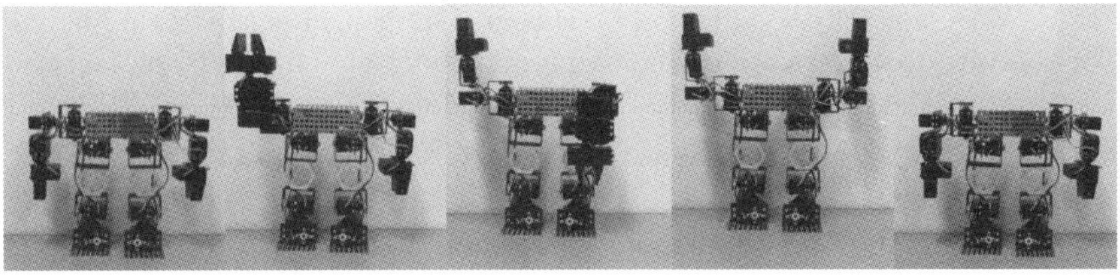

Figure 6: head motion and facial expression based robot programing

6. Conclusion

Alternative user interfaces are offering new ways of interaction that can be exploited in different scenarios other than entertainment, the availability of SDKs offer the opportunity for developing and applying these user interfaces accordingly to cases that can benefit from natural interactions. The implementation presented in this work allowed programming a humanoid robot without needing traditional input devices such as the keyboard and the mouse. In the same manner that repetitive task should be avoided when using the mouse and keyboard, head tracking and clicking using pitch and yaw along eyebrows pull up, can cause discomfort for long periods of using. Even though, for simply programming the humanoid robot none of the users expressed discomfort while moving the servos, however concerns were expressed because using the head as a pointer was a new experience. These inconvenient were addressed when using the hand tracking feature implemented, which was found more natural and no concerns were manifested. Future works will be focused on alternative interaction and voice commands for improving user interaction.

Acknowledgments

The authors thank the Military Nueva Granada University for their support under project ING1188, and undergraduate student Cesar Guerrero from the Multimedia Engineering program.

References

[1] P. Fiorini, "LEGO kits in the lab [robotics education]," *Robotics Automation Magazine, IEEE,* vol. 12, pp. 5-, Dec..

[2] C. N. Thai and M. Paulishen, "Using Robotis Bioloid systems for instructional Robotics," March.

[3] Kyosho, *Manoi,* 2013.

[4] LynxMotion, *LynxMotion Bipeds,* 13.

[5] S. Hakiel, "Delivering ease of use [software development]," *Computing Control Engineering Journal,* vol. 8, no. 2, pp. 81-87, april 1997.

[6] W.-N. Lie, H.-W. Shiu and C. Huang, "3D human pose tracking based on depth camera and dynamic programming optimization," in *Circuits and Systems (ISCAS), 2012 IEEE International Symposium on*, 2012.

[7] Z. Zhang, "Microsoft Kinect Sensor and Its Effect," *MultiMedia, IEEE,* vol. 19, no. 2, pp. 4-10, feb. 2012.

[8] J. Abascal, "Human-computer interaction in assistive technology: from "Patchwork" to "Universal Design"," in *Systems, Man and Cybernetics, 2002 IEEE International Conference on*, 2002.

[9] A. da Gama, T. Chaves, L. Figueiredo and V. Teichrieb, "Poster: Improving motor rehabilitation process through a natural interaction based system using Kinect sensor," in *3D User Interfaces (3DUI), 2012 IEEE Symposium on*, 2012.

[10] M. Tonnis, V. Broy and G. Klinker, "A Survey of Challenges Related to the Design of 3D User Interfaces for Car Drivers," in *3D User Interfaces, 2006. 3DUI 2006. IEEE Symposium on*, 2006.

[11] A. Sherstyuk, D. Vincent, J. J. H. Lui, K. Connolly, K. L. Wang, S. Saiki and T. Cauclell, "Design and Development of a Pose-Based Command Language for Triage Training in Virtual Reality," in *3D User Interfaces, 2007. 3DUI '07. IEEE Symposium on*, 2007.

[12] P. Yanik, J. Manganelli, J. Merino, A. Threatt, J. Brooks, K. Green and I. Walker, "Use of kinect depth data and Growing Neural Gas for gesture based robot control," in *Pervasive Computing Technologies for Healthcare (PervasiveHealth), 2012 6th International Conference on*, 2012.

[13] M. Marinho, A. Geraldes, A. Bo and G. Borges, "Manipulator Control Based on the Dual Quaternion Framework for Intuitive Teleoperation Using Kinect," in *Robotics Symposium and Latin American Robotics Symposium (SBR-LARS), 2012 Brazilian*, 2012.

[14] J. Ekelmann and B. Butka, "Kinect controlled electro-mechanical skeleton," in *Southeastcon, 2012 Proceedings of IEEE*, 2012.

[15] R. El-laithy, J. Huang and M. Yeh, "Study on the use of Microsoft Kinect for robotics applications," in *Position Location and Navigation Symposium (PLANS), 2012 IEEE/ION*, 2012.

[16] W. F. Paul Ekman, Unmasking the face, Malor Books, 2003.

[17] OpenNi, *Open the standard framwork for 3D sensing,* 2013.

[18] E. Suma, B. Lange, A. Rizzo, D. Krum and M. Bolas, "FAAST: The Flexible Action and Articulated Skeleton Toolkit," in *Virtual Reality Conference (VR), 2011 IEEE*, 2011.

[19] Microsoft, *Use the power of Kinect to change de world,* 2013.

[20] J. Ahlberg, *Canide a parameterized face,* 2013.

DYNAMICS OF HUMAN LOWER LIMBS USING CGA DATA AND BSIP PREDICTIVE EQUATIONS

A. BENTO F^{o} [1], V. H. FERNANDES[1], A. H. MAIA[1], E. GARCIA[3], A. FRIZERA[2] and T. BASTOS[2]

[1] *Mechanical Eng. Departments, UFES, Vitoria, ES 29075-910, BRASIL*
antonio.bento@ufes.br
[2] *Electrical Eng. Department, UFES, Vitoria, ES 29075-910, BRASIL*
[3] *CAR - Centro de Automática y Robótica, UPM - CSIC, La Poveda, Madrid, España*

This paper presents the construction of individual dynamic models of human lower limbs using kinematics from Clinical Gait Analysis*(CGA)* data and Body Segment Inertia Parameters*(BSIP)* from predictive equations. It's used human lower limbs *BSIP* from predictive equations and human kinematics from *CGA* data, to get individual *BSIP* and kinematic data. The the human dynamic model is built in a dynamic simulation software environment, to create human body dynamic models, and the *CGA* kinematic data is used as input to study model dynamics and control strategies. This approach would allow the testing and simulation of complex dynamic models without going through the tedious and error prone process of manually calculating equations of motion.
It's shown the use of the strategy presented to collect the data to be used in the simulation of a teen age boy or girl to be used latter in the simulation of a lower limb exoskeleton.

Keywords: CGA, BSIP, YSCS, human walking, gait dynamics, kinematics, exoskeletons, biomechanics

1. Introduction

Great amount of information about kinematics of human walking are being provided, using expert systems based technology equipment and normalized accordion to subjects its body mass and body and leg height.[1] Nevertheless, few information about kinetics (forces and torques) of human walking and about the human dynamic model used in these systems is provided.
It's proposed the human dynamic model construction and simulation through a dynamic simulation software environment. With this scheme one can simulate complex skeletal systems without going through the tedious and error prone process of manually calculating equations of motion; get immediate feedback through plots of any simulation variable; and quickly tweak the model design and control system.

The BSIP were gotten from predictive equations. They correspond to joint centres and to conventional *SCS* and are directly applicable in the conventional *SCS*; and could be obtained for both males and females.[2]
The normalized *CGA* kinetic data is a 50 elements vector[3] representing 2 % stride time angle steps from 0 to 100 %, for each leg joint, which is the time for the leg to describe an entire cycle from heel strike to toe off. This data is interpolated trough 50 *3rd degree* splines in an *OOP Abstract Method Pattern*[4] library which is added to the simulation software package. This normalized *CGA* kinetic data need not to be denormalized as the angles are in *radians.* This *CGA* kinematic data is used as input to the model, to study the model dynamics and control strategies. A three degrees of freedom *DOF* leg model and a lower limb exoskeleton shown in Fig. 2 were built in *YSCS.* The leg joints are: hip flexion-extension, knee and ankle flexion-extension and abduction-adduction; knee flexion; and ankle ;

2. Background

2.1. *The Spring Loaded Inverted Pendulum(SLIP) model*

It has been found in studies[5,6] that the best pattern to represent the race of animals is to bounce, since it is a way to save energy, storing up to 70% of the kinetic energy of flight phase in elastic energy and then use it to take off again.[7]
The *SLIP* model shows that the musculoskeletal system of an animal can be mechanically described as a nonlinear multicomponent spring-mass model, being the simplest possible model for a bouncing system. It's possible to bounce in place or jump ahead, varying the horizontal position. Bouncing in place has a remarkable resemblance to a simple spring-mass model. Jumping ahead at a given speed also involves the leg length as basic parameter of the differential equation model of the spring-mass.

2.2. *Clinical Gait Analysis(CGA)*

It is a human gait evaluation tool that uses expert systems based technologies with which gait analysis can supply information about how human walks and provide answer on why things happen their way.
It's data acquisition process is driven by protocols which ensures the reliability of the data and allow the comparison between different research labs and groups.[8,9] The standardization of the data acquisition process and

representation could be done according to recommendations of the International Society of Biomechanics (ISB).[10–12]
The recent interest in the development of active anthropomorphic exoskeletons, designed through biomechanical analogy with humans, have motivated the use of the configuration and operation of human locomotion from *CGA* data for the design of such devices.
When the intent is to design anthropomorphic lower limb exoskeletons with limb masses and inertias same as in humans, the required joint angles, torques and power for the exoskeleton to perform motion are approximately the same required by a similar sized human to perform the same motion.[13]
In[14] , there are 50 values of joint angles and moments, acquired in 2% time increments which results in 0.24 s which is not sufficient for a precise dynamic modeling of gait and for building a dynamic model to be simulated in Yobotics! Simulation Construction Set.[15]
Nevertheless, using the values from a prospective 5-year study of 16 children study from[16] as interpolation points it could be set up a 49 cubic splines family, which element $s_i^j, i = 1...49$ is shown in Eq.(1) and which spline interpolation is shown in Fig. 1, being a_0, a_1, a_2, a_2 the spline's coefficients and *p[i]* is the pivot of each interpolating interval. These splines can be used for interpolating the joint moments or the joint angles. When interpolating joint angles, the 1st and 2nd derivatives of Eq.(1), leads, respectively, to the angular velocity and acceleration of the joint.

$$s_i^j := a_0^j[i] + a_1^j * x + a_2^j * (x - p[i])^2 + a_3^j * (x - p[i])^3 \tag{1}$$

For a lower limb exoskeleton with 7 degrees of freedom *(DOF)* in each leg there will be a family of more than 340 splines for interpolating all leg joint angles. To illustrate this approach, coefficients values for the s_{10}^j and s_{20}^j hip flex/extension angle splines are shown in Table 1.
The program strategy to build a clean ode for such a great number of polynomials would be better to set it up in *Object Oriented Program - OOP* through an *Abstract Factory Pattern* as showed in.[4]

2.3. *Body Segment Inertia Parameters(BSIP)*

Body Segment Inertia Parameters *BSIP* were first studied[17] in cadavers and, in a latter study[18] demonstrated that body size and moments of inertia are related and that these correlations can be used to develop regression equations for predicting mass distribution characteristics.
Mcconville[18] studied 31 adult males (mean age 27.5 years old, mean weight 80.5 kg, mean stature 1.77 m) and Young[19] studied 46 females (mean age

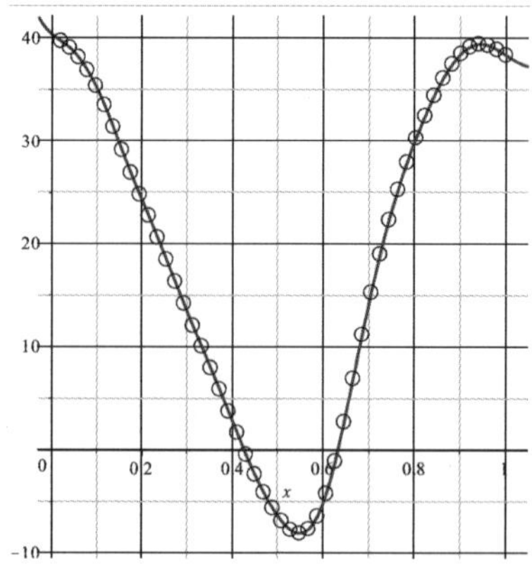

Fig. 1. Spline interpolation for hip flex/extension angle in degrees and intervals of (1/50)% of gait cycle. There is a spline for each interval and a total of 49 spline one for the gait cycle.

Table 1. Typical coefficients values for the s_{10}^{hip} and s_{20}^{hip} hip flexion/extension angle splines, *p[10] = 0.1961* and *p[20] = 0.3922.*

a_i	s_{10}	s_{20}
a_0	+45.3205748594665	45.4911337830222
a_1	-104.893293520992	-106.489377315202
a_2	+159.45179532487	+243.835034131385
a_3	-6155.28615316554	-8962.02253816924

Note: [a] *L: lenght; W: meight; BW: body weight;* X_{COM}, Y_{COM} *and* Z_{COM} *center of mass coordinates; and* r_{ij}*: inertia tensor components.*

31.2 years old, mean weight 63.9 kg, mean stature 1.61 m). Both populations were chosen to represent the entire stature/weight distribution and both studies were performed on living subjects using the same stereophotogrammetric technique. This technique allows the computation of segment volumes through the *3D* reconstruction of surface points. These data are among the few published extensive *BSIP*. These studies provide the *3D* locations of the segment *COM*, the principal moments of inertia and the orientations of the principal axes of inertia with respect to anatomical axes. Dumas et al[2] adjusted these data in order to correspond to joint centres and to conventional *SCS*. Scaling equations were obtained for both males and females that provide *BSIP* which are directly applicable in the conventional *SCS* and do not restrain the position of the centre of mass and the orientation of the principal inertia axes. These adjusted scaling equations are useful for use appropriate *3D BSIP* for research in posture and movement analysis, in dynamic analysis of the human gait and in the design and simulation of exoskeletons.

The adjusted *BSIP* scaling equations, for example, concerning the Thigh of a young woman with weight m_{Body}, can be used as follows. The Thigh length L_{Thigh}, from the Hip Joint Center *(HJC)* to the Knee Joint Center *(KJC)*,[20] can be obtained during *3D* gait analysis, at the same time as the Thigh *SCS* is constructed. In this *SCS* with the origin at the *HJC*, Fig. 3, the *3D* position of the *COM* is:

$$\begin{bmatrix} -0.077 \\ -0.377 \\ -0.090 \end{bmatrix} * L_{Thigh}$$

and the complete inertia tensor is:

$$\begin{bmatrix} 0.31 & 0.07 & -0.02 \\ 0.07 & 0.19 & -0.07 \\ -0.02 & -0.07 & 0.32 \end{bmatrix} * {L_{Thigh}}^2 * 0.123 * m_{Body}.$$

where $0.123 * m_{B}ody$ is the mass of the thigh segment.

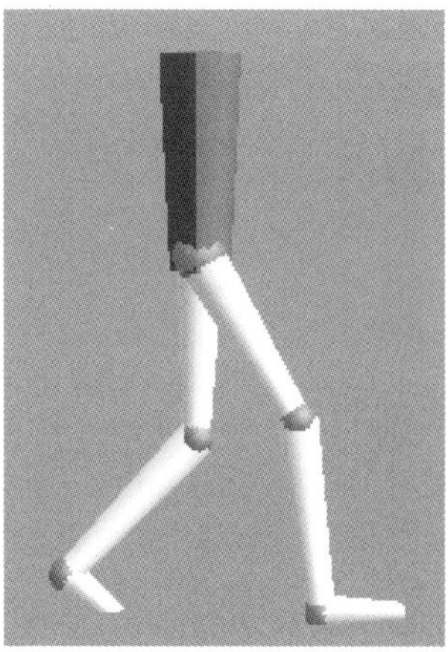

Fig. 2. YSCS exoskeleton model.[15]

3. Discussion

The adjusted *BSIP* scaling equations given in,[2] concerning any human body segment of a young man, can be used directly to access the segment *COM* and inertia tensor expressed in the *ISB* recommended *SCS*, for the standardization in the reporting of kinematic data.
The biomechanical modelling of the human body for movement analysis is classically based on a system of open chain of linked segments. Its possible to have accurate information from kinematics of human walking, using expert systems for *CGA* analysis but its inverse dynamic is not a simple task. The procedure presented in this paper shows how to acquire a subjects gait kinematics and *BSIP* data to realize a dynamic gait analysis, for prostheses and other artificial body parts design and evaluation, and for exoskeleton simulation and design; it's being used to prepare robotic lower limb exoskeletons models, bipeds and agile quadrupeds the *YSCS* environment, allowing the research of leg design for energy efficient applications.

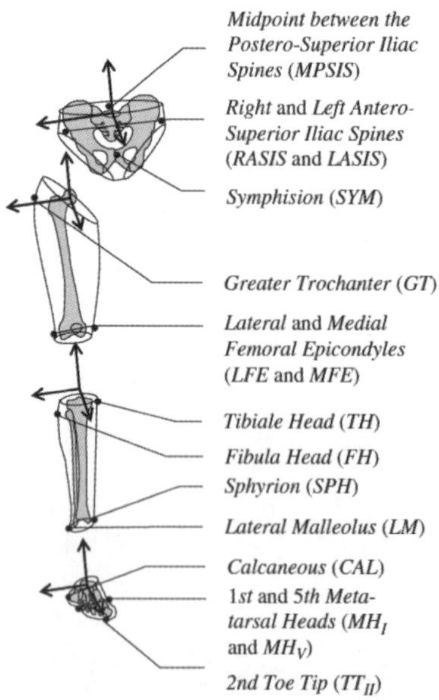

Fig. 3. *SCS* for body segment parameters definition.[2,11]

It was also presented a new efficient and powerful approach to define a subjects gait and inertial data to build his gait dynamic model for com-

Table 2. Body segment inertia properties (*BSIP*) according to Dumas.[2]

Lower limbs	Thigh	Leg	Foot height
L[mm]	379	388	46
L[%H]	24,14%	24,71%	2,93%
W[%BW]	14,60%	4,5%	1,00%
X_{COM}	-7,7%	-4,9%	27,00%
Y_{COM}	-37,7%	-40,4%	-21,80%
Z_{COM}	0,9%	3,1%	3,90%
r_{xx}	31,0%	28,0%	17,00%
r_{yy}	19,0%	10,0%	36,00%
r_{zz}	32,0%	28,0%	35,00%
r_{xy}	7,0%	2,0%	-10,00%
r_{xz}	-2,0%	1,0%	6,00%
r_{yz}	-7,0%	6,0%	-4,00%

paring healthy human gaits with pathologic subject's gait, assessment in occupational and physical therapy, clinical biomechanics, orthosis and prosthesis design, etc.

Also to design anthropomorphic lower limb exoskeletons with limb masses and inertias similar as in humans, the required joint angles, torques and power for the exoskeleton to perform motion can be obtained using the dynamic modelling techniques allowing: a good definition of the power train; the alternatives for power sourcing; the building of better energy efficient exoskeletons, orthesis and prosthesis; and the research of high power to weight density joint actuators.

References

1. A. L. Hof, *Gait & Posture* **4**, 22 (1996).
2. R. Dumas, L. Chaze and J. P. Verriest, *Journal of biomechanics* **40**, 543 (2007).
3. B. W. Stansfield, S. J. Hillman, M. E. Hazlewood, A. A. Lawson, A. M. Mann, I. R. Loudon and J. E. Robb, *Journal of Pediatric Orthopaedics* **21**, p. 403 (2001).
4. E. Freeman and E. Freeman, *Head First Design Patterns*, first edition edn. (OReilly, Octtober 2004).
5. N. C. Heglund, G. A. Cavagna and C. R. Taylor, *Journal of Experimental Biology* **97**, p. 41–56 (1982).
6. G. A. Cavagna, *Exercise and sport sciences reviews* **5**, p. 89–130 (1977).
7. R. Alexander and A. Vernon, *Journal of Zoology* **177**, 265 (1975).
8. A. Ferrari, M. G. Benedetti, E. Pavan, C. Frigo, D. Bettinelli, M. Rabuffetti, P. Crenna and A. Leardini, *Gait & Posture* **28**, 207 (2008).
9. M. R. Pierrynowski and V. Galea, *Gait & Posture* **13**, 193 (2001).

10. G. Wu, F. C. van der Helm, H. E. J. Veeger, M. Makhsous, P. V. Roy, C. Anglin, J. Nagels, A. R. Karduna, K. McQuade, X. Wang *et al.*, *Journal of biomechanics* **38**, 981 (2005).
11. G. Wu and P. R. Cavanagh, *Journal of Biomechanics* **28**, 1257 (1995).
12. R. Baker, *Journal of Biomechanics* **36**, p. 300–302 (2003).
13. A. B. Zoss, H. Kazerooni and A. Chu, *IEEE/ASME Transactions On Mechatronics* **11**, 128 (2006).
14. Kirtley, Cga normative gait database http://www.univie.ac.at/cga/data/index.html, (2006).
15. Yobotics, *www.ihmc.us/groups/scs/* **1** (2012).
16. B. W. Stansfield, S. J. Hillman, M. E. Hazlewood and J. E. Robb, *Gait & Posture* **23**, 288 (2006).
17. C. E. Clauser, *Weight, volume, and center of mass of segments of the human body*, tech. rep., ANTIOCH COLL YELLOW SPRINGS OH (1969).
18. J. T. McConville and C. E. Clauser, Anthropometric assessment of the mass distribution characteristics of the living human body, in *International Ergonomics Association, Congress, 6 th, and Human Factors Society, Annual Meeting, 20 th, College Park, Md*, 1976.
19. J. W. Young, R. F. Chandler, C. C. Snow, K. M. Robinette, G. F. Zehner and M. S. Lofberg, *Anthropometric and mass distribution characteristics of the adult female*, Tech. Rep. FAA-AM-83-1 6, Air Force Aerospace Medical Research Laboratory Wright-Patterson Air Forie Base Ohio 45433 (September 1983).
20. R. Baker, *Journal of Biomechanics* **36**, p. 300–302 (2003).

VARIABLE IMPEDANCE CONTROL OF A PARALLEL ROBOT FOR ANKLE REHABILITATION*

S. Q. XIE AND Y. H. TSOI †

Department of Mechanical Engineering, University Auckland, Auckland, New Zealand

This paper proposes a novel variable impedance controller for a redundantly actuated parallel robot to allow execution of passive ankle rehabilitation exercises while taking into consideration of human biomechanics online. By using a biomechanical model of the human ankle which describes the kinematics and dynamics of the ankle foot structure, the kinematic parameters of the ankle can be estimated online to allow identification of the constrained direction of motion. Impedance adjustment heuristics were also formed by investigating the ankle stiffness over a range of joint displacements through the use of the ankle model developed. The variable impedance controller can prevent application of excessive actuator forces which could threaten the safety of the patient.

1. Introduction

The human ankle is primarily responsible in transferring forces and moment between the lower limb and the ground. It therefore plays an important role in human locomotion and in maintaining the balance of an individual during gait. The ankle can be subjected to large forces and torques and is consequently prone to musculoskeletal injuries. Appropriate rehabilitative treatment must be applied to patients suffering from ankle sprains to ensure complete recovery and prevent recurrent sprains. The treatment regime for ankle sprains involves range of motion (ROM) exercises, muscle strengthening exercises and proprioceptive or balancing exercises [1,2]. Impedance control in particular is very commonly used in the area of rehabilitation [5,7,8]. Variable impedance controller has therefore been proposed to allow adaptation of the robot performance to suit the corresponding environment. To attain additional rehabilitation objectives, [6,11] has based the design of an impedance controller for an elbow rehabilitation device on a skeleton model of the forearm. Variable impedance control has been implemented to reduce the interaction forces between the robotic orthosis and

* This work is supported by New Zealand Bright Future Fellowship

the human arm while also limiting the extension of ligament connecting the radius and ulna. Variable impedance control scheme utilizes knowledge of the ankle joint dynamics and kinematics in adjusting the impedance controller parameters to avoid over exertion of interaction forces in directions of low compliance.

2. Design of Ankle Rehabilitation Robot

An ankle rehabilitation robot has been developed for the purpose of this research. The developed robot consists mainly of four electrical linear actuators connecting a fixed base platform to an end effector platform upon which the patient's foot will be secured [2, 3]. A three dimensional model of the device is shown in Fig. 1. The design of this robot is rather distinct compared to many previously developed devices [1] in the sense that the patient's lower limb actually forms part of the robot kinematical constraint. This has the effect of allowing only anatomically correct motion while also ensuring that the position of the patient's ankle will remain constant during the exercises to permit more precise control of forces and moments applied to it.

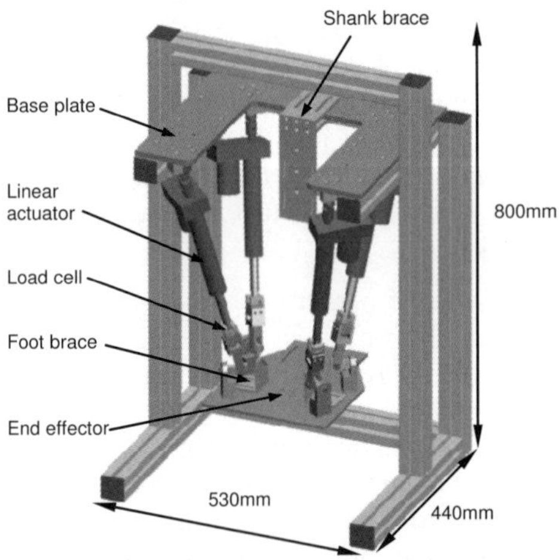

Fig. 1 Three dimensional computer aided design model of the ankle rehabilitation robot designed for this research

Since the rotational range of motion of the ankle and foot structure is rather limited, the disadvantage of limited workspace associated with parallel mechanisms is not particularly detrimental in this application. By considering the normal range of motion at the ankle and the dimensions of the human lower limb, the kinematic parameters of the robot in terms of passive joint locations were selected to provide sufficient workspace to accommodate the ankle range of motion. A singularity analysis carried out during the design of the robot has revealed the presence of singular regions when only three linear actuators are used [7,9]. To remove this singular region, an additional actuator has been included in the final design to prevent rank loss of the Jacobian matrix. The use of redundant actuation has also provided an additional actuation degree of freedom which could be utilized to regulate additional variables such as reaction forces at the ankle.

3. Controller Design

As joint characteristics not only vary with individual but also with joint configuration, it would be best if a variable impedance controller is implemented to maintain good robot performance. Interaction forces between the robot and patient must therefore be kept at low levels during exercises where the foot is being manipulated in a passive manner by the robot.

3.1. *Computed Torque Impedance Control*

The basic impedance controller used can be considered as a computed torque controller with certain proportional and derivative gains, where the proportional gain is the stiffness parameter while the derivative gain is the damping parameter in the manipulator impedance. By considering the robot as a manipulator with three degrees of freedom, the dynamics of the manipulator can be described by (1), where Θ is the task space variable in XYZ Euler angles, M is the inertia matrix, N is the summation of all the centripetal, Coriolis and gravitational forces, $\tau_{\Theta,ext}$ is the patient-robot interaction moment in task space coordinates, J is the manipulator Jacobian relating the actuator velocities, $\dot{l}$ to task space velocities through (2) and F the compressive forces applied at each actuator.

$$M\ddot{\Theta} + N + \tau_{\Theta,ext} = J^T F \tag{1}$$

$$\dot{l} = J\dot{\Theta} \tag{2}$$

However, by assuming that the ankle kinematics model is perfectly accurate and exact kinematics of the ankle is known, the dynamics of the manipulator can be given more precisely as (3), with θ_{as} being the ankle and subtalar joint displacements. Subscript *as* are added to represent the equivalent terms of the robot dynamics when constrained by the ankle kinematics.

$$M_{as}\ddot{\theta}_{as} + N_{as} + \tau_{as,\, ext} = J_{as}{}^T F \tag{3}$$

The impedance controller developed is designed based on the three degrees of freedom dynamic model because the kinematics of the ankle foot structure is not well known.

3.2. *Impedance Adjustment Rules*

To simplify the controller, certain heuristic rules are used to modify the stiffness matrix in the impedance controller. These rules were developed by observing the stiffness of the ankle foot structure which is computed offline throughout certain domain of ankle and subtalar joint displacements. This stiffness can be obtained in the form of a 2×2 matrix describing the change in the moments applied along the ankle and subtalar joints with respect to a change in joint displacements. This is represented in (4). Where K_{as} is the stiffness matrix in the two degree of freedom task space, τ is the revolute joint moment, θ is the displacement of the revolute joint, while subscripts a and s are used to denote quantities relating to the ankle and subtalar joint respectively.

$$K_{as} = \begin{bmatrix} \frac{\Delta\tau_a}{\Delta\theta_a} & \frac{\Delta\tau_a}{\Delta\theta_s} \\ \frac{\Delta\tau_s}{\Delta\theta_a} & \frac{\Delta\tau_s}{\Delta\theta_s} \end{bmatrix} \tag{4}$$

Computation of the stiffness matrix has shown that it is almost symmetrical with near identical values for the off diagonal elements. The values for each of the diagonal element and the off diagonal elements are shown in Fig. 2 as surfaces over a range of ankle and subtalar joint displacements. Analysis of the stiffness matrix element in Fig. 2a shows that an increase in the magnitude of ankle joint displacement will cause an increase in ankle joint stiffness with minimal influence of the subtalar joint displacement, except for when the subtalar displacement is positively large and ankle displacement is negatively large. For the subtalar stiffness, large negative values of subtalar displacement causes very rapid increase in the stiffness values as can be seen in Fig. 2b, while similar increase is also noted when the ankle foot undergoes large dorsiflexion, inversion and adduction.

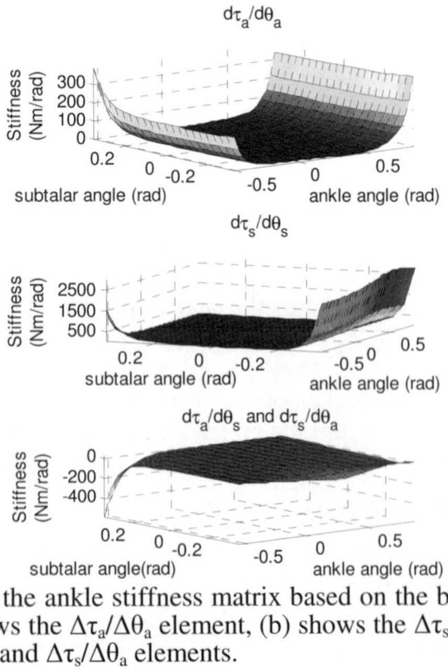

Fig. 2 Elements of the ankle stiffness matrix based on the biomechanical model developed. (a) shows the $\Delta\tau_a/\Delta\theta_a$ element, (b) shows the $\Delta\tau_s/\Delta\theta_s$ element and (c) shows the $\Delta\tau_a/\Delta\theta_s$ and $\Delta\tau_s/\Delta\theta_a$ elements.

Based on the trend of stiffness variation, the domain of joint displacements is divided into several regions. Different heuristics are applied in different regions of the domain to produce multipliers which are used to reduce the magnitude of various elements in the stiffness matrix in the impedance controller. There are altogether four rules used and each rule has the identical form as that shown in (5). μ_i represents the multiplier obtained from rule i, θ is the maximum between the absolute value of both ankle and subtalar displacements and λ_i is the parameter used to determine the how rapidly the stiffness matrix element decays with joint displacement, where $\lambda_i < \lambda_{i+1}$. A higher rule number is applied when the stiffness is changing more rapidly in that region.

$$\text{Rule } i\text{:}\ \mu_i = \frac{2}{1+e^{\lambda_1 \theta}} \tag{5}$$

Variable rules are then developed based on the region defined.

4. Simulation Results

Simulations had been carried out in Simulink to investigate the efficacy of the proposed control scheme. Sinusoidal functions were use as reference trajectories for the desired foot rotational motion. The controller parameters used for the simulation was manually chosen to provide satisfactory performance while limiting oscillations in the motion and force variables. The values of these parameters are provided in table I. In order to create a more realistic scenario, measurement noise was added to both the linear actuator length and actuator force measurements. In order to reduce the effect of measurement noise on the Euler angles and their derivatives which are estimated from the length measurements, third order digital low pass filters were used to attenuate the high frequency component of the actuator length signals. In addition to measurement noise, Coulomb friction was also simulated in the electrical actuators using dead zone discontinuities. The dead zones were set between ±50 N for each actuator.

TABLE I
CONTROLLER PARAMETERS

Parameter	Value
Rule parameters, $[\lambda_1, \lambda_2, \lambda_3, \lambda_4]$	[2, 4, 6, 16]
Stiffness matrix parameters, $[k_{aa}, k_{ss}, k_{as}]$	[100, 50, 0]
Damping matrix parameters, B_2	$\begin{bmatrix} 4 & 0 \\ 0 & 4 \end{bmatrix}$
Joint force controller parameters, $[K_f, \tau]$	$[0.9, 31.8\times 10^{-6}]$

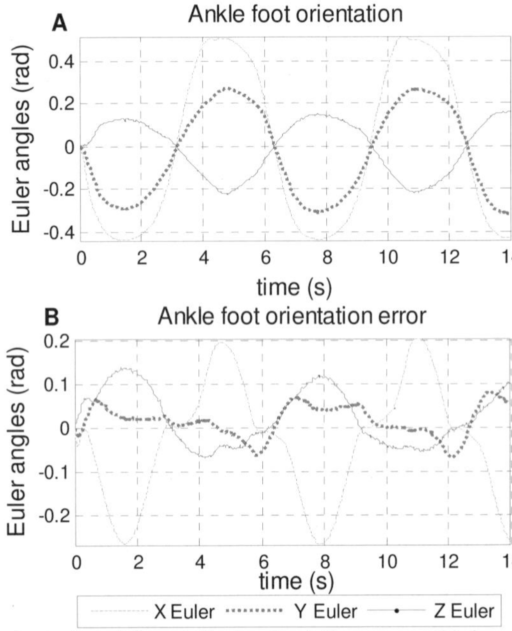

Fig. 3 Foot orientation (a) and foot orientation error (b) in radians as obtained from the simulation. gray lines denote values in the X Euler angle direction, red dotted lines denote values in the Y Euler angle direction while black lines represents values in the Z Euler angle direction.

Fig. 3 shows both the orientation and the orientation error of the foot in Euler angles during the simulated period. It can be seen from the gray lines that the maximum displacement in dorsiflexion (negative direction) is less than that for plantarflexion (positive direction). This is not surprising as the human ankle is expected to be stiffer in dorsiflexion. To compare the performance of the variable impedance controller with that of a constant impedance controller, a performance indicator is required. The performance indicator used in this case is that of a weighted sum of orientation error norm and actuator force norm. This indicator, P is given in (6), with $\Theta_{d,\max}$ being a vector of the maximum magnitudes of Euler angles in the reference trajectory and $F_{\max}$ being a vector of actuator force thresholds which should not be exceeded. These are used to normalize the contribution of each term. Clearly, smaller values of P indicate better controller performance.

$$P = \frac{\Theta_e{}^T \Theta_e}{\Theta_{d,\max}{}^T \Theta_{d,\max}} + \frac{F^T F}{F_{\max}{}^T F_{\max}} \tag{6}$$

Using the simulation results from both the variable and constant impedance controllers, the performance function for each case was evaluated and shown in Fig. 4. It can be seen that the performance of the variable impedance controller (dotted black line) can be considered superior to the constant impedance controller (gray line) for two main reasons. Firstly, there are periods in time when variable impedance produces significantly lower values in the performance function. Secondly, large fluctuations exist in the performance function of the constant impedance controller, indicating oscillations in actuator force or orientation error. Closer inspection of simulation results show that the fluctuations are caused by force oscillations.

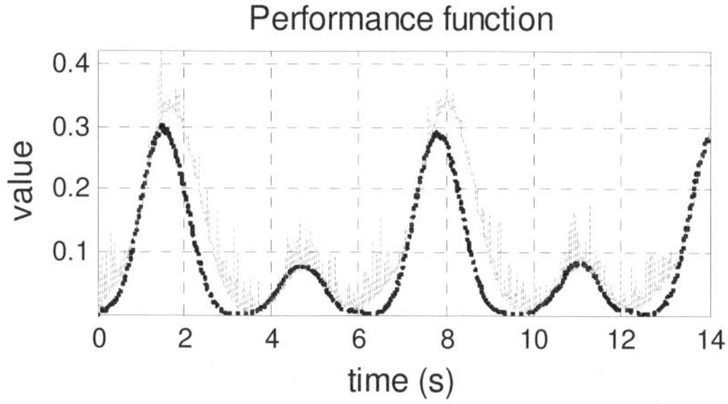

Fig. 4 Performance function values computed from simulations results of variable impedance case (black dotted line) and constant impedance case (gray line). $\Theta_{d,\max} = [2\pi/9 \;\; \pi/12 \;\; \pi/12]^T$ and $F_{\max} = [150 \;\; 150 \;\; 150 \;\; 150]^T$.

5. Conclusions and Future Work

By using a priori information from a biomechanical model of the ankle foot structure, variable impedance adjustment rules were proposed to reduce the stiffness of the impedance controller to maintain a balance between positional accuracy and limitation of actuator forces. The improved performance comes from two areas. Firstly, the desired exercise trajectory can be followed more closely in compliant regions of the ankle. Furthermore, safety of the robot is also improved by preventing application of large forces in stiff configurations of the ankle, thus reducing likelihood of injury. To improve the functionality of the robot, extension to the current work should also be carried out to develop suitable parameter adaptation schemes which can cater for muscle strengthening and proprioceptive exercises. Investigation into methods which can carry out online estimation of the ankle stiffness will also be beneficial as it can reduce the controller's dependency on the biomechanical model used to develop its adjustment rules.

References

1. G. Liu, J. Gao, H. Yue, X. Zhang, and G. Lu, "Design and Kinematics Simulation of Parallel Robots for Ankle Rehabilitation," in *IEEE International Conference on Mechatronics and Automation*, Luoyang, China, 2006, pp. 1109-1113.

2. Tsoi, Y. H., Xie S. Q. ,"Design and Control of a Parallel Robot for Ankle Rehabilitation", International Conference in Mechatornics and Machine Vision in practice, 3-5 December, 2008, Auckland, New Zealand.
3. Andrew M[C]DAID, Yun Ho Tsoi and Shengquan XIE, "MIMO Actuator Force Control of a Parallel Robot for Ankle Rehabilitation", Mechatronics Engineering: Research Development and Education, editors: Maki Habib, ISTE Wiely, 2013
4. J. A. Blaya and H. Herr, "Adaptive control of a variable-impedance ankle-foot orthosis to assist drop-foot gait," *IEEE Transactions on Neural Systems and Rehabilitation Engineering,* vol. 12, pp. 24-31, 2004.
5. Tsoi, Y. H., Xie, S. Q. and Mallinson, G. D., "Joint Force Control of Parallel Robot for Ankle Rehabilitation", the seventh IEEE International Conference on Control and Automation, December 9-11, 2009, Christchurch, New Zealand.
6. Tsoi, Y.H., Xie, S. Q., "Impedance Control of Ankle Rehabilitation Robot", IEEEE International Conference on Robotics and Biomimetics, December 14~17, Bangkok, Thailand, 2008. Cited 7 times by December, 2011. Cited 7 times by 2011.
7. Tsoi, Y. H. and Xie, S. Q., "Trajectory Planning and Optimisation for Ankle Joint Rehabilitation", UKACC International Conference on Control, 2010, 7~10 September, Coventry, UK.
8. Tsoi, Y. H., Xie S. Q., "Design and Control of a Parallel Robot for Ankle Rehabilitation", *International Journal of Intelligent Systems Technologies and Applications* (IJISTA), 2010 - Vol. 8, No.1/2/3/4. pp. 100 - 113.
9. Tsoi, Y. H., and Xie, S. Q., "Online Estimation Algorithm for a Biaxial Ankle Kinematic Model with Configuration Dependent Joint Axes", *Journal of Biomechanical Engineering*, Vol.133, Issue 2, paper number:021005, 11pgs, 2010.
10. Jamwal, P., Xie S. Q., Tsoi, Y. H., and Aw, K. C, "Forward Kinematics Modelling of a Parallel Ankle Rehabilitation Robot Using Modified Fuzzy Inference", *International Journal of Mechanism and Machine Theory*, 45 (11), pp.1537~1554, 2010
11. N. Hogan, "Impedance Control: An Approach to Manipulation: Part III-Applications," *Journal of Dynamic Systems, Measurement, and Control,* vol. 107, pp. 17-24, 1985.

A MOBILE 3D VISION-BASED EMBEDDED SYSTEM FOR ROBUST ESTIMATION AND ANALYSIS OF HUMAN LOCOMOTION

CONG ZONG and PHILIPPE BIDAUD

Université Pierre et Marie Curie, Institut des Systèmes Intelligents et de Robotique, Paris, France
E-mail: cong.zong@isir.upmc.fr
philippe.bidaud@isir.upmc.fr

XAVIER CLADY

Université Pierre et Marie Curie, Centre de Recherche Institut de la vision, Paris, France
E-mail: xavier.clady@upmc.fr

In this paper, we propose and experiment a system for human motion capture which can be incorporated into a mobility aid device such as a walker. This measurement system uses two Kinect 3D vision sensors whose data are mapped after treatment on a physical model of virtual human. This filtering method ensures data consistency as well as the observability of the whole body movement. It is beyond a basis for the analysis of human motor activity and more specifically here of locomotion (gait parameters, posture, joint efforts, energy). We detail the entire process for rebuilding the movement and discuss its performance from the experimental results.

Keywords: Physically-Based Motion capture, human motion analysis, multiple-camera tracking

1. Introduction

The analysis of human locomotion and its changes has led to numerous works by the past. This motor activity can be defined as the combination in time and space of more or less complex movements of different body parts leading to the displacement of the subject. To propel in upright position, the man uses a discrete cycle of unipodal and bipodal supports through the movement of the musculoskeletal system in which it seeks to maintain the controllability of the dynamic equilibrium. Methods for observation and analysis of this function have diversified progressively with technology development.

Most of the techniques used today are based on laboratory devices such as locometers for determining the temporal spatial parameters of gait (walking speed, step length, step duration, the duration of the course and not time propulsion), motion capture systems and posturographic platforms to access respectively the kinematics and the dynamics of the whole body movement. These means are implemented in particular to explore the range of disorders affecting locomotion-temporal spatial parameters of gait, average posture changing and the movements of all body parts.

The development of "holter" technologies for qualitative and quantitative assessment of walking and its disorders is today a challenge for the diagnosis and monitoring of patients at risk.

This paper proposes and explores the performance of a mobile system for capturing and analyzing human motion during walking. The latter consists of two Kinect sensors installed on a walker device. A set of algorithms have been developed for extracting data of the characteristic points of the anatomic envelope. These data are then repositioned on a physical model of virtual human to reproduce the observed motion and be analyzed to find anomalies for instance.

2. Embedded vision-based motion capture system

2.1. *General description*

Two Kinect sensors (see figure 1) $C1$ and $C2$ are used respectively to capture the relative motion of the upper body parts and the feet trajectories with respect to the walker. Wiimote sensors are installed on the wheels of the walker to determine the movement of the walker and then the absolute motion of the person (odometry system[1]).

2.2. *Data extraction*

2.2.1. *Capture of the upper body parts*

The sensor $C1$ is used to capture the upper body part motions. The data extraction method developed in this aim is based on an algorithm combining two steps: 1) Segmentation of the body system; 2) Modeling of the body parts from the sensor data.

A) Segmentation

The objective of the segmentation step is the extraction of the 3D data for each member of the human body. 3D data on the torso, arms and at the

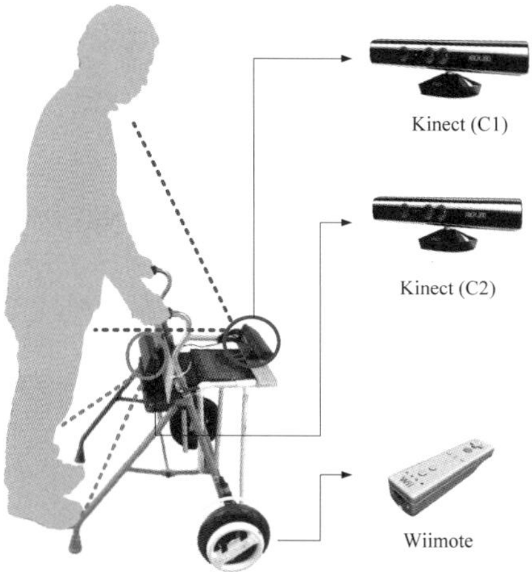

Figure 1. Plate-forme de capture

upper thighs are recorded by the sensor (C1). They are extracted after subtracting the background images using a threshold applied to the depth map. Then, knowing the user anthropomorphic parameters,[2] 3D data of the human body are distributed over the different members, torso, arms, forearms and thighs based on the geodesic distance between different the parts of body during movement. Four geodesic paths are defined between ends of the four members from an origin located near the head (see figure 2).

B) Modeling of the human body parts

After the segmentation phase, we associate a cloud of 3D points to each member in order to estimate the 3D coordinates of the anatomic joints. For that, a cylinder is fitted to the 3D data of each member using iterative Gauss-Newton method. The length of the cylinder is calculated from the anthropomorphic table associated with the human virtual model. The positions of shoulders, wrists and hips joints are detected from the positions of the extreme points of the cylinder axes. The positions of the knee joints are computed from the estimated length of the thighs and the estimated positions of the hip and the directions of the axes of the corresponding cylinders. The elbow positions are inferred as the midpoint between the point in the center of the upper face of the cylinder corresponding to the forearm and of the lower face of the cylinder corresponding to the arm. The

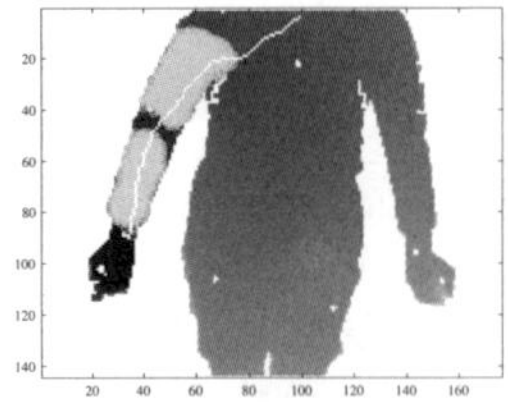

(a)

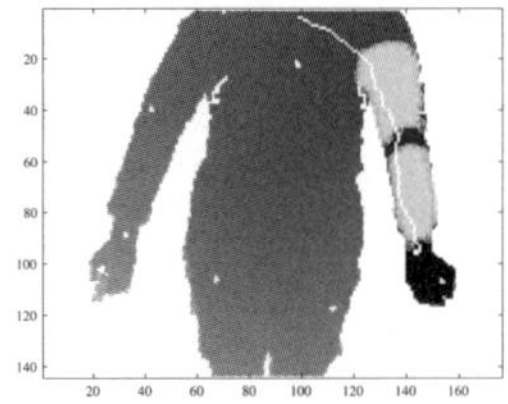

(b)

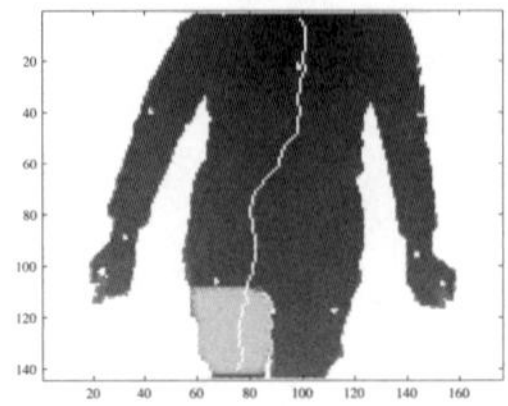

(c)

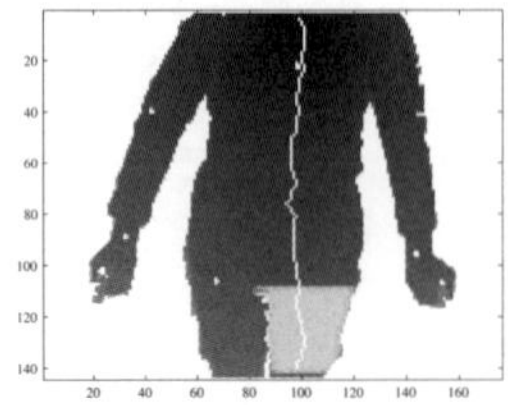

(d)

Figure 2. Different members of the body: (a) the forearm and the right arm (b) the forearm and the left arm (c) the right thigh (d) the left thigh

positions of these anatomic joint points are stacked in the vector $P_h(k)$ at each acquisition period k.

2.3. *Capturing foot positions*

The position of the feet is provided by the sensor data Kinect (C2) directed towards the ground. It is obtained by approximating data of the feet as Gaussian distributions, the centers are considered as a characteristic point of the foot.

(a)

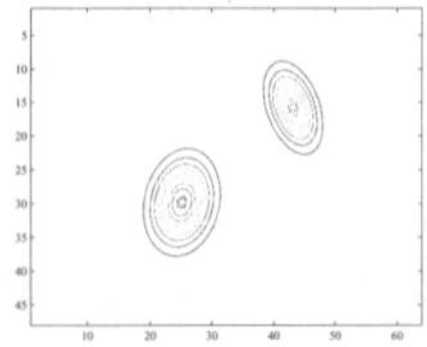

(b)

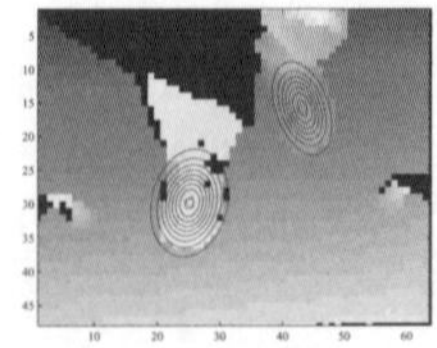

(c)

Figure 3. Example of images capturing the position of the feet

3. Reconstruction of the human motion

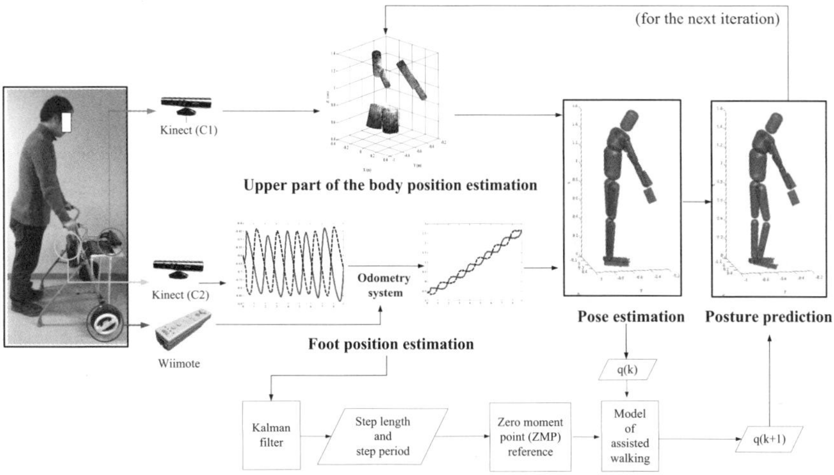

Figure 4. Scheme of system

3.1. *Mapping motion capture data to virtual human model*

To map all of the data collected with the vision system, we use a virtual human model[3] which consists of 19 body segments modeled as rigid bodies (7 for the lower and 12 upper part) and 16 anatomical joints. The virtual human posture is controlled and dynamically animated in a physically consistent way through a Linear Quadratic Program (LQP) controller.[4] At a frequency of 30 Hz, joint key points of the model are driven to the measured positions (see figure 5).

At each period of time, we compute joint motions $V_{measure}$ such as :

$$V_{measure} = \frac{P_{mes}(k) - P_{mod}(k)}{dt} \tag{1}$$

and then are introduced into the objective function of a quadratic program LQP. At each new capture, which seeks to resolve:

$$\begin{aligned} \dot{q}^* &= argmin_{\dot{q}}((J\dot{q} - V_{measure})^T W (J\dot{q} - V_{measure})) \\ &= argmin_{\dot{q}}(\frac{1}{2}\dot{q}^T Q \dot{q} + \dot{q}^T P + \frac{1}{2}C) \\ under\ the\ constraints\ &: \begin{cases} q + \dot{q}dt \leq q_{max} \\ q + \dot{q}dt \geq q_{min} \end{cases} \end{aligned} \tag{2}$$

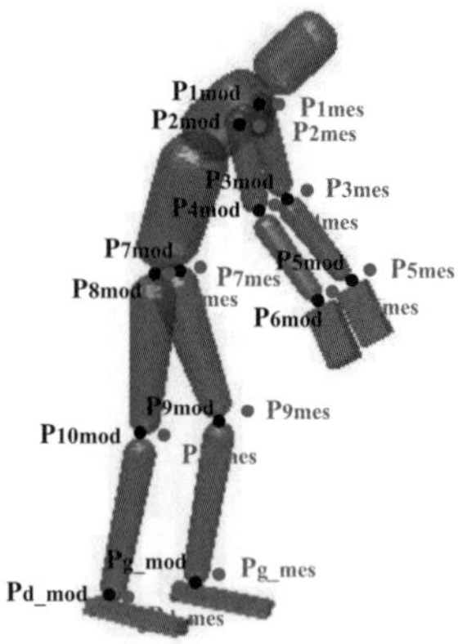

Figure 5. Digital model: the red dots are the measured positions of the points P_I of the joint axis and the black dots are the positions of the analog points on the model P_{mod}

where the C, Q and P matrices are respectively :

$$C = V_{measure}^T W V_{measure} \tag{3}$$

$$Q = J^T W J \tag{4}$$

$$P = -J^T W V_{measure} \tag{5}$$

J is the Jacobian matrix joint points of the virtual human model. W is a diagonal matrix of weights associated with different articulations reflecting motor coordination between the different joints. $W = \begin{pmatrix} w_i & 0 & 0 \\ 0 & w_i & 0 \\ 0 & 0 & w_i \end{pmatrix}$. We define $w_{hand_{right}} = w_{hand_{left}} = 10$ to make that the hands stay in contact with the walker handles. $w_i = 1$ for the other joints. These data are fitted to the current positions of joints estimated from a prediction model.

3.2. *Human posture prediction model*

A biomechanical model of for human walking was proposed by Brubaker[5] to track people from the video stream being processed by a particle filter. It is shown that this approach increases the robustness of the person tracking algorithm, particularly in cases of occlusions. Here we use a model for the human whole-body motion which reflects his dynamic equilibrium when walking to predict the position of the body system at the time of capture and increase the precision of body part fitting process. A Kalman filter is first used to estimate spatio-temporal parameters in the case of assisted walking by a walker: the duration and stride length. This estimation allows to predicts future positions of the feet and therefore the reference trajectory of the Center of Pressure (CoP) from a walking pattern generation using preview control of the Zero-Moment Point (ZMP)[6,7] which coincides with the CoP. It is from this characteristic point that the speed characteristic of the Center of Mass (CoM) and the speed of the foot which is not in contact with the ground (the swing phase) can be calculated. Then, the 3D positions of the joints at the time of capture $k+1$ can thus be predicted using an algorithm which controls the joint positions of virtual mannequin by a LQP program taking into account especially and the constraints imposed by the contacts with the environment (i.e. hands and feet).

4. Experimental results and discussion

In Figure 6, mean values and standard deviations of errors (between a motion capture system Codamotion and the embedded device with two Kinects) are calculated and shown in Table 1. Note that the error values for the upper body parts are around 3 cm and values for the lower body parts are around 6 cm. As stated in,[8] an approximate location of the joints of the body (with the same order of magnitude noise) can achieve good classification accuracy (80%) for a recognition system gait abnormalities.

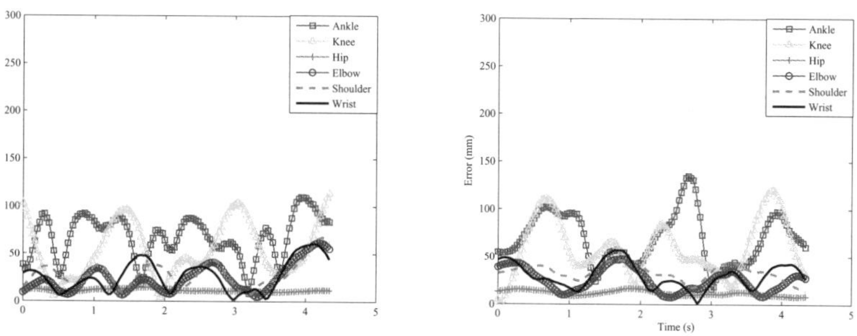

(a) right side

(b) left side

Figure 6. The average errors for different joints

Table 1. The mean values and standard deviations of the errors

Joints		Mean (mm)	Standard deviations (mm)
Ankle	Right	67	22
	Left	69	26
Knee	Right	54	28
	Left	59	29
Hip	Right	11	1
	Left	11	1
Elbow	Right	24	15
	Left	26	9
Shoulder	Right	28	13
	Left	33	8
Wrist	Right	27	16
	Left	29	13

Bibliography

1. J. Gutiérrez, F. V. Medina and M. A. Porta-Gándara, Vertically aligned accelerometer for wheeled vehicle odometry, *Mechatronics* 2010.
2. D. A. Winter, *Biomechanics and motor control of human movement* (John Wiley & Sons,Inc., 2009).
3. R. Mozul, P.-B. Wieber, R. Pissard-Gibollet, L. Boissieux and F. Billet, *HuMAnS*, tech. rep., INRIA, http://www.inrialpes.fr/bipop/software/humans/.
4. J. Salini, S. Barthélemy and P. Bidaud, Lqp controller design for generic whole body motion, in *Proceedings of the 12th International Conference on Climbing and Walking Robots and the Support Technologies for Mobile Machines (CLAWAR)*, (Istanbul, Turkey, 2009).
5. M. A. Brubaker, D. J. Fleet and A. Hertzmann, Physics-based person tracking using the anthropomorphic walker, *International Journal of Computer Vision* 2010.
6. S. Kajita, F. Kanehiro, K. Kaneko, K. Fujiwara, K. Harada, K. Yokoi and H. Hirukawa, Biped walking pattern generation by using preview control of zero-moment point, *IEEE International Conference on Robotics & Automation* 2003.
7. P.-B. Wieber, Trajectory free linear model predictive control for stable walking in the presence of strong perturbations, *IEEE-RAS International Conference on Humanoid Robots* 2006.
8. B. Pogorelc and M. Gams, Discovery of gait anomalies from motion sensor data, in *22nd International Conference on Tools with Artificial Intelligence*, 2010.

A SIMPLE LOAD ESTIMATION OF A PATIENT DURING A STANDING ASSISTANCE MOTION

T. YAMADA and D. CHUGO*

Graduate School of Science and Technology, Kwansei Gakuin University, Sanda, Hyogo, Japan
**E-mail: {yamada, chugo}@chugolab.com*
www.chugolab.com

S. YOKOTA** and H. HASHIMOTO***

***Setsunan University*
Neyagawa, Osaka, Japan
****Advanced Institute of Industrial Technology*
Shinagawa, Tokyo, Japan

In our current research, we have developed a novel standing assistance system for an elderly person. Our developing assistance system is designed compactly for family use. Our system is based on a robotic walker with novel manipulator mechanisms on its body with four parallel linkages. In general, a previous assistive device does not require the patient to use own physical strength and it causes the decreasing the physical strength of elderly people. Therefore, a standing assistance system, which uses a remaining physical strength of a patient, is required. For realizing it, the assistance system needs to estimate the load of a patient during standing motion, however, it is difficult to estimate it without biosensors which should be stuck on the patient as an electromyograph. Thus, in this paper, we propose the load estimation scheme using a human model with measuring information of a patient by our assistance system. Our estimation scheme uses only force sensors and a laser range finder which are equipped on our standing assistance device for measuring a posture and applied force of a patient. Furthermore, our estimation scheme derives the load of each joint by inputting the measuring data to a linkage model, which approximates the human body.

Keywords: Standing Assistance; Walker; Human Linkages Model; Load Estimation.

1. Introduction

In Japan, the population ratio of senior citizen who is 65 years old or more exceeds 23[%] at 2013 and rapid aging in Japanese society will advance in the future.[1] In this aged society, many elderly people cannot perform normal daily household, work related and recreational activities because of decrease in force generating capacity of their body. Today, the 23.5[%] of elderly person who does not stay at the hospital cannot perform daily life without nursing by other people.[2] For increasing their QOL (Quality of Life), they need a domestic assistance system which enables them to perform daily activities alone easily even if their physical strength reduces by aging. Standing motion is the most serious and important operation in daily life for an elderly person who doesn't have enough physical strength[3].[4] In typical bad case, an elderly person who doesn't have enough physical strength will cannot operate standing motion and will fall into the wheelchair life or bedridden life. Furthermore, if once an elderly person falls into such life, the decrease of physical strength will be promoted because he will not use his own physical strength.[5]

In previous works, many researchers developed power assistance devices for standing motion[6].[7] However, these devices assist all necessary power for standing motion and they do not discuss how to use a remaining physical strength of a patient for realizing a standing motion. Thus, we have been developed a novel standing assistance system which uses a remaining physical strength of a patient.[8] Our system is based on a walker which is popular assistance device for an aged person in normal daily life and realizes the standing motion using the support pad which is actuated by manipulators with three degrees of freedom (Fig.1).

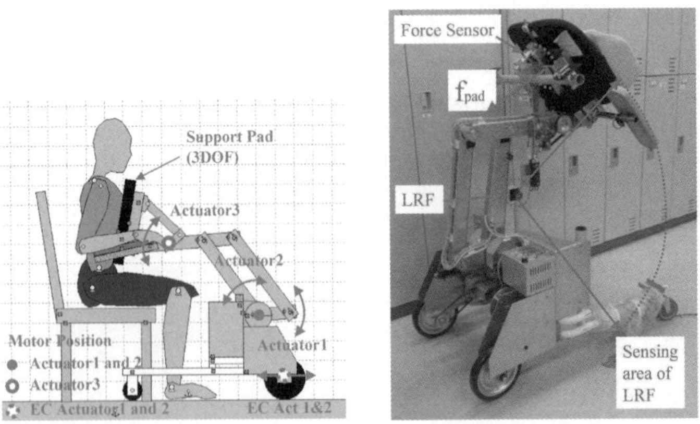

(a) Overview of our system

(b) Our prototype

Fig. 1. Our assistance system

For realizing a standing assistance using a remaining physical strength of a patient, it is important to estimate a load of a patient who uses the standing assistance device because the device is require to coordinate the assistance force according to the load of the patient. However, in general, it is difficult to estimate it without biosensors as an electromyograph and the load estimation with the sensors which are stuck on the patient is not practical. Therefore, in this paper, we propose a real-time load estimation scheme of a patient who uses our assistance system without additional sensors.

2. Proposed Load Estimation Scheme

For estimating the applied load to each joint, we assume the standing motion is symmetrical and we discuss the standing motion as a movement of the linkages model on 2D plane.[9] Using this linkage model, we can derive the traction of each joint and our system can estimate the load of the patient using these data during a standing motion at real time.

In our system, the patient leans on the support pad and puts his arm on the armrest. This means our system applies the assistance force to his chest and his forearm. The direction of these forces is vertical to the support pad and the armrest, respectively. Furthermore, during sitting on the chair, it assists the patient's hip to the perpendicular direction. Considering these conditions, we propose a linkage model which approximates the human body with our assistance device as Fig.2

This model consists of six linkages. The armrest applies the assistance force ($f_{armrest}$) to the center position of the link1 and the support pad applies the force (f_{pad}) to the center position of the link3. The chair applies the force (f_{chair}) to the joint, which connects link3 and link4, only when the patient sits on it. In link1, we can derive the following equations by the balance of applied force and its moment.

$$m_1\ddot{x}_1 = f_{xe} + f_{xarmrest}, \quad m_1\ddot{y}_1 = f_{ye} + f_{yarmrest} - m_1 g \tag{1}$$

$$I_1\ddot{\theta}_1 = f_{xe}\,(y_1 - y_e) + f_{ye}\,(x_1 - x_e) + \tau_e \tag{2}$$

where m_i is mass of the link ($i = 1, \cdots, 6$) and I_i is moment of inertia. (x_i, y_i) is the position of the center of gravity on each link and (x_i, y_i) ($i = a$, k, t, s and e) is the position of each joint. We assume that each linkage is pillar form and its mass is distributed uniformly.

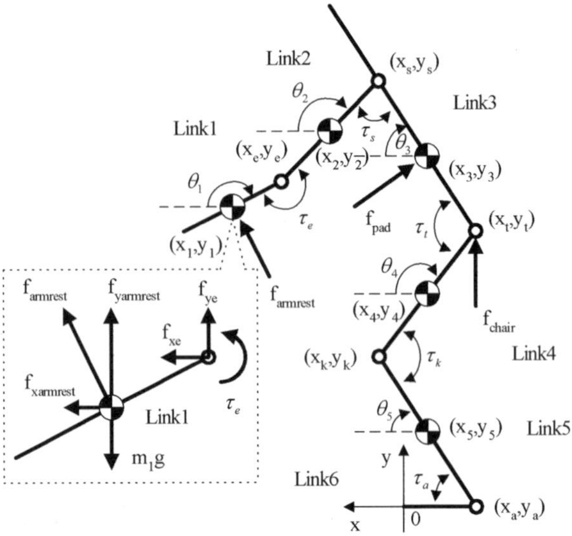

Fig. 2. Human Linkages Model

In this study, we use body parameters which are chosen from a standard body data of Japanese adult male[10] as Table 1. To derive the required body parameters for calculating its moment force, we measure the length of each body segment and the mass of the body on each subject individually.

Table 1. Human Parameters[10]

Number	Link Name	Mass [kg]	Length [m]	Width [m]
Link3	Head	5.9	0.28	0.21
Link3	Trunk	27.2	0.48	0.23
Link3	Hip	18.1	0.23	0.23
Link2	Humerus	4.5	0.39	0.12
Link1	Arm	2.7	0.35	0.08
Link1	Hand	0.5	0.2	0.07
Link4	Femur	9.1	0.61	0.17
Link5	Leg	4.5	0.56	0.16
Link6	Foot	0.8	0.26	0.11

By (1) and (2), the traction output τ_e is derived as (3).

$$\begin{aligned}\tau_e = &(y_e - y_1) f_{xarmrest} + (x_e - x_1) f_{yarmrest} \\ &+m_1 \{(y_e - y_1)\ddot{x}_1 - (x_e - x_1)(\ddot{y}_1 - g)\} + I_1\ddot{\theta}_1\end{aligned} \quad (3)$$

As the same, we can calculate the traction output of each joint as (4)-(7).

$$\begin{aligned}\tau_s = &(y_e - y_s) f_{xarmrest} - (x_e - x_s) f_{yarmrest} + m_2 \{(y_s - y_2)\ddot{x}_2 - (x_s - x_2)(\ddot{y}_2 - g)\} \\ &+m_1 \{(y_e - y_s)\ddot{x}_1 - (x_e - x_s)(\ddot{y}_1 - g)\} + I_2\ddot{\theta}_2 - \tau_e\end{aligned} \quad (4)$$

$$\begin{aligned}\tau_t = &-(y_s - y_t) f_{xarmrest} - (x_s - x_t) f_{yarmrest} + (y_t - y_3) f_{xpad} + (x_t - x_3)(f_{ypad} + f_{ychair}) \\ &+m_3 \{(y_t - y_3)\ddot{x}_3 - (x_t - x_3)(\ddot{y}_3 - g)\} + m_2 \{(y_s - y_t)\ddot{x}_2 - (x_s - x_t)(\ddot{y}_2 - g)\} \\ &+m_1 \{(y_s - y_t)\ddot{x}_1 - (x_s - x_t)(\ddot{y}_1 - g)\} + I_3\ddot{\theta}_3 - \tau_s\end{aligned} \quad (5)$$

$$\begin{aligned}\tau_k = &-(y_t - y_k)(f_{xarmrest} + f_{xpad}) + (x_t - x_k)(f_{yarmrest} + f_{ypad} + f_{ychair}) \\ &+m_4 \{(y_k - y_4)\ddot{x}_4 - (x_k - x_4)(\ddot{y}_4 - g)\} + m_3 \{(y_k - y_t)\ddot{x}_3 - (x_k - x_t)(\ddot{y}_3 - g)\} \\ &+m_2 \{(y_k - y_t)\ddot{x}_2 - (x_k - x_t)(\ddot{y}_2 - g)\} + m_1 \{(y_k - y_t)\ddot{x}_1 - (x_k - x_t)(\ddot{y}_1 - g)\} + I_4\ddot{\theta}_4 - \tau_t\end{aligned} \quad (6)$$

$$\begin{aligned}\tau_a = &-(y_k - y_a)(f_{xarmrest} + f_{xpad}) - (x_k - x_a)(f_{yarmrest} + f_{ypad} + f_{ychair}) \\ &+m_5 \{(y_a - y_5)\ddot{x}_5 - (x_a - x_5)(\ddot{y}_5 - g)\} + m_4 \{(y_a - y_k)\ddot{x}_4 - (x_a - x_k)(\ddot{y}_4 - g)\} \\ &+m_3 \{(y_a - y_k)\ddot{x}_3 - (x_a - x_k)(\ddot{y}_3 - g)\} + m_2 \{(y_a - y_k)\ddot{x}_2 - (x_a - x_k)(\ddot{y}_2 - g)\} \\ &+m_1 \{(y_a - y_k)\ddot{x}_1 - (x_a - x_k)(\ddot{y}_1 - g)\} + I_5\ddot{\theta}_5 - \tau_k\end{aligned} \quad (7)$$

Our system can measure the assistance force ($f_{armrest}$ and f_{pad}) by the force sensors equipped on its support pad and its armrest, and the inclination of a body segment by the laser range finder as Fig.3. Using the laser range finder, our system can estimate an inclination of a femur (Link4) and a leg (Link5). For devivation of them, we use a straight-line approximation technique by a least-squares method. Furthermore, our system can estimate an inclination of a trunk (Link3) easily because the patient leans on the support pad during assistance and this means the inclination of the support pad and one of the trunk are same. We can calculate the position and inclination of the support pad from the kinematical relationship of its parallel linkages.

Using the inclination which is measured by the laser range finder of a body segments and the human body parameters in Table.1, we can derive the position of a body segment easily. Thus, using (5)-(7), we can estimate the load of trunk joint τ_t, the knee joint τ_k and the ankle joint τ_a.

3. Experiments

For verify the effectiveness of our system, three young subjects test our prototype. The subject A is female and the subject B and C are male.

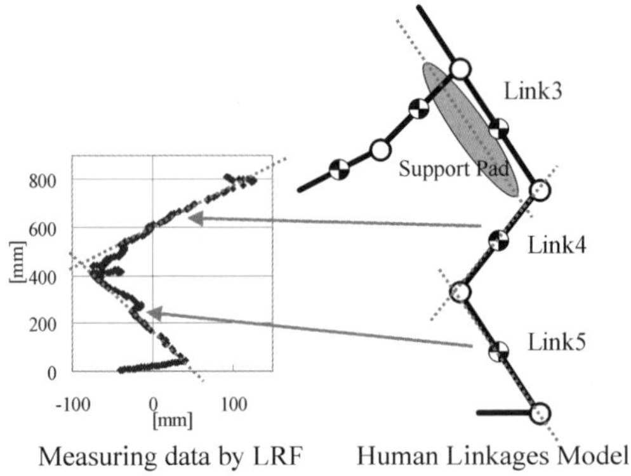

Fig. 3. Posture estimation using the human linkages model

Our prototype assists the subject with two modes. One is the position control mode and the other is force control mode. In general, the force control mode increases the assistance force and applied load of each joint will decrease[11].[12]

Fig.4 shows the standing motion of the subject A with a force mode of our assistance system. The subject stands up at 4[sec] with our system and this standing velocity is natural for young subjects.

Fig.5 shows the estimated applied load of each joint by our proposed scheme. From Fig.5, the applied load of each joint with a force mode is smaller than the load with a position control mode. In general, the applied load will be reduced by the force assistance and the estimated results are consistent. Fig.6 shows the EMG[13] data of a quadriceps muscle which reflects the activity of a waist joint. From these data, we can verify that its activity during the assistance with a force control mode is smaller than its activity with a position control mode. Thus, EMG results are same tendency as the estimated results by our proposed scheme and we can verify that our scheme is useful.

Fig.7 shows the results of all subjects and our proposed scheme works well at other subjects.

4. Conclusion

In this paper, we propose the real time load estimation scheme of the patient who uses the standing assistance device. For realizing it, our system measures the applied force and the posture of the user with the internal

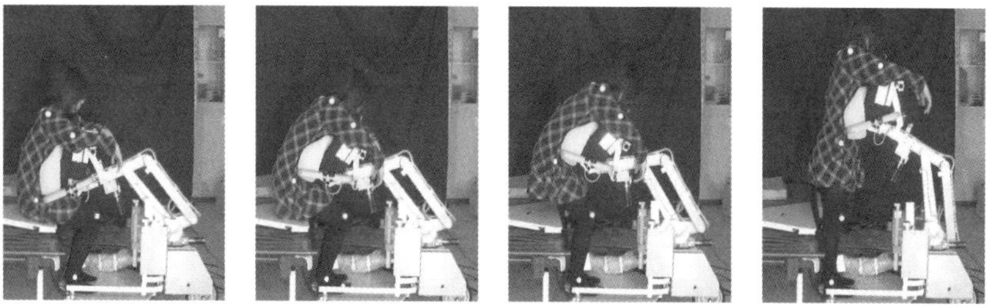

(a) 25[%] (b) 50[%] (c) 75[%] (d) 100[%]

Fig. 4. Standing motion with a force mode of our assistance device. (Subject A)

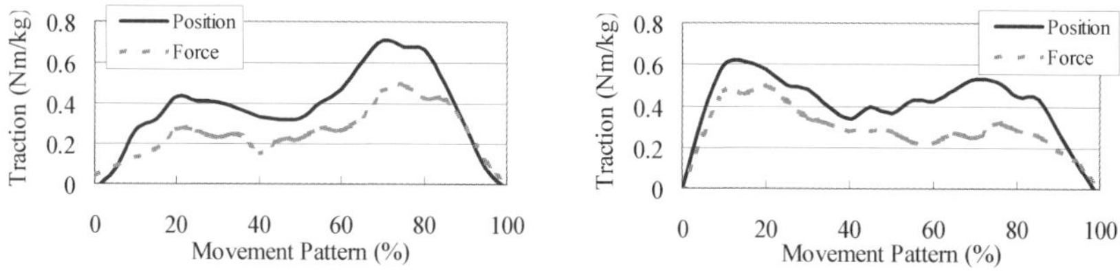

(a) Waist joint (b) Knee joint

Fig. 5. Estimated applied load of each joint during a standing motion. (Subject A)

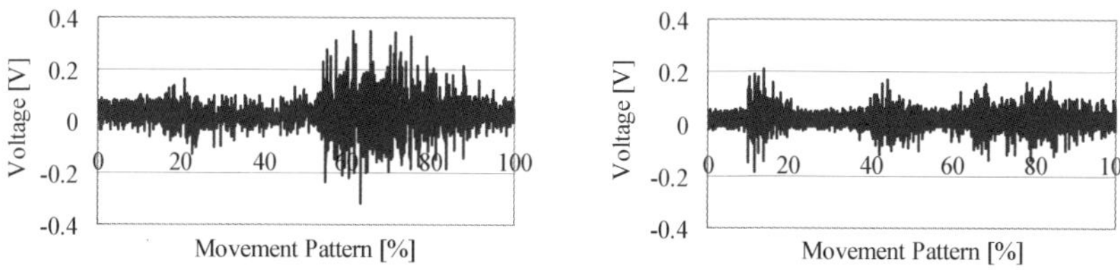

(a) Position control (b) Force control

Fig. 6. EMG of a quadriceps muscle during a standing motion. (Subject A)

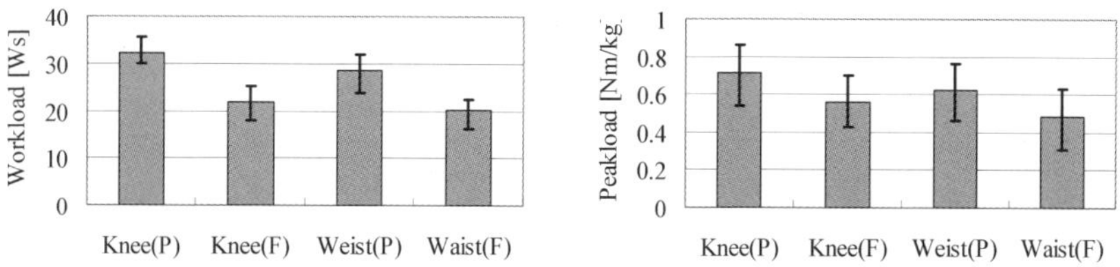

(a) Workload (b) Peak load

Fig. 7. Estimated applied load of each joint during a standing motion. (P) means the position control mode and (F) means the force control mode.

sensors of our assistive device and estimates the load of each joint with the measuring data and the linkage model, which approximates the human body.

In our future work, we will develop the assistance force coordination scheme using the estimated load individually.

Acknowledgments

This research was supported in part by Grant-in-Aid for Young Scientists B (23700676) from Japan Society for the Promotion of Science (JSPS).

References

1. Japan current population estimates as of january 1, 2013 (2013), `http://www.stat.go.jp/data/jinsui/index.htm`.
2. C. Office, *KOUREISHA HAKUSHO (The whitepaper on the aged society, written in Japanese)* (Government of Japan, 2011).
3. N. B. Alexander, A. B. Schultz and D. N. Warwick, *J. of Geometry: MEDICAL SCIENCES* **46**, M91 (1991).
4. M. A. Hughes and M. L. Schenkman, *J. of Rehabilitation Research and Development* **133**, 409 (1996).
5. M. Hirvensalo, T. Rantanen and E. Heikkinen, *J. of the American Geriatric Society* **48**, 493 (2000).
6. K. Nagai, I. Nakanishi and H. Hanabusa, Assistance of self-transfer of patients using a power-assisting device, in *Proc. IEEE Int. Conf. on Robotics and Automation (ICRA'03)*, (Taipei, Taiwan, 2003).
7. A. Funakubo, H. Tanishiro and Y. Fukui, *J. of the Society of Instrument and Control Engineers* **40**, 391 (2001).
8. D. Chugo, W. Matsuoka, J. Songmin and K. Takase, *J. of Robotics and Mechatoronics* **19**, 604 (2007).
9. S. Nuzik, R. Lamb, A. Vansant and S. Hirt, *J. of Physical Therapy* **66**, 1708 (1986).
10. Human body properties database (digital human research center of aist) (2006), `http://www.stat.go.jp/english/data/jinsui/2004np/index.htm`.
11. T. Sugihara, K. Kawabata, H. Kaetsu, H. Asama, K. Kosuge and T. Mishima, Development of a reasonable force sensor for a standing-up and sitting motion support system, in *Proc. of Robotics and Mechatronics Conf. 2004*, (Nagoya, Japan, 2004).
12. D. Chugo, T. Asawa, T. Kitamura, S. Jia and K. Takase, A rehabilitation walker with standing and walking assistance, in *Proc. IEEE/RSJ Int. Conf. on Intelligent Robots and Systems (IROS'08)*, (Nice, France, 2008).
13. H. Kobayashi, Intelligent wireless emg/ecg electrode employing zigbee technology, in *Proc. of SICE Annual Conf. 2011*, (Tokyo, Japan, 2011).

EXOSKELETON FOR ASSISTING HUMAN WALKING

I NYOMAN INDRAWIBAWA

Department of Electrical Engineering, South Westphalia University of Applied Sciences, Luebeker Ring 2, Soest, 59494, Germany

GURVINDER S. VIRK

Faculty of Engineering and Sustainable Development, University of Gävle, SE-80176 Gävle, Sweden

As the ageing society's problems evolve, mobility aid requirements are beginning to gain impetus and detailed studies on humans performing their normal activities for daily living are becoming urgent. The paper presents the design of a lower-body exoskeleton, which can be easily put on and worn, and used for recording human motions. In this way, the mobility requirements for performing daily living tasks are determined. The research involves the design of a 4 degree of freedom (DOF) lower-body prototype system used in acquiring basic motion data from volunteers as they perform basic walking tasks using embedded sensors for recording the hip and knee angles and transmitting them wirelessly. Walking tests with 20 volunteers are carried out after ethical considerations have been addressed; these cover human safety, full informed consent and privacy ensured by presenting the results in an anonymous manner.

1. Introduction

The rapid development in robotics has started to focus on contributing to personal care and health care applications whereas previously, the focus was on manufacturing applications in industrial environments or replacing humans in hazardous situations. This trend has been driven by the global ageing problems and related demands from elderly persons wanting to stay independent in their homes for as long as possible. A major part of the problem is how elderly persons can continue to stay independent and perform basic motion tasks needed for active daily living when their physical capabilities start to diminish. The wheelchair has been used for many years in such situations but has levelled off because its functionality in tight spaces is limited and its use does not satisfy the growing demands of the ageing society. This presents the main challenge for the future: **how the growing demand in the end users can be met, together with the growing end user population.** To meet this paradigm shift, interest in other forms of mobility aids has increased in recent years, and one of the main areas of attention is physical assistant exoskeletons. For this, the solutions are based on external electo-mechanical structures with associated controllers that can be worn by humans and are used to provide physical assistance to perform motions, such as standing or walking. Various systems have been developed and published going back to 1890 and beyond using compressed air for power (see Yagn [1]). Beard et al. [2] published the first patent for a powered gait orthosis in 1991 which was aimed for lower limb paralyzed persons. The assistive power came from mechanical elements such as springs and cables which were driven by the upper limbs of the user. One of the first patents for a computer-controlled exoskeleton was published in 1995, by Durfee et al. [3]; this too was targeted towards lower limb paralyzed persons to help them regain their walking ability. In recent years several companies like Honda [4], and Cyberdyne [5], have developed assistive systems to meet the needs of the elderly whose mobility capabilities are beginning to become impaired due to ageing. These systems are becoming quite sophisticated; for example, the controller for Cyberdyne's HAL suit can sense the muscle activity by using electromyography (EMG) signals and use them to drive the actuators and to support the user in a natural manner to perform various mobility tasks for active daily living. Despite these advances, the systems developed have not been fully commercialised and are still high value items out of the reach of most end users.

2. Biomechanics Study

2.1. *Existing biomechanical knowledge*

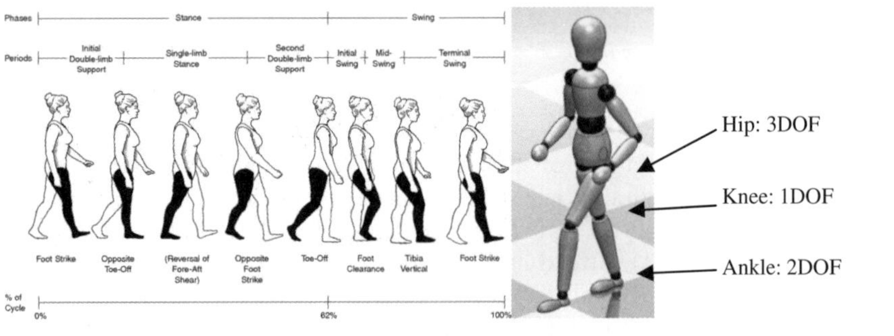

Figure 1: On the left normal human walking cycle [6] and on the right the human lower limb DOF

The focus of the research is to design low-cost assistive exoskeletons which can mimic and support the human during normal motion tasks such as walking and quiet standing. In fact, the basic movements need to be studied much more deeply if the functionality requirements for the needed exoskeletons are to be obtained. The gait cycle on the left of Figure 1 shows the movements for the limbs during normal human walking as they go through the stance (support phase) and into the swing phase with the toe-off, meanwhile the right sketch shows the simplified degrees of freedom (DOFs) that are needed during walking. These DOFs are felt to be the minimum needing adopting in the mechanical configuration concepts aimed at providing natural assistance during human walking without increasing the overall complexity unduly.

2.2. *Biomechanical study using accelerometer*

Accelerometers can measure forces (angular or translational) on a body having known mass [7]. Integrating the measured force gives velocity, and integrating again yields position. Here we plan to use accelerators to determine the hip and knee joint angles during normal human walking (in future this will be increased to include ankle joint angles); hence, 4 sensors are currently used to monitor both legs. 1D accelerators are used to determine the angle data by ensuring correct orientation of the accelerometer on the exoskeleton. The measured data passes to a microcontroller for filtering and processing. The filtration removes the excessive noise in the data by using simple low-pass filters; Figure 4 shows the raw and filtered data from an accelerometer placed at the knee. The data collected and filtered from the accelerators gives good and accurate results.

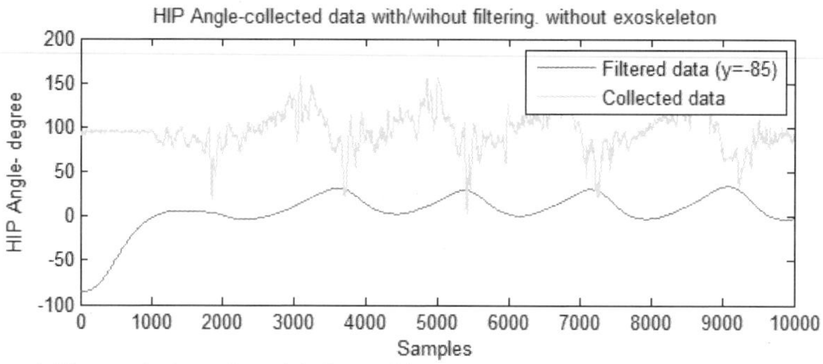

Figure 4: Knee angle data collected during walking: raw and filtered data. The filtered data is offset for clarity

Table 1: Exoskeleton priority demands

Specification	Description	Priority
Primary		
1.Dimensions	Adjustable from 160cm to 190cm in user height	•••
2.Weight	Lightweight: Less than 4kg	••
3.Cost	Affordable: Not exceed €2000	••
4.Safe: actuator power	Limited to 20% of torque needed for human walking	•••••
5.Manufacturability	Easy to manufacture using facilities available in Soest Labs	•••
6.Water proof	Useable outdoors in normal weather	•
7.Look	Appealing aesthetic design	•••
Secondary		
(i) Components	Easy to assemble and disassemble	•••
(ii) Replacement	Available on the market	•••
(iii) Maintenance	Easy to maintain	••
(iv) Manufacturing time	Within 3 months	•••••

Note: Priority level from • (Low priority) to ••••• (High priority)

3. Mechanical design

In order to perform the mechanical design for a wearable lower-body assistive exoskeleton, an overview of the primary and secondary requirements are determined and presented in Table 1. These requirements (fairly obvious to understand) represent the main issues considered in the research and their priorities during the design process.

3.1. *Design section distribution*

The exoskeleton has four basic sections which need to be designed; these are:
1. The connection to the wearer at the waist
2. The thigh section
3. The shin section
4. The ankle/ foot section

Although this paper focuses on the monitoring aspects, the final exoskeleton will be used for providing physical assistance and hence each joint will need actuation and hence the inclusion of the actuators. These have been limited to 20% of the power used in normal human walking ensuring that the exoskeleton will be a low-risk device and hence inherently safe as defined in ISO 13482*. In this way the wearer will be able to overpower the prototype exoskeleton in case of failure as only able bodied persons are planned to use it.

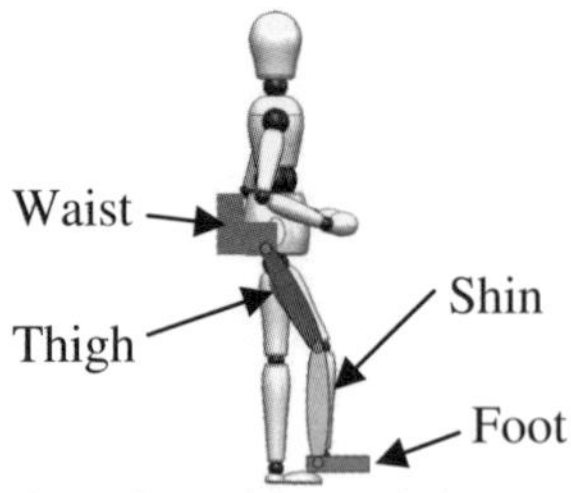

Figure 5: Main sections of the exoskeleton to be designed

Table 2: Specification and HIM for the different parts of the exoskeleton

Section	Specification	HIM
Waist	Provide support for battery and micro-controller Basis of 3 DOF hinge to the thigh Adjustable to fit variable width of user's waist	Belt and soft foam
Thigh	Provide support for the force sensor Adjustable to length of user's joint location Basis of 1 DOF hinge to the shin Provide support for the embedded actuators	Belt
Shin	Adjustable to length of user's joint location Basis of 2 DOF hinge to the user's foot	N/A
Foot	Adjustable to fit user's shoes Provide support for accelerometer sensor	Belt, soft foam, and adjustable pad to fix on shoe

* ISO/FDIS 13482:2013: Robots and robotic devices – Safety requirements for personal care robots

The design of each of the four parts needs to consider the best form of human interface mechanism (HIM) that should be used and appropriate support for embedding the electrical/computing sub-systems.

3.2. *Waist Section*

The waist section has a hinge with 3 DOFs, which provide freedom to rotate the entire leg as well as giving a natural circular walking motion. The HIM of this section uses one locking point with four connection interfaces to the exoskeleton structure, see Figure 6.

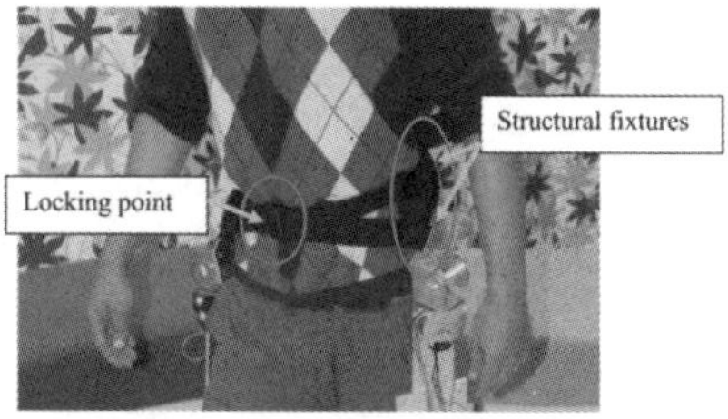

Figure 6: The waist HIM mechanism with one locking point and four fixture points to the structure

3.3. *Thigh section*

The thigh section design includes a curved front fixture allowing for placing of the force sensors. There is a length adjustment feature included using a sliding mechanism permitting the thigh length to be extended by 250mm, and the thigh section provides fixture grounding for two actuators. It also uses a belt with straps and a Velcro locking mechanism to allow easy putting on, taking off and for adjusting the exoskeleton suit.

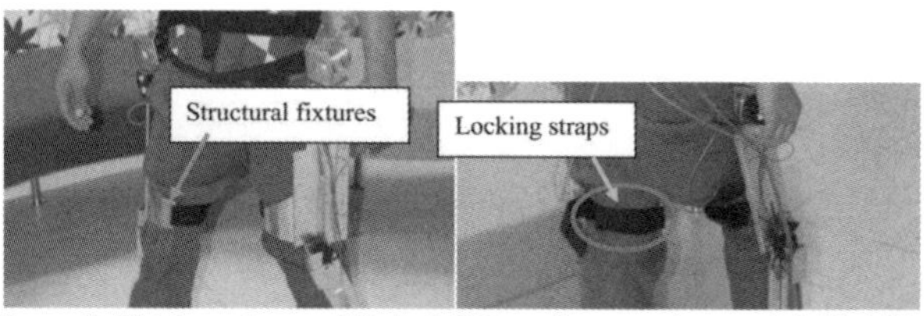

Figure 7: Thigh section and its belt fittings for easy don- and doff-ability

3.4. *Shin section*

The shin section has no HIM since the next connection felt to be needed is on the foot/ankle; however, the shin section is adjustable via a sliding mechanism as shown in Figure 8 for ensuring the comfort of the wearer.

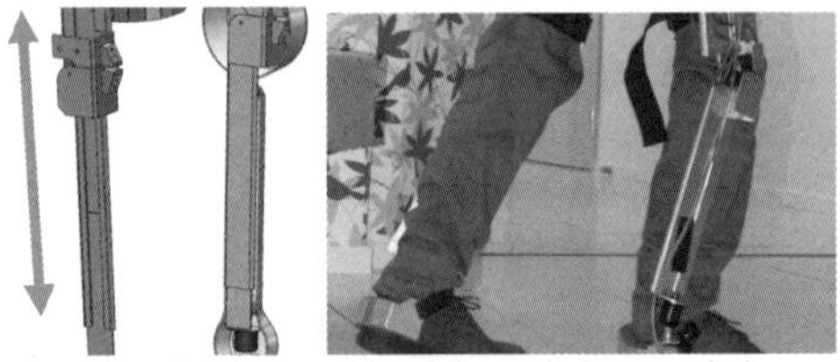

Figure 8: Shin section of the assistive exoskeleton

3.5. *Foot/ankle section*

The foot section implements a ball joint mechanism designed to allow limited angular motion for free rotation in the three axes. This is an advantage, since the foot cannot and should not rotate in all unwanted directions. Normally the human ankle joint rotates mainly in one direction and the ball joint mechanism can mimic such behaviour. To gather angular data, an accelerometer has been used to simplify the design for future work.

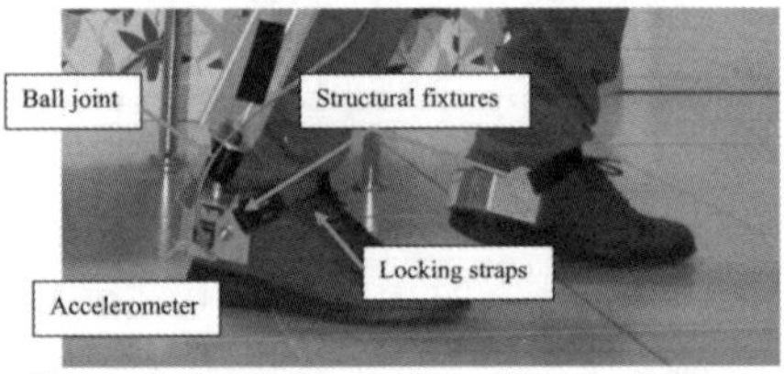

Figure 9: Foot/ankle section of the exoskeleton

4. Electrical Implementation and Results

The exoskeleton uses an Arduino Mega 2560 as the micro-controller board, with a WiFi shield as the communication board. Figure 10 shows the location of the various key elements of the exoskeleton system.

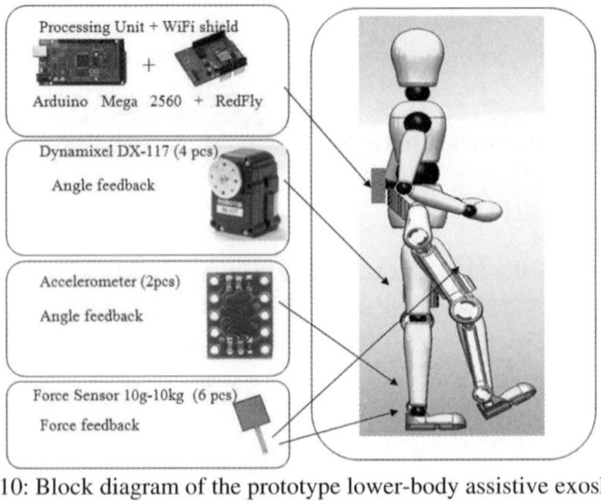

Figure 10: Block diagram of the prototype lower-body assistive exoskeleton

4.1. *Human walking test data*

The data collected from the volunteer walking tests is designed to be stored and analysed using MatLab so that the requirements for designing assistive mobility exoskeletons could be obtained. Figure 11 shows an example of the human walking data collected.

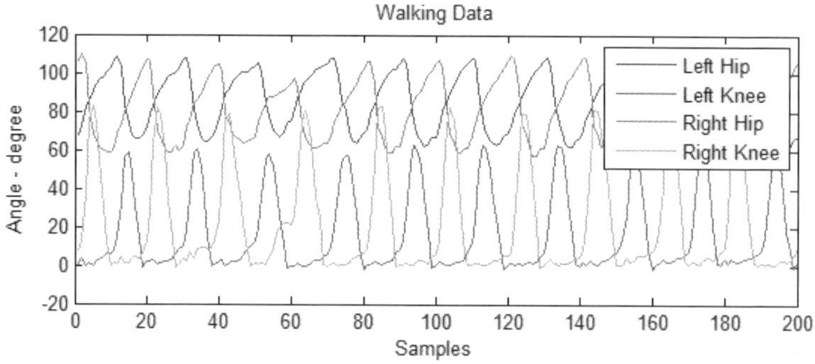

Figure 11: Example of human walking data gathered from servos allocated on the hips and knees

5. Exoskeleton verification tests

5.1. *Ethical protocol for human walking test experiments*

An ethical protocol was developed and adopted for the human volunteer walking experiments ensuring full informed consent was obtained from all test subjects. The ethical protocol included the following issues:

1. Ensuring all volunteers fully understood the planned research and details of the experiments to be carried out
2. Signed agreement to take part in the planned research
3. Signed agreement to use the recorded results in future research and publications in an anonymous manner.

The strategy was first to announce the volunteer walking tests on the University of Gävle's website, use personal contacts at the University as well as in local elderly care homes in Gävle Kommun. For the public announcement, an introductory document was produced and distributed explaining the research aims and the volunteer's requirements. This resulted in 20 volunteers coming forward who signed the ethical consent form and participated in the experimental walking tests. The volunteers wore the exoskeleton suit and performed specified walking tasks so that their individual patterns were recorded directly from the system. These volunteer tests have been designed to help in the design of the controller for the existing exoskeleton or for future developments.

5.2. *Results*

Figure 12 shows some walking volunteers that have performed the tests, and the walking data is currently being analysed.

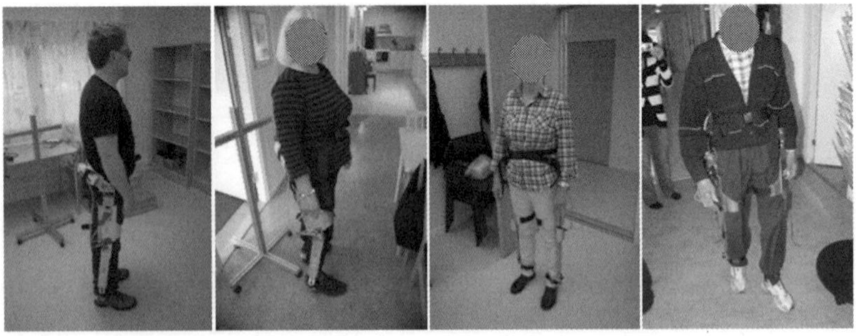

Figure 12: Volunteers wearing the prototype assistive exoskeleton suit

6. Conclusions

The design and development of a prototype exoskeleton has been presented. The focus of the paper has been on the human motion recording aspects although the exoskeleton will be used for providing assistance in the future. The exoskeleton has been used by 20 volunteers whose walking patterns have been recorded and these are being analysed to determine the functionality requirements for designing physical assistance lower-body mobility exoskeletons. Although the design of the sliding mechanism needs some improvements, from the locking side and the size adjustment slider movement, most of the volunteers quoted that the overall exoskeleton design suited their body anatomy well, giving an impression of safety and comfort when worn. The volunteers seemed to have a high hope for such exoskeletons to be produced in an affordable manner for the mass consumer market. Applying EMG sensors as the input controller system can be a future improvement of this project. A better test environment for the volunteers, such as using a special device to control the walking speed, like a controllable walking treadmill will also be adopted in future work.

References

[1] Yagn, N, Apparatus for facilitating walking, running and jumping, USA Patent 440684, 1980.

[2] Beard JE, Spillman FL, Banos O, Showers D, Fussell JM, and Le Blanc G, Powered gait orthosis, USA Patent 5020790, 1991.

[3] Durfee WM and Goldfarb MB, Controlled brake orthosis, USA Patent 5476441, 1995.

[4] Honda US patents: 1) Walking assistance device 7780616 B2 (2010); 2) Walking assist device 8142371 B2 (2012; and 3) Exoskeleton controller 774177B2 (2010),

[5] Cyberdyne, Yoshiykki Sankai, US Patent 2008/0234608 A1, also see information on the internet as URL: www.cyberdyne.jp/english/.

[6] Lim, MR, Huang RC, Wu A, Girardi FP, and Cammisa Jr FP, Evaluation of the elderly patient with an abnormal gait, J Am Acad Orthop Surg; 15:107-117, Feb 2007.

[7] Dimension Engineering, A beginner's guide to accelerometers, [Online]. Available: http://www.dimensionengineering.com/info/accelerometers.

HOPPING MACHINE: A TOOL FOR HUMAN RUNNING SIMULATION AND FOR LEGGED ROBOTS AND EXOSKELETON DESIGN

A. H. MAIA[1], S. LAQUINI JR.[1], A. BENTO F^{o}[1], V. H. B. FERNANDES,[1] and E. GARCIA [2]

[1] *Mechanical Eng. Departments, UFES, Vitoria, ES 29075-910, BRASIL*
antonio.bento@ufes.br
[2] *CAR - Centro de Automática y Robótica, UPM - CSIC, La Poveda, Madrid, España*

This paper presents a *3 DOF (Degrees Of Freedom)* planar hopping machine running in a simulation software environment. The purpose is to use this approach to study the control, simulation and design of legged robots and exoskeletons. The control system allows the machine hopping in place after fall with his leg a little inclined, then hopping forward, then hopping backward and then stop hopping in place. Hopping machines are the most simplified dynamic model capable of representing the elastic behaviour of the legs of mammals and insects in running and hopping. The results of the simulation presented shown that pelvic tilt observed in human running is also observed in the hopping machine running model. This encourages the modelling of anthropomorphic legs for use in human running simulations with hopping machine models and control strategies.

Keywords: running, hopping machine, legged robot, exoskeleton.

1. Introduction

The three *DOF* hopping machine model implemented is showed in Fig. 1. It has a pin joint at the hip and a sliding joint in the leg. The mass and inertia properties, Table 1, are from Raibert.[1] A relatively simple approach in three main parts is used to control the hopping machine dynamic behaviour: supporting the body with a hopping movement; control the attitude of the body at each stance; and position the feet at each flight to keep the balanced movement. This approach can be used to control one legged robots, bipeds and quadrupeds.

The simulation environment is the *YSCS*,[2] a full-featured *OOP* software package for simulations of robots, bio-mechanical systems, and mechanical devices with which it's possible to simulate *N* order rigid body physics;

Table 1. Dimensions and physical properties of Raibert's planar hopping machine.[1]

Parameter	Value	Unit
Overall height	m	0.69
Overall width	m	0.97
Hip height	m	0.5
Total mass	kg	8.6
Unsprung leg mass	kg	0.45
Body moment of inertia	kgm^2	0.52
Leg moment of inertia	kgm^2	0.037
Leg axial motion		
stroke	m	0.25
static force	N	360
Leg sweep motion		
sweep angle	rd	+-0.33
sweep torque	Nm	27
Maximum pressure	kPa	620

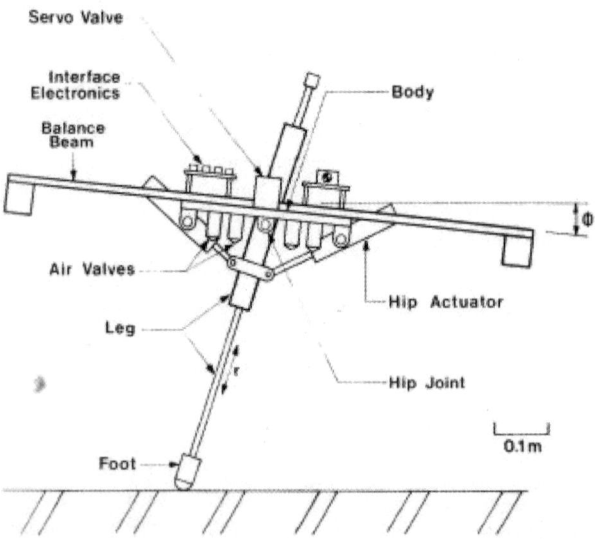

Fig. 1. Raibert's hopping machine; dimensions, inertias and other characteristics are from.[1]

access all joint positions, velocities, and torques; and plot real-time graphs of any variable. The simulation environment, output variables and graphics are specified in the *Simulation* open library; the hopping machine is constructed in the *Robot* library; and the three parts dynamic control is modelled trough a state machine in the *Controller* library.

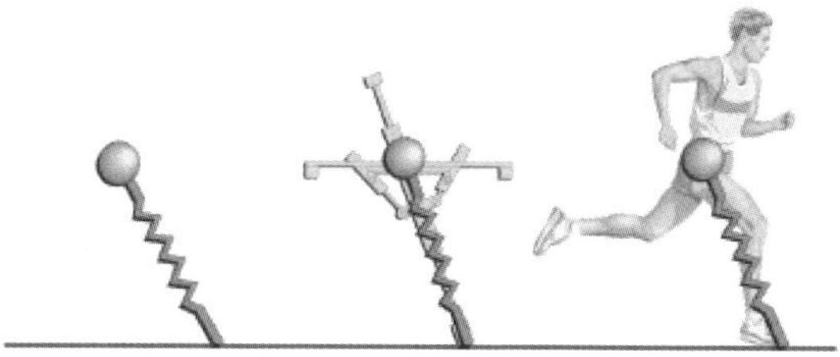

Fig. 2. From left to right: *SLIP* model;[4] Raibert[1] hopping machine ; *SLIP* model illustration in a runner.

In Fig. 3, are presented the pictures from the hopping machine simulation results showing the machine performance. In the simulation, it was showed an unexpected behaviour of the body attitude *nonzero* value during the simulation opposing which was observed by Raibert.[1]

This enabled an opportunity to investigate and compare the model's attitude with the pelvic tilt observed in human running described in the work from Schache.[3] It was found that the model behaviour resembles the human running as expected for the spring loaded inverted pendulum *(SLIP)*[4]model of the hopping machine.

2. Background

2.1. *Human running modelling*

Studies[5–7] have shown that the best pattern to represent the race of animals is to bounce, since it is a way to save energy, storing up to 70% of the kinetic energy of flight phase in elastic energy and then using it to take off again.

The *SLIP* model, Fig. 2, shows that the musculoskeletal system of an animal can be mechanically described as a non-linear multicomponent spring-mass model, being the simplest possible model for a bouncing system. It's possible to bounce in place or jump ahead, varying the horizontal position. Bouncing in place has a remarkable resemblance to a simple spring-mass model.

Using this knowledge, better exoskeleton designs are exploring our understanding of human walking to build energy efficient machines for use as locomotion assisted devices. The lower limb exoskeletons can take the place of a wearer's body function by performing motion like human autonomously. They could realize the tasks such as walking, standing up, etc. and can be used as a function aid apparatus for gait disorder people.[8]

The storage of elastic energy is critical for efficient bipedal locomotion. If the human ankle, during the stance phase of walking, is described in terms of mechanical spring elements, the dynamics in the sagittal plane of a normal ankle can be reproduced at walk speeds slow to moderate, with the energy stored in the spring during passive gait cycle.[9] With numerical simulation it was shown that an exoskeleton using passive elastic devices can substantially reduce muscle strength and metabolic energy to walk.[10] Some successful applications of exoskeletons with elastic energy storage during passive gait cycle are being developed.[?,11]

2.2. *The hopping machine simulation model*

The first simulation is to check if the machine can start hopping and restore equilibrium bouncing vertically, when dropped from a 200 mm height with the leg a inclined 0.1 rd with the body vertical axis. Next the machine walks forward at constant $1.18km/h$ speed and finally a complete simulation with steps: drop down and start bouncing; walk forward at increasing speed; walk forward at constant speed; walk forward at decreasing speed; stop bouncing and repeat backwards. Fig. 3 shows one step sequence: touch down on upper left picture, attitude control from the second to fifth picture, taking off at the bottom left and flight with leg positioning until next touch down at bottom right picture.

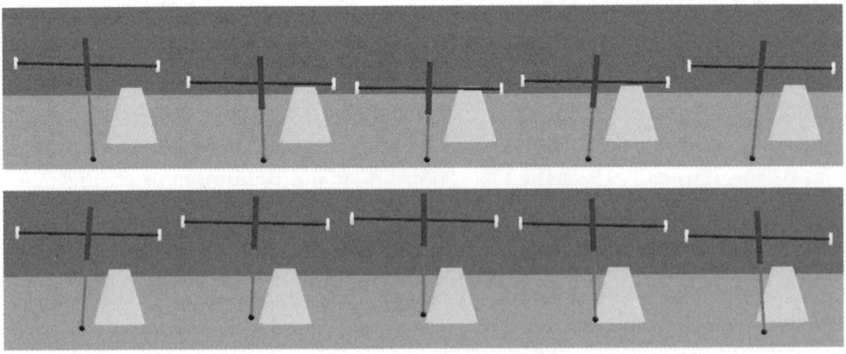

Fig. 3. One step bouncing forward at $0.90km/h$ from left to right and $0.07s$ between frames.

3. Pelvic Tilt Observed from Hopping Forward

The sagittal movements of the pelvis are controlled by gravity, inertia, and the action of the hip flexor and extensor muscles, Fig. 4. The center of gravity is within the pelvis when the body is in the anatomic (upright) position; however, dynamic postural changes can alter the location of the center of mass. The pelvis is inclined forward (anterior) or flexed and moves in a sinusoidal manner with two peaks and two valleys.[12]

In a study[3] it was used a motion analysis system with 14 elite athletes

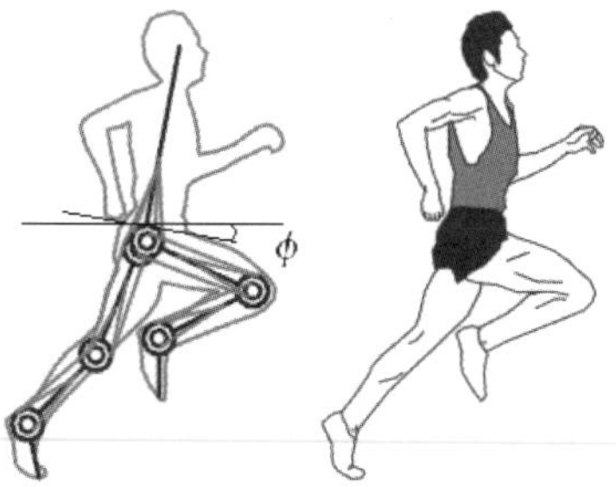

Fig. 4. Human running model illustration.

running on a treadmill at $20km/h$ to assess the relation of anterior pelvic tilt during running to peak hip extension range of motion measured during running and hip extension flexibility measured clinically, using the coordinate system and variables shown in Fig. 5. Results for sagittal plane pelvis

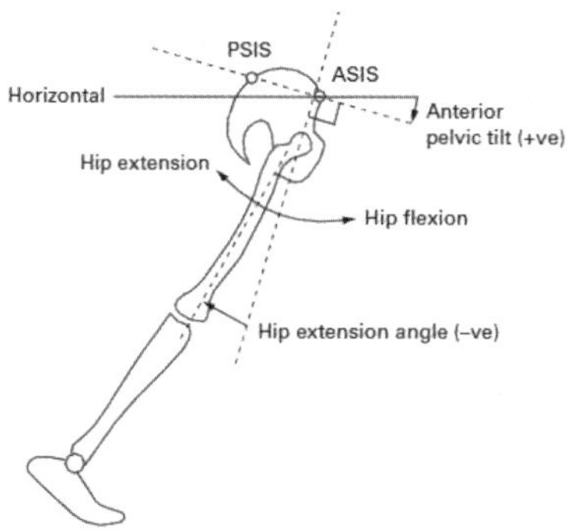

Fig. 5. Diagrammatic representation of the measurement of anterior pelvic tilt and peak hip extension range of motion during running. *PSIS*, posterior superior iliac spine; *ASIS*, anterior superior iliac spine.[3]

and hip angles over the running cycle for a subject is shown in Fig. 6. The anterior pelvic tilt and the hip flexion ranged from approximately -25° to 60° and 10° to 20°, respectively. Toe off marks the single limb stance in which the opposite leg flights to a new stance phase.

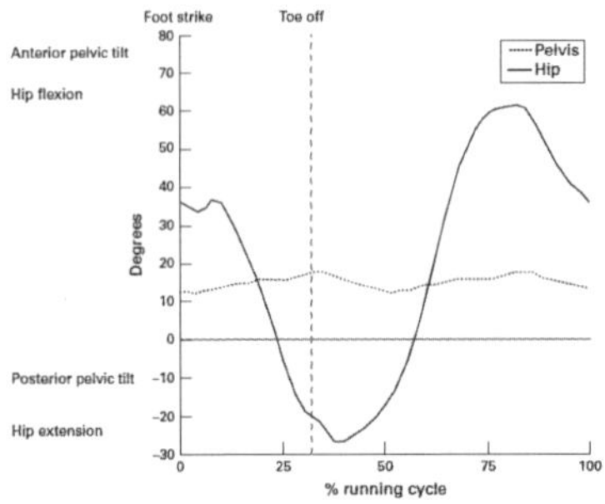

Fig. 6. Sagittal plane pelvis and hip angles over the running cycle for the subject displaying the greatest degree of peak hip extension range of motion running on a treadmill at $20km/h$.[3]

Sagittal plane pelvis and hip angles over the running cycle for the hopping machine running at $1,18km/h$ is shown in Fig. 7. The anterior pelvic tilt and the hip flexion ranged from approximately -3° to 6° and 0.5^o to 1.5^o, respectively. The hopping machine has only one leg and it has to bounce to allow the leg to flight to a new stance phase. The single limb stance observed in a biped, which allows for the opposite leg flight to a new stance phase, is done through a flight phase in the hopping machine.

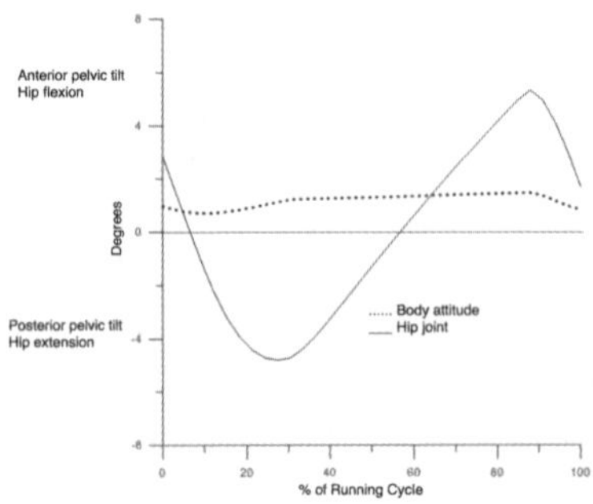

Fig. 7. Sagittal plane pelvis and hip angles over the running cycle for the hopping machine running at 1.18 km/h

4. Discussion

It could be observed a resemblance in the shape of the hip flexion-extension and anterior pelvic tilt curves for the athlete and the hopping machine, as shown in the graphs from Figs. 6 and 7. This resemblance, however, shows only that the behaviours are similar as the athletes maximum and minimum values of hip and pelvic tilt angles are about ten times greater than those from the hopping machine, which, otherwise, could be due to the speed of the athletes have been about 17 times greater than the the hopping machine speed during the data acquisition process. So what could be said is that the simulation results are encouraging to continue the research, improving the hopping machine model.

Although results[13] provide support for the idea that upper body movement is inherently self-tuned, producing stable walking and running even when upper body inertial properties are modified, upper body kinematics are significantly affected by restricting arm swing, which could be due to upper limbs, torso, neck and head inertias and dynamics that react favourably to balance the upper body dynamics.

A leg changes length to propel the body upward and forward, to cushion landings, and to reduce its own moment of inertia and increase its clearance when swung forward. The lengthening and shortening of a leg during these activities is a dynamic action governed by the resonant interaction of leg compliance, body mass, and gravity. The resonant interaction between body mass and springy legs' in the vertical direction has a profound impact on the' behaviour of a running system.[14]

5. Conclusion

One of the motivations to study one-legged robots is to gain a good understanding of system dynamics and extend it to human and animal locomotion.[15]

The procedure presented in this work, shows the use of the *SLIP* model implemented through a hopping machine built in an simulation environment. It could be observed a resemblance in the shape of the hip flexion-extension and anterior pelvic tilt curves for the athlete and the hopping machine. This encourages the searching for improving the model and to use in further studies.

It permits the easily simulation of leg dynamics which allows the study the human gait and it is a powerful tool for lower limb prostheses and other artificial body parts design and evaluation. It also can be used for exoskeleton simulation and design and for agile legged robots simulation and design, the focus in the search of energy efficient applications design.

References

1. M. H. Raibert, *Legged Robots that Balance* (Mit Press, Cambridge, Massachusets, 1986).
2. Yobotics, *www.ihmc.us/groups/scs/* **1** (2012).
3. A. G. Schache, P. D. Blanch and A. T. Murphy, *British journal of sports medicine* **34**, 279 (2000).
4. R. Blickhan, *Journal of Biomechanics* **22**, p. 1217–1227 (1989).
5. R. Alexander and A. Vernon, *Journal of Zoology* **177**, 265 (1975).
6. G. A. Cavagna, *Exercise and sport sciences reviews* **5**, p. 89–130 (1977).
7. N. C. Heglund, G. A. Cavagna and C. R. Taylor, *Journal of Experimental Biology* **97**, p. 41–56 (1982).
8. H. Kawamoto and Y. Sankai, *Advanced Robotics* **19**, 717 (2005).
9. M. L. Palmer, Sagittal plane characterization of normal human ankle function across a range of walking gait speeds, PhD thesis, Massachusetts Institute of Technology, (MIT Press, 2002).
10. A. J. V. D. Bogert, *Biomedical engineering online* **2**, 17 (2003).
11. S. Lee and Y. Sankai, *Nippon Robotto Gakkai Gakujutsu Koenkai Yokoshu* **19**, 1259 (2001).
12. J. Rose and J. G. Gamble, *Human Walking*, 3rd edition edn. (Lippincott Williams & Wilkins, 2006).
13. H. Pontzer, J. H. Holloway, D. A. Raichlen and D. E. Lieberman, *Journal of Experimental Biology* **212**, p. 523–534 (2009).
14. M. H. Raibert, *Ieee Transactions On Systems Man And Cybernetics* **14**, p. 451–463 (1984).
15. A. Sayyad, B. Seth and P. Seshu, *Robotica* **25**, p. 587 a 613 (2007).

DESIGN OF A RECONFIGURABLE WHEELCHAIR WITH A SIT-TO-STAND FACILITY FOR A DISABLED KID

K M GOHER*, M SHAFIQ and A AL-YAHMADI
Department of Mechanical and Industrial Engineering
Sultan Qaboos University, Oman,
**Corresponding author: e-mail: kgoher@squ.edu.om*

The objective of this work is to design and build a reconfigurable wheelchair for the purpose of rehabilitation and self-assistance for disabled 25*kg* child. The proposed work aims at developing a wheelchair with a mechanism that enables the wheelchair user to transform from sit to stand posture and vice versa. This configuration will allow the user to use the wheelchair as a mobility device as well as for rehabilitation purposes without need any external support. The main idea of this project is to allow blood circulation in the legs as sitting for long periods do not allow the blood to circulate normally through the legs and hence inflates the disability. This wheelchair also allows disabled people to stand, complete activities to equate them to normal people and make them capable of assisting themselves without the need of any help or feeling embarrassed. The special feature of this wheelchair is that it has to be of a low cost comparing to other wheelchairs with the same functionality.

1. Introduction

Assistive technology for the mobility impaired includes the wheelchair, lift aids and other devices. However, in most cases; the patient eventually requires assistance to use the device; whether it's someone to push them in a wheelchair, to lift them from the bed to a chair or to the toilet or for guiding them through cluttered areas. With fewer caregivers and more elderly, there is a need for improving these devices to provide them independent assistance (Bostelman). Assistive robotics including wheelchairs is a multidisciplinary work that involves bio-mechanics, mechanical design, electronics, materials science, computer science and robotics (Guo-zheng et al, Li-ning et al). Assistive robots can help patients in order to perform, independently; effective rehabilitation exercises which may lead to fast and better recovery of functions of their muscles (Zhang et al). Many elderly people who can walk are trapped in chairs, which are caused by lack of standing ability. Many others who struggle to stand require a means of rehabilitation and care. With the availability of such facilities; physical rehabilitation and damage prevention and/or necrosis of muscles may be assured (Wang and Jin).

Numbers of powered wheelchair-type aids are developed for elderly and disabled people who cannot walk and have extremely poor dexterity (Yanco,

Hoyer, Katevas, Cooper, Ding and Dicianno). These devices are well suited for people who have little or no mobility. However, they are not appropriate for the elderly with significant cognitive problems. Consequently, the facilities are generally reluctant to permit the elderly residents to use powered wheelchairs (Zhu et al).

The need of transforming human body from a sitting to a standing posture represents a main aspect of human body daily activities. The sit-to-stand task has become a basic test of functional mobility whether measured alone within a laboratory environment (Yamada and Demura) or as part of a larger activity monitoring system for elderly people (Najafi et al). Although there are number of examples of wheelchairs with a sit to stand facility are currently available in the market: XO-202 shown in Figure 1, SuperStand (SS-1) in Figure 2, a conceptual wheelchair design shown in Figure 3 and the LEVO C3 which appears in Figure 4, there is not much research is associated with such designs.

Fig. 1 Karman Healthcare XO-202 power stand-up wheelchair with table

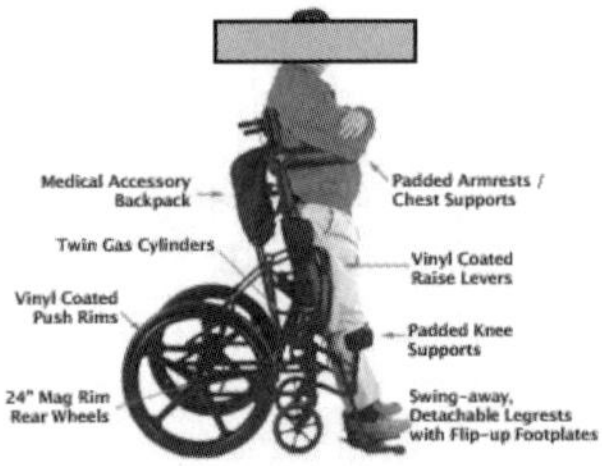

Fig. 2 SuperStand (SS-1) manual standing wheelchair by the Standing Company

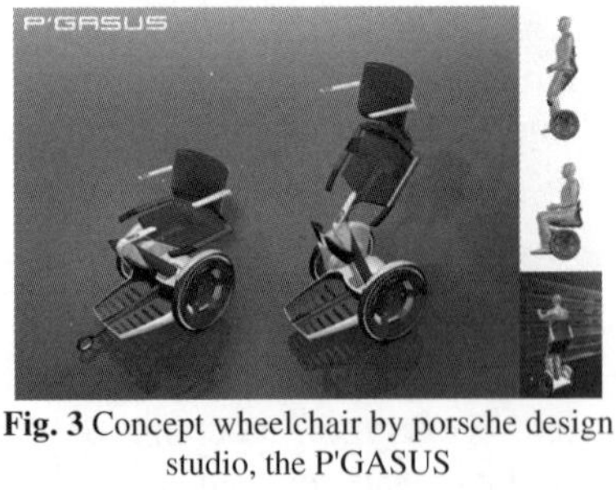

Fig. 3 Concept wheelchair by porsche design studio, the P'GASUS

Fig. 4 LEVO

In this paper; the authors aim at designing and building a reconfigurable wheelchair, with a sit-to-stand mechanism, for the purpose of rehabilitation and self-assistance for a disabled 25 kg child. The proposed design aims at developing a research-based facility for various potential research areas including: studying the interaction between the user and the device, investigation of the kinematics and dynamics of the wheelchair and the human body, studying energy consumption in the wheelchair and the equivalent energy if needed to be offered by the human muscles. The work presented in this paper is arranged as follows: Section II presents and investigates three design alternatives of the wheelchair. Based on the investigations of those design option; a final design is refined in Section III considering all drawbacks identified earlier. In this section; full detailed explanation is provided covering most of the major parts of the wheelchair. Numbers of design safety recommendations are presented in Section IV and the work is concluded in Section V highlighting the main points covered in the research.

2. Proposed Design Options

There are different features and designs for the reconfigurable wheelchairs in the market. The Figures below show two of the design concepts used. These two specific designs had been chosen due to the previously discussed social and economic impacts pacifically. The lift wheelchair has the benefit of reaching higher altitudes even that a healthy standing human can reach, which would relieve some of the patient's inferiority complex. However, the more height the W/C can reach the less stable it is due to the momentum the persons cause. Also this design would suit mostly the lower limbs amputated people who do not need any rehabilitation.

2.1. First Conceptual Design

The first design option as illustrated in Figure 5 is based on the fact that one of the current trends in industry is to minimize the area occupied by a product while manoeuvering. This design mainly simulates an inverted pendulum scenario moving on two wheels. This configuration of the proposed wheelchair allows it to occupy a limited area while standing comparing to currently existing wheelchairs. However, it has the disadvantage of being highly unstable in the standing mode due to the continuous change of the location of the centre of mass of the disabled person which leads in turn to a significant energy consumption at the start of motion.

2.2. Second Conceptual Design

The second design option appears in Figure 6 is a modified version of the previous design where two small cluster wheels are added at the rear side of the wheelchair to gain more stability. Two more pair of motors added to the wheelchair legs to separate them from the wheels in order to provide a reaction force to the patient movements instead of letting the wheel's motors trying to stabilize the wheelchair. This modification will overcomes the problem of instability associated to the inverted pendulum configuration in addition to providing a more safety to the system while changing between sit and stand modes. Both previous design alternatives are equipped with a geared system to change between the chair base and the lower legs of the wheelchair. Although this option offers the same function of the sit-to-stand transformation; it is still suffering from the instability condition while performing the transformation. This design was rejected because of the high energy consumption due to using six motors. In addition; the motors have to be always in the active mode in order to keep the system stable. Furthermore; this configuration is considered to be a double inverted pendulum.

2.3. Third Conceptual Design

In order to overcome the problems associated to the first two designs; an enhanced proposed design is shown in Figure 7. The lower part of wheelchair comprises four wheels and four legs. However, each rear leg is composed of two links as shown in Figure 13. This design offers better stability if compared to previously mentioned ones. In addition it will consume less a relatively power comparing to the other two designs because the motors will be used only while transforming between sit and stand modes. Furthermore; the weight of the person will be distributed among the four legs which mean that only a one quarter of the body's weight will apply the force on the lifting motors. In order to justify the selection of the proposed final design; a summary, shown in Table 1, is presented to show the main differences between the three designs in terms of the working occupied space, power consumption and stability. The data available in this table has been verified based on simulations of the three design options under the same conditions of loading.

Fig. 5 First design option

Fig. 6 Second design option

Fig. 7 Third design option

Table 1: Comparison between the three alternative designs

Comparison	Power consumption	Working space	stability
1st design	high	small	unstable
2nd design	Very high	small	low stability
3rd design	Low	average	stable

3. Final Design Details

The final proposed design, shown in Figure 8, is taking care of all threats and/or problems associated with the previously presented options. Although the design of the joint responsible for changing the orientation of the back seat is designed first as illustrated in Figure 9; this alternative is found to be complicated in terms of manufacturing and the possibility occurrence of improper orientation of the back seat in case of little manufacturing/assembly imprecision. An alternative design of this joint is shown in Figure 10 which relies mainly on using two electrical linear actuators (on both sides of the wheelchair). Detailed explanation of the implementation and the working principle of this hydraulic cylinder are presented in Figure 11.

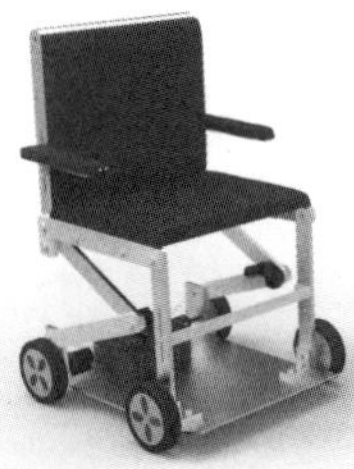

(a) Sitting mode

(b) Standing mode

Fig. 8 Final proposed design with details

3.1. Design of the hydraulic cylinder

As appeared in Figure 10; part 1 is fixed to the back seat and connected to the linear actuator using a revolute joint and the actuator is connected to the seat using a similar joint. This design enables the patient to fully control the back angle as per his/her desire and comfort. However for the time being; this design will be only used to automatically adjust the back seat angle in sit/stand modes. For typical situation; the actuator will be fully extended in the sit mode and the linear distance that the piston rode will move, while changing to the stand position, is equal to the length of part 1 which is 4 cm. The force applied by the linear actuator can be calculated by equating the moments around the pivot point shown in Figure 11. This force is necessary to counteract the patient's pushing reaction force due to his/her back. The range of change the orientation angles of the chair and the lower legs are designed in order to allow the disabled user to full stand; detailed values of those angles are shown in Table 2.

Table 2: Angle Comparison between Sit and Stand Modes

Comparison	Sitting Mode	Standing Mode
Angle 1	90	180
Angle 2	22.45	39.57

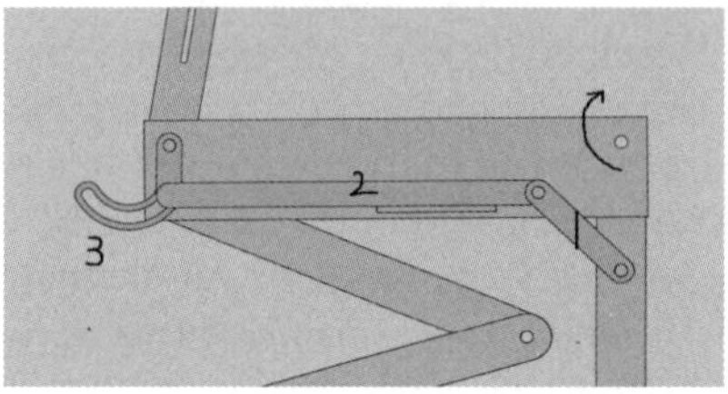

Fig. 9 First design option for back joint

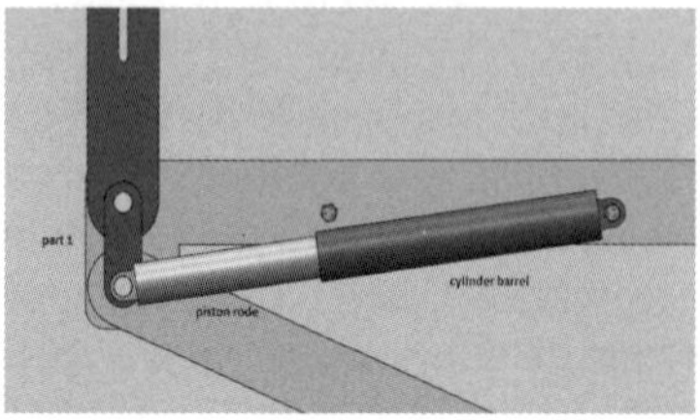

Fig. 10 Linear actuator used to adjust the back seat

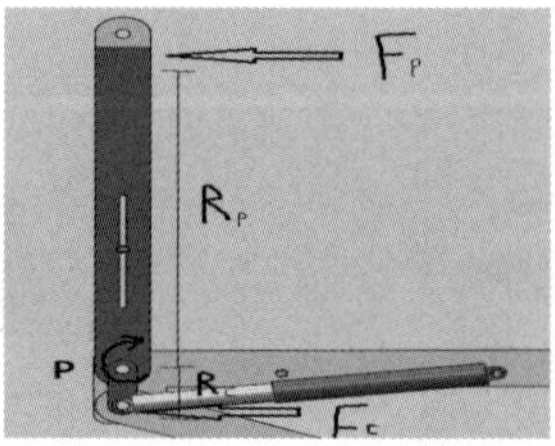

Fig. 11 Moment equation for back seat

Since the linear actuator has to exert a minimum force F_a which is equal to the patient's force F_p; the moment due to the actuator's force has to be equal and opposite to the patient's moment.

3.1.1. Calculation of the patient's force (F_p)

In order to calculate the patient's force F_p; a force sensitive resistor (FSR) is used and placed on a solid surface at a distance equal to R_p from the ground. Then a sample of 6 children, aged between 6-10 years old, is used to calculate the patient's force. Children are advised to lean on the solid surface, one at a time, where his/her body-exerted force is measured by the sensor. The sensor's output is found to be in the range between 0 for maximum force and 1023 N for zero force. To interpret the output readings of the sensor, all measurements have been calibrated and compared with loads with known weights. Table 3 summarizes the details of the children and their associated exerted forces.

Table 3: Force measurement for the user's body

Age (year)	Gender (M/F)	Force applied
9	M	55.01
10	F	54.54
7	M	46.09
7	F	43.69
9	F	39.69
6	F	35.86

Considering the following; F_p= 46.09 N, R_a = 4 cm, R_p = 37 cm, the average force which was used to calculate F_a was calculated from the table to be $F_a = 426.33\ N$ which represents the minimum axial force required to be exerted by each linear actuator for the back seat.

4. Design Safety Facilities

4.1. Location of the entire centre of mass

It is worth to mention in this occasion that the proposed final design will allow the patient to full stand. However, the entire wheelchair and the user body will be inclined with a small angle from the vertical upright position in order to ensure that the entire centre of mass is located slightly towards the rear side of the system which will offer more safety to the user while standing specifically in case of any change in the person posture which will relocate the entire centre of mass.

5. Conclusions

In order to design a reconfigurable wheelchair with a sit to stand facility; number of design options have been proposed and investigated in terms of stability, energy consumption and the occupied workspace. The design which showed better evaluation based on the previous estimators has been refined while ensuring ease of manufacturability, functionality, ease of use and less energy consumption. The proposed design of the wheelchair has the capability to support the weight of a 25kg child and is equipped with a built in mechanism that transform the wheelchair user between sit and stand modes as per his/her desire. The design also has considered number of design safety issues which ensure the safety of the user. Further studies on the developed model are going on including simulation and control as well as the realization of the device.

Acknowledgment

The authors would like to appreciate the financial support received from Sultan Qaboos University towards funding this work.

References

[1] Bostelman, R. and Albus J., (2007), A multipurpose robotic wheelchair and rehabilitation device for the home. Proceedings of the *2007 IEEE/RSJ International Conference on Intelligent Robots and Systems*, pp. 3348–3353.

[2] Cooper R. A., (1995), Intelligent Control of Power Wheelchairs, IEEE Engineering in Medicine and Biology, pp. 423-431.

[3] Cooper, R. A., Boninger, M. L., Spaeth, D. M., Ding, D., Guo, S., Koontz, A. M. and Fitzgerald S. G., (2006), Community Participation, *14*(4), 438–455.

[4] Cooper R. A., Jones D. K., Fitzgerald S., Boninger M. L., and Albright S. J., (2000), Analysis of Position and Isometric Joysticks for Powered Wheelchair Driving, IEEE Transactions on Biomedical Engineering, Vol. 47, No. 7, pp. 902-910.

[5] Dicianno B. E., Spaeth D. M. and Cooper R. A., (2007), Force Control Strategies While Driving Electric Powered WheelchairsWith Isometric and Movement-Sensing Joysticks, IEEE Transactions on Neural Systems and Rehabilitation Engineering, Vol. 15, No. 1, pp. 144-150.

[6] Ding D. and Cooper R. A., (2005), Electric-Powered Wheelchairs, IEEE Control Systems Magazine, pp. 22-34.

[7] Guo-zheng X., Ai-guo S. and Hui-jun L., (2009), System design and control technique of robot-aided rehabilitation, Journal ojClinical Rehabilitative Tissue Engineering Research, vol. 13, no. 4, pp. 717-720.

[8] Hoyer H., (1995), The OMNI wheelchair, Service Robot: An International Journal, Vol.1 No.1, MCB University Press Limited, Bradford, England, pp. 26-29.

[9] Katevas N. I., (1997), The Autonomous Mobile Robot SENARIO: A Sensor-Aided Intelligent Navigation System for Powered Wheelchairs", IEEE Robotics and Automation Magazine, Vol.4, No.4, pp. 60-70.

[10] Najafi B., Aminian L., Ionescu A. P., F. Loew, Bula C. and Robert P., (2003), Ambulatory system for human motion analysis using a kinematic sensor: Monitoring daily physical activity in the elderly, IEEE Transactions on Biomedical Engineering, vol. 50, no. 6, pp. 711–723.

[11] Li-ning S., (2006), Research and Development of Assistive Rehabilitation Robotics, Robot, vol. 28, no. 3, pp. 355-360.

[12] Wang W. and Jin D., (2005), Rehabilitation project development of China, Chinese Journal of Rehabilitation Theory and Practice, vol. 11, no. 3, pp. 161-163.

[13] Yamada T. and Demura S., (2009), Relationships between ground reaction force parameters during a sit-to-stand movement and physical activity and falling risk of the elderly and a comparison of the movement characteristics between the young and the elderly, Archives of Gerontology and Geriatrics, vol. 48, pp. 73–77.

[14] Yanco H. A., (1998), Integrating Robotic Research: A Survey of Robotic Wheelchair Development, AAAI Spring Symposium on Integrating Robotic Research, Stanford University, CA.

[15] Zhang L., Bai D. and Yi L., (2011), Optimization of Walking Assistance Mechanism in Rehabilitation Wheelchair. *Proceedings of the 2011 lEEEIICME International Conference on Complex Medical Engineering*, pp. 621–626.

[16] Zhu, C., Oda, M., Yoshioka, M., Nishikawa, T., Shimazu, S., & Luo, X. (2010), Admittance control based walking support and power assistance of an omnidirectional wheelchair typed robot, Proceedings of the *2010 IEEE International Conference on Robotics and Biomimetics*, pp.381–386.

INTEGRATED DESIGN, MODELLING AND ANALYSIS OF TWO-WHEELED WHEELCHAIR FOR DISABLED

ABQORI AULA, TAREQ M. ALTALMAS, SALMIAH AHMAD, RINI AKMELIAWATI, SHAHRUL NAIM SIDEK

Intelligent Mechatronics System Research Unit, Department of Mechatronics Engineering, International Islamic University Malaysia, Gombak, Selangor 50728, Malaysia

M. O. TOKHI

Department of Automatic Control and Systems Engineering, University of Sheffield, Mappin Street, S1 3JD, United Kingdom

Two-wheeled wheelchairs are considered as highly nonlinear and complex systems. The challenges in handling such complex systems as well as the need for helping disabled people have motivated researchers around the world to do their work in this area with different focuses and objectives. Due to its highly nonlinear characteristics, modeling of this complex system mathematically is a challenging process. This paper presents an alternative way of integrating design and modeling of a two-wheeled wheelchair. The method proposes an integrated design and analysis of two-wheeled wheelchair in which the derivation of physical and mathematical model of the system is done by using Maple, MapleSim software, and the 3D design using Solidworks and MATLAB. Such integrated modelling allows for analysing and visualising the mechanism of the system's model, which is very much required for hardware implementation at a later stage. The obtained model was tested by linear-quadratic regulator and the result showed that the model with proposed design is working well. The designed LQR is able to balance the linearised model of the system.

1. Introduction

During recent years, the number of older adults (>65 years) and disabled people who have injured their extremities because of the wars and traffic accidents, are increasing rapidly. In USA, the elderly people population was expected to be 87 million by 2050 [1]. Hence, many studies have been conducted to help this group of people to be independent in performing their day-to-day activities. However, existing mechanical designs and control algorithms of electric powered wheelchair do not give the users the full ability to act independently in life. For instance, there are wheelchairs that have the property of lifting the chair to reach certain levels - eye to eye level during conversations-, but not many wheelchairs have the ability to work and balance on two wheels. The later feature enables the users to move in narrow spaces and give them capacity to maneuver.

Research on two-wheeled wheelchair is relatively new. Before its rise, there were only researches on two-wheeled mobile robots, also known as self-balancing mobile robots. These two-wheeled mobile robots are technically an extension of research on inverted pendulum on carts. These researches are then further improved for wheelchair application [2].

Previously, some researchers have developed a two-wheeled wheelchair that has the ability to balance on two wheels and lift the seat to reach certain levels. As discussed in [3], the two-wheeled wheelchair was designed using MSC Visual Nastran 4D environment. This design mimics the double inverted pendulum mechanism and they introduced the locking mechanism to protect the user from falling during stationary mode. A new configuration of two-wheeled wheelchair has been designed which also mimics the double inverted pendulum but it has used linear actuator for lifting the seat [4].

This paper proposes an extension of our previous works in [3-4]. In this paper, a new approach, which integrates the improved mechanism and design for balancing and elevating wheelchair's height for a two-wheeled wheelchair is proposed. A suitable control system based on the linear quadratic regulator (LQR) is also proposed to supervise the highly nonlinear, multivariable and unstable characteristic of two-wheeled wheelchair. The system is first modeled mathematically and then virtually in 3D. Due to its very complex nonlinear characteristics, it is difficult, although not impossible, to obtain its nonlinear mathematical representation. Solving the equations, however, is more difficult. This research utilizes computer-aided-engineering software to develop two-wheeled wheelchair's 3D physical model and extract its nonlinear mathematical equations.

The expected outcome from this research is a two-wheeled wheelchair which has the ability to reach certain levels to help the user to do his work independently and to have small turning radius, that is, to give high ability of maneuver in confined spaces, such as at home or in a library.

This paper is organised as follows. Section 2 summarises current commercially available 'smart' wheelchairs in the market. The modelling process of the proposed system as well as the design of LQR controller is discussed in Section 3. The proposed system is then simulated in Section 4. Section 5 concludes this paper and suggests future work.

2. Similar Products

There are quite many 'smart' wheelchairs existing. Independence Technology L.L.C. generated iBOT® Mobility System that has the ability to balance on two wheels. However, it has stopped producing this system [5]. The GENIE Standing Wheelchair was produced by Easycare Products Limited. It has the ability to help the user to stand and reach certain levels but does not have the ability to work on two wheels [6]. Genny™ 2.0 inaugurated a new transportation philosophy in personal mobility that can balance on two wheels but the wheelchair does not have the ability to lift the seat [7].

Two-wheeled people carriers, including Segway™[8], PMP[9-10], and TransBot[11] are considered impractical due to the following limitations:

1. Operable by healthy and able people only;
2. Generally too large to maneuver indoor;
3. Too expensive for most potential users.

Thus, this research proposes a new integrated method of developing the two-wheeled wheelchair to minimize the above-mentioned limitations.

3. System Modeling

3.1. *Physical Modeling*

The proposed two-wheeled wheelchair model is designed within SolidWorks environment. The SolidWorks® CAD software is a mechanical design automation application that give designers the ability to quickly sketch out ideas, experiment with features and dimensions, and produce models and detailed drawings [12]. Figure 1 illustrates the model of the new design in its initial 4-wheeled configuration. The dimensions of the wheelchair are chosen by taking into consideration the ISO 7176-5 standards [13]. The occupied width and length are 63cm and 82cm, respectively. The height (floor to seat) before lifting the chair is 80 cm and after lifting the chair will be 108cm such that the chair can reach its seat level between 80cm to 108cm. The wheels dimensions are the same as those of the commercial wheelchairs in the markets, which are 12.5 inches (= 31.75 cm) for driving wheels and for casters 8 inches (= 20.32 cm) [14]. The actuators that are needed in this design are listed in Table 1. For lifting the seat, a scissor mechanism is used.

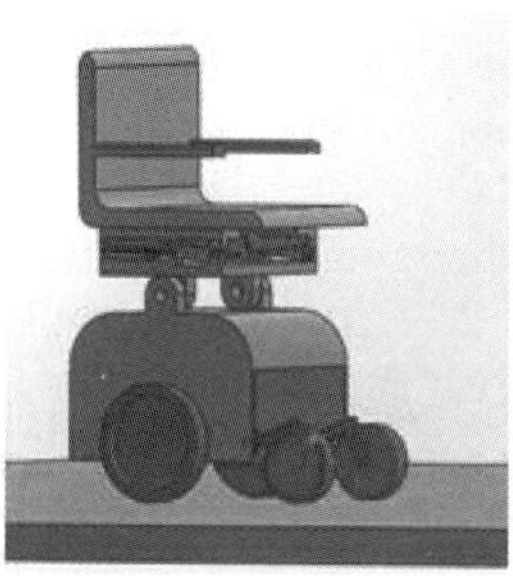

Figure 1. Wheelchair model in SolidWorks

Table 1. The actuators of wheelchair.

Actuator	Number	Purpose
Motor 1	2	To actuate drive wheels
Motor2	1	To lift the casters
Linear actuator 1	1	Scissor mechanism
Linear actuator2	1	Chair tilting

3.1.1. *Converting between Four-wheeled and two-wheeled*

The main goal of this design is to convert the four-wheeled wheelchair to a two-wheeled wheelchair and lifting the seat to certain levels. These two abilities will give the wheelchair users more independence in performing his/her activities and reduce the working space required by the wheelchair. Converting from four-wheeled to two-wheeled consists of several steps. Figure 2 shows the steps of converting process.

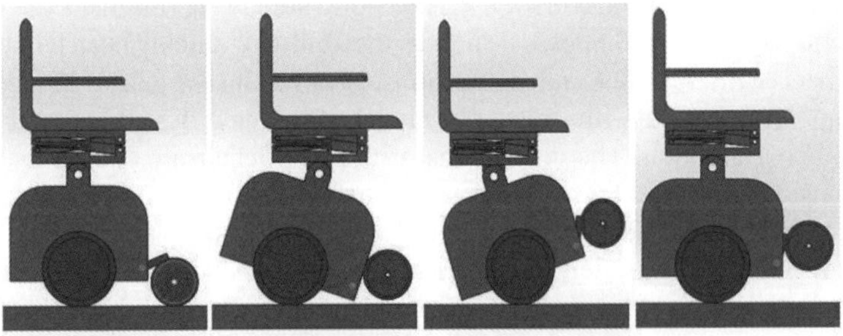

(a) (b) (c) (d)

Figure 2. Converting steps between four-wheeled to two-wheeled

The process is as follows:

1. First, the wheelchair is on the ground on four wheels. After the user sets and gives the instruction to convert to two-wheeled mode, motor 2 will raise the casters Figure 2(a).
2. After the casters leave the ground, the chair may tilt to the front or to the back as in Figure 2(b), 2(c). Tilting mechanism that is applied by linear actuator 2 will help the seat and the user to be in the upright position all the time.
3. After balancing, the wheelchair will be on two-wheeled mode as shown in Figure 2(d).

3.1.2. *Lifting the seat*

For lifting the seat to reach many levels while in the balancing mode, the user can give instruction to raise the seat. It will be raised until reaching the desired level, while at the same time the balancing procedure is active (see Figure 3).

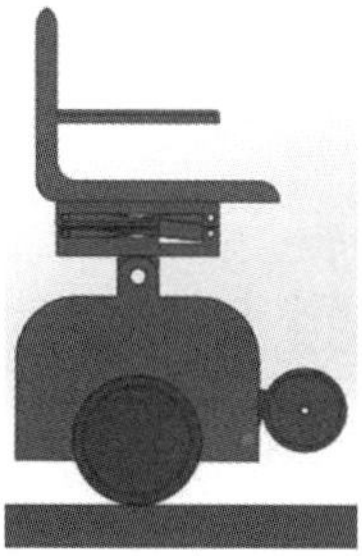

Figure 3. The wheelchair during two-wheeled mode before lifting the seat

3.2. *Mathematical Modeling*

Mechanically, a two-wheeled wheelchair is a nonlinear, multivariable, high order and unstable system. Its two-wheeled mode characteristics are very similar to those of a double inverted pendulum on a mobile platform [15-17]. The diagram of two-wheeled wheelchair and its double inverted pendulum approximation in free-body diagram are shown in Figure 4. Table 2 lists the physical parameters. The system of dynamic equations describing a double inverted pendulum can be derived by Newtonian methods (using free body diagrams) or the Euler-Lagrange method [18]. In this paper, the Euler-Lagrange method is used.

3.3. *The Euler-Lagrange method*

The mathematical model of the design was derived from the dynamics of a double inverted pendulum as shown in Figure 4 using Euler-Lagrange equation, as in Eq. (1).

$$\frac{d}{dt}\left(\frac{\partial L}{\partial q_i}\right)-\frac{\partial L}{\partial q_i}=Q_i\,; \quad i=1,\dots,n \tag{1}$$

where its energy calculation is analysed through Lagrangian dynamics $L=K-P$. Q is a vector of generalized forces (or moments) acting in the direction of generalized coordinates q. There are two forces acting on the system, namely force F on linear motion and torque T on pendulum 2. K is the total kinetic energy of the system, which is the sum of the kinetic energy on the base, on the first link, and on the second link, denoted as $K=K_0+K_1+K_2$.

The total potential energy P is also the sum of the potential energy on the base, on the first link and on the second link, denoted $P = P_0 + P_1 + P_2$.

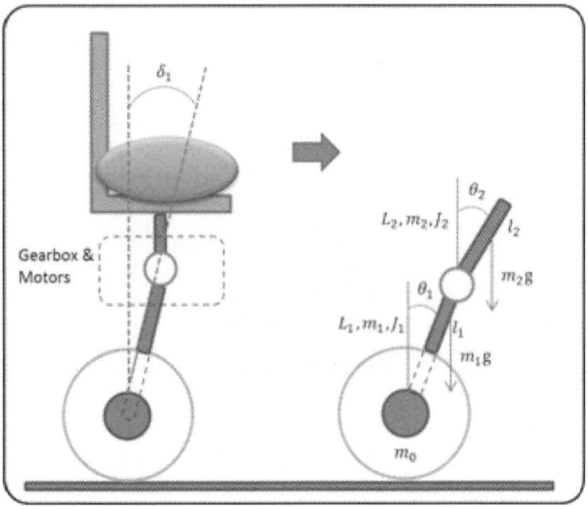

Figure 4. Dynamic similarity to those of a double inverted pendulum when in two-wheel mode

Table 2. Physical parameter definition

Symbol	Definition
m_0	Mass of the base
m_1, m_2	Mass of pendulum 1 and 2
L_1, L_2	Length of pendulum 1 and 2
l_1, l_2	Center of mass of pendulum 1 and 2
θ_1, θ_2	Angular displacement of pendulum 1 and 2
J_0, J_1, J_2	Moment of inertia of wheelbase, pendulum 1 and 2
F	Translational force
T	Rotational force
G	Gravitation force
R	Wheel radius
X	Linear displacement on x-axis

$$K_0 := \left(m_0 + \frac{J_0}{r^2}\right)\dot{x}^2. \tag{2}$$

$$K_1 := \frac{1}{2}m_1\dot{x}^2 + \frac{1}{2}\left(m_1 l_1^2 + J_1\right)\dot{\theta}_1^2 + m_1 l_1 \dot{x}\dot{\theta}_1 \cos\left(\theta_1\right). \tag{3}$$

$$K_2 := \frac{1}{2}m_2\dot{x}^2 + \frac{1}{2}m_2L_1^2\dot{\theta}^2{}_1 + \frac{1}{2}\left(m_2l_2^2 + J_2\right)\dot{\theta}^2{}_2 + m_2L_1\dot{x}\dot{\theta}_1\cos\left(\theta_1\right) \tag{4}$$

$$P := P_0 + P_1 + P_2 = m_1\,g\,l_1\cos\theta_1 + m_2\,g\left(L_1\cos\theta_1 + l_2\cos\theta_2\right). \tag{5}$$

Substituting K and P into Lagrangian L and then substituting L into equation (1), and differentiating it on linear motion and angular motion of first and second pendulum x, θ_1, θ_2, respectively, yield the followings:

$$2\left(m_0 + \frac{J_0}{r^2} + \frac{m_1}{2} + \frac{m_2}{2}\right)\ddot{x} + \left(m_1l_1 + m_2L_1\right)\ddot{\theta}_1\cos\left(\theta_1(t)\right) - \left(m_1l_1 + m_2L_1\right)\dot{\theta}_1^2\sin\left(\theta_1(t)\right) + m_2l_2\ddot{\theta}_2\cos\left(\theta_2(t)\right) - m_2l_2\dot{\theta}_2^2\sin\left(\theta_2(t)\right) = force(t). \tag{6}$$

$$\left(m_1l_1^2 + J_1 + m_2L_1^2\right)\ddot{\theta}_1 + \left(m_1l_1 + m_2L_1\right)\ddot{x}\cos\left(\theta_1(t)\right) - \left(m_1l_1 + m_2L_1\right)\dot{x}\dot{\theta}_1\sin\left(\theta_1(t)\right) + m_2L_1l_2\ddot{\theta}_2\cos\left(\theta_1(t) - \theta_2(t)\right) - m_2L_1l_2\left(\dot{\theta}_1 - \dot{\theta}_2\right)\dot{\theta}_2\sin\left(\theta_1(t) - \theta_2(t)\right) + m_2L_1l_2\dot{\theta}_1\dot{\theta}_2\sin\left(\theta_1(t) - \theta_2(t)\right) - \left(m_1l_1 + m_2L_1\right)g\sin\left(\theta_1(t)\right) = 0. \tag{7}$$

$$\left(m_2l_2^2 + J_2\right)\ddot{\theta}_2 + m_2l_2\ddot{x}\cos\left(\theta_2(t)\right) - m_2l_2\dot{x}\dot{\theta}_2\sin\left(\theta_2(t)\right) + m_2L_1l_2\ddot{\theta}_1\cos\left(\theta_1(t) - \theta_2(t)\right) - m_2L_1l_2\dot{\theta}_1\sin\left(\theta_1(t) - \theta_2(t)\right)\left(\dot{\theta}_1 - \dot{\theta}_2\right) + m_2l_2\dot{x}\dot{\theta}_2\sin\left(\theta_2(t)\right) - m_2L_1l_2\dot{\theta}_1\dot{\theta}_2\sin\left(\theta_1(t) - \theta_2(t)\right) - m_2l_2g\sin\left(\theta_2(t)\right) = torque(t). \tag{8}$$

Equation (6), (7), and (8) represent the dynamic motions in linear displacement x, angular displacement θ_1 and θ_2, of pendulum 1 and 2, respectively. The nonlinearity is highly imminent in these three equations.

3.4. *Linearising the mathematical equations*

In the balanced mode (on two wheels), the two-wheeled wheelchair has similar characteristics as a double-inverted pendulum. Thus, it is represented by a double-inverted pendulum modeled in MapleSim, as shown in Figure 5.

In this design, a simpler modelling which approximates the free-body-diagram of the two-wheeled wheelchair is done in such a way that the extendable part on second link is assumed at its highest length, i.e. fully stretched.

The system has two inputs, a force F driving the system on the translational motion and a torque T acting on the joint of the first and second pendulum. With three measured output x, θ_1, θ_2, the system has six states.

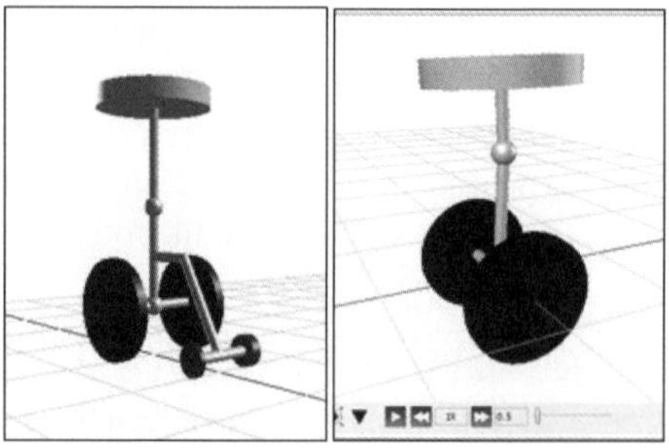

Figure 5. The model is designed in MapleSim

To test the controllability of the newly designed system, a regulator was developed from the linearised version of the model. Linearising the mathematical equations of motion at their equilibrium points and setting the initial condition of θ_1 and θ_2 to 5 and 10 degrees, respectively, the linearised state-space equations were obtained in matrix form by using *Dynamic Systems* and *Linear Algebra* toolboxes of Maple. The linearised state-space matrix $\overline{\mathbf{A}}$, $\overline{\mathbf{B}}$, $\overline{\mathbf{C}}$, $\overline{\mathbf{D}}$ are

$$\begin{bmatrix} 0 & -0.125 & 129.000 & 115.000 & 0 & 0 \\ 0 & 0.284 & -71.500 & -39.200 & 0 & 0 \\ 0 & 1 & 0 & 0 & 0 & 0 \\ 1 & 0 & 0 & 0 & 0 & 0 \\ 0 & -0.088 & 2.770 & 2.460 & 0 & 0 \\ 0 & 0 & 0 & 0 & 1 & 0 \end{bmatrix}, \begin{bmatrix} 0.070 & 26.100 \\ -0.158 & -8.910 \\ 0 & 0 \\ 0 & 0 \\ 0.049 & 0.560 \\ 0 & 0 \end{bmatrix}, \begin{bmatrix} 1 & 0 & 0 & 0 & 0 & 0 \\ 0 & 1 & 0 & 0 & 0 & 0 \\ 0 & 0 & 1 & 0 & 0 & 0 \\ 0 & 0 & 0 & 1 & 0 & 0 \\ 0 & 0 & 0 & 0 & 1 & 0 \\ 0 & 0 & 0 & 0 & 0 & 1 \end{bmatrix}, \begin{bmatrix} 0 & 0 \\ 0 & 0 \\ 0 & 0 \\ 0 & 0 \\ 0 & 0 \\ 0 & 0 \end{bmatrix}$$

The open-loop poles of matrix $\overline{\mathbf{A}}$ are

$$\begin{bmatrix} -9.080 + 0.000\,\mathrm{I} \\ 9.043 + 0.000\,\mathrm{I} \\ 0.160 + 6.197\,\mathrm{I} \\ 0.160 - 6.197\,\mathrm{I} \\ -1.655\,10^{-16} + 8.150\,10^{-9}\,\mathrm{I} \\ -1.655\,10^{-16} - 8.150\,10^{-9}\,\mathrm{I} \end{bmatrix}$$

This clearly shows the instability of the system without controller. Controllability and observability checks on matrix $\overline{\mathbf{A}}$, $\overline{\mathbf{B}}$ and $\overline{\mathbf{C}}$ confirm that the system is controllable and reachable. Choosing weight matrix $\overline{\mathbf{Q}} = \overline{\mathbf{C}}^T\overline{\mathbf{C}}$ and matrix $\overline{\mathbf{R}} = \rho Identity(2)$, the gain matrix $\boldsymbol{K}$ is obtained by solving the algebraic Riccati equation.

4. Simulation

The regulator $\boldsymbol{K}$ is then introduced into the system as shown in Figure 6 to ensure regulation of the system

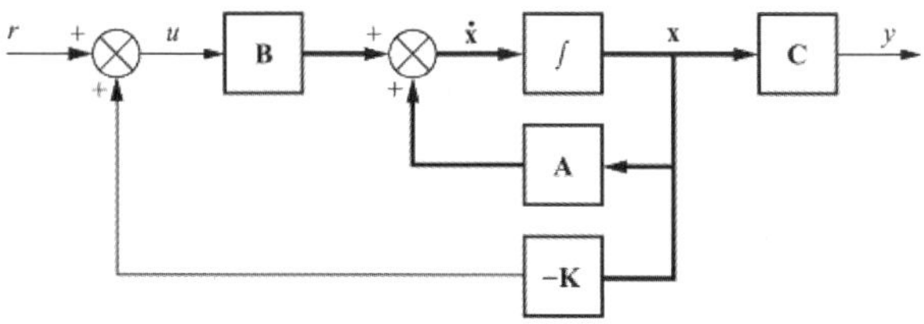

Figure 6. Closed-loop system with LQR controller

The linearised two-wheeled wheelchair in balanced mode together with LQR controller was simulated in MATLAB environment.

Figures 7 and 8 show the simulated linear displacement *x*, and angular displace-ments θ_1 and θ_2. Several results were presented using different values of ρ, which determines how fast the system will converge.

5. Conclusion and Future Work

The newly designed mechanism for two-wheeled wheelchair reveals the similarity of its unstable, multi-variable, nonlinear characteristics with that of a double-inverted pendulum. A linearised model is derived and controlled by an LQR.

Simulations on the closed-loop system show that the designed LQR is able to balance the linearised model of the system.

In the future, the wheelchair's nonlinear characteristics will be preserved and a nonlinear controller will be developed to control the balancing and elevating mechanism. A miniaturized prototype will also be built as a testbed prior to building the actual wheelchair.

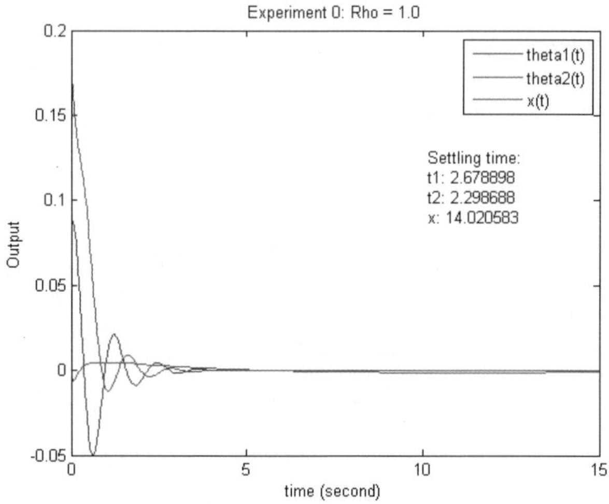

(a)

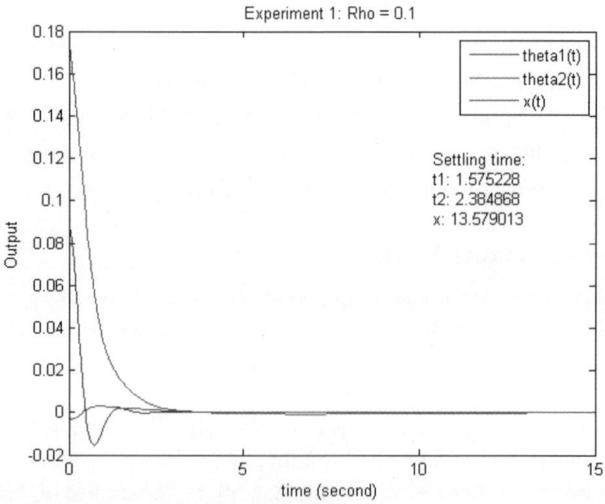

(b)

Figure 7. Simulation results with different values of ρ

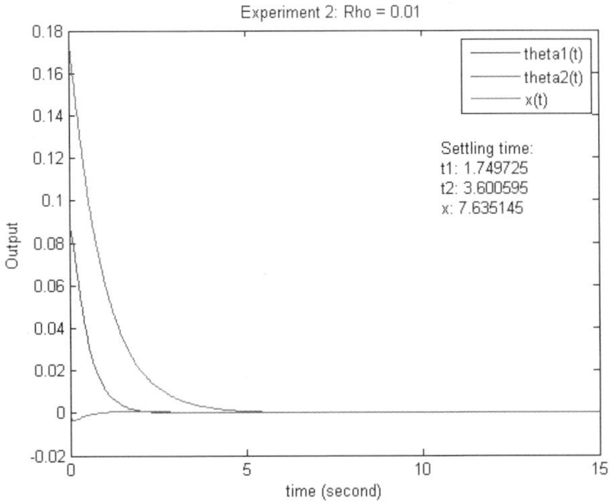

(a)

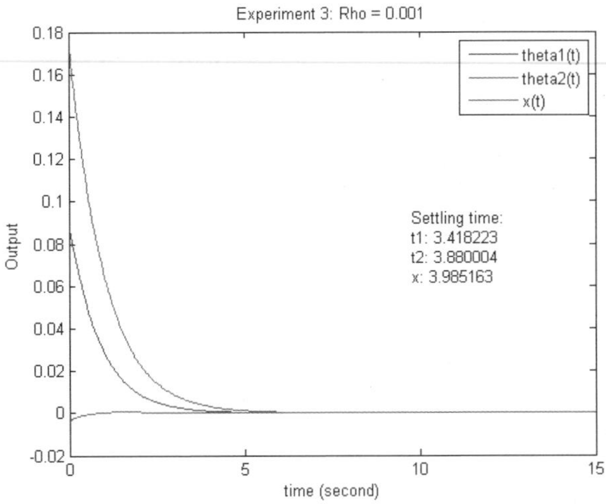

(b)

Figure 8. Simulation results with different values of ρ

References

1. Amol, M. K., Brad, E. D., Rosemarie, C., Diane, M. C., Judith, T. M., Alicia, K., Emily, E. T., Rory A. C.. Demographic Profile of Older Adults Using Wheeled Mobility Devices. *Journal of Aging Research*, (2011).
2. D. Ding and R. A. Cooper, "Electric powered wheelchairs," IEEE Control Systems, vol. 25, no. 2, pp. 22-34, 2005.

3. S. Ahmad and M. O. Tokhi, "Modeling and Control of a Wheelchair on Two Wheels," 2nd Asia International Conference on Modeling &Simulation (AMS 08), Kuala Lumpur, May 2008.
4. Goher, K., Ahmad, S., & Tokhi, M.O.," A new configuration of two wheeled vehicles: Towards a more workspace and motion flexibility," 4th Annual IEEE, Systems Conference, held in San Diego, CA, April 2010.
5. "Independence iBOT Mobility System," Johnson & Johnson Company, [Online]. Available: http://www.ibotnow.com. [Accessed 23 March 2012].
6. Easycare Products Limited, [Online]. Available: http://www.easycareproducts.co.uk/index.htm. [Accessed 4 January 2013].
7. GENNY MOBILITY [Online]. Available: http://www.gennymobility.com/Genny/WhyChooseGenny.aspx. [Accessed 4 January 2013].
8. "Segway - Simply Moving," Segway, Inc., [Online]. Available: http://www.segway.com. [Accessed 23 March 2012].
9. M. Sasaki, N. Yanagihara, O. Matsumoto and K. Komoriya, "Forward and backward motion control of Personal riding-type wheeled Mobile Platform," in Proceedings of IEEE International Conference on Robotics and Automation, 2004.
10. M. Sasaki, N. Yanagihara, O. Matsumoto and K. Komoriya, "Steering control of the personal riding-type wheeled mobile platform (PMP)," in IEEE/RSJ International Conference on Intelligent Robots and Systems, 2005.
11. H. J. Lee, H. W. Kim and S. Jung, "Development of a mobile inverted pendulum robot system as a personal transportation vehicle with two driving modes: TransBot," in World Automation Congress, 2010.
12. Dassault Systèmes SolidWorks Corporation. (1995). Introducing SolidWorks.
13. ISO 7176-5
14. Standard Power Wheelchairs [Online]. Available: http://www.scooterlink.com/departments/wheelchairs/power-wheelchairs/standard-power-wheelchairs.cfm. [Accessed 4 January 2013].
15. Y. Zeng, C. Xia and H. Cao, "A novel LMI based swing-up robust controller for a serial double inverted pendulum," in *IEEE International Conference on Robotics and Biomimetics*, 2009.
16. C. Fu, Y. Liu, W. Tan and J. Zhang, "μ synthesis with regional pole assignment for a double inverted pendulum," in *Chinese Control Conference*, 2010.
17. W. Luhao and S. Zhanshi, "LQR-Fuzzy Control for Double Inverted Pendulum," in *International Conference on Digital Manufacturing and Automation*, 2010.
18. W. Zhong and H. Röck, "Energy and passivity based control of the double inverted pendulum on a cart," in Proceedings of the 2001 IEEE International Conference on Control Applications, 2001.

A COMPARISON OF A PASSIVE AND VARIABLE-DAMPING CONTROLLED LEG PROSTHESIS IN A SIMULATED ENVIRONMENT

JIE ZHAO*, KARSTEN BERNS*, ROBERTO DE SOUZA BAPTISTA† and ANTÔNIO PADILHA L. BÓ†

**Robotics Research Lab, University of Kaiserslautern, Kaiserslautern, 67655, Germany*
**E-mail: {zhao,berns}@informatik.uni-kl.de*

†Automation and Robotics Laboratory, Unversity of Brasília, Brasília, Brazil

This paper presents two control methods for a leg prosthesis: a passive based and a variable-damping based. First, the concept of the mechanical design for the leg prosthesis is described. A simulated environment is used to compare the two control methods. The joint setup and its control scheme, in which a passive or a variable-damping control can be easily implemented, are introduced for the prosthetic leg in the simulated environment. Intelligent prosthesis require a specific adjustment for each gait phase based on kinematic events of knee and ankle joints along a gait cycle. A feasible damping profile is generated based on kinematic and dynamical data of the knee joint along a gait cycle. This data set was extracted from a simulated biped which is 1.8 m in height and 76 kg in weight with 21 degrees of freedom during normal walking. The data show that the damping profile is critical between the toe-off and heel contact events. Tests using the simulated biped with the passive and the variable-damping controlled leg prosthesis during walking enables the comparison of both control approaches.

Keywords: Leg prosthesis; Variable-damping control; Passive control; Rough terrain

Introduction

Amputations of lower limbs, which can occur due to trauma, infections or other diseases, significantly decreases a person's quality of life. Among all basic locomotion affected walking is critical when compared to other movements. This paper addresses the use of new tools in the development of control strategies for artificial legs and presents a comparison between two distinctive control methods, since its merits and drawbacks are still not clear. In the next sections, we start by describing the mechanical design of ongoing leg prosthesis, its working principles and hardware setups. Based on the mechanical setup, a passive control scheme for one joint will be briefly introduced. Next, by parsing captured kinematic and dynamical data from an existed simulated biped, a variable-damping controller is developed for the leg prosthesis. We test both controllers on a simulated human with a leg prosthesis and demonstrate their performances. Results demonstrate both merits and drawbacks of each control method.

Mechanical Design of Leg Prosthesis

The knee module of the ongoing leg prosthesis project is a polycentric knee mechanism with adjustable damping ratios. The main advantages of a polycentric mechanism is a combination of stability at full extension, i.e. in stance phase of gait, and better foot clearance in the swing phase of gait.[1,2] As in various polycentric knee designs, our polycentric mechanism is based on a four bar linkage system.[3] A magnetorheological piston is integrated to the prosthetic knee to provide damping adjustment of the mechanism. Magnetorheological fluids alter its viscosity in the presence of a magnetic field. A piston, or damper, filled with this type of fluid provides a continuously variable damping force controlled by a desired input.[4] In our project we used a commercially available magnetorheological piston from Lord Corporation. In this device, the magnetic field through the magnetorheological fluid is induced by an electric current, which is the control input to adjust the damping coefficient of the system. The prosthetic knee module prototype is currently in the finishing phase of production as shown in Fig. 1. The goal of this project is to investigate different control strategies taking into account human in the loop for above the knee amputees.

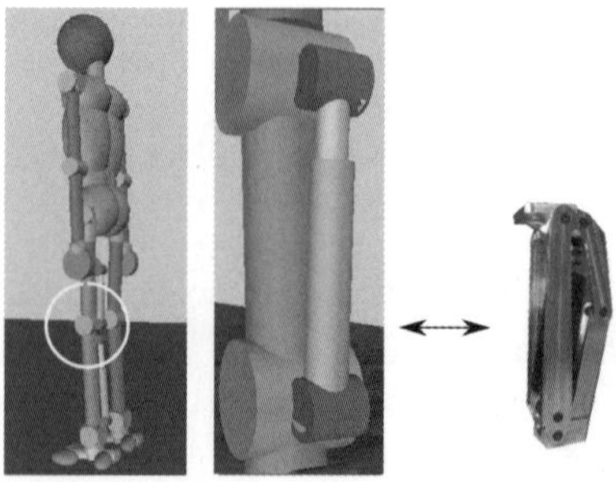

Fig. 1. Mechanical configuration of leg prosthesis in a simulated biped and prototype.

Joint Setup and Control

The main feature of the joint setup is constant stiffness and damping, which makes the joint compliant. Each joint consists of a parallel elastic element and damper, except for prosthetic knee joint that has an additional variable-damping control. Figure 2 presents the torque computation applied to implement a constrained rotational movement of the prosthetic knee and the passive joints at the ankle. Given the constant equilibrium point θ_0 of the

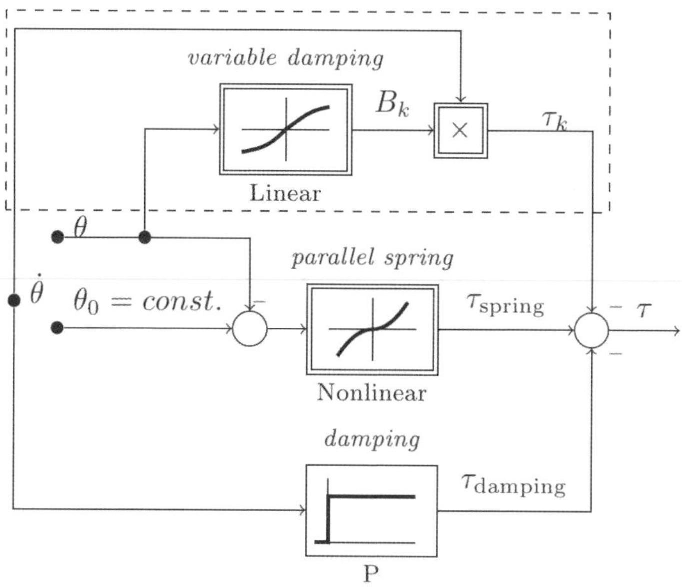

Fig. 2. A variable-damping controlled knee joint. Other two passive joints don't contain elements inside dashed rectangle.

spring and the actual joint angle θ, the resulting torque is computed as

$$\tau_{spring} = \text{sgn}\,(\theta_0 - \theta) \cdot K_{spring} \cdot (\theta_0 - \theta)^2 \,. \tag{1}$$

The combined torque for the joint can be computed as

$$\tau = \tau_{spring} - \tau_{damping} - \tau_k, \tag{2}$$

with K_{spring} being the spring stiffness. Constant damping at each joint is represented by the torque $\tau_{damping}$ which is proportional to the joint's angular velocity while τ_k represents the torque generated by the variable-damping controller at the knee joint if it is activated. Thus, for passive joints the combined torque should not include the last part of Eq. (2) , i.e. τ_k.

Computation of Damping Coefficient at Knee Joint

Kinematic data of the knee joint along gait cycle was obtained simulating a biped in a dynamic environment[a] within the scenario of flat ground walking, as shown in Fig. 3. The power consumption profile at the knee joint after toe-off and before heel strike generally has negative magnitude. Associating this feature to the physical damper, the knee can be modeled as a variable-damper in this phase.

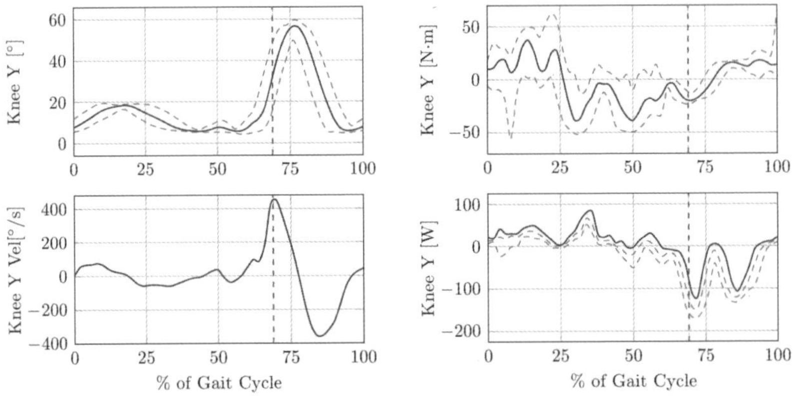

Fig. 3. Knee angle, velocity, torque and power along gait cycle.

The effective damping variable B_k is the ratio between the knee torque τ and knee velocity $\dot{\theta}_k$. By using the data set illustrated in Fig. 3, the damping coefficient can be directly calculated, as shown in Fig. 4.

According to Fig. 4, the damping coefficient can be characterized as a function of knee angle:

$$B_k(\theta_k) = \begin{cases} B_{k_{low}} + \dfrac{B_{k_{up}} - B_{k_{low}}}{\theta_{k_{up}} - \theta_{k_{low}}} \cdot (\theta_k - \theta_{k_{low}}), & \text{if } \theta_k > \theta_{k_{low}} \\ \infty, & \text{otherwise} \end{cases} \quad (3)$$

In Eq. (3) $\theta_{k_{low}}$ and $\theta_{k_{up}}$ represent a range in which knee joint can be modelled as a variable damper whereas $B_{k_{low}}$ and $B_{k_{up}}$ indicate the damping

[a]The detailed control architecture, i.e. biologically inspired control of a dynamically walking bipedal robot can be found in[5]

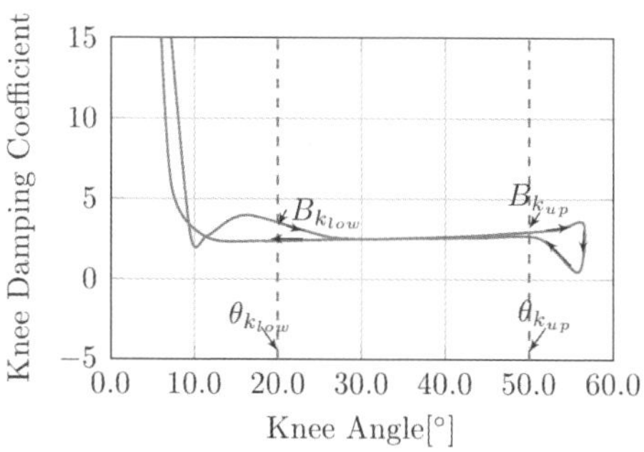

Fig. 4. Knee damping coefficient along the knee angle during gait cycle.

coefficient at $\theta_{k_{low}}$ and $\theta_{k_{up}}$, respectively.

Simulation Results

The variable-damping controlled leg prosthesis consists of one knee joint and ankle joints in both sagittal and frontal plane. Except for the variable-damping controlled knee joint, others joints are passive elements with fixed stiffness and damping. A sliding joint is placed between top of the amputee's stump and the knee joint, whereas another is situated between knee and ankle joint. Those two sliding joints, which are shown in Fig. 1, provide a constant stiffness and lock mechanism that is used to lock knee angle during late swing phase and stance phase.

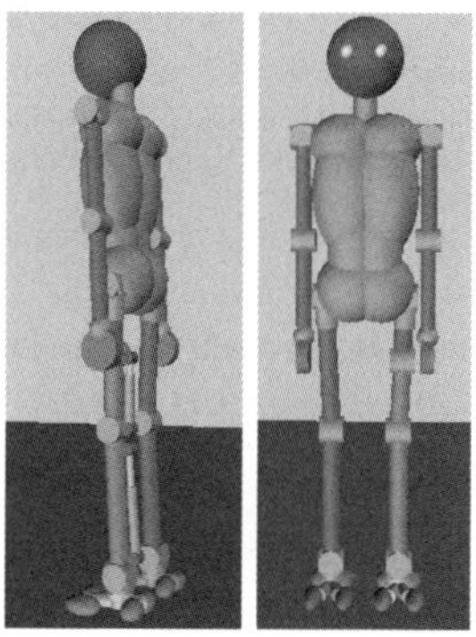

Fig. 5. A simulated biped with a left prosthetic leg.

Meanwhile, the mechanically passive prosthesis consists of three passive joints, e.g. knee joint, ankle joint in sagittal and frontal plane. The stiffness and damping ratio for each joint are predefined and can not be adjusted during the walking phase as shown in Fig. 2.

Walking on the Even Ground

We first simulated the normal walking at a speed of 1.21 m/s on an even platform without the presence of external disturbances. The results allow a detailed evaluation of the calculated damping coefficient for the variable-damping controlled leg prosthesis and the capability of the passive leg prosthesis.

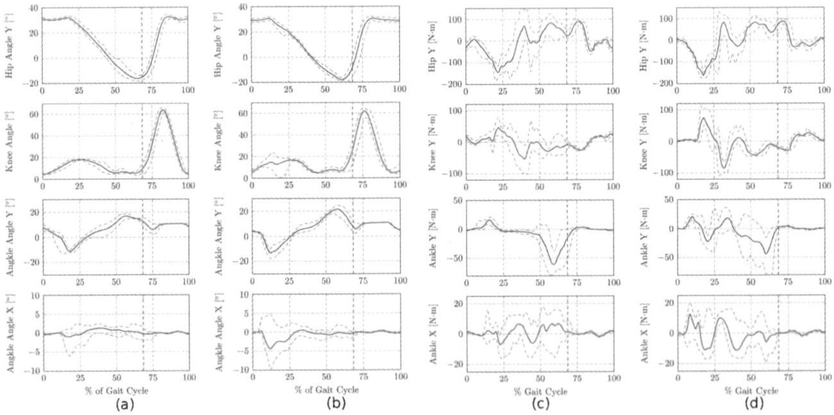

Fig. 6. Joint angles and torques over the course of gait cycle. The solid lines in (a) and (c) represent mean values of passive prosthesis joints over 10 steps of normal walking, whereas those in (b) and (d) show mean values of variable damping controlled prosthesis. The dashed lines mark the minimum and maximum values. The vertical dashed line denotes the transition from stance to swing.

The kinetic data set of both types of prosthesis in this walking scenario is illustrated in Fig. 6. Figures 6 (a) and (b) show the joint angles of passive and variable-damping controlled prosthesis respectively. One can observe that the courses of the kinematic data of the passive controlled joints are actually more stable during gait cycle. Those curves present nearly similar course during the gait cycle, which means that the choice of the damping constant for passive prosthesis is also proper. The joint torques of passive and variable-damping controlled prosthesis are shown in Fig. 6 (c) and (d), respectively. The courses of torque generated in the passive joints have nearly similar profiles as those in variable-damping controlled prosthesis which means they have almost the same performance on energy consumption. However, while the variable-damping controlled prosthesis requires extra mounting of electronic devices and thus more energy consumption, the variable-damping controlled prosthesis exhibits no advantage in this aspect.

Walking on the Rough Terrain

We have also tested walking in scenario of an uneven terrain with the unevenness up to about 3.3 mm. Simulation results show that the variable damping controlled prosthesis is more adaptable to unknown rough terrain than the passive prosthesis. Figures 7 (a) and (b) show the joints angles of passive and variable-damping controlled prosthesis individually. The variances of joint angles of passive controlled prosthesis are clearly larger than those in variable-damping controlled prosthesis. The joint torques of passive and variable-damping controlled prosthesis are shown in Fig. 7 (c) and (d), respectively. The curves representing the passive controlled prosthesis show greater dynamic variations of joint angles along a gait cycle, which means more energy is required for keeping stability.

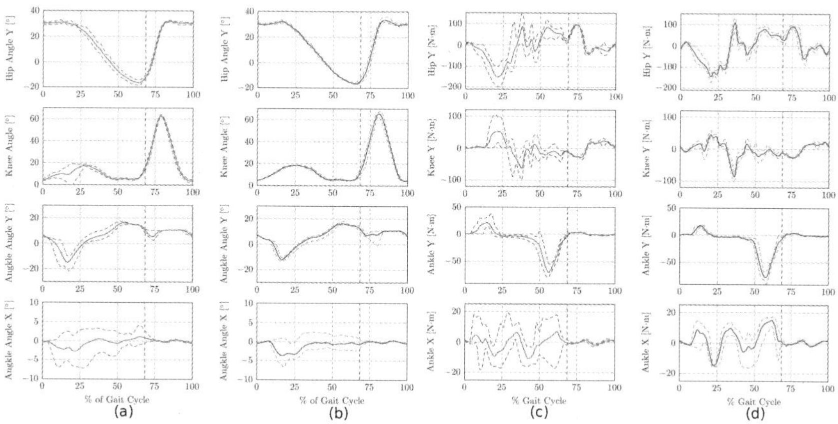

Fig. 7. Joint angles and torques of passive and variable-damping controlled prosthesis of amputees along gait cycle.

Table 1 exhibits standard deviations of the joints in prosthetic legs in walking scenarios of flat ground and uneven terrain respectively. All joint variables are captured from experiment of 10 gaits. One can observe that the passive controlled prosthesis shows more deviations than the variable-damping controlled one. Also, walking on the uneven terrain introduces more deviations on each joint due to unevenness of terrain. This result gives us a clue that using variable-damping controlled prosthesis will improve the stability of walking on unknown rough terrain compared to the passive one.

Joint	Flat Ground		Uneven Terrain	
	passive	variable-damping	passive	variable-damping
Hip Y	1.30°	1.24°	1.52°	1.26°
Knee Y	1.50°	1.34°	1.88°	1.44°
Ankle Y	1.50°	1.40°	1.78°	1.44°
Ankle X	1.32°	1.26°	1.57°	1.30°

Conclusion and Future Works

In this paper we described the development of a variable damping controlled for a lower leg prosthesis using a simulated environment. We then presented a comparison between a passive controlled and a variable-damping controlled prosthesis in different walking scenarios. Simulation results show that there is no obvious difference between two approaches on flat ground walking, but the variable-damping controlled prosthesis may need more energy due to extra electrical instrumentation. On the other hand, the variable-damping controlled prosthesis is more stable than the passive controlled prosthesis on rough terrain walking.

Acknowledgments

The authors thank CAPES[b] and DAAD[c] for the financial support.

References

1. M. P. Greene, "Four Bar Linkage Knee Analysis", Orthotics Prosthetics, 1983.
2. S. Pfeifer, et al., "An actuated transfemoral prosthesis with optimized polycentric knee joint", BioRob, 2012.
3. P. Tang, et al., "Let them walk! Current prosthesis options for leg and foot amputees", J. American College of Surgeons, 2008.
4. F. Li, et al., "The modeling research of Magnetorheological Damper in advanced intelligent prosthesis", CCDC, 2009.
5. T. Luksch, K. Berns, "Control of Bipedal Walking Exploiting Postural Reflexes and Passive Dynamics", ICABB, 2010.

[b] `http://www.capes.gov.br/`
[c] `https://www.daad.de`

SECTION–3
AUTONOMOUS ROBOTS

DODEKAPOD AS UNIVERSAL INTELLIGENT STRUCTURE FOR ADAPTIVE PARALLEL SPATIAL SELF-MOVING MODULAR ROBOTS*

SERGEY SAYAPIN†

Department of Vibration Biomechanics, Institute for Machine Science named after A.A.Blagonravov of the Russian Academy of Science, 4 Maly Kharitonyevsky Pereulok, Moscow, 101990, Russia

ANATOLY KARPENKO

Department of CAD, Bauman Moscow State Technical University, 2-nd Baumanskaya, 5, Moscow, 105005, Russia

DANG XUAN HIEP

Department of CAD, Bauman Moscow State Technical University, 2-nd Baumanskaya, 5, Moscow, 105005, Russia

This article presents a novel concept of a parallel spatial mechanism with twelve degrees of freedom, called Dodekapod robot, as universal intelligent structure for adaptive parallel spatial self-moving modular robots. Dodekapod (from the Greek words dodeka meaning twelve and pod meaning foot or its counterpart leg), unlike hexapod with six degrees of freedom, has twelve bearing rods with linear drives and novel functional capabilities. First time carried out a study of algorithms for movements of Dodekapod robot in the pipes of constant and variable cross-sections. We aim at merging technologies from information technology, special requirements in various branches industry, e.g. methods of transportation, space engineering, swarm robotics, rehabilitation medicine, spatial active vibration protection systems and robotics in order to design adaptive and intelligent robots.

1. Introduction

There are numerous examples of applications of parallel spatial robots (PSR) with 6 degrees of freedom (d.o.f.) (so-called hexapods) in most various fields [1]-[2]. Such robots versus classical robots (so-called serial robots) have the following main advantages: much higher accuracy and rigidity, higher

---

*Work partially supported by the Russian Academy of Science (Programme 15 of basic science research).

†Email: S.Sayapin@rambler.ru.

productivity and reliability, much less sensitive to scaling effect, and ability to manipulate large loads [2]. Compared to the serial robots, the main drawback of hexapods is the smaller workspace [2]. Therefore, some novel designs of PSR with the larger workspace are needed. Dodekapod robots (DR) with 12 d.o.f. is modern stage of development in the field of PSR [3]-[4]. Such robot versus hexapod has the larger workspace. Each of its sides can clamp (unclamp) things with a closed loop surface of various form. Reference [3] shows kinematics analyses of dodekapod manipulator. Currently, many modern multifunctional concepts are needed to expand functional capabilities of DR. Reference [4] shows novel functional capabilities of DR, such as the self-movement, the diagnostics of contact surfaces, and other ones. New functionality of DR is represented below.

2. Description of Dodekapod robot

We present novel concept of the adaptive parallel spatial self-moving modular robot, called Dodekapod robot (DR), where we aim at merging technologies from information technology, special requirements in various branches industry, e.g. rehabilitation medicine, space engineering, methods of transportation, spatial active vibration protection systems and robotics in order to design adaptive and intelligent robots. DR is an octahedral parallel spatial mechanism. Every one of its triangular faces is formed by linear drives that are connected with vertices of octahedron by spherical joints (link balls). This feature opens new functionality of DR versus a hexapod and others parallel spatial robots. The schematic view of the monomodule DR (a) and the simplified structural scheme of the control system (b) are shown in Figure 1. The structural scheme includes maximal number of sensors, radial stops and grippers. This number is dependent on the applications and it may be decreased. DR is executed as the octahedral module 1. All ribs of the octahedron are executed as the rods with the linear drives 2 each of which have the axial force sensor 3, the medial force sensor 4, the relative linear movement sensor 5, and the relative velocity sensor 6. The ends of the adjacent ribs are connected by the spherical joints in the vertices 7 of the octahedral module 1. The vertices 7 contain the radial stops and the middles of the rods contain the grippers (on Figure were not shown) each of which have the temperature sensor 8. The octahedral module 1 has 12 d.o.f., which is a spatial farm as soon as all linear drives 2 are turned off. All vertices 7 have the spatial position sensors 9 which are integrated with the three-axial accelerometers 10. The control system (CS) 11 includes: the computer 12, the software 13 and the digital-to-analog converter (DAC) 14. The inputs of CS 11 are connected to outputs of the analog-to-digital converter (ADC) 15 for the force sensors 3 and 4, ADC 16 for the relative linear movement sensors 5, ADC 17 for the spatial position sensors 9 with the three-axial accelerometers 10, ADC 18 for the relative velocity sensors 6, and ADC 19 for the temperature sensors 8. Outputs of CS 11 are connected to inputs of the software 13 and DAC 14. The outputs DAC 14 are connected to the power amplifier 20 which is connected to each of the linear drives 2.

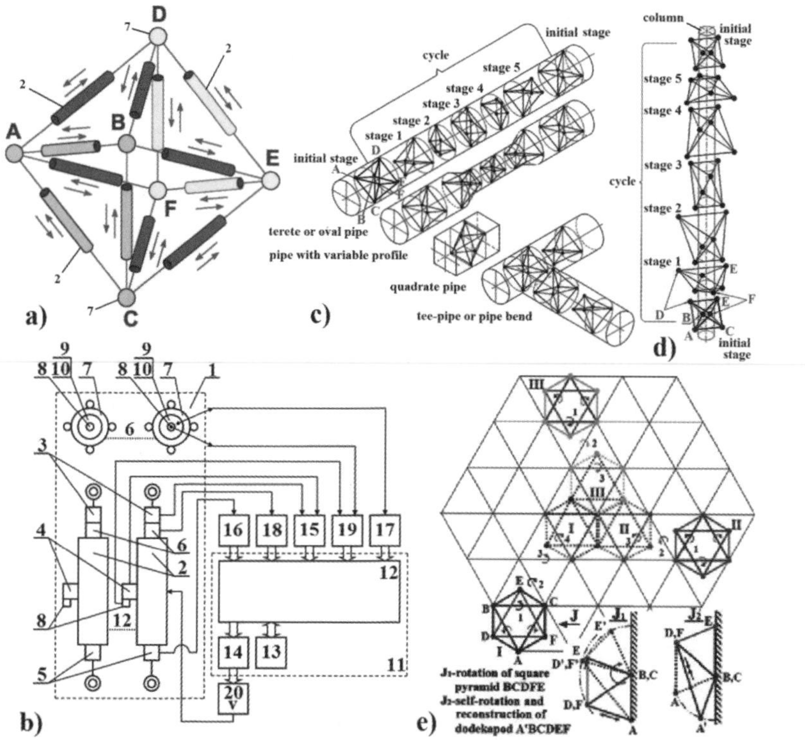

Figure 1. Structure (a, b) and self-traveling of Dodekapod robot in pipes and along column (c, d). Movement of dodekapod robot by rolling and swarm robotics (e).

The octahedral module 1 may be used as a base element not only at the monomodule DR, but also at the multimodule ones. The radial stops and the grippers (on Figure 1 were not shown) provide the transmission of the efforts from linear drives toward the internal and external contact surfaces. The force sensors 3, 4 and temperature sensors 8 provide the operative control of these efforts and temperature in the contact places. The spatial position sensors 9 with three-axial accelerometers 10 provide the operative control of the spatial position of vertices 7 and of vibration along each of axes of rods with linear drives 2. The relative linear movement sensors 5 and the relative velocity sensors 6 of the linear drives 2 register their relative movements and velocities. Before using it we will have to place the octahedral module 1 in the inside or the outside of the closed surface and then carry out the necessary movements depending on requirement. The linear drives 2 and the CS 11 fulfill herewith the coordinated changes of the rib lengths of the octahedral module. As a result the vertices 7 have got spatial movement concerning a base system of coordinates. A geometrical invariability of the octahedral module 1 allows to define the spatial coordinates of all tops 7 as a result of the measurement of the lengths of all rods and to control their spatial movements like as in Stewart's platform [1]. The sensors of the spatial position 9 allow herewith elevating a precision of these measurements. The computer 12 and the software 13 provide the control of real time. The conditions and the algorithms for movements of DR in straight pipes of constant and variable cross-sections (Figure 1) are represented below.

3. The algorithms for the movements of Dodekapod robot in the pipes of constant and variable cross-sections

3.1. *Basic parameters of dodekapod*

The dodekapod (Figure 2, a) contains 6 spherical joints (vertices *A, B, C, D, E, F*) and 12 rods (ribs *AB, BC, AC, DE, EF, DF, AD, AE, BD, BF, CE, CF*) with linear drives.

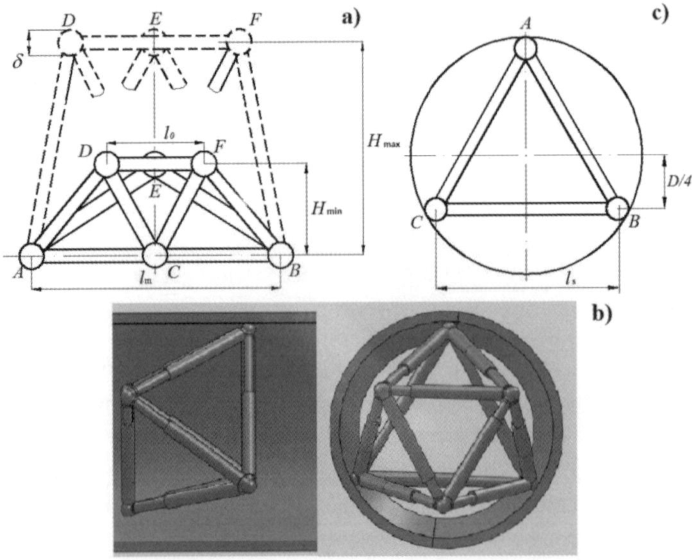

Figure 2. Structure and basic notations of dodekapod in pipe constant cross-sections.

The vertices *A, B, C* and *D, E, F* form two parallel triangular faces (*ABC* and *DEF*). The vertices *A, B, C* and *D, E, F* are dodekapod's stops for a moving in the pipe. When one of the faces in the pipe is fixed, the other face moves by simultaneous changes in the lengths of rods *AD, AE, BD, BF, CE, CF* (Figure 2, b). It is assumed that during the motion dodekapod no slippage between stops and the pipe.

Basic notations (Figure 2, a, c):

l_0, l_m - the minimum and maximum lengths of the rods of dodekapod;

δ - the diameter of the spherical joints of dodekapod;

l_S – the rod lengths of faces *ABC* and *DEF* at the time of contact their vertices with the inner surface of a pipe;

H_{min} - the minimum distance between the faces *ABC* and *DEF* (the length of the side rods reaches the minimum length);

H_{max} - maximum distance between the faces *ABC* and *DEF* (the length of the side rods reaches the maximum length);

h_s - step of the movement of dodekapod, $l_0<l_s<l_m$.

3.2. *The algorithm for the movement of dodekapod in pipe of constant cross-section*

There are a cylindrical pipe with an internal diameter *D* and length *L*. A fixing of the dodekapod in the pipe in its motion provided by alternately fixing the faces (*ABC* or *DEF*) in the pipe by increasing the length of the rods. A fixed face should be an equilateral triangle and perpendicular to the axis of symmetry of the pipe (Figure 2, c). A condition moving of the dodekapod in the pipe: the value l_s is in the range of possible lengths of the rods ($l_0<l_s<l_m$); $l_s = \sqrt{3}D/2$ (from Figure 2, c). As a result we obtain the condition:

$$2l_0/\sqrt{3} \leq D \leq 2l_m/\sqrt{3} \quad (1)$$

The algorithm for the movement of the dodekapod in pipe of constant cross-section (Figure 3):

1. The initial lengths of rods and initial position of the spherical joints are assigned from conditions: $l_{AB} = l_{BC} = l_{CA} = l_{DE} = l_{EF} = l_{DF} = l_s = \frac{\sqrt{3}D}{2}$; $l_{AF} = l_{BF} = l_{BD} = l_{CD} = l_{CE} = l_{AE} = l_m$.
2. The reduced length of rods *AB, BC, CA* from l_s to l_0.
3. The reduced length of rods *AD, AE, BD. BF, CE, CF* from l_m to l_0.
4. The increased length of rods *AB, BC, CA* from l_0 to l_s.
5. The reduced length of rods *DE, EF, DF* from l_s to l_0.

6. The increased length of rods *AD, AE, BD. BF, CE, CF* from l_0. to l_m.
7. The increased length of rods *DE, EF, DF* from l_0 to l_s.
8. A repeat of steps from second to seventh.

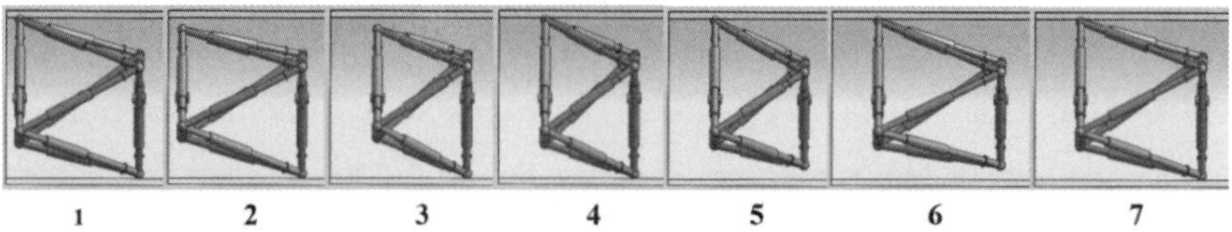

Figure 3. The algorithm for the movement of the dodekapod in pipe of constant cross-section.

3.3. *The algorithm for the movement of dodekapod in pipe of variable cross-section*

Regarded pipe has three sections (Figure 4). Two extreme sections have constant cross section (diameters *D, d* and lengths L_1, L_2), and the middle section has a variable cross-section (length Δ). The complexity of the problem is the possibility of collision of dodekapod with the inner wall of pipe in transition. It is therefore necessary to determine the ratio between the radii of the platform base and the steps of movement.

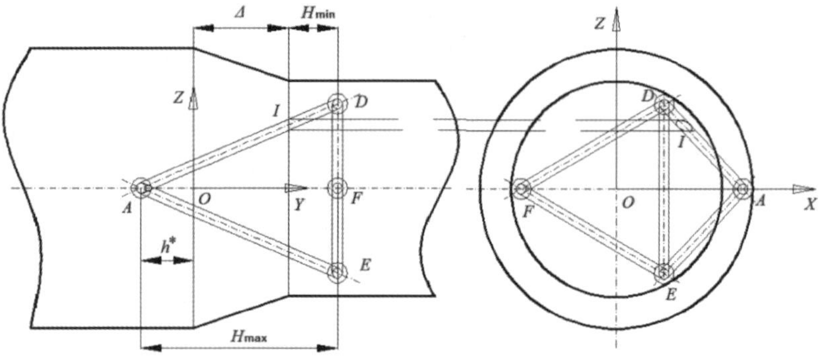

Figure 4. The dodekapod in the middle section of pipe.

The conditions for the movement in the transition: l_s is in the range of possible lengths of the rods $(2l_0/\sqrt{3} \le D,\ \ d \le 2l_m/\sqrt{3})$; $d_{I/OY} < d/2$.
Coordinates of spherical joints *A, D*:

$$A\left(\frac{D}{2}-\frac{\delta}{2}, -h^*, 0\right)\ D\left(\frac{d-\delta}{2}, \Delta + H_{min}, \frac{\sqrt{3}}{4}d - \frac{\delta}{2}\right); \tag{2}$$

$$\overrightarrow{AD} = \left(\frac{d}{4}-\frac{D}{2}, \Delta + H_{min} + h^*, \frac{\sqrt{3}}{4}d - \frac{\delta}{2}\right);$$

$$\overrightarrow{AI} = \frac{h^* + \Delta}{h^* + \Delta + H_{min}} \overrightarrow{AD} = (\frac{h^* + \Delta}{h^* + \Delta + H_{min}} \left(\frac{d}{4} - \frac{D}{2}\right),$$

$$\Delta + h^*, \qquad \frac{h^* + \Delta}{h^* + \Delta + H_{min}} \left(\frac{\sqrt{3}}{4} d - \frac{\delta}{2}\right));$$

$$\overrightarrow{OI} = \overrightarrow{OA} + \overrightarrow{AI} = \frac{h^* + \Delta}{h^* + \Delta + H_{min}} \left(\frac{d}{4} - \frac{D}{2}\right) + \frac{D - \delta}{2},$$

$$\Delta, \frac{h^* + \Delta}{h^* + \Delta + H_{min}} \left(\frac{\sqrt{3}}{4} d - \frac{\delta}{2}\right));$$

$$d_{I/OY} = \sqrt{(\frac{h^*+\Delta}{h^*+\Delta+H_{min}} \left(\frac{d}{4} - \frac{D}{2}\right) + \frac{D-\delta}{2})^2 + (\frac{h^*+\Delta}{h^*+\Delta+H_{min}} \left(\frac{\sqrt{3}}{4} d - \frac{\delta}{2}\right))^2} < \frac{d}{2}; \quad (3)$$

If $h^* \geq 0$ $(H_{min} \leq \Delta)$, then $h^* = \Delta - H_{min}$ and from (3):

$$d_{I/OY} = \sqrt{(\frac{2\Delta - H_{min}}{\Delta} \left(\frac{d}{4} - \frac{D}{2}\right) + \frac{D-\delta}{2})^2 + (\frac{2\Delta - H_{min}}{\Delta} \left(\frac{\sqrt{3}}{4} d - \frac{\delta}{2}\right))^2} < \frac{d}{2}; \quad (4)$$

If $h^* < 0$ $(H_{min} > \Delta)$, then $h^* = H_{min} - \Delta$ and from (3):

$$d_{I/OY} = \sqrt{(\frac{d+2D-4\delta}{8})^2 + (\frac{\sqrt{3}d-2\delta}{8})^2} < \frac{d}{2}. \quad (5)$$

The algorithm for the movement of the dodekapod in pipe of variable cross-section (Figure 5):

1. The movement of the dodekapod near the middle section: the number of steps $N = L_1/h_s$, $h_s = H_{max} - H_{min}$; the length of rods of the faces *ABC* and *DEF* $l_{AB} = l_{DE} = \frac{\sqrt{3}(D-\delta)}{2}$; $l_{AF} = l_m$ - the length of the side rods.
2. The movement of the dodekapod forward with a step $h_{s1} = L_1 - Nh_s$; $l_{AB} = l_{DE} = \frac{\sqrt{3}(D-\delta)}{2}$. The face *DEF* of the dodekapod is installed in leading edge of middle section of pipe (Figure 5, a).
3. If the conditions (2) - (5) are satisfied, then the dodekapod moves through the middle section of the pipe with step $h_{s2} = \Delta + H_{min}$; $l_{AB} = l_{DE} = \frac{\sqrt{3}(d-\delta)}{2}$ (Figure 5, b, c, d).
4. The dodekapod moves in through the extreme section of the pipe with step $h_s = H_{max} - H_{min}$; $l_{AB} = l_{DE} = \frac{\sqrt{3}(d-\delta)}{2}$ (Figure 5, e).

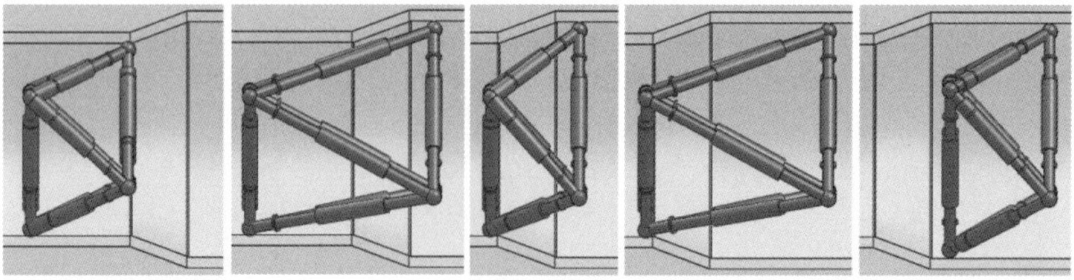

a) b) c) d) e)

Figure 5. The algorithm for the movement of the dodekapod in pipe of variable cross-section.

4. Other Capabilities of dodekapod robot

The novel concept of DR may be used also in many other fields. The examples of applications DR and the new functional capabilities of DR are established, e.g.:

- Possibility of control over of physical-mechanical properties, geometrical shape of contact surface and displacement trajectory.
- Travel of object in pipe with possibility of vibroprotection and positioning.
- Travel of long objects in pipe.
- Possibility of hole drilling in the end of pipe wall.
- Possibility of battering in pipe of wall end.
- Dodekapod robots can connect together, forming some novel mobile structures for various applications.

References

1. Stewart D.: A platform with six degrees of freedom. *J. Proc. Inst. Mech. Eng.,* vol. 180, Part 1, no. 15, pp. 371-386, 1965-1966.
2. Merlet J.-P.: Parallel Robots, 2nd ed. Dordrecht, the Netherlands: Springer, 2006.
3. Bande P., Seibt M., Uhlmann E., Saha S. K., Pao P. V. M.: Kinematics analyses of dodekapod. *J. Mechanism and Machine Theory.*, vol. 40, no.6, pp. 740-756, 2005.
4. Sayapin S. N., Siniov A. V.: Adaptive mobile 3D manipulator and method of organizing displacements and control over physical-mechanical properties, geometrical share of contact surface and displacement trajectory hereby. Invention with a patent of Russian Federation No. 2424893, 2011.

AQUATIC MULTI-ROBOT SYSTEM FOR LAKE CLEANING*

PRANAY AGRAWAL
Department of Mechanical Engineering, Indian Institute of Technology Kanpur, Kanpur, Uttar Pradesh ZIP-208016, India

BISHAKH BHATTACHARYA
Smart Materials, Structures and Systems Lab, Department of Mechanical Engineering, Indian Institute of Technology Kanpur, Kanpur, Uttar Pradesh ZIP-208016, India

This paper discusses the design of a multi-robot system of autonomous aquatic vehicles that can be used for cleaning of lakes and maintenance of fisheries. The traditional method of removing the weeds manually and collecting them along with other surface wastes is highly inefficient and labour intensive. Aimed at automating the entire process, the robots make use of tactile sensors and wireless communication to traverse autonomously and collectively perform cleaning operations such as removing the surface impurities, pumping oxygen into the water, spraying chemicals and distributing food at appropriate locations along with measuring the water quality. A novel algorithm for navigation and waste removal strategy of the multi-robot aquatic system, inspired by insects such as bees and referred to as 'recruitment algorithm', has been proposed. With the help of simulations, supplemented by real environment testing, it has been demonstrated that by the use of a multi-robot system with this strategy, an effective and rapid cleaning of aquatic waste is feasible.

1. Introduction

The condition of the lakes and ponds in majority of the developing countries are abysmal. A lot of money and effort is being spent by the government and non-government organizations on cleaning and maintaining them at regular intervals [1, 2]. Present work aims at developing a multi-robot system of autonomous aquatic vehicles, which can perform various tasks required for the cleaning and maintenance of lakes, ponds and fisheries along with performing secondary functions such as controlled food distribution and fishing.

A few aquatic robots have been built earlier for the purpose of surface cleaning [3, 4]. These works, however, are limited to a few functionalities and are based on single robot system. A few works are available in the open

* Work supported by the Design Program, Indian Institute of Technology Kanpur and UKIERI/MDES/20120045

literature, which have discussed about the modelling and usage of such special purpose robots [5, 6].

Present work aims at developing a more versatile and efficient system by the usage of multiple robots with increased functionality. Activities such as removing surface algae, leaves and twigs, spraying of chemicals at the appropriate locations, pumping oxygen whenever required, checking the water quality as well as deploying payload are planned to be done autonomously. This will save a lot of human effort and provide a sustainable solution to the pervasive problem.

For the purpose of navigation and trash cleaning on the ground, many well-designed algorithms have been developed for both single robotic systems as well as for swarms [7, 8]. However, because of the difference in the dynamic environment, propulsion system, and the difficulty in determining the precise position based on the acceleration and GPS coordinates, these algorithms cannot be directly used on aquatic surfaces. Also, the navigation algorithms developed earlier for autonomous aquatic robots have not been designed with cleaning as an integral part of them [9]. Thus, there is a need for developing a new algorithm for the navigation of these robots and optimizing their performance in waste removal. This paper addresses the issue and presents a viable solution of the problem.

2. Design of the Robotic System

Four important issues considered for designing the aquatic robots are: cost effective solution, robustness, efficacy and durability. Due to the nature of the cleaning work, the vehicle structure is designed such that it can provide high stability, great maneuverability and can easily collect all the waste flowing in between. A pontoon shaped hull works best for this case and fulfills all the hydrostatic, sea-keeping and structural stability criteria [10]. For the purpose of removal and collection of surface waste, a motor driven track-belt system has been designed that can collect the wastes in a section covered with a net. This design provides a simple and effective waste removal and accommodates large amounts of waste within a small space. The lower section of the belt is placed below the water level to easily take out the floating weeds. Moreover, as the collected waste floats on the water surface, the robot does not need to support its weight.

For the prototype (Fig. 1), the hulls were made up of two long PVC pipes supported together with the help of an aluminuim truss. This light and tough

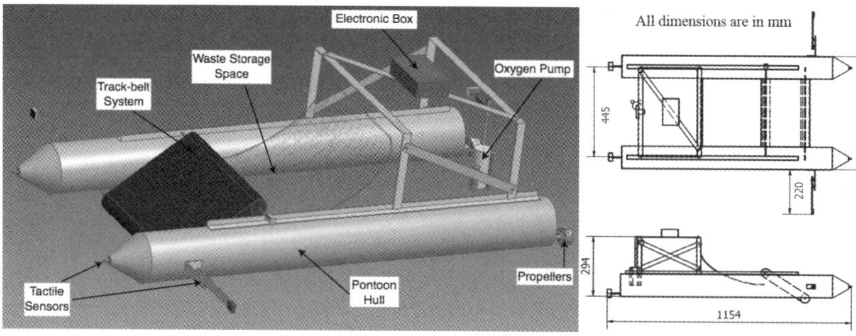

Figure 1. Structure of the robots used for lake cleaning

structure supports the total weight of the system. The specifications of this prototype robot are given in Table 1.

To navigate autonomously in small and constrained spaces, a propulsion system that can provide high mobility and short turning radius is required. Due to its flat front and the central waste collection system, the robot is also required to overcome high form and skin drag, which has been calculated as:

$$Drag = -(18\,v^2 + 0.027\,m\,v^{1.61})$$

where, v is the forward velocity or the robot in m/sec, m is the total mass of the robot and the waste in kg and Drag in Newtons.

Table 1. Specifications of the robot.

Design Parameter	Value
Overall Dimensions	1154mm X 559mm X 294mm
Net Weight without playload	12 kg
Maximum weight with playload	18 kg
Area coverage per hour	760 m^2
Maximum speed at full loading	0.38 m/sec
Maximum forward thrust	3.6 N
Waste removal capacity per run	7.875×10^{-3} m^3
Average battery backup while cleaning	2 hours
Control System	Autonomous + Manual
Drive motors	750 rpm 1.2 N-m torque
Water quality sensors	QRD Photoresistive diode + IR LED
Wirelss module	8 channel Radio Frequency transmitter - receiver

Thus, a propulsion system based on a differential drive mechanism has been designed, where both the propellers are controlled independently. This allows the robots to take a 360° turn on the spot and provides high thrust. Placing the drive motors inside the hull, protects them from water seeping inside its casing.

As the robots are designed to traverse autonomously, they need to sense when they get close to an obstacle or the boundary of the lake. Majority of the sensors such as the ultrasonic sensors, IR sensors, depth ranging etc. that work well on land are either unusable in water bodies or are unable to differentiate between hard rocks and leaves. Tactile sensors form a very good alternative due to their simple design, low cost and their ability to differentiate the collectable wastes from the rigid ones. Measuring the water quality of a lake/fishery is also very important in order to determine where to pump oxygen, detecting regions of contamination and in checking the dissolved salt content, which are essential for the survival of aquatic life [11]. In the current prototype, two tactile sensors are mounted directly onto the nose of the vehicle to detect any obstruction from the front. For the side sensors, a lever mechanism has been used, placing the tactile switch at the lever, fixing one of the ends to the hull and allowing the free end to engage with the obstructions and activate the switch. IR LEDs and photo-resistors are used to get a measure of the total suspended particles in the water.

The robots are primarily designed to work autonomously in groups, but it is also important for them to be remotely controllable whenever the operator wants them to clean up a particular region of a lake or to bring them back to a desired location. The control system thus designed can be toggled between autonomous and manual mode by a remote. The inputs from all the sensors and the wireless receivers are fed into the microcontroller, which processes the information and sends the desired commands to motor drivers for propulsion, waste collection and payload deployment.

3. Navigation Algorithm

The area around a robot has been divided into four regions, separated by radii R_c (Critical radius), R_w (Working Radius) and R_o (Outer Radius). If another robot or boundary of the lake comes inside the R_c of a robot, an obstacle avoidance algorithm [12] is used.

While cleaning a region, if a robot detects waste in its path for five consecutive time units (in seconds), it starts to follow a spiral path in order to collect maximum waste in that region and goes back to a straight-line path if no more waste is detected. In order to achieve the desired spiral path, the 8-bit

Pulse Width Modulation (PWM), which controls the motor voltage, is given by the following signals:

$PWM_{right\ propeller} = \alpha$ and $PWM_{left\ propeller} = \alpha *(1\text{-}e^{-at})$

Here, the value of 'a' controls the distance between successive turnings and α = 255 for an 8-bit PWM signal.

3.1. *Recruitment Algorithm*

A 'Recruitment call' is a wireless signal that a robot sends to all other robots if it finds enough waste in a particular region and wants the other 'free' robots to clean up that particular region. This method of giving recruitment calls has been inspired by insects such as bees and ants [13, 14].

While collecting waste, the robot's trackbelt torque increases, causing a surge in the input current flowing through the motor. This is used to detect whether the robot is collecting waste at that instant.

The algorithms for sending and receiving the recruitment calls have been demonstrated by the help of a pseudo-code described below.

```
/*Sending Recruitment calls*/
if waste is equal to 0{
    time_waste_colected = 0;
    follow a straight path;
    set mode to 'free';
}
if waste > 0{
    set mode to 'busy';
    time_waste_colected = time_waste_colection + 1;
    if time_waste_colected > 5 seconds{
        send recruitment calls;
        execute spiral path;
    }
}

/*Receiving Recruitment calls*/
receive recruitment calls from robots B1, B2.....Bn
if mode = 'free'{
    find the closest robot and set it as B0;
    if distance from B0 is between Rw and Ro
        approach robot B0 for 1 time-step;
}
```

Only the robots between R_w and R_o from the robot giving recruitment calls respond to it as a robot beyond R_o will take a long time and will waste unnecessary energy in getting there. The robots within R_c and R_w are already in close proximity to the robot sending the recruitment calls and therefore ignore the calls. After coming within R_w of the robot giving recruitment calls, the robot starts to follow random paths around it, trying to look for waste.

While approaching the robot which gave the call, if a robot finds enough waste on the way, it goes into the 'busy mode' and ignores the recruitment calls.

4. Simulations

To test the effectiveness and efficiency of the multi-robot system, we have carried out the simulations of a 3-robot system in a lake like environment for varying lake sizes and random waste distribution. The robots have been mathematically modeled as rectangular bodies and the total force (drag and viscous) is calculated as a function of its linear velocity and the amount of waste collected. The simulations have been carried out in C++ using the Enki robotic library and Physics engine, which provides us with the positions and angles of the robots at different instances of time. These position and angles obtained are then used to generate the real-time rendering in MATLAB. The snapshots of a simulation at the beginning and after two hours are shown in Fig. 2. The robots are shown in blue. The green points represent the waste to be removed and the brown region represents a part of the boundary of the lake. As the robot passes over a waste, it is assumed to be removed.

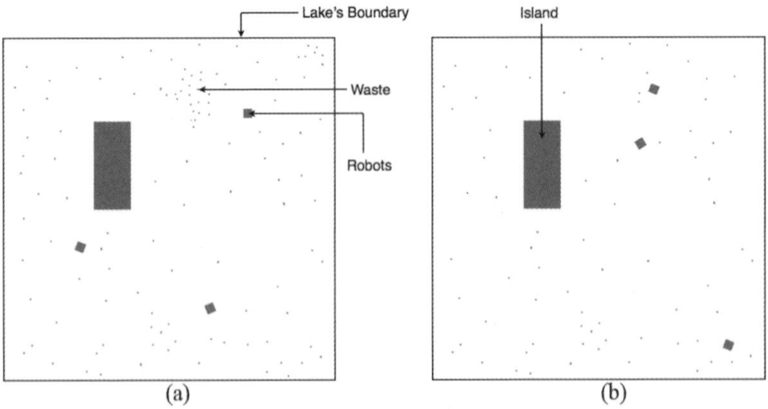

Figure 2. Snapshot of the waste distribution at (a) the start and (b) after two hours

5. Results and Conclusion

This paper has proposed the design of a multi-robot system of autonomous aquatic vehicles that can be used for lake cleaning. As the robots have been designed primarily to be used in developing countries, the cost has been kept below US$ 350 to be economically feasible. Along with the cost as an objective, considering robustness and reliability of the system, a divided hull structure is proposed in which the area between the two hulls has been used for waste removal and collection. The same vehicle has been developed and tested successfully in aquatic condition. For the purpose of efficient waste collection, a novel recruitment algorithm has been developed and simulated for a 3-robot system in a lake like environment.

Though the robots did not have a very high speed, they produced large enough thrust, which was needed to overcome the drag acting on the system. These robots have been designed in such a way that they require least amount of maintenance and human interference, doing most of their work autonomously, but at the same time, having the option to be controlled manually. After the collection area is full, the robots autonomously go back to a defined region on the lake's boundary and dump the waste on the ground.

The simulations performed for a three-robot system with the recruitment algorithm concluded to be successful in removing a minimum of 80% of the randomly distributed waste from a 40,000m^2 lake in a single day. For the simulation shown in Fig 2, though the robots covered just 8% area of the 22,300m^2 lake within two hours, they removed 21% of the total waste, demonstrating the effectiveness of the recruitment strategy.

Apart from the cleaning activities, these robots can also be programmed to perform many other useful functions, such as fishing in a lake. In future, a series of such aquatic robots would be used in a real-time environment to validate their performance.

Acknowledgments

I would like to thank Dr. Roderich Gross for helping me with the swarm algorithm as well as his Ph.D. student Jianning Chan for his support in carrying out the simulation. I also want to thank UKIERI/MDES/20120045 and the Design program IIT Kanpur, for their financial support.

References

1. Ministry of Water Resources, Government of India URL: http://www.worldlakes.org/uploads/Management_of_lakes_in_India_10Mar04.pdf
2. *Department of Ecology, Limnology, University of Lund, Ecology Building, S–223 62 Lund, Sweden* URL: http://www.aseanbiodiversity.info/Abstract/5100 6316.pdf
3. Z. Wang, Y. Liu, H. W. Yip, B Peng, S. Qiao, and S. He "Design and Hydrodynamic Modeling of A Lake Surface Cleaning Robot" AIM 2008.
4. M. A. Abkowitz, "Measurement of hydrodynamic characteristics from ship manoeuvering trials by system identification," SNAME Trans. 88 1980.
5. N. M. P. Kakalis, and Y. Ventikos "Robotic swarm concept for efficient oil spill confrontation" Journal of hazardous materials, 2008
6. M. A. Joordens "Design of a low cost underwater robotic research platform" System of Systems Engineering, 2008
7. Y. Altshuler, A.M. Bruckstein, and I.A. Wagner, "Swarm robotics for a dynamic cleaning problem" Swarm Intelligence Symposium, 2005
8. K. Cheng and P. O. Dasgupta "Dynamic Area Coverage using Faulty Multi-Agent Swarms" Intelligent Agent Technology, 2007
9. M. Caccia, M. Bibuli, R. Bonoa and G. Bruzzone "Basic navigation, guidance and control of an Unmanned Surface Vehicle" Autonomous RobotsNovember 2008, Volume 25, Issue 4, pp 349-365
10. E. K. Boulougouris, A. D. Papanikolaou, Y. L. Corre, F. Ghozlan, O. Turan, N. M. P. Kakalis, Y. Ventikos, D. Fritsch and V. Campos, "Efficient oil spill confrontation by innovative EU-MOP unit" Proceedings of the International Symposium on Maritime Safety, Security and Environmental Protection, Athens, Greece, 2007
11. G. P. Cooper & G. N. Washburn "Relation of Dissolved Oxygen to Winter Mortality of Fish in Michigan Lakes" Transactions of the American Fisheries Society Volume 76, Issue 1.
12. P. Agrawal an internal report on development of aquatic robotic system for Lake cleaning._IITK_ME_SMSS_1_20011-12
13. R. Menzel, K. Geiger, L. Chittka, J. Joerges, J. Kunze and U. Muller "The knowledge base of bee navigation" J. Exp. Biol. 199, 141-146 1996.
14. P. Jevtić, D. A. Gazi, and M. Jamshidi "Building a swarm of robotic bees" World Automation Congress, 2010.

HYBRID SPIRAL DYNAMIC BACTERIAL CHEMOTAXIS OPTIMISATION FOR HYBRID FUZZY LOGIC CONTROL OF A NOVEL TWO WHEELED ROBOTIC VEHICLE

A. M. ALMESHAL[1], K. M. GOHER[2], A. N. K. NASIR[1], M. O. TOKHI[1] and S. A. AGOURI[1]

[1]*Department of Automatic Control and Systems Engineering*
The University of Sheffield, United Kingdom, e-mail: cop10ama@sheffield.ac.uk

[2]*Department of Mechanical and Industrial Engineering*
Sultan Qaboos University, Oman, e-mail: kgoher@squ.edu.om

A hybrid spiral dynamic bacterial chemotaxis (HSDBC) optimisation algorithm is used to optimise the parameters of a fuzzy logic control system of a new structure of two - wheeled robotic vehicle. The vehicle has a novel structure based on double inverted pendulum with a movable payload and 5 degrees of freedom. HSDBC algorithm combines the strengths of bacterial foraging algorithm (BFA) and spiral dynamic algorithm (SDA) for a fast and accurate convergence to the optimum point. This paper demonstrates the efficiency of the HSDBC with fast convergence speed, better accuracy and stability compared to BFA optimisation.

1. Introduction

Control system designs have heightened the need for optimisation algorithms to find the best control design criteria. These criteria can be related control gains, minimum control effort, minimum dissipated energy, and etc. Different kinds of optimisation algorithms are found in the literature inspired by natural phenomena. These include particle swarm optimisation (PSO) developed by Eberhart et al. [1], bacterial foraging algorithm (BFA) developed by Passino et al. [2], spiral dynamic algorithm (SDA) developed by Tamura et al. [3] and hybrid spiral dynamic bacterial chemotaxis algorithm (HSDBC) developed by Nasir et al. [4]. These optimisation algorithms have been developed by researchers to enhance the optimisation process in terms of speed of convergence, avoiding local optima, minimising computation time and accuracy of the solution. In this paper, the HSDBC algorithm is used to optimise a hybrid fuzzy logic controller (FLC) of a double inverted pendulum-like robotic vehicle developed by Al-Meshal et al. [5]. The vehicle has a novel structure with a movable payload and five degrees of freedom (DOF). The design of the vehicle enables manoeuvrability on irregular and inclined terrains as well as in confined spaces. A hybrid fuzzy logic control approach is adopted to control the vehicle with five control loops. This paper demonstrates realisation of HSDBC algorithm to optimise the controller parameters, and simulation results demonstrating performance of the algorithm are presented and discussed.

2. System description

As illustrated in Figures 1 and 2, the vehicle consists of two separate wheels driven by separate DC motors to enable steering in various directions and two pendulums connected by a motor to drive the second link. A payload linear actuator is placed on the upper pendulum to lift up the payload to a demanded height. The vehicle therefore has five degrees of freedom.

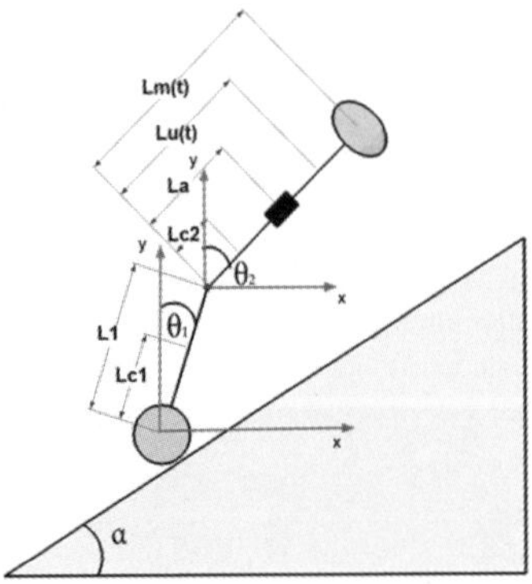

Figure 1. The vehicle on an inclined plane

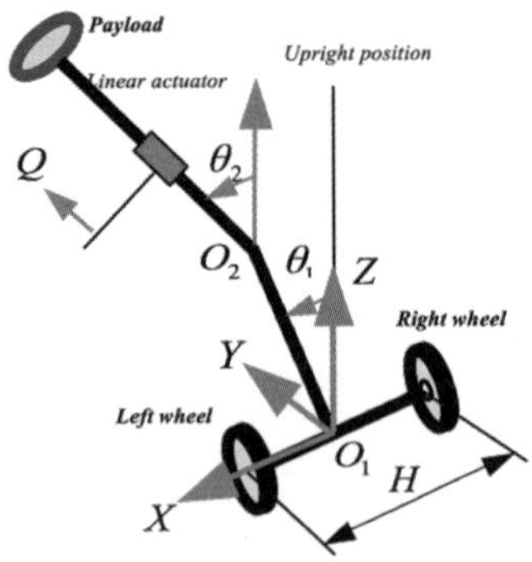

Figure 2. Axonometric diagram of the vehicle

The dynamics of the vehicle were derived using Lagrangian approach, and presented in detail in [1]. The vehicle equations of motion describing the system dynamics are given as follows

$$
\begin{aligned}
&2C_{21}\ddot{\delta}_L + C_{22}\ddot{\delta}_R + C_9\frac{R_w}{2}L_1\ddot{\theta}_1\cos\theta_1 - C_9\frac{R_w}{2}L_1\dot{\theta}_1^2\sin\theta_1 \\
&+\frac{R_w}{2}(C_{10}+C_8Q)\ddot{\theta}_2\cos\theta_2 - \frac{R_w}{2}(C_{10}+C_8Q)\dot{\theta}_2^2\sin\theta_2 \\
&+\frac{R_w}{2}C_8\dot{Q}\dot{\theta}_2\cos\theta_2 = T_L - T_{fL}
\end{aligned}
\qquad (1)
$$

$$2C_{21}\ddot{\delta}_R + C_{22}\ddot{\delta}_L + C_9\frac{R_w}{2}L_1\ddot{\theta}_1\cos\theta_1 - C_9\frac{R_w}{2}L_1\dot{\theta}_1^2\sin\theta_1 + \frac{R_w}{2}(C_{10}+C_8Q)\ddot{\theta}_2\cos\theta_2 - \frac{R_w}{2}(C_{10}+C_8Q)\dot{\theta}_2^2\sin\theta_2 + \frac{R_w}{2}C_8\dot{Q}\dot{\theta}_2\cos\theta_2 = T_R - T_{fR} \tag{2}$$

$$2C_{18}\ddot{\theta}_1 + C_9\frac{R_w}{2}L_1(\ddot{\delta}_L+\ddot{\delta}_R)\cos\theta_1 - C_9\frac{R_w}{2}L_1(\dot{\delta}_L+\dot{\delta}_R)\dot{\theta}_1\sin\theta_1 + 2L_1(C_{10}+C_8Q)\ddot{\theta}_2\cos(\theta_1-\theta_2) - 2L_1(C_{10}+C_8Q)\dot{\theta}_1\dot{\theta}_2\sin(\theta_1-\theta_2) + 2L_1(C_{10}+C_8Q)\dot{\theta}_2^2\sin(\theta_1-\theta_2) + 2L_1C_8\dot{Q}\dot{\theta}_2\cos(\theta_1-\theta_2) + C_9\frac{R_w}{2}L_1\dot{\theta}_1^2(\dot{\delta}_L+\dot{\delta}_R)\sin\theta_1 + 2L_1(C_{10}+C_8Q)\dot{\theta}_1^2\dot{\theta}_2\sin(\theta_1-\theta_2) - C_3g\dot{\theta}_1\sin\theta_1 = 0 \tag{3}$$

$$C_8\ddot{Q} - \frac{1}{2}(C_{12}+2C_8Q)\dot{\theta}_2^2 - C_8\frac{R_w}{2}\dot{\theta}_2(\dot{\delta}_L+\dot{\delta}_R)\cos\theta_2 - 2L_1C_8\dot{\theta}_1\dot{\theta}_2\cos(\theta_1-\theta_2) + C_8g\cos\theta_2 = F_a - F_{fa} \tag{4}$$

$$C_{20}\ddot{\theta}_2 + (C_{12}\dot{Q}+2C_8Q)\dot{\theta}_2 + (C_{12}Q+C_8Q^2)\ddot{\theta}_2 + \frac{R_w}{2}(C_{10}+C_8Q)(\ddot{\delta}_L+\ddot{\delta}_R)\cos\theta_2 - \frac{R_w}{2}(C_{10}+C_8Q)(\dot{\delta}_L+\dot{\delta}_R)\dot{\theta}_2\sin(\theta_2) + C_8\frac{R_w}{2}\dot{Q}(\dot{\delta}_L+\dot{\delta}_R)\cos\theta_2 + 2L_1(C_{10}+C_8Q)\ddot{\theta}_1\cos(\theta_1-\theta_2) - 2L_1(C_{10}+C_8Q)\dot{\theta}_1^2\sin(\theta_1-\theta_2) + 2L_1(C_{10}+C_8Q)\dot{\theta}_1\dot{\theta}_2\sin(\theta_1-\theta_2) + 2C_8L_1\dot{\theta}_1\dot{\theta}_2\cos(\theta_1-\theta_2) + \frac{R_w}{2}(C_{10}+C_8Q)(\dot{\delta}_L+\dot{\delta}_R)\dot{\theta}_2^2\sin\theta_2 - (C_{15}+C_8Q)g\dot{\theta}_2\sin\theta_2 - 2L_1(C_{10}+C_8Q)\dot{\theta}_1\dot{\theta}_2^2\sin(\theta_1-\theta_2) = T_M - T_{FM} - L_dF_d \tag{5}$$

3. Hybrid FLC control strategy

A hybrid fuzzy logic control (FLC) strategy was developed by Almeshal et al. [6] and used to stabilise the vehicle model. Figure 3 shows a block diagram of the closed-loop control system. The system consists of five feedback control loops. Two types of hybrid FLC controllers, namely, a proportional-derivative-like FLC (PD-like FLC) and proportional-derivative plus integral FLC (PD+I FLC), were used to control the robotic vehicle. The inputs to the PD-like FLC are the error signal and the change of error. While the inputs for the PD+I FLC are the error signal, change of error and the sum of previous errors. The FLC is of Mamdani-type with Gaussian membership functions to smooth the outputs with 25 fuzzy rule base.

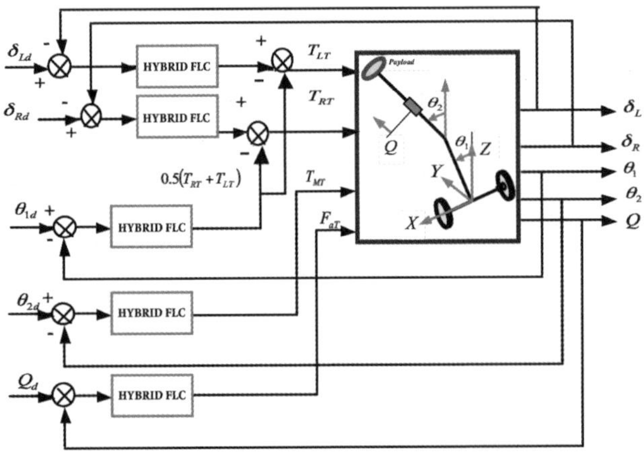

Figure 3. Control system block diagram

4. Hybrid spiral dynamic bacterial chemotaxis algorithm

The hybrid spiral dynamic bacterial chemotaxis (HSDBC) optimisation algorithm for global optimisation was developed by Nasir et al. [4]. The HSDBC algorithm is a hybridisation between the spiral dynamics algorithm (SDA) and the bacterial foraging algorithm (BFA) The BFA algorithm has faster convergence speed to feasible solutions in the defined search space but exhibits oscillations towards the end of the search operation. The SDA algorithm has faster computation time and better accuracy than the BFA algorithm. Furthermore, the SDA algorithm has better stability, due to the spiral steps, when searching towards the optimum point. The HSDBC algorithm combines the strengths of BFA and SDA into a faster, stable and accurate global optimisation algorithm. This is achieved by incorporating the BFA chemotaxis part into the SDA and thus reducing the computational time and retaining the strength and performance of the SDA. The HSDBC parameters are presented in Table 1.

Table 1. HSDBC algorithm parameters

Parameter	***Description***
$\theta_{i,j}$	The angular displacement of the bacteria in the x_i - x_j plane around the origin
r	Spiral radius
m	Number of search point
$k_{\max}$	The maximum number of iterations
N_s	The maximum number of swim per bacteria
$x_i(k)$	Position of bacteria

The HSDBC algorithm is as follows:

Step 0: Preparation

Define the number of search points $m \geq 2$

Define the parameters $0 \leq \theta < 2\pi, 0 < r < 1$ of $S_n(r, \theta)$

Define the maximum number of iterations $k_{\max}$

Define maximum number of swim N_s for bacteria chemotaxis

Set $k = 0$ and $s = 0$

Step 1: Initialisation

Set initial points $x_i(0) \in R^n, i = 1, 2, ..., m$ at the feasible region at random

Define x^* *as* $x^* = x_{ig}(0)$, $i_g = \arg\min_i f(x_i(0)), i = 1, 2, ..., m$

Step 2: Apply bacteria chemotaxis

a-Update x_i as

$$x_i(k+1) = S_n(r,\theta)x_i(k) - (S_n(r,\theta) - I_n)x^* \qquad i = 1, 2, ..., m$$

b- Bacteria swim

i. Check the number of swim for the bacteria i If $s < N_s$, then check fitness the fitness, otherwise set $i=i+1$ and return to step a

ii. Check fitness if *if* $f(x_i(k+1)) < f(x_i(k))$ then update x_i otherwise set $s=N_s$ and return to step a

iii. Update x_i as

$$x_i(k+1) = S_n(r,\theta)x_i(k) - (S_n(r,\theta) - I_n)x^* \qquad i = 1, 2, ..., m$$

Step 3: Updating x^*

$$x^* = x_{ig}(k+1) \ ; \ i_g = \arg\min_i f(x_i(0)) \ ; \ i = 1, 2, ..., m$$

Step 4: Stopping criteria

If $k = k_{\max}$ then stop.

Otherwise $k = k + 1$

Return to step 2

The control system of the vehicle consists of five hybrid FLC controllers with a total of 15 parameters. The performance index of the system is chosen as the minimum mean squared error (MSE). The MSE is calculated for each control loop of the vehicle as follows:

$$MSE = \min\left\{\frac{1}{N}\sum_{i=1}^{N}(e)^2\right\}$$

The objective function is given as the summation of MSE functions of the five control loops. Minimisation of the objective function leads to optimum controller parameters that result in the smallest control-loop errors, and system stability.

5. Constrained HSDBC optimisation

Due to the complexity and coupling nature of the vehicle model, slight changes in the control parameters may affect the stability of the vehicle. Thus, constrained optimisation is needed to define a feasible search space within the stability region of the vehicle model. The optimisation process is thus constrained within the stability region of the system by defining upper and lower boundaries for each control parameter as presented in Table 2.

Table 2. Boundary limits of the controller gain parameters

Parameter	Loop 1	Loop 2	Loop 3	Loop 4	Loop 5
Kp	1.5-2.4	5-6.5	8-12	8-10	30-50
Kd	0.5-1	2.5-4	7.5-9	5-8	10-20
Ki	0.9-1.4	1.5-2	0-0.5	0-0.5	1-10

Penalty-based constrained optimisation approach is used to constrain the HSDBC algorithm. The cost function is penalised whenever a constraint is violated using

$$J(x) = \begin{cases} J(x) & x \in \textit{feasible region} \\ J(x) + P(x) & x \notin \textit{feasible region} \end{cases}$$

where, $J(x)$ is the cost function and $P(x)$ represents the penalty function. The penalty function is added to the cost function to result in a high cost once the constraint is violated, thus eliminating the solution from the feasible solutions region when selecting the minimum cost value in a minimisation problem.

6. Simulation and Results

The system is optimised using two optimisation algorithms, namely BFA and HSDBC for purposes of comparative assessment of performance of the HSDBC algorithm. Tables 3 and 4 show the simulation parameters used for BFA and HSDBC algorithms respectively.

The optimised controller parameters achieved by each optimisation algorithm are presented in Table 5. Table 6 shows the minimum cost function obtained by each algorithm. It is noted clearly that the HSDBC algorithm has found the minimum cost function value of 0.3682

Table 3 BFA parameters

P	**S**	**Nc**	**Ns**	**Nre**	**Ned**	**Ped**	**Sr**
15	20	14	6	2	2	0.25	S/2

Table 4 HSDBC parameters

P	**R**	**Rzw**	**Ns**	**Theta**	**Initial Points**	**Iterations**
15	0.95	0.55	2	$\pi / 4$	10	150

Table 5 Optimised gain values

	Parameter	**BFA**	**HSDBC**
Loop 1	Kp_1	2.0729	2.1566
	Kd_1	0.8572	0.8095
	Ki_1	1.3925	1.2026
Loop 2	Kp_2	6.0155	5.1530
	Kd_2	2.8185	2.6917
	Ki_2	1.6390	1.8754
Loop 3	Kp_3	8.7514	11.4514
	Kd_3	8.1889	8.9946
	Ki_3	0.2449	0.3771
Loop 4	Kp_4	8.9718	9.9903
	Kd_4	5.1315	6.6239
	Ki_4	0.0071	0.0410
Loop 5	Kp_5	49.9646	36.6753
	Kd_5	13.6834	14.3583
	Ki_5	4.0408	5.4203

Table 6 Cost functions

Minimum Cost function value	BFA	HSDBC
J	0.3684	0.3682

Figures 4-9 illustrate the system response with the optimised controller found by BFA and HSDBC algorithms.

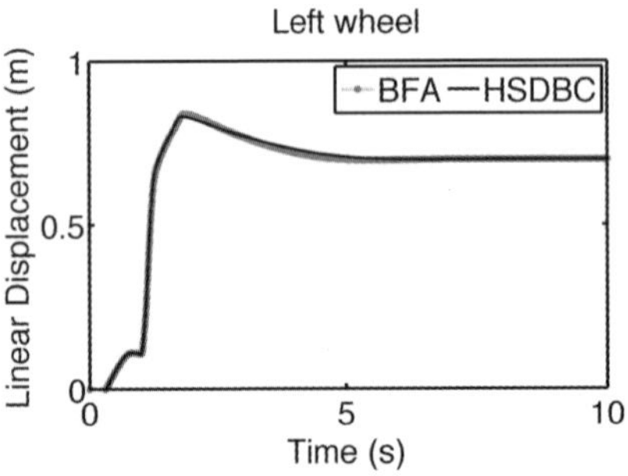

Figure 4 The linear displacement of the left wheel

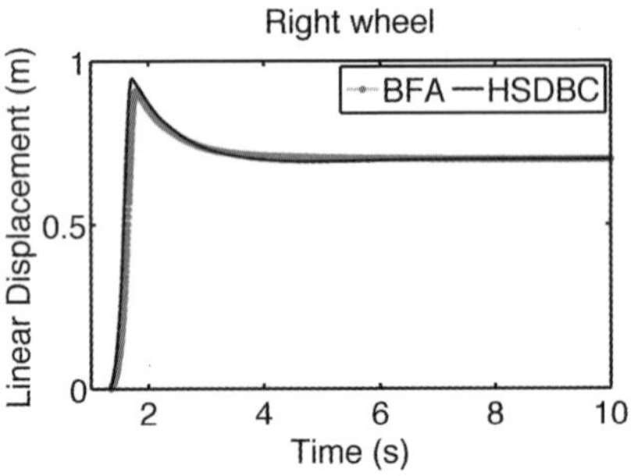

Figure 5 The linear displacement of the right wheel

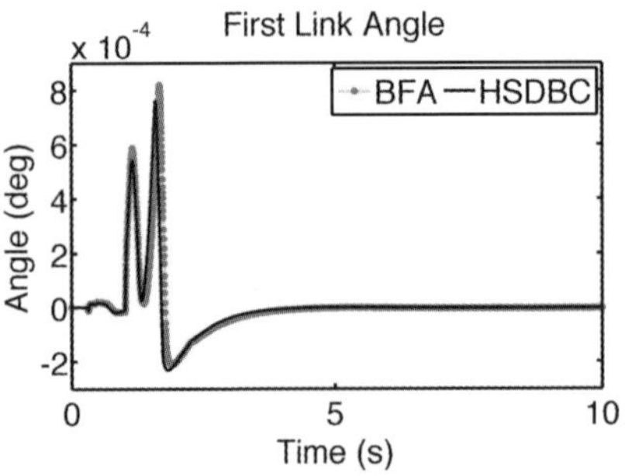

Figure 6 The tilt angle of the first link

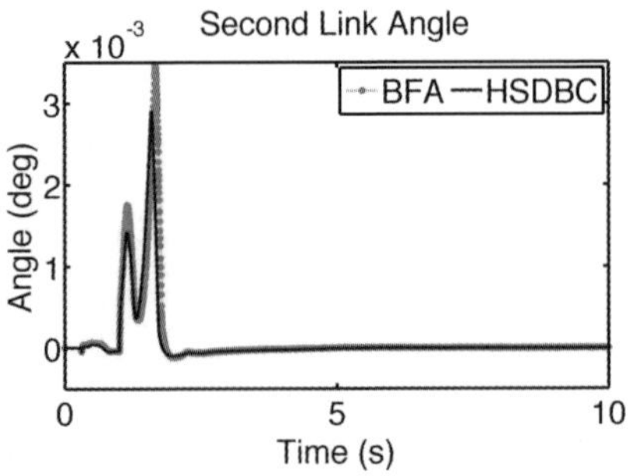

Figure 7 The tilt angle of the second link

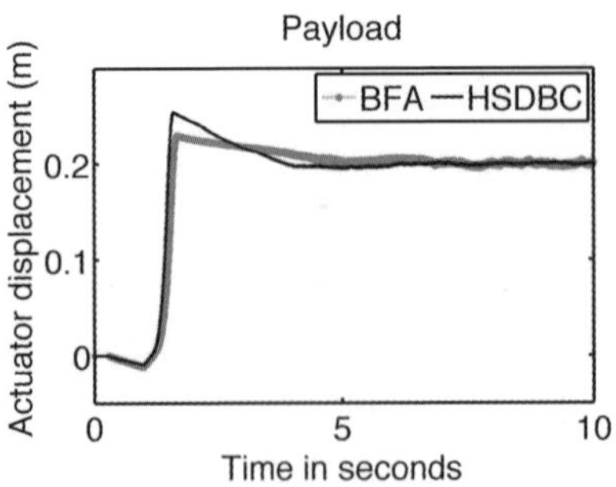

Figure 8 The payload linear actuator displacement

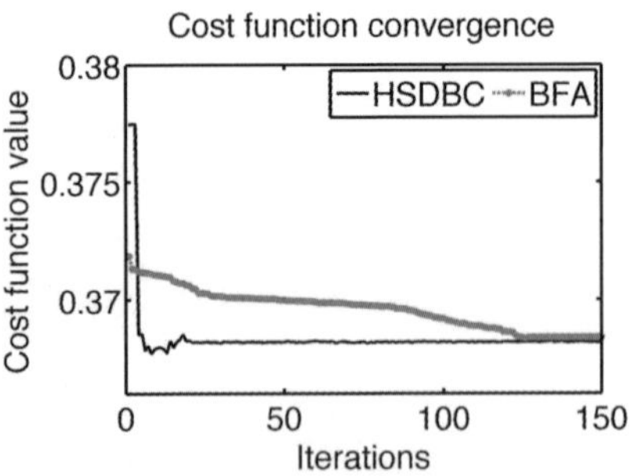

Figure 9 Cost function convergence plot for BFA and HSDBC algorithms

It can be noted that BFA and HSDBC had similar impact on the system response by finding feasible solutions within the system constraints. This is noticed by the decreased percentage overshoot and improved steady state error. However, the HSDBC optimised control system had faster settling time, and this was noticed in the payload linear actuator displacement. Moreover, it had superior performance in minimising the percentage overshoots in the displacement of the left wheel (Figure 4) and pendulum links tilt angles (Figures 5 and 6). Furthermore, HSDBC-optimised control system has clearly improved the settling time of the right wheel and payload actuator displacements (Figures 5 and 8 respectively). The cost function convergence plots are shown in Figure 9. It is noted clearly that the HSDBC optimisation algorithm cost function converged to the minimum value within approximately 25 iterations, while the BFA algorithm needed approximately 126 iterations to settle into best-found minimum values shown in Table 6.

7. Conclusion

This paper has presented application of HSDBC to find the optimum gains for a hybrid fuzzy logic controller of a two-wheeled robotic machine. The hybrid FLC has been optimised using two different approaches; BFA and HSDBC. A comparison of the system performance as been presented has proved that the HSDBC has faster convergence speed, better accuracy and stability.

References

1. Eberhart, R., & Kennedy, J. (1995, October). A new optimizer using particle swarm theory. In *Micro Machine and Human Science, 1995. MHS'95., Proceedings of the Sixth International Symposium on* (pp. 39-43). IEEE.
2. Passino, K. M. (2002). Biomimicry of bacterial foraging for distributed optimization and control. *Control Systems, IEEE*, *22*(3), 52-67.
3. Tamura, K., & Yasuda, K. (2011). Primary study of spiral dynamics inspired optimization. *IEEJ Transactions on Electrical and Electronic Engineering*, *6*(S1), S98-S100.
4. Nasir, A. N. K., Tokhi, M. O., Ghani, N. M., & Ahmad, M. A. (2012, September). A novel hybrid spiral dynamics bacterial chemotaxis algorithm for global optimization with application to controller design. In *Control (CONTROL), 2012 UKACC International Conference on* (pp. 753-758). IEEE.
5. A. M. Al-Meshal, K. M. Goher, M.O. Tokhi, (2011, 8-10 September), Modelling of two-wheeled robotic wheelchair with moving payload. *Proceedings of the 14th International Conference on Climbing and Walking*

Robots and the Support Technologies for Mobile Machines (CLAWAR 2011), Paris, France.

6. Almeshal, A. M., Tokhi, M. O., & Goher, K. M. (2012, September). Robust hybrid fuzzy logic control of a novel two-wheeled robotic vehicle with a movable payload under various operating conditions. In *Control (CONTROL), 2012 UKACC International Conference on* (pp. 747-752). IEEE.

BFA OPTIMISATION OF CONTROL PARAMETERS OF A NEW STRUCTURE TWO WHEELED ROBOT ON INCLINED SURFACE

S. A. AGOURI, M. O. TOKHI, A. M. ALMESHAL
Department of Automatic Control and Systems Engineering, The University of Sheffield, United Kingdom, e-mail: cop09saa@shef.ac.uk

K. M. GOHER
Department of Mechanical and Industrial Engineering College of Engineering, Sultan Qaboos University, Oman

This paper presents an investigation into controller tuning, using quadratic adaptive bacterial foraging algorithm (QABFA) for a two-wheeled robot with an extendable intermediate body (IB). The robot design offers an additional feature to the conventional inverted pendulum on two wheels. The IB of the robot is composed of two co-axial parts connected by a linear actuator and with a payload attached to the end of the upper part. The linear actuator allows the payload to move up and down along the IB of the robot. Considering various positions, speeds and different sizes of a payload, carried by the robot, while maintaining upright balance is the main objective of the current study. The Lagrangian formulation is used to derive the governing equations describing the dynamic behaviour of the system. Simulation results demonstrating the system performance with the QABFA optimised controller are presented and discussed.

1. Introduction

The theory of balancing robot is based on the inverted pendulum model. The wide range of applications and the uniqueness of the technology derived from this unstable system have drawn the interest of many researchers. One of the major nonlinear control problems is the inverted pendulum on a cart. Different control strategies have been implemented for the inverted pendulum on a cart. There are a number of studies done on inverted pendulums on two-wheels. This type of application has emerged from work done on pendulum-carts and rotational pendulums. A prototype of a revolutionary two-wheeled vehicle was built by Grasser et al. [1]. These robots can move on flat and complex terrains, such as sloped and irregular surfaces. A few studies have been conducted in areas related to two-wheeled movement on irregular terrains. [2]. Presented a hill climbing algorithm with regard to an inverted pendulum. Kausar, [3] Presented dynamical modelling and control design of a two-wheeled mobile robot while moving on an irregular terrain. [4] Discussed that inclined slopes will increase the challenge of controlling the inverted pendulum. One aim of this study is to develop a system that is capable of moving on irregular and inclined slopes with a movable dynamic payload. Many researchers have used evolutionary algorithms in controlling an inverted pendulum. For instance [5] used genetic algorithm (GA) to optimize the weighting coefficients of a fuzzy logic controller of a double inverted pendulum system. This paper presents an investigation into controller tuning using an improved bacterial foraging algorithm, namely quadratic adaptive bacterial foraging algorithm (QABFA), introduced by Supriyono [6]. for a two-wheeled robot with an extendable intermediate body moving on an inclined surface. Mathematical model of the system dynamics has been derived and presented by Agouri et al. [7, 8].

2. System Description

The two-wheeled robotic vehicle considered in this work comprises a rod on an axle incorporating two wheels as described in Figure 1. The robot is powered by two DC motors driving the vehicle wheels and a linear actuator connecting the two parts of the IB which allows the attached payload to move up and down according to a pre described motion scenario. A reference Cartesian coordinate frame designated as $OXYZ$ attached to the axle connecting the wheels with its origin located at the vehicle centre point O as shown in Figure 1 is used for the angular and translational motions of the vehicle. The $Z-$ axis points vertically upward, the X axis is parallel and coincides with the wheels axle, and the Y axis is determined according to the right-hand rule in the rectangular coordinate system. The IB is considered balanced if it coincides with the positive Z axis. Partial angular deviation from the Z axis causes an imbalance in the vehicle with a tilt angle θ around the Z axis.

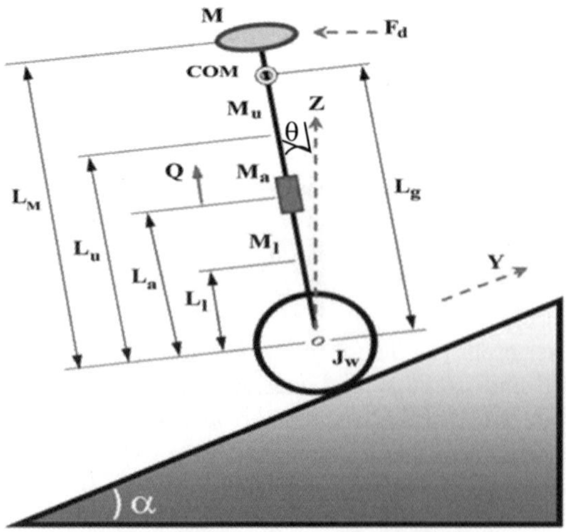

Figure 1. Schematic diagram of the robot

2.1. *Robot Lagrangian dynamics*

In the Lagrange approach, the following dynamic equations can be expressed for n degrees of freedom (DOF) of the system:

$$\frac{dq}{dt}\left(\frac{\partial L}{\partial \dot{q}_i}\right)-\frac{\partial L}{\partial q_i}=Q_i \tag{1}$$

where $L = T - V$ is the Lagrangian function. The generalized coordinates of the system are chosen as $q_i = [Y \quad \theta \quad \mathrm{Q}]^T$ and the generalized force is expressed as $Q_i = [F_c \quad F_d \quad F_a]^T$, Y is the linear displacement of the vehicle, θ is the angular displacement of the IB, Q is the linear displacement of the payload along the IB, F_c is the driving force of the wheels, F_d is an external disturbance force and F_a is the linear actuator pushing force. The system kinetic energy, T_T, and potential energy, V_T, can be expressed as follows:

$$T_T=\frac{1}{2}C_7\dot{y}^2+C_8\dot{Q}^2+\dot{\theta}\dot{y}\cos(\theta+\alpha)\left[C_{10}+M_u(C_5+Q)+M_u(C_6+Q)\right]+ \\ \dot{\theta}^2\left[Q^2C_{12}+C_{13}Q+C_{14}\right] \tag{2}$$

$$V_T=C_{11}y\sin\alpha+\{C_{10}g+M_m g(C_6+Q)+M_u g(C_5+Q)\}\cos\theta+M_C gR_w \tag{3}$$

2.2. *Vehicle dynamic equations*

The three non-linear differential equations describing the robot dynamics alongside the driving forces and the external applied force can be represented as:

$$F_c=\left[C_7\ddot{y}+\dot{C}_{15}+C_{16}Q\right]\ddot{\theta}\cos(\theta+\alpha)-(C_{15}+C_{16}Q)\dot{\theta}^2\sin(\theta+\alpha)+ \\ C_{16}\dot{Q}\dot{\theta}\cos(\theta+\alpha)\ddot{y}+C_{11}\sin\alpha \tag{4}$$

$$F_d=(C_{18}+QC_{16})\ddot{y}\cos(\theta+\alpha)-(C_{18}+QC_{16})\dot{y}\theta\sin(\theta+\alpha)C_{16}\dot{Q}\dot{y}\cos(\theta+\alpha)+ \\ 2\ddot{\theta}(C_{12}Q^2+C_{13}Q+C_{14})+\dot{\theta}(4C_{12}Q+2C_{13}\dot{Q})+C_{19}\dot{y}\dot{\theta}^2\sin(\theta+\alpha)- \\ C_{20}\dot{\theta}\sin\theta \tag{5}$$

$$F_a=2C_8\ddot{Q}-C_{16}\dot{\theta}\dot{y}\cos(\theta+\alpha)-2C_{12}Q\dot{\theta}^2-C_{13}\dot{\theta}^2-C_{16}g\cos\theta \tag{6}$$

3. Control Strategy

Although the system dynamics is strongly coupled , The proposed control scheme consists of three separate control loops as shown in Figure 3. Each individual loop represents a hybrid control scheme; fuzzy PD - like combined with conventional integrator (PD + I FLC). The objective of the first is to stabilize the IB at the vertical upright position. The second controller is utilized to keep the cart wheels within a specified linear position from a predefined reference position while moving on an inclined surface. The third controller is developed to control the linear displacement of the payload along the IB. The system inputs are the driving force F_c , the linear actuator force F_a and the disturbance force F_d.

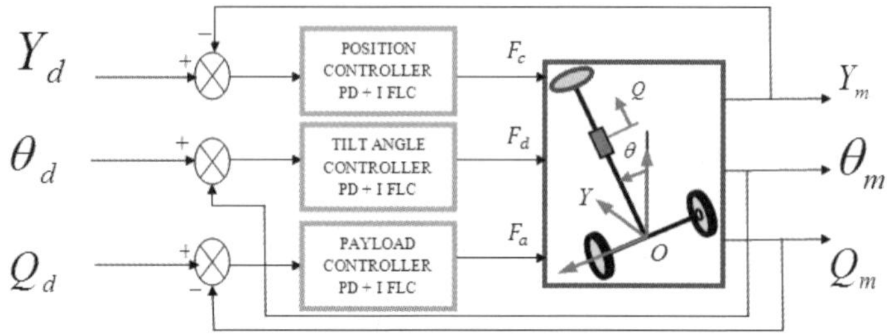

Figure 2. System block diagram

3.1. *Quadratic Adaptive Bacterial Foraging Algorithm*

The QABFA was developed by Supriyono et al. (2010) based on the original BFA (Passino, 2002) by introducing adaptation of the chemotactic step size using a quadratic function. The adaptation scheme is based on changing the chemotactic step size based on the nutrient value. This is achieved by using a quadratic function to find the next chemotactic step size. The approach thus developed speeds up the search operation in finding the optimal solution in a less time compared to the original BFA algorithm. The QABFA step size is calculated using

$$a_q^i = \frac{C_{\max}}{1+\dfrac{b}{d(|J^i|^2+|J^i|)}} \tag{7}$$

where C_{max} = maximum step size, b = tuneable positive scaling factor, d= scaling factor, J^i = nutrient value for every bacterium. Thus, the new adaptive chemotactic step size can be expressed as

$$C_{QABFA}(i) = C(i) \times a_q^i \tag{8}$$

3.2. *Optimisation parameters*

Table 1 shows the simulation parameters used with the QABFA optimisation.

Table 1 simulation parameters

Parameter	Description	Value
P	Dimension of the search space	9
S	Number of bacteria population	40
Nc	Number of chemotactic steps per bacteria	9
Ns	Max. number of swim per bacteria in a constant direction	6
Nre	Reproduction steps	4
Ned	Elimination and dispersal events	2
ped	Probability elimination and dispersal per bacteria	0.25
cmax	Maximum step size	0.1
b	Tuneable positive scaling factor	1
d	Scaling factor	0.045

3.3 Constrained Optimisation

The optimisation process will be constrained within the stability region of the system. Each parameter will have a feasible interval that guarantees the stability of the system within the defined controller gain limits as shown in Table 2.

Table 2 Minimum and maximum parameters

Value	**Kp1**	**Ki1**	**Kd1**	**Kp2**	**Ki2**	**Kd2**	**Kp3**	**Ki3**	**Kd3**
Min	4	0.4	3	4	1	3	10	2	15
Max	5	0.8	4	5	1.3	4	13	3	20

3.4 Objective Function

The performance index of the system is chosen as the minimum mean square error of each loop. The MSE is calculated for each control loop of the vehicle system as follows:

$$\text{Objective \_Function 1} = \min\left[\frac{1}{N}\sum_{i=1}^{N}(\mathrm{Y_d}-\mathrm{Y_m})^2\right] \tag{9}$$

$$\text{Objective \_Function 2} = \min\left[\frac{1}{N}\sum_{i=1}^{N}(\theta_\mathrm{d}-\theta_\mathrm{m})^2\right] \tag{10}$$

$$\text{Objective \_Function 3} = \min\left[\frac{1}{N}\sum_{i=1}^{N}(\mathrm{Q_d}-\mathrm{Q_m})^2\right] \tag{11}$$

The objective function is chosen to be the total of the MSE functions above;

$$\mathrm{J} = \sum_{i=1}^{3} Objective\_function(i) \tag{12}$$

4. Simulation Results

Figures 3-5 compare the results obtained by simulating the manual tuned controller gains with the optimised QABFA controller gains. It is noted that the optimisation algorithm has led to the optimal gain values that successfully stabilised the system with minimum cost function value. The convergence curve of the cost function is shown in Figure 6. The minimum value of the cost function achieved was 0.9. The optimised gain values are shown in Table 3.

Table 3 Optimised gain values

Parameter	Kp1	Ki1	Kd1	Kp2	Ki2	Kd2	Kp3	Ki3	Kd3
Optimal value	4.23	0.74	3.01	4.56	1.03	3.26	11.45	2.65	19.34

Figure 4 shows that the optimised gain values resulted in better system response with a faster settling time and lower percentage overshoot value. The tilt angle response in Figure 5 is noted to have less oscillation and faster settling time with the optimised gain values. On the other hand, the payload displacement shown in Figure 6 is noted to have zero steady state error and lower overshoot but with longer settling time. The cost function presented in Figure 7 shows that the algorithm has converged to the optimal minimum value within 24 iterations. The QBFA has achieved either comparable or better performance, as compared to manual tuning.

5. Conclusion

In this paper PD +I FLC control algorithm with QABFA has been implemented on a two-wheeled robot with an extendable intermediate body. This novel two-wheeled robot design offers an additional degree of freedom in a vertical direction which allows the robot to be utilized for applications such as handling objects at different heights in industrial applications. PD +I FLC based control scheme has been developed and implemented on the system in order to simultaneously achieve upright balance with a limited overall system movement in the inclined surface. The simulation results have demonstrated the potential of the optimisation approach in devising further sophisticated control approaches.

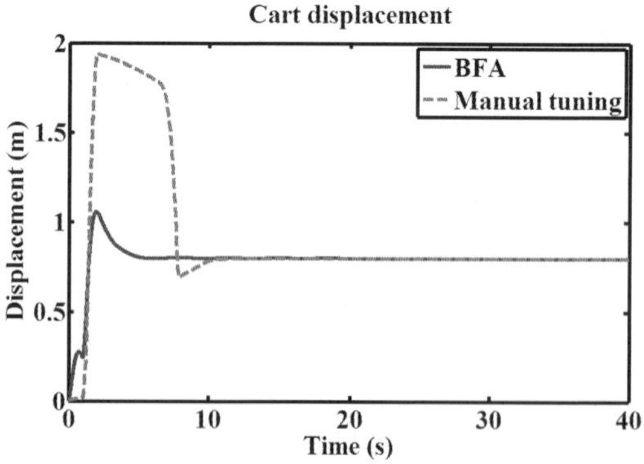

Figure 3. Cart displacement optimised performance compared to manual tuned values

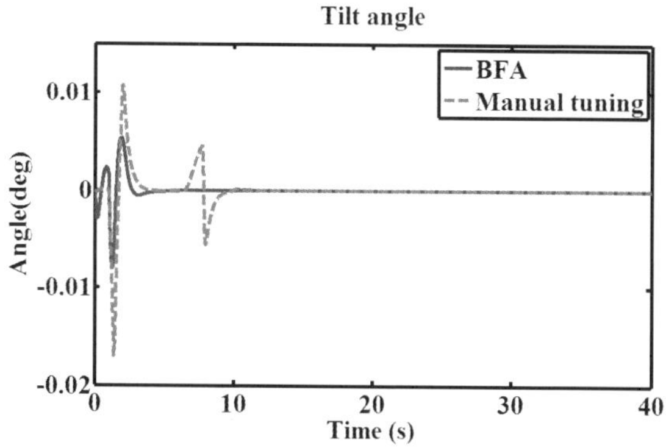

Figure 4. Tilt angle optimised performance compared to manual tuned values

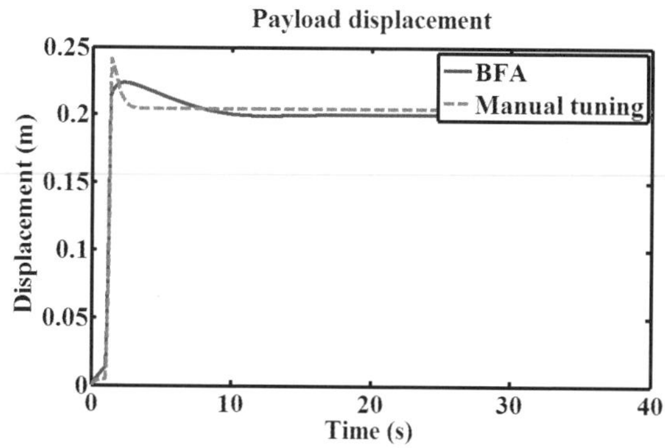

Figure 5. Payload displacement optimised performance compared to manual tuned values

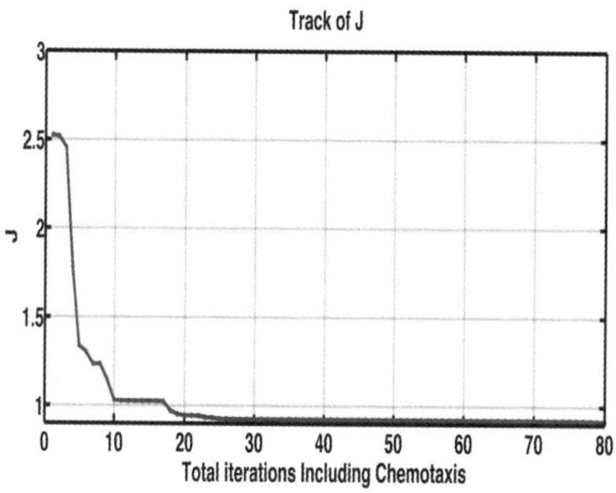

Figure 6. Track of the cost function convergence curve

Appendix

$C_1 = 4l_l + l_l^2 + 4l_l l_u$, $C_2 = 4l_l + 2l_u$, $C_3 = 4l_l^2 + 4l_u^2 + 8l_l l_u$

$C_4 = 4l_l + 4l_u$, $C_5 = 2l_l + l_u$, $C_6 = 2l_l + 2l_u$

$C_7 = M_c + M_l + M_a + M_u + M_m$, $C_8 = \frac{1}{2}(M_u + M_m)$

$C_9 = \frac{1}{2} M_l l_l^2 + \frac{1}{2} J_l + \frac{1}{2} M_a l_a^2 + \frac{1}{2} J_a$, $C_{10} = M_l l_l + M_a l_a$

$C_{11} = M_c g + M_l g + M_a g + M_u g + M_m g$ $C_{12} = \frac{1}{2} M_u + \frac{1}{2} M_m + \frac{1}{2} M_u + \frac{1}{2} M_m$,

$C_{13} = \frac{1}{2} C_2 M_u + \frac{1}{2} C_4 M_u + \frac{1}{2} M_u (C_2 - 2L_g) + \frac{1}{2} M_m (C_4 - 2L_g)$,

$C_{14} = C_9 + \frac{1}{2} C_1 M_u + \frac{1}{2} C_3 M_m + \frac{1}{2} M_u L_g^2 - M_u L_g C_5 + \frac{1}{2} M_u C_1 + \frac{1}{2} M_m L_g^2 + \frac{1}{2} M_m C_3 - M_m C_6 L_g$

$C_{15} = C_{10} + M_u C_5 + M_m C_6$

$C_{17} = M_u C_5 + M_m C_6$

$C_{19} = [C_{10} + M_u (C_5 + Q) + M_m (C_6 + Q)]$

$C_{20} = [C_{10} g + M_u g (C_5 + Q) + M_m (C_6 + Q)]$

$J_m = M_m \{Q^2 + (C_4 - 2L_g) Q + L_g^2 - 2L_g C_6 + C_3\}$

$J_u = M_u \{Q^2 + (C_2 - 2L_g) Q + L_g^2 - 2L_g C_5 + C_1\}$

$C_{16} = M_m + M_u$

$C_{18} = C_{10} + C_{17}$

$L_u = C_5 + Q$

$L_m = C_6 + Q$

$L_u^2 = Q^2 + C_2 Q + C_1$

$L_m^2 = Q^2 + C_4 Q + C_3$

References

1. Grasser, F., A. D'Arrigo, et al. "JOE: a mobile, inverted pendulum." Industrial Electronics, IEEE Transactions on 49(1): 107-114, (2002).
2. Lee, H. and J. Lee. “Hill climbing algorithm of an inverted pendulum”, IEEE., (2009).
3. Kausar, Z., K. Stol, et al. Performance enhancement of a statically unstable Two Wheeled Mobile Robot traversing on an uneven surface. Robotics Automation and Mechatronics (RAM), on IEEE Conference, (2010).
4. Kang .M.S, “Design of a robust controller for the inverted pendulum system” , JIND Sci., Chengji Univ vol 23, no. 2, (2006).
5. Cheng-jun, D., D. Ping, et al. Double inverted pendulum system control strategy based on fuzzy genetic algorithm, IEEE. (2009).
6. Supriyono, H. and M.O. Tokhi, “Bacterial foraging algorithm with adaptable chemotactic step size”, Proceedings of Second International Conference on Computational Intelligence, Communication Systems and Networks 2010 (CICSyN2010), Liverpool, United Kingdom, pp. 72-77, 29-30 July, (2010).
7. Agouri S,Tokhi M O , et al “ Modelling And Control Of A two Wheeled Vehicle With An Extendable Intermediate Body On An Inclined Surfaces”. The 32nd IASTED International Conference on Modelling, Identification and Control, Innsbruck, Austria, 11-13 Feb., (2013).
8. Agouri S Tokhi M O , et al “Dynamic Modelling Of A New Configuration Of Two Wheeled Robotic Machine On An Inclined Surface” Proceedings of the 17th International Conference on Methods and Models in Automation and Robotics, Miedzyzdroje, Poland, 27-30 August., (2012).

SECTION–4
BIOLOGICALLY-INSPIRED SYSTEMS AND SOLUTIONS

COMPLIANT BACKBONES IN MOBILE ROBOTS*

NICOLE I. KERN, RICHARD J. BACHMANN, RYAN J. MICHOLS, RONALD J. TRIOLO, ROGER D. QUINN

Biologically Inspired Robotics Laboratory, Case Western Reserve University, 10900 Euclid Avenue, Cleveland, OH 44106 USA

Mobile robots with body flexibility may exhibit improved performance in speed and agility, especially in maneuvering around obstacles or conforming to small spaces. Body flexibility can also lead to a more natural gait. Two compliant-bodied platforms are being explored. "Cheeter" has been constructed as a hopping robot with a flexible spine. "Caterpillar Whegs" is a wheel-legged robot designed with flexible connections between multiple rigid segments, which may be manipulated to control turning and flexion/extension.

1. Introduction

Robots can provide invaluable support in situations that are unsafe or inaccessible for humans or other animals. Example scenarios include search and rescue, exploration, or maintenance in a hostile environment. A prototypical mobile robot might be composed solely of rigid components, with a box-shaped chassis driven by legs or wheels. This arrangement is simple to fabricate and is easily described by state variables. However, an increase in mechanical flexibility could enhance performance in many applications.

1.1. *Overview of Existing Work*

A variety of mobile robots with body flexibility have been constructed and studied. Snake-inspired robots have been shown to be highly mobile and adapt well to terrain [1, 2]. They can climb with wheels [3] or treads [4], or use their bodies to sidewind, crawl, and climb up and around objects and inside tubes [5]. Patane et al. built a quadruped robot to resemble and interact with rats, and an added spine improved navigation ability in the test space [6]. Park et al. used a waist to improve stability in turning of a quadruped robot [7]. The MIT leg lab simulated gaits on an articulated-spine robot with one front and one rear leg [8]. Takuma, Ikeda, and Masuda showed that gait patterns could be optimized by changing the physical properties of a quadruped robot with a tunable spine [9].

---

* This work was supported by the Department of Veterans Affairs, Rehabilitation Research and Development under Grant B3463R, and by the Department of Defense under Grant PR043074.

2. Flexible-Body Robot Concepts

Two new platforms have been developed in order to experiment with torso flexibility in mobile robots. The first is "Cheeter", a 1 DOF hopping robot with a passive spine. The second is "Caterpillar Whegs" a multiple-segment addition to the Whegs™ family of robots [10]. The field of compliant robots can take obvious inspiration from biology since flexibility, along with dexterity, adds to maneuverability in animals.

2.1. *Cheeter, a Hopping Robot*

The design for the Cheeter torso (Fig. 1) is abstracted from the pronounced torso flexion and extension of a cheetah reaching top speed. In this robot, all functionality unrelated to body elasticity is designed to be as simple as possible; forward propulsion is viewed as a means to test the robustness of the body flexion design under continuous loading.

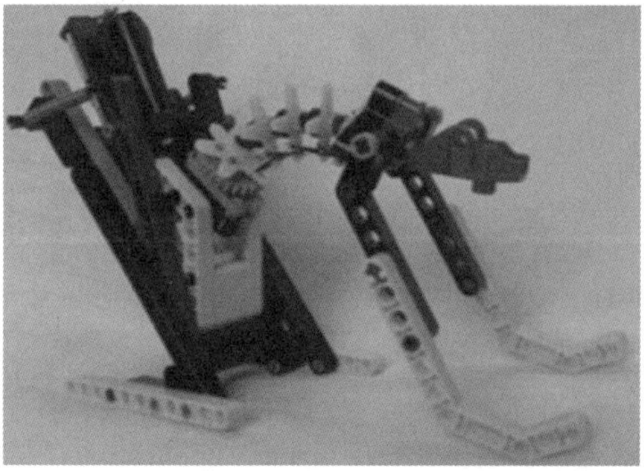

Fig. 1. Cheeter, a simple hopping robot with a flexible spine.

Mobility is achieved by using a cam-driven four-bar linkage to propel the body up and forward with a hopping motion, following in the footsteps of previous robots such as "Grillo" [11] and the miniature jumping robot from EPFL [12]. Favorable linkage lengths were found by iteration to maximize the excursion angle of the ground contact link.

2.2. *Caterpillar Whegs, a Wheel-Legged Robot*

Whegs robots incorporate wheel-legs, which require minimal control to achieve high speeds while having the ability to climb obstacles. Whegs have been specialized for actions such as climbing, jumping, swimming, and flying [13-16]. Previously, Whegs have been built with four to six wheel-legs and a single chassis which was either rigid or had one flexion/extension body joint. Turning has been realized through rack-and-pinion or skid steering.

Caterpillar Whegs (Fig. 2) is designed to add the flexibility of a compliant body to the agility and speed of a wheel-legged robot. The first prototype consists of three segments, each with a drive motor to power one axle with two wheel-legs. A compliant mechanism similar to that in Cheeter provides Caterpillar Whegs with full articulation between segments. Two servo motors use cables to control left/right turning and flexion/extension through bending of two flexible inter-segment connections.

Fig. 2. Caterpillar Whegs with compliant inter-segment connections.

3. Compliant Backbone Structures

In mobile robots, compliant connecting structures must be capable of transmitting force while also enabling flexibility. The added elasticity can be beneficial, but it does increase design complexity. The magnitude of compliance can vary from a small amount of elasticity in an otherwise rigid system to a completely soft-bodied robot. Both Cheeter and Caterpillar Whegs represent a middle ground of compliance since they employ elastic structures as compliant interfaces between relatively larger rigid segments. Including rigid segments offers convenient mounting points for actuators. In the case of the Cheeter robot, the spine could be biased towards flexion to maintain posture (Fig. 3). For both robots, flexion is defined as bending towards an "∩" shape and extension is bending towards a "U" shape.

A flexible middle tube serves to align the Cheeter spine and maintain its general structure. Rubber discs are sized to fit into a hollowed boss in each plastic vertebra. The snug fit between all adjoining parts constrains the system so that each vertebra bends with its neighboring disc and no sliding occurs along the flexible tube. The Caterpillar Whegs compliant connection is simpler since no bending bias is necessary. It comprises rubber bumpers encasing a flexible tube with no vertebrae, as shown in Fig. 3.

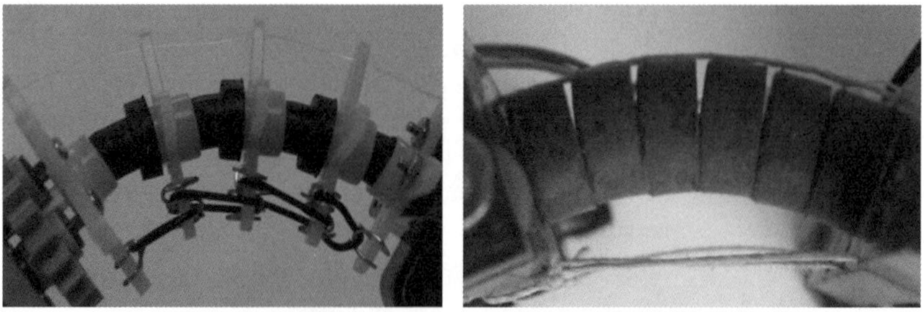

Fig. 3. Close-up of the flexible spine in the Cheeter robot (left) and Caterpillar Whegs (right).

3.1. *Quantified Range of Motion*

In order to control the degree of bending of the body, it is useful to understand how the amount of cable constriction effects movement. For a system that is lightly loaded, relatively short, or controlled by individual cables at each vertebrae, it may be assumed that each disc undergoes the same amount of bending. Fig. 4 illustrates the geometry of a disc of initial length w between vertebrae with distance x from center to control cable. Working under the assumption that the compliant disc material has symmetric stress-strain curves in tension and compression, the neutral axis remains static and located at the centerline (dashed line, Fig. 4).

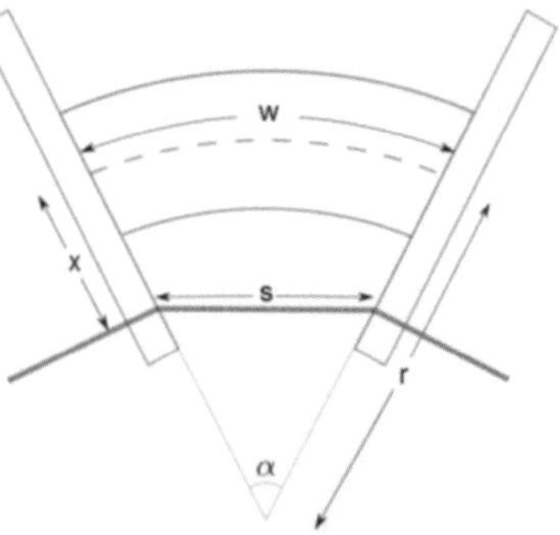

Fig. 4. A disc in bending between two rigid vertebrae. The control cable has been shortened to a new length of s between discs. The bending radius, r, is assumed constant for the length of the spine.

The drawing in Fig. 4 can represent both the alternating disc-vertebrae and the disc-only robotic spine designs. For Cheeter, ignoring the rigid vertebrae, the width w would be the total width of all compliant discs. For Caterpillar Whegs each compliant section would add the same amount of bending. Fig. 3 shows the detail of a backbone section; in this case w = 3.7 cm. The following are variables: the change in cable length ΔL, the bending radius r, and the bending angle per disc α (in radians). The total bending angle of the robot would be α, and the initial cable length through the section would be w. If ΔL, α, or r are assigned a value, given that $w=r\alpha$, variables may be calculated by Equation (1). The Caterpillar Whegs robot uses the same cable control scheme for vertical and horizontal bending. Assuming that the servo motors are capable of lifting the robot, Equation (1) can be used for both the amount of flexion/extension and the amount of turning.

$$\Delta L = w - s = w - (w/\alpha - x)\sqrt{2(1-\cos\alpha)} \tag{1}$$

4. Construction and Behavior

A single drive motor turns the cam that propels the Cheeter robot. The cam mechanism and the four-bar housing are machined from acetal and aluminum, and the remaining robot parts are a colorful construction of flex hose, rubber bumpers, servo horns, springs, and LEGO® pieces. Two AAA batteries power the system and are situated in an attached battery holder above the rear legs of the robot. The robot moves forward at a rate of approximately 3.7 cm/s on carpet, or 5 cm/s on tile. The flexible spine allows more than 90° of bending in all directions and the inner flexible hose is pinched at both extremes to maintain a connection between the two halves of the robot. Throughout hours of run time no catastrophic failures were identified, which validated the flexible spine design and promoted construction of Caterpillar Whegs.

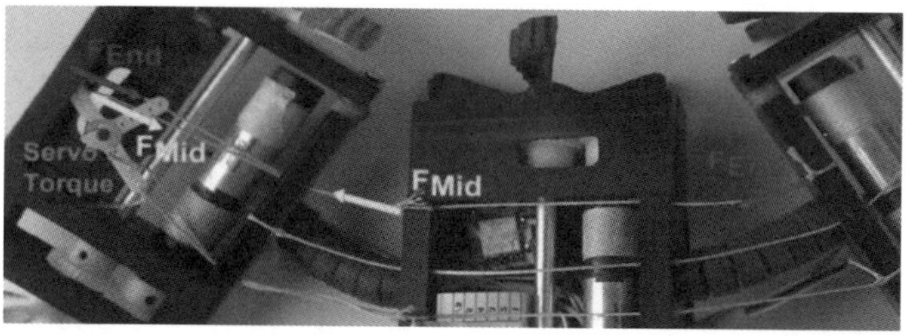

Fig. 5. Illustration of the forces that turn the robot. The moment arm is greater for F_{End} than for F_{Mid} to account for the fact that F_{Mid} shortens the distance between the servo and middle segments, and F_{End} must pick up the slack in the first shortening before causing a length change at the end segment.

In Caterpillar Whegs, the three segment housings are constructed from acetal blocks. The middle segment houses the batteries, motor controller, and receiver, and the outer segments each hold a servo motor to control bending. The front segment contains the flexion/extension servo, and the rear segment controls turning. Fig. 5 depicts the turning control servo in detail; a line of Kevlar thread transmits force individually to each segment for both turning and flexion/extension. Each compliant backbone is connected to segments by a knot at the end of the flexible tubing.

A single motor controller is used for all three drive motors to reduce weight. The drive force from each motor is transmitted to the axle via spur gears, and the gears are currently not enclosed from the environment. The next generation of Caterpillar Whegs robots will feature enclosed weather-proof segment housings to keep out dirt and debris. A baffle or other cover for the threads and inter-segment connections is also desirable. The Caterpillar Whegs robot runs at a maximum speed of 35 cm/s on carpet and 37 cm/s on tile, or approximately an average of 1.4 body lengths per second (Fig. 6). It should be noted that body lengths per second may not be a good metric for a robot with this design since the segments are currently self-driven and any number of segments can be connected together.

Fig. 6. Sequential images of a 37 mm climb. The robot spine is extended, though this height could be achieved with a neutral spine when the wheel-legs optimally contacted the top surface.

The theoretical bending angle of the compliant sections can be calculated from Equation (1) given that the left/right turning threads are 1.3 cm from the centerline and ΔL = .7 cm; $.7=3.7-(3.7/\alpha-1.3)\sqrt{(2*(1-\cos\alpha))}$, which results in α = .52 radians of bending in each compliant section. This value agrees with a protractor measurement of turning. The same calculation for flexion/extension with holes spaced 0.5 cm from the centerline returns an α value of 1.1, though issues with thread slack and servo horn bending constrained flexion/extension to approximately .4 radians.

The active turning radius of the robot is larger than the bending radius for the compliant section since the rigid segments remain straight, and the ground contact points experience less turning. This value was estimated by steering the robot to its extreme in each direction and tracing the resulting paths on the ground (Fig. 7). The amount of turning varied slightly, but the path was close enough to a circle to estimate a 33 cm turning radius. The circular trajectory is entirely a result of sideways bending in the inter-segment connection; no other turning mechanism is needed. The climbing height was also tested; the robot was able to passively climb a rectangular obstacle of 3.5 cm using the wheel-legs alone. A 4 cm obstacle was surmountable with the small amount of added flexion and extension that the servo could produce. Increased flexion and extension, as well as bigger wheel-legs, would improve climbing ability.

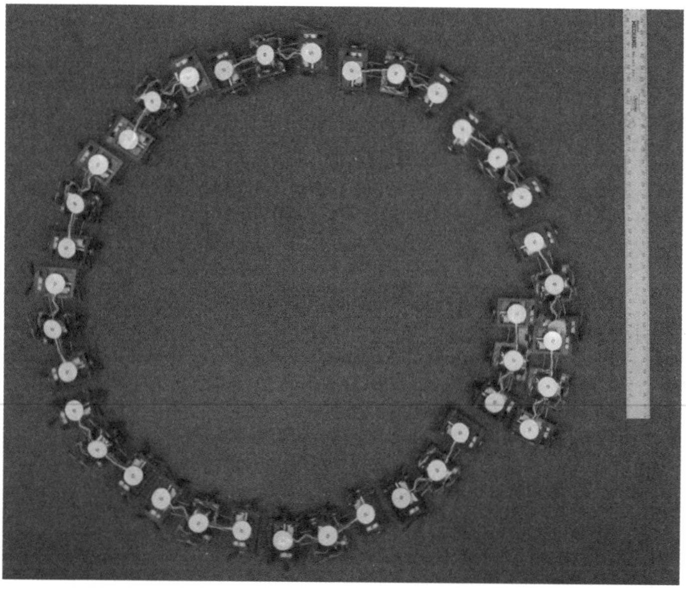

Fig. 7. Overhead view of the Caterpillar Whegs turning path starting from the same initial point and moving forward (upwards) while the body is bent to turn left.

5. Conclusions

The Whegs concept is scalable and the robots may be miniaturized to fit through small spaces, or magnified to negotiate rough terrain and tall obstacles. A multitude of segments would provide traction and power, and a lengthening of inter-segment connections with independent control would increase agility. The size of the segments, the height of the wheel-legs, and the amount of controlled inter-segment bending will define the space in which the robot can navigate.

Mobile robots with compliant backbone structures have the potential to locomote with speed and agility. Turning and climbing can be realized by controlling the relative orientation of connected segments. One consequence of a compliant body is that the robot can conform to its environment, especially while overcoming obstacles, which almost appears life-like. By using designs that minimize the weight and power and maximize flexibility, the robot becomes a more efficient platform for sensors and effectors that may be used to achieve a wide variety of objectives.

References

[1] S. Hirose and A. Morishima, "Design and Control of a Mobile Robot with an Articulated Body," *The International Journal of Robotics Research,* vol. 9, pp. 99-114, April 01, 1990.

[2] E. Shammas, A. Wolf and H. Choset. (2006, 2). Three degrees-of-freedom joint for spatial hyper-redundant robots. *Mechanism and Machine Theory 41(2),* pp. 170-190.

[3] T. Baba, Y. Kameyama, T. Kamegawa and A. Gofuku, "A snake robot propelling inside of a pipe with helical rolling motion," in *Annual Conference of Society of Instrument and Control Engineers (SICE),* Taipei, Taiwan, 2010.

[4] G. Granosik and J. Borenstein, "The OmniTread serpentine robot with pneumatic joint actuation," in *Fifth Int. Workshop on Robot Motion and Control (RoMoCo),* Dymaczewo, Poland, 2005.

[5] C. Wright, A. Johnson, A. Peck, Z. McCord, A. Naaktgeboren, P. Gianfortoni, M. Gonzalez-Rivero, R. Hatton and H. Choset, "Design of a modular snake robot," in *Int. Conf. on Intelligent Robots and Systems (IROS),* San Diego, USA, 2007.

[6] F. Patane, V. Mattoli, C. Laschi, B. Mazzolai, P. Dario, H. Ishii and A. Takanishi, "Biomechatronic design and development of a legged rat robot," in *Int. Conf. on Robotics and Biomimetics (ROBIO),* Sanya, China, 2007.

[7] S. H. Park, D. S. Kim and Y. J. Lee, "Discontinuous spinning gait of a quadruped walking robot with waist-joint," in *Int. Conf. on Intelligent Robots and Systems (IROS),* Edmonton, Canada, 2005.

[8] K. F. Leeser, "Locomotion Experiments on a Planar Quadruped Robot with Articulated Spine." Master's Thesis. MIT, Department of Mechanical Engineering, Feb 1996.

[9] T. Takuma, M. Ikeda and T. Masuda, "Facilitating multi-modal locomotion in a quadruped robot utilizing passive oscillation of the spine structure," in *Int. Conf. on Intelligent Robots and Systems (IROS),* St. Louis, USA, 2010.

[10] R. D. Quinn, J. T. Offi, D. A. Kingsley and R. E. Ritzmann, "Improved mobility through abstracted biological principles," in *Int. Conf. on Intelligent Robots and Systems (IROS),* Lausanne, Switzerland, 2002.

[11] M. Faragalli, I. Sharf and M. Trentini, "Velocity control of a hybrid quadruped bounding robot," in *Intelligent Robots and Systems, 2008. IROS 2008. IEEE/RSJ International Conference on,* 2008.

[12] M. Kovac, M. Fuchs, A. Guignard, J. -. Zufferey and D. Floreano, "A miniature 7g jumping robot," in *Robotics and Automation, 2008. ICRA 2008. IEEE International Conference on,* 2008.

[13] K. A. Daltorio, T. C. Witushynsky, G. D. Wile, L. R. Palmer, A. A. Malek, M. R. Ahmad, L. Southard, S. N. Gorb, R. E. Ritzmann and R. D. Quinn, "A body joint improves vertical to horizontal transitions of a wall-climbing robot," in *IEEE Int. Conf. on Robotics and Automation (ICRA),* Pasadena CA, USA, 2008.

[14] J. M. Morrey, B. Lambrecht, A. D. Horchler, R. E. Ritzmann and R. D. Quinn, "Highly mobile and robust small quadruped robots," in *Int. Conf. on Intelligent Robots and Systems (IROS),* Las Vegas, USA, 2003.

[15] R. Harkins, J. Ward, R. Vaidyanathan, A. S. Boxerbaum and R. D. Quinn, "Design of an autonomous amphibious robot for surf zone operations: Part II - hardware, control implementation and simulation," in *Int. Conf. on Advanced Intelligent Mechatronics (AIM),* Monterey, USA, 2005.

[16] F. J. Boria, R. J. Bachmann, P. G. Ifju, R. D. Quinn, R. Vaidyanathan, C. Perry and J. Wagener, "A sensor platform capable of aerial and terrestrial locomotion," in *Int. Conf. on Intelligent Robots and Systems (IROS),* Edmonton, Canada, 2005.

DEVELOPMENT OF TWO TYPES OF MAINTENANCE ROBOTS FOR A JACKET INSIDE A MIXING TANK

TOMOYA TANAKA, TARO NAKAMURA, DAISUKE SANNOHE, YOUSUKE MORISITA AND MITSUGU TANAKA

Faculty of Science and Engineering, Department of Precision Mechanics
Chuo University
1-13-27 Kasuga, Bunkyo-ku, Tokyo 112-8551, Japan

In this study, we developed two types of maintenance robots using an artificial rubber muscle driven by pneumatics for a jacket inside a mixing tank. One is an excavation robot for cleaning the inside of the jacket, and the other is a polishing robot for inspecting the walls. The excavation robot consists of four units and three joints. This robot mimics the peristaltic crawling motion of an earthworm. On the other hand, the polishing robot consists of five units, a metal brush and two wing parts. This robot polishes the walls inside the jacket. Two types of experiments were performed with the two robots. One of the experiments was a driving test inside the jacket. The other was to eliminate sediment on the local walls.

1. Introduction

When we want to mix some chemicals, we use a mixing tank (Figure 1). When we use the mixing tank, it is important to manage the temperature inside the tank. Temperature management is performed by discharging a refrigerant or a heating medium inside the jacket—the area between the shell (in which chemicals are mixed) and the jacket (which covers the shell).

However, since industrial water or steam or the like is used as the refrigerant or the heating medium, as the mixing tank is used, sediment adheres to or rusts the walls inside the jacket. Thus, the heat-transfer condition into the shell is affected, leading to problems such as some chemical reactions not going to completion in the predetermined time.

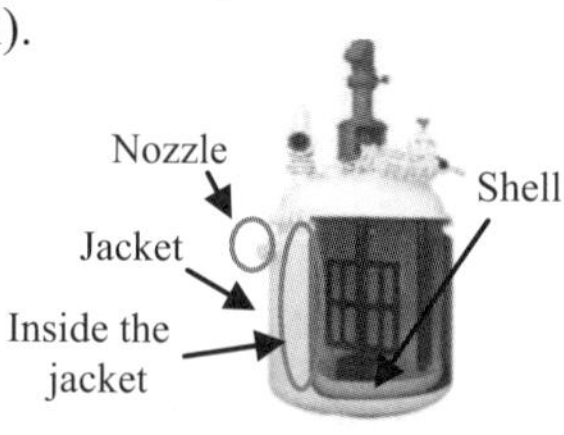

Figure 1. Mixing Tank

Thus, regular cleaning and maintenance inside the jacket is essential. However, cleaning and maintenance methods do not exist at present, because the distance between the walls inside the jacket is very narrow (40–70 mm), and the nozzle leading to the inside the jacket is very narrow (50–60 mm) as well. Thus, it is difficult for a person to clean thoroughly inside the jacket. In addition, there is a robot that can climb between these walls when clean [1], but the robot is very large to climb the narrow space between these sediment-encrusted walls.

In a previous study, we developed an X-shaped peristaltic crawling robot [2] for verifying the inner state of the jacket. Thus, in this study, we developed two types of maintenance robots using an artificial rubber muscle to accommodate the narrow space between the walls.

2. Peristaltic Crawling Pattern of an Earthworm

Figure 2 (a) shows an inner structure of an earthworm's body, and Figure 2 (b) demonstrates locomotion of the earthworm by peristaltic crawling. The earthworm can be moved by a movement of an expansion and extension of a numerous segments (about 110-200). This movement is performed by a two layer muscle; the circular muscle that is arranged inside of a skin circumferentially, and longitudinal muscle that is arranged inside of the circular muscle in axial direction. The earthworm's moving pattern is shown below.

At first, the earthworm expands anterior segments by longitudinal muscles. This expansion is transmitted to a rear segments and the anterior segments is elongated. In this time, the friction is developed between the expansion segments and the ground. By this friction, the elongated segments can get a reaction force of the elongation to the forward. Thus, the earthworm can moves by repetitive the motion of elongation and expansion.

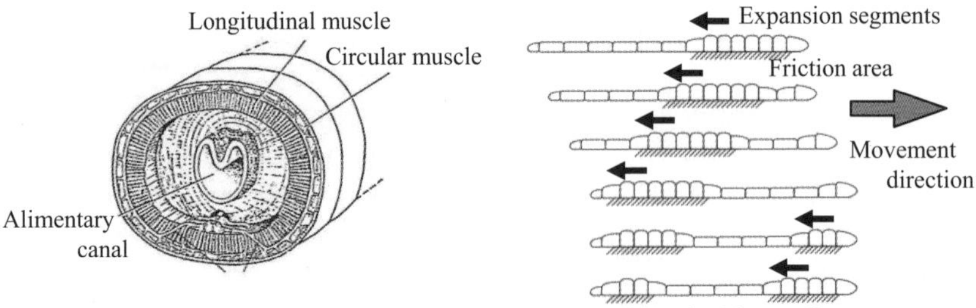

(a) Inner structure of the earthworm's body. (b) Peristaltic crawling.

Figure 2. Earthworm locomotion.

3. Artificial Muscle

Figure 3 shows an overview of the pneumatic artificial muscle. The artificial muscle is composed of micro-carbon fiber and low ammonia natural rubber latex. The carbon fiber has behavior that the fiber does not stretch. So the fiber is placed in the axial direction of the artificial muscle, the artificial muscle does not expands the axial direction. As a result, when air pressure is supplied to the artificial muscle, the artificial muscle contracts in the axial direction and expands in the radial direction.

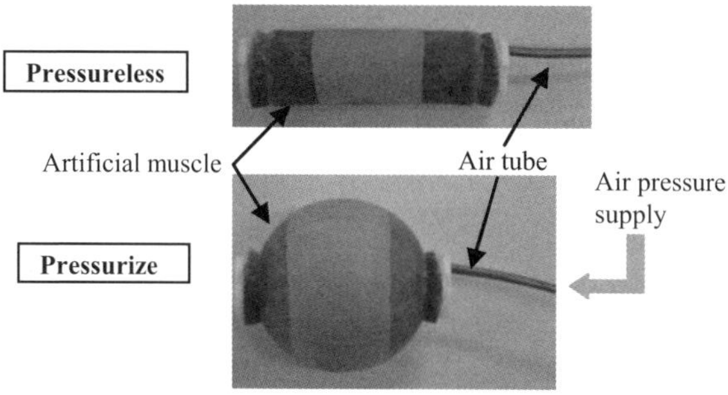

Figure 3. Artificial muscle.

4. Cleaning and Maintenance Method

In this study, maintenance inside the jacket is performed by the following two methods:

(1) First, we approximately eliminate sediment from inside the jacket (discussed in section 4).

(2) Next, we perform an elimination process for sediment and inspect the local walls inside the jacket (discussed in section 6).

5. Excavation Robot for Rough Elimination of Sediment

5.1. *Method for Rough Elimination of Sediment*

Figure 4 demonstrates the method for the approximate elimination of sediment. At first, a left upper nozzle is connected to the right upper nozzle with a horizontal wire. Next, an upper nozzle and a bottom nozzle are connected with a vertical wire, and the vertical wire is fixed to the horizontal wire inside the jacket. The vertical wire is fixed with a brush or the like. Hence, we can approximately eliminate the overall sediment by controlling the horizontal wire by dragging it up and down with the vertical wire.

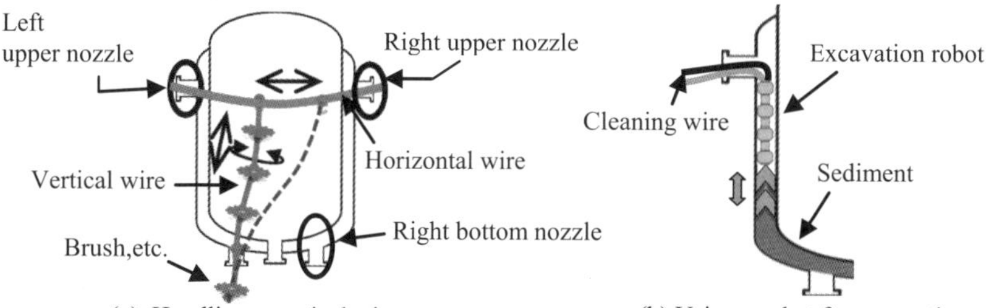

(a) Handling a vertical wire. (b) Using a robot for excavation.

Figure 4. Method for rough elimination of sediment.

However, a large amount of sediment remains inside the jacket; at worst, sediment entirely clogs the area inside the jacket. Therefore, we cannot connect the upper nozzle and the bottom nozzle with the vertical wire in this situation. Therefore, we need to ensure the route of the vertical wire with an excavation robot.

5.2. *Composition of the Excavation Robot*

Figure 5 shows the appearance of the excavation robot. This robot is composed of four units and three joints. An aspect of the joint part is shown in Figure 6. The joint consists of a compression spring, two aluminum plates, and a latex tube. The unit also contains the artificial muscle, a compression spring, two flanges, and bellows (Figure 7). When air is applied into the chamber of the unit, which is an enclosed space, the unit expands in the radial direction and constricts in the axial direction. In contrast, when air is discharged from the chamber, the unit is extended in the axial direction. Thus, the unit mimics the motion of an earthworm segment by repeating extension and expansion. Table 1 shows a specification of the unit.

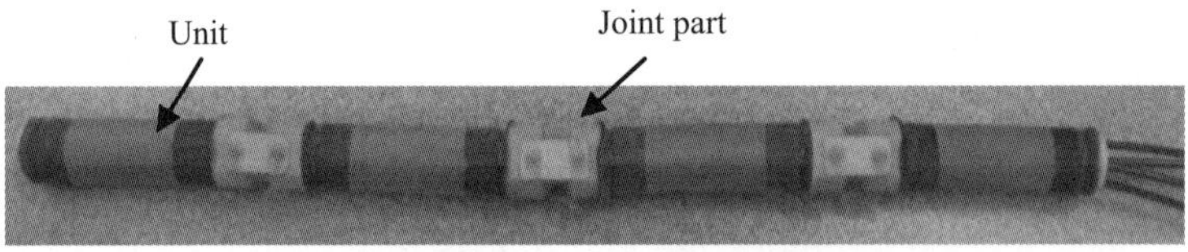

Figure 5. Appearance of robot for excavation.

(a) With a rubber tube. (b) Without a rubber tube.
Figure 6. Aspect of the joint part.

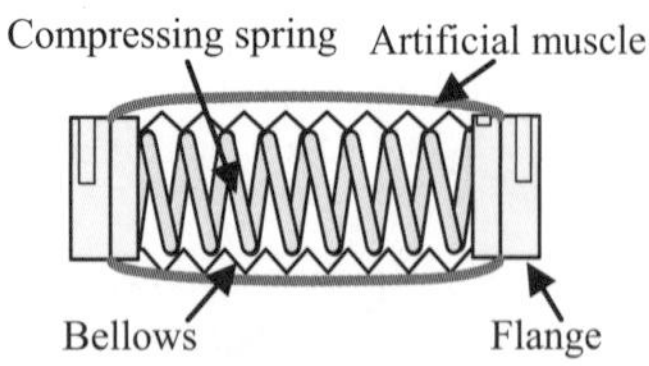

Figure 7. Structure of the unit.

Table 1. The unit's specification when the air pressure of 0.07 MPa is supplied.

Length [mm]	Extension	85.0 [mm]
	Expansion	65.0 [mm]
Width [mm]	Extension	26.0 [mm]
	Expansion	74.0 [mm]

5.3. *Driving Test with the Excavation Robot*

The distance between the two walls of the inside of the jacket at the most predominant mixing tank is 60-70 mm. So we performed a driving test and measured the robot's locomotion speed at between the two walls which distance is 60-70 mm.

5.3.1. *Driving test between the two walls which distance is 60 mm*

In this paragraph, we performed the driving test and measured the robot's locomotion speed at between the two vertical walls which distance is 60 mm. The wall's quality of material is iron.

Figure 8 (a) shows an analysis result of the driving test at between the two vertical walls. And Figure 8 (b) shows an analysis result of the driving test at between the two curved walls. From two figures, we observe the distinctive cyclic waveform of the peristaltic crawling. We also observe that the robot's locomotion speed is 2.58 mm/s at between the two vertical walls, and 2.59 mm/s at between the two curved walls. These results indicate that the robot does not be affected from the wall's form in the driving.

Now when looking a red circle in Figure 6, the displacement of a head part comes down. This reason is the unit's expansion speed is faster than the extension speed. So before the rearward unit is fully extended, the anterior unit grasps the walls (Figure 9).

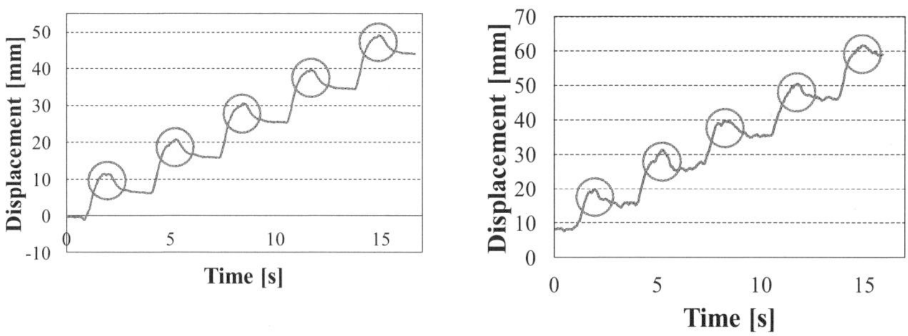

(a) Between the two vertical walls. (b) Between the two curved walls.

Figure 8. Analysis result of the driving test at the distance of 60 mm walls.

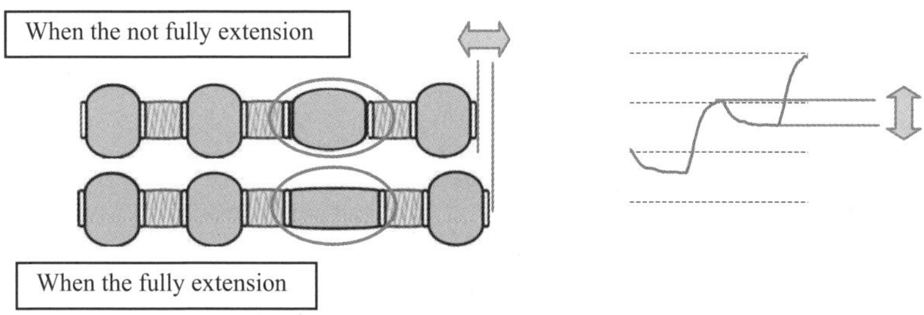

Figure 9. Reason of the coming down of the displacement.

5.3.2. *Driving test between the two walls which distance is 70 mm*

In this paragraph, we performed the driving test and measured the robot's locomotion speed at between the two vertical walls and between the two curved walls with sediments which distance is 70 mm. The wall's quality of material is iron.

Figure 10 (a) shows an analysis result of the driving test at between the two vertical walls. And Figure 10 (b) shows an analysis result of the driving test at between the two curved walls with sediments. From two figures, we observe the distinctive cyclic waveform of the peristaltic crawling. We also observe that the robot's locomotion speed is 4.81 mm/s at between the two vertical walls, and 4.78 mm/s at between the two curved walls with sediments. These results indicate that the robot does not be affected from the sediments in the driving.

In addition, from these results and the results of the driving test at the distance between the walls; 60 mm, the driving speed at the distance between the walls; 70 mm is faster than the driving speed at the distance between the walls; 60 mm. The reason why is that the expansion time at the distance between the walls; 70 mm is longer than at the distance between the walls; 60 mm. Put simply, the amount of expansion at the distance between the walls;70 mm is larger than at the distance between the walls;60 mm. Hence, the rearward unit can become the near fully extension before the anterior unit grasps the walls. But in view Figure 8, the coming down exists yet. So we have to regulate the air pressure in the expansion and break in the expansion time.

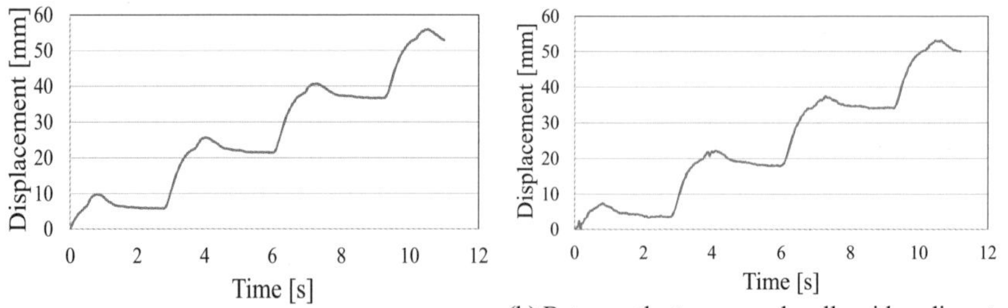

(a) Between the two vertical walls. (b) Between the two curved walls with sediments.

Figure 10. Analysis result of the driving test at the distance of 70 mm walls.

5.4. ***Excavation Test with the Excavation Robot***

In this paragraph, we performed the excavation with the excavation robot and the cleaning wire (Figure 11). Appearance of the test is shown in Figure 12. In this test is performed at the walls distance 70 mm.

Figure 13 shows the result of the test. From this result, we observe that the excavation robot and cleaning wire can excavate the sediments.

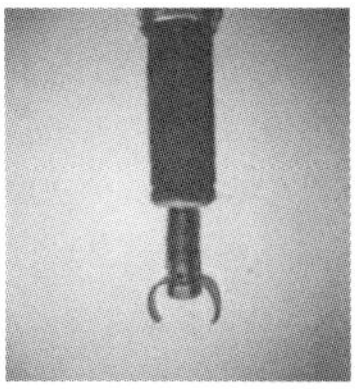

Figure 11. Head part of the cleaning wire.

Figure 12. Appearance of the excavation test.

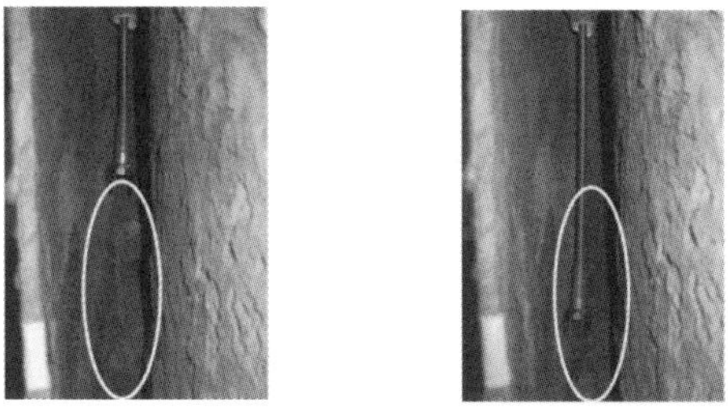

(a) Before the excavation (b) After the excavation

Figure 13. The result of the excavation test.

6. Polishing Robot for an Inspection

6.1. *Method for Local Elimination of Sediment and Inspection*

When we inspect the condition of the local walls, we use the wall thickness as an indicator. However, a problem arises in using this indicator and it is related to the sediment that remains on the walls; thus, using the wall thickness as an inspection criterion is imprecise. Therefore, we need to eliminate the sediment on the local walls, which are the candidate of the examination.

Figure 14 shows the method for eliminating sediment on the local walls. A fixation part grasps the wall on both sides at the place from which the sediment needs to be eliminated. A polishing part is expanded and rolled, thereby eliminating the sediment.

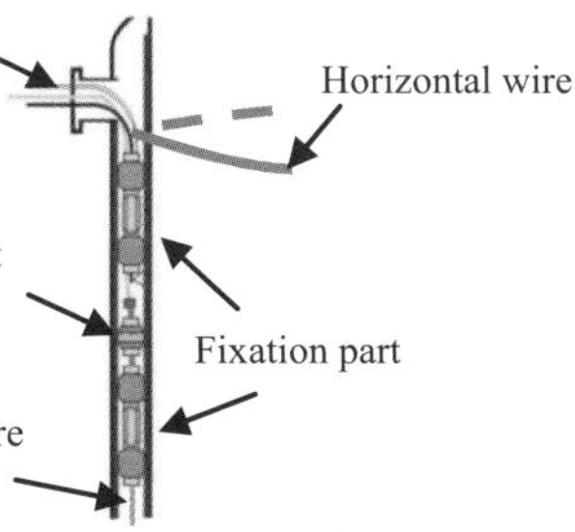

Figure 14. Method for eliminating sediment on the local walls.

6.2. *Composition of Devices for Polishing and Fixation*

Figures 15 and 16 show the parts for polishing and fixation, respectively. The polishing part is loaded with a metal brush. This part can eliminate sediment by self-rotation. The fixation part is loaded on an endoscope and has a wing part to maintain the rotation angle. If the rotation angle cannot be maintained, the condition of the local walls cannot be checked with the endoscope.

Figure 15. Polishing part.

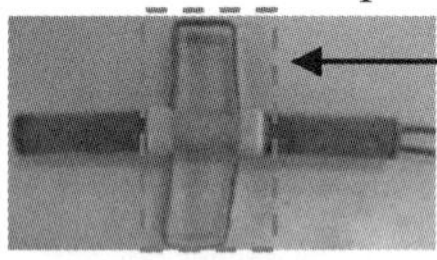

Figure 16. Fixation part.

6.3. *Elimination Test*

We performed an elimination test between the two walls containing the sediment. The distance between the two walls is 70 mm. Figure 17 shows a result of the test. From the figure, we observe that the polishing part can eliminate the sediment clearly.

(a) Before elimination. (b) After elimination.

Figure 17. Result of the polishing test.

7. Conclusion

We developed two types of the maintenance robots for the jacket inside a mixing tank. And we performed some experiments; driving test between the two walls which distance is 60 mm and 70 mm, excavation test, elimination test. Our results were follows.

(1) The excavation robot can drive between the various walls; the curved walls, vertical walls.

(2) The excavation robot doesn't be affected from the sediments when the robot is driving between the curved or vertical walls.

(3) The excavation robot can excavate the sediments.

(4) The polishing robot can polish the sediments on the local walls.

In future, the excavation robot will be improved to increase the maximum locomotion speed.

References

1. A. Ariga, T. Kobayashi, T. Yamaguchi, and S. Hashimoto, *Proc. 2010 IEEE Int. Conf. on Robotics and Biomimetics*, 1507-1512, 2010.
2. D. Sannohe, Y. Morishta, S. Horii and T. Nakamura, *Proc. 2012 4th IEEE RAS & EMBS Int. Conf. on Biomedical Robotics and Biomechatronics*, 1365-1370, 2012.

ADAPTIVE CONTROL OF LEG POSITION FOR HEXAPOD ROBOT BASED ON SOMATOSENSORY FEEDBACK AND ORGANIC COMPUTING PRINCIPLES

A. AL-HOMSY, J. HARTMANN AND E. MAEHLE

Institute of Computer Engineering, University of Luebeck, 23562 Luebeck, Germany
E-mail: alhomsy@iti.uni-luebeck.de www.iti.uni-luebeck.de

Improving robot walking over compliant surfaces is considered an important issue. However, walking on an uneven surface requires controlling the robot's leg position to span a varying distance between the robot's body and the ground. This paper will shed light on the applied decentralized controller approach that enables the robot's leg to adapt its position to be commensurate with the type of walking surface in order to improve walking on compliant surfaces. When the substrate compliance is high (sandy ground), the leg has to correct its position close to the point of contact with the ground. When the substrate compliance is low (gravel ground), the correction of the leg position is low. The novelty of our approach is the evaluation of the local current consumption and the angular position of each leg's joint as somatosensory feedback. Our approach is based on an organic computing architecture and was tested on a low-cost version of the OSCAR walking robot.

Keywords: Organic Computing, Walking Robots, Compliant Surfaces, Gravel Ground, Sandy Ground, Somatosensory Feedback, Control of Leg Position.

1. Introduction

Legged insects have the ability to control the movement of their six legs when they walk on uneven terrain in a fascinating way. Here, the control of the movement of the legs has to cope with the local coordination of the different legs, which has been investigated by Graham [1], and also the control of the leg position, which has been studied intensively by Cruse et al. [2]. This behaviour encourages robotic engineers to benefit from the observed biological behaviour to solve the challenging task of controlling the robot's leg position while walking on uneven terrain and maintaining the coordination between the legs. A related work has been accomplished on the hexapod robot SandBot [3] to enable it to adapt its locomotion on sandy ground. The hexapod robot of Celaya and Porta [4] adapts the height of each foot to the elevation of the ground below it based on a force sensor signal. When the force signal drops below a given threshold, the leg is lowered a bit. In a work closer to the paper presented here,

Palankar and Palmer III [5] describe an approach to control the robot's leg position based on a force threshold controller. Their robot's leg adapts its position also based on a force sensor signal. Giguere et al. [6] introduced an approach to enable their six-legged amphibious robot to identify the environment. The internal sensor information (vertical acceleration, motor current and angle of a particular leg) is used to teach the robot to identify the environment depending on a probabilistic Bayesian classifier.

In this work, an approach will be demonstrated to improve the robot's walk on compliant surfaces by adapting the robot's leg position while walking on highly and weakly compliant surfaces. While previous works mostly rely on additional sensors like force sensors, our approach to control the leg position depends on the evaluation of the consumed current and angular position in the leg's joints. This work has been implemented in the Organic Robot Control Architecture (ORCA) [7] and was tested on the latest version of the hexapod robot OSCAR (Organic Self-Configuring and Adaptive Robot) [8], based on the Bioloid robot kit[1] (see Fig. 1), which uses low-cost off-the-shelf servos.

2. Test Framework

Organic Computing (OC) is aimed at developing technical systems that adapt themselves autonomously and dynamically to the environment, exhibiting self-x properties such as self-adapting, self-configuring, self-protecting, self-organizing, and self-healing [9]. The Organic Robot Control Architecture (ORCA) utilizes the OC principles in its entire layer. It is a modular and hierarchically organized software framework which enables the design of biologically inspired systems [7]. ORCA's architecture consists of two main units, the BCU (Basic Control Unit) and the OCU (Organic Control Unit). BCUs form the backbone of a system and provide sufficient functionality for the system under normal conditions. OCUs are supplemented units to the BCUs. They monitor and observe the BCU and other OCU signals and have the authority to change the BCU's parameters as soon as the BCU's signal shows anomalies, ensuring e.g. the system's function in case of faults.

ORCA has been previously applied to several versions of our hexapod robot OSCAR [10]. In this work, experiments will be conducted on the latest version which is based on the Bioloid robot kit, showing that complex tasks like walking on sandy or gravel ground as well controlling the leg position can be accomplished using low-cost commodity hardware. The OSCAR robot that was used in these experiments (see Fig. 1) has six roughly symmetrically distributed

---

[1] Bioloid Robot Kit, http://www.robotis.com/xe/bioloid_en.

legs which are divided into front, middle and hind legs. Each leg has three joints, which will, as in the previous versions, be called α, β, and γ in the remainder of this paper, with α being closest to, γ furthest away from the body, and β the middle joint. The foot is located at the tip of the leg with a hemispherical shape made of rubber. It is about 1.5 cm in diameter. Each joint is actuated by a Dynamixel AX-12 servo motor, which makes it possible to read, among other parameters, consumed current, the actual servo position and speed. The consumed current is provided as the value at the time of reading. All robot components use two batteries (NiMH3900) as power supply. For testing purposes, the robot is controlled from an external PC via USB. Each leg is now controlled by a set of BCUs that set the target position and read the actual position as well as the consumed current of each servo. The OCU unit for each leg's BCU monitors the consumed current and reacts on detected anomalies.

Figure 1. The test platform: hexapod robot OSCAR. The robot geometry (right), walking on sandy ground (left), gravel ground (middle).

3. Control of Leg Position

Our previous work was focused on analyzing the current consumption and angular position in each servo motor and using it as a somatosensory sensor for the detection of the type of walking terrain [11]. In case the robot walks on hard ground, it detects that the ground is hard as soon as the current consumption I_β in the β-servo motor rises above the maximum predefined threshold (T_{max}) without significant changes in the angular position of the β-servo motor. While in case of walking on sandy ground, the leg will sink and the current consumption in the β-servo motor will rise gradually exceeding the minimum threshold (T_{min}) which is defined as the maximum current consumption for the β-servo motor while the leg moves down without contact with any obstacle. The leg stops sinking into sandy ground as soon as the current consumption in the β-servo motor exceeds the maximum threshold or the leg reaches its maximum position. For more details see [11].

However, walking on compliant surfaces, e.g. walking on sandy or gravel ground, makes the robot's legs sink into the ground and the robot faces great difficulty in walking straight forward. To solve this problem and to improve walking, the position of the legs will be controlled and adapted relative to the substrate compliance based on somatosensory sensing by measuring the current consumption in each β-servo motor only. Each leg's OCU monitors the current consumption and angular position changes in its β-servo motor at the end of the swing phase while the leg moves down. As soon as the β-current consumption exceeds the minimum predefined threshold T_{min}, the OCU records the angular position of the β-servo motor. This position represents the point of contact with the ground. The leg continues moving down until the current consumption exceeds the maximum threshold T_{max} or the leg reaches its maximum position. The leg will sink into the ground depending on the nature of the substrate's compliance. As soon as the leg stops moving down, the OCU sends a signal that the leg has to adapt its position and to return to the position of contact with the ground. The consumed current in the β-servo motor is denoted by I_β. The thresholds T_{max} and T_{min} have been determined by experiment. They are related to the weight of the robot and have the same value for all legs. The following table clarifies our approach.

$I_\beta < T_{min}$	The leg moves down without touching any object and continues its movement down until it reaches its maximal span.
$I_\beta > T_{min}$ && $I_\beta < T_{max}$	The leg touches the ground and the OCU records the β-angular position. The leg is still moving down till the β-current consumption exceeds T_{max} or the leg reaches its maximum span. Then the leg has to adapt its position.
$I_\beta > T_{max}$	The leg stops moving down, the leg has to move back to the point of contact with the ground.

A correction of the leg position to the point of contact with the ground makes the center of mass swing up and down. Moving the center of mass to the bottom side in conjunction with executing the stance phase helps the stance legs to push the robot body forward more effectively than with the center of mass still being up. The center of mass works in this case as a pendulum that lets the robot walk faster and in a more efficient way on highly compliant surfaces. This approach is similar to the stick insect behaviour that is presented by Cruse et al. [2].

4. Experimental Results

To evaluate our approach, several test grounds have been established. The robot has been tested while walking on gravel and sandy ground. The following experiments show the obtained results when the robot walks on low and high compliant surfaces like gravel ground and sandy ground.

4.1. *Walking on Gravel Ground*

In this experiment, the robot walks on a ground covered with small stones with a diameter of about 1.5 cm. The test ground is an indoor area of 200x100 cm. Fig. 2 shows how the front legs adapt their position based on monitoring the β-current consumption and angular position while they move down at the end of the swing phase. The β-current consumption in the front right leg (FR) is monitored while moving the β-servo motor from position 140° to 180°. In the front left leg (FL), the β-current consumption is monitored while the β-servo motor moves from 160° to 120°. Fig. 2 on the left shows how the β-angular position is corrected as soon as the leg stops moving down while walking on gravel ground. For instance, at data point 682, the β-current consumption exceeds T_{min} at the angular position 146° while the leg moves down (the first arrow indicates the point of contact with the ground). The leg continues moving down. At data point 729, the β-current consumption exceeds T_{max} at the angular position 177°. The leg sinks into the gravel ground and in some cases lifts up the robot body height. Now FR corrects its position by moving the β-servo motor from position 177° to 146°, the point of contact with the ground (at the second arrow). The same scenario will repeat itself at the end of each swing phase and the robot body height will vary depending on the type of the walking surface. Fig. 2 on the right shows the same result for FL with its β servo moving from 160° to 120°. The leg will return to the point of contact with the ground as soon as the leg stops moving down.

4.2. *Walking on Sandy Ground*

The setup of the test ground consists of a sandy basin of 200x100 cm. The sand is dry and fine, similar to the sand that is found in sandy deserts. Fig. 3 shows the β-current consumption and the angular position of FR and FL while OSCAR walks on a sandy surface. It is clear that the β-current consumption exceeds T_{min} and rises gradually with significant β-angular position changes until it exceeds T_{max} as shown in Fig. 3. In order to clarify how FR controls its position, we analyze a part of the FR leg curve (from data point 238 to data point 318). FR touches the ground at data point 238 due to the exceeding of T_{min} at β-angular position 151°. At data point 274 the β-current consumption exceeds T_{max} at angular position 170° and the leg stops moving down. Now the leg begins to correct its position and returns to the point of contact with the ground. The leg moves up until the β-servo motor reaches 150°. The same scenario repeats itself at each end of the swing phase while the leg moves down while walking on sandy ground as shown in Fig. 3 on the left. Another example for leg position adaptation during a walk on sandy ground is shown for FL in Fig. 3 on the right. All robot legs will adapt their position in the same way while walking on sandy ground. An adaptive control of the leg position prevents the leg from sinking into sandy ground too deeply and improves the robot's walking on highly compliant surfaces.

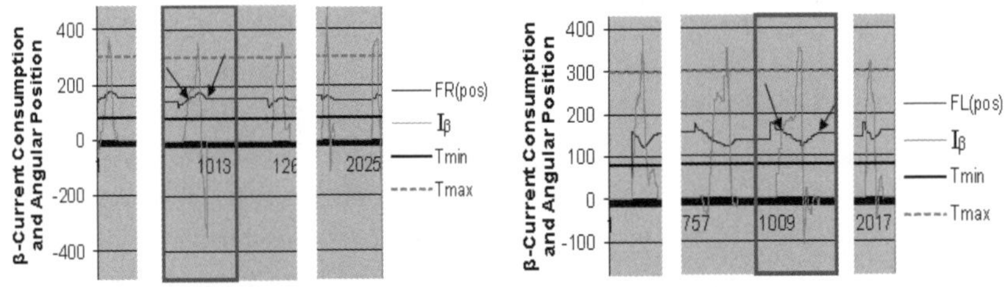

Figure 2. Current consumption and angular position of the β-servo motor in the front legs (FR and FL) at the end of the swing phase while the leg moves down during walking on gravel ground.

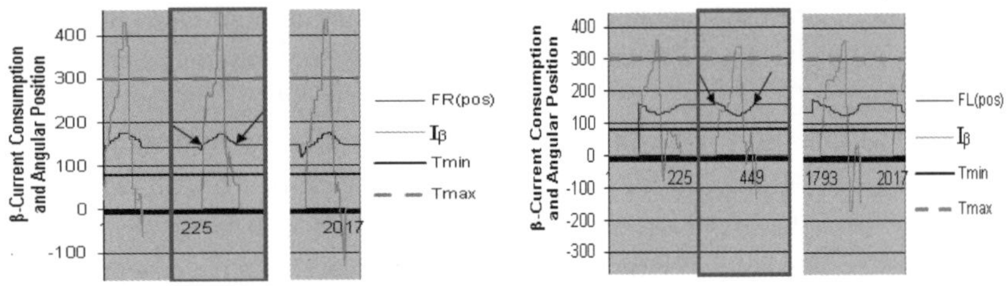

Figure 3. Current consumption and angular position in the β-servo motor in the front legs (FR and FL) at the end of the swing phase while the leg moves down during walking on sandy ground.

4.3. *Walking on Compliant Surfaces with an Amputated Leg*

Research has been previously conducted on adaptive walking for hexapod robots with amputated legs while walking on hard ground, i.e. for OSCAR in [12]. In this paper, leg amputation will be discussed while walking on sandy ground. The amputated leg will be simulated in software, i.e. the amputated leg will be stiffened and moved on top of the body to minimize its effect on the other legs. As soon as the robot loses one of its legs, the robot reconfigures its legs based on the location of the amputated leg to maintain statically stable walking. The neighboring legs of the amputated leg shift their position toward the amputated legs to balance the robots weight while walking.

Without using the adaptive control of the leg position on sandy ground, in our experiments, the robot was not able to move any further when a leg was amputated. With the adaptive control method, the robot is able to continue walking. Fig. 4 shows an experiment, where the middle left leg ML is amputated while the robot walks on sandy ground. Fig 4 (left) shows data obtained from a tilt sensor attached to the robot for reference purposes only. The pitch signal presents the inclination in the walking direction and the yaw signal presents the inclination in lateral direction. Fig 4 (right) shows the step pattern. At data time 650, ML is amputated and the leg loses contact with the ground. The robot reconfigures its legs and continues walking. The robot's tilting will obviously increase leftwards as shown in Fig. 4 (left). This also results in a slightly curved walking.

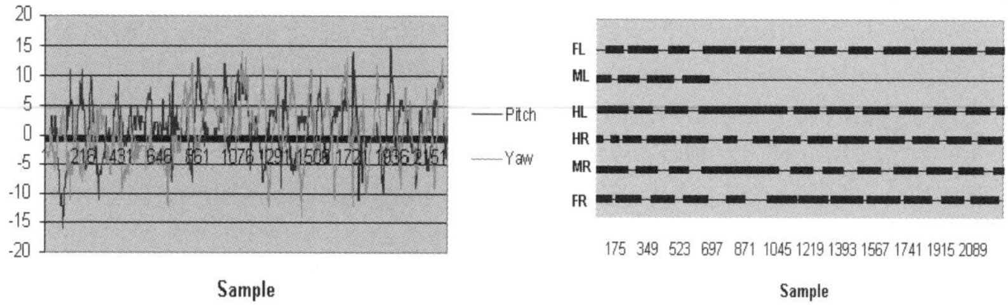

Figure 4. Left: corresponding pitch and yaw readings from a tilt sensor. Right: The step pattern. The black bars present the stance phase and the spaces between the black bars present the swing phase.

5. Conclusion

This paper has focused on improving the locomotion behaviour of a low-cost version of the hexapod robot OSCAR, which is based on the Bioloid robot kit, through adapting the leg's position while walking on different types of terrains. The adaptive control of the leg position approach has been inspired by a stick insect and been implemented on our robot control architecture ORCA. Each leg adapts its position as an elastic system, independent of the other legs, based on monitoring the current consumption and angular position in its middle joint at the end of the swing phase while it is moving down. The presented approach enables the robot to improve the robot locomotion on uneven terrain without using additional sensors, which would increase the complexity of the control system as well as the cost of the robot. In the future, we will focus on the extension of this approach to enable the robot to walk on more complex terrain, i.e. slopy sandy or slopy gravel ground.

Acknowledgments

This work was funded in part by the German Research Foundation (DFG) within priority programme 1183 under grant reference MA 1412/8-2 and the ALBAS University, Syria.

References

1. D. Graham. Pattern and control of walking in insects. *Advances in Insect Physiology*, vol. 18, no. 0065-2806, 31 – 140, (1985).
2. H. Cruse, S. Kühn, S. Park, J. Schmitz. Adaptive control for insect leg position: controller properties depend on substrate compliance. *Journal of Comparative Physiology A*, vol.190, no. 12, 983-991, (2004).
3. C. Li, P. B. Umbanhowar, H. Komsuoglu, D. E. Koditschek and D. I. Goldman. Sensitive dependence of the motion of a legged robot on granular media. *The National Academy of Sciences of the USA*, (2009).
4. E. Celaya, J.M. Porta. Control of a six-legged robot walking on abrupt terrain. *IEEE Robotics and Automation*, vol. 3, 2731 – 2736, (1996).
5. M. Palankar, and L. Palmer III. A force threshold-based position controller for legged locomotion over irregular terrain. *CLAWAR*, 563-570, (2012).
6. P. Giguere, G. Dudek, C. Prahacs and S. Saunderson. Environment identification for a running robot using inertial and actuator cues. *Science and Systems*, Philadelphia, USA, (2006).
7. W. Brockmann, E. Maehle, K.-E. Grosspietsch, N. Rosemann, and B. Jakimovski. ORCA: An Organic Robot Control Architecture. *In: Organic Computing - A Paradigm Shilft for Complex Systems*, 385-397, (2011)
8. A. Al-Homsy, J. Hartmann, and E. Maehle. Inclination detection and adaptive walking for low-cost six-legged walking robots using organic computing principle. *CLAWAR*, 173-183, (2012).
9. C. Müller-Schloer. Organic Computing – On the Feasibility of Controlled Emergence. *IEEE/ACM/IFIP International Conference on Hardware/Software Codesign and System Synthesis*, 2-5, (2004).
10. E. Maehle, W. Brockmann, K.-E. Grosspietsch, A. El Sayed Auf, B. Jakimovski, S. Krannich, M. Litza, R. Maas, and A. Al-Homsy. Application of the Organic Robot Control Architecture ORCA to the Six-legged Walking Robot OSCAR. *In: Organic Computing - A Paradigm Shift for Complex Systems*, 517-530, (2011).
11. A. Al-Homsy, J. Hartmann, and E. Maehle. Slippery and sandy ground detection for hexapod robots based on organic computing principles and somatosensory feedback. *IEEE International Symposium on Robotic and Sensors Environments (ROSE)*, 43-48, (2012).
12. A. El Sayed Auf, F. Mösch, and M. Litza. How the six-legged walking machine OSCAR handles leg amputations. *From Animals to Animats*, 9, 2006.

IMPLEMENTATION OF AN ADJUSTABLE COMPLIANT KNEE JOINT IN A LOWER-LIMB EXOSKELETON

D. SANZ-MERODIO*, M. CESTARI, J.C. AREVALO and E. GARCIA

Centre for Automation and Robotics, CSIC-UPM
La Poveda, 28500 Madrid, Spain
**E-mail: daniel.sanz@csic.es*
http://www.car.upm-csic.es/fsr/egarcia/ATLAS.html

Research on lower-limb exoskeletons and active orthoses is a growing field in service robotics. Nowadays commercial active orthoses present stiff joints but for the intrinsic human-machine interaction task, these actuators need to exhibit compliance. Moreover, they must be powerful enough to move the user limbs while keeping small size for not to bother the user motion and show an aesthetic appearance. In this paper we present the development and main characteristics of a joint prototype with variable stiffness that achieve these requirements. This actuated joint has been implemented in the knee of ATLAS active orthosis. A state machine control scheme takes advantage of the leg dynamics and of the actuator features, achieving a natural, compliant gait without the need of commanding a CGA based pattern. A reduction in the energy expenditure while keeping compliant to accommodate unexpected disturbances is obtained.

Keywords: Exoskeleton; Energy consumption; Variable compliant actuator.

1. Introduction

Lower-limb active orthoses are wearable devices that assist a person with some kind of leg pathology, augmenting the strength and endurance of their legs. In a lower-limb exoskeleton, one of the most important features is safety in human-robot interaction. Thus, changing the stiffness of the joints to adapt to the movements of the user is a need. The stiffness variation can be imitated by software -impedance control techniques-[1] with a stiff actuator, but due to a limited bandwidth it is not possible to absorb shocks, and energy would be dissipated continuously. Usually, users of active orthosis suffer spasmodic movements, that the orthotic device must absorb, which makes a compliant actuator absolutely necessary. A variety of adjustable compliant actuators[2] have been designed due to the demand of actual robotics regarding safe human-robot interaction and energy-efficiency.

Series elastic actuators (SEA) have an inherent mechanical low-pass filter, which makes these actuators safer in human environments. Furthermore, elastic elements can store energy and release it reducing external energy supply.

The actuators used in lower-limb exoskeletons require reduced volume and weight and high power-to-weight ratio. So they usually are designed with a relatively small and light motor and a gearbox with large reduction ratio which provide enough torque and speed output. But this set has the disadvantage of being too rigid. Moreover, these actuators should comply with size, weight and power requirements while providing compliance to the joints.

In this paper, we present a new adjustable compliant actuator which is integrated in the knee joint of the ATLAS active orthosis.[3] It extends along the leg so that it does not protrude. This actuator not only provides elasticity to the joint, it also acts as a joint torque sensor. Its design is based on a stiff conventional actuator in which adjustable compliance has been included by the pivot displacement technique.[4]

The main aim using variable stiffness actuators is to reduce energy consumption by exploiting the natural dynamics of the limbs and the energy stored in the springs. Based on the ideas to reduce energy consumption presented in[5] , a control scheme with discrete stiffness values at different phases is analyzed and implemented.

The requirements for actuators used in lower-limb orthoses are detailed in Section 2. In Section 3, the working principle of the new actuator is presented. The details on the implementation of the new designed actuator in the ATLAS active orthosis and the energy efficient control scheme are presented in section 4 and the experimental results obtained are presented and analyzed in section 5. The main ideas are summarized in section 6.

2. Actuator features required for its implementation in a power active-orthosis

The actuation of the joints in an exoskeleton requires peaks of high torque, about 50Nm at the knee,[6] while allows compliance to the human wearer. Pneumatic actuators with antagonist configuration offer high power output and are able to change stiffness in a biomimetic way. Some rehabilitation devices have been developed with pneumatic actuators,[7] but active orthoses must be portable and nowadays pneumatic and hydraulic actuator needs equipment that is too large and weighted to be feasible. So an efficient electric actuator is needed to drive the knee joint. Also it has to be able to change its stiffness fast enough to accomplish a steady gait at approx. 1 m/s. This kind of actuators must be small enough for not to bother the user and have an aesthetic appearance. Also they have to be as light as possible, for not to increase the inertia moment of the limbs and for being easy to handle, put on and off. For this reason some design prototypes that meet these specifications are being proposed. Some new electric actuators with variable compliance have been designed, as AMASC designed at Carnegie Mellon University,[8] MACCEPA at Vrije Universiteit Brussels,[9] VS-joint and QA-joint mechanisms at the German Aerospace Centre (DLR),[10] and AwAS at the Italian Institute of Technology (IIT),[11] but most of these actuators have been designed for robotic arms, and none of them has been implemented and demonstrated in a portable and wearable active orthosis.

3. Working principle of the variable stiffness actuator

In the first prototype of ATLAS active orthosis, the flexion and extension motion of the hip and the knee joints was driven by electrical brushless flat Maxon motors in combination with Harmonic Drive units. The resulting motor-gearbox set provided enough torque to move the user and showed a large power-to-weight ratio but was too much rigid for a human-machine interaction task. Therefore, based on this configuration for being flat and powerful enough, we have designed our actuator prototype ARES (Actuator with Adjustable Rigidity and Embedded Sensor) that incorporates variable compliance and provides an accurate torque measurement. Figure 1 represents a CAD scheme of our design and the real prototype. The stiffness adjustment is based on the pivot displacement principle in which the actuator position and stiffness control are decoupled achieving more energy efficiency.[4] The torque provided by the motor-gearbox set is transmitted by a slotted bar and a pivot to an elastic device that finally moves the joint. These elastics devices are composed by parallel springs that provide intrinsic compliance to the joint.

Through the slotted bar, the pivot allows vertical displacement driven by a motor which provides stiffness variation. The higher the arm between the center of the motor and the pivot point, the greater rigidity.

A linear encoder has been attached to the pivot point to measure the spring compression, this provides a measure of the torque applied by the relationship:

$$\tau = \frac{2\Delta x K_{equiv} L_{arm}}{cos(\phi)} \tag{1}$$

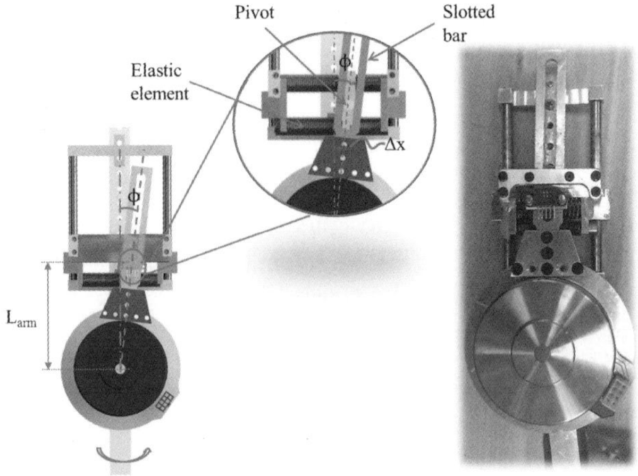

Fig. 1. CAD design of the variable stiffness actuator, principle of operation and real prototype.

where, Δx corresponds to the elastic elements compression, sensed by the encoder, K_{equiv} is the equivalent rigidity of the spring set, angle ϕ is the deflection between the actuator and the joint, and L_{arm} is the length of arm.

Thus, this actuator allows the modification of the stiffness while provides information about the torque applied, allowing compliant force control.

4. Energy-efficient control of locomotion

In our previous work,[5] we proposed some methods to increase energy efficiency, based on the study of high-efficiency passive biped robots. In the design of an energy-efficient control scheme we have focused on three points, due to the characteristics of the actuator:

(1) Utilize the elastic elements of the actuator to store and release energy.
(2) Exploit passive dynamics during swing.
(3) Reduce costs at heel strike collision exploiting the intrinsic impedance.

This new actuator provides intrinsic impedance that the rigid one doesn't, then, to take full advantage of the strategies proposed and achieve a reduction in energy consumption, two parallel strategies are proposed:

a. Modification of the gait pattern to follow based on a force control approach.
b. Design of a state machine controller with different values of stiffness at each state.

The motion control of active orthoses has traditionally been based on the rigid tracking of clinical gait analysis (CGA) reference patterns, typically resulting in high power consumption. The position pattern obtained from the CGA, incorporates the intrinsic compliance that the human joints have. Therefore, instead of following accurately the CGA pattern, we have exploited the inherent compliance of the ARES joint and followed a simplified commanded joint pattern, allowing the joint to adapt to the ground while walking. Figure 2 shows the CGA position pattern followed by the knee and the modified pattern.

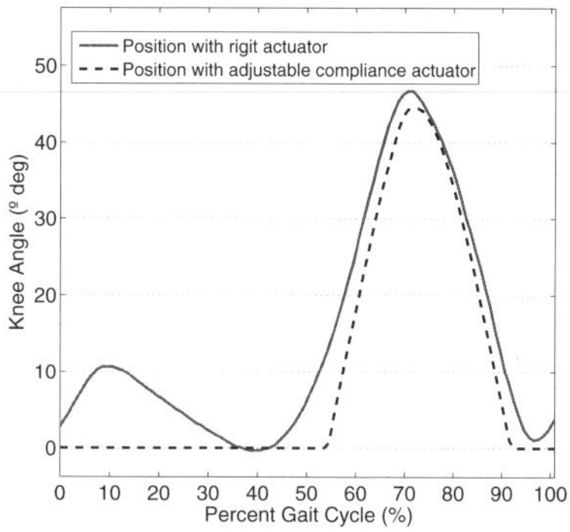

Fig. 2. CGA pattern utilized with the rigid actuator and the pattern implemented in the variable compliant actuator ARES.

During locomotion, a state machine switches between stiffness values for the knee joint. The state machine, see Figure 3, that we have incorporated at the knee joint is based on the study that we have presented in our previous work.[5] Four discrete knee stiffness states are required:

State 1. After toe-off, at the beginning of swing phase, a large stiffness is needed to let the motor lift up the shank.
State 2. When it has reached the highest foot height during swing, zero stiffness allows exploiting the intrinsic dynamics of the limb for swinging it forward and the potential energy stored when the motor changes its direction.

State 3. At the end of swing, the stiffness is increased to prepare for the collision at heel strike but kept low enough to allow the actuator springs to absorb the impact energy, enabling an adequate load response to begin the stance phase.

State 4. At the support phase the motor of the joint is kept blocked and a high stiffness is enforced to support the user weight but still providing some compliance.

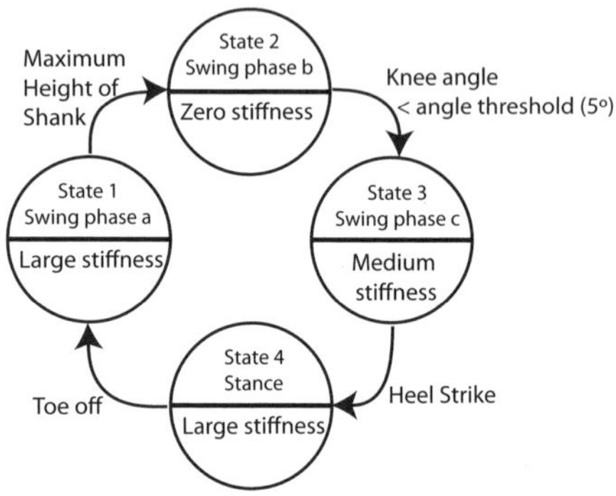

Fig. 3. State machine controller. In each state a range of stiffness and the transition condition are shown.

It is important to remark that due to the characteristics of the variable compliance joint design, the force required to change the stiffness of the joint increases as the torque does. Therefore, to reduce the energy consumed by the motor which changes the stiffness, every modification in stiffness is performed when a reduced torque is being done.

5. Experimental results

The actuator prototype has been incorporated in the knee joint of our ATLAS active orthosis, see Figure 4. The state machine scheme proposed in Section 4 has been implemented but with a variation, due to the fact that with our actuator we cannot achieve zero stiffness. As zero impedance is not achieved, it is necessary to use a control scheme that reduces energy expenditure. Zero torque control reduces the impedance of the joint via software control.

Figure 5 (a) presents the commanded position to the actuator and the real joint position achieved due to the compliance, controlled with the state machine. Notice that even without having set a path that follows the pattern

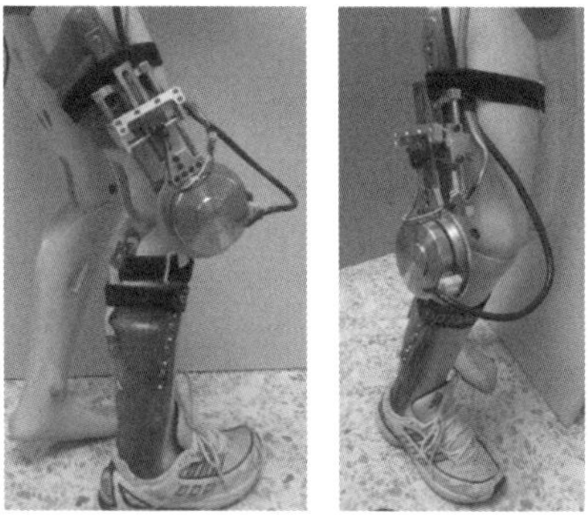

Fig. 4. Implementation of the adjustable compliant actuator.

of hump in the heel strike, it is observed that the load response follows that pattern due to the low impedance implemented that allows the knee to adapt to the ground during the support phase. A comparison with a CGA knee trajectory is shown. The controlled knee stiffness reproduces the natural knee behavior, without commanding a stiff CGA pattern.

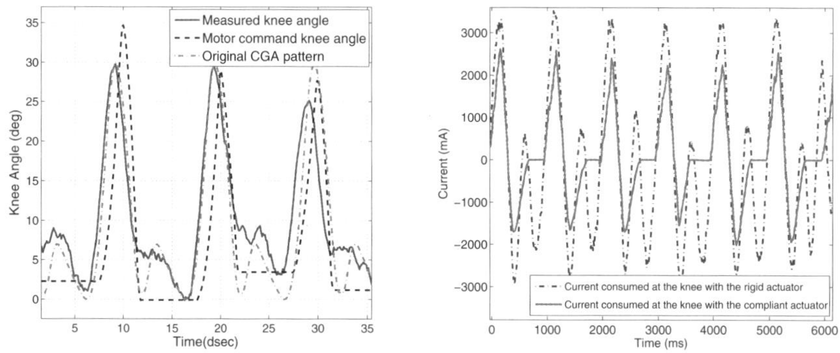

(a) (b)

Fig. 5. (a) CGA pattern utilized with the rigid actuator and the pattern implemented in the variable compliant actuator. (b) Commanded position and the real position reached with the compliant actuator. (c) Current consumption with the rigid actuator and with the variable stiffness actuator.

A comparison of the current consumption in a steady gait with a CGA pattern and rigid actuator and with the new pattern and the compliant actuator is shown in Figure 5 (b). Notice the energy expenditure reduction in each current peak, which corresponds to changes of direction of the trajectory in the joint. This is because the springs accumulate and release energy at each change of direction. During the support phase, no energy is required to block the knee due to the large stiffness, the high reduction ratio in the motor-gear kit and the position pattern proposed. Even so, in Figure 5 (a) it is shown how the knee allows adaptation thanks to the inherent compliance of the joint.

Using variable compliance actuator at the knee joint, a 39% reduction of energy expenditure is achieved.

6. Conclusions

This paper presents promising results on the development of a variable stiffness actuator for the knee joint of the lower-limb exoskeleton ATLAS. The 70% of potential users of an active orthosis suffer spasmodic movements, making absolutely necessary intrinsic compliance in their joints. Besides, the active orthosis actuators must occupy little space and be lightweight to be comfortable, aesthetically pleasing and portable. Therefore, our prototype actuator -ARES- features the demanding characteristics that an active orthosis requires, namely, intrinsic compliance to allow human-machine interaction, small size and low weight.

For the command of the actuator motion in the lower-limb active orthosis a new pattern has been defined and a state machine that switches between different stiffness has been defined to take advantage of the characteristics of the actuator and the dynamics of the legs. Although we have implemented a simpler pattern than that obtained from CGA, the response of the knee joint resembles the CGA data, due to the intrinsic compliance of the actuator, imitating the behavior of a biological joint. Compared to the rigid actuator following a rigid pattern of CGA, the compliant actuator achieves a reduction in energy consumption of 39%. It is observed a reduction in the peak of the current when the direction of the joint changes. An important reduction is achieved keeping the knee motor locked in the support phase, while providing compliance enough to accommodate the body weight.

7. Acknowledgments

This work has been partially funded by the Spanish National Plan for Research, Development and Innovation through grant DPI2010-18702.

References

1. A. Albu-Schaffer and G. Hirzinger, Cartesian impedance control techniques for torque controlled light-weight robots, in *IEEE International Conference on Robotics and Automation, 2002. Proceedings. ICRA '02*, 2002.
2. R. Ham, T. Sugar, B. Vanderborght, K. Hollander and D. Lefeber, *Robotics and Automation Magazine, IEEE* **16**, 81 (2009).
3. D. Sanz-Merodio, M. Cestari, J. C. Arevalo and E. Garcia, A lower-limb exoskeleton for gait assistance in quadriplegia (Guangzhou,China, 2012).
4. L. C. Visser, R. Carloni and S. Stramigioli, Variable stiffness actuators: A port-based analysis and a comparison of energy efficiency, in *Robotics and Automation (ICRA), 2010 IEEE International Conference on*, 2010.
5. D. Sanz-Merodio, M. Cestari, J. Carlos and E. Garcia, *International Journal of Advanced Robotic Systems* , p. 1 (2012).
6. A. M. Dollar and H. Herr, *Robotics, IEEE Transactions on* **24**, p. 144158 (2008).
7. D. P. Ferris, G. S. Sawicki and A. R. Domingo, *Topics in spinal cord injury rehabilitation* **11**, p. 34 (2005).
8. J. W. Hurst, J. E. Chestnutt and A. A. Rizzi, *Robotics, IEEE Transactions on* , 1 (2010).
9. R. Van Ham, B. Vanderborght, M. Van Damme, B. Verrelst and D. Lefeber, *Robotics and Autonomous Systems* **55**, 761 (2007).
10. S. Wolf and G. Hirzinger, A new variable stiffness design: Matching requirements of the next robot generation, in *Robotics and Automation, 2008. ICRA 2008. IEEE International Conference on*, 2008.
11. A. Jafari, N. G. Tsagarakis, B. Vanderborght and D. G. Caldwell, A novel actuator with adjustable stiffness (AwAS), in *Intelligent Robots and Systems (IROS), 2010 IEEE/RSJ International Conference on*, 2010.

NEW PROPULSION SYSTEM WITH PNEUMATIC ARTIFICIAL MUSCLES

IVANKA VENEVA
Institute of Mechanics, Bulgarian Academy of Sciences, "Acad. G. Bonchev" Str. Bl.4. Sofia 1113, Bulgaria, veneva@imbm.bas.bg

The aim of this paper is to present the design of device for control of new propulsion system with pneumatic artificial muscles. The propulsion system can be used for ankle joint articulation, for assisting and rehabilitation in cases of injured ankle-foot complex and all other joints of the human body.

Proposed device for control is composed by microcontroller, generator for muscles contractions and sensor system. During human motion the microcontroller receives the control signals from sensors and modulates ankle joint flexion and extension. The local joint control directly calculates desired pressure levels with a PID position feedback loop and dictates the necessary contractions.

The main goal is to achieve an adaptation of the system and provide the necessary joint torque using position control with feedback.

Key words: Control, artificial muscles, joint articulation, rehabilitation robotics

1. Introduction

Several concepts of pneumatic artificial muscles (PAM) have been developed over time, and commercialized by different companies such as Bridgestone Co. [1], the Shadow Robot Company [2], Merlin Systems Coorporation [3] and Festo [4]. More and more interest for these actuators is growing and several groups all over the world use McKibben like muscles in various robotic and medical applications [5-8].

Actuator technologies that are used in gait therapy devices such as pneumatic and hydraulic actuators can provide the required power with a small device but are impractical as portable devices because they require separate pumps or other air supply. A completely actuated exoskeleton consumes a lot of energy. As energy consumption is an important issue, it remains important to exploit the natural dynamics of human movement by trying to incorporate the unforced motion of a system. The goal of this work is to develop an active lower limb orthosis which combines these two categories. This orthosis should be able to adapt the natural dynamics as a function of the imposed walking motion. For this research a joint actuation with custom made pneumatic artificial muscles and controllable compliance is proposed. It is believed that these pneumatic actuators have characteristics, which can be well adapted, controlled and used for assisting of human planar walking.

2. Methods

The goal of this work is to create a lightweight active orthosis, which is able to exploiting the adaptable passive behavior of the pneumatic artificial muscles in order to maintain adequate stiffness in the joints, reduce energy consumption and control efforts. Stiffness can easily be changed by changing the applied pressures. Research on this topic was done by van der Linde [9] and Wisse [10] through implementation of McKibben's type of pneumatic artificial muscles.

A pneumatic artificial muscle is essentially a volume, enclosed by a reinforced membrane, that expands radially and contracts axially when inflated with pressurized air. Hereby the muscle generates a unidirectional pulling force along the longitudinal axis. When neglecting the membrane's material deformation and the low inertial muscle properties, the generated force F is expressed:

$$F = -p\frac{dV}{dl} \tag{1}$$

where p is the gauge pressure inside the muscle, dV - enclosed muscle volume changes and dl - actuator length changes. The volume of the actuator increases with decreasing length until a maximum volume is reached.

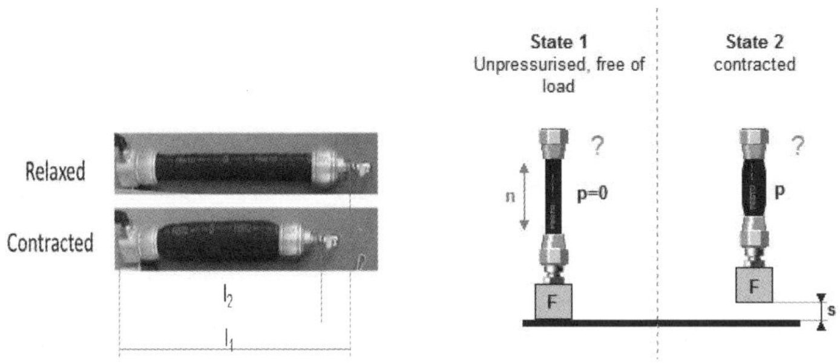

Figure 1. McKibben muscle. The volume of the contracted actuator increases with decreasing length

In Figure 1 the concept of the McKibben muscle is given. It contains a rubber inner tube which will expand when inflated, while a braided sleeving transfers tension. Inherent to this design are dry friction between the netting and the inner tube and deformation of the rubber tube. Depending on the geometry and type of the membrane, the specific force characteristic alters. At maximum contraction, forces become zero, and at low contraction these forces can be very high. Typical working pressure values range from 1 to 5 bar and more. Due to a threshold of pressure which depends on the rubber characteristics, these muscles do not function properly at low pressures. For the practical application, contractions will be bounded somewhere between 5 and 35 %.

Many actuator materials and devices have been put forth as "artificial muscles." Pneumatics and hydraulics (including soft-bodied actuators such as the "McKibben Muscle" can imitate much of the performance of natural muscle and have a shape and feel similar to natural muscle, but they are noisy, difficult to control, and require a separate pump to provide the fluid energy.

In this paper we propose an adaptive device for ankle joint actuation with custom made pneumatic muscles. The device for actuation, data acquisition and control of active ankle-foot orthosis recently proposed by Veneva [11] was used as a basis. We use pneumatic membranes for registering foot contacts during normal level walking and for joint actuation.

The purpose was to confirm our new dynamic control scheme and to confirm that by harnessing the stored energy in the pneumatic element, motor and energy requirements were significantly reduced. Thus, we designed a pneumatically powered, controlled ankle-foot orthosis as a tool for rehabilitation and studying human locomotor adaptation.

3. Description of the system

Ankle-foot orthosis is a system with one degree of freedom which foot segment is connected to the shank segment by a rotational joint. Two identical artificial pneumatic muscles are attached laterally to the ankle (Fig. 2). The pneumatic muscles are lightweight custom made and can produce high power outputs. They are made from latex tubing surrounded by a braided polyester shell. Inflating the tubing causes the shell to expand radially and shorten axially. The tubing is positioned into two end fittings which close the muscle and provide tubing to inflate and deflate the enclosed volume. Due to its specific design, the PAM can easily work at pressures as low as 20 mbar. Muscle contraction can be more than 40 %, depending on its original dimensions. The muscle prototype has a weight of about 100 gr while it can generate forces up to 1 kN. A several test was performed, at which a muscle moves up and down a load of 1 kg by a slow varying gauge pressure between 1 and 3 bars.

The *Control algorithm* is based on the biomechanical interpretation of the locomotion [11, 12]. Within a given walking cycle, four distinct positions were used corresponding to the phases: *heel strike, stance, toe-off and swing*. During the swing phase, where the clearance of the toe is released, the system must actively adjust the flexion of the orthosis and keep this position till the heel strike appears. Thus the ankle torque has to be modulated from cycle-to-cycle throughout the duration of a particular gait phase. This algorithm works well and will be used in the new system with pneumatic artificial muscles.

Four pressure membranes and tactile sensors (TR1, TR2 and TL1, TL2) are incorporated under the heel and the toes in the shoes of both legs. During each gait cycle, by pressuring the membranes, under the influence of weight we push air to the muscles. Inflating the tubing causes the shell to expand radially and shorten axially thus generating muscle contractions. During each gait cycle, by pressing the corresponding tactile sensors the electrical valves are open and the air comes out decreasing the tubing, thus generating muscle relaxation. Thereby ankle torque of one leg is modulated during the particular gait phase by pushing the pressure membranes of the opposite leg.

In this way the required flexion of the ankle joint is realized by pneumatic muscles contraction using pneumatic membranes for registering foot contacts and for joint actuation. Thus we propose an adaptive actuation of the joint harnessing the stored energy in the pneumatic element.

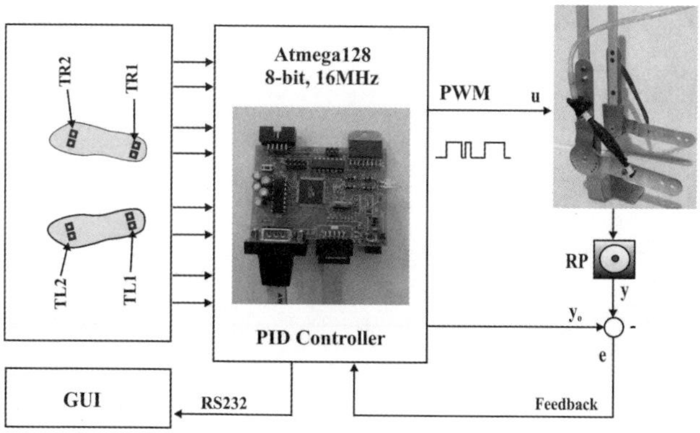

Figure 2. Control system with ankle-foot orthoses and pneumatic artificial muscles

The control module is based on the device for actuation, data acquisition and control of active ankle-foot orthosis, recently proposed by Veneva [11]. It has been realised using microcontroller ATmega128 (*Atmel Co.*). The microcontroller receives the diagnostic information about the system from the sensors and generates the signals to the valves for opening or closing. The PWM channel is connected to the driver to control the direction, speed and torque of the actuator by varying the duty cycle of the PWM output. Control signals are received in real time from sensors. The tactile sensors and a rotary potentiometer measure ankle joint position and send signals to the microcontroller. A Proportional-Integral-Derivative control with feedback was used to estimate the trajectory of the foot and positioning the actuated foot segment of AFO when the foot rotates about the ankle. During each gait cycle a microcontroller estimates forward speed and modulates swing phase flexion and extension in order to assure automatic adaptation of the joint torque.

4. Ankle joint setup

The leg consists of three parts: lower leg, upper leg and foot (Fig. 3a). The length of the i-th link is l_i, its mass is m_i and the moment of inertia about its centre of mass G_i is I_i. While the foot is in contact with the ground, this model has one degree of freedom which is represented by the relative knee angle θ.

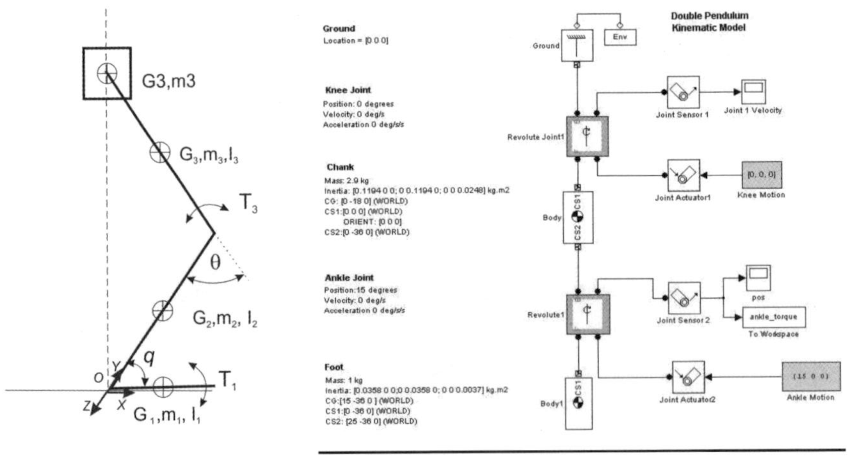

a) b)

Figure 3. a) Shematic overview of the lower limb; b) Kinematic model of the personalized AFO in SimMechanics MATLAB

Different mechanisms will be used for the hip, knee and ankle joint actuation. Here we are going to discuss the model of the ankle joint actuation. The dynamic model of the system is in the swing phase and it assumes that the shank is inertially fixed (Fig. 3b).

Ankle-foot orthosis is built of two segments - shank and foot. The foot segment is connected to the shank segment by a rotational (hinge) joint with a single rotational degree of freedom which is represented by the ankle angle q.

To determine the kinematic expressions of the joint system, an orthogonal XY-coordinate system is defined. The X-axis is aligned with the floor, while the vertical Y-axis is attached to the shank and the axis Z represents the ankle joint axis. To position the foot, we enforce the appropriate angle between the shank and the foot. We simulate the model in *Inverse Dynamics mode* in SimMechanics MATLAB to compute the joint torque required to rotate the foot in desired position [12]. During the simulation the geometry of the orthosis is presented as a double pendulum. The joint torque is given by following expression:

(2) $$T = T_d - T_c - T_g,$$

(3) $$T_d = (J_c + md^2)\ddot{q} + k\dot{q} + mgd\sin q,$$

where T_d is the driving torque; T_c – the torque caused by the friction; T_g - torque caused by the gravity; J_c is the foot body inertia moment; q – generalized coordinate; d - distance between the origin O and the centre of mass of foot segment; m - sum of masses of the foot and orthosis foot segment.

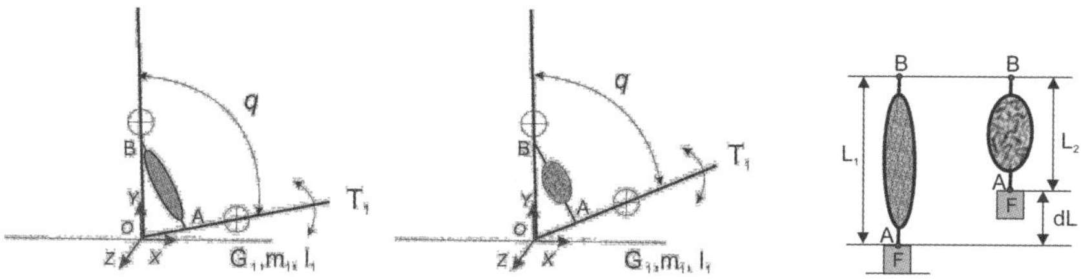

Figure 4. Ankle joint with pneumatic muscle generated pulling force: relaxed (left), contracted (right)

The essential parameters to be determined during the design process of the joint (Fig. 4) are L_1 - the length of muscle when relaxed; L_2 - the length of muscle when contracted; r_1 - the distance between the origin O and the point A - the muscle attachment of the foot segment; r_2- the distance between the origin O and the point B - the muscle attachment of the shank segment; q - the ankle angle between the vector OA and OB, (counter-clockwise is positive).

The contracted pneumatic artificial muscle generates a pulling force along the longitudinal axis.

(4) $$\tau = r_1 F \sin q$$

(5) $$q = f(L, p)$$

The model can be personalized using the physical parameters of the patient and estimate the length of muscle and point of attachment A and B, the pulling force required to rotate the foot about the ankle and the minimum and maximum angle of joint rotation. The foot parameters are known from the conventional anthropometric tables.

5. Results and Discussion

In order to test the control algorithm and system functionalities a laboratory model with two hinge joints and attached pneumatic muscles was designed. A healthy subject equipped with special shoes with four pressure membranes and tactile sensors mounted under the heel and the toes part of the insole performs different trials of slow and normal level walking. Lower limbs movement was measured during walking using signals from sensors. A potentiometer is mounted on the hinge joint, coinciding with the axis of the ankle joint. Hinged joints are attached laterally of both ankles. The footswitches placed under the foot beneath the heel and the toe have been used to detect in real time the precise moments of heel strike (when the foot first touches the floor) and toe-off (when it takes off).

The sensors work together to detect walking over one given interval of time and to collect the following parameters: ankle joint angles, foot (heel and toe) contacts and foot velocities.

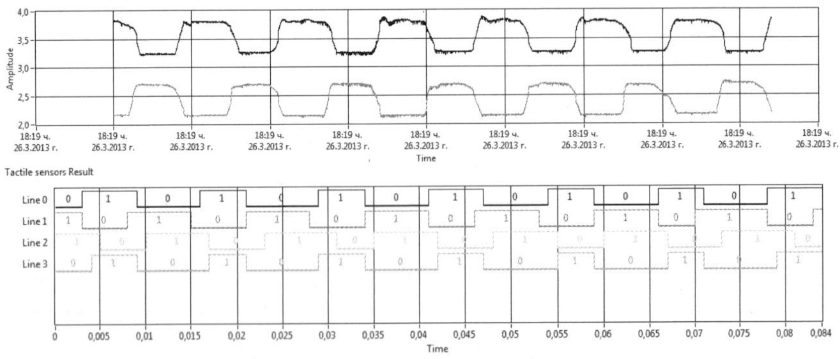

Figure 5. Visualization of human motion data (in LabView): a) Ankle angle rotation (potentiometer data in volts); b) Tactile sensors results

For visualization of the signals a LabView virtual instrument has developed. The data from sensors were collected with multifunctional (DAQ) module (NI-USB-6211, National Instruments and LabVIEW). The graphic on Fig.5a shows the kinematics of the ankle. Ankle angle rotation (potentiometer data in volts) is shown with the presence of peaks during the flexion or negative peaks during toe-off and extension. The range of the measured angles correspond to the rotation of 30° and obtained potentiometer signals are with sensitivity of 4 mV/deg. Signals line0 and line1 are recorded digital signals from the switches mounted under the heel and the toes part of the left leg insole while signals from the right leg are line2 and line3 (Fig. 5b). The first one transition of line0 signal from 1 to 0 shows the heel strike component (left stance) and the second one transition of line1 signal from 0 to 1 shows the toe-off component (for the left leg). The values of the signals were detected in milliseconds.

It is obvious that the algorithm is applied in the same way for both legs. The pneumatic muscles have a high power to weight ratio and can be coupled directly without complex gearing mechanism. Due to the compressibility of air, a joint actuated with these pneumatic actuators shows a compliant behaviour, which can be positively employed to reduce shock effects. Joint compliance can be adapted while controlling position which enhances the possibilities of exploitation of natural dynamics.

6. Conclusion

The presented propulsion system for control of active ankle-foot orthosis integrates adaptive pneumatic system with artificial muscles and biomechanics based algorithms. The pneumatic system with artificial muscles is automatically modulated in order to optimize the heel-to-forefoot transition during the stance or the swing phase of walking. The data from the sensors are used in every step from the control algorithm.

The proposed control device can be used in the new propulsion system with pneumatic artificial muscles to control the orthosis functionalities for ankle joint articulation, for assisting and lower limb rehabilitation.

References

1. Inoue, K. [1987]. Rubbertuators and applications for robotics, Proccedings of the 4th International Symposium on Robotics Research, pp. 57-63.
2. Shadow Robot Company [2003]. Design of a dextrous hand for advanced CLAWAR applications, Proc. of the 6th Int. Conf. on Climbing and Walking Robots and the Support Technologies for Mobile Machines, Catania, Italy, pp. 691-698.
3. Merlin Systems Coorporation [2003]. Merlin actuators: Automation,
4. Festo [2004]. Fluidic muscle MAS, Festo Brochure Fluidic Muscle.
5. Raparelli, T., Mattiazzo, G., Mauro, S. and Velardocchia, M. [2000]. Design and development of a pneumatic anthropomorphic hand, Journal of Robotic Systems 17(1): 1-15.

6. Pomiers, P. [2003]. Modular robot arm based on pneumatic artificial rubber muscles (PARM), Proc. of the 6th Int. Conf. on Climbing and Walking Robots and the Support Technologies for Mobile Machines, Catania, Italy, pp. 879-886.
7. Klute, K. G., Czerniecki, J. M. and Hannaford, B. [2002]. Artificial muscles: Actuators for biorobotic systems, The International Journal of Robotics Research 21(4): 295-309.
8. Daerden, F., Verrelst, B., Lefeber, D. and Kool, P. [1999]. Controlling motion and compliance with folded pneumatic artificial muscles, Proc. of the 2nd Int. Symposium on Climbing and Walking Robots, Portsmouth, UK, pp. 667-677.
9. Van der Linde, R. Q. [1999]. Design, analysis and control of a low power joint for walking robots, by phasic activation of mckibben muscles., IEEE Transactions on Robotics and Automation 15(4): 599-604.
10. Wisse, M. [2001]. Biologically inspired biped design, Symposium of the Int. Soc. For Postural and Gait Research, Maastricht, The Netherlands, pp. 900-903.
11. Veneva I., Device for Control of Active Ankle-Foot Orthosis and Monitoring System for Gait Analysis, (2010), Journal of Theoretical and Applied Mechanics, vol. 40, No. 4, pp. 81–92
12. Veneva I., Intelligent device for control of active ankle-foot orthosis, Proceedings of the 7th IASTED Int. Conf. on Biomedical Engineering "BioMed 2010", Innsbruck, Austria, 2010, pp. 100-105

DEVELOPMENT OF A WALL CLIMBING ROBOT USING THE MOBILE MECHANISM OF TRAVELING WAVES PROPAGATION

YUTAKA MIZOTA, KAZUTOSHI TAKAHASHI, YUSUKE GOTO AND TARO NAKAMURA

Faculty of Science and Engineering, Department of Precision Mechanics Chuo University 1-13-27 Kasuga, Bunkyou-ku, Tokyo 112-8551, Japan

Recently, there have been many studies for developing robots that perform tasks instead of humans. In this study, we focus on tasks to be performed at high places above the ground, which can be dangerous and costly. Therefore, a robot having the ability to climb a wall is preferred for an automatic maintenance system, instead of humans. This paper reports the development of a wall climbing robot, for which two new mechanisms have been adopted: a mobile mechanism using traveling wave propagation and an adhesion mechanism using a centrifugal fan. This mobile mechanism modeled the locomotion mechanism of a snail. This mobile mechanism has a large contact area and is extremely stable. Therefore, it can move stably on both smooth and rough surfaces. In this study, we developed a mobile mechanism that uses a traveling wave propagation system. In addition, we developed a wall climbing robot using this mobile mechanism, and conducted climbing experiments on various walls. We confirmed that this robot can climb the walls in the tests.

1. Introduction

Recently, there have been many studies that use robots to perform dangerous tasks instead of humans. The focus of these studies has been on tasks in high places such as bank buildings, atomic power plants, and so on. The tasks in bank buildings include cleaning, maintenance, and inspection. These tasks are usually performed by highly skilled workers. However, these tasks are dangerous and costly. Therefore, a robot which has the ability to climb a wall is expected to be useful as an automatic maintenance system instead of relying on humans.

Various types of wall climbing robots have been developed worldwide. Wall climbing robots consist of a mobile mechanism and an adhesion mechanism. The adhesion mechanism includes magnetic adhesion [1] and vacuum adhesion [2], while the mobile mechanism includes the wheeled mechanism [2] and crawler mechanism [1]. Different adhesion and mobile mechanisms can be combined to realize different types of wall climbing robots.

The authors developed traveling wave propagation type mobile mechanisms that consisted of several slider cranks [3], but their mechanisms require complicated structures and are bulky. Accordingly, in this study, we developed a traveling wave propagation type mobile mechanism using a screw cylinder. So we developed a wall climbing robot using this mobile mechanism and conducted climbing experiments on various walls. We confirmed that this robot can climb a variety of walls.

2. Locomotion Mechanism of a Snail

A snail moves by propagating traveling waves. By extending and contracting muscles, a small part of its body leaves the ground surface, and as it sends the aperture between the body and surface in the forward direction like a wave, the entire body moves in the forward direction [4]. Figure 1 shows the state of the traveling wave during the forward motion for the snail shown in Fig. 1.

The advantage of this locomotion is its high stability. There are three reasons for this: First, the contact area is large so the load is spread over a wider surface. Second, since the area in transition is small, the change in the load of the contact area is also small. Finally, because it moves only by a small aperture, there is little chance that it will slip off. Therefore, the locomotion of a snail is advantageous in unstable locations, such as walls. Therefore, we adopted this traveling wave propagation as the locomotion mechanism for our wall climbing robot.

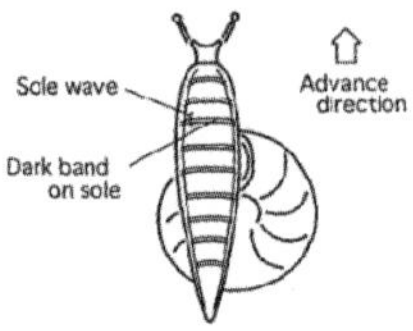

Figure 1. Locomotion mechanism of a snail [5]

3. Overview of the Wall Climbing Robot

An overview of the wall climbing robot using the mobile mechanisms of the traveling wave propagation and a centrifugal fan is shown in Fig. 2. The specifications of the robot are shown in Table 1. This robot is controlled by the proportional control system and is able to move by itself by loading the battery.

Characteristics of this robot are as follows. First, it has extremely stable movement because the contact area is large. Second, this robot can change directions without scraping the wall because minor frictional forces are absorbed by a sponge rubber, which is affixed to the back of the spring leaf. Third, because the contact area does not rotate, the sensors can easily be loaded at the contact area. Finally, this robot can adhere to both smooth and rough surfaces.

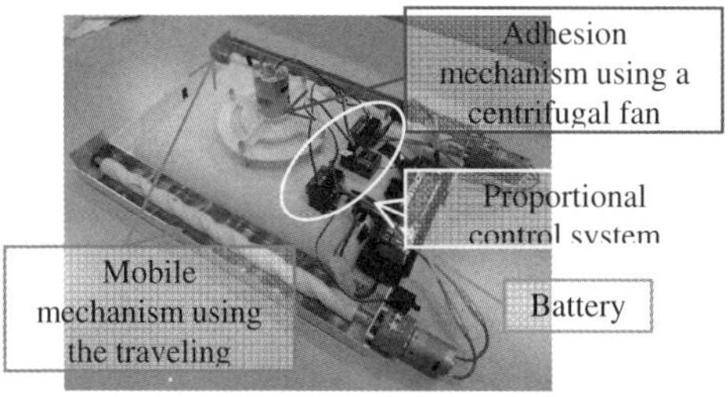

Figure 2. Overview of the wall climbing robot

Table 1 Specifications of the robot

Length	460 [mm]
Width	300 [mm]
Height	130 [mm]
Weight	1.55 [kg]

4. Mobile Mechanism Using the Traveling Wave Propagation

4.1. *Outline of the Mobile Mechanism*

The outline of the mobile mechanism using the traveling propagation is shown in Fig. 3. The spring leaf is put under the screw cylinder and compressed in the anteroposterior direction. Therefore, the spring leaf follows the wave shape of the screw cylinder. The spring leaf is supported on both sides by guide rails. The sponge rubber, which is thicker than the guide rail, is affixed to the back of the spring leaf. Therefore, waves make contact with the ground and the wall, and its operation is similar to a snail's gastropod. By rotating the screw cylinder using a motor, the waves are propagated to the axial direction. The test model of the mobile mechanism is shown in Fig. 4.

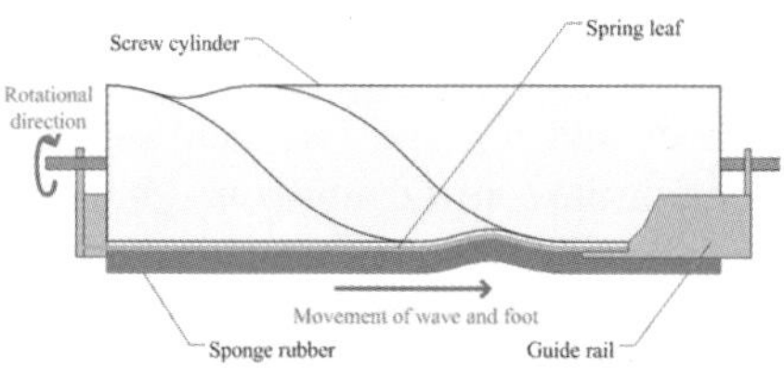

Figure 3. Outline of the mobile mechanism

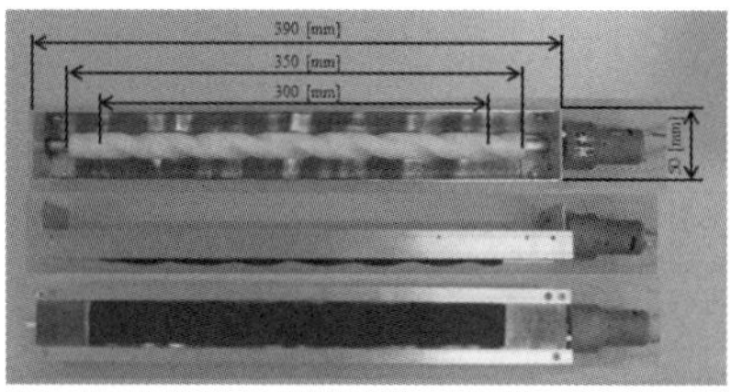

Figure 4. Test model of the mobile mechanism

4.2. *The Theoretical Values of the Horizontal Velocity*

The horizontal velocity of this mobile mechanism depends on the shape of the wave. The distance moved when a wave has passed is determined by the difference in the gauge length that includes waves and the gauge length that does not include waves. Therefore, on the basis of this difference, the horizontal velocity is determined.

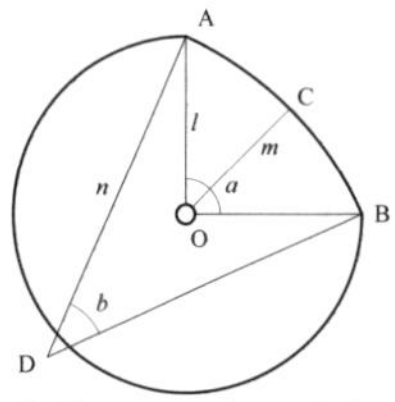

Figure 5. Cross section of the screw cylinder

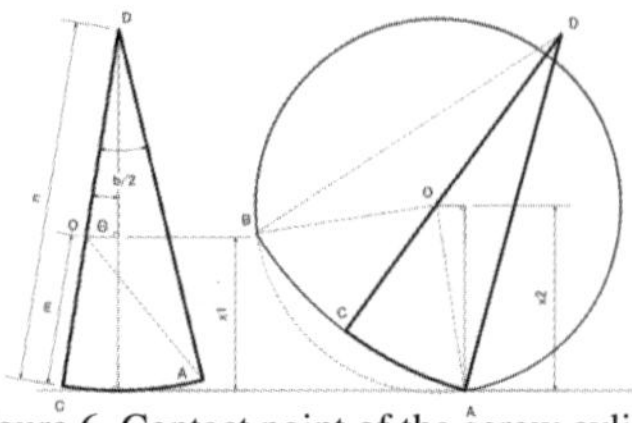

Figure 6. Contact point of the screw cylinder and a wave for each angle of rotation

The cross section of a screw cylinder is shown in Fig. 5. The shape of a wave is expressed in the function using this figure. In this figure, A, B, C, D, and O are the vertexes; *l*, *m*, and *n* are the lengths of the segments; and *a* and *b* are angles. ABD is the circular sector with a radius of n and central angle of *b*. The cross section of the circular sector is determined using lines *l* and *m*, and angle *a*.

$$n = \frac{l^2 + m^2 + 2lm\cos\left(\pi - \frac{a}{2}\right)}{2l\cos\left(\pi - \frac{a}{2}\right) + 2m}. \tag{1}$$

$$b = 2\sin^{-1}\left\{\frac{l\sin\left(\pi - \frac{a}{2}\right)}{n}\right\}. \tag{2}$$

The contact point of the screw cylinder and a wave for each angle of rotation is shown in Fig. 6. The cross section of the screw cylinder is symmetric around the line OC. Therefore, the state in which the wave has a tangent at point C (with which it is in contact) is defined as the initial condition. We determine the distance from the center of the screw cylinder to the wave when the angle of rotation of the screw cylinder is a variable number θ.

$$0 \le \theta \le \frac{b}{2}\ (x = x_1),$$

$$x_1 = n - (n - m)\cos\theta. \tag{3}$$

$$\frac{b}{2} \le \theta \le \frac{a}{2}\ (x = x_2),$$

$$x_2 = l\cos(\frac{a}{2} - \theta). \tag{4}$$

The length of an axial directional line Y, which is from the starting point of a wave to the end point, is given by the number of turns p, the length of the screw cylinder L, and the circular sector of the angle a.

$$Y = \frac{La}{2\pi p}. \tag{5}$$

The angle of rotation, which is from the starting point of a wave to the end point of a wave, is angle a. The length, which extends from the cross section of the state (where segment OC is perpendicular to the surface) to the cross section of the arbitrary state, is shown by angle θ.

$$0 \le y \le \frac{Y}{2} \quad \left(0 \le \theta \le \frac{a}{2}\right),$$

$$y = \frac{Y}{a}\theta. \tag{6}$$

From the above, the shape of a wave is shown as follows.

$$S = 2\int_0^{\frac{a}{2}} \sqrt{x'^{2}_{(\theta)} + y'^{2}_{(\theta)}}\, d\theta. \tag{7}$$

By substituting Eq. (3), Eq. (4), and Eq. (6) in Eq. (7), we obtain the length of the shape of a wave. Therefore, the distance moved when a wave has passed is determined as follows.

$$f = S - Y. \tag{8}$$

From the above, when the number of rotations of the screw cylinder is q and the number of threads of the screw cylinder is r, the theoretical value of the horizontal velocity V is determined as follows.

$$V = fqr. \tag{9}$$

Because the horizontal velocity is amplified by the sponge rubber, which is affixed to the back of the spring leaf, the theoretical value of the horizontal velocity V is multiplied by the amplification coefficient T. Therefore, the corrected horizontal velocity V' is determined as follows.

$$V' = Tfqr. \tag{10}$$

4.3. *Experimental Test of the Mobile Mechanism*

We carried out an experiment of the mobile mechanism. In this experiment, the test model in the case where the sponge rubber is not affixed to the back of the spring leaf is moved on the sponge rubber. The voltage applied to the test model was every 1.0 V from 3.0 V to 12.0 V using the stabilized power supply. We measured the number of rotations of the screw cylinder and the time taken for it to move when the test model moved 20 mm for each of the voltages. In this experiment, we examined the changes in the horizontal velocity due to the number of threads of the screw cylinders. We used two types of screw cylinder that were a single spiral and a double spiral. The experimental result is shown in Fig. 7. This graph shows the relationship between the number of screw cylinders and horizontal velocity of the test model. This result proved that the horizontal velocity depends on the number of threads. Therefore, we concluded that the double spiral is best for this mobile mechanism.

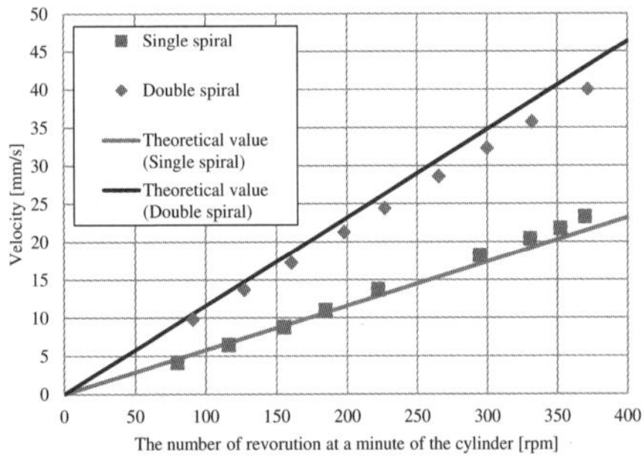

Figure 7. Relationship between number of rotations of the screw cylinder and horizontal velicity

5. Adhesion Mechanism Using a Centrifugal Fan

An outline of the adhesion mechanism using a centrifugal fan is shown in Fig. 8. With this adhesion mechanism, the negative pressure area between the robot and the wall was generated by a centrifugal fan because the pressure of the influent air is greater than the effluent air. Therefore, the adhesion force is generated from the difference in the atmospheric pressure. Although the suction cup cannot adhere to rough surfaces such as brick and concrete, this adhesion mechanism can adhere to rough surfaces.

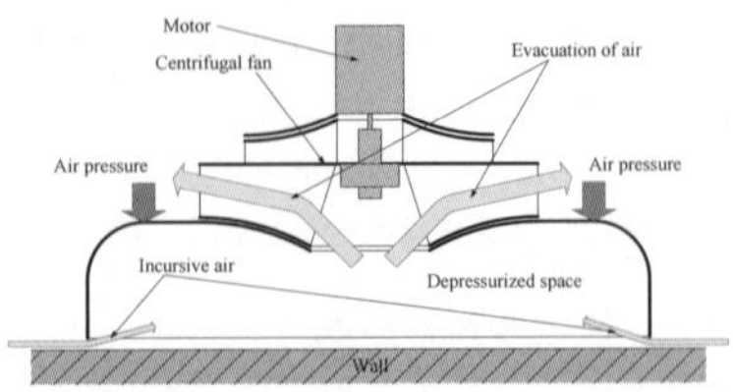

Figure 8. Outline of the adhesion mechanism using a centrifugal fan

6. Wall Climbing Experiments

6.1. *Experiments on the Vertical Surfaces*

The climbing velocity of this robot was measured on vertical surfaces. The experimental conditions were as follows: (1) smooth horizontal surface, (2) smooth vertical surface, (3) vertical glass surface, and (4) vertical concrete surface. In this experiment, the mobile mechanism and adhesion mechanism are supplied with a 7.2 V DC from a stabilized power supply.

The experimental results are shown in Fig. 9. This graph shows the climbing velocity of the robot under various conditions. The images of the wall climbing robot while climbing the vertical surfaces are shown in Fig. 10. From this result, the climbing velocity is observed to be slower than the horizontal velocity, because this robot is subjected to opposing vertical gravitational force. On the concrete surface, the climbing velocity decreases due to sliding between the mobile mechanism and the concrete surface. This is because the friction force is reduced by powder at points in which the surface of the wall is exfoliated and formed.

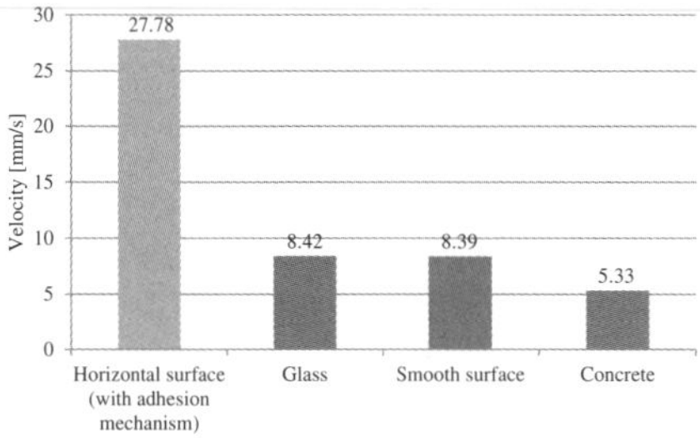

Figure 9. Difference in the climbing velocity on the vertical surfaces

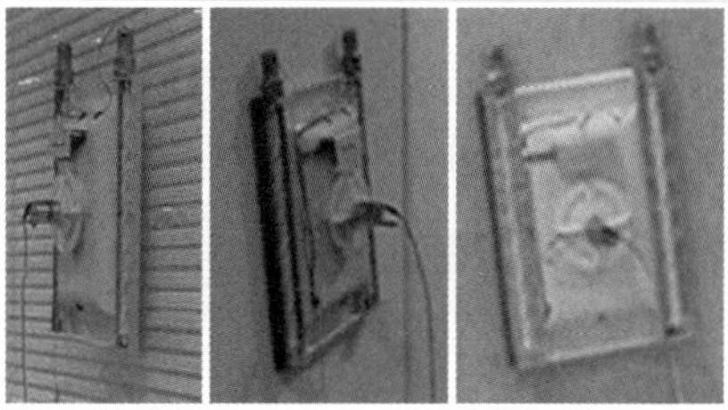
Glass Smooth surface Concrete

Figure 10. Image showing movement along the vertical surfaces

6.2. *Experiments on the Ceiling Surface*

We examined the horizontal velocity of the wall climbing robot on a ceiling surface made of gypsum board. In this experiment, the mobile and adhesion mechanisms are supplied with a 7.2 V DC voltage by a stabilized power supply.

The experimental results are shown in Fig. 11. The image of the wall climbing robot while moving along the ceiling surface is shown in Fig. 12. This graph shows the velocity of the robot in the horizontal, ceiling surface, and vertical surface. These results confirmed that the horizontal velocity on the ceiling surface is faster than the climbing velocity. We believed the reason for this is that there is no gravitational influence on the direction of motion as opposed to the case when climbing on vertical surfaces.

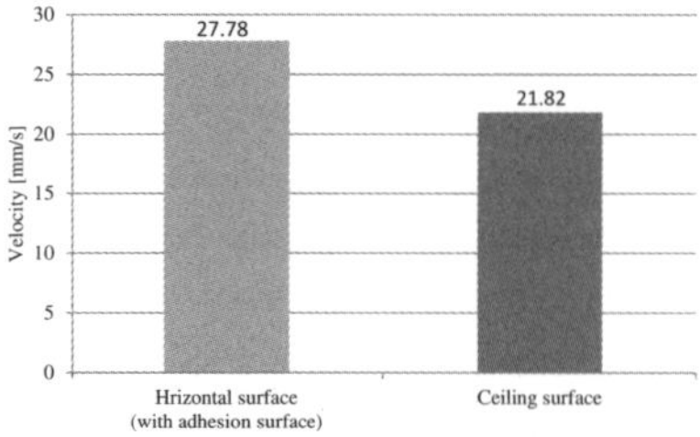

Figure 11. Difference in the climbing velocity on the ceiling surface

Figure 12. Image showing movement along the ceiling surface

7. Conclusion and Future Work

Our conclusions are as follows:

1. We developed a mobile mechanism that has a system that incorporates traveling wave propagation by observing a snail's gastropod.
2. We modeled this mobile mechanism numerically.
3. We examined the specifications of the mobile mechanism. In addition, we made a comparison between the modeling data and the experimental results for a mobile mechanism. We found that our experimental results were almost the same as those obtained numerically.
4. We demonstrated the climbing experiment on vertical and ceiling surfaces, confirming that our proposed robot could move along these surfaces.

Hereafter, to improve the horizontal velocity and climbing velocity, we should improve the mobile mechanism. And we plan to include in our design the ability of the wall climbing robot to transfer from the wall to the ceiling.

References

1. Weimin Shen, "Permanent Magnetic System Design for the Wall-climbing Robot," in Proc. of the IEEE International Conference on Mechatronics & Automation, 2005
2. T. Miyake, H. Ishihara M. Yoshimura "Basic Studies on Wet Adhesion System for Wall Climbing Robots," in Proc. of the IEEE/RSJ International Conference on Intelligent Robots and Systems, 2007
3. T. Nakamura, T. Hayakawa, "Development Of Omnidirectional Wall-climbin Robot By Using Travering Wave Propagation," Clawar, 2011, pp. 165-172
4. R. Fujihara, H. Morikawa, S. Kobayashi, "The Mechanism of Pedal Locomotion of Gastropod," The Japan Society of Mechanical Engineers, JAPAN, vol. 67 No. 658 (2001-6)
5. R. Fujiwara, H. Morikawa, Y. Hukuya, H. Sakai, S. Kobayashi, "Pedal-Like Locomotion Mechanism Modeled on Pedal Crawling of Terrestrial Gastropod," The Japan Society of Mechanical Engineers, Japan, Vol. 70 No. 695 (2004-7)

BIO-INSPIRED ELBOW IMPEDANCE MODULATION USING A COMPLIANT TECHNICAL JOINT DRIVE

S. ANNUNZIATA* and J. PASKARBEIT

Research Group Biomechatronics, University of Bielefeld,
Bielefeld, Germany
** E-mail: salvatore.annunziata@uni-bielefeld.de*
www.elan.uni-bielefeld.de

A. SCHNEIDER

Faculty of Engineering and Mathematics, University of Applied Sciences,
Bielefeld, Germany

In recent years the robotics community has focused its interest on the control of the impedance of compliant actuators to increase safety and to make the interaction with humans more natural. Biological systems are elastic and are able to modulate joint impedance while keeping stability through co-activation and reciprocal activation of muscles. In order to implement a bio-equivalent, technical drive actuation system for prosthetics, a bio-inspired position and stiffness control strategy has been implemented and connected to a technical model of an elbow joint. The sum of all muscle–torques actuating the joint represents the net-torque that should be generated in the technical elbow to realize the desired motion. This net-torque is transmitted to a miniaturized lightweight joint drive with inherent serial elasticity and controlled with a speed–torque control cascade. The impedance range of the biological musculoskeletal system is evaluated in simulation and compared to the range obtained when the technical drive is acting instead of its biological counterpart. The impedance range of the technical drive using biological controllers is equivalent to that achieved in the biological example. Simulation results of the biologically controlled drive for different load situations show that the system successfully modulates impedance both in static cases and during movements while keeping stability.

Keywords: Bio-inspired control; impedance control; muscles; prosthetics; orthotics.

1. Introduction

The development of robot joints ensuring safety during interaction tasks is of growing interest in robotic research. Modulation of joint impedance, lightweight construction and the introduction of inherent elasticity in the joint are considered to be the key elements for safe interaction. Biological systems are able to regulate joint impedance through antagonistic muscle co-activation due to the muscle–intrinsic nonlinear elasticity.[1] In this work, the impedance of the human elbow joint is evaluated in simulation with the objective of realizing a bio-equivalent technical actuation system for robotics, prosthetics and orthotics.

The work is structured as follows. In Sec. 2, a simplified model of the human arm, represented by a hinge joint moved by two muscles, is introduced. A bio-inspired position and stiffness control scheme provides the activations for the muscles which in turn generate the necessary net-torque at the joint (Sec. 3). Section 4 describes the miniaturized lightweight joint drive and the control approach based on a cascade of speed-torque controllers. Impedance modulation is analysed and the controlled joint drive is tested in a numerical simulation using Matlab/Simulink for different load situations (Sec. 5). Simulation results confirm the ability of the combined bio-inspired control scheme with the technical drive to simultaneously regulate the position and, with respect to different external perturbations, successfully modulate the joint impedance.

2. Biological system modeling

A miniaturized, lightweight, joint drive actuator with inherent elasticity is controlled to perform movements and react to external perturbations similarly to the behaviour of the human elbow joint. A bio-inspired control scheme connected to a model of the biological setup reproduces a (co-)contraction behaviour of the muscles. The output of this biological controller is the net-torque of the antagonistic muscles. This net-torque is used as input for the technical drive system. Figure 1(a) depicts the actuation system for the elbow joint and the superimposed bio-equivalent case with two muscles. To realize this equivalence, a transfer of biological properties to technical systems is necessary.

Two muscles, modeled using the Hill-type muscle model[2] and connected in an antagonistic setup with a pulley, represent a simplified version of the elbow joint. Figure 1(b) shows the joint stiffness (solid lines) and damping (dashed lines) plotted over the joint angle α in the case of equal levels of activation for both muscles (co-activation) and slow movements (see Ref. 3 for details). The muscle parameters adopted in this work are based on measurements of elbow muscles from Ref. 4. Values of maximum isometric force, maximum shortening velocity and curvature constant in the force–

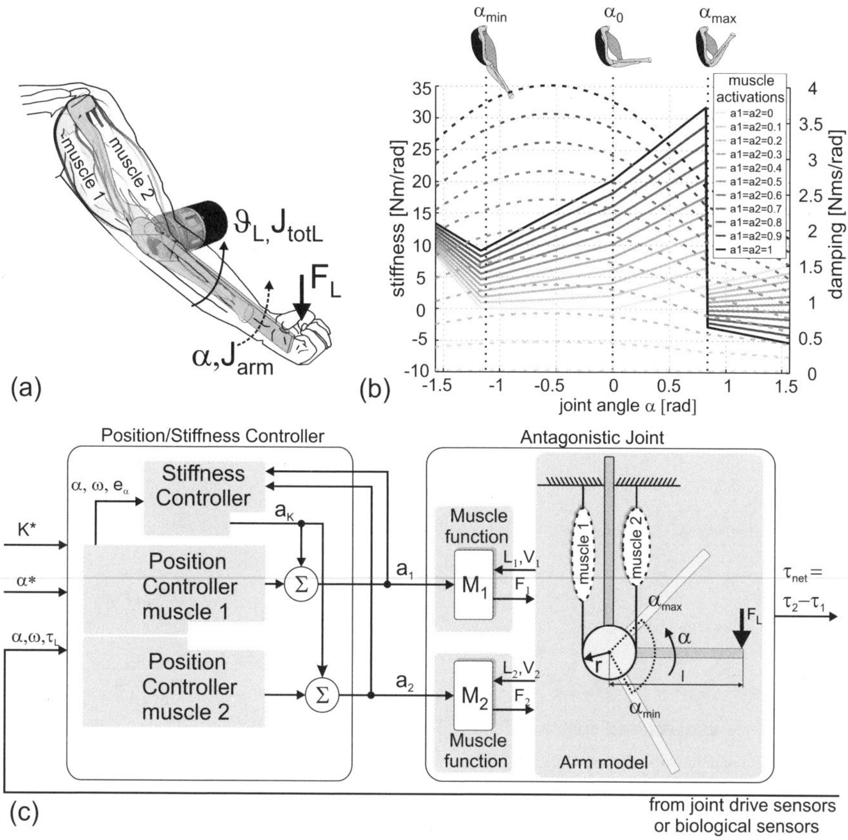

Fig. 1. (a) Simplified depiction of the human elbow joint with two antagonistic muscles and a joint drive with a lever arm representing a forearm prosthesis. (b) Joint stiffness (solid lines) and damping (dashed lines) plotted over joint angle for different co-activations ($a_1 = a_2$). (c) Schematic view of the simultaneous position/stiffness controller acting on the antagonistic joint setup.

velocity muscle curve are selected based on the values of elbow joint stiffness and damping experimentally measured in humans (see Ref. 5).

3. Bio-inspired control scheme

A bio-inspired control scheme was designed based on physiological models and properties of biological systems. Figure 1(c) shows its schematic design. A detailed description can be found in Ref. 3. The two antagonistic muscles drive a pulley in a hinge joint and receive the activation signals from a position controller and a stiffness controller. A stiffness computation block calculates the current joint stiffness based on the muscle activations, joint position and velocity. The stiffness controller activates all muscles with the same signal (co-activation). The position controller, which is responsible for generating the necessary torque in order to reach the desired position, acts on one muscle at a time (reciprocal activation). A further compliance adaptation algorithm, based on the position error e_α, modulates the stiffness set-point, providing the system with the ability to react to external forces while keeping joint stability. The complete system provides the net-torque τ_{net} (sum of the torques of all muscles actuating the joint) that should be generated at the elbow to realize the desired stiffness K^* and motion α^*.

4. Technical joint system description

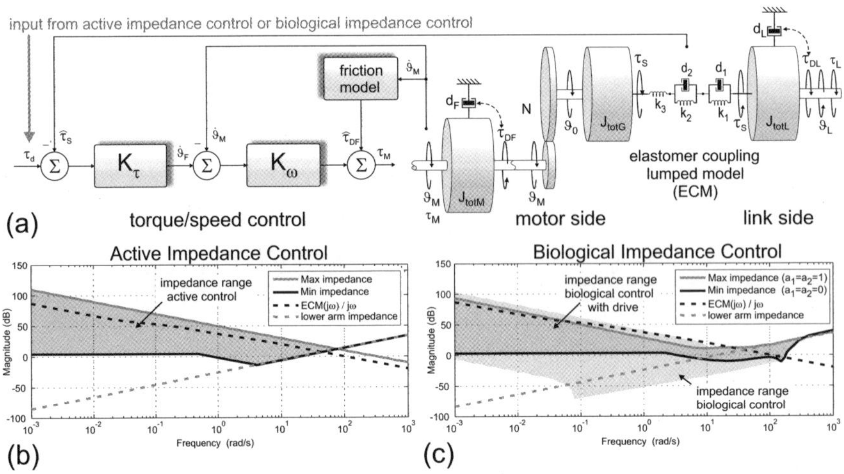

Fig. 2. (a) Schematic view of the torque/speed cascade control acting on the joint drive with the elastomer coupling lumped model. (b) Impedance modulation range when an active control of the joint compliance is realized with a proportional position controller P. (c) Impedance modulation range for a bio-inspired control based on antagonistic muscles.

The physical joint drive used in this work consists of a brushless DC motor, a harmonic drive gear, a sensorized elastomer coupling (providing an inherent elasticity[6]) and an output shaft with a load. Rotary encoders for measuring the position of the output shaft and the torsion angle of the elastomer are also included. The elastomer coupling is approximated with a second order linear model. This model is used to estimate the torques acting on the output side based on the torsion measurement. For the technical data of the rotatory drive and the parameters of the model see Ref. 6.

The control scheme (Fig. 2(a)) is based on a cascade feedback control of motor speed and torque with additional motor friction compensation. The motor speed controller is a PI and was tuned in order to reach the desired velocity with a settling time of $9\,ms$ with an overshoot of 20%. The torque controller feedback loop is closed on the estimation of the torque $\hat{\tau}_S$ obtained by measuring the torsion ϑ_S of the elastomer coupling. The torque controller adopted in this work is also a PI and its design plays a crucial role for the stability of the joint drive. Stability considerations set limitations on the control parameter values. Taking into account those considerations, the controller has been tuned in order to reach the desired torque in $0.5\,sec$ without overshooting. The bandwidth of the torque controller loop is $60\,rad/sec$.

5. Bio-inspired impedance modulation

The mechanical impedance represents the measure of how much the system resists to movements when perturbed. High impedance corresponds to high stiffness while low impedance results in a more compliant behaviour of the system under perturbation.

The classical way to regulate the impedance of a technical drive, starting from the scheme of Fig. 2(a), is to close an outer position loop with a proportional controller (P), which acts as desired stiffness (active impedance control).[7]

To provide an estimation of the impedance range, an impedance analysis has been performed in Matlab/Simulink® (The MathWorks Inc. Natick, MA, USA). The Linear Analysis Tool was adopted for the linearization of the non-linear system around an operating point. The bode diagram, representing the impedance frequency response, was obtained considering as input of the entire system the load torque and as output the joint speed.

The impedance range obtained for different values of P is reported in Fig. 2(b). For low frequencies the actuator is able to reproduce the desired stiffness while for higher frequencies the impedance matches the load (lower arm) inertia. The impedance of the elastomer coupling is shown as black dashed line.

Another approach is to take advantage of the virtual antagonistic muscle setup and the bio-inspired control strategy as shown in Fig. 1(c). The theoretically achievable impedance range using this approach is shown in Fig. 2(c) in light gray. However, if the biological control scheme is used to control the technical drive by providing as input τ_{d} to the torque controller the τ_{net} of Fig. 1(c), the achievable impedance range is reduced. The dark gray area in Fig. 2(c) represents the range of achievable impedance when the drive has to reproduce the biological behaviour. The friction in the output (for example due to ball bearings) reduces the achievable minimum impedance (see black line) while the maximum torque that the actuator can generate reduces the maximum impedance (dark gray line). The achievable impedance range of the technical drive using biological controllers is equivalent to that achieved in the biological example. The light gray area below the minimum impedance (Fig. 2(c)) which can be theoretically generated by the biological system does not seem to play an important role in real world scenarios.

An important result concerning the stability of the system can be drawn from the impedance plot described above. The presented system is passive, and therefore stable, for the entire frequency range for which the rendered impedance is lower than the pure elastomer coupling impedance ($ECM(jw)/jw$) (see Ref. 7 for more details).

5.1. *Simulation results*

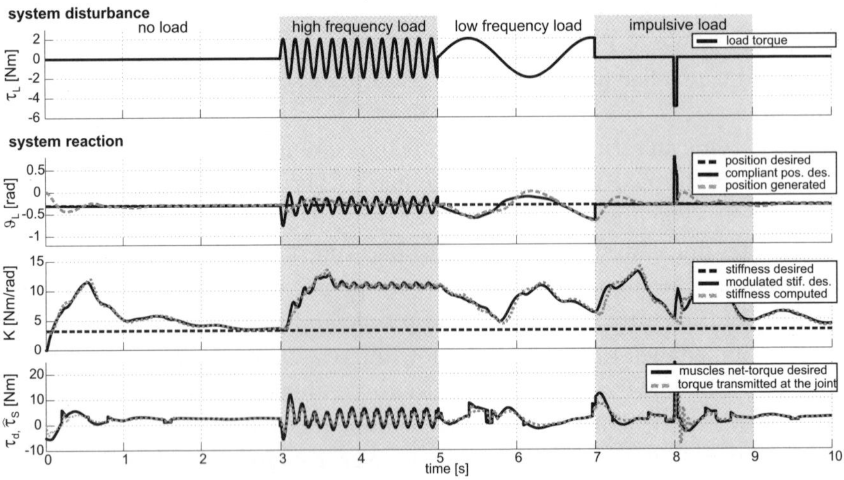

Fig. 3. Simulation results of the position and stiffness control with bio-inspired impedance modulation in presence of different load situations. Depending on the perturbation, joint stiffness adapts producing a stabilizing effect at joint level.

Figure 3 depicts the simulation results of the position control and impedance modulation with respect to different external perturbations when the joint is commanded to reach a desired position $\alpha^* = \vartheta_{\mathrm{L}}{}^* = -0.3\,rad$ and a desired stiffness $K^* = 3\,Nm/rad$.

The bio-inspired control mechanism acting on the muscles and working in combination with the technical drive is able to simultaneously regulate the position and, in dependence of the perturbation frequency, adapt the joint impedance. For rapid perturbations stiffness increases so as reducing the effect of the external torque on the lever, while for low frequency perturbations the joint is compliant. It is proven that humans show the same behaviour as they increase voluntary muscle activation to move faster or to move against a load.[8]

The last row in the figure depicts the net-torque (solid black line) obtained from the bio-inspired control scheme, which represents the desired torque for the technical drive, and the effective torque transmitted at the joint (dashed gray line). To generate the desired biological behaviour, the torque controller which acts on the drive has to make sure that these signals are the same. Besides a short transitory, the desired net-torque is reached thus ensuring the robotic joint to achieve both desired position and stiffness performances.

6. Conclusions

A bio-inspired control scheme featuring co-activation and reciprocal activation combined with virtual antagonistic muscles in a hinge joint setup has been tested on a lightweight miniaturised joint drive. The underlying objective was to construct a robotic limb for prosthetics and orthotics able to reproduce the human arm behaviour. The used actuator possesses an inherent serial elastic element in form of a sensorized elastomer coupling, which provides the ability to store energy and to attenuate shocks due to unexpected impacts. The measurement of the torsion in the coupling is used for the estimation of the torque acting at the output. A speed–torque cascade control is responsible for controlling the drive thus producing the necessary motion to implement the desired bio-equivalent behaviour.

To evaluate the advantages and limitations of such a bio-equivalent robotic approach, the impedance modulation range was analysed, first when the joint drive is controlled with a classical active impedance control approach, and then when the bio-inspired control algorithm is adopted. For the sake of the analysis, constraints like maximum torque generated by the drive and friction were also considered. The two impedance ranges obtained in the analysis were comparable. Nevertheless, the advantage to use the biological approach is evident when the system is subject to perturbation thanks to the inherent ability to modulate the impedance. The biologically controlled drive was tested for different load situations. Results show that the system successfully reaches the desired position while modulating the impedance. Furthermore, as long as the rendered impedance lies below the intrinsic stiffness of the pure elastomer coupling, the system is stable due to its passivity properties.

The present study was conducted based on computer simulations. Future work will be the implementation and test of the speed and torque controller cascades on the real joint drive. In addition, the bio-inspired control laws, as proposed in this work, will be also implemented and used in combination with the controlled technical joint drive to generate bio-equivalent movements.

Acknowledgments

This work has been supported by the Federal Ministry of Education and Research (BMBF) within the BIONA program (ELAN-project to A.S.), within the Excellence Initiative of the German Research Foundation (DFG Center of Excellence "Cognitive Interaction Technology", EXC277) and by an EU-FP7 grant (ICT-2009.2.1, No. 270182 to A.S.).

References

1. N. Hogan, *Transactions on Automatic Control* **29**, 681 (1984).
2. F. E. Zajac, *Critical Reviews in Biomedical Engineering* **17**, 359 (1989).
3. S. Annunziata and A. Schneider, *Applied Bionics and Biomechanics* **9**, 249 (2012).
4. W. M. Murray, T. S. Buchanan and S. L. Delp, *Journal of Biomechanics* **33**, 943 (2000).
5. D. Bennett, J. Hollerbach, Y. Xu and I. Hunter, *Experimental Brain Research* **88**, 433 (1992).
6. J. Paskarbeit, S. Annunziata and A. Schneider, A resilient robotic actuator based on an integrated sensorized elastomer coupling., in *Proceedings of the 16th International Conference on Climbing and Walking Robots*, (Sydney, Australia, Clawar 2013).
7. H. Vallery, J. Veneman, E. van Asseldonk, R. Ekkelenkamp, M. Buss and H. van der Kooij, *Robotics Automation Magazine, IEEE* **15**, 60 (2008).
8. T. E. Milner, *Experimental Brain Research* **93**, 177 (1993).

A RESILIENT ROBOTIC ACTUATOR BASED ON AN INTEGRATED SENSORIZED ELASTOMER COUPLING

J. PASKARBEIT* and S. ANNUNZIATA

Research Group Biomechatronics, University of Bielefeld,
Bielefeld, Germany
**E-mail: jan.paskarbeit@uni-bielefeld.de*
www.elan.uni-bielefeld.de

A. SCHNEIDER

Faculty of Engineering and Mathematics, University of Applied Sciences,
Bielefeld, Germany

Whereas in classical robotics the goal is to construct stiff actuators in order to achieve a high precision, current research also focusses on compliant drives. These can be used in bioinspired robotic systems as well as in classical robotic scenarios where the focus lies on safety – for example in human-machine interactions. Due to the elasticity of the compliant drives, the forces during impacts can be reduced compared to stiff systems. Usually, the compliance is achieved by pure control or by integration of steel springs. In this work, an approach is presented that utilizes a sensorized elastomer coupling. The advantages over the mentioned systems is the inherent damping due to the elastomer properties and the possibility to build smaller and therefore lighter systems as compared to steel spring based approaches. A disadvantage is the need for a more complex model in order to derive the acting torque from torsion measurements. Based on experimental data, a lumped physical model has been developed that is able to estimate the torques with a high accuracy, yet simple enough to be implemented on a microcontroller.

Keywords: Bioinspired robotics; compliant actuators; system modeling; serial elastic actuator.

1. Introduction/Drive-Layout

For bioinspired robotics – especially for walking robots – a crucial requirement is the application of actuators with a high torque/weight ratio. Although still defeated by the outstanding overload capabilities of real muscles,[2] modern actuators are already in the same magnitude regarding their nominal power as compared to muscles. The rotatory drive presented in this work (see Fig. 1 (a), (d)), has been developed for the application in the six-legged walking robot HECTOR.[5–7] The weight of the robot is mainly governed by the combined weight of the actuators. Due to the sprawled leg posture, the torques the actuators must be able to exert in order to lift the robot are approximately proportional to the weight of the robot. Therefore, the main focus during the actuator development was to achieve a high torque/weight ratio. However, also for the application in robotic arms, a high torque/weight ratio is important since the actuators at the base have to hold the weight of the subsequent elements which is dominated by the weight of the integrated actuators. Also regarding safety during human-machine interaction, a low moved mass reduces the danger of injuries in case of impacts.

Another premise during the development was the integrability of the control and drive electronics. This reduces the computational load for the main control unit and considerably simplifies the wiring. To integrate the electronics into the mechanical drive frame, its power dissipation must be sufficiently low. Therefore, up-to-date power MOSFETS with a very low on-resistance have been chosen that enable a continous operation without the need for active cooling (see Fig. 1 (b)).

The drive is actuated by a BLDC motor in combination with a lightweight harmonic drive gearbox. The motor commutation is accomplished using an absolute angle encoder that is placed directly on the control electronics. In order to simplify the application of the drives e.g. in closed kinematic chains, a serial elastic element has been introduced into the drive. The classical way to achieve compliance is to integrate a torque sensor between the output of the gearbox and the output of the drive and to mimic elasticity by control.[4] An advantage of this approach is the ability to change the elasticity online. The disadvantage of this virtual elasticity is the need for controllers that are fast enough to cope with high torque peaks. For example during collisions of a connected lever arm with the environment, high torque peaks may occur that could be harmful for the gearbox. Especially the inbuilt harmonic drive type gearbox is sensitive to torque peaks. These disadvantages can be eluded by the integration of an inherently elastic element that is able to damp high torque peaks. Just as the torque sensor, the elastic element is placed between the gearbox and the output of the drive. By measuring the torsion of the elastic element, the acting torques can be estimated using a model of the elastic element. Using this torque estimation, the elasticity of the drive can be varied and even suppressed by control. In contrast to the purely controlled elasticity, the inherent elasticity is also maintained in cases of outage, control errors etc.

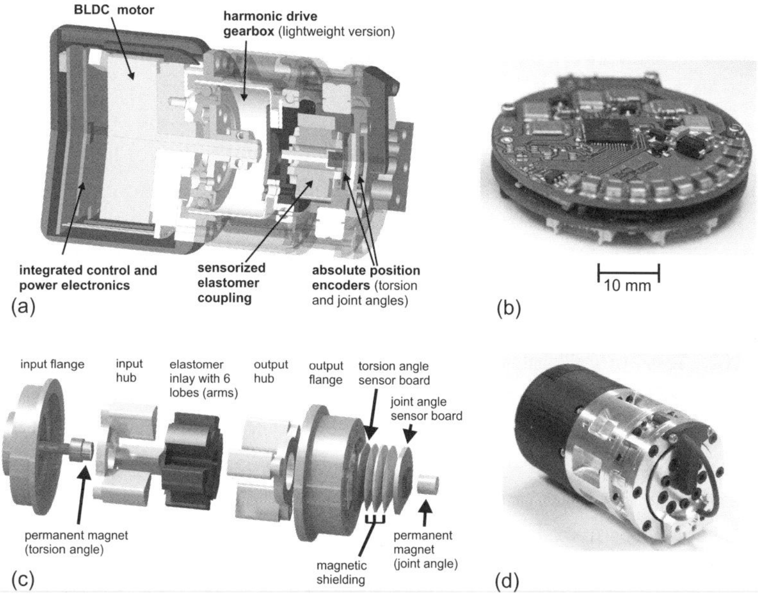

Fig. 1. (a) Section view of the compliant rotatory drive for actuation of robotic joints. The self-contained drive includes power and control electronics, a BLDC motor, a lightweight Harmonic Drive gearbox, a novel, sensorized elastomer coupling and integrated position encoders. Technical data: length: ~90 mm, diameter: ~50 mm, max torque: ~15 Nm, weight: ~0.39 kg, power/weight: ~170 W/kg. (b) Image of the power- and control-electronics stack. (c) Exploded view of the elastomer coupling. (d) Image of the assembled actuator.

2. Integrable, Sensorized Elastomer Coupling

To maintain a lightweight and small-sized construction a new sensorized elastomer coupling based on the principle of jaw couplings was designed and integrated into the drive (see Fig. 1 (c) for an explosion view). A prototype of the coupling is depicted in Fig. 2 (a). The coupling consists of an star-shaped elastomer inlay that has six lobes and two metal hubs of which each has three teeth. The teeth of these hubs alternatingly gear into the spaces between the elastic lobes of the inlay. A torque that is exerted between the two hubs results in a compression of three elastic lobes. An opposed torque causes a compression of the other three lobes. The maximum torsion and therefore the maximum torque is restricted by the strain the used material is able to withstand without plastic deformation. For the first prototype,

a rubber with a shore hardness of 94 A was used (SPP1838/81; Kraiburg, Waldkraiburg, Germany). This material is able to withstand strains up to 60 %. To reduce the load for the compressed elastomer lobes, the lobes have been bonded to the metal teeth. In this way, when loaded, half of the lobes are stretched and the other half is compressed. This can reduce the strain in the compressed lobes, thus increasing the maximum applicable torque. In addition, the geometry of the inlay and the metal hubs can be optimized in order to show a certain torsion-strain behaviour. This can further reduce the strain in the elastomer lobes. In Fig. 2 (c), the torsion-strain-dependency for the compressed and the stretched elastomer lobes is depicted. The underlying data has been obtained via a finite element simulation (MSC Software Corporation, Santa Ana, USA). According to this data, the coupling can be twisted up to 9.5° before overstrain is reached.
To measure the torsion angle of the elastomer coupling and therefore to derive the acting load, an angular encoder has been integrated into the coupling as shown in Fig. 1 (c). For this encoder, a Hall-effect based IC that measures the angular position of a moving diametrically magnetised permanent magnet was chosen due to its small size and the high angular resolution of 12 bit. The magnet must be mechanically connected to one side of the coupling (in this case the input hub) and the sensor IC must be connected to the other side (the output hub). To guarantee a precise and absolute joint angle measurement (output position) another angular encoder is placed back-to-back to the first sensor. The magnet associated to the joint angle measurement is connected to the housing of the drive [not shown in Fig. 1 (c)]. Since the magnetic field of the torsion-measurement-magnet also influences the position-measurement-sensor and vice versa, a magnetic shielding, consisting of high-permeability sheets, was placed between the sensors.
The main advantages of this design in comparison to steel spring based approaches[8] is the high integrability since the need for support structures is minimal. Furthermore, steel springs have a very low attenuation, whereas elastomer-based couplings reduce potentially harmful undamped resilience in the output due to their comparably high damping characteristics. To overcome this drawback, a tunable friction has been proposed in combination with steel springs.[3] This approach, however, is hard to miniaturize and shows a lower power/weight ratio.

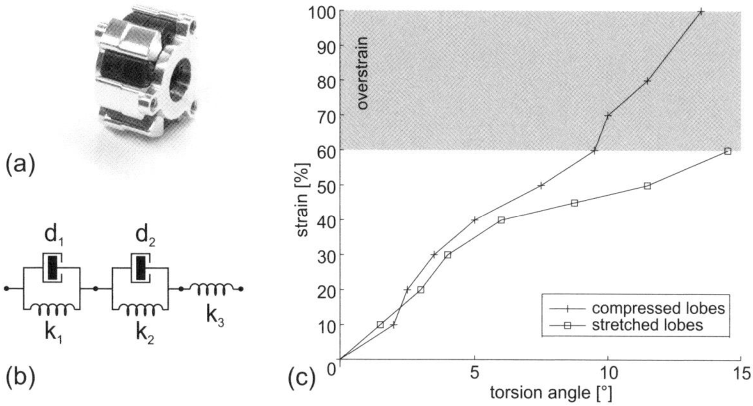

Fig. 2. (a) Image of the elastomer coupling prototype. (b) Lumped equivalent network used for torque estimation. (c) Torsion-strain dependency for compressed and stretched elastomer lobes.

3. Torque Estimation

An advantage of steel spring based serial elastic elements over elastomer based elements is the easy computability of the acting torque. A simple linear spring model is sufficient for the estimation. Due to the desirable damping characteristic of the elastomer, a higher order model must be used. For this purpose, different lumped models have been tested. Although quite simple, good results have been obtained for a model built up of a serial spring and two parallel spring-damper-subsystems in series (see Fig. 2 (b)). When provided with the measured elastomer torsion, the model estimates the torque acting on the elastomer inlay with a best fit of 84 %. This fit has been obtained with the following parameters:

$$\begin{aligned} d_1 &= 2.172 \frac{N\,m\,s}{deg} \\ d_2 &= 13.46 \frac{N\,m\,s}{deg} \\ k_1 &= 3.803 \frac{N\,m}{deg} \\ k_2 &= 0.564 \frac{N\,m}{deg} \\ k_3 &= 1.737 \frac{N\,m}{deg} \end{aligned} \tag{1}$$

The corresponding continuous time transfer function of this second order system is:

$$\frac{T(s)}{\Theta_s(s)} = \frac{a_1\, s^2 + b_1\, s + c_1}{a_2\, s^2 + b_2\, s + c_2} \tag{2}$$

with the parameters

$$\begin{aligned} a_1 &= d_1\, d_2\, k_3 \\ b_1 &= d_2\, k_1 + d_1\, k_2 \\ c_1 &= k_1\, k_2\, k_3 \\ a_2 &= d_1\, d_2 \\ b_2 &= d_2\, k_1 + d_1\, k_2 + d_1\, k_3 + d_2\, k_3 \\ c_2 &= k_1\, k_2 + k_1\, k_3 + k_2\, k_3 \quad . \end{aligned} \tag{3}$$

As a consequence of its simplicity, the corresponding discrete time transfer function can be implemented on the embedded microcontroller and therefore be used in torque control applications.[1]

Experimental results are depicted in Fig. 3 (a) and (b). They show the hysteresis curves of the estimated torque plotted over the torsion for high (a) and low (b) frequency load changes. In the insets, a section of the measured and estimated load data is plotted over time. As can be seen, the model is able to reproduce the elastic and the damping properties of the elastomer coupling with good results for different frequencies.

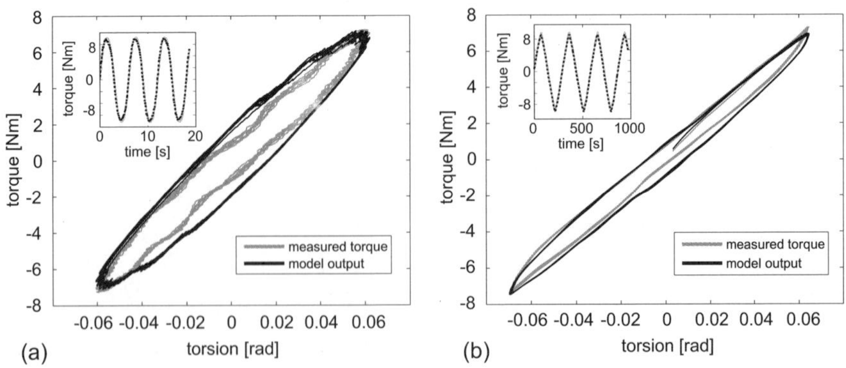

Fig. 3. (a) Hysteresis curve for high frequency load changes. (b) Hysteresis curve for low frequency load changes.

4. Discussion

In this paper an actuator for robotics applications has been presented that features a novel sensorized elastomer coupling. Compared to previous implementations of serial elastic elements using steel springs, the elastomer based coupling has inherent damping properties that increase the safety e.g. during human-machine interaction. Since the need for support structures is considerably reduced, also the mass and the size of the actuator could be reduced. Due to the reduced inertia, the inherent elasticity and the damping, impacts are less dangerous – both for the robot (and especially the gearboxes of the integrated drives) as well as for contact objects.
Using an angular encoder, the torsion of the coupling can be sensed. Based on a lumped equivalent network of springs and dampers, a second order model of the elastomer coupling has been developed that is able to reproduce the behaviour of the coupling for load changes of different frequencies. In combination with the angular torsion sensor, the model is able to predict the acting load. The simplicity of the model and the corresponding transfer function also allow for an implementation on the microcontroller embedded in the actuator, thus allowing for decentral torque control.

5. Acknowledgments

This work has been supported by the Federal Ministry of Education and Research (BMBF) within the BIONA programme (ELAN-project to A.S.), within the Excellence Initiative of the German Research Foundation (DFG Center of Excellence "Cognitive Interaction Technology", EXC277) and by an EU-FP7 grant (ICT-2009.2.1, No. 270182 to A.S.).

References

1. S. Annunziata, J. Paskarbeit, and A. Schneider. Bio-inspired elbow impedance modulation using a compliant technical joint drive. In *Proceedings of the 16th International Conference on Climbing and Walking Robots*, 2013.
2. R. K. Josephson. Contraction dynamics and power output of skeletal muscle. *Annu Rev Physiol*, 55:527–46, 1993.
3. M. Laffranchi, N. Tsagarakis, and D.G. Caldwell. A compact compliant actuator (compact) with variable physical damping. In *Robotics and Automation (ICRA), IEEE International Conference on*, pages 4644–4650, 9-13 May 2011 2011.
4. M. Mason. Compliance and force control for computer-controlled manipulators. *IEEE Trans on Systems, Man, and Cybernetics*, 11(6):418–432, 1981.
5. J. Paskarbeit, J. Schmitz, M. Schilling, and A. Schneider. Design of an insectoid robot as a versatile carrier for bioinspired sensors. In *Proceedings of the*

13th International Conference on Climbing and Walking Robots, pages 173–180. World Scientific, 2010.
6. A. Schneider, J. Paskarbeit, M. Schaeffersmann, and J. Schmitz. Biomechatronics for embodied intelligence of an insectoid robot. In *ICIRA (2)*, volume 7102 of *Lecture Notes in Computer Science*, pages 1–11. Springer, 2011.
7. A. Schneider, J. Paskarbeit, M. Schaeffersmann, and J. Schmitz. Hector, a new hexapod robot platform with increased mobility - control approach, design and communication. In *Advances in Autonomous Mini Robots*, pages 249–264. Springer Berlin Heidelberg, 2012.
8. N.G. Tsagarakis, Matteo Laffranchi, Bram Vanderborght, and D.G. Caldwell. A compact soft actuator unit for small scale human friendly robots. In *Robotics and Automation, 2009. ICRA '09. IEEE International Conference on*, pages 4356–4362, May 2009.

DERIVATION OF MATHEMATICAL MODELS OF THE PERISTALTIC CRAWLING ROBOT FOR MAINTENANCE OF A MIXING TANK

YOSUKE MORISHITA, DAISUKE SANNOHE, TATSUYA OSAWA, TOMOYA TANAKA AND TARO NAKAMURA

Faculty of Science and Engineering, Department of Precision Mechanics
Chuo University 1-13-27 Kasuga, Bunkyou-ku, Tokyo 112-8551, Japan

Mixing tank is a device which is used in a variety of chemicals to stir. When using this apparatus, the controlling temperature in the tank is important. So flow a heat medium or a refrigerant to the internal of the jacket to control it. Although a large amount of deposits are formed in the inner wall of jacket, there is no way to clean up or to inspect inside of the jacket because it's too narrow space and entrance. Therefore, we developed peristaltic crawling robot for vertical narrow space, and realized the locomotion in 50[mm] wide space. In addition to that, we simulate in this robot and do experiment to compare the results of simulation. Also, to achieve driving on curved narrow space in mock-up of the mixing tank.

1. Introduction

Mixing tanks have become useful in various chemical processing such as the production of electronic materials, remedies, and food. Traditionally, mixing tanks have a dual structure comprising an inside container (called a shell) and an outside container (called a jacket), as shown in Fig.1. In this structure, the interior temperatures of the mixing tanks are adjusted by injecting a heat medium or refrigerant into the interspace of two containers. However, a large amount of excrescences are formed on the inner wall of the jacket. Currently, there is no way to maintain the inside jacket of this device because the intake of the medium and the width of the shell and jacket are designed to be narrow (of the order of 40–60 mm), and it is not easy to clean them by hand. In addition, there have been few studies involving robots that can move in narrow spaces.

A previous study of a robot that can move between two planes, was conducted by Ariga, et al [1], who reported a wheel robot with pantograph-type structure that can move between the interspace of two walls with a spacing of above 270 mm, but no robot that can adapts to such a narrow space has been developed till.

Therefore, we have been developing a peristaltic crawling robot [2]-[3] that realizes its movement by using pneumatic artificial rubber muscles for the purpose of performing inspection the inner pipe. The advantage of using peristaltic crawling technique by pneumatic artificial muscles is as follows:

1) It adapts to width changes of narrow spaces caused by excrescences.

2) It can become slim when passing the intake and has enough thick grasp the wall when driving.

We developed the robot and experimented on it to confirm the basic characteristics. In this paper, we report to build the model of the robot to do the automatic route selection, and do experiment to compare the target tracks.

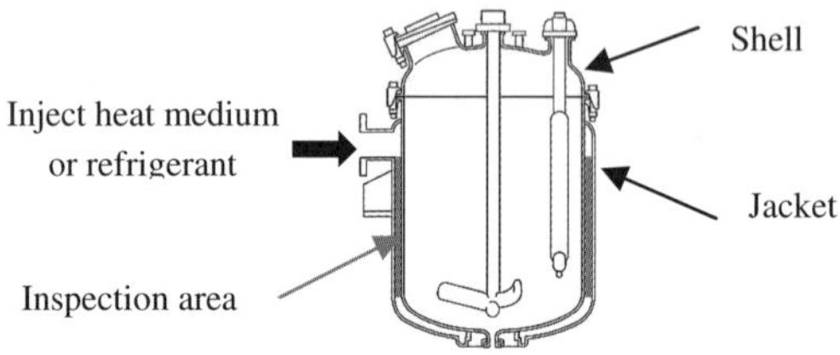

Fig.1 Cross-sectional view of a mixing tank

2. X-Shaped Peristaltic Crawling Robot

Fig.2 shows the appearance of the developed robot. This robot consists of two parts: an artificial muscle part and an X-shaped joint part. The Artificial muscle part is made of pneumatic artificial rubber muscle and is used to perform peristaltic crawling for realizing movement. The X-shaped joint part consists of two components and one servomotor. It is used to change the direction. The motion pattern of the robot is shown in Fig.3.

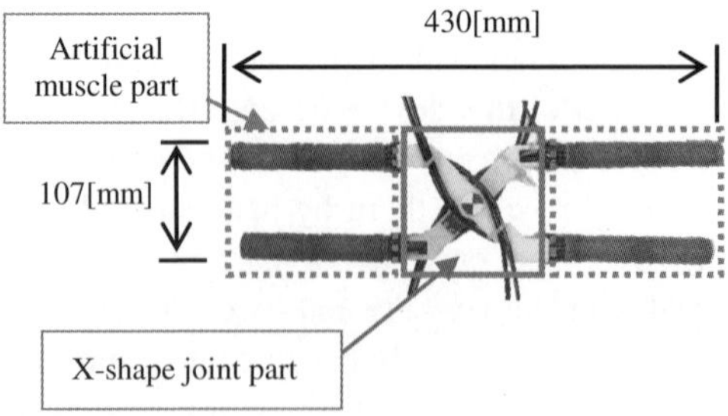

Fig.2 Appearance of the X-shaped peristaltic crawling robot

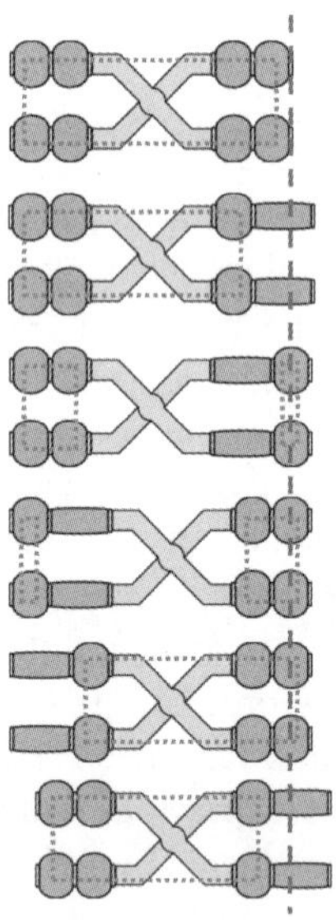

Fig.3 Motion pattern

3. Modeling of the Robot and Simulation

The developed robot is modeled to evaluate its position and attitude, and a simulation analyses its movement over given routes in 2-dimensional space.

3.1. *Setup coordinates*

The robot and its setup coordinates are shown in Fig.4. The robot can be divided into the artificial muscle part and the X-shaped joint part modeled as a slide joint and a rotational joint, respectively. The rotational angle, anterior segment, and posterior segment are paid particular attention the coordinates of each unit are set as Σ_0–$\Sigma_{(A,B)(R,L)i}$ shown in Fig.5. l_i is a distance between each point, θ_a is the rotational angle at Σ_0, and θ_b is the angle between the two parts of rotating parts. Further, Σ_0 is the world coordinate system.

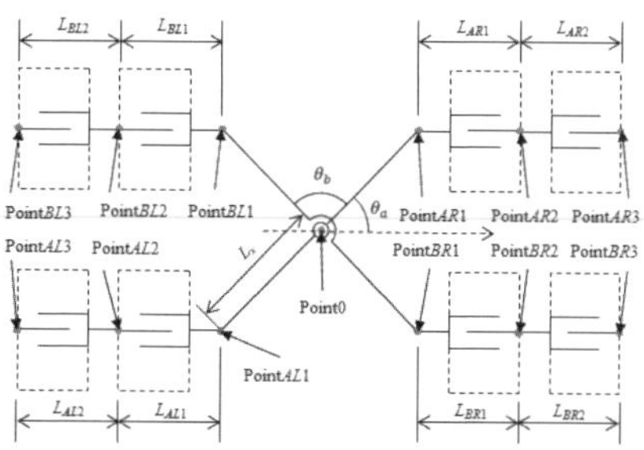

Fig.4 Encoding of the robot

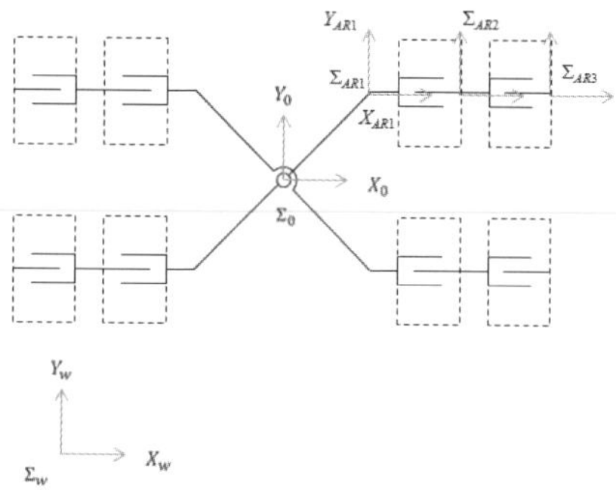

Fig.5 The setup of coordinates of the robot

3.2. *Homogeneous transformation matrix of the robot*

The attitude of the robot observed from the basic unit in each motion is derived. It is transformed to world coordinate systems to derive a homogeneous transformation matrix of the robot. Then, the transformation matrixes of the robot are shown in (1) (2) when $i = 1$.

$$ {}^{0}T_{A(R,L)1} = \left(\begin{pmatrix} c\theta_a & -s\theta_a & L_x \\ s\theta_a & c\theta_a & 0 \\ 0 & 0 & 1 \end{pmatrix}, \begin{pmatrix} -c\theta_a & s\theta_a & L_x \\ -s\theta_a & -c\theta_a & 0 \\ 0 & 0 & 1 \end{pmatrix} \right) \tag{1} $$

$$ {}^{0}T_{B(R,L)1} = \left(\begin{pmatrix} c\theta_{ab} & -s\theta_{ab} & L_x \\ s\theta_{ab} & c\theta_{ab} & 0 \\ 0 & 0 & 1 \end{pmatrix}, \begin{pmatrix} -c\theta_{ab} & s\theta_{ab} & L_x \\ -s\theta_{ab} & -c\theta_{ab} & 0 \\ 0 & 0 & 1 \end{pmatrix} \right) \tag{2} $$

Here, $\theta_{ab} = \theta_a + \theta_b$. Where the other point are obtained from the transformation matrix shown in (3) (4),

$$^{(A,B)(R,L)1}T_{(A,B)(R,L)i+1} = \Pi_{k=1}^{i} \, ^{(A,B)(R,L)k}T_{(A,B)(R,L)k+1} \tag{3}$$

$$^{(A,B)(R,L)i}T_{(A,B)(R,L)i+1} = \begin{pmatrix} 1 & 0 & \operatorname{sgn}\left(x_{(R,L)}\right)L_{(A,B)(R,L)} \\ 0 & 1 & 0 \\ 0 & 0 & 1 \end{pmatrix} \tag{4}$$

$x_{(R,L)}$ is the x coordinate of point$(A,B)(R,L)1$. In addition, as shown in Fig.5, the transformation matrix that gives the relationship between the world coordinates Σ_w and Σ_0 is shown in (5). Coordinates of it is fixed in the world coordinate system as shown in Fig.5. Here, ϕ is the angle between the robot and the world coordinate.

$$^{w}T_0 = \begin{pmatrix} \cos\varphi & -\sin\varphi & x_j \\ \sin\varphi & \cos\varphi & y_j \\ 0 & 0 & 1 \end{pmatrix} \tag{5}$$

4. Simulation and Experimental Results about the Running Experiment between the Planes and Discussion

We set a route and had the robot follow it. Fig.6 shows the analysis results of the track. I consider about it, there is no problem of robot move in the vertical direction. However, at the time of the move in the horizontal direction, it's moving down gradually while. Because when this robot moves in a horizontal position, there is a moment to grasp only the first unit and the rear unit. While the robot is deflected by gravity at that time, the center of the robot drops a little, transition to the next operation in that state. As a result, the position of the robot is going down as more and more shown by the waveform. In addition, the moving distance of the robot in the horizontal direction have not reached the goal, it is because the length elongation of the traveling direction are absorbed by the deflection.

5. Experimental Results about the Running Experiment between the Curved Surfaces and Discussion

We improved the robot to able to drive between the curved surfaces. The improved robot is shown in Fig.7. The body of the robot must be bent to drive between curved surfaces. So we attached the parts as shown in Fig.8. This is not bent in the vertical direction, but it is possible to bend in the horizontal direction along the wall. Fig.9 to 11 shows the experimental apparatus, the state of the experiment and the analysis results of track.

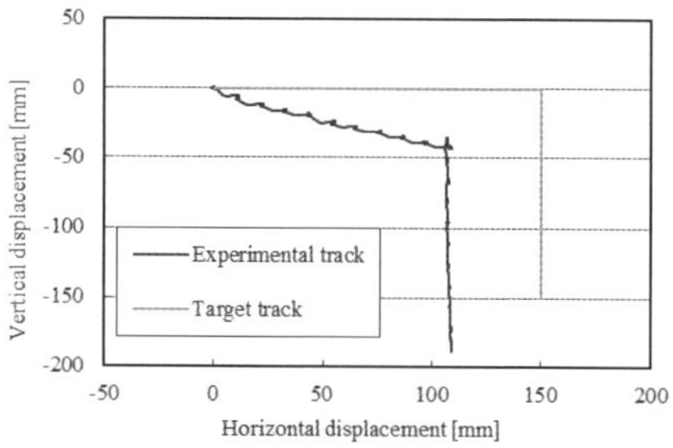

Fig.6 Experimental result

Looking at the Fig.11, is progressing while falling downward than when driving between the planes. This is because, the length of the robot is increased by the parts, and greater torque is applied. As a result, the body of the robot is bent and the position of the robot is going down.

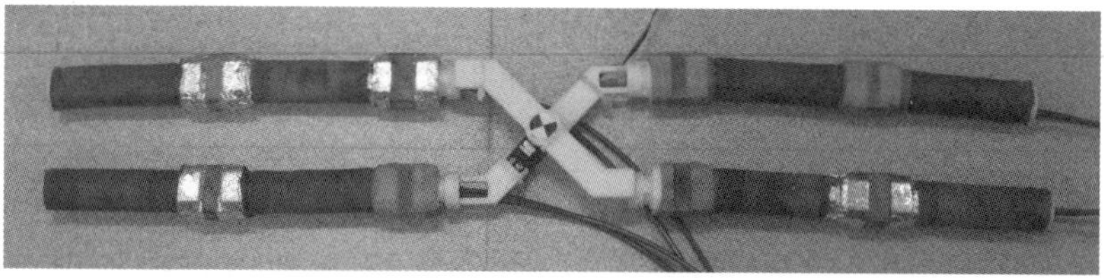

Fig.7 Appearance of the improved robot

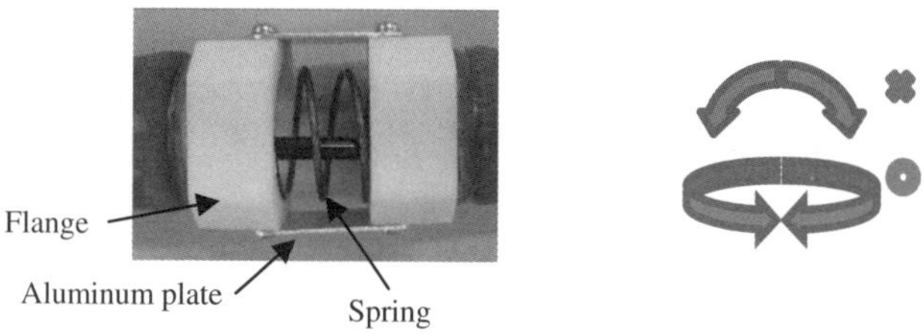

Fig.8 The structure of bending only in a certain direction

Fig.9 The mock-up of the mixing tank

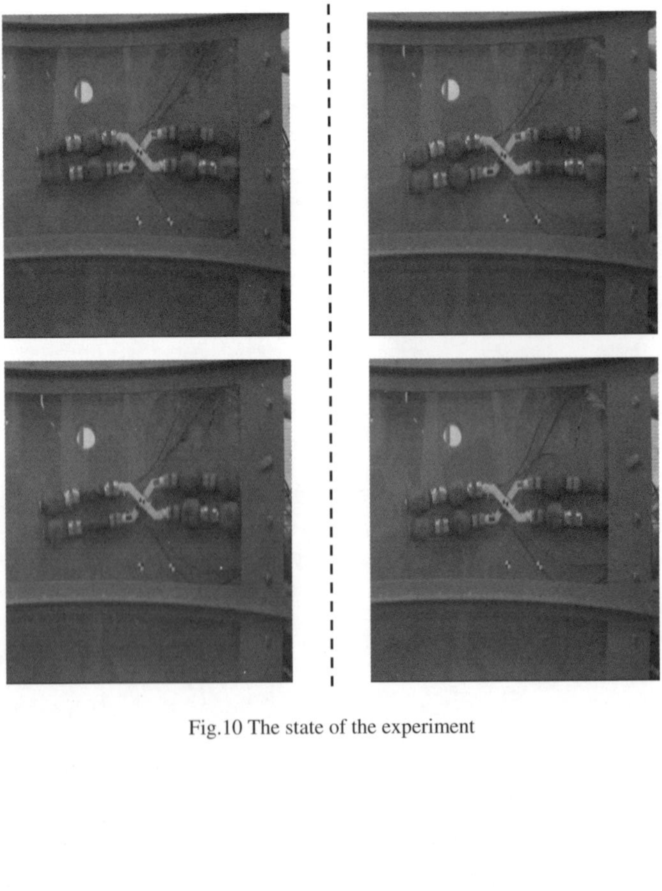

Fig.10 The state of the experiment

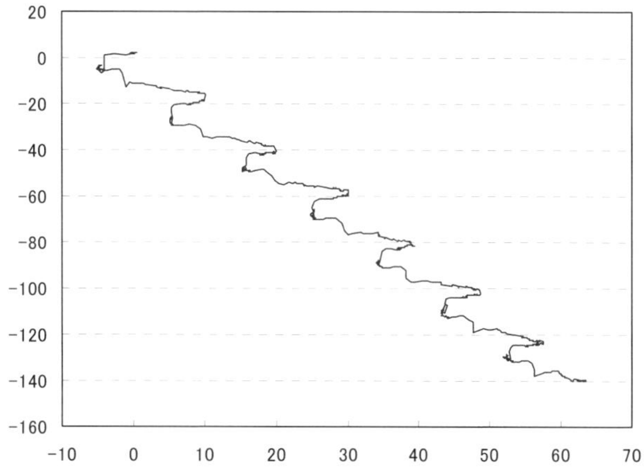

Fig.11 Experimental result

6. Conclusions and Future Work

6.1. *Conclusions*

We built a model of the peristaltic crawling robot using pneumatic artificial muscles for simulation. And we experiment on the robot between planes and curved surface. From the experimental results, we considered about the causes to go downward.

6.2. *Future Work*

In this paper, the robot would gradually drop while driving. So we should solve this problem. Then I thought in order to overcome this difficulty, we attempt to attach the parts as shown in Fig.12. This item will be able to support to reduce the deflection of the robot. This is not bent in the vertical direction, but it is possible to bend in the horizontal direction along the wall like the parts of Fig.8. By passing it through the hole in the center of the robot, we can solve the problem.

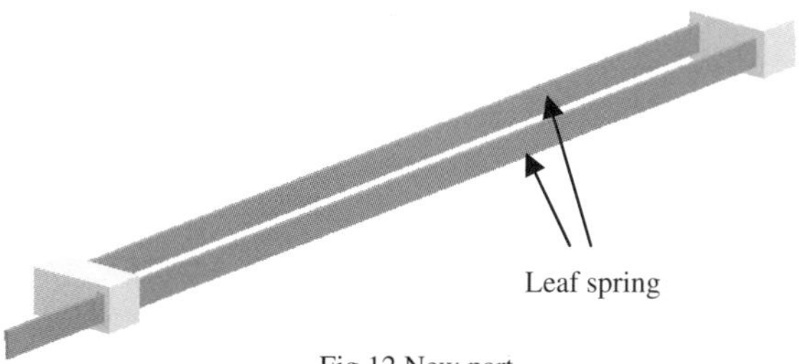

Fig.12 New part

References

1. A. Ariga, T. Kobayashi, T. Yamaguchi, and S. Hashimoto, "Wall Climbing Robot in Narrow Space with Pantograph-type Structure," Proceedings of the 2010 IEEE International Conference on Robotics and Automation (ICRA 2008), pp.3028-3033, 2008.
2. T. Nakamura and T. Iwanaga, "Locomotion Strategy for a Peristaltic Crawling Robot in a 2-Dimensional Space," Proc. IEEE International Conference on Robotics and Automotion (ICRA 2008), 2008.
3. K. Adachi, M. Yokojima, Y. Hidaka, and T. Nakamura, "Development of multistage type endoscopic robot based on peristaltic crawling for inspecting the small intestine," Proceedings of IEEE Advanced Intelligent Mechatronics (AIM 2011), pp.904-909, 2011.

IMPROVEMENT OF ENERGY CONSUMPTION BY MOVEMENT OF CENTER OF ROTATION OF JOINT

KAZUTOSHI TANAKA*, SATOSHI NISHIKAWA*,** and YASUO KUNIYOSHI*

*Graduate School of Information Science and Technology,
The University of Tokyo, Tokyo, 113-8656, Japan
**Research fellow of the Japan Society for the Promotion of Science
E-mail: {tanaka, nisikawa, kuniyosh} @isi.imi.i.u-tokyo.ac.jp,
www.isi.imi.i.u-tokyo.ac.jp

Robots mimicking the structure of the musculoskeletal systems have been developed in order to realize high physical performance or to reveal the contribution of the musculoskeletal systems to body movements. The movement of center of rotation (CoR) of biological joints have been focused on in various research fields. However, to date, the isolated contribution of the joint displaceable center of rotation (JDCR) on physical performance has not yet been analyzed. Revealing this contribution is expected to provide guidelines for the design methodology of robots. In this paper, we compared two models in order to investigate the effect of JDCR on physical performance. The first model has two virtual links to represent the joint including the JDCR (JDCR model). The second model represents the joint with fixed CoR (JFCR model). We evaluated the two models by analyzing energy consumption during reaching motions. In addition, we examined energy consumption while the end effector traced a circular trajectory using extended models (eJDCR model and eJFCR model). The results showed that the energy consumption of models which include JDCR lower than models which does not under all tested conditions. This study suggested that a joint including JDCR can improve energy consumption.

Keywords: Bio-inspired robot, Movement of center of rotation of joint

1. Introduction

Musculoskeletal structure of the animal body contributes to the improvement of physical performance in the dynamic movements such as walking and climbing. In recent years, numerous musculoskeletal robots have been developed to realize robots which exhibit outstanding physical performance or to better understand the contribution of the animal musculoskeletal systems toward adaptive motions [1, 2]. Most joints in these robots have been hinge joints or ball joints, so that their center of rotation (CoR) is fixed.

However, CoR of biological joints moves by rolling and sliding. Thus, joints with displaceable center of rotation (JDCR) have been widely studied in fields such as biomechanics [3, 4], prosthesis [5] and exoskeletal robots [6]. In these studies, the contribution of joint movements, including the JDCR, to physical performance have been suggested. However, it is very difficult to isolate the effect of the JDCR because many factors interact with each other in the animal's musculoskeletal structure. Revealing the isolated contribution of the JDCR provides important clues for the design of future robots so they can for realize the better physical performance.

In recent years, bio-inspired joints including the JDCR have been developed. Hosoda *et al.* developed anthropomorphic open joints which is stronger against impact than ball bearings [7]. Etoundi *et al.* proposed a bio-inspired joint mimicking the curve profiles of bones in the human knee joint, and indicated that the proposed joint has superior performance to a hinge joint in terms of its stiffness and compactness [8]. Hamon and Aoustin proposed a joint model representing the complexity in the human knee joint, and indicated that the model has the advantage of energy consumption of walking [9]. These studies indicated advantages of bio-inspired joints including the JDCR. However, the isolated contribution of JDCR has not yet been revealed. Therefore, we investigated its isolated contribution by modeling joint movements with JDCR.

2. Numerical analysis using two joint models

In this section, we examine physical performance of two planar models with numerical analysis in order to evaluate the isolated contribution of JDCR. One of the models is modeled as a joint with JDCR and another as a joint without it. In order to investigate how the two models affect the physical performance, we measured their energy consumption ratios in a reaching motion not only as a simple but as an essential movement. We use energy consumption to measure physical performance because energy consumption can be defined as the efficiency of a movement independent of the type of actuator used and type of driven system.

2.1. *Model of joints including JDCR*

Two models to evaluate contribution of JDCR and their body parameters are shown in Fig. 1. The first model representing the JDCR, called JDCR model, has three degrees of freedom. The second model representing fixed CoR in joints, called JFCR model, has one degree of freedom. Link 1 and

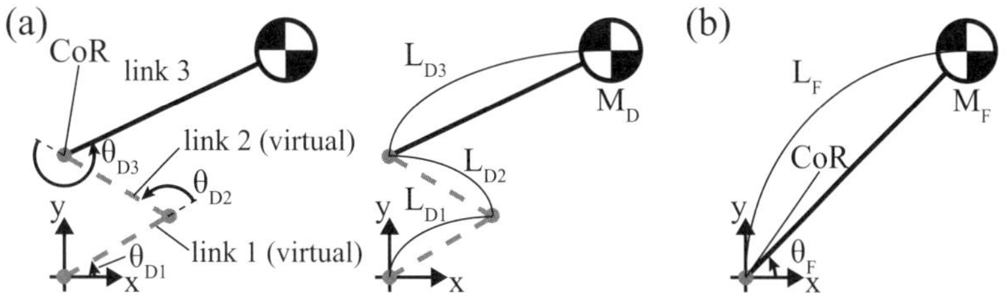

Fig. 1. (a) JDCR model. (b) JFCR model.

link 2 in the JDCR model are massless virtual links. The planar movement of the joint between link 2 and link 3 corresponds with the movement of CoR. The position of the CoR is given by

$$\mathrm{x_{CoR}} = \begin{bmatrix} L_{\mathrm{D1}} cos(\theta_{\mathrm{D1}}) + L_{\mathrm{D2}} cos(\theta_{\mathrm{D1}} + \theta_{\mathrm{D2}}) \\ L_{\mathrm{D1}} sin(\theta_{\mathrm{D1}}) + L_{\mathrm{D2}} sin(\theta_{\mathrm{D1}} + \theta_{\mathrm{D2}}) \end{bmatrix}. \tag{1}$$

Eq. 1 shows that the CoR can move freely in an area without constraint when $L_{\mathrm{D1}} = L_{\mathrm{D2}}$ so that this model can simulate most movements of the CoR. Therefore, we set $L_{\mathrm{D1}} = L_{\mathrm{D2}}$ in the simulations.

2.2. *Setup*

These models were evaluated under the following conditions:

(1) The positions, velocity and acceleration of their end effectors are the same at both initial and final states of the reaching motion.
(2) Body parameters of links are the same, that is, $M_{\mathrm{D}} = M_{\mathrm{F}}$, $L_{\mathrm{D3}} = L_{\mathrm{F}}$.
(3) Reaching times T are the same.

JDCR model was considered advantageous when energy consumption of the JDCR model was less than the JFCR model.

In order to investigate how the two models affect physical performance in a reaching task, we calculated their energy consumption using MATLAB (MathsWorks, Inc.). The simulation parameters are shown in Table 1. In both initial and final state, we set that angular velocity and angular acceleration of each joint in them equals zero. In the initial state, we set $x_{\mathrm{COR}}(0) = [0\ 0]^T$ and the position of the end effector $x_{\mathrm{E}}(0) = [L_{\mathrm{F}}\ 0]^T$. Therefore, the parameters were, $\theta_{\mathrm{D2}}(0) = \pi$ [rad], $\theta_{\mathrm{D3}}(0) = \pi - \theta_{\mathrm{D1}}(0)$ [rad] and $\theta_{\mathrm{F}}(0) = 0$ [rad] respectively in the initial state. In the final state,

Table 1. Simulation parameters.

L_{D1} [m]	L_{D2} [m]	L_{D3} [m]	L_F [m]	M_D [kg]	M_F [kg]	T [s]
0.3	0.3	1.0	1.0	1.0	1.0	1.0

$\theta_F(T)$ was varied from $\frac{1}{10}\pi$ to π by $\frac{1}{10}\pi$ [rad] as the goal position of the end effectors, and $\theta_{D1}(t)$, $\theta_{D2}(t)$ and $\theta_{D3}(t)$ were calculated so that the positions of the end effectors for both models were fixed. We selected the $\theta_{D1}(t)$ values in initial and final states which yielded the lowest energy consumption. Each $\theta(t)$ in eJDCR model is calculated by assumption as,

$$\theta(t) = \sum_i^5 a_i t^i, \tag{2}$$

where a_i $(i = 0, ..., 5)$ is a constant. Each joint torque is obtained from the trajectories with inverse dynamics. Energy consumption E is calculated as,

$$E = \sum_i \int \tau_i \dot{\theta}_i dt, \tag{3}$$

where τ_i and $\dot{\theta}_i$, denote the torque and angular velocity of the i-th (i = 1, ..., 3 or 1) joint, respectively. In addition, The energy consumption ratios r_E are defined as,

$$r_E = \frac{E_D}{E_F}, \tag{4}$$

where E_D and E_F are energy consumption of JDCR model and JFCR model, respectively. We interpreted that the JDCR model is superior to JFCR model in terms of energy consumption when $r_E < 1$.

2.3. *Result*

The results of the analysis indicate the following three points. First, the JDCR model is superior to the JFCR model under all tested conditions in terms of energy consumption (Fig. 2 (a)). Second, r_E values decrease as $\theta_F(T)$ values increase. This means that the JDCR model becomes increasingly advantages as the distance to the goal of reaching motion increases. Third, $\theta_{D1}(t)$ and $\theta_{D3}(t)$ show little changes during reaching, and $\theta_{D1}(0)$ and $\theta_F(T)$ are in the following linear relationship:

$$\theta_{D1}(0) = \frac{1}{2}\theta_F(\mathrm{T}) + \pi. \tag{5}$$

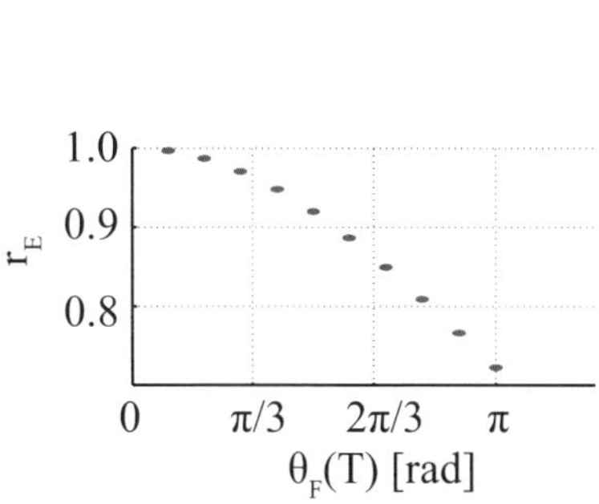

(a)

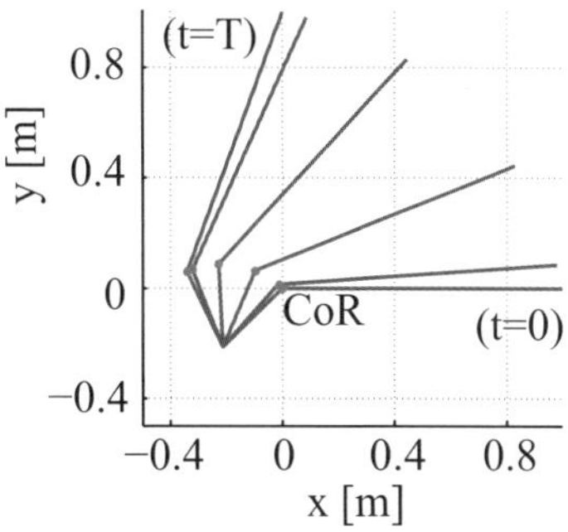

(b)

Fig. 2. (a) Energy consumption ratios of JDCR model to JFCR model. (b) Images from a reaching of JDCR model taken at 0.2-s intervals.

As a result, the CoR in the JDCR model takes on a circular arc trajectory (Fig. 2 (b)). We consider that these points are caused by distance displaced by their masses.

3. Numerical analysis using two extended joint models

In this section, we examine the physical performance of two extended joint models in order to evaluate movements of a body which includes the JDCR model. In order to investigate how the two extended models affect physical performance, we measured their energy consumption ratios while the end effector traced a circular trajectory.

3.1. *Extended model of joints including JDCR*

Two extended models include JDCR model and JFCR model. Their body parameters are shown in Fig. 3. The first model is the extended JDCR model, called eJDCR model, which includes the JDCR model. The second model is the extended JFCR model, called eJFCR model, which includes the JFCR model. Link 2 and link 3 in the eJDCR model are massless virtual links. The planar movements of the joint between link 3 and link 4 corresponds with movements of CoR in the JDCR model. In addition, we set $L_{D2} = L_{D3}$ in the simulations.

3.2. *Setup*

These models were evaluated under the following conditions:

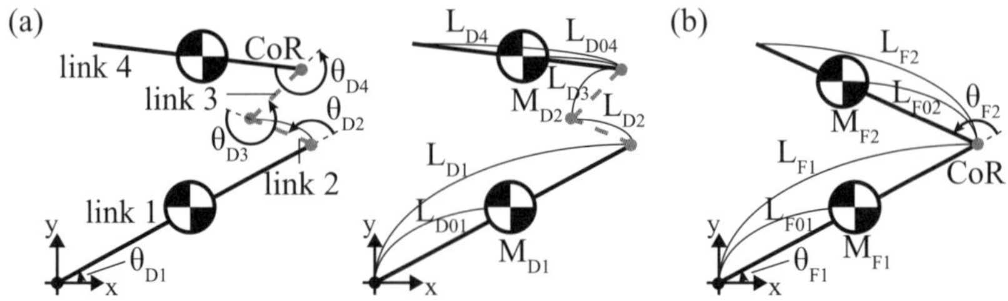

Fig. 3. (a) eJDCR model. (b) eJFCR model.

Table 2. Simulation parameters.

L_{D1} [m]	L_{D2} [m]	L_{D3} [m]	L_{D4} [m]	L_{F1} [m]
1.0	0.3	0.3	1.0	1.0
L_{F2} [m]	M_{D1} [kg]	M_{D2} [kg]	M_{F1} [kg]	M_{F2} [kg]
1.0	1.0	1.0	1.0	1.0

(1) The positions, velocity and acceleration of their end effectors are always the same in the motion.
(2) Body parameters of links are the same, that is, $M_{\mathrm{D1}} = M_{\mathrm{F1}}$, $M_{\mathrm{D2}} = M_{\mathrm{F2}}$, $L_{\mathrm{D1}} = L_{\mathrm{F1}}$, $L_{\mathrm{D4}} = L_{\mathrm{F2}}$, and we set $L_{\mathrm{D01}} = L_{\mathrm{F01}} = \frac{1}{2}L_{\mathrm{D1}}$, $L_{\mathrm{D04}} = L_{\mathrm{F02}} = \frac{1}{2}L_{\mathrm{D1}}$.
(3) In eJDCR model, joints except the joint corresponding with CoR do not move, that is, $\theta_{\mathrm{D2}} = \phi_2$, $\theta_{\mathrm{D4}} = \phi_4$, where ϕ_2 and ϕ_4 are constants.

eJDCR model was considered advantageous when energy consumption of the eJDCR model was less than the eJFCR model. The simulation parameters are shown in Table 2. We set the position of the end effector as,

$$\mathrm{x_E(t)} = \mathrm{R}\begin{bmatrix} cos\omega t \\ sin\omega t \end{bmatrix} + \begin{bmatrix} 0 \\ d \end{bmatrix}, \tag{6}$$

where R, ω and d are constants. We set tracking time $T = 2\pi\omega^{-1}$. θ_{D1} and θ_{D4} are obtained with inverse kinematics. Each joint torque is obtained from the trajectories with inverse dynamics, and energy consumption is calculated based on Eq. 3. In the simulation, other parameters were set to the following: $R = 0.15$ [m], $\omega = 1.0$ [rad·s^{-1}], $\phi_2 = \frac{3}{2}\pi$ and $\phi_4 = \frac{3}{2}\pi$. In addition, d was varied from 0.9 to 1.8 by 0.1 [m]. Movements of eJDCR are shown in Fig. 4 (a).

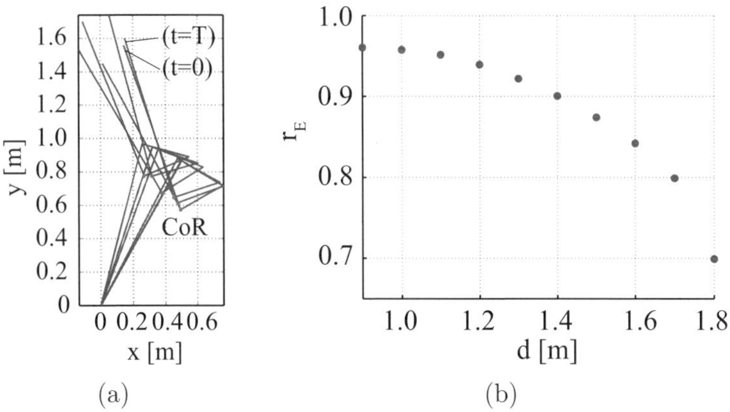

Fig. 4. (a) Images from tracking of eJDCR model taken at 0.3-s intervals ($d = 1.6$ [m]). (b) Energy consumption ratios of eJDCR model to eJFCR model.

3.3. *Results*

The results of the analysis indicate the following three points. First, the eJDCR model is superior to the eJFCR model under all tested conditions in terms of energy consumption (Fig. 4 (b)). Second, r_{E} values decrease as d values increase. This means that the JDCR model becomes increasingly advantages as the end effector moves more far from the body. Third, distance displaced by M_{D1} and M_{D1} decreases as the end effector moves more far from the body, and this distance correlates closely with energy consumption.

4. Conclusion

In this paper, we compared JDCR model with JFCR model and eJDCR model with eJFCR model in order to investigate the isolated effect of JDCR on physical performance. In numerical analysis using JDCR model and JFCR model, we compared energy consumption in a reaching task. Results of this show that JDCR model is superior to the JFCR model in terms of energy consumption, especially when reaching for a far goal. In addition, we compared energy consumption in a tracking task using eJDCR model and eJFCR model to examine physical performance of a body including JDCR in its joint. Results of this show that eJDCR model has an advantage in movements that the end effector moves far from the body. We consider that these results are caused by distance displaced by their masses. These results suggest that joints including the JDCR can improve the physical performance of robots. These joints in walking robots have advantages particularly because robots move the foot far from their body during walking.

It is suggested that joints including the JDCR can improve the physical performance of robots by results of numerical analysis in this paper. We thereby obtained guidelines for the future design of musculoskeletal robots. Our future work is to examine the design of joints including JDCR. We will also develop such joints and verify its physical performance in future research.

References

1. I. Mizuuchi, T. Yoshikai, Y. Sodeyama, Y. Nakanishi, A. Miyadera, T. Yamamoto, T. Niemela, M. Hayashi, J. Urata, Y. Namiki and others, Development of musculoskeletal humanoid kotaro, in *Proc. of IEEE International Conference on Robotics and Automation*, pp.82–87, (2006).
2. R. Niiyama and Y. Kuniyoshi, Design principle based on maximum output force profile for a musculoskeletal robot, *Industrial Robot: An International Journal*, vol.37, No. 3, pp.250–255, (2010).
3. P. G. Grant, Biomechanical significance of the instantaneous center of rotation: the human temporomandibular joint, *Journal of biomechanics*, vol. 6, No.2, pp. 109–113, (1973).
4. G. L . Smidt, Biomechanical analysis of knee flexion and extension, *Journal of Biomechanics*, vol. 6, No. 1, pp. 79–92, (1973).
5. L. E. Amigo, A. Casals and J. Amat, Design of a 3-DoF joint system with dynamic servo-adaptation in orthotic applications, in *Proc. of IEEE International Conference on Robotics and Automation*, pp.3700–3705, (2011).
6. F. H. Gunston, Polycentric knee arthroplasty prosthetic simulation of normal knee movement, *Journal of Bone & Joint Surgery*, vol. 53, No. 2, pp. 272–277, (1971).
7. K. Hosoda, Y. Sakaguchi, H. Takayama, and T. Takuma, Pneumatic-driven jumping robot with anthropomorphic muscular skeleton structure, *Autonomous Robots*, vol. 28, No. 3, pp. 307–316, (2010).
8. A. C. Etoundi, V. Ravi and S. C. Stuart, A bio-inspired condylar hinge joint for mobile robots, in *Proc. of IEEE/RSJ International Conference on Intelligent Robots and Systems*, pp.4042–4047, (2011).
9. A. Hamon and Y. Aoustin, Study of different structures of the knee joint for a planar bipedal robot, in *Proc. of IEEE-RAS International Conference on Humanoids*, pp.113–120, (2009).

A SIMPLIFIED VARIABLE ADMITTANCE CONTROLLER BASED ON A VIRTUAL AGONIST-ANTAGONIST MECHANISM FOR ROBOT JOINT CONTROL

XIAOFENG XIONG, FLORENTIN WÖRGÖTTER and PORAMATE MANOONPONG

the Bernstein Center for Computational Neuroscience, Institute of Physics III, University of Göttingen, Göttingen, 37077, Germany
E-mail: (xiong, worgott, poramate)@physik3.gwdg.de.

In this paper, we propose a simplified variable admittance controller applied to robot joint control. It is based on a virtual agonist-antagonist mechanism (VAAM) consisting of contractile and parallel elements (CEs and PEs). "Virtual" here means that every joint physically actuated by a standard servo motor can produce variably compliant motions as if it were driven by a pair of agonist and antagonist muscles. This makes it different from variable stiffness actuators (VSAs) with mechanically bulky and complex mechanisms. Moreover, the controller differs from other conventional PID admittance and variable admittance controllers since it only relies on force sensing at the end effector of robot rather than complex force/torque sensing of every joint. We have successfully implemented the controller on our hexapod robot which enables it to perform variable compliant behaviors, thereby reducing contact force and preventing leg damages when it is imposed with static or dynamical perturbation.

Keywords: Muscle model, variable admittance control, hexapod robot

1. Introduction

A friendly interaction with environments is a prerequisite of emerging robotic applications so as to achieve safety, adaptivity and compliance. This can be done by developing actuators[1] and controllers[2] with variable admittance/impedance, which have been applied to safe human-robot interactions[3] and adaptive assist devices.[4] Generally, variable admittance/impedance actuators are characterized by mechanically sophisticated structures while variable admittance/impedance controllers rely on force/torque sensing as inputs. In this paper, we propose a virtual agonist-antagonist mechanism (VAAM)[b] for a simplified variable admittance controller. This novel controller endows the system with easily changing compliance by tuning stiffness and damper coefficients $[K, D]$ of the parallel elements (PEs) of the VAAM. This makes it more suitable for joint control of small robots in comparison with mechanically bulky and complex variable stiffness actuators (VSAs).[6–8] Moreover, this novel controller depends only on force sensing at the end effector rather than complex torque/force sensing at each joint as PID admittance controllers.[9] It also does not include desired states like standard variable admittance controllers.[4]

2. A Virtual Angonist-Antagonist Mechanism (VAAM)

In Fig. 1(a), a joint, physically actuated by a standard servo motor, is driven by the VAAM (i.e., $M1$ and $M2$). $M1$ and $M2$ mimic the function of agonist and antagonist muscles when confronted with an external load F_{ext}. The joint motions are excited by F_{ext} at the shank. O is the resting position of the shank. $M1$ and $M2$ consist of parallel and contractile elements (PE and CE). Each PE is modelled by a spring-damper system (see Fig. 1(b)). The passive forces $F^P_{(1,2)}$ created by $PE_{(1,2)}$ can be represented as:

$$F^P_{(1,2)} = K_{(1,2)}(l^P_{(1,2)} - l_0) + D_{(1,2)}(v^P_{(1,2)} - v_0), \tag{1}$$

where $l^P_{(1,2)}$ are the lengths of $PE_{(1,2)}$. $v^P_{(1,2)}$ are their velocities, and their initial values $[l_0, v_0]$ are equal to $[0.085, 0]$. $K_{(1,2)}$ and $D_{(1,2)}$ are stiffness and damper coefficients of $PE_{(1,2)}$, respectively. Here, we set $K_1 = K_2 = K$ and $D_1 = D_2 = D$.

The active force produced by the VAAM is the product of its activation α and length-velocity function $F(l_C, v_C)$. The active forces $F^C_{(1,2)}$ generated by $CE_{(1,2)}$ are represented as:

$$F^C_{(1,2)} = \alpha_{(1,2)} F(l^C_{(1,2)}, v^C_{(1,2)}), \tag{2}$$

where $l^C_{(1,2)}$ are the lengths of $CE_{(1,2)}$, and $v^C_{(1,2)}$ are their velocities. $\alpha_{(1,2)}$ are the functions of the $CE_{(1,2)}$ activation.

The total forces F^T_1 and F^T_2 are the sum of the passive and active forces produced by $M1$ and $M2$ respectively (derived from Eqs.(1) and (2)),

[b]In this paper the joint, physically actuated by a standard servo motor, is driven by a virtual agonist-antagonist mechanism. Its contractile and parallel elements (CEs and PEs) are simulated.

$$F_1^T = \underbrace{K(l_1^P - l_0) + Dv_1^P}_{F_1^P} + \underbrace{\alpha_1 F(l_1^C, v_1^C)}_{F_1^C}, \tag{3}$$

$$F_2^T = \underbrace{K(l_2^P - l_0) + Dv_2^P}_{F_2^P} + \underbrace{\alpha_2 F(l_2^C, v_2^C)}_{F_2^C}. \tag{4}$$

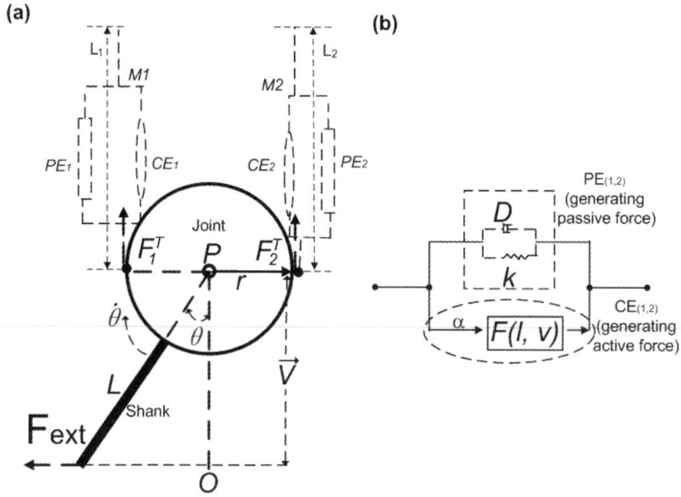

Fig. 1. The virtual agonist-antagonist mechanism (VAAM). (a) A joint driven by the VAAM. It consists of an agonist mechanism $M1$ and an antagonist mechanism $M2$ with lengths L_1 and L_2. Each of them consists of contractile and parallel elements (CE and PE). After applying force F_{ext} via a shank with length L to a joint P with the radius r, the joint angle θ changes. (b) The agonist and antagonist mechanisms $M1$ and $M2$. Their PEs (i.e. $PE_{(1,2)}$) producing passive forces are modelled as spring-damper mechanisms. Their CEs (i.e. $CE_{(1,2)}$) yielding active forces depend on their activation α and length-velocity function $F(l_C, v_C)$.

The antagonist mechanism $M2$ resists the extension of joint angle θ when it is excited by F_{ext}. Simultaneously, the agonist mechanism $M1$ produces opposing force against $M2$. Therefore, the direction of F_1^T is clockwise when the direction of F_2^T is counter-clockwise (see Fig. 1). According to the right-hand rule, the torque $\tau(F_2^T)$ direction points outward from the page. The torque $\tau(F_1^T)$ and $\tau(F_{ext})$ directions point into the page. We assume the torques pointing into the page to be positive. Then the torques $\tau(F_1^T)$,$\tau(F_2^T)$ and $\tau(F_{ext})$ acting on the joint P can be represented as:

$$\tau(F_1^T) = F_1^T r = (K(l_1^P - l_0) + Dv_1^P + \alpha_1 F(l_1^C, v_1^C))r, \tag{5}$$

$$\tau(F_2^T) = -F_2^T r = -(K(l_2^P - l_0) + Dv_2^P + \alpha_2 F(l_2^C, v_2^C))r, \tag{6}$$

$$\tau(F_{ext}) = F_{ext}\vec{V} = F_{ext}(L + r)\cos(\theta), \tag{7}$$

where $\vec{V}$ is the displacement vector of F_{ext} relative to the joint P. r is the radius of the joint P.

When applying Euler's laws of motion to the rotations of the joint P (see Fig. 1(a)), the net torque $\sum \tau$ acting on the joint P is equal to the product of its moment of inertia I and angular acceleration $\ddot{\theta}$,

$$I\ddot{\theta} = \sum \tau = \tau(F_{ext}) + \tau(F_1^T) + \tau(F_2^T). \tag{8}$$

Substituting Eqs.(5), (6) and (7) into Eq.(8), we obtain:

$$\begin{aligned} I\ddot{\theta} = {} & F_{ext}(L + r)\cos(\theta) \\ & + r[(K(l_1^P - l_0) + Dv_1^P + \alpha_1 F(l_1^C, v_1^C)) \\ & - (K(l_2^P - l_0) + Dv_2^P + \alpha_2 F(l_2^P, v_2^P))]. \end{aligned} \tag{9}$$

The lengths (i.e. $L_{(1,2)}$) of $M1$ and $M2$ are equal to the lengths of their CE and PE (i.e., $L_1 = l_1^P = l_1^C, L_2 = l_2^P = l_2^C$). In Fig. 1, $M1$ is shortening when $M2$ is lengthening. $\Delta L_{(1,2)}$ are the displacements of $M1$ and $M2$, i.e. $-(\Delta L_1) = \Delta L_2$. Here we postulate the relationship between their displacements $\Delta L_{(1,2)}$ and the joint angle θ,

$$-(\Delta L_1) = \Delta L_2 = \theta r = l_2^P - l_0 = -(l_1^P - l_0), \tag{10}$$

The relationship between velocities $\Delta \dot{L}_{(1,2)}$ and the joint velocity $\dot{\theta}$ can be represented as:

$$-\Delta\dot{L}_1 = \Delta\dot{L}_2 = \dot{\theta} r = v_2^P = -v_1^P. \tag{11}$$

Note that the initial values of $[v_1^P, v_2^P]$ are equal to zero. Substituting Eqs.(10) and (11) into Eq.(9), we obtain:

$$\begin{aligned} I\ddot{\theta} = {} & F_{ext}(L + r)\cos(\theta) \\ & + r[\underbrace{(\alpha_1 F(\phi_1, \phi_2) - \alpha_2 F(\phi_3, \phi_4))}_{\text{active force}} \\ & - \underbrace{(2K\theta r + 2D\dot{\theta} r)}_{\text{passive force}}], \end{aligned} \tag{12}$$

Equation (12) consists of the active and passive forces generated by $CE_{(1,2)}$ and $PE_{(1,2)}$, respectively.

Many physiological experiments have shown that the passive properties of muscles play a key role in animal joint stability when confronted with external perturbations.[10] Inspired by this finding for robotic joint control, we set the $M1$ and $M2$ activations α_1, α_2 from Eq.(12) to zero. Thus, we obtain:

$$I\ddot{\theta} = F_{ext}\vec{V} - 2r^2(K\theta + D\dot{\theta}), \vec{V} = (L + r)\cos(\theta). \tag{13}$$

According to Eq.(13), the VAAM relies on its passive properties (i.e. the parallel elements (PEs)) resulting in a variable admittance controller.

The variable admittance control based on the VAAM is applied to the hexapod robot AMOS-II (see Manoonpong et.al[11] for more detail) having three joints (TC-, CTr-, FTi- joints[a]) at each leg (see Fig.2 (a)). A pair of the virtual PEs excited by vertical loads is implemented on the CTr- and FTi- joints of AMOS-II, since the TC- joints of AMOS-II only allow for horizontal movements. Without force/torque sensing at each joint, some computational simplifications need to be made for the displacement vectors $\vec{V}_{(1,2)}$ at the joints $J_{(1,2)}$ caused by the external force F_{ext} . According to the leg kinematic configuration, the displacement vector $\vec{V_1}$ of the FTi- joint can be represented by (derived from Eq. (7)):

$$\vec{V_1} = (L_F + r)cos(\theta_1). \tag{14}$$

The displacement vector $\vec{V_2}$ of the CTr- joint, on which F_{ext} indirectly acts can be approximated by :

$$\vec{V_2} = (L_C + r)cos(\theta_2) + \vec{V_1}. \tag{15}$$

Note that similar simplified computations of displacement vectors can also be made for legs with complex structures.

The legs can move vertically when they are excited by F_{ext}, which are detected by the force sensors installed in the legs. Here the link lengths L_F and L_C are set as: $L_F = 0.065(m), L_C = 0.11(m)$. r is the joint radius, which is equal to $0.01(m)$.

The rotation matrix can be represented as (derived from Eq.(13)):

$$\ddot{\theta}_{(2\times2)}I_{(2\times1)} = F_{ext}\vec{V}_{(2\times1)} - 2r^2(\theta_{(2\times2)}K_{(2\times1)} + \dot{\theta}_{(2\times2)}D_{(2\times1)}), \tag{16}$$

[a]The thoraco-coxal (TC-) joint enables forward and backward movements, the coxa-trochanteral (CTr-) joint enables elevation and depression of the leg, and the femur-tibia (FTi-) joint enables extension and flexion of tibia.

where the acceleration matrix $\ddot{\theta}_{(2\times2)}$ and inertia matrix $I_{(2\times1)}$ can be written as:

$$\ddot{\theta}_{(2\times2)} = \begin{bmatrix} \ddot{\theta}_1 & 0 \\ 0 & \ddot{\theta}_2 \end{bmatrix}, I_{(2\times1)} = \begin{bmatrix} 5.0\times10^{-7} \\ 5.0\times10^{-7} \end{bmatrix}. \quad (17)$$

The displacement vector matrix $\vec{V}_{(2\times1)}$ is:

$$\vec{V}_{(2\times1)} = \begin{bmatrix} \vec{V}_1 \\ \vec{V}_2 \end{bmatrix}, \begin{matrix} \vec{V}_1 = (L_F + r)cos(\theta_1) \\ \vec{V}_2 = (L_C + r)cos(\theta_2) + \vec{V}_1 \end{matrix}. \quad (18)$$

The joint angle matrix $\theta_{(2\times2)}$, stiffness coefficient matrix $K_{(2\times1)}$, velocity matrix $\dot{\theta}_{(2\times2)}$ and damper coefficient matrix $D_{(2\times1)}$ are:

$$\theta_{(2\times2)} = \begin{bmatrix} \theta_1 & 0 \\ 0 & \theta_2 \end{bmatrix}, K_{(2\times1)} = \begin{bmatrix} K \\ K \end{bmatrix}, \dot{\theta}_{(2\times2)} = \begin{bmatrix} \dot{\theta}_1 & 0 \\ 0 & \dot{\theta}_2 \end{bmatrix} D_{(2\times1)} = \begin{bmatrix} D \\ D \end{bmatrix}. \quad (19)$$

3. Experimental results

In this setup (see Fig. 2 (b)), AMOS-II was placed between supporters (black wheels) having the total height of 18 cm (i.e. $L_s = 18(cm)$). Then a static load (i.e., a whiteboard) was placed on top of AMOS-II. It carried the full load, since its height (i.e. $22cm$) is higher than that of the supporters. The video clips of this experiment can be seen at http://www.manoonpong.com/CLAWAR2013/E1.wmv . When the FTi- and CTr- joints of AMOS-II are driven by the virtual PEs, it can automatically adapts its body height. Therefore, the legs with the virtual PEs suffer less contact force when AMOS-II is imposed with a static load. In contrast, without the virtual PEs, AMOS-II have to resist the load when the passive springs installed in the legs cannot be compressed anymore.

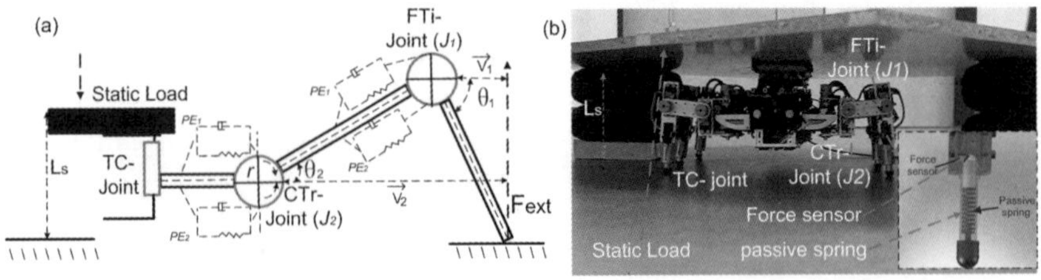

Fig. 2. Experimental model and setup. (a) AMOS-II leg with a simplified variable admittance controller (see Eq.(16)). The joints $J_{(1,2)}$ are driven by a pair of the virtual PEs, respectively. (b) AMOS-II is imposed with a static load (i.e. a white board).

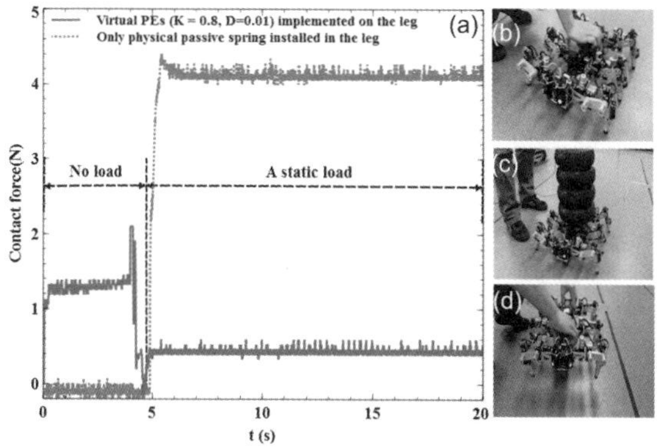

Fig. 3. (a) Comparison of the contact force arising with a static load. The force is measured at a foot contact sensor. (b) Hand pushing. (c) Placing a heavy load with a total weight of $5.72kg$. (d) Sudden Dropping (i.e. about ten centimetres).

Figures 3 (b)-(d) show experiments when AMOS-II was imposed with dynamical perturbations, like hand pushing, heavy load and sudden dropping. The video clips of these experiments can be seen at http://www.manoonpong.com/CLAWAR2013/E2.wmv . In hand pushing conditions, AMOS-II can produce 'softer' (i.e., $D = 0.001$) and 'stiffer' (i.e., $D = 0.1$) behaviors, by tuning the damper coefficients D in Eq.(16). However, these setups (i.e., $D = 0.001$ and $D = 0.1$) make AMOS-II unstable and suffer more contact force. Therefore, intermediately 'soft' (i.e., $D = 0.01$) was applied to AMOS-II when it was imposed with a heavy load and suddenly dropped. Overall, the PEs of the VAAM enable AMOS-II to perform variable compliant behaviors by tuning the damper coefficients D of its PEs. As a consequence, they make AMOS-II reduce contact force and prevent leg damages when it is imposed with static and dynamical perturbations. All stiffness coefficients K of its PEs were set to 0.8.

4. Conclusion

We presented a simplified variable admittance controller based on a virtual agonist-antagonist mechanism (VAAM). Two characteristics of this novel controller are: (1) through the simple tuning, this simplified variable controller can generate variable compliant behaviors without complex physical spring and damper mechanisms, e.g. variable stiffness actuators (VSAs), and (2) it also doesn't rely on complex force/torque sensing but rather simple force sensing at the end effector. The variable compliant behaviors are thereby produced by tuning the stiffness and damper coefficients (K, D) of the PEs. These two characteristics enable robots to reduce contact force and generate variably 'soft'/'stiff' behaviors when receiving perturbations. In future work, the modular neural networks[11] will be used to control the CEs of the VAAM enabling AMOS-II to traverse different terrains with natural movements.

Acknowledgements: This research was supported by the Emmy Noether Program (DFG, MA4464/3-1), the Federal Ministry of Education and Research(BMBF) by a grant to the Bernstein Center for Computational Neuroscience II Göttingen (01GQ1005A, project D1), and European Communitys Seventh Framework Programme FP7/2007-2013 (Specific Programme Cooperation, Theme 3, Information and Communication Technologies) under grant agreement no. 270273, Xperience. We thank Martin Biehl, Guanjiao Ren and Frank Hesse for their discussions.

References

1. R. Ham, T. Sugar, B. Vanderborght, K. Hollander and D. Lefeber, *Robotics Automation Magazine, IEEE* **16**, 81 (september 2009).
2. C. Yang, G. Ganesh, S. Haddadin, S. Parusel, A. Albu-Schäffer and E. Burdet, *Robotics, IEEE Transactions on* **27**, 918 (oct. 2011).
3. S. Parusel, S. Haddadin and A. Albu-Schäffer, Modular state-based behavior control for safe human-robot interaction: A lightweight control architecture for a lightweight robot, in *Robotics and Automation (ICRA), 2011 IEEE International Conference on*, may 2011.
4. A. Lecours, B. Mayer-St-Onge and C. Gosselin, Variable admittance control of a four-degree-of-freedom intelligent assist device, in *Robotics and Automation (ICRA), 2012 IEEE International Conference on*, may 2012.
5. A. Albu-Schäffer, O. Eiberger, M. Grebenstein, S. Haddadin, C. Ott, T. Wimböck, S. Wolf and G. Hirzinger, *RAM, IEEE* **15**, 20 (september 2008).
6. M. Catalano, G. Grioli, M. Garabini, F. Bonomo, M. Mancinit, N. Tsagarakis and A. Bicchi, Vsa-cubebot: A modular variable stiffness platform for multiple degrees of freedom robots, in *Robotics and Automation (ICRA), 2011 IEEE International Conference on*, may 2011.
7. O. Eiberger, S. Haddadin, M. Weis, A. Albu-Schäffer and G. Hirzinger, On joint design with intrinsic variable compliance: derivation of the dlr qa-joint, in *Robotics and Automation (ICRA), 2010 IEEE International Conference on*, may 2010.
8. J. Hurst, J. Chestnutt and A. Rizzi, *Robotics, IEEE Transactions on* **26**, 597 (aug. 2010).
9. W. Yu, J. Rosen and X. Li, Pid admittance control for an upper limb exoskeleton, in *American Control Conference (ACC), 2011*, 29 2011-july 1 2011.
10. S. Sponberg and R. J. Full, *Journal of Experimental Biology* **211**, 433 (2008).
11. P. Manoonpong, U. Parlitz and F. Wörgötter, *Front. in Neural Circuits* **7** (2013).

SECTION–5

HMI, INSPECTION AND LEARNING

DETERMINATION OF TRAJECTORIES USING NON-INVASIVE BCI TECHNIQUES IN 3D ENVIRONMENTS

TEODORO GARCÍA-EGEA[A, B], CARLOS ALBERTO DÍAZ-HERNÁNDEZ[A], JUAN LÓPEZ-CORONADO[A] AND JOSÉ L. CONTRERAS-VIDAL[C, D, E]

[A] *Department of Automation and Systems Engineering, Polytechnical University of Cartagena, Campus Muralla del Mar, Cartagena, SPAIN*

[B] *Department of Polytechnic Sciences, Polytechnic Faculty, UCAM Catholic University of San Anthony of Murcia, SPAIN*

[C] *Fischell Department of Bioengineering, University of Maryland, College Park, Maryland, USA*

[D] *Department of Kinesiology, University of Maryland, College Park, Maryland, USA*

[E] *Laboratory for Non-invasive Brain-Machine Interface Systems, Department of Electrical and Computer Engineering, University of Houston, Houston, TX 77204, USA*

tgarcia@pdi.ucam.edu; ca.al.di.he@gmail.com; jl.coronado@upct.es; jlcontreras-vidal@uh.edu

The feasibility of continuous decoding of self-initiated, self-selected hand movements to three-dimensional (3D) spatial targets from scalp electroencephalography (EEG) using linear decoders has been recently demonstrated. In this paper, we show that it is possible to train linear classifiers to decode hand movement direction to eight 3D spatial targets, in both planning and movement windows, using only the fluctuations in the amplitude of smoothed low-frequency signals from high-density scalp EEG. Taken together these results support the design of brain-computer interfaces (BCI) based on non-invasive scalp EEG signals and suggest that the current perception of the limits of EEG as a source signal for BCI applications merits further examination.

1. Introduction

Recent studies on neural decoding of movement or movement intention based on electroencephalography (EEG) signals have shown that EEG signals contains useful motor information that can be extracted and classified [6, 17, 3, 4, 10, 11]. These studies showed that decoding of movement direction or movement trajectories can be accomplished from the low-pass filtered fluctuations in the amplitude (i.e., amplitude modulation, or AM) of the EEG signals or from the spectral modulations in the delta (0.1- 4Hz) band. Moreover, The feasibility of continuous decoding of self-initiated, self-selected hand movements to three-dimensional (3D) spatial targets from scalp electroencephalography (EEG) using linear decoders has been recently demonstrated [3]. In general, neural decoders can be designed to predict state estimates such as discrete classifications of multiple internal states such as intended movement direction to spatial targets [6, 17] or to predict continuous time variables of endpoint hand trajectories [3, 10] or angular kinematics [11] that could serve as reference signals to an upper or lower limb prosthetic device, computer cursor, spelling device, or virtual keyboard - that is, a brain-computer interface (BCI) system. One objective of BCI research is to develop systems capable of decoding neural representations of natural movement planning and execution. Understanding the limits of scalp EEG with respect to the number of degrees of freedom, complexity, reliability, speed, and level of intuitive control that can be achieved are important topics in current BCI research.

Typically, BCI systems communicate a user's intent by converting the individual's brain signals into functional outputs for control or communication. Because this type of communication does not depend on peripheral nerves and muscles, the development of non-invasive BCI systems has focused on applications for people who are severely paralyzed that restore communication and interaction with their environment. Unfortunately, most EEG-based BCI systems require weeks to months of training before subjects can achieve some level of brain control and are only capable of slow communication rates. Communication rate, of course, determines how quickly and accurately buttons, icons and targets can be selected by the end-user of a BCI. Training time and bit rate are therefore primary determinants of acceptance of BCI technology by potential users.

The limited performance of current EEG-based BCIs may be related to their choice of input feature spaces (e.g., P300, sensorimotor rhythms), and/or their use of stereotypic and nonspecific neural signals (not directly correlated with intended action) to trigger predefined system outputs. However, recent decoding studies suggest that the smoothed (< 4Hz) amplitudes of EEG signals across the scalp contain detailed information about the representation of movement at the macro-scale that could lead to improvements in communication rate and training time.

In this regard, Waldert et al [17] investigated EEG activity during self-chosen targeted hand movements and showed that the hand movement direction can be inferred, on a single-trial basis, using the low-pass (3Hz) filtered neuronal activity measured over 20 sensors distributed over the scalp. Hammon et al [6], using high-density EEG and a set of eight different feature vectors, showed that it is feasible to decode intended (e.g., during motor planning) hand reaching to spatial targets (one out of four targets arranged in 2D) and hand movement direction during actual reaches (one out of three targets in 3D) from approximately 140-150 EEG channels. Waldert et al [17] compared decoding accuracy (DA) and Shannon's mutual information between the decoded and the measured movement direction (defined as decoded information, DI, in bits) in the above studies. However, it is not clear how an increasing number of possible reaching targets would affect decoding accuracy, and therefore the optimal number of targets for maximizing both DA and DI is unknown. Thus, in this study we investigated DA and DI during self-initiated, self-selected, unconstrained, hand reaching movements to eight targets in 3D space using linear discriminant analysis (LDA) of EEG derived features.

2. Materials and methods

Experimental procedure. The Institutional Review Board of the University of Maryland at College Park approved the experimental protocol, which follows that reported in Bradberry et al. [3]. After giving informed consent, five healthy, right-handed subjects sat upright in a chair and executed self-initiated, center-out reaches to self-selected push-button targets near eye level and arranged in 3D space. The distance from the center position to each of the targets was ~22cm. We instructed subjects to attempt to make uniformly distributed random selections of the eight targets without counting. Subjects were instructed to fixate a LED on the center target throughout data collection and to only blink when their hand was resting at the center target (See Figure 1 of Bradberry et al [3], for an illustration of the pointing apparatus). For each subject, the experiment concluded after each target was acquired at least 10 times.

Data collection. A 64-sensor Electro-Cap was placed on the heads of subjects according to the extended International 10–20 system (ear-linked referenced). Continuous EEG signals were sampled at 1000 Hz, band-pass filtered from 0.5 to 100 Hz, and notch filtered at 60 Hz. Horizontal and vertical electro ocular activity was measured with bipolar sensor montages. The EEG data were decimated from 1 kHz to 100 Hz by applying a low-pass, anti-aliasing filter with a cut-off frequency of 40 Hz and then down sampling by a factor of 10. Fingertip position was sampled at 100 Hz using a motion-sensing system that tracked an infrared LED secured to the index fingertip.

Artifact detection and removal. To remove physiological and non-physiological artifactual components from the EEG data, we used Independent Component Analysis (ICA) running under EEGLab [5]. We used the option 'reject component by map' in the EEGLab toolbox by visually-inspecting and selecting those ICs that showed scalp map and spectral features known to represent eye blinks and/or muscle artifacts [12]. The remaining components were then transformed back into cleaned EEG time series.

Signal pre-processing. The cleaned EEG data were 1) sorted by target direction, 2) aligned to movement onset (MO). (MO was defined as the time point when the subject released his/her fingertip from the home switch. The end of the movement was considered the time that the subject pressed the target button), 3) segmented (-0.2 to 1.2s), 4) smoothed (zero-phase shift, 6th-order, 18Hz Butterworth low-pass filter (LPF)), 5) differentiated (Matlab's diff command), 6) re-sampled to a fixed median length of 1.4 sec (that is, 120 samples) comprising -0.2 to 1.2s, and 7) filtered again (zero-phase shift, 10th-order, 3 Hz Butterworth LPF) to extract the amplitude information within the delta band. Importantly, the re-sampling operation preserved the pattern information (i.e., fluctuations in the amplitude of the EEG signals during the chosen analysis interval) while providing feature input vectors (n=120) of constant length for training and classification.

Neural decoding method. We used linear discriminant analysis with regularization (rLDA) implemented by Ye et al (2006) on a single-trial basis to perform classification of time domain signal patterns in the delta band. rLDA is a stable and regularized algorithm because small variations in the training set do not considerably affect classification performance. Moreover, the regularization property provides a good generalization performance and high robustness with respect to outliers (Ji & Ye, 2008). The regularization parameter (λ) was optimized from the following value set (λ = [10-10 10-7 0.1 10 100 104 10-7]. As inputs to the rLDA classifier, we used the windowed and re-sampled time–varying EEG signals gathered from four different time periods: 1) the pre-movement period (200ms of data prior to movement onset), 2) from -200ms to +375ms, 3) -200ms to +700ms, and 4) -200ms to 1.4 s; all aligned to MO.

Decoder validation. The percentage of correctly decoded trials, termed decoding accuracy (DA), was used to quantify the performance of the rLDA classifier. For each subject, the DA was calculated using a 10-fold cross-validation procedure, where the set of trials used for training the RLDA (90%) and the set of trials used for decoding (10%) were mutually exclusive. In addition, we computed the decoded information (DI) as in [17] to quantify the amount of information about movement direction contained in the extracted

EEG signals. Specifically, the Shannon mutual information between the decoded and real hand movement directions was estimated (see Eq. 2 of [17]) and the function DI(DA) was assessed assuming equal probabilities for correct predictions in each class and equal distributions of false predictions across all classes (see Eqs. 3-8 of [17]).

3. Results

Figure 1 depicts the individual and the mean decoding accuracy (DA) across the five subjects for time-domain (i.e., smoothed EEG amplitudes) feature vectors representing the plan or pre-movement (200ms before movement onset) period and the period that included the 'plan' + 'move' windows (-200ms to 1.2 s aligned to MO). Figure 1A shows that the mean (S.D.) DA for the pre-movement period was low (21±4.4) but well above chance level (12.5%), whereas the mean DA for the full-length vector was 62.5% ± 17.67.

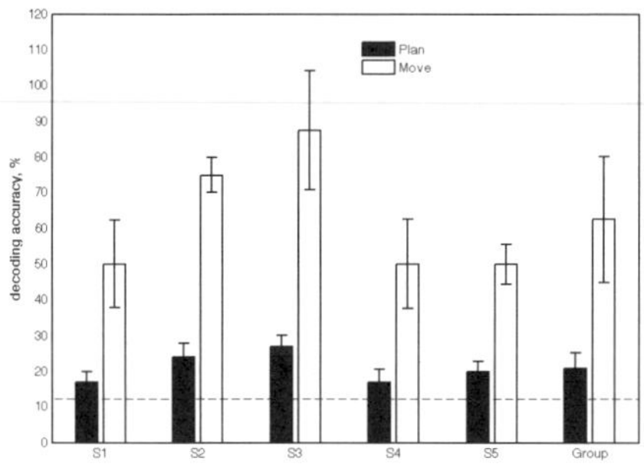

Figure 1. Decoding accuracy (%) per subject and across the group (N=5). The horizontal dotted line represents chance (12.5% for eight targets). 'Plan' corresponds to the pre-movement period (-200ms – MO), whereas "Move" corresponds to the full-length movement window (MO – 1.2 s).

Figure 2 shows the DA as a function of the time window used for classification. It provides an insight into the targeted movement information that is carried in various time windows used by the decoder. Overall, DA increased with the number of samples used for classification, but the pre-movement period already contained target information resulting in DA scores above chance.

Figure 3 shows the confusion matrix for the outcome of the classifier across all subjects. The confusion matrix has a complex structure and allows us to understand how and where the misclassifications and the correct predictions occur. Rows indicate the correct class label, and columns indicate the predicted class label. Thus, each cell shows what fraction of examples from the row class was predicted to belong to the column class. From Fig. 3 it is noted that targets 4 and 5 had the highest misclassification rates (60%); target 4 corresponded to reaches to the nearest contralateral target in the top left corner, whereas target 5 was located in the far ipsilateral right corner. The best decoded targets were Target 1 (100% classification rate; reaches nearest the top-right corner), Target 2 and 7 (80% classification rate; hand reaches to the nearest middle-right corner and far contralateral middle-bottom corner), and Targets 3, 6, and 8 (60% classification rates; contralateral medium-left, ipsilateral middle-bottom, and contralateral medium-left, respectively).

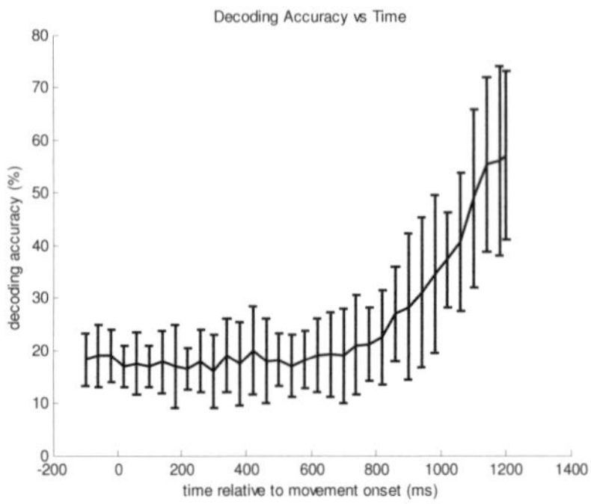

Figure 2. Evolution of decoding accuracies (DA) using delta band (0.5-3 Hz) low-pass filtered EEG as a function of epoch length (aligned to MO). The graph shows DA of the following epoch sizes: 200ms immediately before the time indicated in the abscissa as time 0 (MO), and the cumulative time windows up to 1200ms immediately following MO. The chance level is 12.5% for 8 targets.

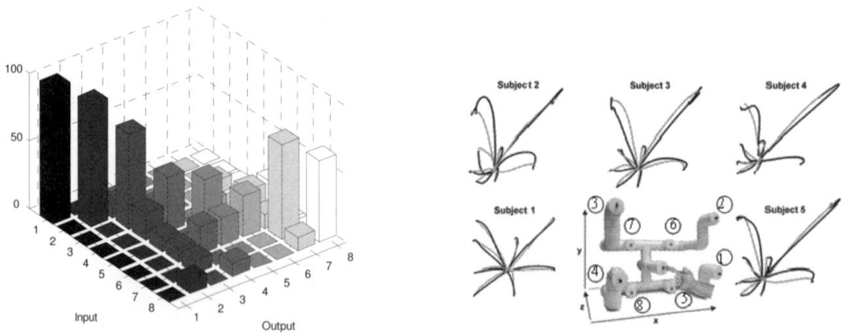

Figure 3. (Left) Confusion matrix for the neural classifier: Gray bars relate to target predicted in each case (in %). Each label in the Input represents the instances in an actual class (true target), whereas the labels in the Output represent the predicted class (assigned target) by the classifier. (Right) Experimental setup and mean finger paths. The reaching apparatus is shown in the middle along with the Cartesian coordinate system. Mean finger paths for center-to-target (black) and target-to-center (gray, not included in analysis) movements exhibited movement variability among subjects. Reproduced with permission from [3].

4. Discussion

In this paper we demonstrate the feasibility of decoding hand movement directions to eight 3D spatial targets using the smoothed (0.5-3 Hz) time-varying amplitudes of high-density EEG signals acquired across the scalp. The decoding accuracy was comparable to that reported in both invasive (electrocorticograph or ECoG) and non-invasive EEG decoding studies (See Table 1). Surprisingly, our results show decoding accuracies comparable to those from an ECoG study [2] that also involved reaches to 4 or 8 spatial targets. In order to compare our results with other decoding studies using a different number of targets or recording method, we estimated the Decoded Information (DI) for our experiment as in [17] and included our results in the modified Figure 7 of [17] shown here as Figure 4. That movement can be decoded from EEG to the same extent as ECoG is supported by ECoG decoding studies based on low-frequency (smoothed) local field potentials, also termed local motor potential (LMP) or movement-related potentials (MRP) [13, 2, 1]. These low frequencies are likely to suffer less attenuation due to the skull [16] and would have higher signal-to-noise ratios (SNRs) than higher frequency bands.

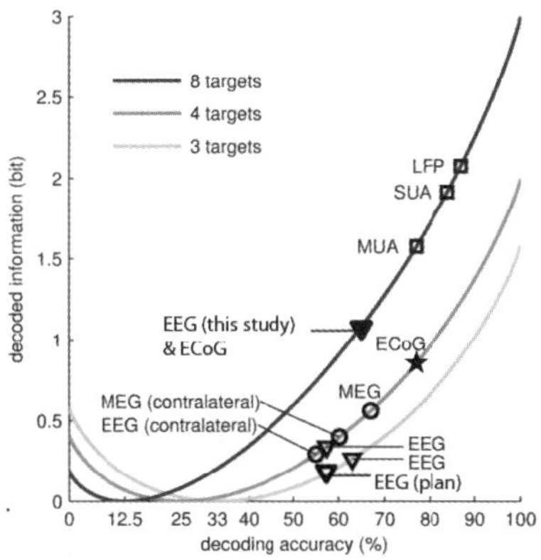

Figure 4. Comparison of different recording techniques with respect to DA and DI about movement direction for center-out reaching experiments. The gray curves reflect the dependency of DI on DA under certain simplifying assumptions (Eq. 2 of [17]) for 3, 4, or 8 targets. Symbols: open squares; star [2]; circles [17](contralateral: only sensors above contra-lateral motor areas were used); triangles – open [6]; filled triangle – this study (62.5% DA) and [6] (63%). Adapted and revised, with permission, from Figure 7 [17].

Which EEG signal component in the time or frequency domain carries the most information about hand movement direction?

Most decoding studies based on analog (e.g., LFP or ECoG) recordings have generally focused on the identification of the optimal frequency band(s) for spectral-power-based decoding [2, 9, 15, 14, 7, 20]; with the underlying assumption that spatially-resolved gamma band frequencies are critical for movement decoding. However, some of the above studies have evaluated decoding from movement-related potentials (time-domain) and from spectral amplitude modulations (frequency-domain) in very low frequencies and in the high gamma band. Specifically, Ball et al. [2] reported high decoding accuracy of arm movement direction based on amplitude modulation both in the delta band (<2 Hz) and in the high gamma band (52–128 Hz), yielding considerably higher decoding accuracy than the alpha, beta and low gamma bands.

Ince et al. [7] examined decoding of movement target direction based on single unit activity (SUA) and LFP signals, and reported that the decoding accuracies from simultaneously recorded SUA and LFP signals were similar; whereas the directional information varied with the LFP frequency sub-band, being greater in low (0.3–4 Hz) and high (48–200 Hz) frequency bands than in intermediate frequency bands. They attributed the high decoding accuracy from LFP signals to the spatial organization of the LFP signals over the recorded areas (primary motor and premotor cortices). Morever, Zhuang et al. [20] showed decoding of 3D reach-to-grasp movements from LFP signals based on delta and gamma band information.

Taken together, the above EEG, ECoG and LFP studies and our current results indicate that information about movement direction is contained in both the delta and the high gamma bands, and that scalp EEG may be a suitable source signal for decoding movement intent in BCI applications.

TABLE I. COMPARISON WITH OTHER STUDIES OF MOVEMENT DIRECTION DECODING

Author	Modality (Sensor density) / Target Number	Decoding Accuracy (DA)
Hammon et al. (2008)	EEG (140-150 electrodes) / 3 Targets	58.9%
Hammon et al. (2008)	EEG (140-150 electrodes) / 4 Targets	57.3%
Waldert et al, (2008)	EEG (20 contralateral electrodes) / 4 Targets	55%
Ball et al. (2009)	ECoG (112 electrodes) / 8 Targets	63%
Present Study	**EEG (55 electrodes) / 8 targets**	**62.5%**

Acknowledgments

This research was supported in part by the US National Science Foundation (NSF) Award #IIS-1219321, SENECA (Murcia Regional Research Foundation) Award InCamFut_15401/PI/2010, and the Spanish Research Foundation (DTP_MICINN-CICYT) Award TEC2009-10639-C04-03.

References

1. Acharya S, M.S. Fifer, H.L. Benz, N.E. Crone, N.V. Thakor (2010) Electrocorticographic amplitude predicts finger positions during slow grasping motions of the hand," J. Neural Eng. 7: 264–275, 2010.
2. Ball, T., Schulze-Bonhage, A., Aertsen, A., Mehring, C., 2009. Differential representation of arm movement direction in relation to cortical anatomy and function. J. Neural. Eng. 6 (1), 016006.
3. Bradberry TJ, Gentili RJ, Contreras-Vidal JL (2010). Reconstructing three-dimensional hand movements from non-invasive electroencephalographic signals. J Neurosci., 3;30(9):3432-7.
4. Bradberry TJ, Gentili RJ, Contreras-Vidal JL (2011). Fast attainment of computer cursor control with noninvasively acquired brain signals. Journal of Neural Engineering, 8(3): 036010, PMID: 21493978.
5. Delorme A, Makeig S (2004). "EEGLAB: An open source toolbox for analysis of single-trial EEG dynamicss," J. Neurosci. Methods, 134(2):9-21.
6. Hammon, P.S., Makeig, S., Poizner, H., Todorov, E., de Sa, V.R., 2008. Predicting reaching targets from human EEG. IEEE Signal Proc Mag 25 (1), 69–77.
7. Ince NF, Gupta R, Arica S, Tewfik AH, Ashe J, Pellizzer G. (2010). High accuracy decoding of movement target direction in non-human primates based on common spatial patterns of local field potentials. PLoS One., 5(12): e14384.
8. Ji S, Ye J. (2008). Generalized linear discriminant analysis: a unified framework and efficient model selection. IEEE Trans Neural Netw. 19(10):1768-82.
9. Leuthardt EC, Miller KJ, Schalk G, Rao RP and Ojemann JG. (2006). Electrocorticography-based brain computer interface—the Seattle experience," IEEE Trans. Neural Syst. Rehabil. Eng. 14: 194–8.
10. Lv J, Li Y, Gu Z (2010) Decoding hand movement velocity from electroencephalographic signals during a drawing task. Biomed Eng Online, 9:64.
11. Presacco A, Goodman R, Forrester L, and Contreras-Vidal JL (2011). Neural decoding of walking from non-invasive, high-density electroencephalographic (EEG) signals. Journal of Neurophysiology, 106(4): 1875-1887, PMID: 21768121.
12. Rong F, Contreras-Vidal JL (2006), Magnetoencephalographic artifact identification and automatic removal based on independent component analysis and categorization approaches, J Neurosci Meth, 157:337-54.
13. Schalk G, Kubanek J, Miller KJ, Anderson NR, Leuthardt EC, et al. (2007). Decoding two-dimensional movement trajectories using electrocorticographic signals in humans. J. Neural Eng. 4: 264–275.
14. Schalk G, Miller KJ, Anderson NR, Wilson JA, Smyth MD, Ojemann JG, Moran DW, Wolpaw JR, and Leuthardt EC. (2008). Two-dimensional

movement control using electrocorticographic signals in humans," J. Neural Eng., 5:75-84.

15. Shenoy P, Miller KJ, Ojemann JG, and Rao RP. (2008). Generalized features for electrocorticographic BCIs. IEEE Trans. Biomed. Eng. 55: 273–80.
16. Voytek B, Secundo L, Bidet-Caudet A, Scabini D, Stiver SI, Gean AD, Manley GT, Knight RT (2009). Hemicraniectomy: A New Model for Human Electrophysiology with High Spatio-temporal Resolution, J Cog Neurosci, in press.
17. Waldert, S., Preissl, H., Demandt, E., Braun, C., Birbaumer, N., Aertsen, A., Mehring, C., (2008). Hand movement direction decoded from MEG and EEG. J. Neurosci. 28 (4), 1000–1008.
18. Vidaurre C, Blankertz B. (2010). Towards a cure for BCI illiteracy, Brain Topogr. 23(2):194-8.
19. Ye J, Xiong T, Li Q, Janardan R, Bi J, Cherkassky V, Kambhamettu C (2006). Efficient model selection for regularized linear discriminant analysis. CIK'06, Nov 5-11, 2006, Arlington, VA (ACM 1-59593-433-2/06/0011).
20. Zhuang J, Truccolo W, Vargas-Irwin C, Donoghue JP (2010). Decoding 3D reach and grasp kinematics from high-frequency local field potentials in primate primary motor cortex. IEEE Trans. Biomed. Eng. 57(7): 1774-84.

DEVELOPMENT OF A PERISTALTIC CRAWLING INSPECTION ROBOT WITH PNEUMATIC ARTIFICIAL MUSCLES FOR A 25A ELBOW PIPE

TATSUYA KISHI, TARO NAKAMURA, MEGUMI IKEUCHI

Faculty of Science and Engineering, Department of Precision Mechanics
Chuo University,
1-13-27 Kasuga, Bunkyo-ku, Tokyo 112-8551, Japan

In this study, we developed a peristaltic crawling robot that can pass smoothly through a 25A 90-degree elbow pipe, whose radius of curvature equals its inside diameter. The actuator of the robot is composed of artificial rubber muscles. The robot performs earthworm-like locomotion by supplying air pressure to the artificial muscles. To enable smooth passage through the elbow pipe, we designed the robot with low friction and devised a variable friction unit with an aluminum sheet. This robot passes through the elbow pipe while retaining speed and maximum friction force as in a straight pipe.

1. Introduction

The objective of this study is the development of an in-pipe inspection robot that can inspect a complex network of 25A gas pipes.

The 25A nominal-diameter gas pipe (inside diameter is approximately 28 mm) is a small-diameter gas pipe that requires regular inspection. 25A gas pipes are used in houses and between gas meters and the main pipe, which lies under the road. In addition, many of these pipes are 90-degree elbow pipes, whose radius of curvature equals its inside diameter (R_c = 1.0ID). The inspection of 25A gas pipes involves checking the health of pipes, preventing illegal constructions, locating pipes, and identifying areas of water leakage.

Currently, 25A gas pipes are inspected with an endoscope. However, owing to the limitations of the endoscope, a number of pipes remain uninspected. When probing complicated and long piping, the endoscope eventually deflects and fails to penetrate further. To investigate complicated piping systems, in-pipe self-propelled inspection robots are required. Recently, various in-pipe robots have been developed for pipe inspection. However, they have some drawbacks. For instance, a snake-like robot [1] needs space that is wider than itself. Vibration mechanisms [2] cannot go backwards and vertical movements. Wheel-type robots [3]-[5] have been developed, but their bodies are difficult to downsize because of their complex structure. Consequently, these robots cannot pass through narrow pipes or the elbow pipes (R_c = 1.0ID). Passing through the elbow pipe is rendered difficult by steps and extra friction.

In this study, we developed a peristaltic crawling robot for the in-pipe inspection of 25A gas pipes, including 90-degree elbow pipes (R_c = 1.0 ID). Peristaltic crawling enables the robot to enter narrow spaces. In addition, the robot can be rendered low-friction by an aluminum sheet, enabling smooth passage through elbow pipes.

2. Peristaltic Crawling Pattern of an Earthworm

Peristaltic crawling is used by life forms such as earthworms and inchworms for moving forward. The robot developed in this paper emulates the earthworm's motion.

An earthworm consists of 110–200 segments. Figure 1 (a) shows the structure of an individual segment. The earthworm has two muscle layers: a circular muscle that is arranged inside of a skin circumferentially, and a longitudinal muscle that is arranged inside the circular muscle in the axial direction. The former elongates and the latter contracts the segments in the axial direction [6]. Figure 1 (b) shows the earthworm's peristaltic crawling pattern. The earthworm contracts and expands as follows. First, the earthworm contracts an anterior segment using the longitudinal muscles. Next, this contraction is transmitted to rear segments while the anterior segments are elongated by the circular muscles simultaneously. Finally, the contraction segments generate frictional force between the segments and a surface. Because of this frictional force, a reaction force to elongate contraction segments is generated.

Thus, the earthworm can move because of the repetition of thick-short and thin-long motion.

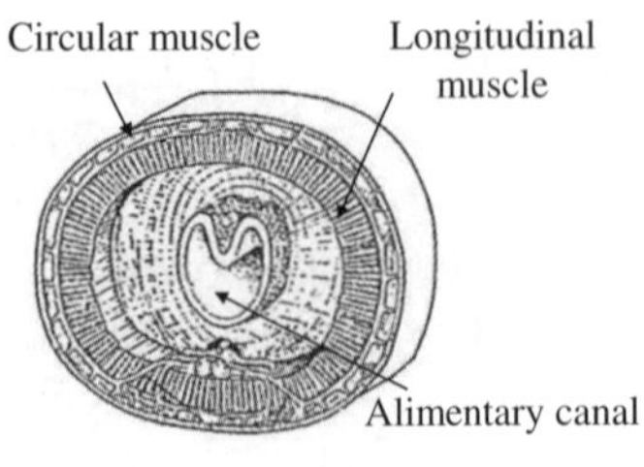

(a) Inner structure [6].

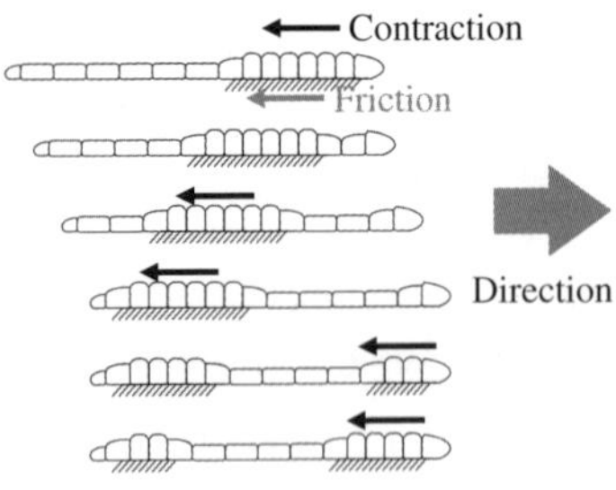

(b) Peristaltic crawling.

Figure 1. Peristaltic crawling pattern of an earthworm.

3. Straight-Fiber Artificial Muscle

We use a straight-fiber artificial muscle as the actuator for the robot. Figure 2 shows an overview of the artificial muscle, which is made of low-ammonia natural-rubber latex and micro-carbon fiber. This construction offers the following advantages: lightness, low price, flexible, and high output. The materials are formed in a tube, and a micro-carbon sheet is inserted into the tube in the long-axial direction. The range of the sheet extends when the rubber elongates [7].

This actuator is controlled by air pressure. When air is supplied to the actuator, the artificial muscle contracts in the axial direction and expands in the radial direction because the film expands but the carbon sheet barely does.

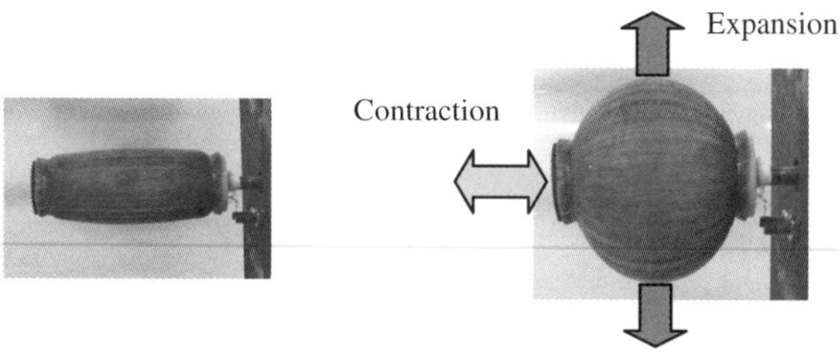

(a) Unpressurized. (b) Pressurize.

Figure 2. Straight-fiber artificial Muscle.

4. Peristaltic Crawling Robot

4.1. *Peristaltic Crawling robot*

Figure 3 is an image of the robot, developed by emulating an earthworm's motion. This robot has six units, five joints, and a head section. This robot weighs 98 g and length 480 mm when fully extended. The robot is coated with aluminum sheet to reduce friction and step of the robot so that the robot can pass through elbow pipe smoothly. Friction and step are the main obstructions faced by a robot passing through an elbow pipe. Figure 4 is an image of the robot without aluminum sheet. This robot is hollow and equipped with an endoscope (Figure 5).

The structure of a unit is shown in Figure 6, and its specifications are given in Table 1. When air is inhaled by the chamber, the artificial muscle expands radially and contracts in the axial direction. When air is discharged from the chamber, the artificial muscle extends in the axial direction. In this way, by repeated the extension and contraction, an earthworm's motion can be emulated.

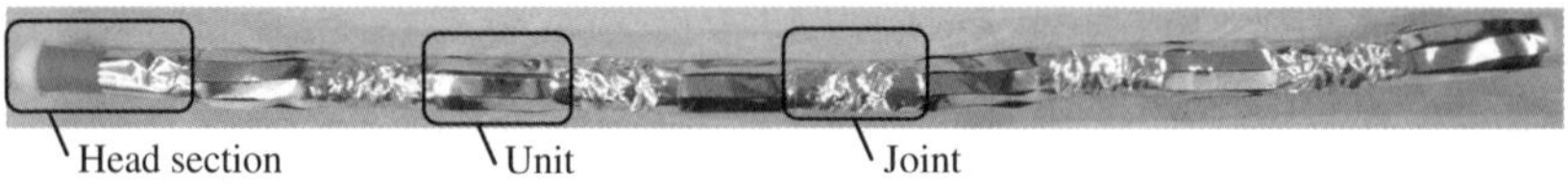

Figure 3. Peristaltic crawling robot.

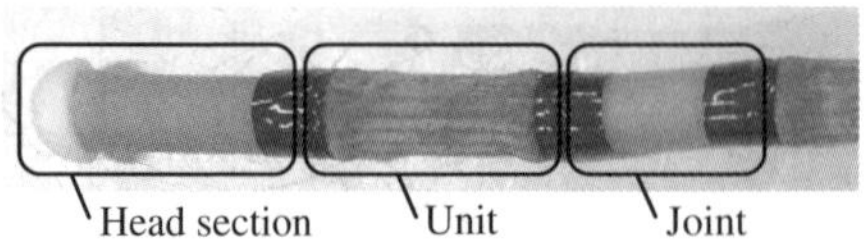

Figure 4. Without aluminum sheet.

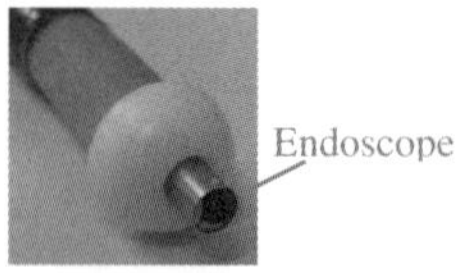

Figure 5. Head section with an endoscope.

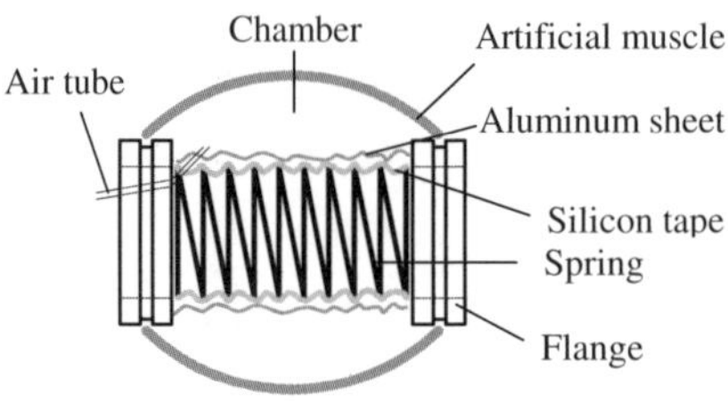

Figure 6. Structure of the unit.

Table 1. Unit specifications.

Weight [g]		6.8
Length [mm]	Extension	46.6
	Contraction	35.6
Width [mm]	Extension	15.4
	Contraction	37.7

4.2. *Variable Friction Unit*

The robot was modified into a low-friction device so that it could pass smoothly through an elbow. However, doing so reduces the maximum friction force and speed of the robot in a straight pipe. Therefore, we developed a variable friction unit that can pass smoothly through an elbow while retaining maximum friction force and speed of the robot in a straight pipe, as shown in Fig. 7. As shown in the figure, the aluminum sheet is cut in the axial direction, so the artificial muscle is exposed when the unit contracts so that the robot develops friction between itself and the pipe.

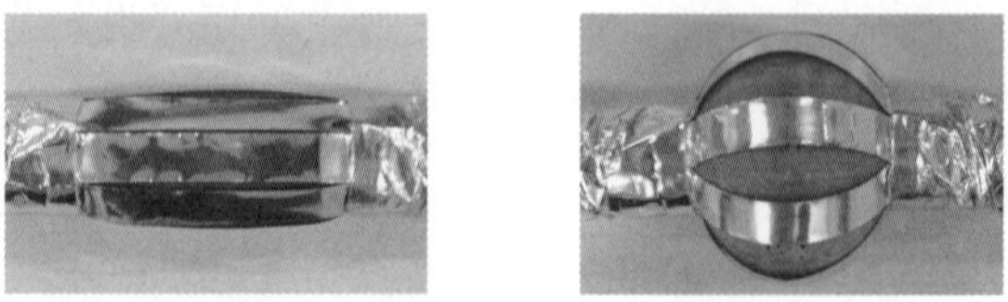

(a) Extension. (b) Contraction.

Figure 7. Variable friction unit.

4.3. *Method of Operation*

The robot is controlled by a H8-3052F microcomputer. Digital signals output from the H8 microcomputer are converted into analog signals using a digital-analog converter. These analog signals are then input to a proportional solenoid valve. The air pressure that is supplied to the robot via a manifold from a compressor is proportional to the input voltage. Thus, the units are free to extend and contract independently.

5. Motion Patterns

By altering the extension or contraction of each unit, several motion patterns can be achieved. Motion patterns consist of a wavelength, propagation speed, and waves. Here, wavelength is the number of units extended in the axial direction. The propagation speed is the number of units propagated in the rear. Here l, s, n, and N are the wavelength, propagation speed, number of waves, and number of units, respectively. Henceforth, we identify the basic motion patterns as motion l-s-n. For example, motion 4-1-1 (wavelength–propagation speed–number of waves) is shown in Figure 8. Each of the parameters must satisfy the following equations:

$$(l+s)\times n < N. \quad (1)$$

$$N-l > s. \quad (2)$$

$$l \geq s. \quad (3)$$

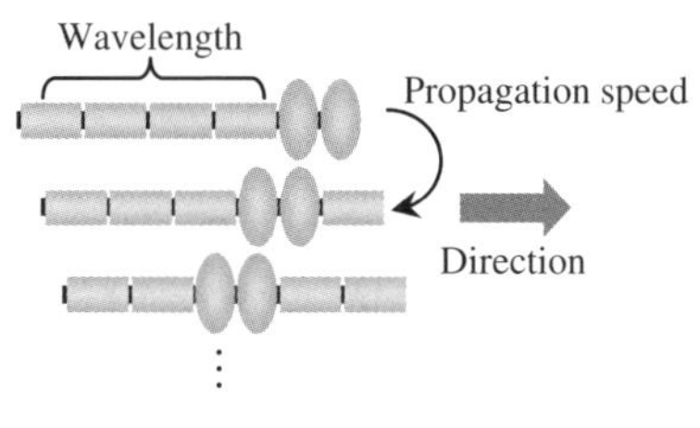

Figure 8. Motion pattern 4-1-1.

6. Experimental Results and Discussion

6.1. *Maximum Friction Force*

To verify that maximum friction force is not affected by aluminum sheet significantly, we compared the maximum friction force of the robot coated with aluminum sheet to that without aluminum sheet.

The maximum friction force of the developed robot was measured using the experimental setup shown in Figure 9. The number of contracted units is varied from one to six. The maximum friction force is measured by a push-pull scale. First, the units of the robot were contracted in a horizontal 25A acrylic pipe and cut off from the electrical supply to maintain contraction. Then, we connected the robot to the push-pull scale by a string and pulled the 25A acrylic pipe along a guide in a straight line. As soon as the 25A acrylic pipe slips from the robot, the friction force is the maximum static friction force. An air pressure of 0.1 MPa was applied to the unit.

The experimental results are shown in Table 2. We observe that the maximum friction force of the aluminum sheet-coated robot does not decrease more than 10% compared to its uncoated counterpart. On the contraction of two units, the robot broke because it was unable to withstand the load.

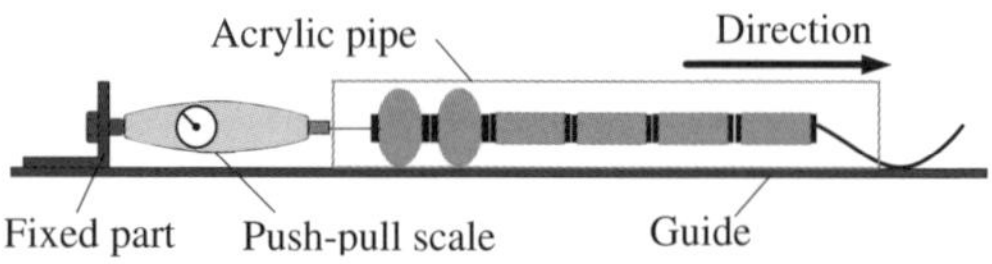

Figure 9. Experimental setup for measuring frictional force.

Table 2. Maximum friction force in a 25A acrylic pipe.

The number of contracting units		1	2
Maximum friction force [N]	Without aluminum sheet	61	(80)
	With aluminum sheet	56	(78)

6.2. *Locomotion Speed*

We measured the locomotion speed of the robot undertaking Motion 4-1-1. Motion 4-1-1 is the motion pattern that gets highest locomotion speed of the robot in a 25A acrylic pipe by experiment. These experiments were performed using a 25A acrylic pipe. In addition, an air pressure of 0.1 MPa was applied to the unit.

The experimental results are shown in Figure 10. The results suggest that aluminum sheet exerts no significant impact on locomotion speed.

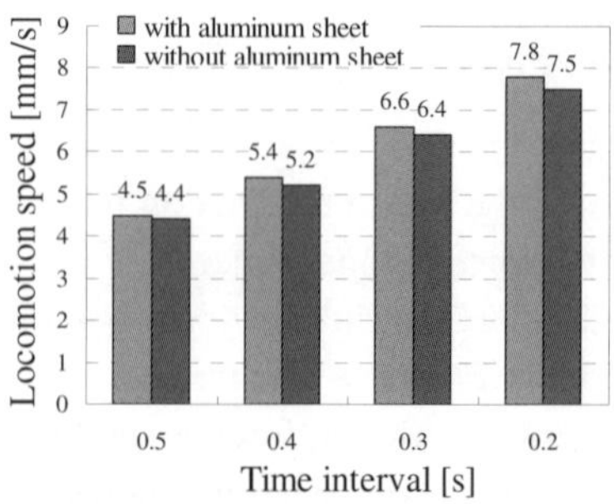

Figure 10. Locomotion speed in a 25A acrylic pipe.

6.3. *Locomotion Speed in a 90-degree elbow pipe*

To confirm that the aluminum sheet-coated robot can pass smoothly through the elbow pipe, we measured its locomotion speed through the 90-degree elbow pipe (R_c = 1.0 ID). These experiments were performed using Motion 4-1-1 with a time interval of 0.2 s. In addition, an air pressure of 0.1 MPa was applied to the unit.

The experimental results are shown in Figure 11. The locomotion speeds of the aluminum sheet-coated and uncoated robot are 6.8 and 4.1 mm/s, respectively. Therefore, the aluminum sheet-coated robot passes through the elbow pipe approximately 1.7 times faster than the uncoated robot. Thus, we confirmed that the aluminum sheet-coated robot can pass through an elbow pipe smoothly. Thus, the proposed approach leads to reduced friction and step of the robot and is thus effective in enabling the robot to smoothly pass through an elbow pipe.

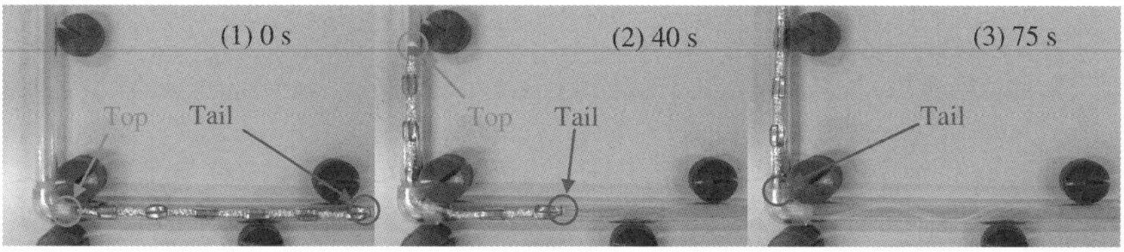

(a) With aluminum sheet.

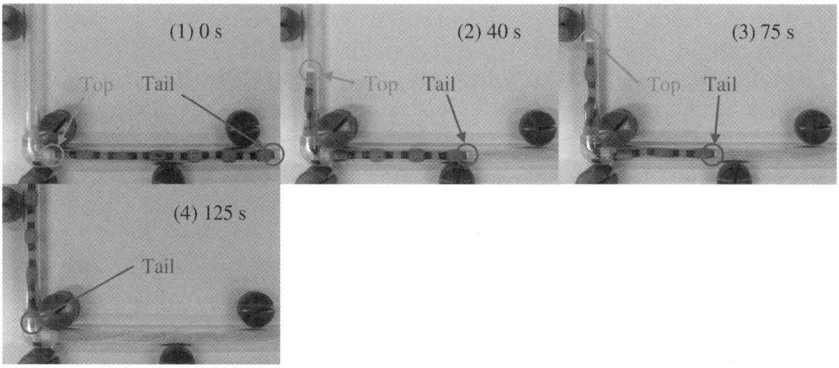

(b) Without aluminum sheet.

Figure 11. Passage time in a 25A 90-degree acrylic elbow pipe.

7. Conclusion

We developed a peristaltic crawling robot for the in-pipe inspection of 25A gas pipes that can pass through the 25A 90-degree elbow pipes (R_c = 1.0 ID). There is no report of robots that can pass through the elbow pipes except the robot. The robot passed through the elbow pipes (R_c = 1.0 ID) 1.7 times faster than its without aluminum sheet while retaining maximum friction force and speed in a straight pipe by reducing friction, and devising the variable friction unit with aluminum sheet.

In future, to use the robot in a real-life environment, we should enable the robot that can pass through continuous elbow pipes. Moreover, we should enable the robot that can travel long distances.

References

1. Akira Kuwada, Kodai Tsujino, Koichi Suzumori, and Takefumi Kanda "Intelligent Actuators Realizing Snake –like Robot for Pipe Inspection," Proc. IEEE International Symposium on Micro-nano Mechatronics and Human Science, pp. 1-6, 2006
2. Masashi Konyo, Kazunari Hatazaki, Kazuya Isaki, and Satoshi Tadokoro, "Development of an Active Scopecamera Driven by Ciliary Vibration Mechanism," Proc. of the 12th ROBOTICS symposia, pp.460-465, 2007
3. Peng Li, Shugen Ma, Bin Li, and Yuechao Wang, "Development of an Adaptive Mobile Robot for In-pipe Inspection Task," Proc. IEEE International Conference on Mechatronics and Automation, pp. 3622-3627, 2007
4. Tokuji Okada and Tsuyoshi Sanemori, "MOGER: A Vehicle Study and Realization for In-pipe Inspection Tasks," IEEE JOURNAL OF ROBOTICS AND AUTOMATION, VOL. RA-3, NO. 6. DECEMBER 1987
5. Amir H. Heidari, M. Mehrandezh, R. Paranjape and H. Najjaran, "Dynamic Analysis and Human Analogous Control of a Pipe Crawling Robot," Proc. IEEE/RSJ International Conference on Intelligent Robots and Systems, pp. 733-740, 2009
6. H. Sugi, "Evolusion of muscle motion," The University of Tokyo Press, p. 72, 1977 (in Japanese).
7. T. Nakamura, N. Saga, and K. Yaegashi, "Development of Pneumatic Artificial Muscle Based on Biomechanical Characteristics," Proceedings of IEEE International Conference on Industrial Technology (ICIT 2003), pp. 729-734.

EXAMINATION OF SURFACE FEATURE ANALYSIS AND TERRAIN TRAVERSABILITY FOR A WALL-CLIMBING ROBOT

D. SCHMIDT*, M. JUNG and K. BERNS

Robotics Research Lab, University of Kaiserslautern
D-67663 Kaiserslautern, Germany
**E-mail: dschmidt@cs.uni-kl.de*
http://agrosy.cs.uni-kl.de/cromsci/

Terrain traversability is a common problem in outdoor robotics. The main task is to determine if the current or upcoming terrain can be overcome by the locomotion system of the robot. But, this problem increases in terms of climbing robots which do not only have a locomotion, but also an adhesion system which has to be considered. This paper addresses these aspects and presents an approach to identify and analyze important surface features and to estimate the system's behavior. Supervised learning techniques are used to classify surfaces into different categories depending on their traversability. As test-platform, a wall-climbing robot using negative pressure adhesion and an omnidirectional drive system is considered which is able to drive on flat concrete buildings.

Keywords: climbing robot, surface features, feature extraction, supervised learning, support vector machine

1. Introduction

Whether a terrain is traversable by a robot or not generally depends on two aspects: The condition or characteristics of the surface and the robot's ability to overcome this surface with its locomotion system. Commonly, the terrain is either classified into different categories like concrete, gravel or sand[1] or it is analyzed based on aspects like inclination or roughness.[2,3] Basis of this is the environmental perception via laser range sensors, cameras or other sensors. The goal is to adapt vehicle motion to this information and e. g. slow the robot down or evade specific regions. In theory, it is also possible to determine the needed amount of sensor accuracy based on vehicle settings and to guarantee operability of the system depending on its parameters and the expected environmental features.[4]

In the field of climbing robots, the traversability additionally relies on the adhesion system and its ability to handle different terrains. In the present case, the climbing robot CROMSCI (figure 1) using an omnidirectional drive system in combination with negative pressure adhesion has been considered.[5] This robot has been developed to inspect concrete buildings like bridges or dams area-wide and semi-autonomously. Although the negative pressure system is highly sophisticated with optimized control components and seven adhesion chambers, it may still fail on certain surfaces.

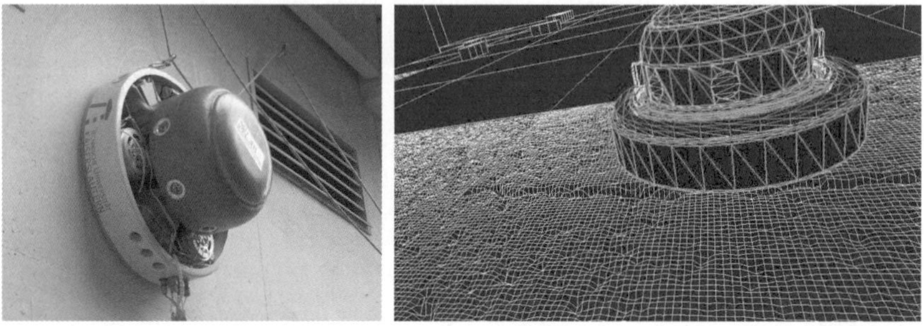

Fig. 1. Real prototype (left) and wireframe view of simulated environment (right).

This paper presents ways to analyze the upcoming terrain by a coupling of surface parameters and reactions of the system by using techniques of supervised learning. Figure 2 illustrates the concept and the internal components which are used for feature extraction and traversability estimation. The offline training has to be executed beforehand based on training examples which include a depth image and a rating of the current robot adhesion to determine the influence of the surface on the adhesion system. The goal is to predict the reaction of the adhesion system on the upcoming terrain and to determine, if it can be overcome by the robot or not.

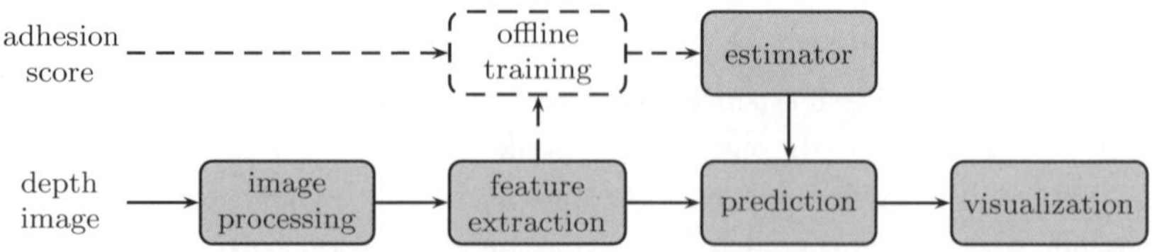

Fig. 2. Components of the traversability estimation including offline elements (dashed).

Next section 2 presents surface parameters and steps of feature extraction. Section 3 contains the surface analysis itself as well as approaches of supervised learning. A comparison of these methods is given in section 4, section 5 sums up the findings of this work.

2. Feature Extraction and Surface Parameters

The feature extraction itself depends on the devices for environmental perception. Here, a stereo camera system is considered which will be placed at a height of about 50 cm at the robot chassis with a pitching angle of 35°. This sensor setup has been created in simulation (figure 3) due to two reasons: At first, it is not possible to use the current version of CROMSCI for sufficient training examples since its hardware is still not reliable enough for long-term applications. Additionally, it was not clear if it is possible at all to acquire useful information out of 3D surface data for traversability information and what image resolution and accuracy would be needed.

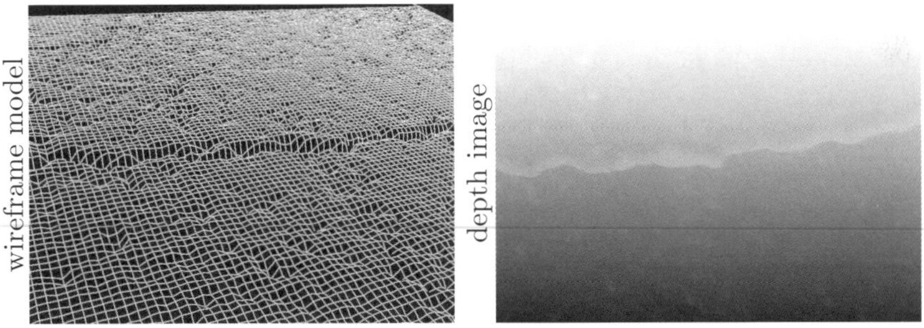

Fig. 3. Basic surface mesh (left) and acquired depth image (right) for feature extraction.

The *image processing* part of figure 2 consists of the height calculation of each pixel based in the acquired depth image. This step also includes a linear interpolation for areas which are e. g. hidden behind superstructures and a perspective transformation into a top-view of the perceived terrain.

For an analytic description of the 3D surface (figure 3) ten different parameters according to upcoming standard ISO 25178-2[6] are calculated: Statistical parameters like *arithmetic mean height* S_a and *root mean square height* S_q describe significant height deviations like cracks or holes, the *skewness* S_{sk} determines the symmetry of the height distribution, and *kurtosis* S_{ku} indicates the presence of either peak or valley defects. The topographical properties are defined by *root mean square gradient* S_{dq} and *developed area ratio* S_{dr} to differentiate surfaces with a similar average roughness. The volume is analyzed via *peak material volume* V_{mp} and *core material volume* V_{mc} indicating the clinging capability, the *core void volume* V_{vc} provides information about the resulting void volume, and the *dale void volume* V_{vv} indicates the remaining void volume after clinging. Additionally, three other characteristics are taken into account: S_v and S_p indicate the *deepest valley* and *highest peak* respectively, S_z is the *maximum height*.

3. Surface Analysis and Supervised Learning

Since it is not useful to calculate the surface parameters for the complete area they are determined for ring masks corresponding to the outer circular sealing area (figure 4). This mask has been chosen because the outer sealing is the most important connection between the adhesion system of the robot and the ground. The idea is to calculate the surface parameters of these masks beforehand and to measure the reaction of the adhesion system when the robot reaches this area. Of course, it is also possible to apply other shapes like rectangles, pie slices (representing chamber areas) or spokes (sealing areas). The main challenge is to find a meaningful rating of the analyzed area to get a suitable rating of its characteristics. Experiments have shown that the circular shape in combination with a so-called adhesion score delivers the most significant information, so this is considered here.

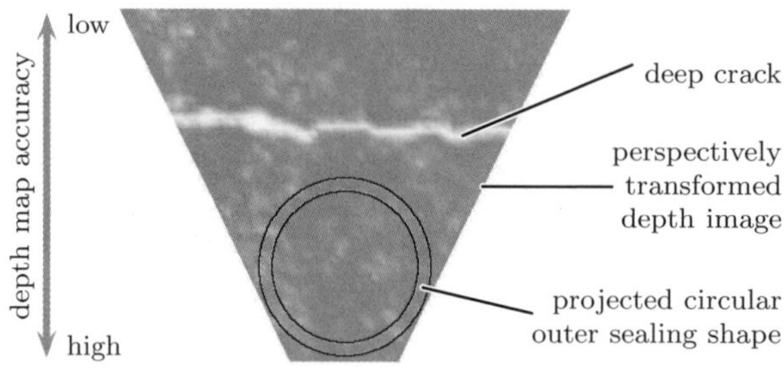

Fig. 4. Top-view of camera field-of-view with circular shape for feature analysis.

In the present case this adhesion score function $S_A \in [0,1]$ evaluates downforce F_z and point of downforce (x_F, y_F) of the adhesion system[7] according to equation 1. Here, d^{max} denotes a maximal distance between point of downforce to robot center, F_z^{min} and F_z^{max} are thresholds for minimal and maximal forces. The final training sets consist of the surface features of one circular area calculated beforehand and the corresponding adhesion score when the negative pressure chambers reach this area.

$$S_A = \max\left(0, \min\left(1, \max\left(\frac{\sqrt{x_F^2 + y_F^2}}{d^{max}}, \frac{F_z^{max} - F_z}{F_z^{max} - F_z^{min}}\right)\right)\right) \tag{1}$$

The goal is, again, a prediction of the reaction of the adhesion system based on the extracted features. Therefore, an *estimator* (compare figure 2) needs to be set up which is able to evaluate the current depth map in driving direction of the robot and to determine whether the upcoming terrain can be overcome by the robot or if it would drop off.

For the given application of feature classification, methods of supervised learning seem to be most effective. In total, three different machine learning approaches have been used and evaluated here:

- **Linear Regression** assumes a linear function $f(\vec{x})$ according to equation 2 with inputs $\vec{x} = (x_1, x_2, ..., x_p)$. The goal is to estimate the unknown coefficients β_j based on the feature vectors $\vec{x}_i$ and desired outputs y_i. The most popular method is *least squares* which minimizes the residual sum of squares[8] over all N training sets.

$$f(\vec{x}) = \beta_0 + \sum_{j=1}^{p} x_j \cdot \beta_j \tag{2}$$

- **Decision Trees** are binary trees in which each internal node corresponds to a test of one feature[8] – e. g. if $S_a > 0$ or not. Therefore, each branch represents an outcome of the test separating the remaining set (subtree) in the best way. Finally, each leaf node holds a class label which represents the prediction. A popular algorithm for the construction of a decision tree is CART using the gini coefficient as measure for a suitable division, as it is also applied here.
- **Support Vector Machines** create one or more hyperplanes in a more-dimensional space.[9] In cases of non-linear functions the input vectors need to be transformed into higher dimensions via kernel functions K to allow a linear separation in this space. The final decision function (equation 3) uses N resulting support vectors $\vec{x}_i$, coefficients α_i, y_i, and intercept β_0 to classify the input vector $\vec{x}$:

$$f(\vec{x}) = \sum_{i=1}^{N} \alpha_i \cdot y_i \cdot K(\vec{x}_i, \vec{x}) + \beta_0 \tag{3}$$

The implementation has been done using *scikit-learn*[a] which is a Python module providing machine learning algorithms. In a first step, several terrain types in form of surface patches have been created. To identify general features the training sets use normed surfaces, whereas the test sets use more realistic structures. Then, the simulated robot collects surface features and the corresponding adhesion scores during motion, serving as the desired outputs of machine learning. A total number of 3 300 datasets was recorded with a depth image resolution of 480x360 pixels.

---

[a] `http://scikit-learn.org/`

4. Experimental Results and Application

For comparison of the introduced learning methods several experiments were executed. Figure 5 illustrates the results once for the training data (left) and for the datasets for testing (right) by applying the trained regressions. For comparison, two different indicators are considered here: The *mean-square error* (MSE) should be close to zero if the prediction nearly fits the ground truth value S_A from the data sets. Additionally, the R^2 score is used which is a statistical measure of how well the regression line fits the given data. This score should be around one in an optimal case.

In terms of *linear regression* third order polynomials of each feature delivered best results, although the test examples are still to scattered which makes this method inapplicable for the given task. The *decision tree regression* delivers very promising results in the training phase (second row in figure 5) with a maximal tree depth of five. But, the resulting scores for prediction of S_A of the test data are much worse which can be seen in a mean-square error (MSE) of 0.123 compared to 0.015 in the training phase. The *support vector machine* delivers best results for the test data as it can be seen in the MSE and R^2 values in the last line of figure 5. Basis of this method was a grid search to optimize the tuning parameters and a cross-validation, which splits the training data into k subsamples of equal size, one for training and one for validation.

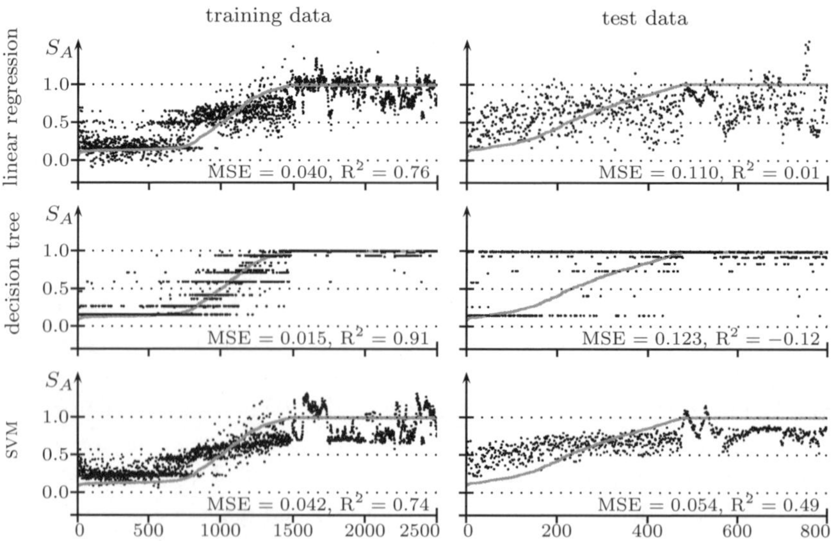

Fig. 5. Predicted (dots) and real (gray curve) outcomes of S_A using different regressors in terms of 2 500 training data (left) and 800 sets for validation (right).

Beside the presented regression methods also *decision tree classification* and *support vector classification* were tested. Here, the adhesion score has been divided into the three classes *safe* ($S_A \leq 0.25$), *drop-off* ($S_A > 0.99$) and *risky* (else). Both classificators delivered very good results in prediction of a drop-off, but the overall precision rate lies only at 0.57 in case of the decision tree classification and 0.40 for using support vectors.

Since the results seemed goal-leading but not sufficient so far, 500 additional training sets were used to test if the results could be improved by more datasets. The resulting predicted values S_A were expanded and shifted via function T (equation 4) to get more significant values.

$$T(S_A) = \max\left(0, \min\left(1, 4 \cdot S_A - 1.5\right)\right) \tag{4}$$

The result for the scaled SVM regressor can be seen in figure 6. Again, the predicted values and the ground truth of the 800 test samples are shown. Finally, a mean-square error of 0.032 and a R^2 score of 0.71 could be achieved which is a significant improvement compared to figure 5.

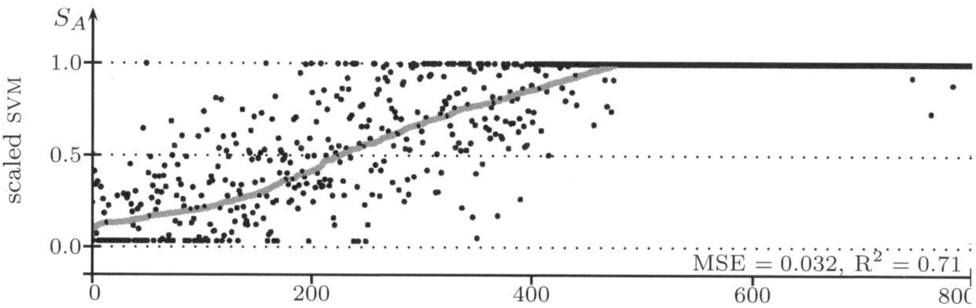

Fig. 6. Scaled validation results of predicted (dots) and real (gray curve) test data of the final support vector machine regressor.

It can be stated that each of the examined methods performs rather well on the training datasets, but none of the examined methods could achieve good results for the untrained datasets used for testing. Among the regression approaches, the support vector machines especially in combination with the scaling transformation delivered the best results. The scatter plots of the decision tree regressor e. g. show that this method predicts the state riskier than it really is. Based on these results further experiments and trainings seem to be necessary. On the one hand the training surfaces were constructed fine-grained structures whereas the test environment is coarser. On the other hand it seems that the camera resolution might be too low during data acquisition.

5. Conclusion

The presented paper aimed at aspects of terrain traversability in the range of climbing robots. Starting from a depth map captured by a simulated stereo camera, several surface features were extracted that distinguish among varying terrain structures. The traversability information are acquired using methods of supervised learning based on training sets. To determine the best supervised learning method, an investigation of a couple of different approaches was undertaken. It turned out that a modified support vector regressor reached a high prediction quality although there is still space for further improvements.

Next steps are the implementation of an improved rating value to characterize the adhesion state and the testing of further surface characteristics. Finally, more statistical data as well as training data from the real robot and stereo vision system are needed to bring this approach into application.

References

1. M. Castelnovi, R. Arkin and T. R. Collins, Reactive speed control system based on terrain roughness detection, in *Proceedings of the International Conference on Robotics and Automation (ICRA)*, (Barcelona, Spain, 2005).
2. T. Braun, H. Bitsch and K. Berns, Visual Terrain Traversability Estimation using a Combined Slope/Elevation Model, in *Proceedings of the German Conference on Artificial Intelligence (KI)*, (Kaiserslautern, Germany, 2008).
3. S. Thrun, M. Montemerlo and A. Aron, Probabilistic terrain analysis for high-speed desert driving, in *Robotics Science and Systems Conference*, (Philadelphia, USA, 2006).
4. A. Kelly and A. Stentz, Rough terrain autonomous mobility - Part 1: A theoretical analysis of requirements, in *Journal of Autonomous Robots*, (2) (Springer, 1998) pp. 129–161.
5. D. Schmidt, C. Hillenbrand and K. Berns, Omnidirectional locomotion and traction control of the wheel-driven, wall-climbing robot, cromsci, in *Robotica Journal*, (7) (Cambridge Univ Press, 2011) pp. 991–1003.
6. T. Torims, J. Vilcans, M. Zarins, I. Strazdina and A. Ratkus, Implications of the new iso surface roughness standards on production enterprises, in *Annals of DAAAM and Proceedings*, (DAAAM International Vienna, 2010)
7. D. Schmidt and K. Berns, Risk prediction of a behavior-based adhesion control network for online safety analysis of wall-climbing robots, in *Proceeding of the 9th International Conference on Informatics in Control, Automation and Robotics (ICINCO)*, (Rome, Italy, 2012).
8. T. Hastie, R. Tibshirani and J. Friedman, *The Elements of Statistical Learning: Data Mining, Inference and Prediction* (2001).
9. H. Yu and S. Kim, Svm tutorial - classification, regression and ranking, in *Handbook of Natural Computing*, (Springer, 2012)

DEVELOPMENT OF HUMANOID ROBOT TEACHING SYSTEM BASED ON A RGB-D SENSOR*

CHUNTAO LENG†

School of Mechanical Engineering, Shanghai Jiao Tong University, Shanghai, China
Engineering Training Center, Shanghai Jiao Tong University, Shanghai, China

QIXIN CAO

School of Mechanical Engineering, Shanghai Jiao Tong University, Shanghai, China

BO FANG

Engineering Training Center, Shanghai Jiao Tong University, Shanghai, China

YANG YANG

School of Mechanical Engineering, Shanghai Jiao Tong University, Shanghai, China

ZHEN HUANG

School of Mechanical Engineering, Shanghai Jiao Tong University, Shanghai, China

A 17 degree-of-freedom humanoid robot is developed, and a humanoid robot teaching system based on RGB-D sensor is proposed. Through the analysis of the robot's joint configuration, the transformation formula from the Cartesian coordinate space to the joint space was deduced. Using Microsoft Kinect, the human body skeleton is mapped to the humanoid robot joint. To smooth data without causing delay, a Velocity-based Selective Mean Filter (VSMF) was proposed. The teaching system was applied on the small humanoid robot and the experiment results verified the effectiveness.

1. Introduction

The Robot Teaching is the important area of robotics research, and it is widely used in practice, such as Welding robot teaching [1], Handling robot teaching [2]. Currently, robot teaching is mainly based on the simulation software [3]. It is necessary to manually specify the joint angle of the key points of the motion sequences, and to achieve the smooth movement of the robot by

---

* This work is supported by the National High Technology Research and Development Program of China (No.2012AA041401, 2012AA041403)

† Email: ctleng@sjtu.edu.cn

interpolation method. However, this method is very complicated in practical applications, and it needs higher skill requirements for the operator.

With the development of computer vision technology, researchers have realized the real-time acquisition of human motion model. Especially after the release of the Microsoft Kinect sensor, there is great progress for the real-time modeling of body posture based on RGB-D sensor. When this body posture real-time modeling techniques are applied, we can quickly obtain the three-dimensional coordinates of the major joints of the human body [4].

The key point for robot teaching based on RGB-D sensor is to identify the mapping relationship between the collected joint data and the real robot joint. There are some related researches. The literature [5] presented a system which enabled a human user to teach a robot how to grasp an object. The literature [6] captured Human postures using Kinect sensor and controlled the humanoid robot. But the noisy data was not dealt with, which affect robot teaching accuracy.

In this paper, we developed a small humanoid robot, called BioRobo, and built a robot teaching system based on RGB-D sensor. According to the configuration of the robot, we obtain the three-dimensional coordinates of the body joint, and achieve the conversion from the Cartesian coordinate space to the articulation space. VSMF filtering was proceeded for the noise of the sensor data, and a stable joint angle data can be obtained. Finally, the proposed method was applied to the small humanoid robot, and achieved good results.

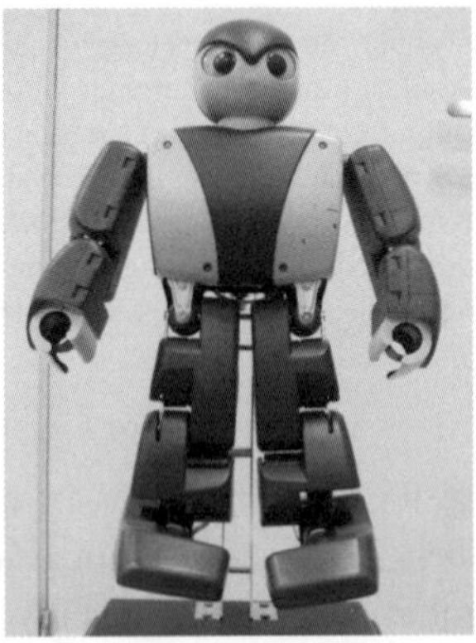

Figure 1. BioRobo

2. BioRobo robot system

BioRobo robot is configured with 17 degrees of freedom, which is shown in Figure 1. In order to improve the real-time performance, stability and scalability of small humanoid robot control system, we designed a distributed robot control system in hardware. Based on STM32-core processing module, we build the robot master controller module, the sensor interface module, the acceleration sensor module and the camera module. We created a digital bus servo control system and used a field programmable processor array as the core of robot motion control module.

3. Introduction to the teaching system

Robot teaching system consists of a RGB-D sensor, computer, humanoid robot. Kinect is applied in this system as the RGB-D sensor (Figure 2). Kinect is developed by Microsoft, which contains a low cost RGB camera, a standard CMOS camera and a depth camera which can collect the depth information. When Kinect collected the controller's body gestures, we can calculate and obtain the joint data. After mapping these joint data to the robot, the robot can follow the controller's movement.

Figure 2. Microsoft Kinect

4. Humanoid Robot Teaching Method

Because of the difficulty of small humanoid robot motion teaching and the complexity of procedures, we proposed a method based on the RGB-D sensor. According to human body action, we mapped the body joint angle to the robot joint angle. After DOF cutting and Velocity-based Selective Mean Filter, the users can teach robot by their body motions. We also introduced the collision detection to improve the safety of robot teaching. After a period of use in many areas, the experiments of small humanoid robot BIOROBO has got satisfactory results.

4.1. *Robot joint configuration and calculation for joint space parameters*

The general humanoid robot manipulators have six or seven degrees of freedom. In this paper, we take the typical service robot as example. There are seven degrees of freedom for each arm, which are shown in Figure 3.

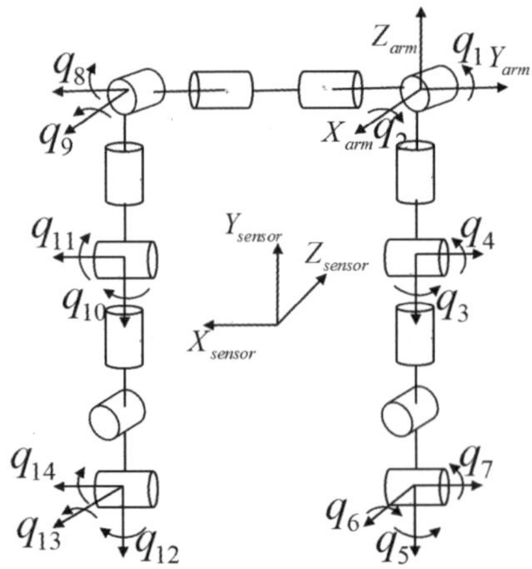

Figure 3 Typical humanoid robot joint configuration

Using Kinect sensor and OpenNI drive, we can obtain three-dimensional coordinates of the human head, neck, shoulder, elbow, wrist, waist, hip, knee, ankle joint, and etc., total 15 joints. With the three-dimensional coordinates of the shoulder, elbow and wrist, it is possible to obtain the rotational angle of four joints of the arm. Similarly, with the three-dimensional coordinates of the hip, knee and ankle, it is possible to obtain the rotational angle of four joints of the leg.

Take the left arm as example. The sensor coordinate system, the robot arm coordinate system, the zero position and the rotation direction of each joint is shown in Figure 3. Set the coordinates of the shoulder, the elbow and the wrist joint in the sensor coordinate system are $P_{sh} = [x_{sh}, y_{sh}, z_{sh}]^T$, $P_{el} = [x_{el}, y_{el}, z_{el}]^T$, $P_{wr} = [x_{wr}, y_{wr}, z_{wr}]^T$. Next, we will calculate the joint angle q_1, q_2, q_3, q_4.

The vector pointing from the shoulder to the elbow is $V_{up} = [x_{up}, y_{up}, z_{up}]^T = P_{el} - P_{sh}$, and the vector pointing from the elbow to the wrist is $V_{low} = [x_{low}, y_{low}, z_{low}]^T = P_{wr} - P_{el}$.

According to the robot kinematics and Figure 4, we can obtain Eq. (1) and Eq. (2).

$$q_1 = a\tan 2(z_{up}, y_{up}) \tag{1}$$

$$q_2 = a\sin(x_{up}) / \|V_{up}\| \tag{2}$$

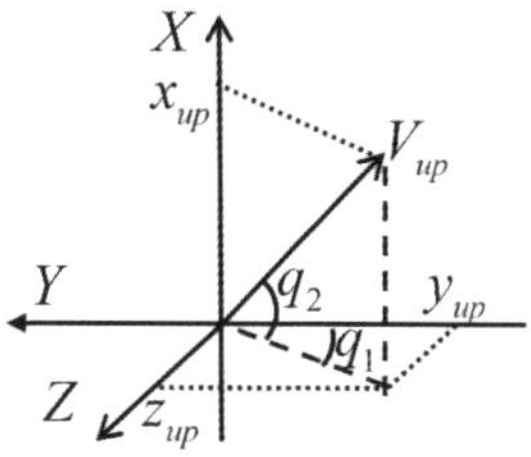

Figure 4. The first two joint diagram of upper arm

Next, we calculate q_3 and q_4. q_3 is the angle between the vector (N) and the initial position. N is the normal vector of the plane defined by the upper arm and forearm. According to the definition above, $N = V_{up} \times V_{low}$, set the normal vector of the initial position is N_0, then we can get the q_3 as follows:

$$q_3 = a\cos(\frac{N_0 \cdot N}{\|N_0\| \cdot \|N\|}) \tag{3}$$

According to the coordinate system defined in Figure 3, $N_0 = [1,0,0,1]^T$ in the initial position. When the first two joints of the left arm is rotated, N_0 changes accordingly in the sensor coordinate system, and the Transformation matrix is:

$$T_{nor} = \begin{bmatrix} 1 & 0 & 0 & 0 \\ 0 & \cos(q_1) & -\sin(q_1) & 0 \\ 0 & \sin(q_1) & \cos(q_1) & 0 \\ 0 & 0 & 0 & 1 \end{bmatrix} \cdot \begin{bmatrix} \cos(q_2) & -\sin(q_2) & 0 & 0 \\ \sin(q_2) & \cos(q_2) & 0 & 0 \\ 0 & 0 & 1 & 0 \\ 0 & 0 & 0 & 1 \end{bmatrix} \tag{4}$$

Therefore, the coordinate of N_0 in the sensor coordinate system is $N_0 = T_{nor} \cdot N_0$. Substituting this value into Eq. (3), we can get the third joint angle. According to the equation, the value of q_3 ranges from 0 to π, but this value may be negative in the actual case. In order to deal with this situation, we can determine the angle is positive or negative according to the direction of the vector. In order to determine the direction of the vector in the sensor coordinate system, redefine N_0 and N, $N_0 = [1,0,0,1]^T$, $N = inv(T_{nor}) \cdot N$. Then we can obtain Eq. (5).

$$q_3 = a\cos(\frac{N_0 \cdot (inv(T_{nor}) \cdot N)}{\|N_0\| \cdot \|(inv(T_{nor}) \cdot N)\|}) \cdot sign((N_0 \times (inv(T_{nor}) \cdot N)).y) \tag{5}$$

The fourth joint angle q_4 is the angle between the upper arm V_{up} and the forearm V_{low}, which is

$$q_4 = a\cos(\frac{V_{up} \cdot V_{low}}{\|V_{up}\| \cdot \|V_{low}\|}) \tag{6}$$

4.2. *Data filtering and smoothing*

Due to the presence of measurement noise, the calculation of the angle of each joint must be error. For example, when the human body maintains a posture unchanged, the joint angle measured persist fluctuations. In order to smooth the data, the commonly used method is mean filter. But the mean filter often brings lag problem. Therefore, we propose a new method, which called Velocity-based Selective Mean Filter (VSMF). Before processing the mean filter on the current frame data, according to the historical movement speed, judge the human is in rapid movement period or relative stationary period. If the human is in the relative stationary period, we do the mean filter on the frame data. Otherwise, we do not do the mean filter. The pseudo code of the algorithm is shown in Table 1 below:

Table 1. Pseudo code of VSMF

1. Setting the measured data is *data*, the result after VSMF is *dataF*. Setting the width of the filtering window is *L*, the speed threshold is *T*, the initialization stabilization time $t = 0$;
2. The data of current frame is *data(i)*

```
if abs(data(i)-data(i-1))<T
    t++;
else
    t=0;
end
if t>L
    dataF(i)=mean(data(i-L:i))
else
    dataF(i)=data(i);
end
```

5. Experiments

In order to test the effect of VSMF, we analyze the data of fourth joint of left arm. In Figure 5, the data is processed by mean filter and VSMF respectively. The abscissa represents the data serial number, and ordinate represents the fourth joint angle. As can be seen from the figure, the fourth joints experienced a period of rapid movement between the two relatively stable periods. During the relatively stable period, the performance of the VSMF is almost the same with the mean filter. Both of them can effectively reduce the data shock. But during the fast movement, the mean filter method results a certain lag, and the VSMF method can quickly follow the change of data.

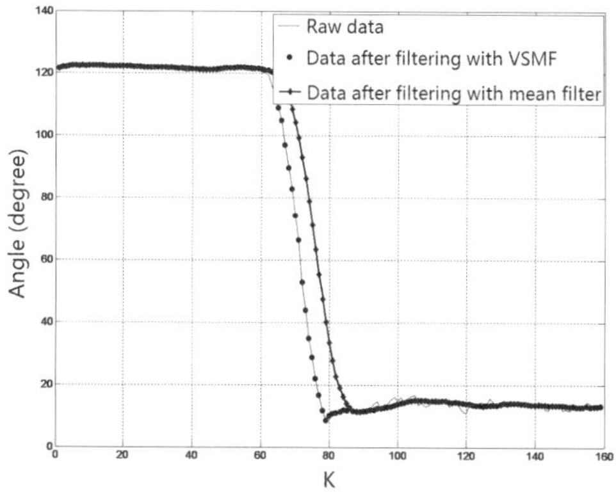

Figure 5 comparison between VSMF and mean filter

In order to verify the effectiveness of VSMF method, we performed experiments on BioRobo developed in our lab. There are three degrees of freedom for the robot arm and leg. Compared to the previous model, BioROBO has no torsional freedom of the upper arm and forearm, and the freedom of wrist. So we ignore these degrees of freedom in mapping. Using the robot teaching method proposed above, the real robot can be controlled following the human gesture. Figure 6 shows the photos of experimental scene. The experimental results show that the robot can correctly follow human action.

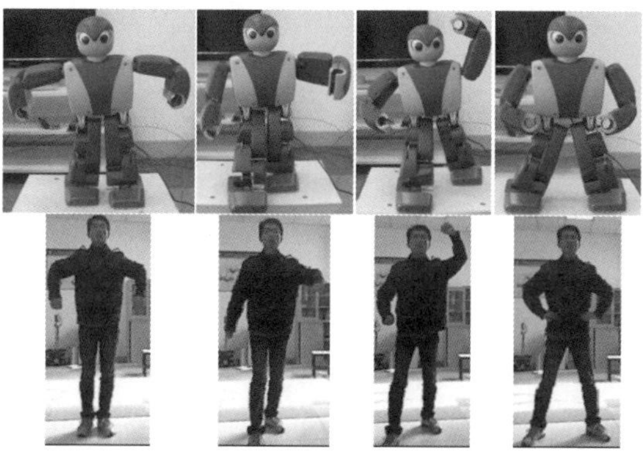

Figure 6. Humanoid robot teaching for BioRobo

6. Conclusion

A humanoid robot teaching system based on the RGB-D sensor is proposed in this paper. We use Microsoft Kinect to get the human body skeleton and map it to the humanoid robot, also propose the VSMF method for data filtering and smoothing. Finally, the effectiveness of the proposed method is verified by experiments on BioRobo. Without loss of generality, the method can also be applied to other similar humanoid robot platform with more degrees of freedom.

References

1. GONG Ye-fei, LI Xin-de, DAI Xian-zhong, et al. A weld robot off-line programming system integrated with virtual structured-light sensor [J]. Transactions of the China Welding Institution, 32(4): 17-20 (2011).
2. LIU Xiang-quan, YUN Chao, ZHANG Zhi-qiang, et al. Development of mixed-connection robot palletizer based on PMAC [J]. Journal of Machine Design, 26(5): 49-52 (2009).
3. GAN Zhi-gang, Xiao Nan-feng. Design and implementation of virtual teaching system for humanoid robot[J]. Journal of South China University of Technology (Natural Science Edition), 36(1): 18-24 (2008).
4. Shotton J, Fitzgibbon A, Cook M, *et al.* Real-Time human pose recognition in parts from single depth images [C]. IEEE Conference on Computer Vision and Pattern Recognition, USA: IEEE, pp1297-1304 (2011).
5. Tien Cuong KIEU, Kimitoshi YAMAZAKI, Ryo HANAI, Kei OKADA, Masayuki INABA, "Grasp Observation and Reproduction by Humanoid Robots Using Color Camera and 3D Sensor", IEEE/SICE International Symposium on System Integration (SII), pp 808-813 (2011)
6. RR Igorevich, EP Ismoilovich, et al, "Behavioral synchronization of human and humanoid robot", 8th International Conference on Ubiquitous Robots and Ambient Intelligence (URAI), pp655-660, Nov. 23-26 (2011)

SECTION–6

INNOVATIVE DESIGN OF CLAWAR

DEVELOPMENT OF WORM-RACK DRIVEN CYLINDRICAL CRAWLER UNIT

JUN-YA NAGASE

Department of Mechanical and Systems Engineering, Ryukoku University, 1-5 Yokotani, Seta Oe-cho, Otsu, Shiga 520-2194, Japan

KOICHI SUZUMORI

Division of Manufacturing System and Physical Science, Okayama University, 3-1-1 Tsushimanaka, Okayama, Okayama, 700-8530, Japan

NORIHIKO SAGA

Department of Human System Interaction, Kwansei Gakuin University 2-1 Gakuen, Sanda, Hyogo, 669-1337, Japan

In Japan, since the Great Earthquake, attention has surrounded development of disaster victim relief activity using robots. Development of a robot that is adaptive to various environments in irregular places, rubble, and confined areas is needed. A wheel movement type robot, a robot in the shape of a snake, a crawler type robot and a multi-leg robot are all described in the relevant literature as rescue robots that are currently studied. Although these robots are high efficiency and/or high ground-covering ability, they need comparatively large space to move. In this study, a cylindrical crawler robot based on worm-rack mechanism, that doesn't need large space to move and has high ground-covering ability, is proposed. In this report, structure, driven mechanism, design, prototype, and experimental evaluation are reported.

1. Introduction

The Great Hanshin-Awaji Earthquake and the September 11, 2001 attacks triggered the development of rescue robots for relief and search in disaster scenarios. It is strongly demanded in particular that robots be used to conduct relief and search efforts in places that are inaccessible to humans. For that purpose, various rescue robots have so far been developed, such as wheeled robots [1], multi-legged robots [2], snake-like robots [3], and a peristaltic crawling robot [4]. Among them, the tracked crawler robot [5] can travel by traversing uneven ground flexibly with its crawler belt attached to the ground surface firmly in conditions that afford a large ground contact area. It is particularly advantageous in its potential to break through irregular areas such as rubble heaps. However, conventional crawler robots have difficulty in travelling in confined spaces such as under rubble or inside a thin pipe, because of their complicated structure and mechanism constraints.

A flat-body crawler robot [6] has been developed with a shape that allows it to move around inside a space of low height, such as under rubble. However it still cannot travel inside confined spaces or in a thin pipe. A crawler robot having multiple conventional crawler mechanisms in axial symmetry [7] has been developed for travelling in a pipe, but its complicated structure and numerous actuators do not permit it to travel in a pipe with a small bore. In addition, it cannot move across flat ground.

This study is therefore aimed at implementation of a robot with high running performance in confined spaces such as those under rubble or in a thin pipe. A small, lean, and simple crawler robot has been developed, with multiple crawlers configured in axial symmetry at equal intervals on a cylindrical frame and driven by a single motor through a worm rack mechanism. This report describes the structure, driving principle, design, and prototype evaluation of the proposed cylindrical crawler unit.

2. Mechanism

2.1. *Basic structure*

Figure 1 depicts a structural drawing of the cylindrical crawler unit proposed in this study. This robot is a cylindrical robot comprising a geared motor, crawler belts made of silicone rubber, and a worm and a frame made of resin.
Six crawler belts are spaced at equal intervals around the longitudinal axis of the cylindrical frame. Each crawler belt is wound around the frame, of which both ends are connected to produce a loop. In addition, teeth having a rake angle corresponding to the lead angle of the worm tooth are formed on the outer surface of the crawler belt. The corner of those teeth forms a curved surface, which allows the crawler belt to expand the contact area with the ground when they are in contact, so that enhanced driving force is generated.

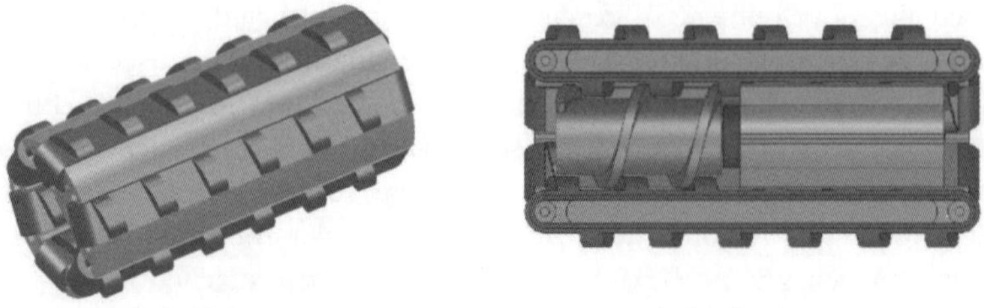

(a) Outview (b) Section

Fig.1. Structural drawing of the cylindrical crawler robot proposed in this study.

A pair of rollers is furnished for each crawler belt at both ends of the frame to suppress the energy loss by friction of the crawler belt. Furthermore, notches on the teeth of a crawler belt, decrease the normal force from the belt to the roller and reduce the frictional force. The gearmotor is mounted as stationary inside and coaxially to the frame. The worm is attached to the motor shaft, and is positioned to engage with each of the six crawler belts.

Its driving principle is described. First, the motor fixed to the frame is driven and the worm is rotated via the gear. Then, the rotating worm provides all six crawler belts engaged with propulsion to the axial direction to kick the contact surface. This is the mechanism by which the robot moves forward. Moreover, inverse rotation of the motor enables backward movement.

This robot has a simple and compact cylindrical crawler mechanism in which multiple crawler belts in axial symmetry to a cylindrical frame are driven by a single actuator via a single worm. Therefore it is suitable for travelling in confined spaces such as those under rubble or in a thin pipe, which has persisted as a difficult task to date. The features of this cylindrical crawler robot can be summarized as follows:

(i) simple structure driven by the worm rack mechanism, facilitating considerable reduction in size compared to the conventional cylindrical crawler structure,
(ii) suitable for travelling in confined spaces, such as those under rubble or in a thin pipe,
(iii) operable even with top-bottom or left-right sides of the robot in contact with walls at the same time by virtue of its multiple crawler belts in axial symmetry, and
(iv) easily retrievable from a sudden fall at a level difference.

Because robots are strongly demanded for rescue operations at disaster scenes that are inaccessible to humans, implementation of such a robot is an extremely important subject. As a solution to this issue, making the most of the crawler mechanism of this study, which allows travel under rubble or in a thin pipe employing the features of the high running performance of the conventional crawler mechanism, is considered to contribute greatly to relief and search activities at earthquake disaster sites and at nuclear plants. In addition, the proposed robot mechanism has such a simple structure that substantial size reduction is anticipated.

2.2. *Extend to 2 degrees of freedom of motion*

Multiple crawler units connected in series in the articulated structure allows turning by bending joints between units. However, conventional articulated structure [5-8] spends many time and high energy loss in order to enter into the narrow space, because it should readjust each joint angle several times. Therefore in this study, the structure that can move the crawler unit mentioned above to lateral direction was devised. The articulated structure, which is composed of such a structure, can reduce the time loss and energy loss, because it can conduct various motion such as lateral movement, rotation on the spot and diagonally forward movement as well as backward and forward.

As a conventional crawler robot that can move laterally, there is Omni-Crawler [8]. It is difficult to downsize such a crawler, because this crawler structure has at least two geared motors. One geared motor is for moving backward and forward, and another one is for moving laterally. On the other hand, the number of geared motor of our proposed crawler has only one. Therefore our proposed crawler can also enter into the narrow space because it is easy to downsize the crawler.

Section of the structure of our devised crawler unit is shown in Fig.2. This crawler unit has two kinds of mode. Fig.2 (a) shows "backward and forward mode", and Fig.2 (b) shows "lateral mode". These modes are changed by the small linear actuator that is arranged behind of geared motor. Because the length of the linear actuator is much shorter than that of the geared motor, it is easy to downsize the body. Driven mechanism of the crawler unit is as follows.

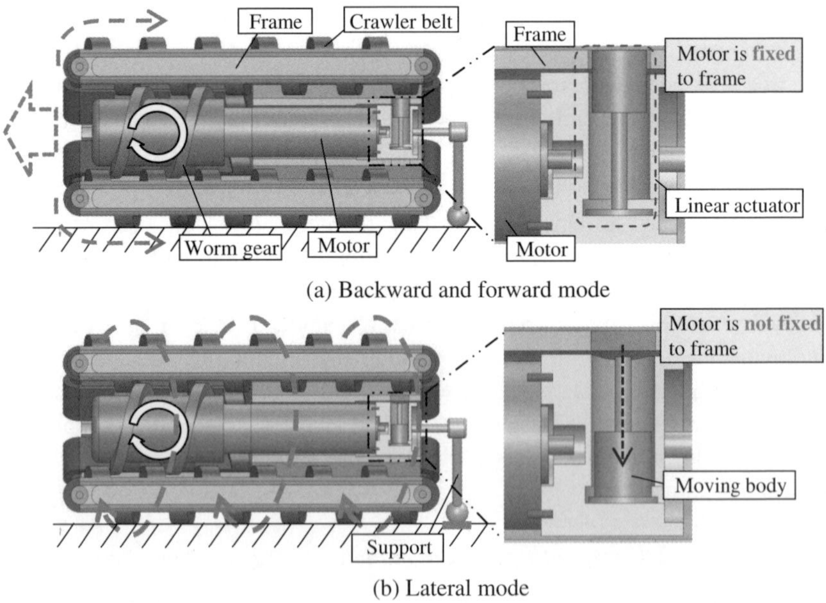

Fig.2 Structure and driven mechanism of our proposed crawler unit

In "backward and forward mode" of Fig.2 (a), the geared motor is fixed to the frame by the moving body of the linear actuator through the attachment. This structure is equivalence to the structure of Fig.1. Therefore, after the geared motor fixed to the frame is driven and the worm is rotated via the gear, the rotating worm provides all six crawler belts engaged with propulsion to the axial direction to kick the contact surface. On the other hand, in "lateral mode" of Fig.2 (b), the geared motor is not fixed to the frame, because the moving body of linear actuator is down. The support rod is attached to the motor through the attachment as shown Fig.5 (2). It is slightly longer than outer radius of the crawler unit. Therefore, when the geared motor is driven, the reaction force for the force generated between worm and crawler belt is transmitted to the surface between the tip of support rod and the ground. And then, the crawler unit rotates to lateral direction by radius force generated between worm gear and crawler belt tooth. As mentioned above, our proposed crawler unit can move laterally as well as backward and forward by single geared motor and single small linear actuator. Moreover, it can also move to omni-direction by connecting multiple units in series.

3. Design

This chapter describes the design of the proposed crawler unit. Because the crawler belt of this crawler unit is made of rubber, insufficient tooth thickness might cause large elastic deformation in belt toothing at the time of engagement with the worm, which degrades the driving efficiency. However, an overly thick tooth might increase the energy loss by friction with the frame or the roller. For that reason, this study adopted 8 mm by trial-and-error as the minimum tooth thickness that does not cause decreased driving efficiency by distortion of belt toothing.

In addition, an overly small face width of a crawler belt might bring about high stress in the crawler belt toothing, such that large distortion might occur for

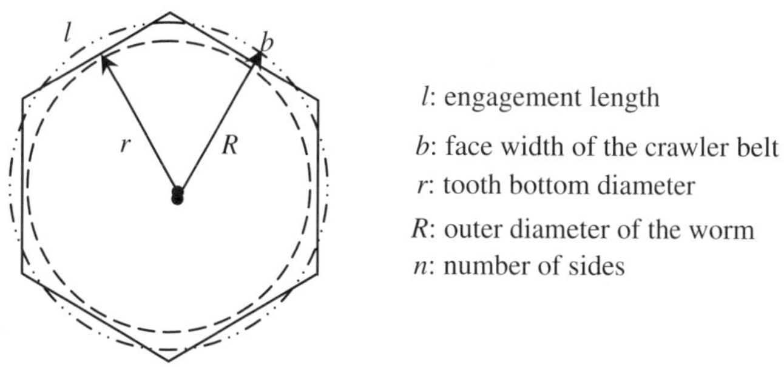

Fig.3 Schematic diagram of the proposed robot in an axial view, indicating the spatial relationship of the worm and the crawler-belt toothing.

the reason described above. Therefore, it is necessary to design a device to make the contact length as long as possible to secure the strength of crawler belt toothing. Furthermore, because the tip face of the crawler belt toothing delivers driving force to the ground surface, the contact area should be as large as possible to assure frictional force. Consequently, the crawler belt maximum engagement length was examined using the geometric spatial relationship of the worm and the crawler belt. The face width was determined. Figure 3 presents a schematic diagram of the proposed robot in an axial view, indicating the spatial relationship of the worm and the crawler-belt toothing, where n, b, l, r, and R respectively denote the number and the face width of the crawler belt, the engagement length, the tooth bottom diameter, and the outer diameter of the worm. Geometric spatial relations such as the number of crawler belts, the engagement depth, and the tooth bottom diameter of the worm in Fig. 3 derive the face width b_{max} when the engagement length l becomes the maximum with the outer diameter of the worm being R and the number of crawler belts n as in the following equation:

$$b_{\max} = 2R\sin\left(\frac{\pi}{n}\right) \tag{1}$$

The tooth bottom diameter was determined as the maximum so that it would not interfere with the crawler belt tip. That is, when an n-gon is circumscribed to a circle of radius r in Fig. 3, the maximum radius rmax can be calculated using the following equation:

$$r_{\max} = \frac{b}{2\tan(\pi / n)} \tag{2}$$

The traveling speed of the robot V is obtained as

$$V = \frac{pN}{i} \tag{3}$$

where the rotating speed of the motor N, the reduction ratio of the gear i, and the pitch of the worm tooth p. The target traveling speed of the robot was set to 100 mm/s in this study. Therefore, a gear with tooth pitch of 16.3 mm and a reduction ratio of 67 was employed, and a motor of a rotating speed of 30,000 rpm was adopted according to Eq. (3).

4. Prototype end experiment

Based on the design described above, a crawler unit was prototyped. The production procedure of the crawler belt made of silicone rubber by mold forming is described first. Liquid rubber (KE-1600; Shin-Etsu Chemical Co.,

Ltd.) was poured into a mold, with a lid on, and was left untouched for 60 min in a 70 ℃ thermostatic chamber. Then, after taking out the rubber fabricated by solidification, liquid rubber was applied to both ends, which were bonded and left untouched for 60 minutes in the thermostatic chamber again for complete adhesion. The crawler belt was finished. The frame and worm made of ABS resin were molded with a 3D printer (Dimension Elite; Stratasys, Inc.). The employed gearmotor was a brushless DC motor (GP13A SL; Maxon Motor AG). Figure 4 exhibits the prototype of the cylindrical crawler unit. Table 1 presents its specifications. The prototyped robot dimensions are total length of 100 mm, an outermost diameter of 47 mm, and mass of 0.13 kg.

Fig.4. Prototype of cylindrical crawler robot

Table 1 Specification of proposed crawler.

Total length	110mm
Outside diameter	50mm
Total weight	0.16kg
Number of crawler belt	6
Material of crawler belt	Silicone rubber
Pitch of worm gear	16.3mm
Lead angle of worm gear	26mm
Length of support	35mm
Motor	6W GP13A SL (Maxon motor)

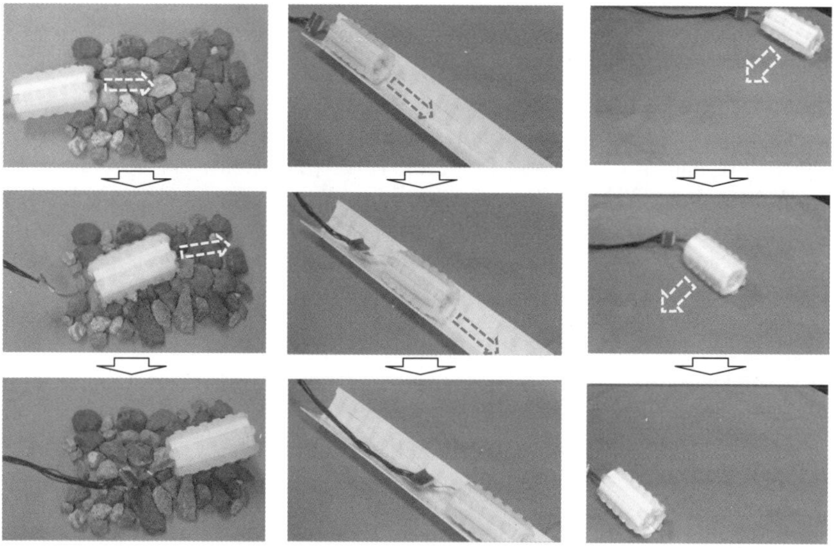

(a) Forward on rubble (b) Forward on v-groove (c) Lateral motion

Fig.5 Experiment result

Next, a performance test of the prototyped crawler unit was carried out. In this test, two kinds of running condition, that is, on rubble, inside pipe, are set. Then we examined whether the crawler unit could move on uneven ground and in confined spaces. Moreover we examined whether it could move laterally. The experimental result is shown in Fig.5. As a result, the crawler unit traveled by traversing uneven ground such as on rubbles with its crawler belt attached to the ground surface firmly. It cloud also propelled inside a pipe. Furthermore, the crawler unit could move laterally by the rotation itself.

5. Conclusion

This paper describes a worm-rack-driven cylindrical crawler unit developed as a rescue robot that can travel in confined spaces such as those under rubble or in a thin pipe, and can move laterally by the rotation itself. This study is concluded as follows.

(i) A worm-rack-driven cylindrical crawler unit was developed, following different mechanisms from the driving principle of the conventional crawler, for use as a rescue robot that can travel in confined spaces or in a thin pipe. This is a simple and compact crawler unit that propels a robot by driving multiple crawler belts with a single motor.

(ii) In the performance test, the crawler unit traveled by traversing on rubbles and propelled inside a pipe. Furthermore, the crawler unit could move laterally by the rotation.

References

1. P. Li, S. Ma, B. Li and Y. Wang, Proc. of 2008 IEEE/RSJ International Conference on Intelligent Robots and Systems, 2643 (2008).
2. K. Ito and F. Matsuno, Proc. of IEEE Int. Conf. on Robotics and Automation, 3392 (2002).
3. S. Hirose, H. Yamada, IEEE Robotics & Automation Magazine, 88 (2009).
4. N. Saga and T. Nakamura, Intelligent autonomous system, Vol.8, No.1, 85 (2004).
5. K. Nagatani, D. Endo and K. Yoshida, Proc. of International Conference on Robotics and Automation, 2752 (2007).
6. I. Fukumoto and A. Sato, Japanese Unexamined Patent Application Publication No. 2008-213671, in japanese.
7. H. Miyamoto, Japanese Unexamined Patent Application Publication No. 2002-220049, in japanese.
8. K. Tadakuma, R. Tadakuma and K. Nagatani, Proc. of IEEE/RSJ International Conference on Intelligent Robots and Systems, 2422 (2008).

UNCONVENTIONAL FIVE-LEGGED ROBOT FOR AGILE LOCOMOTION

Marek WĄSIK, Mikołaj WASIELICA, Piotr SKRZYPCZYŃSKI*
Institute of Control and Information Engineering, Poznan University of Technology
ul. Piotrowo 3A, PL-60-965 Poznań
**E-mail: piotr.skrzypczynski@put.poznan.pl*

This paper presents the unconventional five-legged walking robot. The machine has radial symmetry, and its legs are equally distributed around the body. This results in omnidirectionality – the ability to change the walking direction without the need of rotation. The main objective of the presented research was to test the concept of a five-legged, radially symmetric robot and to demonstrate the motion agility of such an unique configuration.

Keywords: Walking robot, Five-legged, Kinematics, Gaits

1. Introduction

The ability to traverse rough terrain and clear major obstacles makes legged robots useful in patrol, search and rescue applications.[2,8] In order to have good obstacle negotiation properties a walking robot should be able to change its posture while walking – that is to lift or to lower the trunk with regard to the ground, which provides the ground clearance when necessary, but also allows the robot to reach farther with its legs, in order to overcome a ditch or a curb. A very useful property is also the omnidirectionality of motion. Many legged robots have the ability to walk backward or sideways (like crabs do), but usually these walking modes are slower than the normal forward walking, because the kinematic workspace of the legs can be used only partially, due to some mechanical limitations. Statically stable gaits are safer in uneven terrain, and robots using mostly statically stable gaits do not need such complicated control algorithms and exteroceptive sensing as robots performing dynamic walking.[2] The smallest number of legs that allows static walking is four, but the quadruped crawl gait, in which only one leg is moved at time is intrinsically slow.[4] The speed of a walking robot depends on the number of legs the robot has. If a statically stable gait is considered a hexapod can achieve higher speed than a quadruped because it can use the tripod gait, in which two sets of three legs each are moved repeatedly. However, a hexapod configuration usually needs more actuators and results in a bigger, heavier machine. If agile motion in confined spaces is considered, e.g. to enter collapsed building structures, it is also important for the robot to be compact and lightweight.

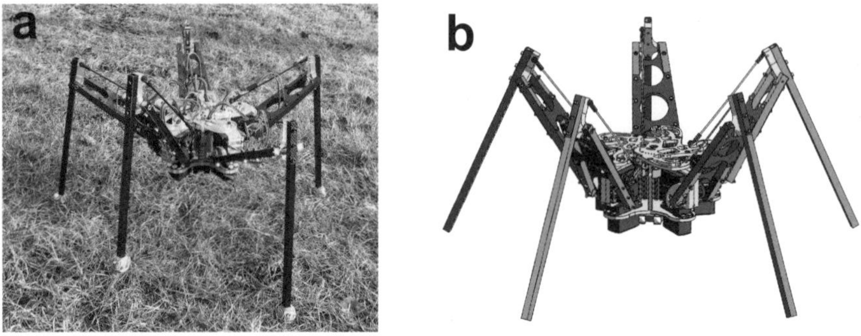

Fig. 1. PentOpiliones, the five-legged robot: traversing natural terrain (a), and in a CAD drawing (b)

Considering the above-mentioned requirements and limitations, and the properties of known walking robot configurations, we conceived an unconventional, five-legged walking robot. In Nature all land animals and insects have a bilateral body plan and an even number of legs. However, there exist species of sea organisms that have radial symmetry, and an odd number of limbs. One example of such an animal is the starfish, which has fivefold symmetry and five legs. Robots with odd number of legs are rarely described in the literature, two examples are given in Refs. 3 and 7. However, as shown in[5] optimization of the distribution of the legs around a robot's body can result in a design, which is more efficient from the viewpoint of mass reduction and energy consumption. Five-legged gaits were also studied in the context of six-legged robots with a leg failure, and it was demonstrated that such a robot can detach a leg that is failed (e.g. due to a blocked joint) and continue the mission using five legs.[6]

This paper describes in details the design of the five-legged robot prototype, named PentOpiliones (Fig. 1a), which is a proof-of-concept for a bigger machine intended mostly for search and rescue missions. The paper focuses on the advantages of the five-legged, radially symmetrical configuration with regard to the agility of motion, and its remainder is structured as follows: section 2 presents the design of the robot's mechanical structure and its control system, while section 3 provides description of the kinematics and describes the gaits implemented in the robot. Experimental evaluation of the robot's motion agility is summarized in section 4. The paper is concluded in section 5 with some final remarks and comments on possible improvements to the design.

2. Design of the robot

The dimensions of the PentOpiliones are 45×30 cm (diameter×height), and its mass is 2.05 kg. The legs have insect-like structure, and each of them has three degrees of freedom (DOF). The first two joints of a leg are placed very close to each other, which results in a design similar to a single joint of the fourth kinematic class. The two main segments of a leg (femur and tibia) have length of 150 and 250 mm, respectively. The exact dimensions and the kinematic structure of the robot are given in Fig. 2a.

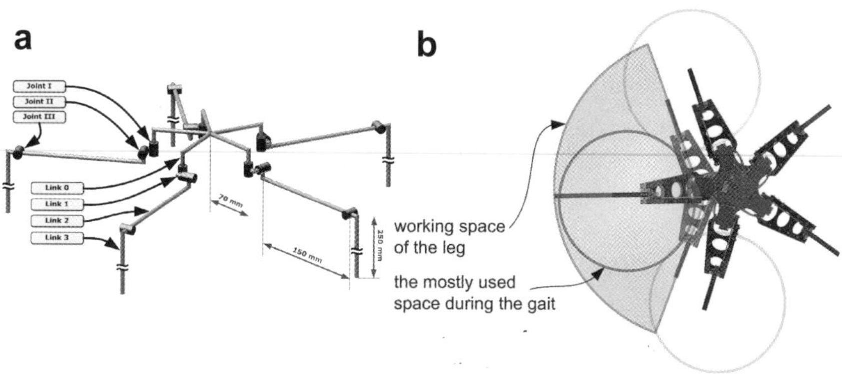

Fig. 2. Kinematics scheme of the symmetrical five-legged robot (a), and the horizontal projection of leg's workspace (b)

To drive the robot's joints we use standard servomotors, which are like those often applied in remote-controlled model toys. The servos used have a torque of about 1 Nm, a speed of 0.15 sec/60° with no load, a mass of 55 gram and metal gears. All servos are mounted in the trunk, to keep the legs as lightweight as possible. Legs are made from carbon fiber square tubes, which are reinforced with a carbon fiber sheet. Other main mechanical components of the robot are made from polycarbonate that is easier in treatment (Fig. 1b).

The robot control system is based on the STM32f4 microcontroller with the ARM core Cortex-M4. It is clocked with the frequency of 168 MHz with a hardware FPU, and has a rich set of peripherals. All components are connected to the on-board microcontroller, which controls the behavior of the robot. The robot is equipped with a number of internal and exteroceptive sensors that can be used depending on the task at hand. There is an inertial measurements unit (IMU) containing a gyroscope, an accelerometer, and a compass. Using measurements from these sensors the microcontroller can estimate the attitude of the robot's body with regard to the gravity vector and the magnetic north vector.

On each of the five sides of the robot's trunk there are mounted infrared distance sensors (Sharp GP2D12), which allow to detect obstacles in the range of 150 cm, and ambient light sensors, which can detect the direction of the external lighting source. There are also sensors which measure external environment parameters: temperature, humidity and air pressure. The controller's board contains some unused connectors, which allow to connect additional modules, such as GPS, laser scanner (e.g. Hokuyo URG-04LX) or a CCD camera. All 15 servomotors are connected directly to the main microcontroller, and are controlled using standard PWM signals. To communicate with an external computer the robot uses a wireless Bluetooth module. The PentOpiliones robot can be powered from an on board battery or an external power supply. The robot has a high-current Lithium-Polymer rechargeable battery, which is located in its trunk and lasts for about 90 minutes of walking.

3. Gait generation

The legs of the robot are regularly fixed on the circumference of the circle, with the angular separation of 72° (Fig. 2b). Therefore, the robot has radial symmetry and is omnidirectional. This means that it can change the walking direction without the necessity of rotation, and can walk in each direction with the same easiness and speed.

The basic gait pattern implemented in the PentOpiliones robot always uses three legs to support the machine. These supporting legs are moved backward, while the two other legs are carried over to the front, regardless of the walking direction. With this pattern of motion for a 5-legged machine the swing phase has to be 1.5 times shorter than the stance phase. It is beneficial, because the legs being in the swing phase are not loaded, and can be moved quicker. Always two legs on the opposite sides of the robot are in the swing phase, but they are not synchronized. While one of these legs is in the middle of the swing phase, the other one is finishing the motion, then it touches the ground, and another leg begins the swing phase (Fig. 3a). In other words, two neighboring legs are always shifted 144° in the phase.

During the gait programming we considered the limitations of the

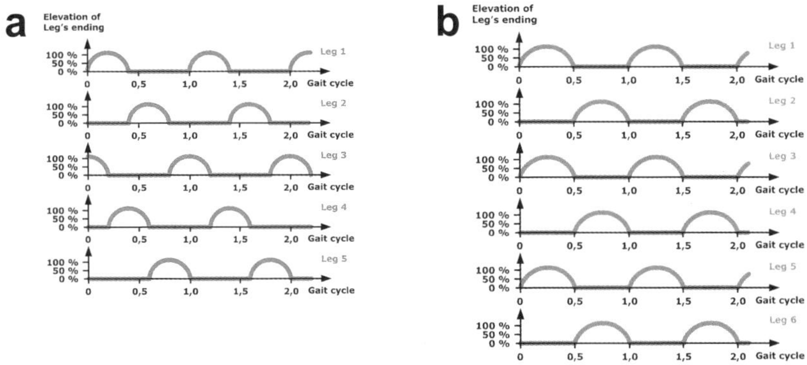

Fig. 3. Phases of gait cycles of a five-legged (a) and a six-legged (b) robot

workspace of the robot's legs, as well as the possibility of collisions between the legs. In typical hexapod robots, which have the legs placed symmetrically on both sides of the trunk, like our Messor robot,[8] there is a problem with the workspaces of the legs that cover each other. Considering a tripod gait (Fig. 3b), if one of the middle legs is transferred to the front, the front leg is in the support phase, and thus it is moved to the back, which easily can cause a collision between these legs, and in practice limits the usable workspaces of the legs involved. The hexapod robot should have bigger distance between legs, which makes the machine bigger and less maneuverable, or it has to take shorter steps, which decreases the speed. In the configuration of legs used in the PentOpiliones this problem does not exist, because the workspaces of the legs fill evenly the space around the trunk (Fig. 2b). As a result, the robot can take longer steps without a risk of collision between the legs.

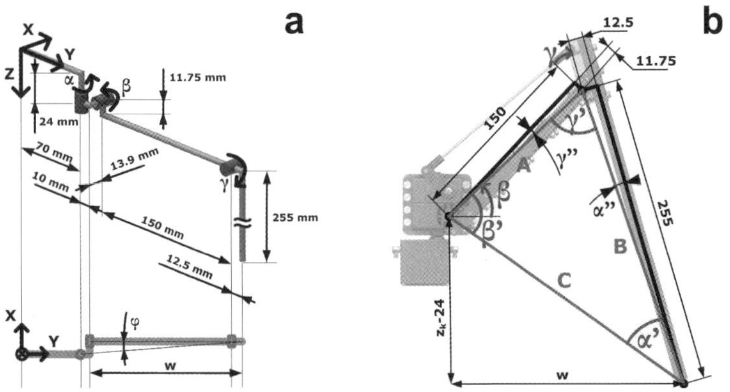

Fig. 4. Kinematics of a leg (a), femur and tibia geometry (b)

The tips of the robot's legs have a point-like contact with the ground, and each leg can be treated like a 3-DOF manipulator. Therefore, the problem of robot gait generation can be splitted up to define the coordinates of the leg's tips, and then to calculate the joint angles from the inverse kinematics of a leg. The inverse kinematics is computed from the geometric relations. Considering the kinematics of a leg and its dimensions (Fig. 4a) we compute α:

$$\alpha = \arctan \frac{x_k}{y_k - 70} - \arctan \frac{13.9}{w + 10}, \tag{1}$$

where x_k and y_k are the coordinates of the leg's tip, and $w = -10 + \sqrt{-13.9^2 + x_k^2 + (y_k - 70)^2}$. Angles in the remaining two joints can be computed considering Fig. 4b:

$$\beta = \arccos\left(\frac{A^2 + C^2 - B^2}{2AC}\right) + \gamma'' - \arctan \frac{z_k - 24}{w}, \tag{2}$$

$$\gamma = 90^\circ - \arccos\left(\frac{A^2 + B^2 - C^2}{2AB}\right), \tag{3}$$

where z_k is the vertical position of the leg's tip, $A = 150.46$, $B = 255.31$, $C = \sqrt{w^2 + (z_k - 24)^2}$, and $\gamma'' = \arctan \frac{11.75}{150} = 4.48^\circ$.

Assume that many times per second there is given a small vector $\mathbf{q}_i$, which defines the robot movement. Each leg being in the stance phase is then moved by this vector, contributing to the displacement vector $\mathbf{z}$ of the whole body. Each leg has its neutral point of support – a point in the middle of the leg's workspace, where the leg's tip is located in the starting position. At the beginning of the swing phase, the controller remembers the desired direction of walking, and calculates the coordinates of the point in which the moving leg should be put down. At the swing phase the leg's tip is moving with a constant velocity along a semi-circular trajectory to the desired point. This point is located at the distance equal to the half of the step's length from the neutral point of support, towards the direction of walking (Fig. 5). Using this gait control strategy we can instantaneously change the walking direction of the robot, and it is not necessary to wait until the end of the current step. This feature increase the agility of motion with regard to a typical hexapod.

The gait generation algorithm is characterized by a number of parameters, which can be set during motion in order to change the posture of the robot. We can modify the elevation at which the tips of the legs are raised during motion. The legs can be raised higher while walking on rugged terrain, or they can be raised lower, to enable a shorter gait cycle (bigger

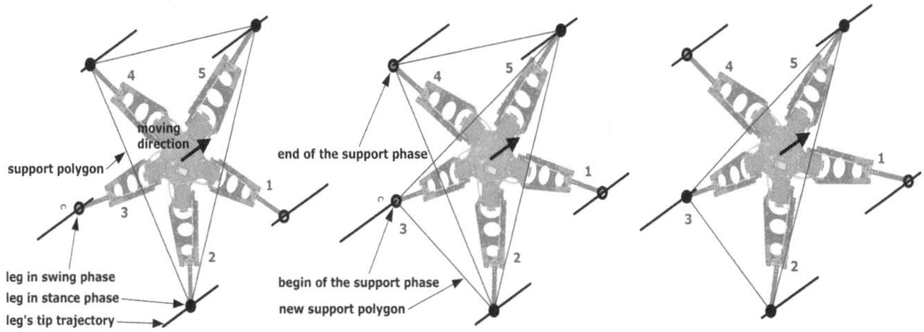

Fig. 5. Subsequent phases of the five-legged gait with the support polygon indicated

speed). We can also freely set the height of the trunk above the ground. When the trunk is high, the robot can walk over obstacles, but the trunk can be also lowered, to increase stability or to walk under an obstacle (e.g. under a car). Another parameter which can be changed is the length of the step. Longer steps increase the speed of walking, while shorter steps give more stability. Next parameter that modifies the posture of the robot is the distance between the center of the robot and the neutral point of support. This point should be set in the center of leg's workspace to ensure a big enough range of leg movements. Nevertheless, it can be moved closer to the center of the robot, to allow the machine to walk through narrow gaps, or can be moved farther, to increase the stability.

4. Experimental results

The PentOpiliones robot was tested outdoors and indoors, also during public events. The indoor experiments on a rocky terrain mockup were aimed at demonstrating qualitatively the motion agility. Figure 6 illustrates the possibility to negotiate obstacles in the teleoperation mode, and to traverse uneven terrain.

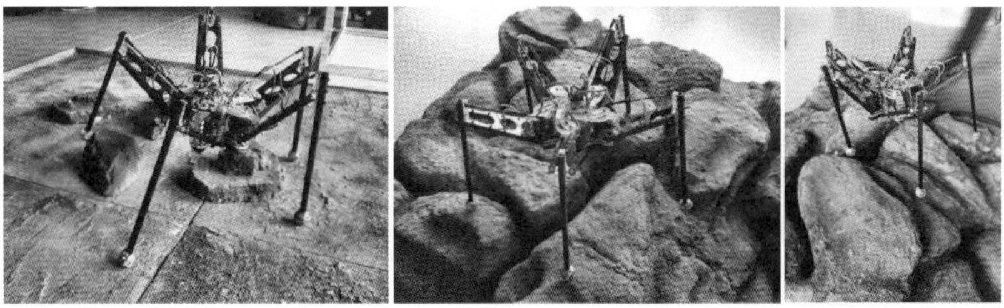

Fig. 6. Indoor experiments – negotiating obstacles on terrain mockups

The maximal speed of walking was 20 cm/s, the maximal ground clearance was 30 cm, while the trunk can be raised up to 42 cm (e.g. when carrying a camera for inspection). The span of the tips of the legs can change from 29 to 70 cm, giving the robot a great flexibility in negotiating various obstacles. The performance of the robot is shown in the accompanying video available at: http://lrm.cie.put.poznan.pl/clawar2013.wmv.

5. Final remarks

This paper characterizes the five-legged, symmetrical robot prototype. The PentOpiliones robot was successfully tested indoors and outdoors, demonstrating good ability to overcome obstacles, to traverse various types of terrain, and to change its posture while in motion. Future research plans include integration of new exteroceptive sensors, such as the omnidirectional vision system, which should extend the autonomy of the robot. Moreover, we are interested in automatic synthesis of efficient gaits for the PentOpiliones robot using the bio-inspired approach introduced in Ref. 1.

References

1. Belter D., Skrzypczyński P. (2010) A Biologically Inspired Approach to Feasible Gait Learning for a Hexapod Robot, Int. Journal of Applied Mathematics and Computer Science, 20(1), 69–84.
2. Belter D., Skrzypczyński P. (2011) Rough Terrain Mapping and Classification for Foothold Selection in a Walking Robot, Journal of Field Robotics, 28(4), 497–528.
3. Besari A., Zamri R., Prabuwono A., Kuswadi S. (2009) The Study on Optimal Gait for Five-Legged Robot with Reinforcement Learning, Intelligent Robotics and Applications, LNCS Vol. 5928, Springer, 1170–1175.
4. Gonzalez de Santos P., Galvez J. A., Estremera J., Garcia E. (2003) SILO4 – a True Walking Robot for the Comparative Study of Walking Machine Techniques. IEEE Robotics & Automation Magazine, 10(4), 23–32.
5. Gonzalez de Santos P., Garcia E., Estremera J. (2007) Improving Walking-Robot Performances by Optimizing Leg Distribution, Autonomous Robots, 23(4), 247–258.
6. Mostafa K., Tsai C.-S. Her I. (2010) Alternative Gaits for Multiped Robots with Leg Failures to Retain Maneuverability, Int. Journal of Advanced Robotic Systems, 7(4), 33–40.
7. Pac M., Popa D. (2010) Kinematic Analysis of a Five-Legged Mobile Robot, Proc. ASME IDETC/CIE2010, Montreal, 1427–1435.
8. Walas K., Belter D. (2011) Messor – Versatile Walking Robot for Search and Rescue Missions, Journal of Automation, Mobile Robotics & Intelligent Systems, 5(2), 28–34.

CLIMBING ROBOT EQUIPPED WITH A POSTURAL ADJUSTMENT MECHANISM FOR CONICAL POLES*

YASUHIKO ISHIGURE†

Marutomi Seiko Co., Ltd., Kurachi Aza-Ikuda 3147-7
Seki, 501-3936, Japan

HARUHISA KAWASAKI,
TAICHI KATO, KATUYUKI HIRAI, NOBUYUKI IINUMA

Department of Human and Information Systems, Gifu University Yanagido 1-1,
Gifu, 501-1193, Japan

SATOSHI UEKI

Department of Mechanical Systems, Toyota National College of Technology,
Toyota, 471-8525, Japan

A climbing robot with a postural adjustment mechanism for conical poles is presented. The climbing method driven by servomotors with a warm-wheel reduction mechanism can rest on a tree by using its own weight without any energy expenditure. To realize both straight climbing and spiral climbing for conical poles, a postural adjustment mechanism is needed to move the steering mechanisms of the active wheels smoothly. We present the design of the robot's two-link arm mechanism with 1 DOF as the postural adjustment mechanism. It was demonstrated experimentally that the climbing robot, equipped with four active wheels, can straight climb and spiral climb conical poles by using the proposed postural adjustment mechanism.

1. Introduction

Robots that can climb poles are under development and are expected to be used in the inside/outside maintenance of buildings, observations of disaster scenes from a height, pruning trees, and more.

As an alternative, we developed and analyzed a climbing method [1, 2]. The climbing mechanism is equipped with a servomotor and a warm-wheel reduction mechanism which has non-back-drivability. It can stay on a tree using its own weight, since the center of its mass is located outside of the tree. A climbing

* This work was supported in part by the NEDO project (P03040) and the NARO Program for the Promotion of Basic and Applied Researches for Innovations in Bio-oriented Industry.

robot that uses its own weight, with one active wheel and two passive wheels, was described in 2010 [3, 4]. However, the climbing speed of the robot is low, since the robot can slip off when an active wheel slips. These robots can only climb straight up; they cannot climb in a spiral. Therefore, it is difficult to observe omnidirectionally and, for example, to prune trees after climbing up. Applications of climbing robots will be expanded if they can climb not only straight up but also spirally [2].

Most trees are conical thicker at the bottom and gradually becoming thinner with increasing height. This is called the stem taper in the logging/timber industry. When a climbing robot using its own weight climbs straight up a conical pole, the pole can get stuck in between the robot's wheels because the robot tilts downward as it climbs the conical pole. Moreover, it can be difficult to control the steering angle of the active wheels during the climb, even if the robot has not gotten stuck in the tree.

This paper proposes a postural adjustment mechanism to be used with a climbing robot that has four wheels. It consists of two 1-degree of freedom (DOF) two-link arm mechanisms, one located on the top side and the other underneath, to make the robot lightweight. The results of field experiments conducted in a forest are also presented.

2. Climbing robot using its own weight

2.1. *Mechanism of climbing robot using its own weight*

It is desirable that a climbing robot has more than two wheels, since it is quite possible for a wheel to slip out of place; more wheels will contribute to stable climbing. As shown in Figure 1, our climbing robot is equipped with four active wheels. Two wheels are located on top (numbers 1 and 2) and two wheels are located on the bottom (numbers 3 and 4). Each wheel is placed at the same angle $\pm \pi / 4$ rad relative to the $\boldsymbol{x}$-axis when viewed from the top of the cylinder. The active wheel is driven by a DC servomotor via a warm-wheel reduction mechanism. The warm-wheel reduction mechanism, which has non-back-drivability, allows the robot to be at rest without any energy expenditure when the input electric current for the servomotor is zero.

We ensured that the center of mass of the robot is located outside of the cylinder by placing the control system at the side of the down-side active wheels. Each of the four active wheels has a steering system to control the steering angle. The steering system is driven by a DC servomotor via a warm-wheel reduction mechanism.

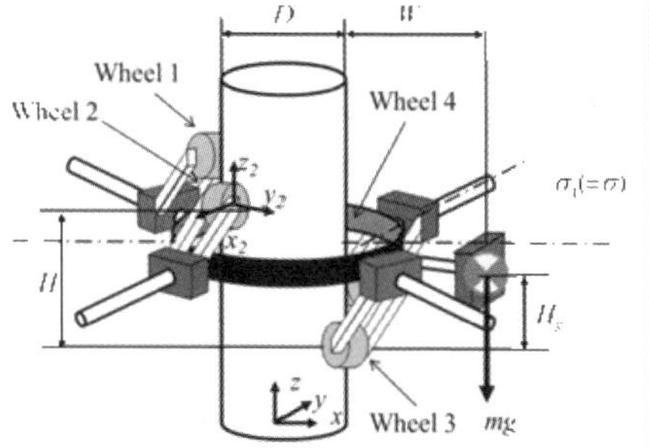

Fig. 1 Climbing robot equipped with four active wheel

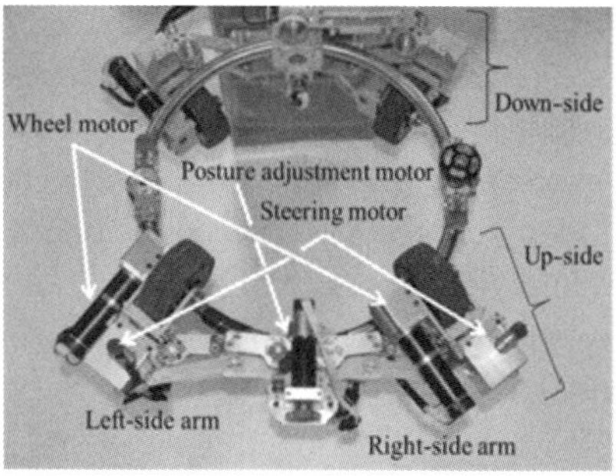

Fig. 2 Postural adjustment mechanism

3. Postural adjustment mechanism

3.1. *Mechanical structure*

A postural adjustment mechanism is necessary to help keep the weight of the robot down. As shown in Figure 2, we designed a postural adjustment mechanism that adjusts the posture of robot by moving two 1-DOF two-link arm mechanisms. The arms are located at the left side and right side such that the active wheel attached to the end link moves to the center of the pole. The arm mechanisms are placed at the top and bottom sides. The tilt of the robot is adjusted by controlling the servomotor, which moves the active wheels to the center of the pole and shortens the distance between the opposite wheels. The robot can rest without expending any energy, since the motor drives the arm joint via the warm-wheel reduction mechanism, which has non-back-drivability.
In this mechanism design using one degree of freedom, the active wheels move along the reference line (the dashed line in Figure 3), and it is desirable

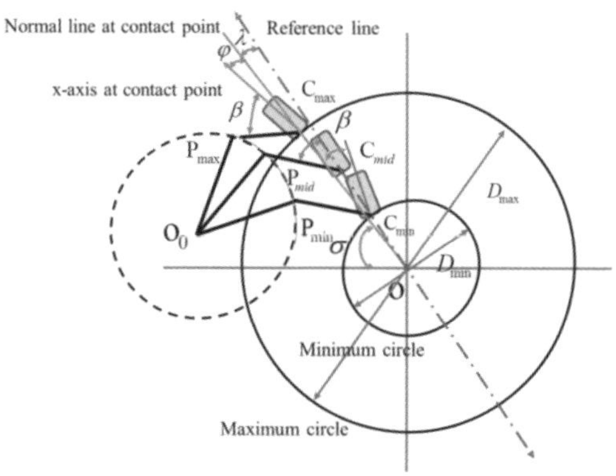

Fig. 3 Postural adjustment mechanism

that the posture of the wheel relative to the reference line at the contact point on the pole is not so large. In the figure, O is the origin of the base coordinate system fixed on the conical pole, O_0 is the origin of the arm base coordinate system, C_{max} is a contact point when the diameter of the pole is maximum (D_{max}), C_{min} is a contact point when the diameter of the pole is minimum (D_{min}), C_{mid} is a center point between C_{max} and C_{min}, P_i (i=*max,min,mid*) is a position vector of the end point of link 1 at C_i, and β is the mounting angle between the wheel and link 2.

3.2. *Design method*

The posture adjustment by the 1-DOF arm mechanism has a limit because the active wheel has three degrees of freedom in two-dimensional space. Hence, we defined a design guide as follows:

1. To maintain a wheel configuration that can resist an external force, a gap angle λ, which is defined as an angle between the reference line and the normal direction at the contact point, is less than about ± 5 deg as targeted.
2. To ensure a contact region of a wheel while avoiding uneven contact, a wheel posture φ, which is defined as an angle between the wheel travelling direction and the normal direction of the pole surface, is less than about ± 5 deg as targeted.
3. To reduce the mechanism's weight and to reduce the required torque at the arm joint, the total length of arms 1 and 2 is made as short as possible.
4. The 1st link is driven by the motor, and the 2nd link is driven by a four-linkage closed loop mechanism.

In order to satisfy these design guides, the length of the link mechanism was determined as follows:

a. On the contact points C_{max} and C_{min}, the wheel posture φ, the gap angle λ, and the origin position of the arm base coordinate system O_0 are given.
b. The mounting angle between the active wheel and link 2, β, is given, the end point of link 1 corresponding to C_i (i=*max,min,mid*), P_i, is calculated, and a relational expression of the length l_2 of link 2 is derived, which satisfies that an intersection of the perpendicular bisector between P_{max} and P_{mid} with the perpendicular bisector between P_{min} and P_{mid} coincides with the origin position of the arm base coordinate system O_0.
c. The length l_1 of link 1 is calculated, and β is derived by seeking the minimum of the total length of the arm, $l = l_1 + l_2$.
d. The angle q_1 of link 1 which passes through P_i, and the angle q_2 of link 2

which passes through C_i are calculated for $i=max,min,mid$, and the length of the four-linkage mechanism is determined.

e. The wheel posture and the gap angle are derived with respect to the diameter of the pole. If these are within the acceptable ranges, then this process is finished. If it is not so, then the process is returned back to (a) after adjusting the position of the origin of the arm base coordinate system O_0.

3.3. *Numerical Calculation*

The relation between all link lengths and the mounting angle of the prototype climbing robot was derived using the Maple software program (Cybernet Systems Co., Ltd.).

Figure 4 shows the relation between the mounting angle and the total arm length and the relation between β and the absolute maximum $|\varphi|_{max}$ and $|\lambda|_{max}$ for φ and λ. It is hard to set both the total arm length and the maximum angle change best because the relation between them is conflicting relation. We therefore used $\beta=\pi/10$ rad, $l_1=98.4$ mm, and $l_2=86.9$ mm in the developed robot design. In this case, $|\varphi|_{max}=0.058$rad and $|\lambda|_{max}=0.086$rad.

The changes of the wheel posture φ and the gap angle λ with respect to the pole diameter are shown in Figure 5. Although the wheel posture varies according to the diameter of the pole, the design specification is almost satisfied.

This design method requires trial and error calculations. However, this is a useful design method for a 1-DOF mechanism that can generate three degrees of freedom planar motion of active wheels.

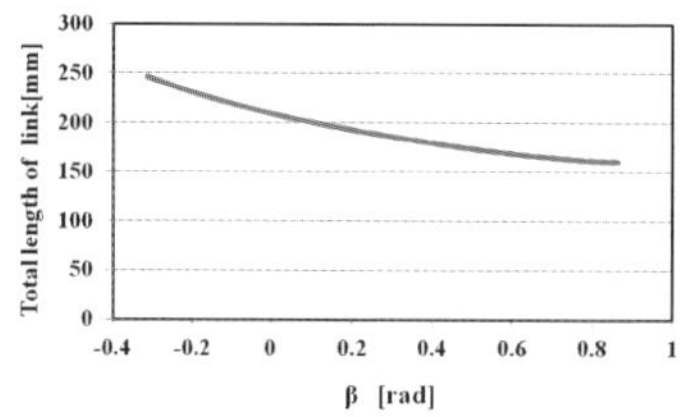

(1) Total link length

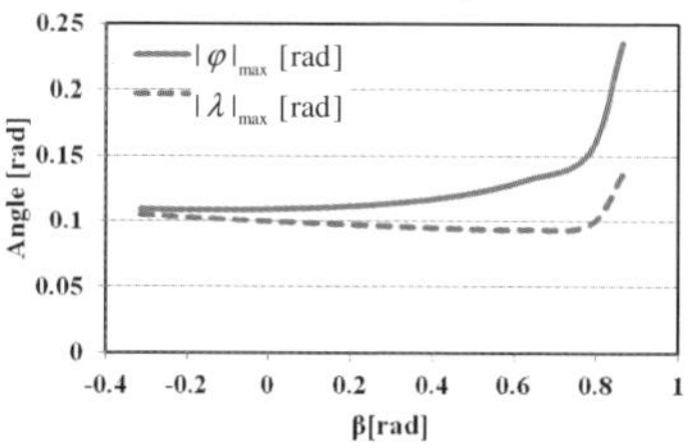

(2) Maximum posture variations of wheel

Fig. 4 Total link length and posture variations

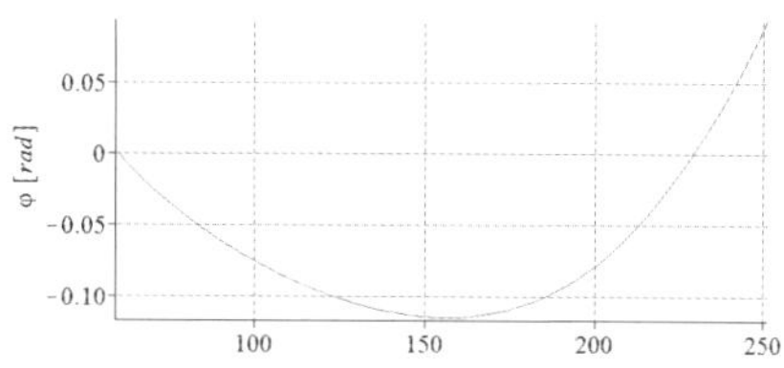

(1) Deviance angle of contact point

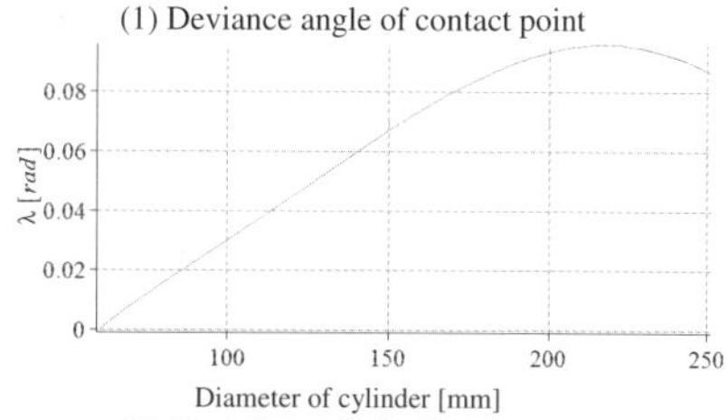

(2) Variations of wheel posture

Fig. 5 Variations of wheel posture and deviance angle of contact point

3.4. *Control method*

The tilt of the climbing robot is measured by the posture sensor. In order to match the central axis of the robot to that of the tree, one of the two arms (called the main arm) is implemented a posture PID control, and the other (called the slave arm) is implemented a joint-angle PID control in which the desired value is the arm joint angle of the main arm. That is, the posture PID control for the up-side arm is:

$$u_m(t) = k_{mP}(\theta_d - \theta(t)) - k_{mD}\dot{\theta}(t) + k_{mI}\int_0^t(\theta_d - \theta(t))dt \tag{8}$$

where $u(t)$ is the control input, k_{mP}, k_{mD}, k_{mI} are the proportional feedback gain, differential feedback gain, and integral feedback gain, respectively, and $\theta_d, \theta(t)$ are the desired posture angle and the measured posture angle, respectively. The actual desired posture angle is set to zero. The down-side arm is the slave arm, and the implemented joint-angle PID control law is expressed as follows:

$$u_s(t) = k_{sP}(q_m(t) - q_s(t)) - k_{sD}\dot{q}(t) + k_{sI}\int_s^t(q_m(t) - q_s(t))dt \tag{9}$$

where $q(t)$ is the angle of the 1st joint, subscript m is the main arm, and subscript s is the slave arm; the desired joint angle of the slave arm is set to be the joint angle of the main arm.

4. Experimental evaluation

4.1. *Developed climbing robot*

We developed a climbing robot to evaluate the effect of the posture adjustment mechanism. The robot can move up and down in both straight and spiral manners while the posture of the robot is regulated by the posture adjustment mechanism. There are four active wheels on the tip of the arm mechanisms which are driven by four 20-watt DC servomotors. Each DC motor is equipped with a rotary encoder and a planetary gear mechanism and coupled to a warm-wheel reduction mechanism. In addition, each active wheel has a steering mechanism driven by a 3-watt DC motor coupled to the warm-wheel reduction mechanism. Thus, the robot can move up and down spirally by the steering angle control.

A battery power supply and a control system are attached to the robot such that the center of the mass of the robot is located outside of the cylindrical pole. The control system consists of an FPGA-based control circuit and motor drive circuits. The robot is remote-controlled through a wireless LAN. The mass of the climbing robot, m, is 13.0 kg; the diameter of the active wheel is 0.075 m, and the outer dimensions of the robot are W = -0.005 m, H = 0.158 m.

4.2. *Experiment on a tree*

In generally, cedar and hinoki trees in the man-made forests in Japan have a tapering characteristic feature in which the diameter of the tree is reduced by approx. 0.01 m with each 1-m increase in height from the base of the tree [5]. To confirm the effect of the posture adjustment mechanism, we tested the climbing robot as shown in Figure 6 in an experimental forest. The tree was a hinoki with the breast-height diameter 0.155 m. The motion profile of the robot for the test was comprised of six motions: straight climbing upward for approx. 1.25 m, wheel steering with the steering angle 1.36 rad, spiral climbing for approx. 0.5 m, wheel steering with the steering angle -1.36 rad, straight moving down the tree, then stopping with posture adjustment. Figure 7 shows the experiment results for the position, velocity and posture of the robot.

Interval 1 is the straight climbing, intervals 2 and 4 are the wheel steering, interval 3 is the spiral climbing, and interval 5 is straight moving down. The velocity PI control was adopted for active wheels at moving up and down. The control gains of each wheel control system were the same. In Figure 7, the label "Desired" indicates the targeted value and "Wheel" shows the mean of the positions and velocities of the four active wheels, which were calculated from the rotation angle of the wheel motor and the diameter of the wheel. The Wheel position was almost equal to that of Desired, and the position error was approx. 0.1 m at the target climbing height 1.25 m, which occurred in association with sliding of the wheel. The Wheel velocity almost converged to the target velocity, 0.25 m/s. The posture of the robot changed not a little at the interval of straight climbing, due to the taper of the tree and the irregularity of the tree surface. However, the posture error was less than ± 0.02 rad during the steering interval and less than ± 0.04 rad during the spiral climbing; these values are within the tolerance level for wheel steering and pruning. These experimental results showed that the climbing robot with the posture adjustment mechanism can move up and down a tapered cylindrical column.

Fig. 6 Field test of the climbing robot

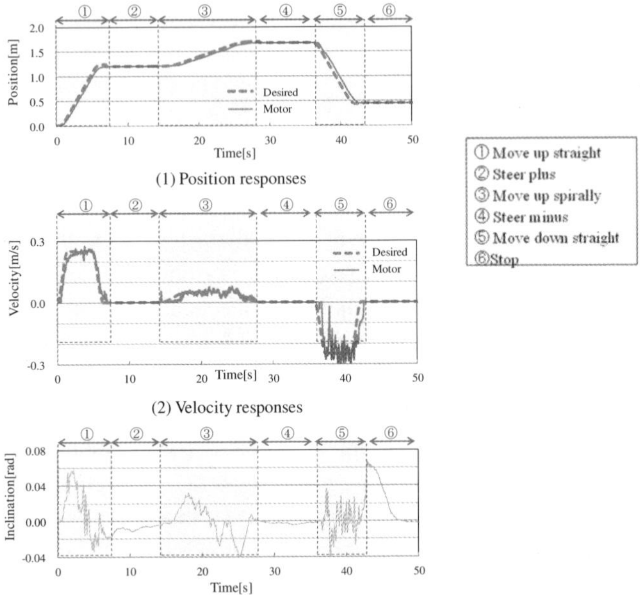

(1) Position responses

(2) Velocity responses

(3) Posture responses

Fig. 7 The climbing robot's position, velocity and inclination responses to moving upward and downward (Left)

5. Conclusion

We have developed a novel climbing robot equipped with a posture adjustment mechanism consisting of two 1-DOF arms with joints located at the up-side and down-side. The robot can move straight up and down and spirally and can steer the active wheels in case of not only straight cylindrical poles but also conical poles. The robot can also remain stationary by using its own weight: no energy expenditure is needed to do so. The design method of the posture adjustment mechanism has also been presented. The robot’s straight climbing velocity was 0.25 m/s, which is the fastest velocity among the existing climbing robots. We will evaluate the robot’s ability to prune trees in the near feature.

References

1. H. Kawasaki, S. Murakami, H. Kachi, and S. Ueki, *Proc. of SICE Annual Conference 2008*. p. 160 (2008).
2. H. Kawasaki, S. Murakaki, K. Koganamaru, W. Chonnaparamuty, Y. Ishigure, and S. Ueki, *Proc. of CLAWAR 2010*. p. 455 (2010).
3. J. C. Fauroux and J. Morillon, *An International Journal.* **37:3**, p. 297 (2010)
4. A. Sadeghi, H. Moradi, and M. N. Ahmadabadi, *Robotica*. **30** (2011).
5. Website of Gifu Prefecture Reseach Institute for Forests, http://www.cc.rd.pref.gifu.jp/forest/shiyou/hosori.html (in Japanese).

CAMINANTE: A PLATFORM FOR SENSITIVE WALKING

VADIM CHERNYAK, ENNIO CLARETTI, STEPHEN S. NESTINGER, and *EDUARDO TORRES-JARA

Sensitive Robotics Laboratory, Worcester Polytechnic Institute, Worcester, MA 01609, USA
** E-mail: etorresj@cs.wpi.edu*

This paper presents the development of *Caminante*, a walking robotic platform with the characteristics necessary to develop *Sensitive Walking*. *Sensitive Walking* makes extensive use of tactile feedback from sensors on the robot's feet to determine the robot's actions, effectively adapting to variations in the terrain. This approach is inspired by biological legged locomotion that utilizes skin sensors to robustly walk over most terrains by feeling the real-time interaction of the feet with the ground. The challenges and design requirements considered while building *Caminante* are presented in this paper.

Keywords: Legged Robots; Tactile Sensors; Sensitive Walking.

1. Introduction

Legged locomotion is important in robotics because of its adaptability to varying terrain conditions. Legs work well indoors on flat floors and stairs, as well as outdoors on flat roads and rocky terrain. Moreover, about half of the Earth's landmass is currently inaccessible by wheeled or tracked vehicles.[1] However, in general, state-of-the-art legged robots are still incapable of ensuring stability while traversing irregular terrain. Emphasis has been

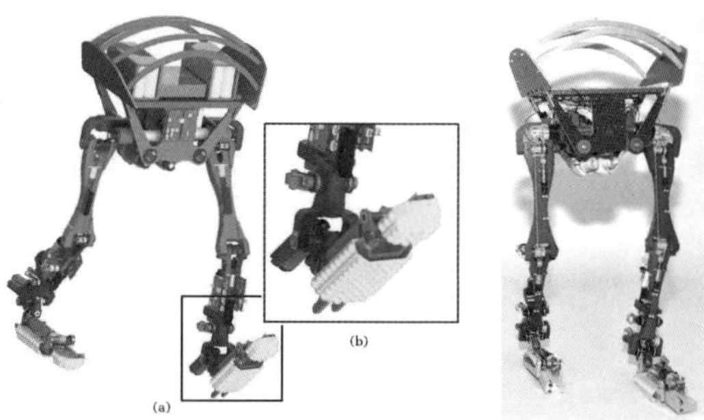

Fig. 1. **Left.** (a) CAD model. (b) Tactile sensors. **Right.** Robot *Caminante*

given to the stability and balance,[2–4] dynamics,[5,6] and gait generation[7–9] of the robots, which has given the field deep insight on how to develop walking robots. Limited attention has been given to sensing the interaction between the foot and the terrain;[10,11] it is not common to equip a robotic foot with a large tactile sensor array.[12–16] Contact information becomes relevant on irregular terrain, which is where legged locomotion is more effective than wheeled. For instance, human feet are covered by a large array of tactile sensors that provide information about the geometry of the floor (an edge, a flat surface), the force distribution throughout the foot (concentrated in the toe or heel), the stability of the contact (is it slippery?), and the presence of an obstacle in the trajectory (the toe touched a step). Contact information makes human walking adaptive.

The advantages of using contact information has been demonstrated in robotic manipulation by modern approaches[17–19] such as Sensitive Manipulation.[20–22] Inspired by sensitive manipulation, we are developing *Sensitive Walking*, a new approach to robotic walking that uses tactile feedback to effectively adapt the robot's response to the possible variations in the terrain. To develop and evaluate this approach, the bipedal robot *Caminante* (Figure 1) has been designed and fabricated with the required characteristics. This paper describes the challenges faced in the design and realization of this robot.

2. Robot Design

Sensitive Walking assumes that there is an array of tactile sensors on the robot's feets to provide information about the physical interaction with the surface in particular and the environment in general. The following sections describe the design challenges associated with developing *Caminante* to evaluate *Sensitive Walking*.

2.1. *Design Challenges*

A *major challenge* to implement *Sensitive Walking* is creating a platform with foot sensors that have the right characteristics for the task. In the human foot, this is solved using tactile feedback from the skin. The characteristics of these tactile sensors are crucial for the implementation of this approach. The tactile sensors need to be highly sensitive to normal and lateral forces, detect the contact independent of local surface geometry, provide a soft interface to the contact, and reduce slippage with a high friction coefficient. A *second challenge* was designing feet that allow the robot to take advantage of its sensors. A *third design challenge* for this type of robot was to give *Caminante* the capability of gently coming in contact with the environment on each step.

2.2. *Tactile Sensor Design*

Compliant tactile sensors[23] were invented for developing *Sensitive Manipulation* in the robots Obrero[21] and GoBot.[22] These types of sensors have the required properties (Section 2.1) for *Sensitive Walking*. In these sensors, a rubber dome is deformed by applied normal and lateral forces. The force value is estimated by measuring the deformation of the dome using an optical array (See Figure 2). The operational range and sensitivity of these sensors depend on the material and thickness of the dome. For this design, the sensors handle the robot weight distribution along the foot. Another characteristic of these sensors is that they do not get damaged when they are saturated. This is an important characteristic because when the foot initially contacts a sticking point (a small stone) the force will temporarily be concentrated at that point, saturating the sensor. This condition is handled by the sensor while remaining operational. The domes are molded silicone rubber, which enables dense sensor arrays. An example of the molds is shown in Figure 2. The geometry of the foot is discussed in Section 2.3. The challenge of creating the molds is having all the sensor on the same fabric to facilitate assembly. Each foot of the *Caminante* has a 95 sensor array that are sampled at 100Hz using 12 bit analog-to-digital converters with an SPI interface.

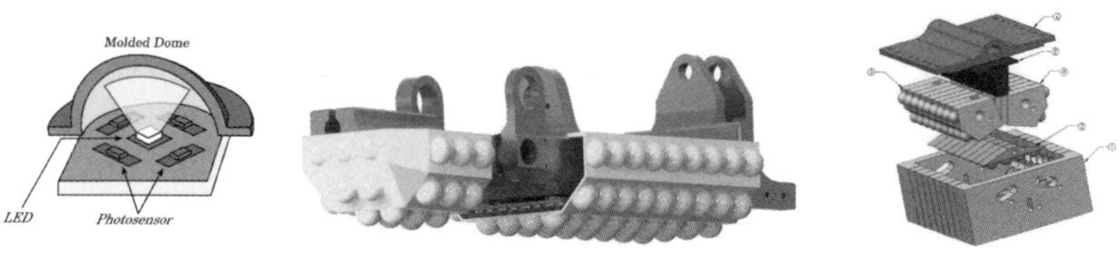

Fig. 2. **Left**. Compliant tactile sensor.[23] **Middle**. Sensors mounted onto the foot. **Right**. Sensor mold.

2.3. *Foot Design*

The foot design is of great importance because the success of *Sensitive Walking* depends on covering, as much as possible, the foot surface with tactile sensors. Moreover, the motion of the foot enables the capability of sensing by coming in contact with the world. Figure 2 shows a CAD model of the foot, which has an articulated toe and is covered by tactile sensors. The sensors are placed on the sole of the foot, at a 45 degree angle to the bottom of the foot and at a perpendicular angle to the foot. Additional sensors fully encapsulate the sides and bottome of the big toe. This arrangement allows for the measurement of forces acting on the sole of the foot while the robot is walking. The sensors on the side and front of the foot allows for the detection of floor imperfections and obstacles. The articulated toe, covered with tactile sensors, enables smooth walking. During a step, the tactile sensors on the back of the foot detect when there no force is applied. Therefore, the back of the foot can be lifted while the toe remains in contact with the floor, providing a smooth heel-to-toe transition. The articulated toe has a degree of compliance in its actuation, which makes the foot more adaptable when stepping on irregular terrain.

2.4. *Mechanical Design: Legs and Body*

To implement *Sensitive Walking*, the feet need to gently come into contact with the environment. With *Caminante*, gentle contact is achieved by using joint compliance. Extensive work has been done to implement compliance in walking robots.[24] One of the successful mechanisms is **series elastic actuators[25] (SEAs)**, which uses a spring in series with the motor to measure the force and reduce the mechanical impedance. An example of the type of SEA[26] used in *Caminante* is shown in Figure 5. In addition to the compliance, the design of *Caminante* was governed by considering dexterity. Dexterity is essential in allowing natural walking motion with regard to shifting the weight of the robot between legs. Hence, the legs were designed with 5 degrees to mimic the major degrees of freedom in a human leg. The length of each leg joint was chosen to mimic a toddler. Figure 3 gives an explanation of every joint location and direction, the D-H parameters of one leg, and the range of motion of each individual joint. Motors torques were determined based on dynamic calculations under maximum load conditions. The calculations were validated by a dynamic model that simulated a walking gait using all twelve joints. The motor torques for a $2km/h$ walk can be seen in Figure 4. The simulator used was V-REP

The robot uses eight $36V$ $90W$ brushless four pole DC motors at the four hip joints, two knee joints and two of the four ankle joints. The remaining two ankle joints and big toe joints are powered using $36V$ $12W$ brushless DC motors. High energy density compression springs with a stiff-

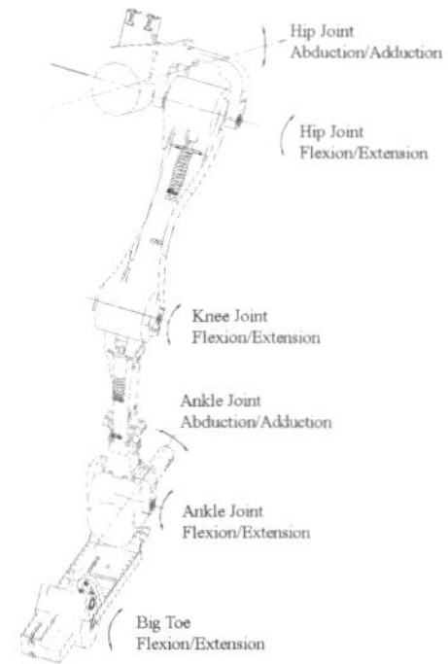

θ	a	d	α
$\theta_1(t)$	0	0	$\pi/2$
$\theta_2(t)$	0.21	0.08	0
$\theta_3(t)$	0.22	0	$\pi/2$
$\theta_4(t)$	0.02	0	$-\pi/2$
$\theta_5(t)$	0.03	0.14	0
$\theta_6(t)$	0.02	0	0

Joint	Max	Min
Hip Abd./Add.	40°	-30°
Hip Flex./Ext.	95°	-95°
Knee Flex./Ext.	95°	0°
Ankle Flex./Ext.	20°	-60°
Ankle Abd./Add.	90°	-90°
Toe Fle./Ext.	85°	-20°

Fig. 3. Leg joints. Denavit-Hartenberg parameters $[m, radians]$. Range of motion [°].

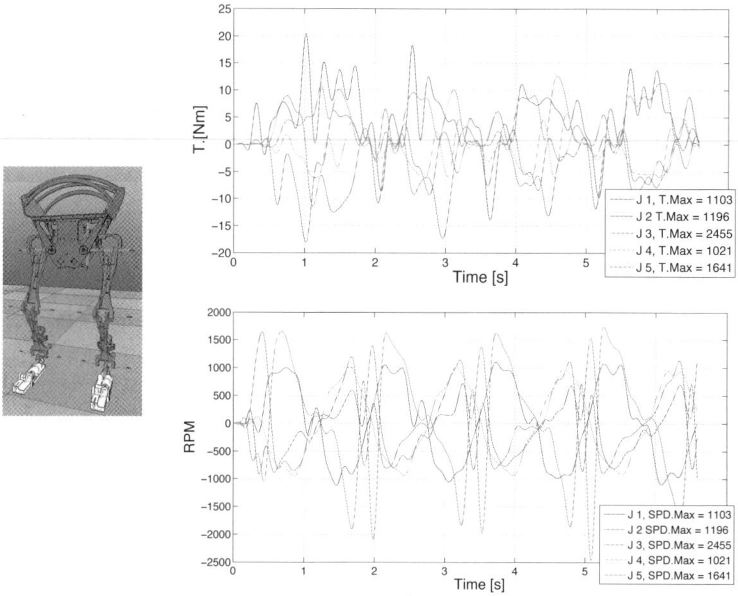

Fig. 4. **Left.** Model on Simulator V-REP. **Right.** Joint torques $2km/h$ walk.

ness of $24000N/m$ for the top four joints, $6900N/m$ for the ankle abduction/adduction, and $6600N/m$ for the big toe joint were selected based off the expected joint torque, motor torque, and pulley diameter. The relationship between expected Forces on each joint and the resultant deflection was derived. Based on this derivation, springs were chosen that provide minimal deflection during dynamic load and up to 20° when in contact with the environment. This serves as adequate compliance for rough terrain. The robot's transmission mechanism uses a cable drive system for the purpose of minimizing backlash, placing actuators in a convenient location, and minimizing weight. Dyneema cable was selected for its low stretch, low weight, high flexibility, and high strength properties. To fully take advantage of legged motion, a knee locking mechanism was designed into each leg, as is done with passive walking mechanisms.[27]The locking knee removes the need to completely control the swing of the leg and engages when the upper and lower leg links are collinear. Instead, the leg can passively swing and catch the robot. The locking mechanism also handles the correct angular position of the leg. The body of the robot was built with carbon fiber birch composite due to its high strength to weight ratio and high rigidity.

2.5. *System Integration*

Figure 5 shows the overall system architecture of *Caminante*. The robot has an on-board computer that takes in data, via USB, from the tactile sensor array, the four brushless motor controllers, and the inertial measurement unit. An LPC2148 microcontroller per foot provides the information from it's respective tactile array. Sets of three motors are managed with a single microcontroller. Each microcontroller reads six potentiometers: three for joint position feedback and three for force feedback (SEA).

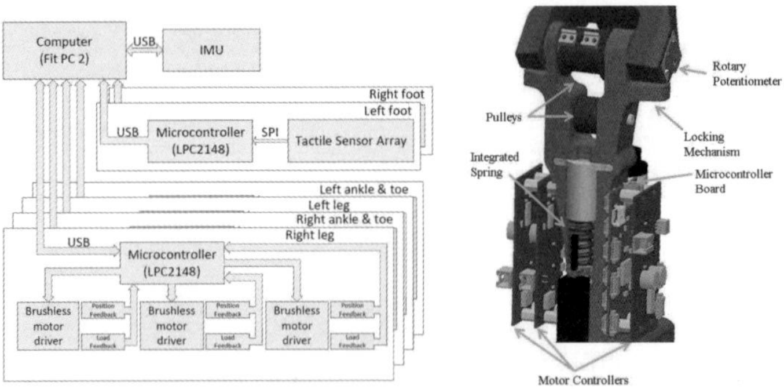

Fig. 5. **Left.** Hardware architecture. **Right.** Knee detail.

3. Building Caminante

Caminante weight 13 kilograms, stands 1 meter tall, and is entirely untethered (See Figure 1). Experimental validation focused on the output of the foot tactile sensor and the assembly of the mechanical design.

3.1. *Tactile Sensor Response*

Figure 6 shows three frames of a heel-to-toe walking gait performed by *Caminante*. As the robot transitions its weight from being fully planted on the ground to lifting the foot, the sensors detect the shift in force. The sensor response will be used to control the motion of the robot.

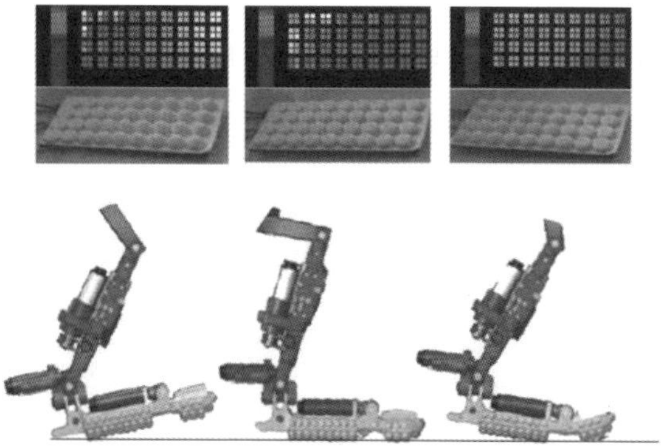

Fig. 6. Typical sensor output in a heel-to-toe walk. As the domes are pressed, the sensors detect the change on the forces applied.

4. Conclusion

This paper presents the design of *Caminante*, a robotic walking platform developed to demonstrate *Sensitive Walking*. To meet the necessary requirements for *Sensitive Walking*, *Caminante* employs compliant tactile sensors and SEAs to reliably sense foot forces and gently come into contact with the environment. The foot sensors enhance real-time walking by allowing *Caminante* to effectively adapt to variations in the terrain. This paper discussed the design challenges and considerations for the sensors and the SEAs used in *Caminante*. Motor selection was verified via dynamic simulation of the robot in V-REP, which provided the necessary joint speeds and torques for a $2km/h$ walk. The use of the foot tactile sensors to sense temporal weight distribution across the foot during a step was verified using a basic heel-to-toe walking step. The sensor's response demonstrated the required sensitivity to normal and lateral forces, a soft contact interface, and a high friction coefficient.

References

1. M. Raibert, Dynamic legged robots for rough terrain, in *Proc. IEEE-RAS Int. Conf. Hum. Robots*, 2010.

2. M. Vukobratovic and J. Stepanenko, *J. Math. Biosci.* **15**, 1 (1972).
3. H. Herr and M. B. Popovic, *J. Exp. Biol.* **211**, 467 (2008).
4. E. C. Whitman and C. G. Atkeson, *Proc. IEEE Int. Conf. Hum. Robots* , 210 (2010).
5. A. Hofmann, M. Popovic and H. Herr, *Proc. IEEE Int. Conf. Robot. Autom.* , 4423 (2009).
6. M. Vukobratovic, H. Herr, B. Borovac, M. Rakovic, M. B. Popovic, A. Hofmann, M. Jovanovic and V. Potkonjak, *Int. J. Hum. Robot.* **5**, 639 (2008).
7. M. B. Popovic, A. Goswami and H. Herr, *Int. J. Robotics Research* **24**, 1013 (2005).
8. B. Ugurlu and A. Kawamura, *IEEE Trans. Robotics* **28**, 1406 (2012).
9. F. Asano, Z.-W. Luo and M. Yamakita, *IEEE Trans. Robotics* **21**, 754 (2005).
10. A. Ananthanarayanan, S. Foong and S. Kim, *Proc. IEEE Int. Conf. Robot. Autom.* , 1398(May 2012).
11. M. Y. Chuah, M. Estrada and S. Kim, *Proc. IEEE/RSJ Int. Conf. Intelligent Robots and Systems* , 1963(Oct. 2012).
12. K. Byl and R. Tedrake, Approximate optimal control of the compass gait on rough terrain, in *ICRA*, 2008.
13. M. Raibert, K. Blankespoor, G. Nelson, R. Playter and the BigDog Team, Bigdog, the rough–terrain quadruped robot, in *Proc. of the 17th World Congress*, 2008.
14. K. Suwanratchatamanee, M. Matsumoto and S. Hashimoto, Balance control of robot and human-robot interaction with haptic sensing foots, in *HSI*, 2009.
15. Honda, Asimo (2011) key specifications (2011).
16. A. Kalamdani, C. Messom and M. Siegel, *IEEE Instrumentation Measurement Magazine* **10**, 46 (october 2007).
17. J. Romano, K. Hsiao, G. Niemeyer, S. Chitta and K. J. Kuchenbecker, *IEEE Transactions on Robotics* **27**, 1067 (December 2011).
18. RoboSKIN Project http://www.roboskin.eu/index.php.
19. A. Dollar, L. Jentoft, J. Gao and R. Howe, *Autonomous Robots* **28**, 65 (2010).
20. E. Torres-Jara, Sensitive manipulation, PhD thesis, MIT CSAIL2007.
21. L. Natale and E. Torres-Jara, A sensitive approach to grasping, in *Sixth international workshop on Epigenetic Robotics*, (Paris, France, 2006).
22. E. Torres-Jara and G. Gomez, Fine sensitive manipulation, in *Australasian Conference on Robot. Autom.*, (Canberra, Australia, 2008).
23. E. Torres-Jara, I. Vasilescu and R. Coral, *A Soft Touch: Compliant tactile sensors for sensitive manipulation*, Tech. Rep. MIT-CSAIL-TR-2006-014, MIT (March 2006).
24. G. A. Pratt, *Integrative and Comparative Biology* **42**, 174 (2002).
25. G. Pratt and M. Williamson, Series elastic actuators, in *Proc. IEEE/RSJ Int. Conf. on Intelligent Robots and Systems*, 1995.
26. E. Torres-Jara and J. Banks(March 2004).
27. S. Collins, A. Ruina, R. Tedrake and M. Wisse, *Science* **307**, p. 1082 (2005).

LAYERED BODY FOR FLEXIBLE MONO-TREAD MOBILE TRACK

T. HAJI* and S. ARAKI

Matsue College of Technology, Matsue, Shimane, JAPAN,
**E-mail: haji@matsue-ct.jp*

T. KINUGASA, T. ITO and K. YOSHIDA,

Okayama University of Science, Okayama, JAPAN,

H. AMANO,

National Research Institute of Fire and Disaster, Tokyo, JAPAN

R. HAYASHI,

Kagoshima University, Kagoshima, JAPAN,

K. TOKUDA,

Wakayama University, Wakayama JAPAN,

M. IRIBE,

Osaka Electro-Communication University, Osaka, JAPAN

Robot technology is expected to become applicable to missions on rough terrain, such as rescue activities, environmental investigations, and planetary exploration. Tracked vehicles are effective against such environments because the contact pressure of the vehicle can be distributed more widely. In order to improve mobility, new mechanisms such as a serpentine-type tracked vehicle have been proposed. We previously proposed the flexible mono-tread mobile track (FMT). The first prototype called WORMY has a mechanical problem, i.e., track belt interference and derailing, when it moves flexing. This paper proposes and validates a new design strategy of the FMT adopting layered structure of its body. The layered structure reduces space between vertebrae, which can solve the derailing problem. A new prototype adopted the layered structure is developed in order to validate its mobility. We also evaluate effectiveness of passive retro-flexion against obstacles.

Keywords: Flexible Mono-tread Mobile Track; Tracked vehicle; Flexible layered structure.

1. Introduction

Robot technology is expected to become applicable to missions on rough terrain, such as rescue activities, environmental investigations, and planetary exploration. Tracked vehicles are effective against such environments because the contact pressure of the vehicle can be distributed more widely. In order to improve mobility, new mechanisms such as a serpentine-type tracked vehicle – i.e., articulated multi-tracked vehicles composed of several serially connected mono-tread,[1] dual-tread[2–5] or multi-tread[6,7] tracked vehicles and steerable mono-tread tracked vehicles with bendable[8–14] – have been proposed.

Fig. 1. FMT with vertebral structure

We previously proposed the flexible mono-tread mobile track (FMT).[15] The FMT has a single track that can flex in three dimensions. Its fundamental mobility, e.g., wall climb-ability, ditch traffic-ability and step climb-ability, was demonstrated by the prototypes RT02-WORMY[15] (Fig.1, left). WORMY was composed of six vertebrae connected by five vertebral disks made of rubber cylinder or ball joints with a coil spring. WORMY has a mechanical problem, i.e., track belt interference and derailing, when it moves flexing (Fig.1, right).

This paper proposes and validates a new design strategy of the FMT adopting a layered structure of its body. The layered structure reduces space between vertebrae and realizes smooth flexion, which can solve the derailing problem. A new prototype adopted the layered structure is developed in order to validate its mobility through some experimental tests.

2. Modification From Vertebral Structure to Layered Structure

Figure 2 is a commercial slatted conveyer belt using as a track belt. The belt is composed of track shoes connected each other by "D" shape pins. The shoe has a top plate and two float preventing tabs. A pitch of the shoe, 38 mm, is defined by distance between two pins. The belt slides on a guide rail constrained between the plate and the tab. If there is open space wider than the tab along the guide rail for flexion, then the belt, especially the float preventing tab, interferes or derails through the open space of the guide rail. Moreover, the larger vertebrae are and the shorter vertebral disks

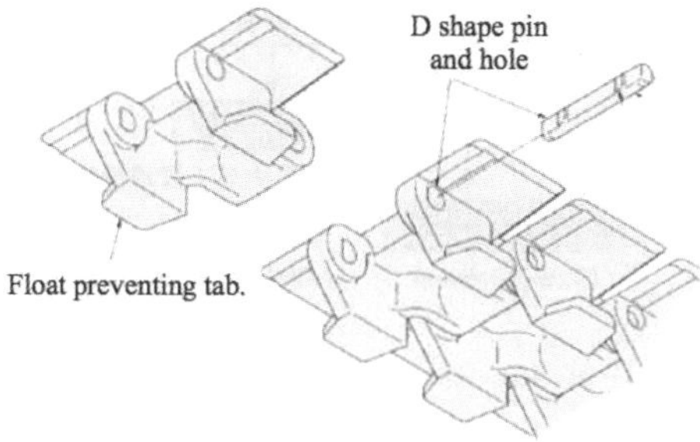

Fig. 2. Track shoe (slatted belt)

are, the larger the slippage between the belt and the ground is, because the flexed posture turns a polygon from a smooth arc. The slippage results in increase of driving resistance and ground damage.

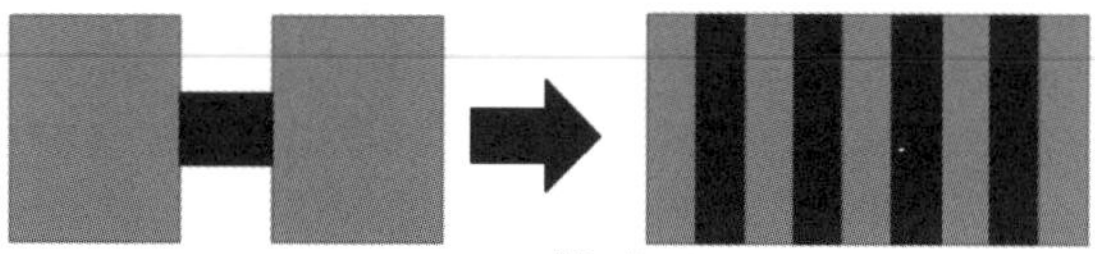

Fig. 3. Modification image from vertebral structure to layered structure

Fig. 4. RT06 adopted a layered body

We adopted layered structure instead of the vertebral structure. Figure 3 shows an image of a part of the vertebral structure and the layered structure. Concretely, the body of FMT consists of a layer of MDF plates (medium density fiberboard, thickness: 5.5 mm) and rubber sponge plates (thickness 10 mm). The thickness of the both plate are thinner than the tab, thus the open space becomes narrow and the body flexion approaches the arc.

Figure 4 represents a new prototype of the FMT called "RT06" (left) with the layered structure (right). In the right figure, black layers are made from rubber sponge and brawn layers are from MDF. A side view is shown in the below figure of Fig. 4. From the side view, the vehicle is composed of a drive segment (the left hand side), a control segment located at the center of the body, an idler segment (the right end) and the layered segments.

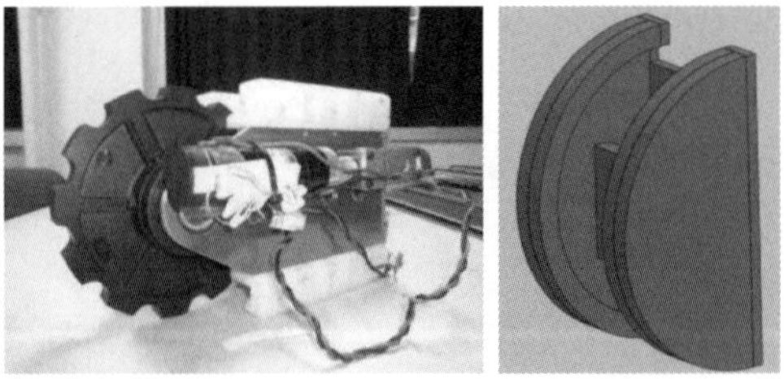

Fig. 5. Drive segment and idler segment

The drive segment (Fig.5, left) has two motors for driving the track belt and flexing in lateral direction. Flexion in vertical direction called retro-flexion was passive, i.e., an actuator for the retro-flexion implemented to WORMY was removed. The control segment has a control device called TPIP 2 board made by Sanriz Automation Co. Ltd. and a two channels motor driver, of which arrangement can be compatible in the layered segments. The idler segment (Fig.5, right) is composed of a semi-circular guide rail made of MDF. The layered segments are the layer of 45 rubber sponge plates and 47 MDF plates. Total length of RT06, 1.2 m, and the height, 0.2 m are the same as WORMY, however, the weight of the idler segment, 200 g is a tenth of that of WORMY and the layered segments are slightly lighter. As a result, the total weight of RT06 became 9 kg (3 kg lighter than WORMY). RT06 is teleoperated by the control device, TPIP2,[17] and uses Li-Fe batteries for drive (13.2 V) and control (6.6V).

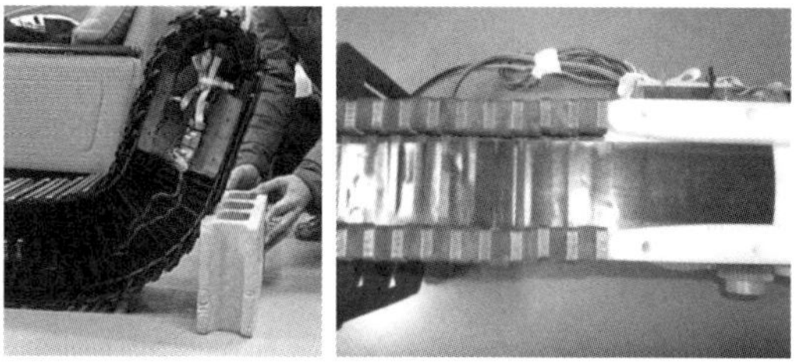

Fig. 6. Drive segment and idler segment

The rubber sponge layer is too soft to prevent over flexing around relatively heavy segments such as the drive segments or the control segments as shown in Fig.6 (left). Hence we adopted vinyl tape to prevent over flexion as shown in Fig.6 (right). The vinyl tape gives additional stiffness to the layer.

3. Experiment without Active Retro-flextion

The prototype RT06 was used in the experiment. The purpose of the experiment was to validate the layered structure. We tested fundamental performance of RT06 without the active retro-flexion. The prototype was able to run and turn on the flat surface without the interference nor derailing.

Next, step climbing performance was tested as shown in Fig.7. Radius of wheel or sprocket decides climb-ability of tracked or wheeled vehicle against a step. Because the climb-able height of the step is lower than the radius basically, mobile robots moving on a rough terrain adopted sub-mechanism such as paddle, sub-track and articulation. On the other hand, the FMT can climb the step of a half of its length by using the retro-flexion. The retro-flexed posture becomes a part of circle of which radius is much larger than its sprocket. From Fig.7, the prototype could climb over the step 480 mm high. The procedure of the climbing is as follows: RT06 reaches to the step and touches the wall to begin with, next, the head of RT06 lifts along the wall passively and adjusts, and then it climbs over the step. We can confirm that the prototype flexes continuously and smoothly and adjusts its body to shape of the step, even if the FMT does not use any actuators for the retro-flexion.

Finally, stairway climbing performance was tested. A step of the stairway is 295 mm deep and 195 mm high, which is usual stairway in office buildings under the Japanese regulation. In addition, the step has slight overhang, i.e., its side wall inclines 10 degrees from the vertical position.

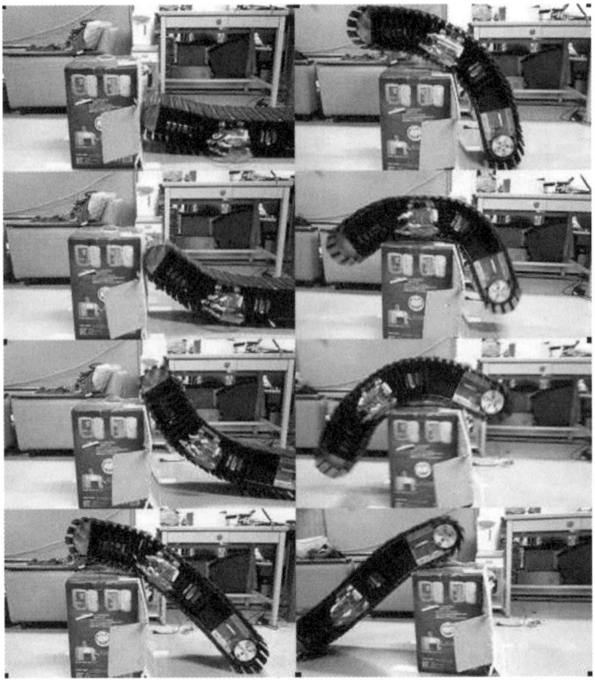

Fig. 7. Step climb-ability

RT06 cannot retro-flex actively, thus it cannot lift the head (the idler segment) when it tries to climb as shown in Fig.8. As a result, RT06 could not climb the stairway up. There would still exist the same problem even if RT06 overcame the first step of the stairway, because the head droops until it reaches to the next step. We restrict angle of the passive retroflexion in order to avoid drooping, then RT06 can climb the stairway as shown in Fig.9. If the FMT retro-flexes actively such as RT02-WORMY and RT04-NAGA, then it can lift its head, adapt to shape of steps and climb a stairway. Therefore, the active retro-flexion should be adopted to the FMT.

4. Conclusion

In the paper, we proposed a new layered structure of a flexible mono-tread mobile track in order to overcome mechanical problem of the previous prototype of the FMT. A new prototype of the FMT was developed using alternately layered rubber sponge and MDF plates, and its fundamental

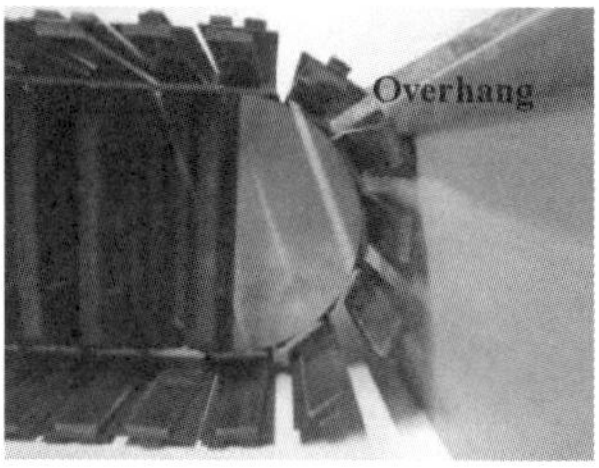

Fig. 8. Stuck in overhang

Fig. 9. Stairway climb-ability

performance were evaluated through some experimental tests. The results are summarized as follows:

- The layered structure reduces possibility of derailing.

- The stiffness of the layer next to heavy segments should be relatively hard.
- Smooth and continuous flexion realizes step climb-ability using passive retro-flexion.
- The new prototype could climb up and down a step 480 mm high and stairs.

It is impossible to climb an overhung step using the passive retro-flexion only, because the head of RT06 gets stuck in the overhung step. However the passive retro-flexion realize adaptability against rough terrain, which gives wider contacting surface to the ground, and thus mobility could have improvement. Although the retro-flexion should be actuated, alternate use of passive and active retro-flexion might be recommended.

Acknowledgment

This work was supported in part by the Ministry of Education, Culture, Sports, Science and Technology of Japan through a Financial Assistance Program (No. S1201041 and S1101036). We received generous support for the teleoperation control device (TPIP board[17]) of our prototypes from Sanriz Automation Co., Ltd.

References

1. Y. Tanaka, et al., *Proc. of SSRR 2006* **TUE-AM1** (2006).
2. T.Takayama, S.Hirose, *J. of Robotics and Mechatronics* **15(1)** 61/69 (2003).
3. M. Arai, et al., *Proc. of IROS2004* 52/57(2004).
4. M. Arai, et al., *J. of Field Robotics* **25(1)** 31/65 (2008).
5. T. Kamegawa and F. Matsuno, *J. of RSJ* **25(7)** 1074/1081 (2007).
6. K. Osuka and H. Kitajima, *J. of Robotics and Mechatronics* **15(5)** 561/570 (2003).
7. J. Borenstein, A. Borrell, *Proc. of IEEE ICRA 2008* 1766/1767 (2008).
8. H. E. Best, *US Patent* **No.3548962** (1970).
9. V. H. Ames, *US Patent* **No.3565198** (1971).
10. J. C. Stancy, Jr, *US Patent* **No.4453611** (1980).
11. N. Iwatsuka, and K. Sadakane, *Japan Patent* **No.60-80573** (1985).
12. T. Fukuda, et al., *Trans. of JSME* **C 60(569)** (1994).
13. H. Schempf, *Robotics Res.* **Springer Verlag Berlin** 453/465 (2003).
14. Automatika, Inc. (AURORA) website:
    http://www.automatika.com/products_aurora.htm
15. T. Kinugasa, et al., *Proc. of IEEE/RSJ IROS 2008* 1642/1647 (2008).
16. T. Kinugasa, et al., *Proc. of SSRR2011* 86/91 (2011).
17. TPIP board, http://www.sanritz.co.jp/product_domain02_tpip_sys.html

SURFACE ADAPTATION ROBOT FOR DEFECT DETECTION BY PERFORMING CONTINUOUSLY AN ULTRASOUND WHEEL PROBE

HERNANDO LEON-RODRIGUEZ

Industrial Engineering Department, Nueva Granada Military University, Bogota, Colombia, ZIP: 111101, E-mail: hernando.leon@unimilitar.edu.co

This paper describes the design of an umbilical-free mobile non-destructive testing (NDT) climbing robot for industrial applications that is capable of inspecting oil tanks, pipelines, petro-chemical tanks, bridges, railways, etc. The mechanical design is based on being adapted itself to the surface (flat, convex and concave along a pipe or tank length and width). It can makes transitions between surfaces that have an angle up to sixty degrees between them. The robot is a wheeled mobile robot with magnetic adhesion forces optimised with flux concentrators. It is carrying a NDT wheel probe in order to detect weld defects, wall thickness and corrosion. The robot has an on-board the motion control system, the NDT flaw detector and the high frequency card; the entirely system can be controlled remotely from ground via wireless communication.

1. Introduction

Application areas for automated and robotic non-destructive testing (NDT) are in the inspection of critical infrastructure on offshore oil platforms, nuclear power plant, shipyards, petrochemical and other storage tanks, aircraft, buildings, pipelines, bridges and railways [1, 2, 3]. Access to test sites on these structures is obtained by constructing scaffolding or abseiling down from the top. Erecting scaffolding is expensive and hence testing the structural integrity of these structures is an expensive and time consuming activity.

An increasing interest in the development of special climbing robots has been seen in recent years [5, 6]. This has been motivated by a desire to increase operation efficiency in the performance of dangerous and hazardous tasks by reducing the turn-around time in both planned and unplanned outages. These capabilities in climbing robots would not only allow them to replace human workers in the carrying out of these dangerous duties but would also eliminate costly scaffolding. One of the most interesting applications of a climbing robot involves steel tank inspection. For example, thousands of storage tanks in oil refineries have to be inspected manually to prevent linkage or potential catastrophe.

A prototype NDT robot is described in this paper which uses permanent magnets, with an air gap of 5 mm to avoid studs, small obstacles, full contact with the surface, etc. The robot is able to attach itself and be adapted in the most usual shape in the petro-chemical industry like flat, convex and concave ferrous surfaces. It is able to change surfaces up 60 degrees while maintaining an adhesion force and climb on vertical surfaces. The robot is designed to scan along the weld lines with continuous motion. It is carry and deploys a dry contact ultrasonic wheel probe to detect weld defects, wall thickness and corrosion. The ultrasound detector is carried on board the robot; it is completely umbilical free with all control and NDT data acquisition on-board. The NDT inspector is able to control the robot wirelessly via a PC laptop computer and set all parameters from the ground. The robot saves all data on-board, so in case of loss of communication with the ground, it keeps on collecting data until the communication is restored. It can be adapted in different surface curvatures with a minimum diameter of 1 metre. It can be attached to the surface from inside or outside and inspect for defects or wall thickness. Its mass is 5 kg which makes its deployment very easy and its batteries last in 2 hours of continuous scanning.

2. Design of the Ultrasound NDT Robot

The NDT robot is been designed and built as shown in Figure 1, to have three hinged sections, with three groups of wheels to drive the robot, and one support wheel in the front of free motion. The two sets of wheels in each half of the robot are driven differentially by four DC geared motors.

Figure 1: Prototype of a wall climbing robot performing ultrasound wheel probe for NDT inspection.

The main requirement is to keep the NDT probe in permanent contact with the surface at all times i.e. on concave and convex surfaces and when moving across and along any pipeline as show in figure 2 and figure 7. The robot can travel around the pipeline surfaces both internally and externally and also when it is travelling along the pipeline. With a given overall size of the robot, when the robot is working on concave surfaces, the air gap between the magnets and surface in the sections design is much bigger than in the convex surface i.e. d1 > d2. As a consequence, the adhesion force on convex surfaces is greatly increased because of the reduction of the air gap.

The three section design not only enables the robot to transfer between sharp angled adjoining surfaces, but also improves its climbing ability on curvatures; in consequence, this design satisfies this requirement. [6, 7] The robot moves at low constant speed to scan for weld defects, thickness and corrosion. The robot speed can be modified by the operator to suit the inspection requirements. Its maximums speeds is 10 m/min, it is speed can be improved a lot, but considering that the speed´s inspection should be very low (1 m/min) because of NDT devices.

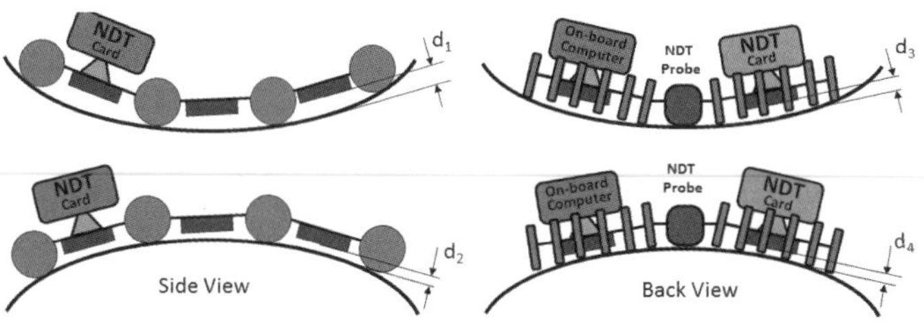

Figure 2: Robot adaptation to surface curvatures in tank or pipeline.

The robot weights 5 kg and it has the ability to carry nearly 6 kg extra payload in a vertical climbing. It is a high power wheeled design robot with capability to travel over any weld obstacle and provide the option to make angular transitions between surfaces of 0 to 60°. It is driven by four powerful 15V DC motors with an output torque of 1000Ncm on each motor. They transmit the power to each set of wheels by connecting them with a steel chain, providing a big pulling force and highly secure transmission.

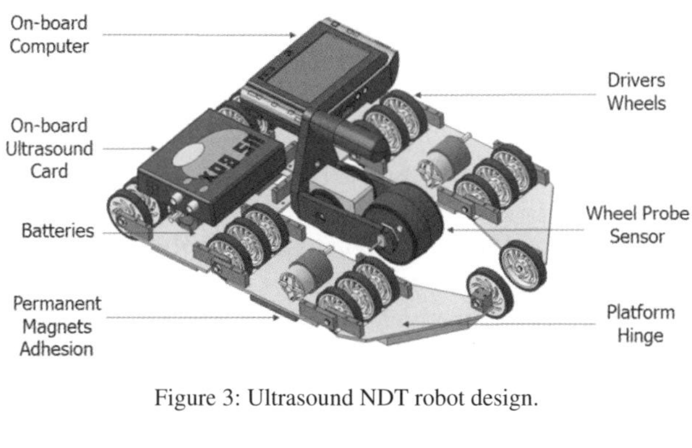

Figure 3: Ultrasound NDT robot design.

The robot has on-board all controllers including two motor control modules, small computer system, one NDT card, one ultrasound NDT probe and batteries. See figure 3. All cards are connected to the computer via USB and the on-board computer is connected to the ground via Wi-Fi interface; so that, the operator can control the robot and the NDT devices via Ethernet network. All the on-board modules with standard USB interface are plugged to a USB hub and then connected with the computer.

The operator can access the on-board computer via Wi-Fi wireless communication to just supervise the inspection or to control the robot manually. Furthermore, all the on-board devices are powered by a high capacity lithium-ion Polymer (Li-Po) battery pack, so that the robot is totally umbilical free. The NDT application and graphic user interface (GUI) are placed on-board, this guarantees that all information is safe and the operator does not need to be worry about losing connection or interference from other devices. See figure 4 for a schematic of the control system.

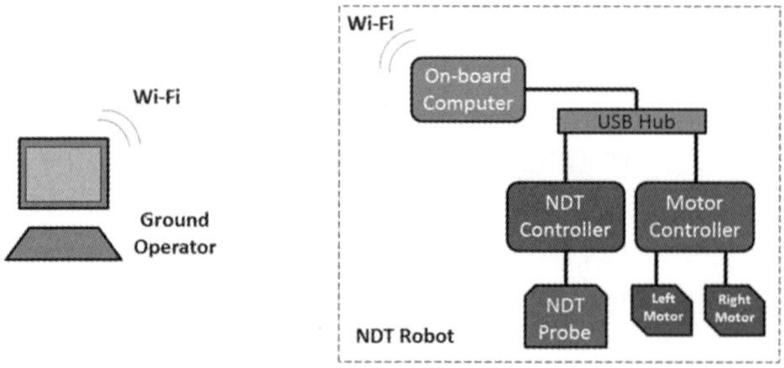

Figure 4: Schematic diagram of the Control System - NDT robot.

3. Optimization of Magnetic Adhesion

The wall climbing robot uses magnetic adhesion force in order to be attached in a ferrous steel wall. The big optimization in the adhesion force Arrays was split the permanent magnets into three hinge modules of the robot, called back, middle and front magnetic blocks, see figure 5. In addition, to make even stronger the magnetic force behind all set of magnets were been placed a main flux concentrator plates on each hinge.

The adhesion force has been calculated for all three blocks independently by using finite element magnetostatic analysis. The largest block of ten magnets (shown in figure 5 and figure 6) was first modelled in ANSYS design modeller and then imported into ANSYS Magnetostatic analysis. Like all the Finite element methods procedure, the meshing of the magnets was carried out and the boundary conditions were defined and simulated. The results of the simulations were verified by experimental results to validate the simulation setup and boundary conditions. [4]

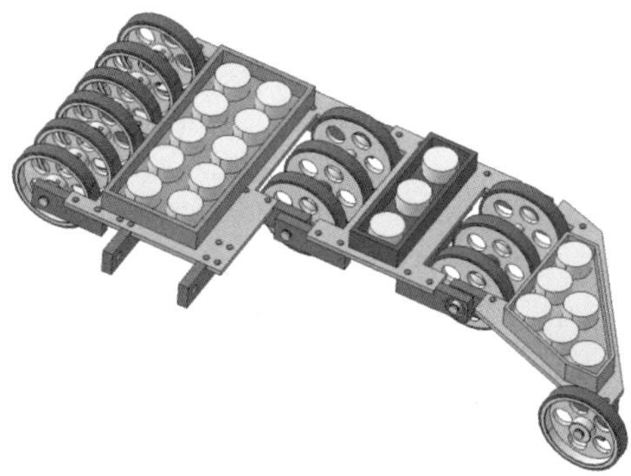

Figure 5: Half section of robot showing the three blocks of permanent magnets – Bottom view.

The ANSYS Magnetostatic Analysis is able to analyse different magnetic properties of the blocks. It includes flux density, field intensity, force simulation, torque, energy and magnetic flux. The same arrangement of blocks of permanent magnets are placed on both sides of the robot, but for the ANSYS analysis we have evaluated just one side, so the total adhesion forces available is double. [4]

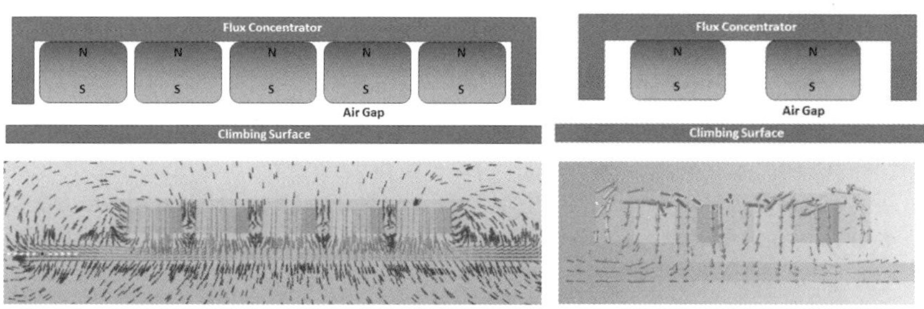

Figure 6: Arrangement of block magnets and magnetic flux line analysis for each block - side and long view.

Figure 6 shows a conceptual design of the biggest magnetic block and the results of finite element analysis of how the flux lines around the block are facing the flux concentrator from north poles of two magnets and finished these flux lines cross into the south pole after passing through the wall to complete the magnetic circuit.

These magnetic blocks show in figure 5 are placed in the back of the robot where the highest attraction force is required. The magnetic block is composed by the magnetic flux concentrator with 5mm thickness of mild steel, two groups of neodymium magnets of N35 with 20 mm diameter and 10 mm height (ten in total) with minimum technical requirement of 5 mm air gap between climbing surface (usually steel) and the robot. If the air gap is increased, the magnetic adhesion decreases, but this air gap is necessary to avoid obstacles.

The figure 7 shows the auto-adaptation of the robot in flat concave and convex surface. It also shows the continuous inspection of weld defect in all surfaces.

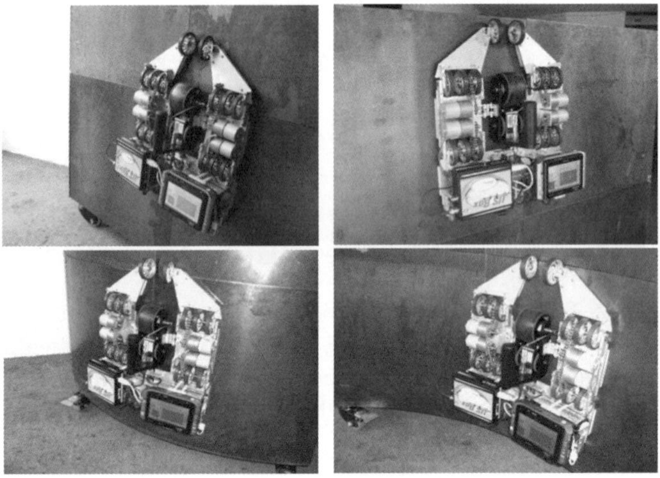

Figure 7: Auto-adaptation of climbing robot in different shapes.

4. NDT Performance and Results

The robot carries a single wheel probe to test weld defects, thickness or corrosion. The single element wheel probe is provided by Sonatest Ltd. It uses rubber technology for near-dry Ultrasonic coupling. The probe is designed to operate between 1MHz and 10MHz and is able to run in a low pressure tyre of 58mm diameter. It can be operated with or without an optical position encoder, placed in the back of the wheel probe.

The project had achieved some manual NDT test with the purpose of understand some pre-defined defects in different conditions like depth, size and shape. The results are presented considering visible pre-defined defects of three different test pieces of 50x50x200mm size; on this document only test piece 1 are presented. The figures 8 and 9 show the trials that were carry out in manual NDT inspection over a mock-ups to analyse the different types of defects and echoes sending back from the probe; as a results, the NDT ultrasonic sensor identify them as expecting, different size in a constant distance between them.

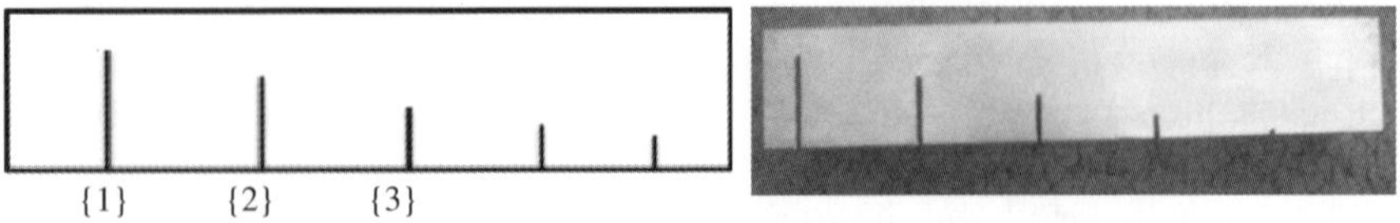

Figure 8: Test piece 1 with transversal cuts, simulated weld defects.

The result of manual evaluation has a clear picture of how the ultrasonic responds on each pre-defined defect, where, doing a comparison between the five graphs, it is possible to observe how the signal get bigger on the first cut, and it begins to become smaller until the last defect. See figure 9.

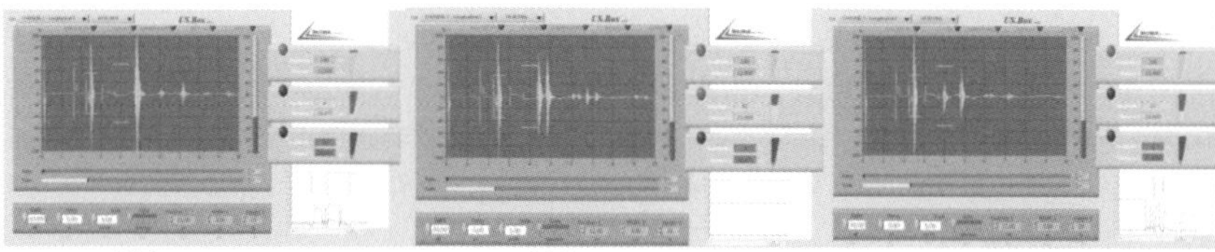

{1} {2} {3}

Figure 9: Result of manual NDT analysis in test piece 1.

The figure 10 is showing the results of the automated mode inspection of the NDT robot over the mock-up flat surface performing in a continuous motion and detections. It's clearly seen in the NDT picture where the wheel probe recognises the different by increasing the size of curve-plot.

The obtained results from the automated inspection are the measurement of any detectable change on the surface varies provided by the A-scan readings. From these measurements can be seen that the manual inspection resulted in Figure 9 present noise on the detected signals from the surface, due to the non-constant pressure over the wheel probe during the inspection. It is worth noting that the noise is considerably reduced when using the automated inspection due to the adhesion forces applied in the target surface.

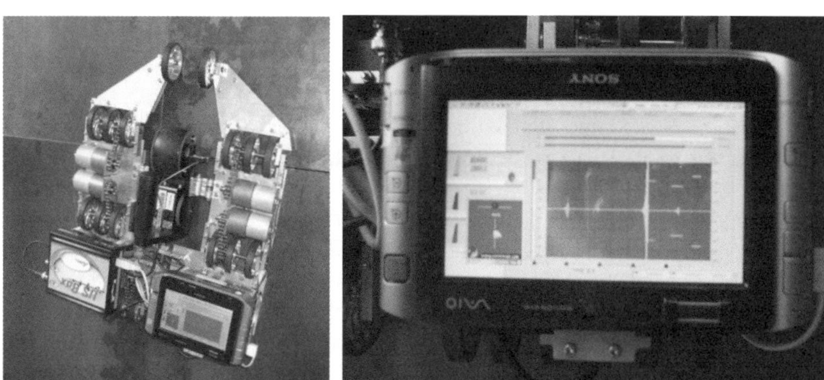

Figure 10: Result of automated NDT analysis in flat wall mock-up.

5. Conclusions

The lightweight NDT robot has been designed with the capability of climbing on steel walls carrying a payload of up to 6 kg. The high payload capability, considering the small side of the robot, has been obtained by using neodymium magnets, magnetic flux concentrators to increase adhesion forces and a mechanical design that keeps the magnet arrays at a more or less constant distance from the surface even when moving on large surface curvatures and shapes. It is able to climb on curved surfaces with excellent manoeuvrability, and can transfer between angled surfaces up to sixty degree.

The robot has perform in automated mode an ultrasound NDT wheels probe, with well defects detection and thickness measurement, it has a clear picture of the defect detection based on the constant force applied over the wheel probe.

References

[1] Shang J., Bridge B., Sattar T.P., Mondal S. and Brenner A., "Development of a Climbing Robot for Weld Inspection", *Proceedings of the 10th International Conference on Climbing and Walking Robots and Supporting Technologies for Mobile Machines, CLAWAR 2007*, Singapore, 2007.

[2] Bridge, B., Sattar T.P., Leon H., "Climbing Robot Cell For Fast And Flexible Manufacture Of Large Scale Structures", *22nd International Conference,CAD/CAM Robotics and Factories of the Future (CARS & FOF 2006)*, pp 584-597, Eds Narayanan, Gokul Kumar, Janardhan Reddy, Kuppan, ISBN Narosa Publishing House, ISBN 13:978-81-7319-792-5, ISBN; 10:81-7319-792-X, 19-22 July 2006.

[3] Sattar T.P., Zhao Z., Feng J., Bridge B., Mondal S., Chen S., "Internal In-service Inspection of the floor and walls of Oil, Petroleum and Chemical Storage Tanks with a Mobile Robot", *5th International Conference on Climbing and Walking Robots and the Support Technologies for Mobile Machines*, pp 947-954, Edited by Philipe Bidaud and Faiz Ben Amar, ISBN 1 86058 380 6, Professional Engineering Publishing Ltd. UK, 2002.

[4] Hussain S., Sattar T. Salinas E. Parameters Analysis and Design Framework for Magnetic Adhesion wall Climbing Wheeled Robot, *International Journal of intelligent System and applications, 2012*

[5] Shang, Jianzhong; Sattar, Tariq; Chen Shuwo; Bridge Bryan, "Design of a Climbing Robot for Inspecting Aircraft Wings and Fuselage", *Industrial Robot: An International Journal*, vol. 34, No. 6, 2007, Emerald Group Publishing Limited, ISSN 0143-991X, 2008.

[6] Fernández R, González E, Feliú V, González A., A Wall Climbing Robot for Tank Inspection. An Autonomous Prototype, *36th Annual Conference on IEEE Industrial Electronics Society*, pages 1424 – 1429, ISSN: 1553-572X, 2010.

[7] F. Tâche, W. Fischer, G. Caprari, R. Siegwart, R. Moser, and F. Mondada, "Magnebike: A magnetic wheeled robot with high mobility for inspecting complex-shaped structures," *Journal of Field Robotics*, vol. 26, pp. 453-476, 2009.

OPTIMAL DESIGN OF A MAGNETIC ADHESION FOR CLIMBING ROBOTS

PETER WARD*, DIKAI LIU, KEN WALDRON, MAHDI HASAN

Centre for Autonomous Systems,
University of Technology, Sydney, NSW, Australia
**E-mail: Peter.K.Ward@student.uts.edu.au*

Designing a magnetic adhesion system for climbing robots requires careful selection of design parameters to achieve a feasible solution. There are many considerations which must be taken into account, such as, size constraints for the intended environment and robot configuration, the maximum load that can be supported by the climbing robot, and the expected air gaps during operation. With consideration of the design challenges, an optimal design for a magnetic adhesion system is presented. Based on the optimal design a prototype footpad has been constructed for use on an inchworm climbing robot and experimental results are presented.

Keywords: optimal design; magnetic adhesion; climbing robot

1. Introduction

Although adhesion is one of the most challenging problems in designing a climbing robot, it is generally accepted that magnetic adhesion is the most suitable for ferromagnetic structures.[1,2] Magnetic adhesion can be split into two types, being electromagnetic and permanent magnetic adhesion. Permanent magnetic solutions are considered the most suitable with the primary benefit of not requiring power for active adhesion, hence, are inherently power-fail safe. This is essential for industrial applications where safety tethers may not be possible in accessing difficult to reach areas. Many forms of adhesion are not suitable for surfaces which are rough, uneven, and contain layers of paint which may be peeled from the underlying structure. Furthermore, permanent magnets allow safe adhesion to the structure rather than the exterior layers unlike other forms of adhesion.

While there are several novel adhesion systems which utilise permanent magnets such as Electro-Permanent Magnets,[3,4] Magnetic-Switchable-Devices[5] and others,[6–8] this paper begins the design process by first considering what an optimal design for an adhesion system means. The term footpad is used to refer to the integration of multiple adhesion modules, and analyses the design problems in integrating adhesion modules to form an optimal footpad design.

2. Footpad Design Objectives and Parameters

The primary design challenge is maximising performance while minimising weight. The performance is determined by the five modes of failure for the footpad as identified in Fig. 1. The modes include failure through excessive pull force F_p, slip force F_s, bending moments M_x and M_y, and torsion M_z due to the load imposed by the robot and payload. The design variables chosen to maximise the performance criteria include the radius of the footpad R_{pad}, the magnet diameter D_m, the magnet thickness T_m, and the number of magnets n.

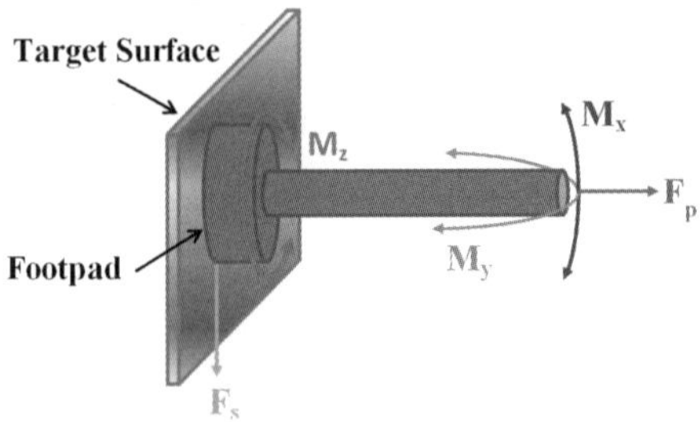

Fig. 1. Loads imposed by climbing robot on footpad while attached to vertical surface

This paper makes several assumptions. All magnets have the same radius from the footpad centre and are equally spaced around the circumference of the footpad. The magnet is fixed to a large sector gear which is rotated from the surface through use of a motor and pinion gear as seen in Fig. 2. An inchworm configuration has been chosen and optimised to provide greatest manoeuvrability for the intended design applications.[9] Each end of the inchworm robot will use identical footpads.

The fixed design parameters include the weight of the footpad base plate W_b, the height of footpad H_p, the height of the frame H_f, the thickness of the frame T_f, the density of aluminium ρ_{Alum}, the density of N52 grade Neodymium ρ_{NdFeB}, the length of the robot L_r, the weight of the robot W_r, and the coefficient of static friction μ_s, the weight of the magnet housing W_h. The weight of the magnet housing includes the weight of the pinion gear, sector gear, motor and axle. Other design parameters include the weight of the magnet $W_m(D_m, T_m)$, the weight of the frame $W_f(R_{pad})$, and the weight of the footpad $W_{pad}(R_{pad}, W_m, W_f)$, and the pull force of the magnet $F_m(D_m, T_m)$.

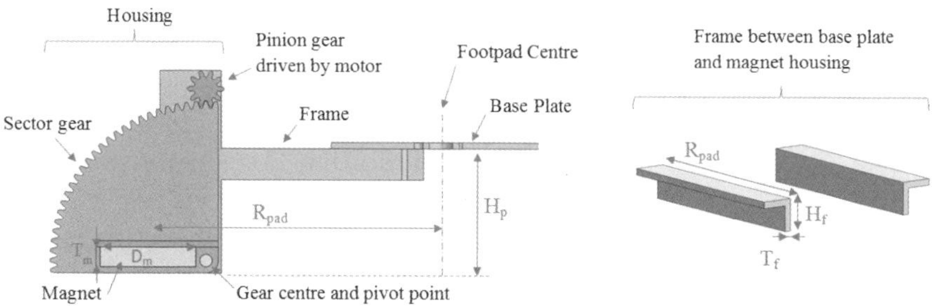

Fig. 2. Components which make up footpad and climbing robot

The pull force, F_m, of the magnet is determined by D_m, T_m and the air gap between magnet and ferromagnetic surface. The assumed magnetic solution uses a protective housing with a separation distance of 2mm from the surface. Taking into account the expected air gaps due to roughness and paint in the intended environment, the footpad is designed for a worst case scenario with a 5mm air gap. Figure 3 shows the force model through testing for an N52 Neodymium magnet with a 5mm air gap.

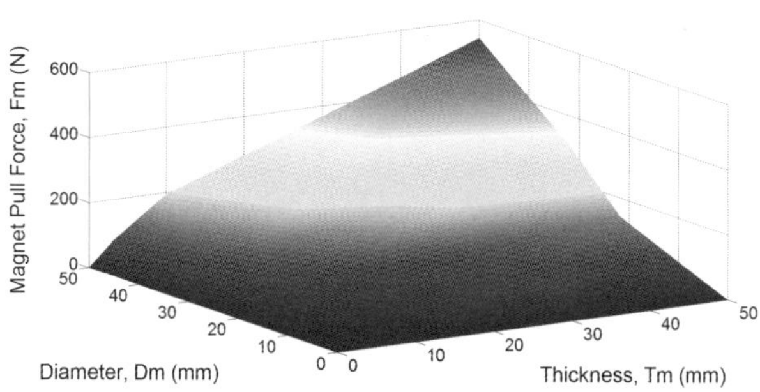

Fig. 3. Force Model $F_m(D_m, T_m)$, for N52 NdFeB with 5mm air gap

A design problem now exists where we must trade off performance criteria to reach a feasible weight. This requires careful selection of the footpad radius, magnet diameter, thickness, and number of magnets.

3. Optimal Design Model

An optimal design model has been developed for a footpad in order to minimise the weight and maximise the performance criteria. The footpad weight is normalised by the maximum load the robot is capable of sustaining, given its specified joint torques. Hence the normalised weight $N(W_{pad})$ approaches a maximum value of one as the weight approaches zero. The performance criteria are normalised to estimates of the applied load in a worst case scenario i.e. F_p is normalised to $N(F_p)$, using an estimate of the weight for the entire robotic system. Multiplying the normalised performance and weight values together results in a value we wish to maximise. The function is converted to a minimisation problem by multiplying by negative one, the objective function is then expressed as follows.

$$\underset{R_{pad},D_m,T_m,n}{\text{Minimise:}} \quad -N(F_p) \times N(F_s) \times N(M_x) \times N(M_y) \times N(M_z) \times N(W_{pad}) \tag{1}$$

$$F_p = n \times F_m(D_m, T_m) \tag{2}$$

$$F_s = n \times F_m(D_m, T_m) \times \mu_s \tag{3}$$

$$M_x = \sum_{i=0}^{n-1} F_m(D_m, T_m) \times R_{pad} \times \cos(\frac{2 \times \pi}{n} \times i) \tag{4}$$

$$M_y = \sum_{i=0}^{n-1} F_m(D_m, T_m) \times R_{pad} \times \sin(\frac{2 \times \pi}{n} \times i) \tag{5}$$

$$M_z = n \times F_m(D_m, T_m) \times R_{pad} \times \mu_s \tag{6}$$

$$W_{pad} = (n \times [W_m(D_m, T_m) + W_h + W_f(R_{pad})] + W_b) \tag{7}$$

Where,

$$W_m(D_m, T_m) = \pi \times (D_m/2)^2 \times H_m \times \rho_m \tag{8}$$

$$W_f(R_{pad}) = 4 \times \rho_a \times T_f \times H_f \times R_{pad} \tag{9}$$

3.1. *Design Constraints*

In order for the robot to perform a 360 degree plane transition as seen in Fig. 4, the upper limit of the footpad radius is restricted by the length of members in the robot chain. Furthermore, the lower limit of the footpad radius is constrained by the minimum footpad size for a given magnet diameter. Hence, Eqn. 10 prevents an intersection of the magnets and ensures the construction of the footpad is feasible, see Fig. 4.

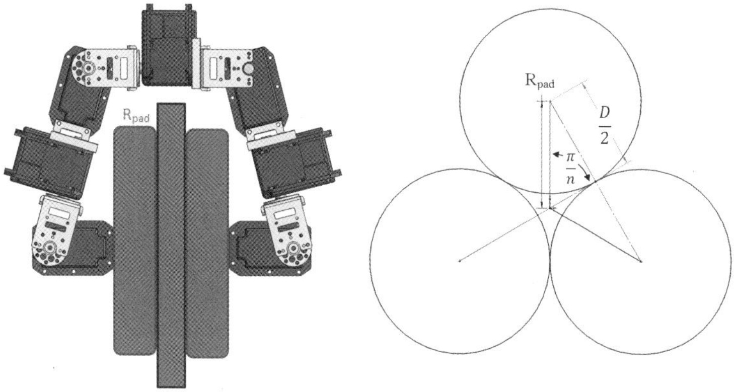

Fig. 4. Geometric constraints for footpad radius being limited by 360 degree plane transfer and the minimum locations of magnet

Equation 11 limits the footpad weight by the maximum payload W_{max}, that the climbing robot can support. This has been determined by considering the motor torque capacity for the inchworm robot. The pull force is constrained in Eqn. 12 to ensure that the total system weight can be supported while inverted.
The slip force is constrained in Eqn. 13 to ensure sufficient adhesion to overcome the weight of the robot when on a vertical surface.
A constraint on the bending moments in the x direction (Eqn. 14) and y direction (Eq. 15) ensures that the footpad can support the climbing robot in cantilevered positions (Eq. 17) for either orientation.
Equation 16 ensures that the pad will not rotate when the robot is horizontal to and parallel to a vertical surface, Eq. 18.
The set of constraints become:

$$0.15m \geq R_{pad} \geq \frac{D_m}{2 \times \sin(\pi/n)} \tag{10}$$

$$W_{max} \geq W_{pad} \tag{11}$$

$$F_p \geq (2 \times W_{pad} + W_R) \tag{12}$$

$$F_s \geq (2 \times W_{pad} + W_R) \tag{13}$$

$$M_x \geq M_R \tag{14}$$

$$M_y \geq M_R \tag{15}$$

$$M_z \geq M_T \tag{16}$$

Where,

$$M_R = (H_p + L_r) \times W_{pad} + (H_p + \frac{L_r}{2}) \times W_r \tag{17}$$

$$M_T = \frac{L_r}{2} \times W_r \times (\frac{H_p}{2} + L_r) \times W_{pad} \tag{18}$$

3.2. *Boundary Conditions and Application Parameters*

Boundary conditions have been used to narrow the search for an optimal solution. The number of magnets range from $n = 1 : 6$, the foot pad radius R_{pad}=0:100, magnet diameter D_m=0:50mm, magnet thickness T_m=0:50mm. The fixed design parameters are $H_p = 0.03m$, $L_r = 0.567m$, and $W_R = 1.2kg$, $W_{max} = 0.6kg$, $W_h = 0.110kg$, $W_b = 0.080kg$, $\rho_{Alum} = 2700kgm^{-3}$, $\rho_{NdFeB} = 7501kgm^{-3}$, $\mu_s = 0.6$. These values have all been converted to units of newtons and newton-metres where required before being used.

3.3. *Optimal Design Results*

The optimisation problem finds multiple local minima. Due to the limited boundary conditions and large step size, it is feasible to use an exhaustive search to guarantee the solution for global minimum. Using a step size of 1 unit for each design variable, the optimally designed footpad was found with a radius of 100mm, using three magnets with a 42mm diameter magnet and a 3mm thickness. Due to availability a magnet with diameter 31.75mm and thickness 6.35mm was selected. For the given magnet dimensions, the new footpad radius was determined to be 97mm with the following results: F_p=240.0N, F_s=144.0N, M_x=15.5Nm, M_y=13.4Nm, M_z=14.0Nm, W_{pad}=560g.

4. Construction and Experimental Results

Based on the optimisation results, a footpad of 612g was constructed as seen in Fig. 5(a). The robot system can be seen in Fig. 5(b)(c) to verify the geometric constraints and the ability to support the robot in cantilever. The footpad is seen in Fig. 5(d) supporting more than 2.5x factor of safety in bending moment with a 0mm air gap. Figure 6 compares the theoretical pull forces and bending moments to experimental results for two thicknesses of sheet steel. To achieve full adhesion capacity a minimum of 2.93mm of sheet steel is required to prevent magnetic saturation. The test rig is constructed from 1.5mm sheet steel, hence lower adhesion was expected.

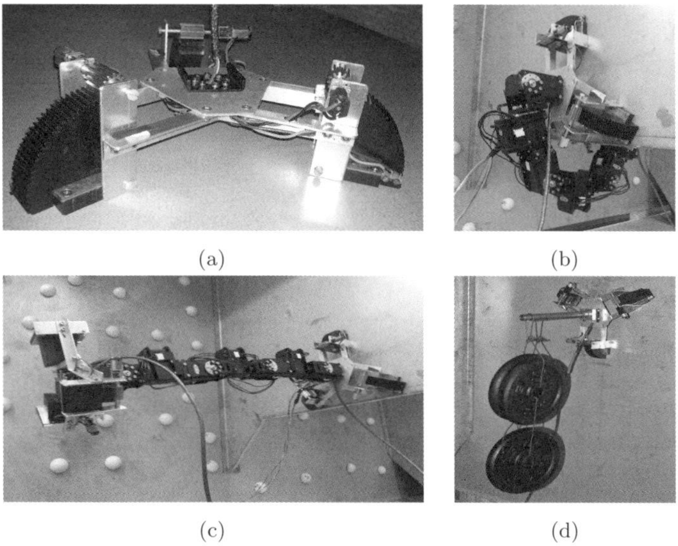

Fig. 5. (a) Construction of optimally based footpad (b) Verification of geometric constraints in 360 degree plate transition (c) Footpad supporting complete inch-worm robot in cantilever (d) Footpad supporting 18.4 Nm load

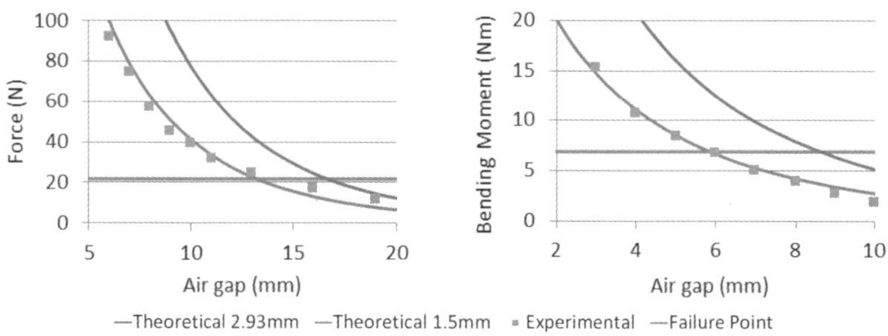

Fig. 6. Comparison of theoretical results for 2.93mm sheet steel and 1.5mm sheet steel, experimental results, and failure point for pull force and bending moment in x direction

5. Conclusion

Several design issues were addressed in designing a footpad intended for use on climbing robots. A generic optimisation model which can be fitted to different application requirements has been formulated. In the process a novel magnetic footpad has been constructed based on the optimisation model. Experimental results support the design model and satisfy the design constraints.

6. Acknowledgements

This work is supported by the Australian Research Council (ARC) Linkage Grant (LP100200750), the NSW Roads and Maritime Services, the Centre for Autonomous Systems (CAS) at the University of Technology, Sydney. The authors would like to thank the climbing robot team for their support.

References

1. M. F. Silva and J. A. T. Machado, *Climbing and Walking Robots* (In-Tech, March 2010), ch. A Survey of Technologies and Applications for Climbing Robots Locomotion and Adhesion, pp. 1–22. ISBN 978-953-307-030-8.
2. J. Xiao and A. Sadegh, *Climbing and Walking Robots: towards New Applications* (InTech, 2007), ch. City-Climber: A New Generation Wall-climbing Robots, p. 546. ISBN 978-3-902613-16-5.
3. A. N. Knaian, Electropermanent magnetic conectors and actuators: Devices and their aplication in programable matter, PhD thesis, Electrical Engineering and Computer Science, (Massachusetts Institute of Technology, 2010).
4. D. L. Peter K. Ward, Design of a high capacity electro permanent magnetic adhesion for climbing robots, in *Proceedings of the 2012 IEEE International Conference on Robotics and Biomimetrics*, ed. H. H. Hong Zhang (IEEE, Guangzhou, China, December 2012).
5. F. Rochat, P. Schoeneich, M. Bonani, S. Magnenat, F. Mondada and T. M. O. M. H. V. G. S. Hannes Bleuler Fujimoto, Hideo, Design of magnetic switchable device (msd) and applications in climbing robot, in *Climbing and Walking Robots and the Support Technologies for Mobile Machines.The 13th International Conference on*, CLAWAR 2010 (World Scientific, Nagoya, Japan, Sep 2010).
6. F. Ta andche, F. Pomerleau, W. Fischer, G. Caprari, F. Mondada, R. Moser and R. Siegwart, Magnebike: Compact magnetic wheeled robot for power plant inspection, in *Applied Robotics for the Power Industry. 1st International Conference on*, CARPI (Montreal, Canada, 2010).
7. B. B. S. Mondal, T.P. Sattar, Tofd inspection of v-groove butt welds on the hull of a container ship with a magnetically adhering wall climbing robot, in *5th International Conference on Climbing and Walking Robots and the Support Technologies for Mobile Machines*, eds. P. Bidaud and F. B. Amar (Professional Engineering Publishing Limited, 2002).
8. X. D. Z. Y. S. H. G. X. W. Yan, L. Shuliang, Development and application of wall-climbing robots, in *Proceedings of the 1999 IEEE International Conference on Robotics and Automation*, (Detroit, Michigan, USA, 1999).
9. K. W. David Pagano, Dikai Liu, A method for optimal design of an inchworm climbing robot, in *Proceedings of the 2012 IEEE International Conference on Robotics and Biomimetrics*, ed. H. H. Hong Zhang (IEEE, Guangzhou, China, December 2012).

COMPLIANT FOOT SYSTEM DESIGN FOR BIPEDAL ROBOT WALKING OVER UNEVEN TERRAIN*

NING WU†, BOON-HWA TAN, CHEE-MENG CHEW, AUN-NEOW POO

Mechanical Engineering, National University of Singapore,
9 Engineering Drive 1, EA-03-06, Singapore, 117575

This paper presents a new foot system for biped walking on uneven terrain and its design process. Stabilization of contact states between the foot and the ground and proper landing on unknown terrain are the criteria that ensure stable walking motion on uneven terrain. Generally, the conventional rigid and flat foot could not maintain level when the robot is walking on uneven terrain. In addition, the impulsive force exerted during landing on rough terrain must be suppressed. The authors proposed a point-contact type foot with hydraulic fluid balance mechanism. The size of the proposed foot mechanism is 160 mm x 277 mm and its weight is 1.6 kg. The foot system consists of four contact points each of which is equipped with a force sensing resistor (FSR) to detect the landing state. The foot can generate a support polygon on uneven terrain by using three or four contact points. The main advantages of the proposed foot system are that it enables the stabilization of the contact state and absorption of the impact when the swing foot lands.

1. Introduction

Uneven terrain walking for biped robot is one of the most challenging tasks in current bipedal walking research. In order to generate a stable bipedal walking motion on uneven terrain, some researchers have studied the motion pattern generation methods while other researchers have researched on real-time stability control methods [1, 6, 8, and 9]. During single support phase, most of these methods have assumed that the contact state of the foot is supported by four contact points. However, this assumption is not applicable for a bipedal robot that is walking on uneven terrain. As a bipedal robot moves its Center of Mass (COM) during single support phase, the contact state between the foot and the ground determines the walking stability for subsequent walking cycle. For a bipedal robot that is equipped with rigid and flat feet, it is challenging for the robot to maintain its foot contact with the rough terrain because the foot changes its contact state easily and randomly. In order to achieve stable walking motion on uneven terrain, it is desirable for the biped robots to maintain four-point-

† Work corresponding author. Email: wuning@nus.edu.sg.

contact during the single support phase. It is because this will provide large support polygon for the biped. The paper by Hashimoto, et al [3] described a new foot system, WS-1 (Waseda Shoes - No.1) that is able to maintain four points contact at the contact state. However, due to improper sensors mounting, landing state detection is not very accurate. Hence, Hashimoto et al. has developed another improved biped foot system, WS-1R (Waseda Shoes - No. 1 Refined) [4, 5] which can maintain large support polygons on uneven terrain. Nevertheless, this design could not deal with concave surface where the large support polygon could not be maintained. Also, this design is heavier (1.9 kg) than conventional rigid and flat feet. Heavy foot would reduce the swing speed of the swinging leg and reduce the stability of the robot. Sano and Yamada [7] have proposed a new point-contact type foot with springs (PCFS). This proposed foot could adapt to rough terrain by minimizing the impact force and disturbance. However, this design is not strong enough to be mounted on a robot which has a weight of more than 50kg.

In this paper, a novel foot design is proposed to achieve stable walking motion on uneven terrain. This new foot design is a combination of shock absorbing mechanism, landing surface detection mechanism and stabilization mechanism of supporting leg and landing leg. The proposed foot system is equipped with Bang-bang controller to activate the foot system mechanism. Based on the Pascal's Law, the proposed foot system would maintain four-point-contact with respect to the contacting terrain and balance by itself. This design does not only simplify the controller for uneven terrain walking motion but also increase the stability of walking motion.

This paper is organized as follows. Section 2 illustrates the foot system design and its working principle. Section 3 discusses the walking tests and key considerations for the application of the proposed foot system. Section 4 shows the experimental results with the proposed foot system. Lastly, a conclusion is made to summarize the work done and the future development.

2. Proposed Foot Design and Working Principle

In this section, the design process and the working principle for the proposed foot system is discussed in detail. In order to achieve stable walking motion on uneven terrain, stable landing state should be provided support exchange. Also, the ZMP of the robot should be maintained within the support polygon of the foot to ensure the supporting foot does not rotate or slip with respect to the ground. A bipedal robot with flat and rigid foot would have difficulty to maintain the ZMP within the support polygon when the ground is uneven as the support polygon may become smaller on uneven terrain.

The novel foot system (Figure 1) is light, compact, simple and with fast response. Its principle is based on Pascal's Law which states that if pressure is exerted at any point within a confined incompressible fluid, the pressure will be transmitted equally in all directions throughout the fluid so that the pressure of the fluid remains the same [2].

The proposed foot system is attached with four hydraulic cylinders with a stroke length of 25mm. They are interconnected by polyurethane tubes such that fluid exchange can be enabled among them. The proposed foot system is "locked" (maintain current position and orientation) when all the four contact points are in contact with uneven terrain and the ZMP is near to the center of the foot. The foot system is said to be "locked" when the fluid exchange is stopped. Under normal circumstances the foot will be maintained at a particular orientation.

Three solenoid valves are used to 'lock' or 'unlock' the fluid exchange between front and back, left and right. Four force sensing resistors (FSR) are mounted at the four contact points to detect the landing state. By using the principle of Pascal's Law, when one of the contact points is in contact with the terrain, fluid exchange would be triggered until all the four contact points exert the same pressure to the contact terrain. At this moment, a locking mechanism is enabled instantaneously to stop the fluid exchange among the cylinders such that the four contact points are maintained at the position where ZMP is near the center of the foot as shown in Figure 3. The proposed foot system is able to adapt to terrain with seven degree inclination as well. The working flow chart of the proposed foot system is summarized as in Figure 2.

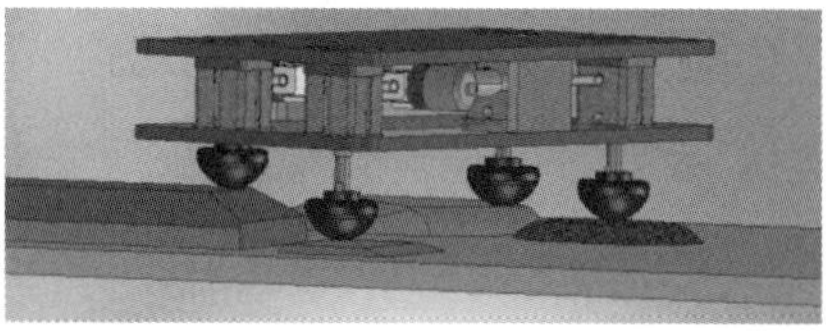
(a) CAD illustration

(b) Final Assembly

Figure 1: Proposed Foot System in CAD (a) and final assembly (b) respectively

2.1 Locking Mechanism

Locking mechanism is the most important feature of the proposed foot system. This mechanism should able to sustain landing impact and then maintain the locking function during walking motion. The foot system would become heavy if the locking mechanism is complicated. Heavy foot would reduce the swing leg velocity and hence reduce the gait velocity. Hence, a simple but robust locking mechanism is required.

The locking mechanism must be locked instantaneously in arbitrary positions that the landing state could adapt to random uneveness. Also, in order to prevent the testing robot from toppling, the locking mechanism must be triggered right before swinging leg leaves the contact surface, regardless of contact state conditions. The locking mechanism is based on the bang–bang controller as follows.

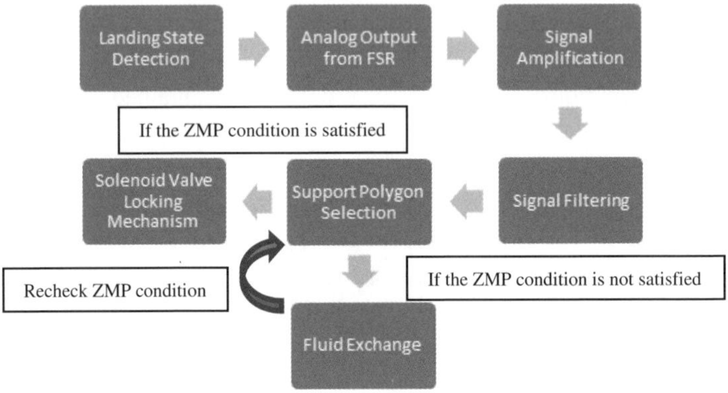

Figure 2: Working flow chart of the Proposed Foot System

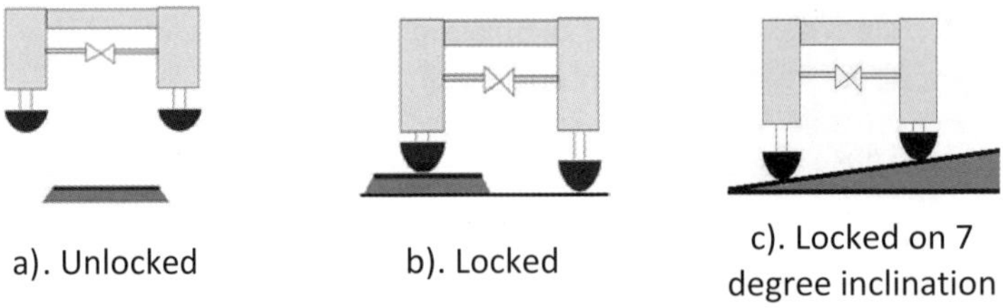

Figure 3: Locking and unlocking Conditions

$$u = \begin{cases} -V & if\ y < U_L \\ +V & if\ y > U_H \\ 0 & if\ U_L \leq y \leq U_H \end{cases} \quad (1)$$

where u is the control output, V is a constant output; U_L is the lower boundary where the ZMP is close to the center of the foot and within the user predefined stability margin whereas U_H is the upper boundary where the ZMP is outside the user predefined stability margin but within the foot support polygon. The constant output V refers to the 'locked (negative)' and 'unlocked (positive)' state of the solenoid valves. The solenoid is in idle state if there is no contact state detected or the foot is not in contact with ground.

2.2 Stability Index Estimation

The ZMP is used as the stability index to indicate the walking motion stability. For the case of rigid and flat foot on uneven terrain, the ZMP is difficult to be estimated because the contact state changes easily. Since the proposed foot system has only four contact points, the ZMP can be estimated easily by measuring the reaction force that exerted on each contact point. From the magnitude of reaction forces and the positions of the contact points, the position of ZMP $p = [p_x \ \ p_y]$ can be determined via the equation below:

$$p_i = \frac{\sum_{i=1}^{4} f_i p_i}{\sum_{i=1}^{4} f_i} \qquad (2)$$

where f_i ($i = 1,2,3,4$) is the vertical reaction force that exerted on each contact point and $p_i = [p_{x_i} \ p_{y_i}]$ is the two dimensional position vector of each contact point.

3. Walking Tests and Key Considerations

In order to justify the feasibility and the functionality of the proposed foot system, the authors have chosen the ZMP stability criterions as the stability indicator. This criterion imposes the constraint that the ZMP must be situated inside the foot support polygon to ensures the foot does not rotate along one of its edges and free of slippage. A bipedal robot is stable if the ZMP is inside the foot support polygon. Then, given the criterions, the authors have conducted two sets of different experiments as follow:

a) Comparison of ZMP criterions when the robot is walking on an obstacle (15 cm height) with and without the new proposed foot.

b) Comparison of ZMP criterions when the robot is walking on a slope (7 degree) motion with and without the new proposed foot.

The proposed foot was tested on ASLAN which is a life-sized humanoid robot developed in National University of Singapore (NUS). The step time for the testing humanoid robot is 1 step per second and the walking speed is 20mm/step. Besides, the timing for locking mechanism is vital. The locking mechanism must be done within single step cycle. The flat foot system on ASLAN has a weight of 2kg whereas the proposed foot system has a weight of 1.6kg.

Since the testing humanoid robot is around 60kg, the hydraulic cylinders and solenoid valves selected must be able to sustain this payload. For safety, the foot system is designed such that it can withstand 70kg pay load. According to the experimental results by Hashimoto, et al. [3], for uneven terrain with fluctuation more than 20mm, it is difficult for biped walking robots to walk stably on the uneven terrain by solely maintaining the ZMP position at the center of foot. Hence, the proposed foot system is designed to adapt uneven terrain with fluctuation less than 20mm. For fluctuation more than 20mm, more advanced controllers need to be added at the top of the proposed foot system.

4. Experimental Results

In order to increase the adaptability of the proposed foot system on an obstacle, the authors have implemented landing pattern with dorsiflexion and plantar flexion. Given this landing pattern, the testing robot could walk stably on an obstacle with a height of 15mm. Figure 5 and 6 show the variation of COM in sagittal direction (X_{ZMP}) and lateral direction (Y_{ZMP}) respectively when the robot is walking on the obstacle for three consecutive walking cycles. Based on the graph above, the X_{ZMP} and Y_{ZMP} of the robot with the proposed foot are maintained near to the ZMP reference. This implies that the robot with the proposed foot has achieved higher walking stability. Also, during each single support phase, the variation of ZMP is less for the case with the proposed foot. For the case with flat foot, the robot falls backward at the time of 3.3s (marked as a red cross in the figures). In the steady state, X_{ZMP} and Y_{ZMP} are maintained at a steady value until the end of single support phase. The steady ZMP value implies that the walking motion is subjected to less oscillation.

Equipped with the proposed foot system, the testing humanoid robot could walk stably on a slope with gradient of 7 degree. Figure 7 and figure 8 show the variation of X_{ZMP} and Y_{ZMP} respectively when the robot is walking on the obstacle for three consecutive walking cycles. Based on the graph above, the X_{ZMP} and Y_{ZMP} of the bipedal robot with the proposed foot are maintained near to the ZMP reference. The bipedal robot requires at most 0.6s to enter the steady state. During steady state, the X_{ZMP} and Y_{ZMP} are maintained at the center of the supporting ankle until the end of single support period of the swinging leg. The steady ZMP value implies that the walking motion is subjected to less oscillation. Also, during each single support phase, the variation of ZMP is less for the case with the proposed foot. The Red Cross indicates the falling state of the testing humanoid robot with flat foot.

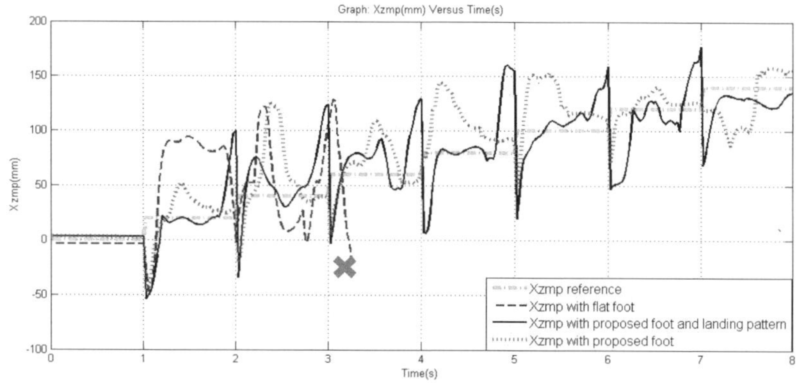

Figure 5: The Variation of X_{ZMP} (mm) when the robot is walking on an obstacle (with and without the proposed foot)

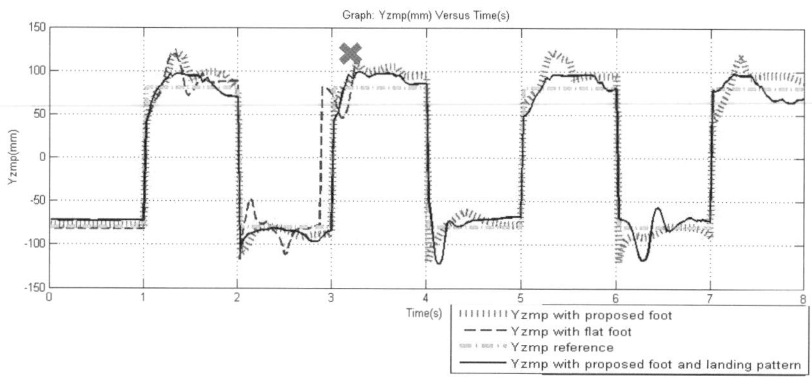

Figure 6: The Variation of Y_{ZMP} (mm) when the robot is walking on an obstacle (with and without the proposed foot)

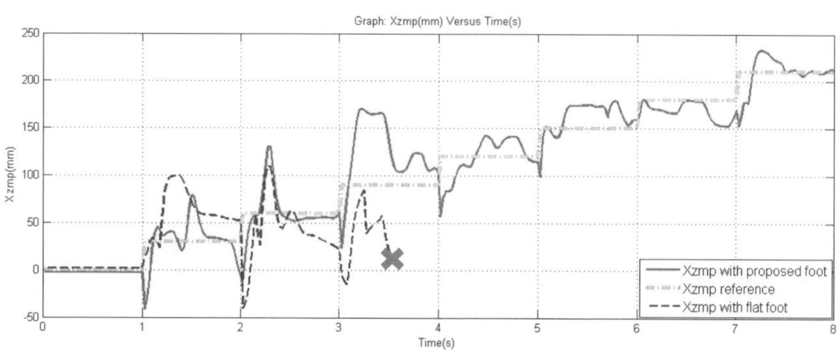

Figure 7: The Variation of X_{ZMP} (mm) when the robot is walking on the slope (with and without the proposed foot)

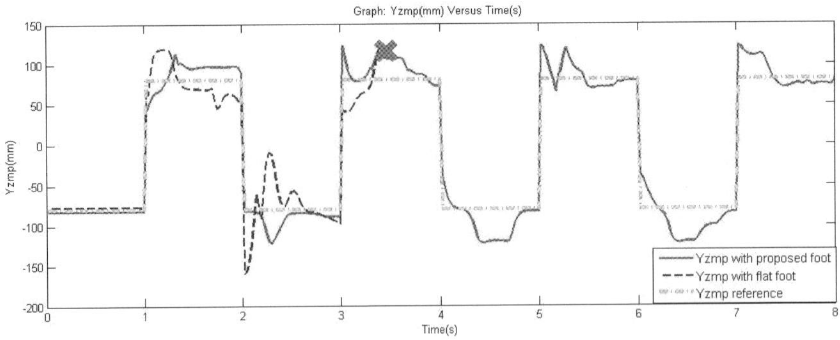

Figure 8: The Variation of Y_{ZMP} (mm) when the robot is walking on the slope (with and without the proposed foot)

5. Conclusion

To achieve stable bipedal walking motion on rough terrain, it is necessary to stabilize the contact state between the foot and the ground. Hence, the authors have presented a new foot system that has the following advantages which includes stabilization of contact state, estimation of the ZMP position, absorption of landing impact and faster response in achieving stable state on an obstacle if compared with flat foot system. The effectiveness of the proposed foot system design has been verified experimentally. The proposed foot system enables the testing bipedal robot to walk forward faster, walk stably on an obstacle and walk steadily on a slope. Current foot design has four contact points. However, it is not easy to maintain four contact points simultaneously especially on the uneven terrain. It can achieve three contacts most of the time and this cut the support polygon by half.

References

1. Takanishi, T.Takeya, H.Karaki, M.Kumeta, and I.Kato, “A Control Method for Dynamic Walking under Unknown External Force,” Proc. of the IEEE/RSJ IROS 1990, pp.795-801, Tsuchiura, Japan, July, 1990.
2. Bloomfield, Louis (2006). How Things Work: The Physics of Everyday Life (Third Edition). John Wiley & Sons. pp. 153. ISBN 0-471-46886-X.
3. Hashimoto, K.; Sugahara, Y.; Ohta, A.; Sunazuka, H.; Tanaka, C.; Kawase, M.; Lim, H.; Takanishi, A.; , "Realization of Stable Biped Walking on Public Road with New Biped Foot System Adaptable to Uneven Terrain," Biomedical Robotics and Biomechatronics, 2006. BioRob 2006.

The First IEEE/RAS-EMBS International Conference on , vol., no., pp. 226-231, 20-22 Feb. 2006

4. K. Hashimoto, Y. Sugahara, H. O. Lim and A. Takanishi, “Development of New Foot System Adaptable to Uneven Terrain for All Biped Robots,” Proc. of the 17th CISM-IFToMM Symposium on Robot Design, Dynamics and Control (ROMANSY2008), pp. 391-398, 2008.
5. K. Hashimoto, Y. Sugahara, H. O. Lim and A. Takanishi, “New Biped Foot System Adaptable to Uneven Terrain,” Journal of Robotics and Mechatronics, Vol. 18, No. 3, pp. 271-277, 2006.
6. S. Kagami, et al., “Online 3D Vision, Motion Planning and Bipedal Locomotion Control Coupling System of Humanoid Robot: H7,” Proc. of the IEEE/RSJ IROS 2002, pp. 2557-2562, Lausanne, Switzerland, October, 2002.
7. S. Sano, M. Yamada, N. Uchiyama, and S. Takagi, ”Point-Contact Type Foot with Springs and Posture Control for Biped Walking on Rough Terrain”, Proc. the 10th IEEE Int. Workshop on Advanced Motion Control, vol.2, pp.480-485, 2008.
8. Yamaguchi, J.; Takanishi, A.; Kato, I.; , "Experimental development of a foot mechanism with shock absorbing material for acquisition of landing surface position information and stabilization of dynamic biped walking," Robotics and Automation, 1995. Proceedings. 1995 IEEE International Conference on, vol.3, no., pp.2892-2899 vol.3, 21-27 May 1995
9. Y. Okumura, T. Tawara, K. Endo, T. Furuta, and M. Shimizu, “Realtime ZMP Compensation for Biped Walking Robot using Adaptive Inertia Force Control,” Proc. of the IEEE/RSJ IROS 2003, pp. 335-339, Las Vegas, USA, October, 2003

NEW DESIGN OF PERISTALTIC CRAWLING ROBOT WITH AN EARTHWORM MUSCULAR STRUCTURE

HIKARU NOZAKI

Department of Science and Technology, Kwansei Gakuin University, 2-1 Gakuen Sanda, Hyogo 669-1337, Japan

SATOSHI TESEN

Department of Science and Technology, Kwansei Gakuin University, 2-1 Gakuen Sanda, Hyogo 669-1337, Japan

NORIHIKO SAGA

Department of Science and Technology, Kwansei Gakuin University, 2-1 Gakuen Sanda, Hyogo 669-1337, Japan

HIROKI DOBASHI

Department of Science and Technology, Kwansei Gakuin University, 2-1 Gakuen Sanda, Hyogo 669-1337, Japan

JUN-YA NAGASE

Department of Mechanical and Systems Engineering, Ryukoku University, 1-5 Otsu, Shiga, 520-2194, Japan

In recent years, in consideration of the danger of secondary disaster in disaster areas, the development of rescue robots has been anticipated. We have devoted attention to the mechanism of peristalsis performed by earthworms and developed a robot using the mechanism based on the following three points. First, earthworms require no large spaces in which to move. Second, the posture of earthworms is stable. Finally, the migratory mechanism of earthworms is simple. Devoting attention to earthworm muscle structures of two kinds, we developed a robot that recreates the earthworm muscle motion using actuators of two kinds. The salient feature of this mechanism is that the robot is always able to move a fixed distance, even in a narrow space using linear actuators. Herein, we propose the robot mechanism and examine the robot motion patterns.

1. Introduction

In recent years, rescue operations using robots have been attracting attention [1]. To support rescue operations, we have devoted our attention to peristalsis, which is performed by earthworms as a means of moving in narrow spaces. Worms have muscles of two layers. One layer is parallel to the body lengthwise direction. The other muscles surround the body. Using these muscles, earthworms expand segments and contact walls. This is called retentive motion. Earthworms progress by making a regressive wave using these two muscle layers. Earthworms can progress if some chink exists on a surface to offer some traction. Therefore, earthworms require minimal space for movement. In addition, earthworms can move whichever face of segments touches the ground and can maintain a stable posture without tumbling. Moreover, the movement mechanism is simple: earthworms move only by segment contraction. Some peristaltic drive systems have been adapted to crawling robots such as pneumatic actuators, magnetic fluids, and shape memory alloys (SMA) [2]–[4]. In many mechanisms, lengthwise and circumferential expansions of the body are done using one actuator. However, that presents a problem. Expansion in the lengthwise direction is related to the migration length. Therefore, in narrow spaces, because contraction of a circuit direction becomes small, the migration length also becomes short. To resolve this problem, we propose a new mechanism that always moves a fixed distance using linear actuators. Furthermore, we propose some moving patterns that correspond to various environments.

2. New Robot Mechanisms

2.1. *Mechanisms of Robot Segments*

A robot we propose consists of multiple segments as shown in Figure 1. In this section, we present the mechanism of a segment. We propose a mechanism that uses two kinds of actuators to separate circumferential extension for retention and lengthwise extension for movement as shown in Figure 2. Therefore, we can solve the problem above that migration length is short. We use servo motors for retention and use linear actuators for movement. The movement distance of the robot is influenced by the stretching of linear actuators, so this robot always moves a fixed distance by this mechanism.

2.2. *Robot Composition*

In this section, we show the entire robot composition. Parts which make up the robot are classifiable into four groups. We designate these parts as the “Robot head”, “Segment”, “Input position of battery and microcomputer”, and “Joint”. The installation image of each part is shown in Figure 1. A microcomputer is installed in the robot center. Two segments are installed at each right and left. Therefore, this robot can move both forward and backward.

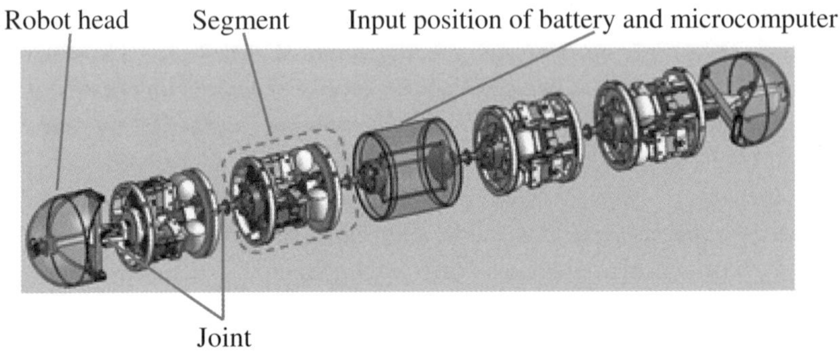

Figure 1. Overall view of the new robot.

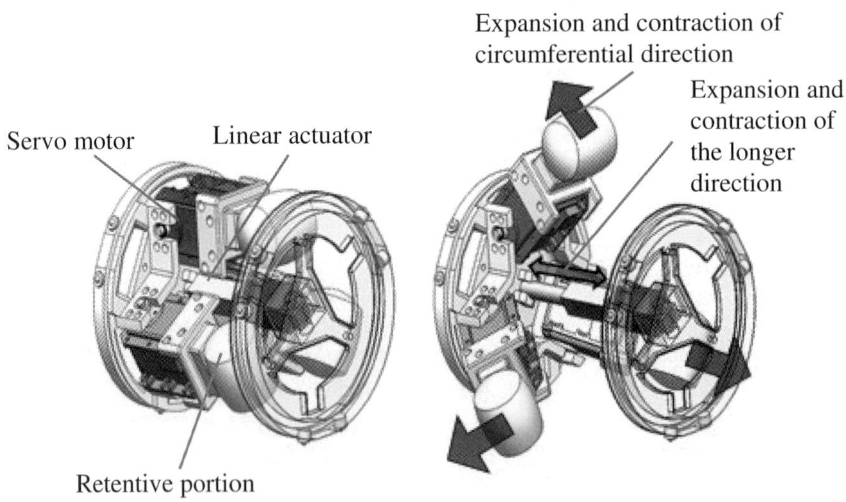

Figure 2. Segmental structure of the peristaltic crawling robot.

3. Proposed Moving Patterns

3.1. *Two Moving Patterns*

We consider two useful moving patterns using actuators of two kinds. First, we consider the "Fastest Mode". In this mode, only the head and tail segments are used for retention. The other segments are used for movement. Second, we consider "Stable Mode". This mode forms a retrogressive wave by working all segments separately. We use a simple two-dimensional model of the robot to represent moving patterns of the robot like Figure 3 and explain movement of actuators. The head segment is defined as segment 1. The other segments are defined as segment 2, segment 3, and segment 4 sequentially from the head as shown in Figure3. The black dot show servo motors. Fundamentally, there are three servo motors in one segment, but there are two servo motors shown in Figure 3 for model simplification.

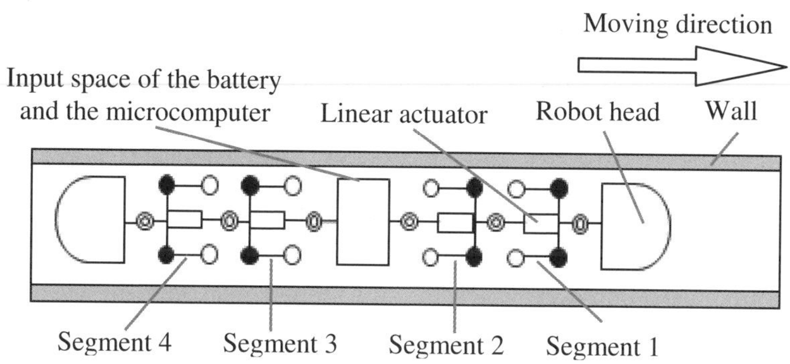

○: Retentive portion
●: Servo motor
◎: Joint

Figure 3. Simplified model of the robot.

3.2. *Fastest Mode*

We define a moving pattern by which the robot can move the fastest as "Fastest Mode". This mode is shown in Figure 4. First, only segment 4 is retained and all the linear actuators expand. Second, segment 4 stops retention, only segment 1 is retained and all linear actuators contract. The robot moves by these movements. All the linear actuators always function simultaneously. Therefore, the robot can move without being affected by the volume of space. The situation of an experiment of Fastest Mode is shown in Figure 4. The robot ran in a pipe of 200 mm in diameter. It took 40 s for the robot to move 150 mm.

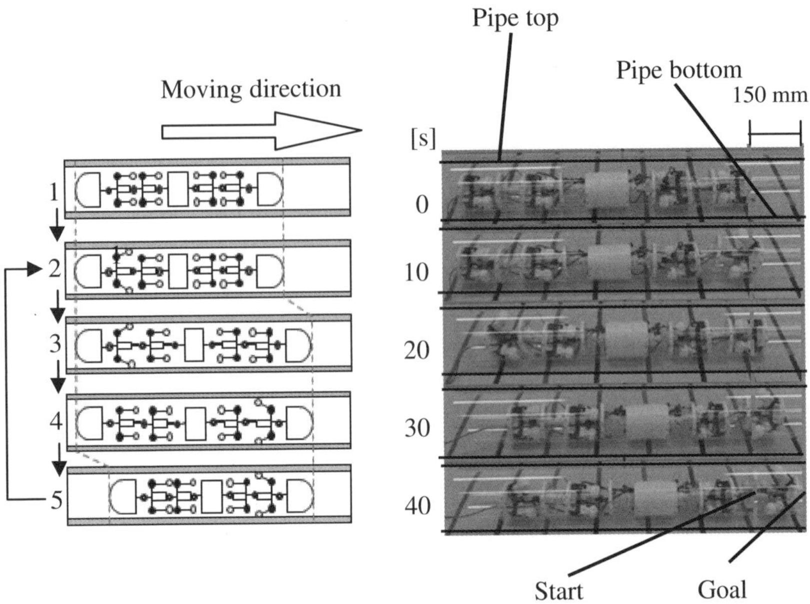

Figure 4. Fastest mode motion of the robot and moving experiment.

3.3. *Stable Mode*

We define a moving pattern which forms a retrogression wave by moving all segments separately as “Stable Mode”. This mode is shown in Figure 5. This is a moving pattern resembling earthworm peristalsis All segments are retained except the segment which stretches. This is the mode in which the stability of the body is the highest. However, because the linear actuator to elongate is one per operation, it takes much greater time than the Fastest Mode to move in this pattern. The arrangement of an experiment of Stable Mode is shown in Figure 5. The robot ran in a pipe of 200 mm in diameter. It took 240 s for the robot to move 150 mm.

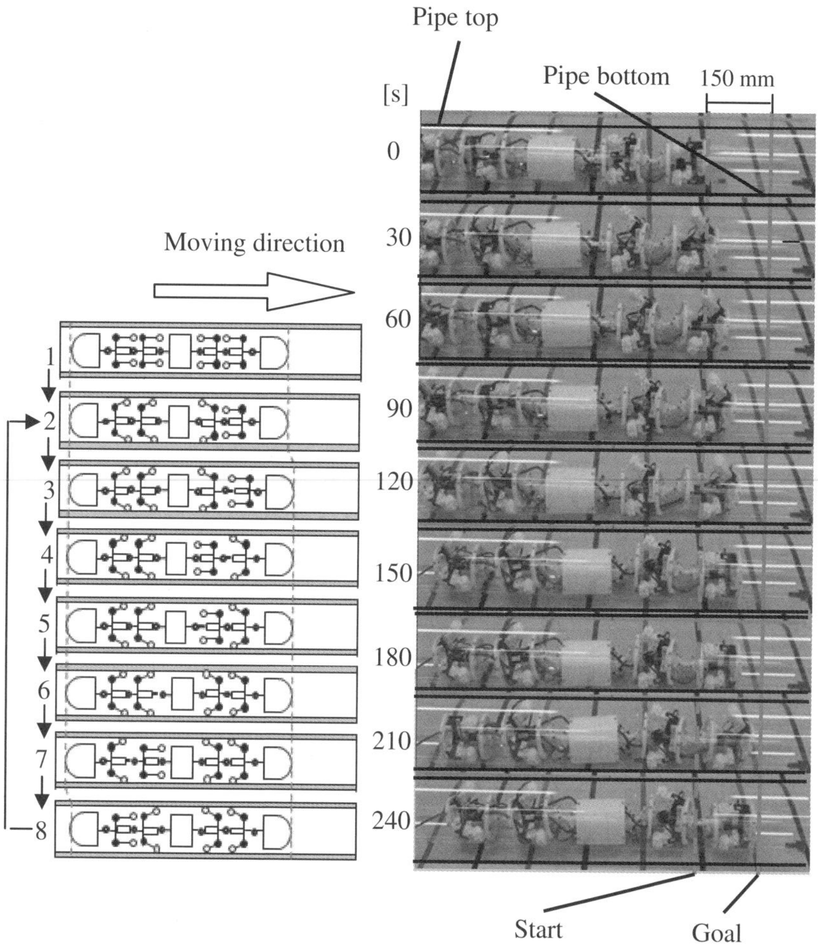

Figure 5. Stable mode of the robot and moving experiment

4. Conclusion

We encountered a problem with a robot we produced in earlier work: the robot cannot always move a fixed distance. In this work, we examined a mechanism using actuators of two kinds to solve the problem. Moreover, we defined two moving patterns for movement in various environments, "Fastest Mode" and "Stable Mode", and we conducted experiments of operation in each mode. We shall examine the course-decision algorithm used for the robot head as future work.

References

1. S. Tadokoro : "Present Situation and Future Prospects of Rescue Robots", *J. IEICE*, **Vol.92, No.3**,203-208, (2009).
2. T. Noritsugu and M. Kubota : "Development of In-Pipe Mobile Robot using Pneumatic Soft-Actuator (in Japanese)", *Journal of Robotics Society of Japan*, **Vol.18, No.6**, 73-80, (2000) .
3. N. Saga and T. Nakamura : "Elucidation of Propulsive force of micro-robot using magnetic fluid", *Journal of Applied Physics*, **Vol.91, No.10, Parts 2 and 3**, 7003-7005, (2002).
4. C. Onal, R. Wood and D. Rus : "An Origami-Inspired Approach to Worm Robots", *Transactions on Mechatronics*, 1-9, (2012).

HEX-PIDERIX: A SIX-LEGGED WALKING CLIMBING ROBOT TO PERFORM INSPECTION TASKS ON VERTICAL SURFACES*

XOCHITL YAMILE SANDOVAL-CASTRO †, MARIO A. GRACIA-MURILLO, JONNY PAUL ZAVALA-DE PAZ AND EDUARDO CASTILLO-CASTANEDA†

†*Instituto Politécnico Nacional, Centro de Investigación en Ciencia Aplicada y Tecnología Avanzada, Unidad Querétaro, Cerro Blanco 141, Colinas del Cimatario, C.P. 76090 Querétaro, México. E-mail:* xsandovalc1100@alumno.ipn.mx

This paper presents the kinematic analysis of the climbing walking robot Hex-piderix. It has 3DOF in each of the six legs, a suction cup for climbing vertical surfaces on each limb tip and a locomotion algorithm. The authors proposed a computer vision system that uses image analysis for inspection employing a mounted camera. It proposes an invariant method for illumination to detect cracks on surfaces. Imagesare improved dimming the light and calculating a threshold by binarizing it using the Otsu method. Finally, morphological operations are applied to highlight the crack.

1. Introduction

There are many tasks in which an employee would be exposed to a very high risk. Therefore several tools and equipment is being developed to exempt men from performing remote or dangerous operations, [1], [2]. Teleoperated robots are used in several fields due to their advantage in the security standpoint. They prevent the user from working in hazardous environments, [3].

Hex-piderix is a 7DOF serial robot developed to safely inspect cracks in bridges. It takes pictures of them to analyze their length, width and depth (the reliability of the data collected by the robot depends on the operator's training). The robotic system consists of three parts: A specially designed crane to move the robot, a robotic mechanism and a vision system, [4].

Wheeled robots have also been designed to detect cracks inside tunnels, where the robot must keep a parallel distance with the wall, [5]. These robots can only be used on flat surfaces whereas legged robots are easily handled in a variety of terrains. ROBUG IIs [6], is a quadruped robot that can walk on rough terrains, climb vertical surfaces an can automatically change trajectories from ground to wall.

* This work is supported by CONACYT.

Legged robots can be classified as: Bipeds, quadrupeds, hexapods and octopods. Most climbing robots are hexapods or quadrupeds; the former provide a better static stability. They can perform staticgaits by supporting the robot's body on five legs at any time, while quadrupeds can only walk steadily when supported by at least three legs. This feature makes hexapods much more stable than quadrupeds, since they can use a bigger support polygon, [7].
In order to develop dynamic models and control algorithms for a legged robot it is important to calculate the kinematics. This can be achieved by using homogeneous transformation matrices, [8], [9], [10].

2. Kinematics model

2.1. *Geometric Description*

The basic morphology proposed for Hex-piderix is shown in Fig.1. The framerepresents the origin and the {Q} indicates the local frame. The robot consists of six identical legs distributed symmetrically around the robot's thorax. Each limb has 3DOF that consist of three links: Coxa, femur, and tibia, connected by rotational joints. The tip of the tibia has a suction cup to adhere to vertical surfaces.

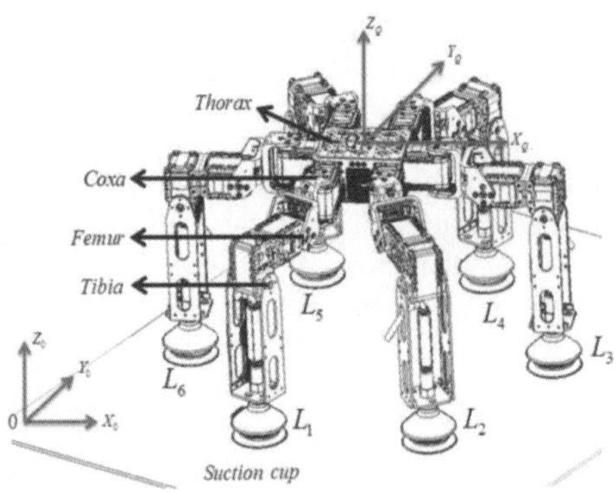

Figure 1. Robot architecture.

2.2. *Position, Velocity, and Acceleration Analysis*

The mechanism's symmetry simplifies the position analysis. The purpose of forward kinematics analysis is to find the thorax's pose in relation to the overall frame for a given set of joint angles$q_{i,j}$, where i represents the limb number $(i = 1, \dots, 6)$ and j is the joint number $(j = 1,2,3)$. In order to determine the robot's pose it is assumed that three legs are touching the ground $(L_1, L_3 \; and \; L_5)$. Coxa links represented by P_i form an isosceles triangle composed by base a and sides b. The local frame $\{Q\}$ is located at the triangle's centroid, its axis X_Q is directed to P_1[11], illustrated in Fig.2 (a).

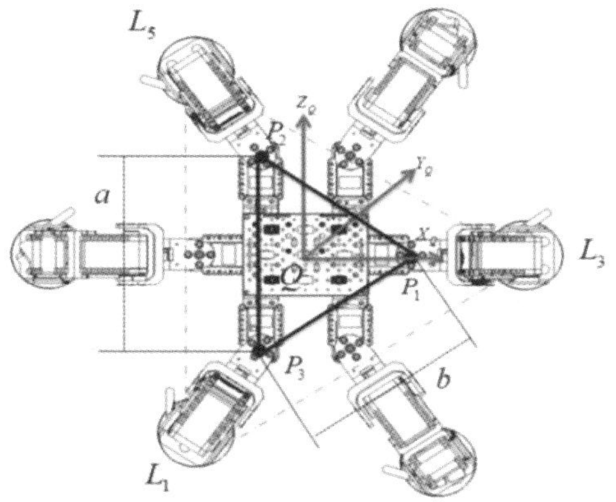

Figure 2. (a) Prism formed by coxa links.(b) i-th kinematic chain robot

Inverse kinematics consists in determining the joints angles, $q_{i,1}$, $q_{i,2}$ and $q_{i,3}$, starting from the given position of the tip.The problem is solved for each leg separatelyso it can be formulatedin a geometric way (Fig. 3(a)), taking into account the femur and tibia always rotate in parallel axes.

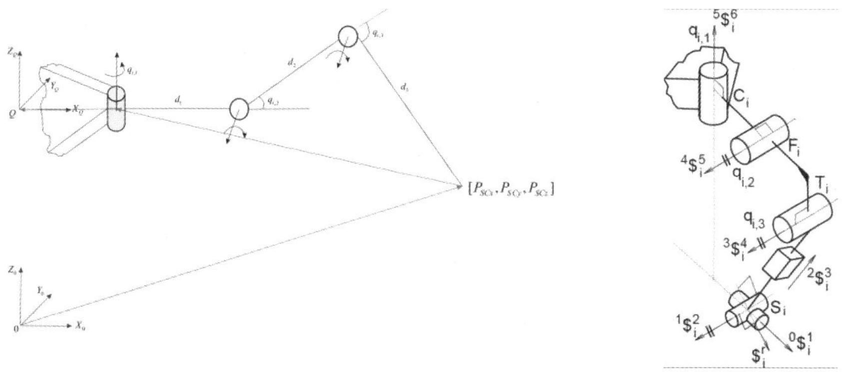

Figure 3. (a) Configuration leg mechanism. (b) i-th kinematic chain robot

The position equations are given by:

$$q_{i,1} = \tan^{-1}\left(\frac{P_{SCy}}{P_{SCx}}\right) \tag{1}$$

$$q_{i,2} = \tan^{-1}\left(c \pm \sqrt{a^2 + b^2 - c}\right) - \tan^{-1}\left({}^{a}/_{b}\right) \tag{2}$$

Where,

$$a = 2d_2\left(\sqrt{P_{SCx}^2 + P_{SCy}^2} - d_1\right)$$

$$b = 2P_{SCz} * d_2$$

$$c = \left(\sqrt{P_{SCx}^2 + P_{SCy}^2} - d_1\right)^2 + P_{SCz}^2 + d_2^2 - d_3^2$$

$$q_{i,3} = \cos^{-1}\left(\frac{\left(\sqrt{P_{SCx}^2+P_{SCy}^2}-d_1\right)^2+P_{SCz}^2-d_2^2-d_3^2}{2*d_2*d_3}\right) \quad (3)$$

The velocity and acceleration analysis are solved by using the screw theory. As an approach to the study of the dynamics of the robot, the manipulator is assumed as a parallel mechanism. Its limbs are modeled as shown in Fig.3(b). The forward velocity analysis consists in finding the velocity state of the thorax V_0 in relation to the global reference frame given the joint rate velocities, ${}_a\omega_b^i$. The velocity state V_0 can be calculated without knowing the ratio of passive velocities, which represents an advantage over other models. Kinematic model equations can be reviewed in [12].

3. Locomotion for hexapod robots

Locomotion is the ability of autonomous movement [13]. Many insects adopt a tripod gait which allows them to walk fast with a stable posture [14]. Others can change their locomotion strategy to walk on steep or vertical surfaces [15].
In this project the tripod gait was implemented, to do this the robot must have three limbs touching the ground forming a polygon as shown in Fig.4 (a). The center of gravity projection must be included to ensure stability. Fig.4 (b) describes the locomotion sequence.To ensure that the robot will not fall, six legs touch the groundin a time instant.

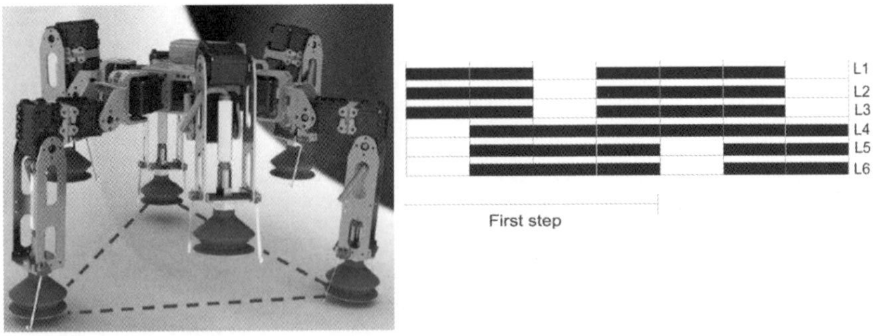

Figure 4. (a) Stability polygon, (b) Tripod gait.

4. Attachment mechanism

To avoid falling while moving over any surface (flat or steep), legged animals generate vertical groundreaction forces equivalent at least to gravity. This is a problem on steep and vertical terrains because vertical forces make slipping easier. There are two classes of attachment mechanisms that can potentially solve this problem: Interlocking and bondingcontain [16].
Physical processes utilized for bonding in climbing animals include *suction* [17], [18] *chemical adhesion* [19], *capillary adhesion* [18], *and Van der Waals adhesion* [20]. Animals climbing with bonding mechanisms can run on vertical or inverted surfaces on a wide variety of substrate materials.

4.1. *Integration of the Bounding Mechanism*

The most frequent approach to guarantee the robot's adhesion to a surface is to use the suction force. This operating principle allows it to climb over arbitrary surfaces made of diverse materials, and can be implemented by using different strategies. The vacuum can be generated through the Ventruli Principle [21]. A vacuum system has been adapted to the Hex-piderix to allow it to climb vertical surfaces. This system was set in each limb, which consists in a bellow suction cup and a vacuum generator. The electrovalve uses 6 bar of pressure provided by a compressor. The suction cup has a 55mm diameter with a maximum clamping force of 106N. The vacuum system (Fig. 5) is controlled by the PLC CPX-4DE.

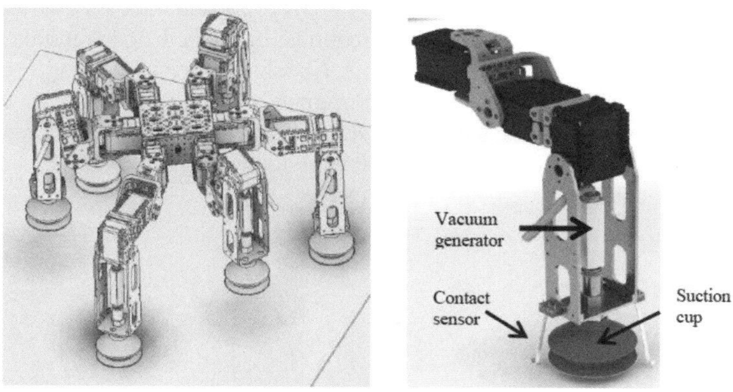

Figure 5. Attachment of Bounding Mechanism.

5. Visual inspection

5.1. *Image Acquisition*

Images were acquired with aWebcam (WEC SOMIKON-360), mounted on the Hex-piderix as shown in Fig. 6. Ten images were chosen to test the methods proposed for enhancement and detection.

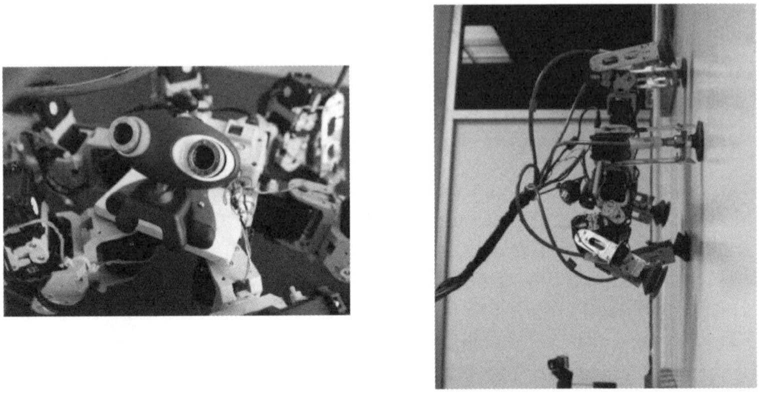

Figure 6. Camera mounted on the thorax of Hex-piderix.

Most of them have pixel intensity variations, Fig. 7(a), because image segmentation is complex. Due to its complexity a method must be considered in the pre-processing stage to increase the contrast between the background cracks and to correct the image illumination. Fig. 7(b) shows the binary image obtained by applying the Otsu method [22] to the original image. In its center we can see part of the crack, but on the sides the crack is lost. These crack details cannot be seen because the levels of gray are lower in the sides preventing us from detecting the cracks located outside the image.

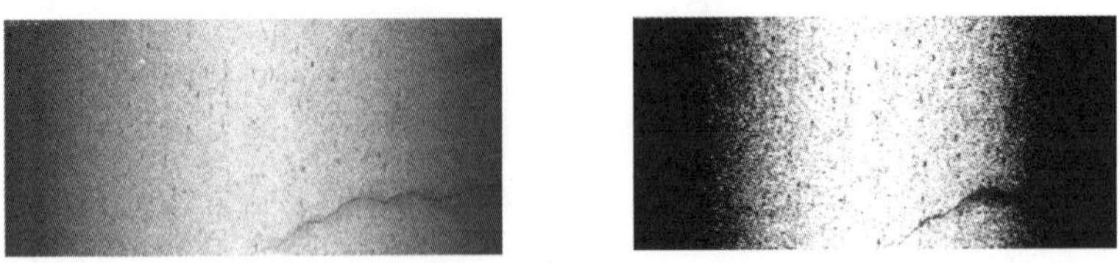

Figure 7. (a) Gray scale image. (b) Binary image, Otsu method aplied on Fig. 6(a).

5.2. *Proposed Method*

The enhancement process is one of the most important for optimal results in vision systems. Three significant characteristics of crack images are: The pixels have low gray levels, histogram is distributed in a narrow range in the levels of gray scale, and lighting in the picture is not uniform. First a logarithmic transformation is applied to the image in order to expand the range of gray levels and clarify it. A logarithmic transformation is defined by the following equation.

$$s = c.\log(1 + r) \quad (4)$$

To correct the illumination an illumination pattern extraction is realized based on the level of pixel intensity on each of the image's columns. In Fig. 8(a) the original points are plotted in green, and the adjusted data is represented with the red curve. The adjusted data is the trend in the illumination of the image, while the original data also contains the details thereof. The result of the adjusted curve is a vector $f(x)$ of size 1xN. Subsequently it creates patterns of light as aMxN matrix where each row is formed by the elements of the vector $f(x)$(Fig. 8 (b)).

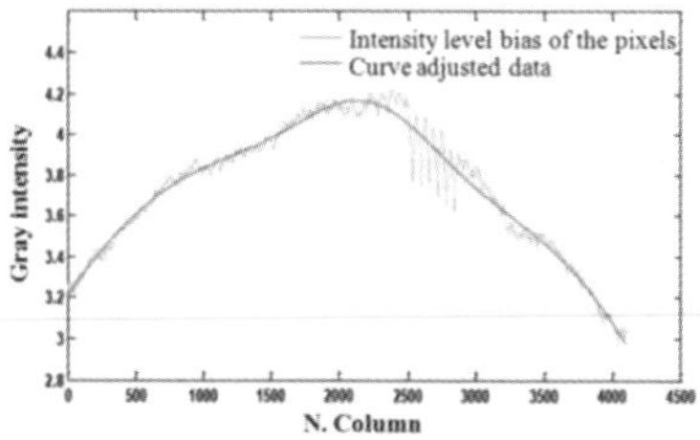

Figure 8. (a) Data adjusted with a Gaussian function. (b) Illumination pattern of the image.

Wavelet transform [23], was applied in two dimensional discrete images with corrected illumination and a series of morphological operations in order to retain only the particles with an area greater than 10 pixels (ref. Fig. 9 (a)). Fig. 9 (b) shows the skeleton of the crack after performing morphological operations on the binary image.

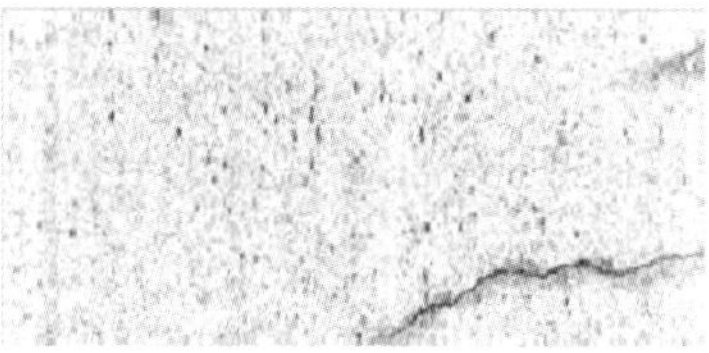

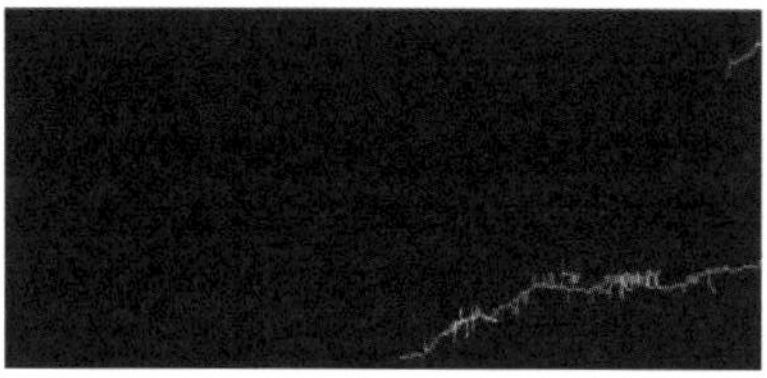

Figure 9. (a) Sum of the images after the wavelet transformation for windows and logarithmic transformation. (b) The skeleton of the crack.

6. Conclusion

Hex-piderix, a climbing walking robot was designed using a suction system for its motion. Kinematics analysis were performed for position, withoutincluding the bellows suction cup. A tripod gait was defined and implemented using a PLC to activate the electrovalves.

Hex-piderix is an excellent experimental basis for projects that belong to the field of mobile robotics. It can be used as a prototype in inspection applications on vertical surfaces, which allows crack detection using image analysis. An efficient method to correct the illumination and to increase image contrast was implemented. The crack is extracted by morphological operations.
To complete the analysis we conducted statistical comparisons of cracks extraction processes using images with and without illumination. This method is executed on-line for real-time inspection of surface cracks and it automatically integrates an algorithm for crack's classification.

Acknowledgments

This work has been supported by the grant *ConsejoNacional de Ciencia y Tecnología*CONACYT.

References

1. H. R. Everet. Sensor for Mobile Robots -Theory and Application, A.K. Peters Ltd. Wellesley, Massachusetts (1995).
2. J. Jones Mobile Robots. Inspiration to Implementation. A. K. Peters Ltd. Wellesley, Massachuset (1993).
3. J. A. Fernández. Manipulador virtual teleoperado, Instituto Militar de Ingeniería, Río de Janeiro (2002).
4. Oh Je-Keun, Jang Giho, Oh Semith, and Lee Jeong Ho. "Bridge inspection robot with machine vision". *Automation in Construction.* Vol. 18. Issue 7, pp 929-941(2009).
5. JuSeung-Nam, Jang Jae-Ho, and Han Chang-Soo. "Auto inspection system using a mobile robot for detecting concret cracks in a tunnel". *Automation in Construction.* Vol. 16. Issue 3, pp 255-261 (2007).
6. AudeBillard and Jan IjspeertAuke. "Biologically inspired neural controllers for motor control in a quadruped robot". *In IJCNN. International Joint Conference on Neural Network*(2000).
7. P. Gonzalez de Santos, E. Garcia and J. Estremera. *Quadrupedal locomotion: an Introduction to the control of four-legged robots.* Verlag, London: Springer, pp. 3-32 (2006).
8. A. Sintov, T. Avramovich and A. Shapiro. "Design and motion planning of an autonomous climbing robot with claws." *Robotics and Autonomous System*, vol.59, pp.1008-1019, June (2011).
9. S. Shekhar Roy, A. Kumar Singh and D. Kumar Pratihar. "Estimation of optimal feet forces and joint torques for on-line control of six-legged robot."*Robotics and Computer-Integrated Manufacturing*, vol.27, pp.910-917, March (2011).
10. M.C. García-López, E. Gorrostieta-Hurtado, E. Vargas-Soto, J.M. Ramos-Arreguín, A. Sotomayor-Olmedo and J.C. Moya-Morales. "Kinematic

analysis for trajectory generation in one leg fo a hexapod robot." *Procediathecnology*, vol.3, pp.342-350, (2012).
11. J. Gallardo–Alvarado, C.R. Aguilar–Nájera, L. Casique–Rosas, L. Pérez–González, J.M. Rico–Martínez, "Solving the kinematics and dynamics of a modular spatial hyper–redundant manipulator by means of screw theory." Multibody Syst. Dyn. vol20, pp. 307–325(2008).
12. Xochitl Yamile Sandoval-Castro, Mario Garcia-Murillo, Luis Alberto Perez-Resendiz and Eduardo Castillo-Castañeda (2013). Kinematics of Hex-Piderix - A Six-Legged Robot - Using Screw Theory, *International Journal of Advanced Robotic Systems,*Houxiang Zhang, Shengyong Chen (Ed.), ISBN: 1729-8806, InTech, DOI: 10.5772/53796. Available from: http://www.intechopen.com/journals/international_journal_of_advanced_robotic_systems/kinematics-of-hex-piderix-a-six-legged-robot-using-screw-theory
13. A. Campbell Neil, and Jane B. Reece. Capacidad de movimiento autónomo. Biología. Séptima Edición. Editorial medica panamericana (2007).
14. Andrew A. Biewener "Animal Locomotion". Oxford University Press. ISBN 0-19-850022-X (2003).
15. Tobias Seidl and RüdigerWehner. Walking on inclines: how do desert ants monitor slope and step length. *Frontiers in Zoology*. Vol. 5 doi:10.1186/1742-9994-5-8 (2008).
16. M. Cartmill, Functional Vertebrate Morphology, M. Hildebrandt, D. M. Bramble, K. F. Liem, D. B. Wake, Eds. TheBelknap Press of Harvard University Press, Cambridge, MA, pp. 430 (1985).
17. W. Kier, A. Smith, *The structure and adhesive mechanism of octopus suckers*, Integr. Comp. Biol. 42 (2002).
18. G. Hanna, W. J. P. Barnes, Adhesion and detachment of the toe pads of tree frogs, Journal of Experimental Biology 155, (1991).
19. P. Flammang, J. Ribesse, M. Jangoux, Biomechanics of adhesion in sea cucumber Cuvierian tubules (Echinodermata, Holothuriodea), Integr. Comp. Biol. 42, (2002).
20. K. Autumn, Properties, principles principles, and parameters of the gecko adhesive system, in Biological Adhesives A. Smith, J. Callow, Eds. Springer Verlag, Berlin Heidelberg (2006).
21. Yang Li, Man tian Li, and Li ning Sun. Design and passable ability of transitions analysis of a six legged wall-climbing robot. In Proc. of the IEEE Int. Conf. on Mechatronics and Aut., pp. 80-804, Harbin, China (2007).
22. N. Otsu, "A threshold selection method from gray level histograms", IEEE Trans. Syst. Man Cybern, vol. 9, pp. 62-66, January 1979.
23. S. Mallat, "A Wavelet Tour of Signal Processing", nd Edition, Academic Press, 2009.

SECTION–7

LOCOMOTION

REINFORCEMENT LEARNING OF BIPEDAL LATERAL BEHAVIOUR AND STABILITY CONTROL WITH ANKLE-ROLL ACTIVATION

BERNHARD HENGST

School of Computer Science and Engineering, University of New South Wales, Sydney, NSW 2052, Australia
E-mail: bernhardh@cse.unsw.edu.au

Nimble bipeds are expected to execute a rich repertoire of movement behaviours and switch between them seamlessly without falling over. A good example demanding this responsiveness is playing a game of soccer. This paper explores a less well studied approach to biped locomotion using model-based multi-goal reinforcement learning. We apply this approach to learn the side-to-side movement of a small humanoid robot in simulation. By on-line learning one transition function, it is possible to modify the behaviour of the robot varying the cost function. Behaviours include: changing support feet at different frequencies; standing upright; standing on either foot; and switching between these behaviours, subject to the imposed constraints. The behaviours are stable over much of the experienced phase domain. They respond optimally to disturbances and when switching.

Keywords: Machine Learning; Artificial Intelligence; Reinforcement Learning; Robotics; Bipedal Locomotion; System Identification.

1. Introduction

Future bipedal robots are expected to be more agile, exhibiting a variety of behaviours. It is difficult to hand-code these behaviours and to transition seamlessly between them.

In this paper we take a reinforcement learning (RL) approach to learning and controlling behaviour. We apply this approach in simulation, specifically to tackle the problem of lateral (frontal or coronal-plane) behaviour to aid bipedal locomotion. The novelty is in a combination of:

- A random action routine to learn the system state transition function on-line rather than the action-value function $Q(s,a)$ as is more usual in RL.[1] We do not rely on a (linear) inverted pendulum model, or any other model, but perform automatic system identification.

- Using the reward function to craft multiple behaviours such as side-to-side rocking at different frequencies, standing still and balancing on one leg.
- Seamlessly switching between behaviours. The behaviour response to switching is theoretically guaranteed to be optimal.
- Stability of the behaviours to disturbances. The response to any disturbance is an optimal trajectory back to the continuous cycle or state behaviour required to minimise overall cost.

In the rest of this paper we will briefly review related research, present our method and approach, and provide results showing the system behaviour under changing goals, and with disturbances. We conclude with a discussion and planned future work.

2. Related Work

We briefly mention some prominent research on bipedal walking.

Passive-dynamic walkers, originally developed by McGeer[2] based on a "rimless wheel" model, were improved by others and powered to walk on level ground.[3] These machines have no explicit controllers yet exhibit humanlike motions, but are limited in their repertoire of behaviour.

Another research direction is the control of the Zero Moment Point to stay within the support polygon. Exemplified by the work of Kajita, et al,[4] the use of a (linear) inverted pendulum model allows 3D bipedal gaits to be developed. This control-theoretic approach relies on *preview control*[5] - a needed more responsive feed-forward method.

Much research has been devoted to planar bipedal locomotion. An extensive exposition is given by Westervelt et al.[6] Planar research has been extended to 3D.[7] Grizzle reminded researchers that even the simplest bipedal locomotion is challenging, let alone aperiodic walks, non-flat ground, etc.[8] These approaches follow the control system methodology where system identification is manual and specified using differential equations.

Despite some early promises for RL, and while the literature on bipedal walking is extensive, there has been relatively few RL approaches to bipedal locomotion. Exceptions include frontal plane control using an actuated passive walker;[9] point-feet placement;[10–12] temporally extended actions applied to swing foot placement underpinned by semi-MDP theory.[10]

In this paper we propose to exploit the properties of RL to model non-linear, discrete and continuous dynamics and to learn optimal control for several different behaviours via reward functions.

3. Method

The general method we employ has the following steps:

(1) Construct and constrain the system, specifying joints, links, etc
(2) Define a Markov state and available control actions
(3) Learn the state transition function through exploration
(4) Define cost functions to achieve various behaviours
(5) Learn the optimal action-value Q function for each cost function
(6) Execute the behaviours by switching between optimal policies

We now describe the application of the above method to learn lateral bipedal motion behaviours.

3.1. *Construct and Constrain*

For real or simulated systems, this involves a specification of both the physical and software constraints. In our case we model a 23 DoF Aldebaran Robotics Nao humanoid robot with the Open Dynamics Engine (ODE) physics simulator (see Figure 1).

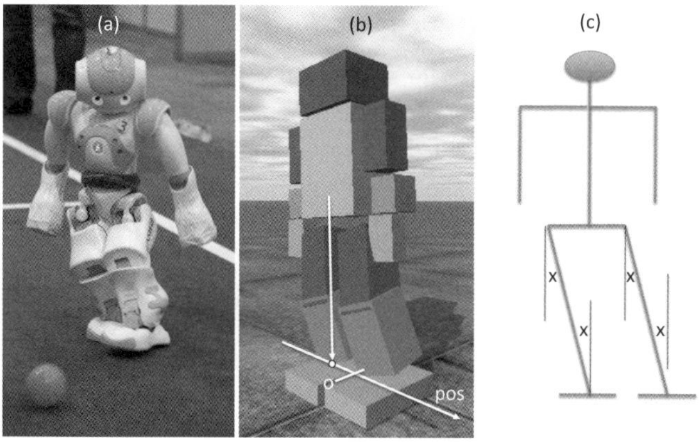

Fig. 1. (a) the Aldebaran Nao Robot, (b) ODE simulation showing the *pos* variable, and (c) the hip and ankle roll - they are equal, except for an additional control action described in Section 3.2.

To aid lateral movements we set the hip and ankle rolls to follow in direct proportion the perpendicular projection of the centre-of-mass (CoM) of the torso unto the line through the feet CoMs with the origin at the half-way point. We call this variable *pos* (see Figure 1). The support foot is determined by the sign of *pos*. The swing foot starts to lift immediately on change of support foot and moves up and down under constant acceleration with a period determined manually.

3.2. *Specify State and Actions*

RL is defined by the tuple $\langle S, A, T, R\rangle$ where S is the set of states, A the set of actions, T a stochastic transition function, and R a stochastic reward or cost function.[1] One would expect that rigid body physics could be modelled by point distributions, but even the ODE simulator generates noise due to the random nature of collision detection.

We define the lateral dynamic state of the Nao by the triple (pos, vel, act), where *pos* is a continuous variable defined in Section 3.1, *vel* is the velocity of *pos*, and *act* is the discrete control action applied at the last time-step to model the motor delay (of the simulator) of 2ms.

The three discrete actions A are $\{-\delta, 0, +\delta\}$, where δ is an additional amount of roll added to the ankle joints. In our application $\delta = 0.045$ radians. These actions may accelerate the robot to the left or right.

3.3. *Learning the Transition Function*

At each time-step we record the transition between states in S given an action from A. Since the sub-state space (pos, vel) is continuous we model the transitions at discrete grid-point values and use radial basis function approximation to generalise the transition function for off-sample states.

We explore and learn the transition function through a random action routine without lifting the swing-foot foot (but see discussion). Actions are generated at random using Equation 1 to adequately explore actions that persist.

$$action = \begin{cases} \text{random action from } A & 10\% \text{ of the time at random} \\ lastAction & \text{otherwise} \end{cases} \tag{1}$$

A portion for the learned transition function is depicted in Figure 2. Depending on the resolution of the grid, $resPos$ and $resVel$, learning may take several hours on the simulator.

3.4. *Define the Cost Function*

The cost function specifies the immediate rewards in RL (cost being the negative of reward). In our implementation we specify three components:

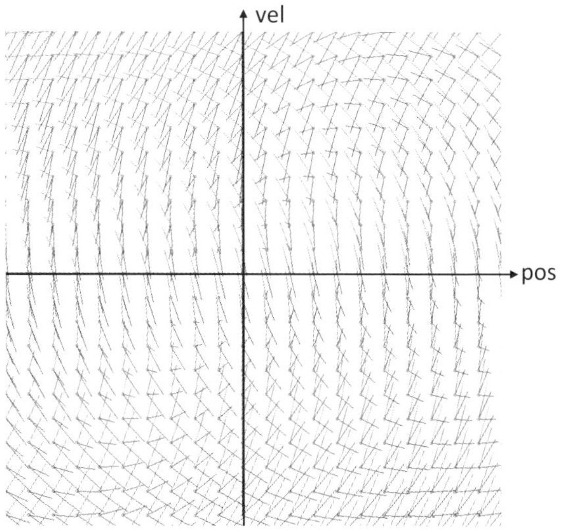

Fig. 2. Part of the transition function centred at ($pos = 0, vel = 0$) learned on-line showing the next state transition from grid-points for each of the three ankle roll actions.

the cost of taking an action; the cost when reaching a desired set of goal states in the (pos, vel) space; and the cost of switching actions (to minimise motor usage). For example, the reward function for the robot to stand still is specified by Equation 2.

$$reward = \begin{cases} 50 & \text{if } |pos| < resPos \text{ and } |vel| < resVel \\ -2 & \text{if } action \neq lastAction \\ -1 & \text{otherwise} \end{cases} \tag{2}$$

Rocking from side-to-side can be similarly specified by rewarding a velocity value at the time of switching the support foot. The magnitude of the velocity will determine the period of the rock. To stand on the left leg we reward states near (CoM-left-foot, 0).

3.5. *Learning the Optimal Control Policy*

We use standard RL policy iteration (PI) on the discrete Q action-value function.[1] The optimal function, $Q^*(s, a)$, is the future cost of taking action a in state s and following the optimal policy thereafter. The only subtly is the handling of the temporal difference backups as the next state for transitions from the grid-points is unlikely to be another grid point in the continuous state-space. However, it is straight forward to estimate the Q

functions for the next states by calculating a radial-basis weighted average value from neighbouring grid-points.[13] Assuming the function approximator is well behaved, the control policy is guaranteed to be optimal, a result that follows directly from MDP theory.[1] The Q values for the cost functions in Section 3.4 are learned and stored separately. Since the hard work is already done in Section 3.3, PI takes about a minute on a standard PC to learn six different policies: stand still; stand on one of each the two legs; and three sideways rocking motions at different frequencies.

3.6. *Execution*

The control policy is obtained from the optimal Q function, i.e. the optimal action $= \text{argmax}_a Q^*(state, a)$. Control is a matter of calculating the optimal action on-line at each time-step from the stored Q values given the current state and activating the motors accordingly.

4. Results and Discussion

After implementing all the steps in Section 3 we test the various behaviours by switching between them and observing the rendered robot. We also plot the evolution of the *pos* and *vel* state as a time series and with phase portraits.

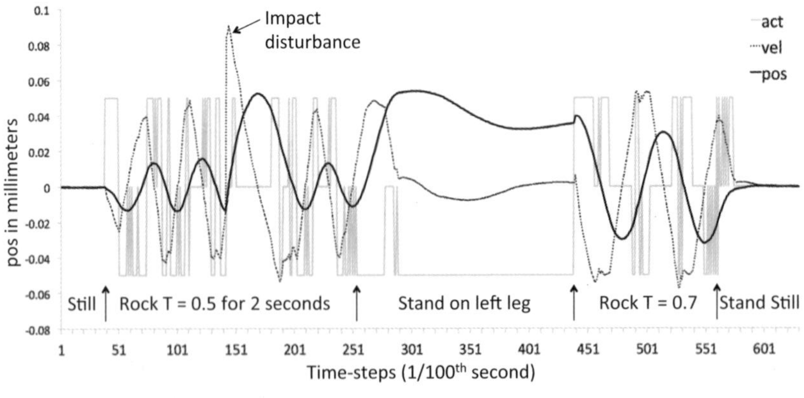

Fig. 3. Time series showing the value of *pos*, *vel* and the *action* for the series of behaviours described in Section 4. Note the quick recovery after the sideway impact disturbance and the smooth transitions between behaviours.

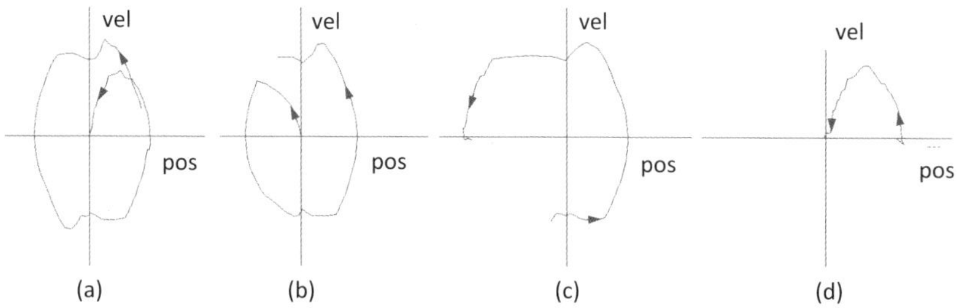

Fig. 4. Example phase portraits, switching between policies: (a) T=0.7sec rock to standing upright, (b) standing upright to T=0.7sec rock, (c) T=0.7sec rock to standing on right leg, (d) standing on left leg to standing upright.

The combination of switching and durations for the behaviours is infinite, so we show results for an example series of behaviours, namely: from a standing start, execute a rock with period 0.5 seconds for 2 second, then stand on the left leg for 2 seconds, followed by a rock of period 0.7 seconds for 1 second, and back to the standing upright position. During the 2 second rock we impart a sideways impact force (delivered by shooting a cannon-ball at the machine). The results are plotted in Figures 3. Figure 4 shows phase portraits switching between several behaviours. Both Figures show stable behaviour for each policy and when switching.

If the transition function in Section 3.3 is instead learned while lifting the swing foot, the RL will use the replacement of the foot to increase the acceleration of the body. This is a consequence of maximising future reward. In practice this may cause the robot to thump its feet, increasing wear and tear. For this reason we settle for learning without lifting the foot. The downside is that introducing the lifting action subsequent to learning, introduces what appears as a disturbance to the natural cycle and hence the kink in the limit cycles seen in the phase portraits in Figure 4.

The sideways rock was combined with an open loop forward (sagittal) and a sideways gait. Both performed well for small step-sizes, but there is much do to. With a RL sagittal gait under our belt[14] the next step is to combine and synchronise the two orthogonal motions.

Another step is to adapt the simulated results to the real Nao. An advantage of our approach is that the transition function learned on the simulator can be used as a starting point and adjusted on the real robot hopefully with less training time. The actions (in our case ankle rolls) are discrete. A fruitful development would to introduce continuous actions by function fitting and interpolation.

5. Summary

In summary, we have presented a promising approach to tackle more versatile stable biped locomotion, generating multiple behaviours from the one learned transition function, and seamlessly and optimally switching between the behaviours.

References

1. R. S. Sutton and A. G. Barto, *Reinforcement Learning: An Introduction* (MIT Press, Cambridge, Massachusetts, 1998).
2. T. McGeer, Passive dynamic walking, *I. J. Robotic Res.* **9**, 62 (1990).
3. S. Collins, A. Ruina, R. Tedrake and M. Wisse, Efficient bipedal robots based on passive-dynamic walkers, *Science* **307**, 1082 (2005).
4. S. Kajita, F. Kanehiro, K. Kaneko, K. Fujiwara, K. Harada, K. Yokoi and H. Hirukawa, Biped walking pattern generation by using preview control of zero-moment point, *ICRA'03* , 1620 (2003).
5. T. B. Sheridan, Three models of preview control, *IEEE Transactions on Human Factors in Electronics* **7**, 91(June 1966).
6. E. R. Westervelt, J. W. Grizzle, C. Chevallereau, J. H. Choi and B. Morris, *Feedback Control of Dynamic Bipedal Robot Locomotion* (CRC Press, Boca Raton, 2007).
7. A. Ames and R. Gregg, Stably extending two-dimensional bipedal walking to three dimensions, *American Control Conference* , 2848 (2007).
8. J. W. Grizzle, C. Chevallereau, A. D. Ames and R. W. Sinnet, 3d bipedal robotic walking: Models feedback control, and open problems, *8th IFAC Symposium on Nonlinear Control Systems* (2010).
9. R. Tedrake, Stochastic policy gradient reinforcement learning on a simple 3d biped, in *Proc. of the 10th Int. Conf. on Intelligent Robots and Systems*, 2004.
10. J. Morimoto, G. Cheng, C. Atkeson and G. Zeglin, A simple reinforcement learning algorithm for biped walking, in *Robotics and Automation, 2004. Proceedings. ICRA '04. 2004 IEEE International Conference on*, april-1 may 2004.
11. S. Wang and J. Braaksma, Reinforcement learning control for biped robot walking on uneven surfaces, in *Proceedings of the 2006 International Joint Conference on Neural Networks*, 2006.
12. J. Morimoto, J. Nakanishi, G. Endo, G. Cheng, C. G. Atkeson and G. Zeglin, Poincaré-map-based reinforcement learning for biped walking., in *ICRA'05*, 2005.
13. B. Hengst, On-line model-based continuous state reinforcement learning using background knowledge, *Twenty-Fifth Australasian Joint Conference on Artificial Intelligence (AI12)* (2012).
14. B. Hengst, M. Lange and B. White, Learning ankle-tilt and foot-placement control for flat-footed bipedal balancing and walking, *11th IEEE-RAS International Conference on Humanoid Robots* (2011).

OBSTACLE/GAP DETECTION AND TERRAIN CLASSIFICATION OF WALKING ROBOTS BASED ON A 2D LASER RANGE FINDER

PATRICK KESPER, EDUARD GRINKE, FRANK HESSE,
FLORENTIN WÖRGÖTTER, and PORAMATE MANOONPONG*

Faculty of Physics, Third Institute of Physics - Biophysics
Georg-August-Universität Göttingen,
Friedrich-Hund-Platz 1, 37077 Göttingen, Germany
**E-mail: poramate@physik3.gwdg.de*

This paper utilizes a 2D laser range finder (LRF) to determine the behavior of a walking robot. The LRF provides information for 1) obstacle/gap detection as well as 2) terrain classification. The obstacle/gap detection is based on an edge detection with increased robustness and accuracy due to customized pre and post processing. Its output is used to drive obstacle/gap avoidance behavior or climbing behavior, depending on the height of obstacles or the depth of gaps. The terrain classification employs terrain roughness to select a proper gait with respect to the current terrain. As a result, the combination of these methods enables the robot to decide if obstacles and gaps can be climbed up/down or have to be avoided while at the same time a terrain specific gait can be chosen.

Keywords: Legged locomotion, Autonomous robots, Climbing, Gap avoidance

1. Introduction

The evaluation of environmental information is the basis to enable autonomous (walking) robots to successfully navigate through complex environments. To tackle this problem different approaches have been proposed. For example, Labecki et al.[1] used a laser range finder (LRF) to build a height map of the environment. Maier et al.[2] combined monocular vision and LRF readings for humanoid robot navigation. Wooden et al.[3] utilized a LRF and stereo vision in combination with complex navigation and perception algorithms. In contrast to these works, here we use a LRF based simple control algorithm without further sensor modalities for obstacle/gap detection and terrain classification. In this paper we will show that such a simple control algorithm is sufficient to enable a hexapod robot to successfully traverse difficult terrains. Furthermore this algorithm can serve as a basis for more complex control structures.

2. Materials and Methods

2.1. *The Hexapod Robot AMOSII*

AMOS II is a six-legged walking robot[4] (see Fig. 1) used throughout this work. At the front of the robot, a Hokuyo URG-04LX-UG01 laser range finder (LRF) is mounted on a small tower-like mounting device (Fig. 1). We selected this LRF since it is lightweight, independent from light conditions and needs only little power.[5] The LRF is mounted at a height of $h = 22.5$ cm, its horizontal angle β is set to 25° (see Fig. 1) for the work at hand. We found this to be the optimal trade off between decreasing look ahead distances for higher angles and reduced accuracy and robustness for lower angles.[5] Besides the LRF, two ultrasonic sensors are attached at the front part of the robot. These sensors are used for climbing behavior[4] which is triggered by the presented obstacle/gap detection algorithm (described below).

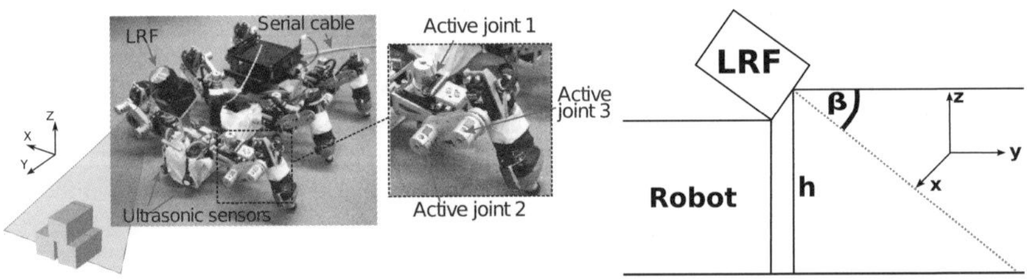

Fig. 1. Left: The walking robot AMOS II with the LRF (overall picture and zoom of leg with three active joints). An illustration of the LRF function is depicted. Right: Sketch of the LRF setup.

2.2. *Obstacle/Gap Detection*

In the work at hand, we propose an obstacle/gap detection algorithm consisting of three steps. First the raw laser scan data is preprocessed to remove measurement artifacts, then edges of objects are detected based on height differences (see Fig. 1, Left) and in the postprocessing edges are combined to objects. The details of each step are provided in the following subsections.

Preprocessing: In a first step the LRF readings are transformed from spherical into cartesian coordinates. Then, in order to increase robustness of the edge detection, outliers and measurement artifacts such as mixed pixels[5] are removed. A simple exclusion algorithm based on the deviations of x and z coordinates (see Fig. 1, Right) between neighboring points is used (see Algorithm 2.1, Left).

Edge Detection Algorithm: In this work, we use a fixed value thresholding algorithm for edge detection.[6] This algorithm is based on the fact that obstacles/gaps correspond to height differences in the x-z plane of the 2D laser data.[6] By applying a threshold to the height difference between two data points, edges can be detected.

Postprocessing: Here, neighboring edges with a similar mean height are joined as they have high probability to correspond to only one entity, e.g., a rough object. For the remaining edges average (z_{avg}), maximum (z_{max}) and minimum (z_{min}) values are calculated and compared with thresholds defining obstacles and gaps (see Sec. 2.4).

This algorithm can be applied to detect different types of obstacles including walls and objects having sufficient height different to the ground. It has been successfully tested to detect, e.g., different box sizes.

2.3. *Terrain Classification based on Roughness Estimation*

Obtaining an estimation of the surface roughness enables walking robots to select a terrain specific gait, thereby performing effective locomotion. We introduce a criterion for roughness estimation based on the amplitude and the rate of height changes of the LRF data. These features are represented in the root-mean-square value $R_q = \sqrt{\frac{1}{n}\sum_{i=1}^{n} d_i^2}$. It evaluates the deviation d_i of each data point from the average height of the calculated edges. The roughness estimation value can be used to select a gait depending on terrain surface roughness. However, to enhance the robustness of the gait selection, R_q is filtered with a low pass filter.

2.4. *Behavior Control for Complex Environments*

The processed LRF data provides obstacle/gap and roughness information about the environment used to decide which action the robot should perform. The behavior control based on these data is summarized in Algorithm 2.1, Right. First the minimum and maximum height values are evaluated to detect non crossable objects/gaps. In case of the average height being between the positive and negative ground threshold, no climbing is initiated since this situation is considered as obstacle/gap free. Thus, the robot continues walking forward. On the other hand, if the average height

Algorithm 2.1 Left: Pseudo code for preprocessing LRF data. Right: Pseudo code of behavior control.

```
for all p_i in dataPoints do
    transform(p_i)
    Δx = |p_i.x − p_{i+1}.x|
    Δz = |p_i.z − p_{i+1}.z|
    if Δx > minX AND
        Δz < maxZ then
        save(p_i)
    else
        remove(p_i)
    end if
end for
```

```
if z_min < minHeight then
    turn()
else
    if z_max > maxHeight then
        startObstacleAvoidance()
    else
        if z_avg > ground then
            if dist < distThreshold then
                startClimbing()
            else
                startWalkingForward()
            end if
        end if
        if z_avg < −ground then
            startClimbing()
        else
            startWalkingForward()
        end if
    end if
end if
```

is above the positive ground threshold (i.e., climbable obstacle) or below the negative ground threshold (i.e., climbable lower step), climbing behavior is activated. During climbing behavior the front part of the robot is tilted upwards/downwards. These motions are controlled by the ultrasonic sensors-driven backbone joint (see Goldschmidt et al.[4] for more details). While climbing, the LRF does not produce reliable measurements and thus its values are not used. During locomotion the gait can additionally be changed according to the roughness R_q of the terrain.

3. Experimental Results

3.1. *Behavior Control based on Obstacle/Gap Detection*

The behavior control using LRF-based obstacle/gap detection was investigated on a track consisting of a ground floor and two elevated platforms. The robot started walking towards the first platform (stage 1, Fig. 2). It continued walking until it was near enough to the platform. It then detected the platform as a climbable obstacle since the average height (z_{avg}, Fig. 2) was above a positive ground threshold. The decreasing distance to the platform can be seen in the increasing outputs of the ultrasonic sensors

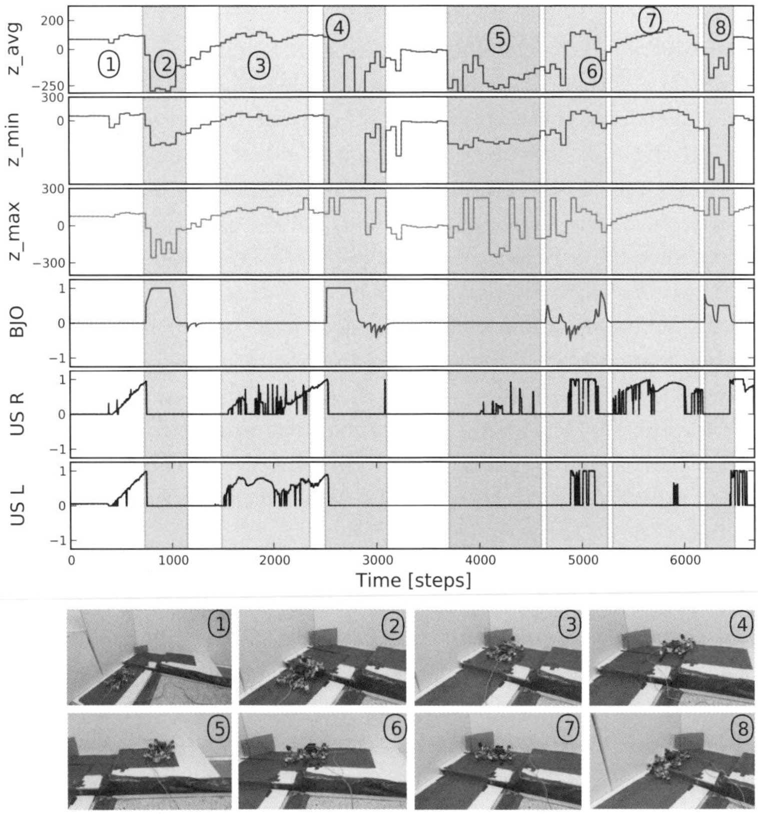

Fig. 2. From top to bottom: Average height of the LRF data (z_{avg}), minimum height in LRF data (z_{min}), maximum height in LRF data (z_{max}), backbone joint signal (BJO) and left and right ultrasonic sensors (US_L, US_R). All height values are given in mm. If the backbone joint signal is positive, the joint is tilted upwards and downwards for negative values. The ultrasonic signals represent the distance to an obstacle, where a higher value is obtained for nearer obstacles (zero represents no obstacle in field of view). Within the green areas the robot avoided obstacles, while in the blue area a gap was avoided by turning. Within the red areas, the climbing routine was enabled. In these areas the output of the LRF was ignored.

(US_L, US_R, Fig. 2). As these values exceeded a threshold, the robot started to climb (stage 2). The progress of the climbing routine is represented by the backbone joint motion. At the beginning the front part of AMOS II moved up to reach the edge of the obstacle, indicated by an increase in the backbone joint signal (BJO, Fig. 2). Afterwards the front part bent down to support the climbing of the main body, hence the backbone joint signal decreased to negative values until the climbing finished. While walking on the first platform the robot detected the wall in front of it. The height of the wall was correctly classified as too high to climb (z_{max}, Fig. 2) and the robot then turned (i.e., obstacle avoidance, stage 3). At some point the wall vanished from the field of view, the robot thus made some forward steps until the wall appeared again in its field of view. This can be seen in the ultrasonic sensor outputs oscillating around the distance threshold value. After avoiding the wall, the robot climbed onto the second platform (stage 4), showing a behavior similar to the first climbing procedure. On top of the second platform only the LRF was able to detect the gap limiting the platform. Its minimum value z_{min} decreased below the gap threshold. The robot turned until facing the first platform again (stage 5). At this point the robot climbed down, as the average height z_{avg} was between gap and ground thresholds. Again the climbing routine can be recognized by the backbone joint signal (stage 6). After climbing down to the first platform the robot walked forwards, avoided the wall directly in front (stage 7) and then climbed down to the ground (stage 8).

This experiment shows that the behavior control based on LRF data proposed in Sec. 2.4 allows the robot to autonomously locomote on the complex terrain. We recommend readers to also see the supplementary video of this experiment at http://manoonpong.com/CLAWAR2013/suppl.wmv.

3.2. *Behavior Control based on Terrain Classification*

To evaluate the terrain classification using LRF data as described in Sec. 2.3, an experimental track was built, where the robot has to traverse three areas with significant differences in roughness. The starting point was on a flat wooden platform (area 1). From there the robot had to continue to walk over fine (area 2) and coarse-grained gravel (area 3). The gait of AMOS II can be controlled via one control input.[4] In this experiment this input was determined by the result of the terrain classification (R_q). Figure 3 shows the sensor data of the robot while traversing the different terrains. As expected R_q was very small on the wooden platform (area 1), thus a fast walking pattern was chosen (tripod gait).

When arriving at the second area the filtered R_q increased to values between 5 and 25. This resulted in a gait change, from a tripod gait to a wave gait, which is slower and better suited for fine gravel.

In front of the third area the filtered R_q value increased above 25. This led to a change of the wave gait to a tetrapod gait, which enabled the robot to perform faster and proper walking on uneven terrain, like the coarse-grained gravel. The successful completion of this track indicates that the terrain classification provided by the LRF can be used to choose gaits according to the terrain the robot approaches. We recommend readers to also see the supplementary video of this experiment at http://manoonpong.com/CLAWAR2013/suppl.wmv.

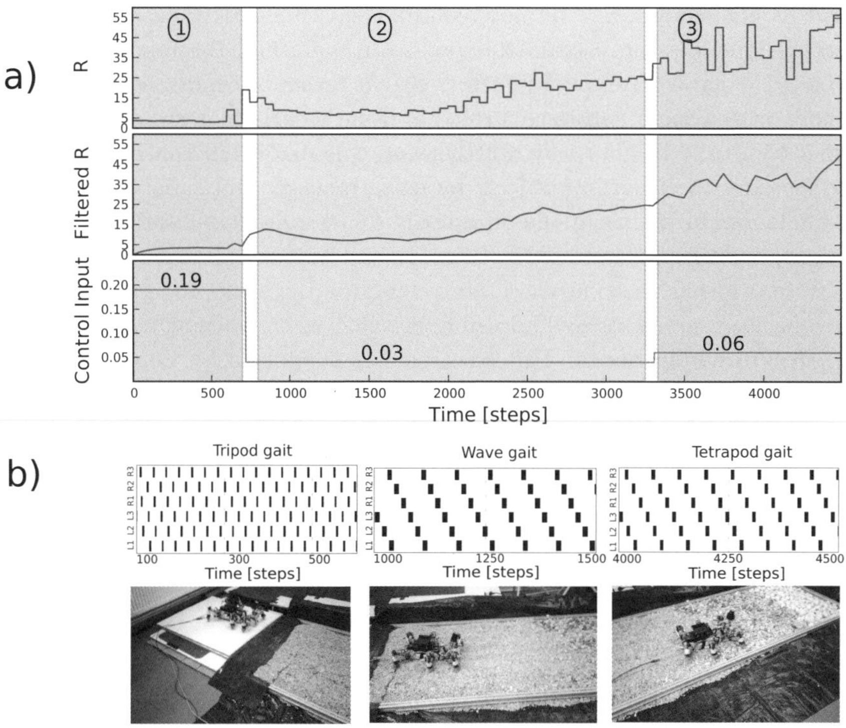

Fig. 3. Signals and pictures of the robot while traversing different terrains. a) The first two rows show the raw and filtered results of the terrain classification (R_q) followed by the control input which determines the gait of the robot. b) The gaits used by AMOS II shown as gait diagrams (black swing phase, white stance phase) and below the places in the track where these gaits are used.

4. Conclusion

We showed that a 2D LRF can be used for obstacle/gap detection. This allows the hexapod robot to climb up and down platforms surrounded by walls and deep gaps. Furthermore, terrain classification based on LRF data can be used to adapt the gait of the robot to different terrains, like flat ground and different grained gravels.

However there are still some aspects which can be improved. Basically, the LRF can provide position and width information about obstacles in its field of view. This information could be used to determine the direction of turning when facing several obstacles with different heights or provide a basis for more advanced behavior control. With this enhancement the robot will be able to navigate in more complex environments than the comparatively simple setup used in this paper. In particular the robot would be able to judge if it would be better, e.g. in terms of energy or danger, to climb or to avoid an obstacle instead of using a decision based on the maximal climbable heights only. Furthermore the algorithm can be extended to measure the length of objects by using memory and simple navigation techniques. In addition, the thresholds for terrain classification and the corresponding gaits were set by hand. In real world applications, where the terrain is usually not known, a more elaborate threshold and gait selection can be necessary. This could possibly be achieved by learning techniques.[7,8]

Acknowledgments: This research was supported by Emmy Noether grant MA4464/3-1 of the Deutsche Forschungsgemeinschaft, the Bernstein Center for Computational Neuroscience II Göttingen (BCCN grant 01GQ1005A, project D1) and the Bernstein Focus Neurotechnology Göttingen (project 3B, 01GQ0811).

References

1. P. Labecki, D. Rosinski and P. Skrzypczynski, *Transactions of the ASME, Journal of Basic Engineering* (2011).
2. D. Maier, M. Bennewitz and C. Stachniss, Self-supervised obstacle detection for humanoid navigation using monocular vision and sparse laser data, in *Proc. IEEE Int. on Robotics and Automation (ICRA)*, 2011.
3. D. Wooden, M. Malchano, K. Blankespoor, A. Howardy, A. Rizzi and M. Raibert, Autonomous navigation for bigdog, in *Proc. IEEE Int. on Robotics and Automation (ICRA)*, 2010.
4. D. Goldschmidt, F. Hesse, F. Wörgötter and P. Manoonpong, Biologically inspired reactive climbing behavior of hexapod robots, in *Proc. IEEE Int. Conf. on Intelligent Robots and Systems (IROS2012)*, 2012.
5. Y. Okubo, C. Ye and J. Borenstein, Characterization of the Hokuyo URG-04LX laser rangefinder for mobile robot obstacle negotiation, in *Proc. SPIE*, 2009.
6. G. A. Borges and M.-J. Aldon, *J. Intell. Robot. Syst.* **40**, 267 (2004).
7. S. Steingrube, M. Timme, F. Wörgötter and P. Manoonpong, *Nature Physics* (2010).
8. P. Manoonpong, C. Kolodziejski, F. Wörgötter and J. Morimoto, *Advances in Complex Systems* (2013).

MOTION CONTROL FOR THE 6-LEGGED ROBOT IN EXTREME CONDITIONS*

YURY F. GOLUBEV, VICTOR V. KORIANOV,
VLADIMIR E. PAVLOVSKY, ALEXEY V. PANCHENKO
Keldysh Institute of Applied Mathematics
Russian Academy of Sciences Moscow 125047, Russia

This paper deals with the problem of motion design for six legged mobile robot when it is overcoming complicated obstacles. The previous stage of development consisted of overcoming the combination of vertical column standing near a high shelf, two shelves connected with a narrow horizontal beam, two-sided vertical corner, and stepladder. In this paper the problem of overcoming obstacles of high degree instability is discussed. As an example of such obstacle a big enough balls, which can freely roll along the horizontal supporting plane, was chosen. The possibility of control for autonomous six-legged robot climbing from supporting horizontal plane to the massive ball, climbing from the ball to the high shelf and climbing from one ball to another one was demonstrated. The motion is executed with the help of dry friction forces only. The motion of the body and legs is executed by imposing servo-constraints in the form of adaptive step cycles of legs and the required geometrical structure of body motion. The results of 3D computer simulation of the controlled robot dynamics are presented. The electromechanical prototype is developed and presented.

1. Introduction

Up today many experimental multi-legged robots are created to investigate new locomotion possibilities [1]. At the moment the most advanced robot of that type is the BigDog aimed for load transportation along the open rough terrain [2]. However the problem exists on creating the climbing robot with capability of motion in condition of windfall terrain and along the surfaces of complicated geometrical forms.

In the paper the problem of motion design for six-legged mobile robot when it is overcoming complicated obstacles is considered. The aim of investigation is to study the capabilities of a walking robot in extreme conditions.

Note, walking robots are able potentially to move on structures of considerable height using the Coulomb friction forces. In this situation the robot

* This work is supported by Russian Foundation for Basic Research (#13-01-00184).

can use only the dry friction force at points of contacts of feet with the support surface. The complexity of algorithms of motion design is caused by the fact that the reaction of the support surface at the points of contact with the surface should not go outside the limits of the friction cones whose axes can have arbitrary directions in the space. In [3] efficient algorithms for constructing the motion of a six-legged robot able to overcome a combination of obstacles in the form of a vertical column standing near a shelf, for climbing the roof of a vertical corner along its walls, as well as an algorithm of passing from one shelf to another along a narrow horizontal beam, for climbing along a stepladder, for stabilization the six legged robot on a big enough ball, which can freely roll along the horizontal supporting plane were presented.

In certain extreme situations, the ability of a walking robot to overcome obstacles may considerably depend on the ability of the robot to provide efficiently the dynamic stability when there is no static stability. In [3] the dynamic stability is provided by a balancing motion of middle robot legs when, on a narrow beam, the robot has only two supporting points.

The solution to the problem of robot passage from a shelf to a massive ball, which can roll on a horizontal plane, the following acceleration and deceleration of the ball by special motions of the robot on the ball, a passage of the robot from the ball to another shelf, was presented in [4]. The problem of climbing of the insectomorphic robot onto the freely movable ball from a supporting horizontal plane was solved in [5]. Some approach to the solution of the problem of motion design along the unsteady surface can be obtained by investigating possibility of robot climbing from one movable ball to another. In this paper a set of typical methods that allow the robot to execute pointed maneuvers under a given friction coefficient are proposed. Typical motion elements are given using geometric constraints (servo-constraints) on the trajectories of feet movements and the trajectories of motion of special points of the body.

The efficiency of the proposed methods is confirmed by means of three-dimensional computer simulation [6] of the complete dynamics of balls together with the robot having 24 degrees of freedom. The results of calculations illustrating the operation of the described algorithms for motion construction are presented.

The motion is executed with the help of dry friction forces only. The motion of the body and legs is executed by imposing servo-constraints in the form of adaptive step cycles of legs and the required geometrical structure of body motion. Asymptotically stable prescribed motion of legs with respect to the body and the entire system is provided by a PD-controller. The results of 3D computer simulation of the controlled robot dynamics are presented. The electromechanical prototype is developed. Results of experiments with a robot prototype will be presented as well.

2. Results of computer simulation

Computer simulation of full robot dynamics was performed to evaluate the efficiency and robustness of control algorithms of a robot. This simulation was performed with the help of the program complex "Universal Mechanism" [6]. The robot's motion, as a result of computer simulation, is fully determined by initial conditions, electrical voltage sending to leg's angles drives, weights of mechanical system's elements and reactions of a supporting surface. Digital integration of the differential equations of motion was performed. The results of computer simulation were obtained with using a specially designed virtual environment that allows to experiment with a model of robot dynamics in the same way as it can be done in physical experiments. All aspects connected with a stable motion of the robot especially at non regular stages of motion were worked out. They includes problems of robot's falling down from the obstacle, non-regular contacts of the obstacles and elements of the robot, mutual intersections of legs, right choice of supporting points and so on. The results of the simulations show the effectiveness of the developed methods of robot control and realizability of the proposed scheme of robot motion.

Figure 1 shows fragments of the preparation stage of robot's motion when it climbs onto the free ball which can roll along the horizontal plane without slippage. The basic goal of the preparation stage is to locate the body in the tangential direction to the ball in the vertical plane. To make the support more reliable, the front feet should be located near the top of the ball. To make the grasp of the ball after the initial push more reliable, the support points of the middle legs on the ball should be a sufficient distance away from the support points of the front legs [5]. To impart the desired initial velocity to the body and the ball, the support points of the rear legs on the plane should lie on the longitudinal axis of the body. The capability of performing the preparation stage is ensured by the kinematic and dynamic properties of the robot-ball system [5]. Based on the theorem on variation of angular momentum of a mechanical "robot-ball" system with respect to the point where the ball touches the horizontal plane, conditions are found under which this maneuver is possible. For example the initial velocity v_0 of the ball should satisfy the relations

$$v_0 \geq 2P\sqrt{gR},\ P = \ln\frac{1+p}{1+p\sin\varphi_0},\ p = \frac{mRr}{J+(M+m)R^2},$$

where g is the acceleration of gravity, R — ball's radius, φ_0 — the angle initial position of robot's center of mass relative to the ball, M, m — masses of the ball and the robot respectively, J — central momentum of inertia of the ball relative to a horizontal axis, r — distance from robot's center of mass to the ball's center.

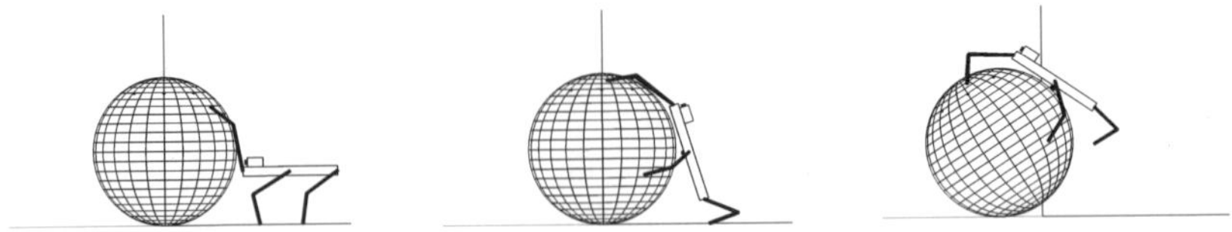

Figure 1. Climbing onto the free rolling ball.

Estimate $\tilde{R}$ is obtained for the maximum radius of the ball that the robot can climb. Let friction coefficient is $f = 0$. Then,

$$\tilde{R} = \frac{l_2 + 2\hat{h} + \sqrt{5l_1^2 - (2l_2 - \hat{h})^2}}{5},$$

where l_1, l_2 are lengths of a leg links and $\hat{h}$ is the maximal possible height of front legs attachment points lift above the support plane. This value of radius $\tilde{R}$ is obtained when the link l_2 lies horizontally and the foot rests on the top point of the ball (Fig. 2, a). This is quite natural, because, in the absence of friction, this is the only point to hang on without forcing the ball to slip. It is not difficult to see that the maximum of $\tilde{R}$ is attained when $\hat{h} = 2(l_1 + l_2)$ and is given by $\max_{\hat{h}}(\tilde{R}) = l_2 + l_1$.

(a) (b)

Figure 2. Maximum ball radius.

Let $f = 1$. Then $\tilde{R} = \hat{h} + l_1 + l_2$. This value of radius $\tilde{R}$ corresponds to the case where the robot with the front legs directed upward reaches the ball point lying at the height of ball center (Fig. 2, b). In the given case, this point belongs to the friction cone with the vertex located at support point on the ball and the vertex angle $\pi/4$. Taking the limit value $\hat{h} = 2(l_1 + l_2)$, we obtain

$\max_{\hat{h}}(\tilde{R}) = 3(l_2 + l_1)$. As can be seen, the presence of friction greatly affects the maximum radius of the ball and facilitates climbing the ball.

When the robot has reached the top of the ball then the possibility of a horizontal displacement of the system robot-ball exists. To produce an increment of velocity the robot has to make reciprocal movements of body of short duration in the desired direction. The amplitude of these movements depends on difference between of real velocity of ball's center and desired value of velocity [4]. The acceleration stages can be performed both on each gait period and in several periods. For a sufficiently large number of acceleration stages, the ball velocity asymptotically approaches the desired value.

When the robot goes on the moving ball by tripod gait there are two main problems. One of them is that the center of mass of the robot displaces from its right vertical position due to asymmetric arrangement of legs in their transfer phase. This problem is solved by appropriate displacement of robot's body. Another problem is connected with the motion of support points together with the ball. To compensate this motion the future supporting points are placed on the intersection of the ball and an inclined plane. The angle of inclination depends on the ball velocity [4].

Horizontal motion of the system "robot-ball" can be accomplished when the ball has reached the desired position on the support plane. For example it may be vertical shelf with horizontal upper platform. Fig. 3 presents fragments of motion for the case when robot passes from the ball to the shelf. The main difficulty of the problem consists in the fact that in the course of the passage, the ball may roll away from the shelf (releasing constraint) or it may start rolling along the shelf under an insufficient friction force between the ball and the vertical wall of the shelf. If the robot press the ball actively to the shelf, then the friction force, arising at the point of contact of the ball with the vertical wall of the shelf and bounded by the friction cone, counteracts the roll of the ball along the vertical wall of the shelf. This circumstance can be used by the robot in order to get safely to the upper platform of the shelf. The pressing of the ball to the shelf is regulated by the fact that the projection of the center of mass of the robot on the horizontal plane during the passage from the ball to the shelf is displaced in the required way towards the shelf from the point at which the ball touches the bottom ground. During the passage of the robot from the ball to the shelf the pressure of legs on the ball should be, if possible, symmetric with respect to the vertical plane passing through the ball center and containing the direction of the future ball motion. So the "gallop" gait is used. The maneuver of passing from the ball to the shelf can be executed successfully both in the case when the ball stops at a small distance to the shelf and in the case when the ball has a small velocity in approaching the shelf. In a considerable collision between the ball and the shelf, the ball gains the reverse velocity, and the robot goes on the ball in the reverse direction and can execute a new attempt in order to approach the shelf.

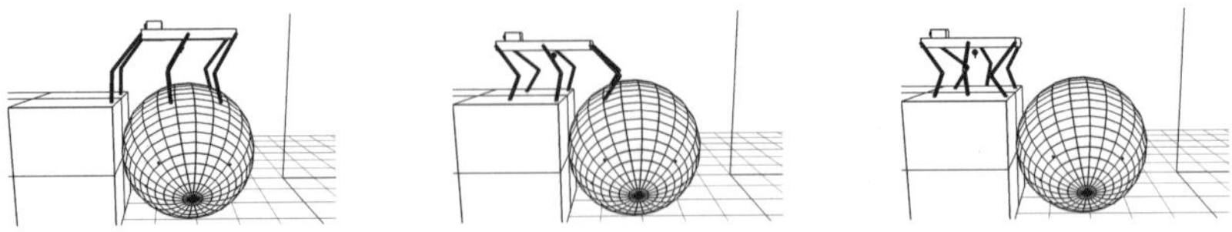

Figure 3. Climbing from the ball to the shelf.

A more difficult problem concerns the robot climbing from one movable ball to another movable ball. Both of balls are identical and can freely roll along the horizontal plane without slippage. It is supposed that at the initial stage balls touch one another. They are at the rest and robot stays on the top of one of them. The difficult is that, while robot is going to climb, both balls begin to move due to the gravity force momentum. The gravity force momentum originates by the displacement of the robot's center of mass from its equilibrium position. Dry friction forces exist in a contact point of balls. These forces decelerate the motion of balls and facilitate the procedure of climbing. Fig. 4 presents fragments of the robot climbing from one ball to another one.

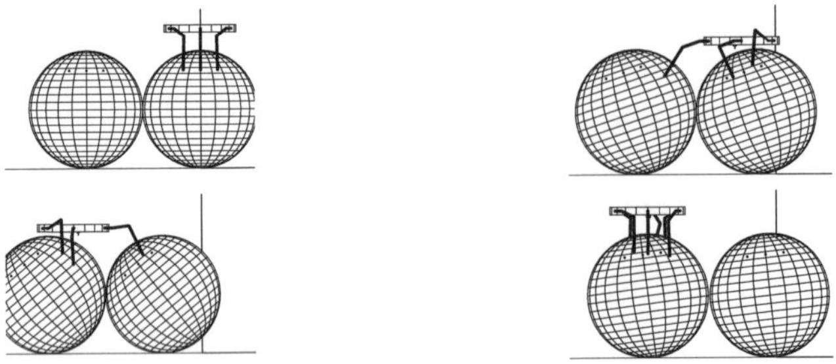

Figure 4. Climbing from the ball to another one.

The body of robot keeps its horizontal orientation during the motion. So the body can't go down lower then balls diameter. The robot has to move faster than the balls, because there is a problem of reaching the support points on the ball under the condition that the robot's body does not touch balls.

At the beginning of climbing the robot has to squat. It allows him to reach support points not only on one ball but on both balls. Balls may separate one from another if the robot dispose support points on both balls. It is necessary to make an additional displacement of support feet together with the target ball to prevent this separation.

Modeling showed efficiency of the designed algorithm of control. The elaborated control algorithm is rather theoretically simple and doesn't demand huge computing capacities. Generalization of algorithm of creation of basic trajectories for devices with different kinematics of feet and with arbitrary arrangement of points of fastening of feet is possible. Relying on results of numerical experiments, it is possible to draw a conclusion on necessary dynamic properties of drive mechanisms in robot hinges.

3. Realization of the prototype

Working off of the described control algorithm on rigorous computer simulation and on the real model of the HexMini robot (Fig. 5) is fulfilled.

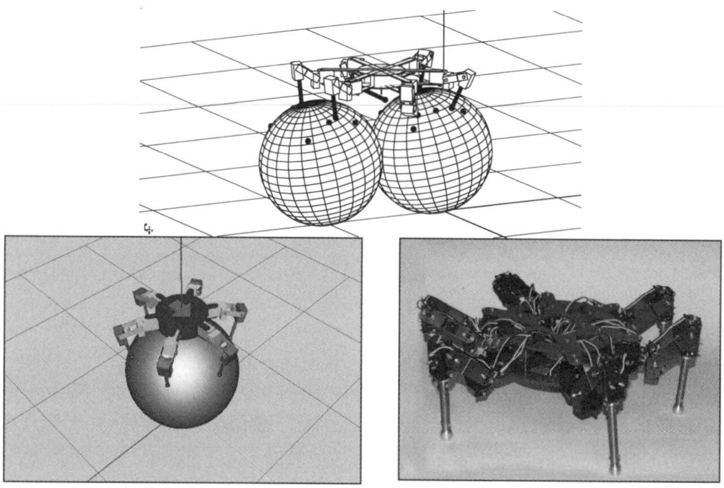

Figure 5. The HexMini robot.

Diameter of the robot prototype is about 50 cm, the equipped weight of 4 kg. Presumably, the robot will operate with a ball to overcome it and to move on a top [7]. The results will be demonstrated.

4. Conclusion

Typical solutions of new problems on climbing different obstacles by an insectomorphic robot with the help of dry friction forces have been obtained. The execution of proposed maneuvers is hampered by the risk of robot falling and of mutual crossing of legs because of the narrowness of the support polygon on the corresponding surface. The proposed algorithms of stabilization and organization of purposeful motion of the robot on obstacles were tested by means of computer simulation in a program environment implementing interaction of a complete 3D dynamic model of the robot and objects surrounding it with due account of Coulomb friction forces and models of electromechanical drives at robot joints. It was shown that the proposed program motion ensuring active stabilization of the motion of the entire system can be implemented for the friction coefficient at support points not exceeding 1.

References

1. http://en.wikipedia.org/wiki/Hexapod_(robotics)
2. http://www.bostondynamics.com/robot_bigdog.html
3. Yu. F. Golubev and V. V. Koryanov, "Motion Design for an Insectomorphic Robot on Unstable Obstacles," in Proceedings of 11th International Conference CLAWAR-2008, Coimbra, Portugal, 2008 (World Scientific Publishing Co. Pte., 2008) pp. 654–661.
4. Yu.F. Golubev and V.V. Koryanov. Motion Control for an Insectomorphic Robot on a Movable Ball. Pleiades Publishing, Inc., Journal of Computer and System Sciences International. Vol. 48, No. 5, 2009. Pp. 801-814.
5. Yu.F. Golubev and V.V. Koryanov. Insectomorphic Robot Climbing a Freely Rolling Ball. Pleiades Publishing, Ltd., Journal of Computer and System Sciences International. Vol. 49, No. 6, 2010. Pp. 1009-1019.
6. http://www.umlab.ru
7. V.E.Pavlovsky, A.V.Panchenko. Research of dynamics of the small 6-legged robot in a program complex "Universal Mechanism". // Proc. of the International conference "An Extreme Robotics – 2011", on November 23-25, 2011, SPb, CNII RTK, CD, p. 122-125. (In Russian).

DEVELOPMENT OF HEXAPOD ROBOT SUPPORTED MECHANICALLY USING PNEUMATIC RUBBER ARTIFICIAL MUSCLES

H. TOMORI AND Y. HIRATA

Department of Precision Mechanics, Faculty of Science and Engineering, Chuo University, 1-13-27 Kasuga, Bunkyo-ku, Tokyo, 112-8551, Japan

T. NAKAMURA AND H. OSUMI

Chuo University, 1-13-27 Kasuga, Bunkyo-ku, Tokyo, 112-8551, Japan

Recently, the development of mobile robots is actively pursued. These robots are expected to have high mobility, allowing them to work in various scenarios such as rescue and exploration. Therefore, we focused on the hexapod robot for stable operation in an irregular terrain. We developed a leg using the parallel linkage mechanism and pneumatic rubber artificial muscles, and constructed a model of the legs. The model reproduced the experimental result. Then, we developed a hexapod robot which can support its weight mechanically using these legs, even without a power supply. Also, we demonstrated that our robot could walk. Furthermore, we confirmed experimentally that the robot supported loads that were about 4 times its weight.

1. Introduction

Recently, development of mobile robots is actively pursued. These robots are expected to have high mobility, allowing them to work in various scenarios such as rescue and exploration. However, these mobile robots are mainly wheel type or crawler track type robots [1][2]. It is difficult for these types of robots to move an irregular terrain.

Therefore, we focused on the hexapod robot for stable operation in an irregular terrain. As the existing research of hexapod robots, Arai developed the hexapod robot using motors [3]. And Nonami developed the hexapod robot using hydraulic actuators [4]. However, in these cases, the output power of the motor is low and the hydraulic actuator is heavy. In addition, neither actuator can support the robot without power supply.

To solve this problem, we focused on developing a pneumatic rubber artificial muscle. This actuator is light weight and has high output power when compared with motor and hydraulic actuators. In this study, we use a straight-fiber-type artificial muscle [5]-[7], which does not extend more than the natural length. Therefore, we developed a leg that can support the robot mechanically using this artificial muscle, even without power supply. We also developed a hexapod robot using this leg. Furthermore, we experimentally obtained the relationship between the load and displacement of the robot to investigate the weight capacity.

2. Straight-Fiber-Type Artificial Muscle

In this study, we use the straight-fiber-type artificial muscle shown in Figure 1, and the specification of the artificial muscle is shown in Table 1. It has a tubular shape and is made of natural rubber latex. The muscle structure includes a carbon fiber layer in the tube along the direction of its length. By this carbon fiber, this actuator is not extended more than natural length.

According to this structure, it contracts in the direction of its length and expands in the radial direction using air pressure. The contractile force is then used as an actuator. This artificial muscle is flexible, lightweight, and has high contractile force.

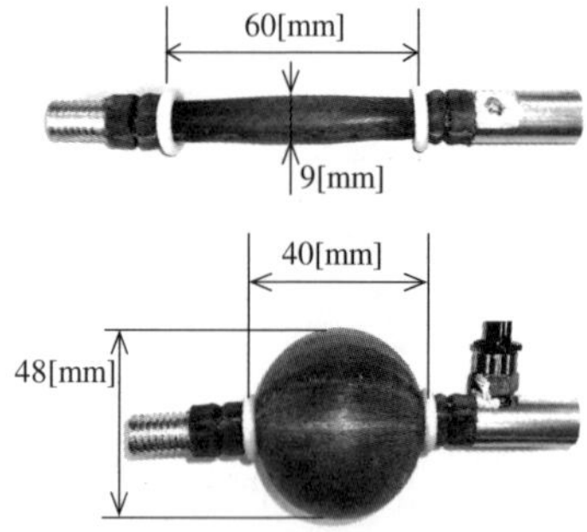

Figure 1. Schematic of fiber-double-layer-type artificial muscle.

Table 1. Specification of an artificial muscle.

Length l [mm]	60.0
Diameter d_0 [mm]	9.0
Thickness t [mm]	0.13
Elastic modulus K [MPa]	5.5

3. Structure of the Leg

The proposed structure of the leg is shown in Figure 2 and the motion of the leg is shown in Figure 3. This leg uses the parallel linkage mechanism, and the pneumatic rubber artificial muscles are arranged on the diagonal line of this parallel linkage mechanism. In Figure 2, the artificial muscle 1 is used to raise the leg, and artificial muscles 2 are used to lower the leg. With this mechanism, the leg can independently control the height and stiffness.

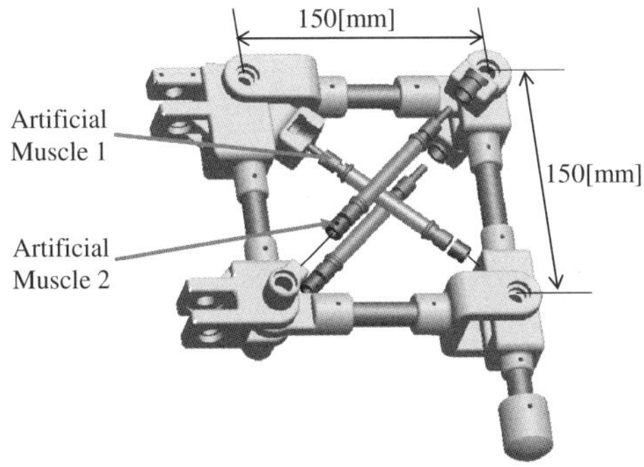

Figure 2. Proposed structure of the leg.

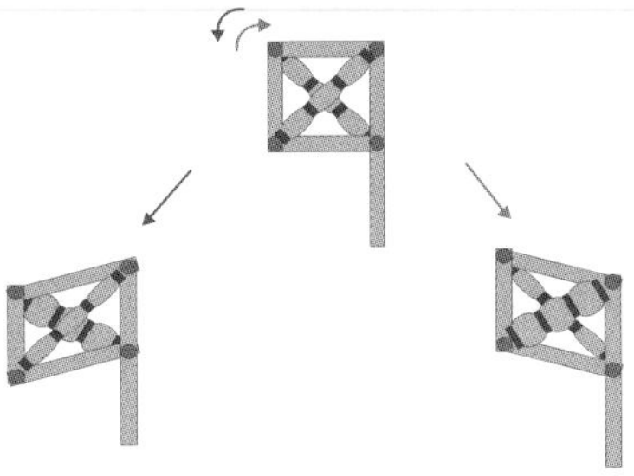

Figure 3. Driving of the leg.

4. Model of the Leg

In this chapter, we constructed a model of the leg, which includes the mechanical equilibrium model of the artificial muscle [5].

4.1. *Mechanical Equilibrium Model of the Artificial Muscle*

In this section, we describe the mechanical equilibrium model that treats the static characteristics of the artificial muscle. Since the artificial muscles are highly nonlinear in contraction, contractile force, and pressure, they are difficult to control. Therefore, we used a mechanical equilibrium model to control the artificial muscle. Through this mechanical equilibrium model, the contractile force of an artificial muscle can be calculated from the pressure P and amount of contraction x.

$$F(x,P)=\frac{PG_3(\phi_0)-G_1(\phi_0)}{G_2(\phi_0)} \tag{1}$$

$$G_1(\phi_0)=\frac{2Kt}{d_0}\left[\frac{l_0}{d_0}\right]^2\left[\frac{\sin\phi_0-\phi_0\cos\phi_0}{\phi_0^{\,2}}\right] \tag{2}$$

$$G_2(\phi_0)=\frac{M\tan\phi_0}{d_0nb} \tag{3}$$

$$G_3(\phi_0)=\left[\frac{l_0}{d_0}\right]^2\left[\frac{\phi_0-\sin\phi_0\cos\phi_0}{\phi_0^{\,2}}\right]$$
$$+2\frac{l_0}{d_0}\frac{\sin\phi_0}{\phi_0}-\frac{\pi d_0M}{4nb}\tan\phi_0 \tag{4}$$

$$\phi_0(x)=\frac{2\alpha l_0^{\,1.5}x^{0.5}}{(l_0-x)^2+\alpha^2l_0x}\quad(\alpha=1.1) \tag{5}$$

Where α is the constant that approximates the relation between the amount of contraction and the diameter, l_0 is the initial length, d_0 is the initial diameter, K is the elastic coefficient, t is the thickness of the rubber, M is the coefficient of the fiber of artificial muscle, n is the number of fibers, and b is the width of the fiber.

4.2. *Displacement of the Leg*

Simple figure of the leg is shown in Figure 4. From this figure, displacement of the leg X_c is expressed by Eq. (6).

$$X_c=b\sin\theta \tag{6}$$

$$\theta=\frac{\pi}{2}-\phi \tag{7}$$

$$\phi=\pi-\zeta \tag{8}$$

$$\cos\zeta=\frac{a^2+b^2-(d-x)^2}{2ab} \tag{9}$$

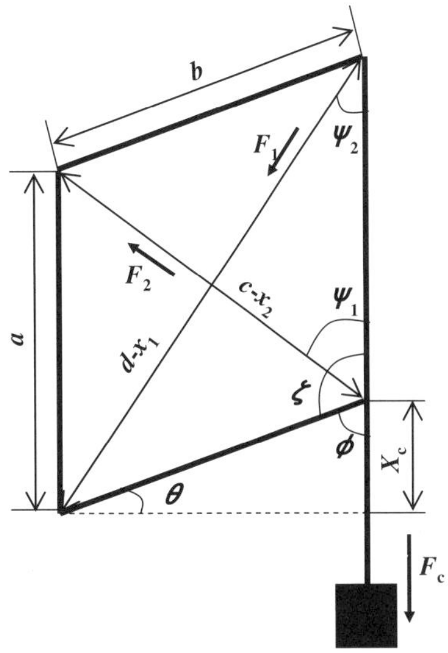

Figure 4. Simple figure of the leg.

4.3. *Force of the Leg*

From Figure 4, force of the leg F_c is expressed by Eq. (10).

$$F_c = 2F_1 \cos\psi_2 - F_2 \cos\psi_1 \tag{10}$$

$$\cos\psi_1 = \frac{\left(\frac{c}{2}\right)^2 + a^2 - \left(\frac{d-x}{2}\right)^2}{ac} \tag{11}$$

$$\cos\psi_2 = \frac{\left(\frac{d-x}{2}\right)^2 + a^2 - \left(\frac{c}{2}\right)^2}{a(d-x)} \tag{12}$$

5. Load Characteristic Experiment of the leg

In this chapter, we obtained relationship between displacement of the leg, force of the leg, and pressure of the artificial muscle. Figure 5 shows experimental results and theoretical lines using the model.

From these figures, experimental results reproduced theoretical lines. Thus the model of the leg can be used for control the leg.

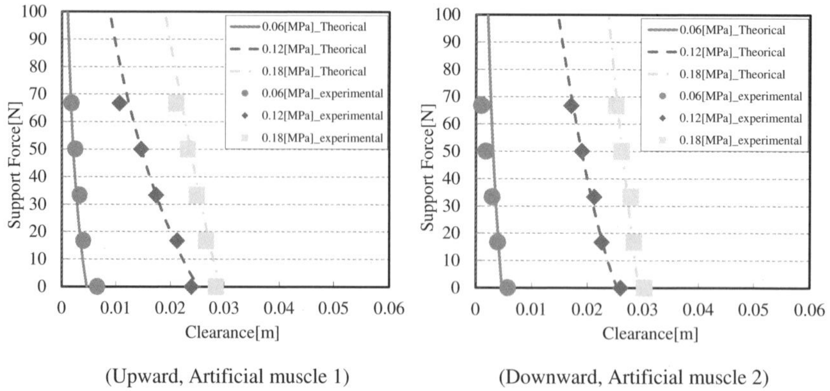

Figure 5. Load characteristic of the leg.

6. Development of the Hexapod Robot

We developed the hexapod robot using legs, as explained in Chapter 3. Figure 6 shows the robot and its simplified schematic. In Figure 6, the foreleg, hind leg, and middle leg of the other side are joined by a wire, and the wire is pulled by the pair of artificial muscles. Thus, the robot walks with an alternating tripod gait, as shown in Figure 7.

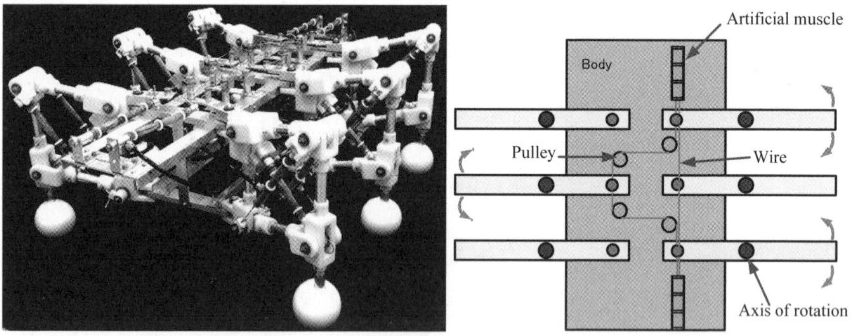

Figure 6. Hexapod robot using artificial muscles and a simplified schematic of the robot.

Table 2. Specification of the robot

Weight	73.5[N]
Width	570[mm]
Length	820[mm]
Height	280[mm]

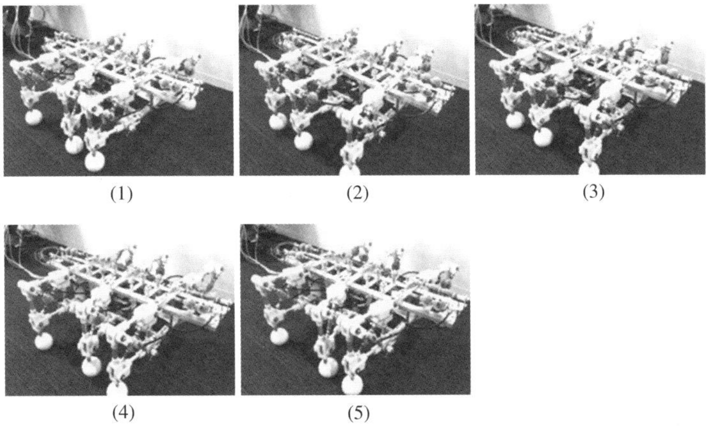

Figure 7. Motion of the hexapod robot.

7. Load Experiment of the Robot

To investigate the weight capacity, we experimentally obtained the relationship between the load and displacement of the robot. In this experiment, the robot supported its weight using 3 legs. A pressure of 0.05 [MPa] was applied to artificial muscle 1, and 0.12 [MPa] to artificial muscle 2. Also, other legs were raised by 25 [mm] as idling legs. Then, we measured the displacement of the body in the vertical direction as the loads increased.

Figure 8 shows the experimental result. From this figure, the robot was able to support the load of 300 [N] floating idling legs. Therefore, we confirmed that the robot supported a weight that is about 4 times its own weight.

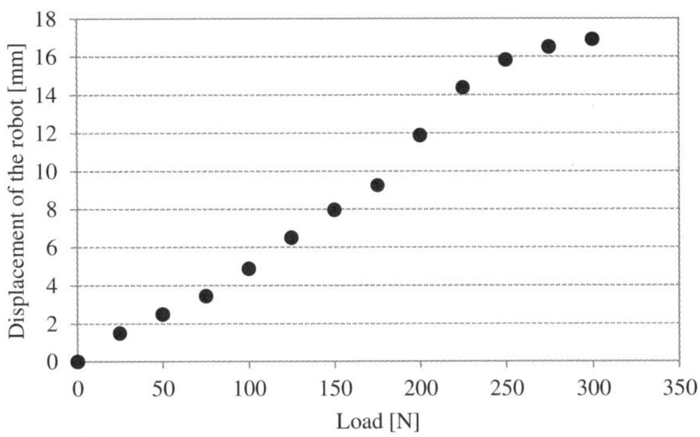

Figure 8. Relationship between load and displacement of the robot.

8. Conclusion

In this paper, we developed a leg using pneumatic rubber artificial muscles and constructed a model of a leg. The model reproduced the experimental result. Then, we developed a hexapod robot using these legs and realized walking of the robot. Furthermore, we confirmed experimentally that the robot supported about 4 times its own weight.

In the future, we plan to add an air compressor and controller to the robot.

References

1. N. Ushimi and K. Tsuruta, "High Speed Control for Two-Wheeled Mobile Robots," *JSME annual meeting*, Vol. 2007, No.7, pp. 275-276, 2007.
2. K. Hoshi, Y. Fukuoka, N. Minakata and Y. Imai, "Remote Control of a Crawler Robot Running Swiftly on Irregular Terrains using its Humanoid Upper Body," *The 27th Annual Conference of the Robotics Society of Japan*, Vol. 27, Page. ROMBUNNO. 3Q1-07, 2009.
3. T. Takubo, T. Yoshioka, T. Arai, Y. Mae and K. Ohara, "Leg-Wheel Hybrid Locomotion for Multi-Legged Robot," *Trans. Japan Soc. Mech. Eng., Series C*, Vol. 75 No. 759, pp. 2996-3004, 2009.
4. S. Imamura and K. Nonami, "Walking Control of Hydraulically Actuated Hexapod Robot with Kick Motion COMET-IV," *JSME Robotics and Mechatronics Conference 2010*, Vol. 2010, Page. ROMBUNNO. 1A1-B20, 2010.
5. T. Nakamura and H. Shinohara, "Position and Force Control Based on Mathematical Models of Pneumatic Artificial Muscles Reinforced by Straight Glass Fibers," Proceedings of IEEE International Conference on Robotics and Automation (ICRA 2007), pp. 4361-4366, 2007.
6. Nakamura, T., "Experimental Comparisons between McKibben type Artificial Muscles and Straight Fibers Type Artificial Muscles," SPIE International Conference on Smart Structures, Devices and Systems III, 2006.
7. Tomori, H. and Nakamura, T., "Theoretical Comparison of McKibben-Type Artificial Muscle and Novel Straight-Fiber-Type Artificial Muscle," International Journal of Automation Technology (IJAT), Vol. 5, No. 4, pp. 544-550, 2011.

LOCAL REFLEX GENERATION FOR OBSTACLE NEGOTIATION IN QUADRUPEDAL LOCOMOTION

MICHELE FOCCHI [°] [*], VICTOR BARASUOL[†], IOANNIS HAVOUTIS[*], JONAS BUCHLI[+], CLAUDIO SEMINI[*] and DARWIN G. CALDWELL[*]

[*]*Dept. of Advanced Robotics, Istituto Italiano di Tecnologia (IIT),*
[°]*E-mail: michele.focchi@iit.it, http://www.iit.it/en/advanced-robotics.html*
[†]*Federal University of Santa Catarina (UFSC),*
[+]*Agile & Dexterous Robotics Lab (ADRL), ETH Zurich*

Legged robots that dynamically locomote through rough terrain need to constantly handle unpredicted collisions (e.g. foot stumbling due to an obstacle) due to the unstructured nature of the environment. If these disturbances are strong enough they can cause errors in the robot's trunk that are difficult to control with a common feedback-based controller, imposing a serious risk to the overall system stability. The impulsive nature of such disturbances demands a very short reaction time, especially in case of dynamic gaits (trot, gallop, etc.). A quick reaction becomes increasingly crucial when the robot is deprived of reliable visual feedback (e.g. smoky areas or thick vegetation) or when an accurate map of the environment is not available. In this paper we propose a *local elevator reflex* which enables the robot to reactively overcome high obstacles. The reflex is implemented and experimentally evaluated on the hydraulic quadruped - *HyQ*. We demonstrate the feasibility and effectiveness of our approach showing that the robot is able to step over a platform of $11cm$ height (14% of the leg length) without prior knowledge of the terrain.

Keywords: Reflex, Quadrupedal locomotion, Control.

1. Introduction

Natural terrain is often not level, may be slippery and provide sparse footholds. One way to approach locomotion is to perform footstep planning, assuming that a reasonable map of the environment is available. However, this approach is valid for slow locomotion and has limited applicability under rapidly changing conditions. In the latter case fast reactions are mandatory for maintaining stability. Biological studies have shown that animals use local reflexes to cope with terrain irregularities.[1] This way, animals initially retract the foot and then try swinging over the obstacle. This is particularly useful in situations where the leg is obstructed by an unexpected or unperceived elevation of the terrain that causes a frontal impact, making the robot stumble.

Early implementations of reflexes in robotics date back to the 80's with Bekey's work,[2] where reflex activity consisted of relating joint motion to sensory patterns following certain rules. Park et al.[3] proposed a reflex method for a legged robot to deal with slippery surfaces. More recently a reflex behaviour has been effectively implemented for the bipedal robot *Mabel.* Mabel has been shown to step up on platforms that are as high as 12% of its leg length.[4]

In this paper we propose the implementation of a reflex strategy called the *local elevator reflex*, for the hydraulic quadruped robot HyQ[5] (Fig. 1(right)). This reflex is implemented as a kinematic modification of the reference trajectory of the foot of the swing leg. An advantage of our reflex approach is that it depends on a small set of parameters that have physical meaning and can be directly linked to the locomotion parameters. This way the robot is capable of regulating the *strength* of the reflex to the locomotion situation at hand.

2. Reflex implementation

HyQ's trotting involves periodic motion generation (ellipses) for the feet together with a controller for body posture stabilization. The feet trajectories are generated by non-linear oscillators (CPGs), as described in.[6] The following section describes the generation of the reflex trajectory $\mathbf{P}_{rfx}$ for one foot, that is added as an offset to the reference $\mathbf{P}_{ref}$ generated by the CPG. $\mathbf{P}_{rfx}$ is a function of intuitive physical parameters: the reflex duration t_{rfx}, the maximum retraction r_{max}, the desired angle of retraction θ_0 and the duration t_{ret} of the retraction impulse. These parameters can be easily expressed as a function of the gait parameters. In particular the reflex duration can be varied according to the swing phase duration while the maximum retraction can be linked to the step height (Fig. 2(left)).

The reflex trajectory is generated in a coordinate frame, the *foot frame*, that is aligned with the trunk frame (Fig. 1(left)). In this derivation we are generating trajectories that lie in the $X-Z$ plane of this frame, (where x is aligned with the direction of motion of the robot) but this can be easily extended to any orientation. The reflex can only be triggered when the leg is in the air (swing phase), when the CPG phase angle $\alpha > 90°$ (Fig. 2(left)). This is because for $\alpha < 90°$ the reflex trajectory is cancelled out by the CPG trajectory, that is at that time descending. The CPG phase angle α represents the angle between the desired foot position vector and the x axis in the $X-Z$ plane of the foot frame. The swing phase starts at the lift-off and finishes at the touchdown event. The touchdown is detected

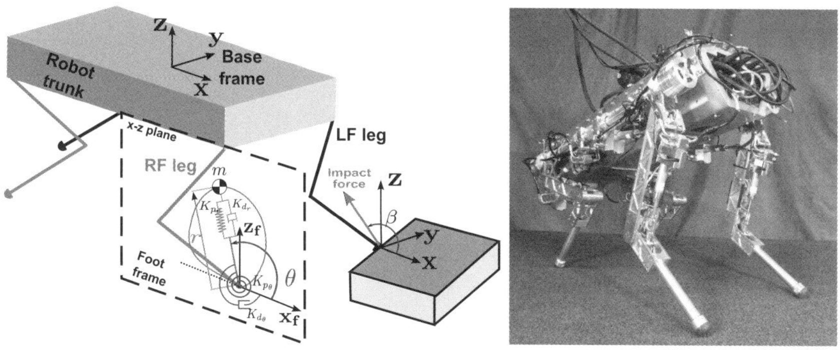

Fig. 1. (**Left**) The reflex trajectory, if restricted to the $X - Z$ plane, can be described with radial r and angular θ coordinates. The angle β of the impact force (see left-front leg) is expressed in the $X - Z$ plane. (**Right**) Picture of the HyQ robot.

by sensing the *ground reaction forces* (GRFs). The GRFs are estimated using the torque sensors at the joints and the Jacobian transpose of the leg kinematics. In the HyQ motion generation framework there is an adaptation mechanism which "cuts" the reference ellipses when a ground impact is detected (touchdown). This adaptation mechanism is very important to enhance robot stability when walking on rough terrain.[6] The touchdown event determines the beginning of the stance phase in which the periodic motion becomes a motion along a line parallel to the ground. This results in the leg pushing backward in the horizontal direction and thereby propelling the robot forward.

Fig. 2 (left) shows three different touchdown events and the corresponding modification of the elliptic trajectories. It also shows the areas where the reflex can be activated (ascendant semi-ellipse) and the touchdown detected (descendant semi-ellipse). These two areas should not overlap to prevent undesired activation of the reflex during stance phase. Once an impact is detected at the end-effector, during the reflex-active phase, the vector of the impact force measured at the end-effector is projected into the $X - Z$ plane to evaluate if this is a frontal impact. Then the reflex is triggered if the angle β about the x-axis (Fig. 1(right)) is such that $\beta > 120°$ or $\beta < -150°$. This range was experimentally evaluated and can be tuned according to the level of responsiveness that is desired from the robot.

After the reflex is triggered, a fast retracting motion for the foot is generated. This enables the foot to move away from the obstacle, prevents foot *trapping* and results into higher step clearance. The foot is then driven in a circular trajectory overcoming the obstacle (*swing over* motion). Since

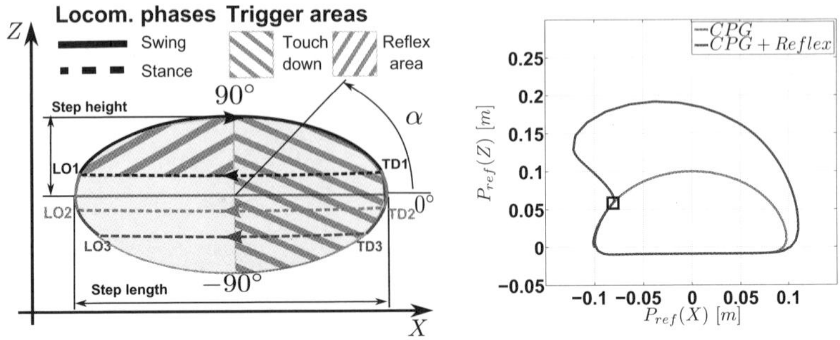

Fig. 2. (**Left**) Areas of the CPG cycle where a possible impact can trigger the limb reflex or the touchdown. α is the phase angle of the CPG trajectory. TD_i are different touch-down events that result in different stance trajectories. LO_i are the correspondent lift-off events. (**Right**) Periodic foot (clockwise) trajectory (red), and the same trajectory modified by the reflex action (blue) for overcoming a frontal obstacle. The black square indicates the collision point.

we constrain the trajectory to lie on a plane, we can implement this behavior by looking at the dynamics of a controlled unitary mass expressed in polar coordinates:

$$\ddot{\theta} + K_{d_\theta}\dot{\theta} + K_{p_\theta}\theta = 0, \qquad \ddot{r} + K_{d_r}\dot{r} + K_{p_r} = F_r(t) \tag{1}$$

where r is the radius and θ is the angle about the X axis as depicted in Fig. 1 (left). Linear and rotational virtual springs and dampers $K_{p_r}, K_{d_r}, K_{p_\theta}, K_{d_\theta}$ attract the state variables back to the origin. Since the equations in (1) are decoupled, it is possible to define different dynamics for the radial and angular motions, respectively.

Since we initially want to have a retraction, we set an external input $F_r(t)$ only in the radial equation. This retraction force is applied only for a limited time t_{ret} (square pulse).

$$F_r(t) = \begin{cases} 0 & t < 0 \\ |Fr| & 0 > t > t_{ret} \end{cases} \tag{2}$$

After the time t_{ret} has elapsed, the trajectory will only move towards the origin under the effect of the virtual elements. The acceleration is extracted from (1) and by double integration the reflex trajectory, $(\theta(t), r(t))$, can be obtained in polar coordinates. Before starting the integration, it is important to set the initial conditions $\theta(0) = \theta_0, r(0) = 0$ for the state variables to have the force F_r injected in the desired direction. After transforming to Cartesian coordinates ($x = rcos(\theta), y = 0, z = rsin(\theta)$) the reflex trajectory $\mathbf{P}_{rfx}$ is obtained.

Fig. 2 (right) shows the effect of the addition of the reflex to the CPG trajectory. The shape of the reflex trajectory can be tuned by choosing the settling time for the angular t_{rfx_θ} and radial t_{rfx_r} dynamics. In particular these dynamics are described by the eigenvalues λ_i of the associated homogeneous linear systems.

$$\lambda^2 + K_d\lambda + K_p = 0, \qquad \lambda_{1,2} = \frac{-K_d \pm \sqrt{K_d^2 - 4K_p}}{2} \tag{3}$$

For the sake of simplicity, we decided to set $t_{rfx_r} = t_{rfx_\theta}$ by choosing the same gains K_d, K_p for θ and r. K_d is chosen to place the eigenvalues at $\lambda_i = 4.6/t_{rfx}$[7] in order to obtain the desired settling time t_{rfx}, while K_p is chosen to have λ_i as real coincident roots (critical damping):

$$K_d = 9.2/t_{rfx}, \quad K_p = K_d^2/4. \tag{4}$$

An inverse dynamics algorithm is used to compute the torque necessary to perform the reference trajectory alleviating the work of the feedback joint position controller and achieving higher control bandwidth.[8] As a matter of fact a high control bandwidth is mandatory to track the kinematic modification of the reflex because its duration will be constrained in a fraction of the time interval in which the leg is in swing phase.

Finally the value of $|Fr|$ can be set to achieve a desired maximum retraction r_{max}. To express r_{max} we investigate the analytical solution of the radial velocity $\dot{r}(t)$ finding the time instant t_{max} in which it crosses zero. We must underline that the instant where maximum retraction occurs, t_{max}, is not the instant when the pulse ends ($t = t_{ret}$) because at t_{ret} the unity mass has a non-zero instantaneous velocity that must be decelerated. The time response of the velocity to a square pulse input for a critically damped 2nd order system, is:[7]

$$\dot{r}(t) = \frac{|F_r|}{K_p}(t\lambda^2 e^{t\lambda} - \mathbf{1}_{t>t_{ret}}(t)(t - t_{ret})\lambda^2 e^{-(t-t_{ret})\lambda}) \tag{5}$$

where $\mathbf{1}(t)$ is an indicator function. After some manipulation the time instant where $\dot{r}(t_{max} = 0)$ can be found:

$$t_{max} = \frac{-t_{ret}\lambda^2 e^{t_{ret}\lambda}}{\lambda^2(1 - e^{t_{ret}\lambda})} \tag{6}$$

by putting (6) in the time response of the radius $r(t)$ we obtain the relationship:

$$r_{max} = r(t_{max}) = \tfrac{|F_r|}{K_p}\underbrace{(-(1 + t_{max}\lambda)e^{-t_{max}\lambda} + (t_{max} - t_{ret})\lambda e^{-(t_{max}-t_{ret})\lambda})}_{A} \tag{7}$$

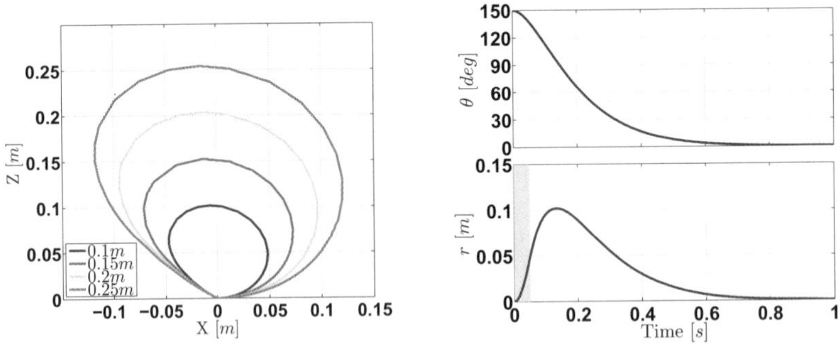

Fig. 3. Simulation. (**Left**) Reflex trajectory (clockwise) in the $X - Z$ plane for different maximum retraction values. (**Right**) Time response of the polar coordinates ($t_{rfx} = 0.5s$, $r_{max} = 0.1m$ and $\theta_0 = 150°$). The shaded area highlights the interval during which the retraction impulse is acting.

Then the value of $|Fr|$ can be linked to the desired maximum retraction:

$$|\mathbf{F}_r| = \frac{r_{max} K_p}{A}. \tag{8}$$

Figure 3 (left) shows an $X - Z$ plot of the reflex trajectory for different retraction parameters, i.e. varying r_{max}. After the initial fast retraction it can be seen that the *swing over motion* is due to the dynamics of θ that, starting from θ_0, is moving towards the origin. Fig. 3 (right) shows the time response of the polar coordinates for a settling time $t_{rfx} = 0.5s$, $r_{max} = 0.1m$, and $\theta_0 = 150°$. The figure shows r and θ have the same settling time.

3. Experiments

In this section we present some experimental results that demonstrate the effectiveness of the proposed reflex algorithm, when the robot is trotting and stepping up on a $11cm$ high platform. The forward speed has been set to $0.1m/s$ while the step height has been set to $9cm$ making the reflex action necessary to overcome the step. The step length was set to $5cm$ and throughout our test we have used a cycle period frequency of $1.7Hz$. The max retraction r_{max} was set to $13cm$, and the desired angle of retraction θ_0 to $135°$. The value of the settling time t_{rfx} was computed in real-time depending on which was the value of α at the moment of the reflex trigger. In particular, smaller t_{rfx} (faster reflex trajectories) were obtained if the trigger occurred close to the apex $\alpha = 90°$. This is important for assuring the synchronization of the foot with the CPG.

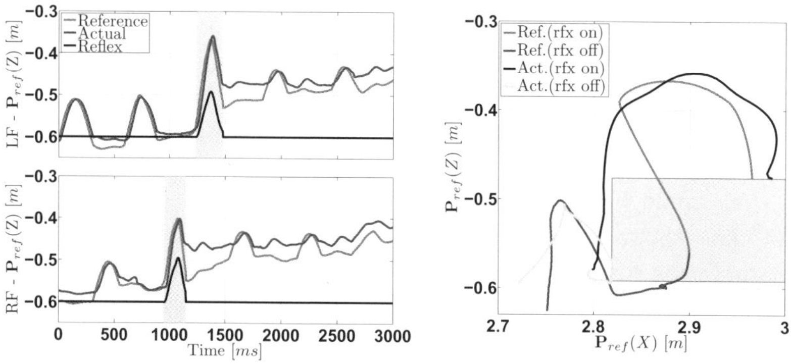

Fig. 4. Experimental results. (**Left**) Step up. Vertical (Z) component of the foot position for the left-front (upper plot) and right front (lower plot) leg. Red plots are the reference trajectories while blue plots the actual ones. The reflex component of the reference trajectory (black) is offset of -0.6 m for display purposes only. The grey area shows when the reflex is active. (**Right**) Foot trajectory in the $X - Z$ plane in world coordinates when the reflex is triggered for the left-front leg.

Fig. 4 (left) presents the vertical components (Z) of the reference and actual foot trajectory during the step up when the robot is trotting towards the platform. The reflex is effectively triggered after the frontal impact of the feet with the platform, initially for the right-front (RF) leg and $250ms$ later for the left-front (LF) leg. Then when each foot lands on the platform a new touchdown condition is detected. The shadow area shows when the reflex is active. The foot trajectories are mapped into an inertial frame fixed to the ground at the point where the robot base is when the trot starts. This helps in estimating the real obstacle height. The small oscillations in the desired trajectory are artifacts due to the estimation errors in the robot odometry. Fig. 4 (right) is a plot in the $X - Z$ plane that shows how the reflex modifies the trajectory of the left foot after the frontal impact with the platform. The trajectory is mapped into an inertial frame fixed with the ground. The figure shows that after the first impact the reflex was not triggered as the reference trajectory had already overpassed the apex ($\alpha = 90°$). However the reflex is successfully triggered in the following CPG cycle and the leg successfully overcomes the obstacle. This demonstrates the robustness of the proposed approach in the worst situation in which the foot hits the obstacle when it is almost at the end of the cycle ($\alpha < 90°$). The experiment has been successfully repeated on a set of 10 trials, 5 of them with an obstacle height of 8 cm and a CPG step height of 7 cm, while the other 5 with an obstacle height of 11 cm and a CPG step height of 9 cm.

4. Conclusions

In this paper we proposed the implementation of a *local elevator reflex*, intended to improve robot locomotion robustness when hitting high obstacles, that would otherwise cause the robot to stumble. The underlying idea is that the leg should "*give in*" instead of trying to command a trajectory that would be infeasible. As a matter of fact, if the kinematic plan is not changed, tracking the original trajectory would cause the injection of destabilizing forces. The generation of the trajectory can be linked to gait parameters that enable us to adapt the reflex shape and duration to different locomotion situations. From Fig. 4 (left) it can be seen that the cutting of the ellipses causes a reduction of the available step height when the touchdown occurs on top of the platform, this in turn reduces the step length and so the locomotion speed.

Therefore future work considers the development of a more sophisticated adaptation mechanism, that has the purpose to move the origin of the ellipse in order to adapt to the new terrain elevation and recover the locomotion capabilities.

Acknowledgments: Jonas Buchli is supported by a Swiss National Science Foundation professorship. The authors would like to thank CAPES for the scholarship granted to V. Barasuol (Grant Procs. 6463-11-8).

References

1. K. Espenschied, R. Quinn, R. Beer and H. Chiel, Biologically based distributed control and local reflexes improve rough terrain locomotion in a hexapod robot *Robotics and autonomous systems* **18** (Elsevier, 1996).
2. G. Bekey and R. Tomovic, Robot control by reflex actions, in *Proc. IEEE Int. Conf. Robotics and Automation*, 1986.
3. J. H. Park and O. Kwon, Reflex control of biped robot locomotion on a slippery surface, in *Proc. ICRA Robotics and Automation IEEE Int. Conf*, 2001.
4. H.-W. Park, K. Sreenath, A. Ramezani and J. W. Grizzle, Switching control design for accommodating large step-down disturbances in bipedal robot walking, in *Proc. IEEE Int Robotics and Automation (ICRA) Conf*, 2012.
5. C. Semini, HyQ – design and development of a hydraulically actuated quadruped robot, PhD thesis, Istituto Italiano di Tecnologia (IIT)2010.
6. V. Barasuol, J. Buchli, C. Semini, M. Frigerio, E. R. De Pieri and D. G. Caldwell, A reactive controller framework for quadrupedal locomotion on challenging terrain, in *2013 IEEE International Conference on Robotics and Automation (ICRA)*, may 2013.
7. G. Franklin, *Feedback Control of Dynamic Systems*, 3rd edn. (Addison-Wesley Longman Publishing, Boston, 1993).
8. J. Buchli, M. Kalakrishnan, M. Mistry, P. Pastor and S. Schaal, Compliant quadruped locomotion over rough terrain, in *Proceedings of IEEE/RSJ International Conference on Intelligent Robots and Systems (IROS)*, 2009.

EVENT DRIVEN GROUND-IMPEDANCE IDENTIFICATION FOR LEGGED ROBOTS

JUAN CARLOS AREVALO, MANUEL CESTARI, DANIEL SANZ-MERODIO and ELENA GARCIA

Centre for Automation and Robotics, UPM CSIC
Carretera Campo Real km.0200, Arganda del Rey, Spain 28500,
E-mail: juan.arevalo@car.upm-csic.es

Legged robots, in order to increase their adaptability to complex terrain, need to have information about the contact properties of the terrain they need to negotiate. This information, in principle, is not available a priori. In such case, it is necessary to extract the contact properties on line. This work is motivated by this need. To solve this problem we propose to use a fist order identification method to this task. To do this, taking advantage of a leg prototype structure, we develop an event driven system identification approach to extract the contact properties of the ground. A series of experiments conducted on two different types of terrain indicate good convergence and low error on the identification. In both types of ground, the identification is performed within a single step.

Keywords: Legged robots, Impedance Identification, Contact Modeling

1. Introduction

Legged robots are a good choice for missions over complex terrain such as disaster scenarios, extraterrestrial exploration, etc., because of their superior adaptability to these terrains. Over the years, several legged robots have been built. However, very few are capable of the agile locomotion required for the aforementioned missions.[1] Moreover, whenever a robot is in contact with its environment, the controlled robot's dynamic behavior should be chosen according to the environment which it will be in contact with.[2] Thus, improving the adaptability to rough terrain requires the use of control algorithms that account for contact forces, or to know beforehand the properties of the terrain in order to be able to predict them.

In principle, there is no reason why this information should be available beforehand. Nonetheless, extracting the properties directly from the terrain is a challenge because, in general, to identify an unknown system it is necessary to excite all the frequency lines of interest.[3] In the case of walking robots this is not possible to achieve, because the situation calls for specially designed signals (typically a chirp signal or a sine wave), which are not possible to implement in a walking robot without falling. Nevertheless, if one assumes a certain contact model from start it is possible to use a real time method of identification like the recursive least sqares (RLS) or the windowed least squares method (WLS) to the task.

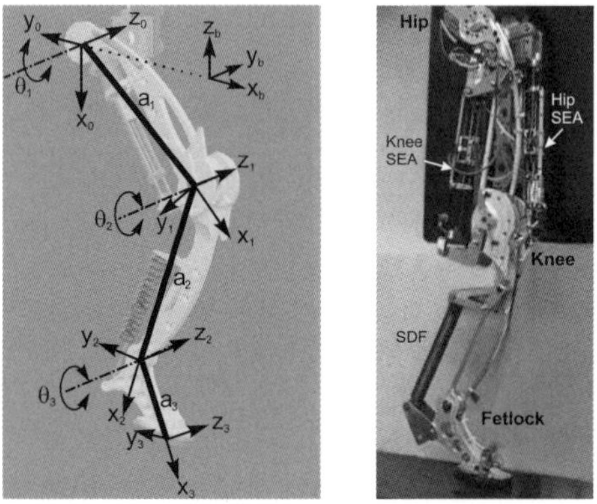

Fig. 1. The HADE Leg Prototype. Kinematics (left). Real prototype (right)

In this work we propose a method based on the RLS algorithm to identify the parameters of the ground contact model. To develop this method, we take advantage of the structural properties of the HADE leg prototype (see Fig. 1). This prototype is a planar 3 DOF leg directly driven by two Series Elastic Actuators (SEA);[4] one at the hip and other at the knee, while the fetlock joint is passively driven by a spring resembling the superficial digital flexor in horses.[1,5]

This work is organized as follows: Section 2 provides a summary of the identification method and the contact model used; Section 3 presents the identification algorithm; Section 4 presents the experimental setup and results; and finally, in Section 5 the conclusions are stated.

2. Impedance Identification Method and Contact Model

To identify the impedance of the ground it is first necessary to choose a contact model, one that is particularly easy to implement and that gives a good estimate of the energy loss and the impact and contact force is the linear spring-dashpot model[2,6] which equation is the following:

$$F_c = k\delta + b\dot{\delta} \quad (1)$$

where F_c is the contact force, δ and $\dot{\delta}$ are the penetration in the ground and its derivative, while k and b are the stiffness and damping coefficients of the ground.

The next step is to choose an identification method, which in our case is the RLS method[7,8] because it offers good tracking when used in similar applications.[9] This method considers the system to be an autoregressive moving average (ARMA) process. To apply this method, it is necessary to convert equation (1) into an autoregressive form, i. e. to express it in the following form:

$$y_t = \varphi^T \theta \quad (2)$$

where $\varphi^T = [-y_{t-1}, -y_{t-2}, ..., -y_{t-n}, u_t, ...u_{t-m}]$ and $\theta = [a_1, a_2, ..., a_n, b_0, .., b_m]^T$. Where u_{t-i} and y_{t-i} are the input and the output of the system at instant $t-i$ and $t-j$ respectively; while b_i and a_j are the parameters of the model, with $i = 1, 2, ..., m$ and $j = 0, 1, 2, ..., n$.

To represent the contact model as an ARMA process, one alternative is to choose the contact force F_c as the input of the model and the penetration δ as the output. This yields:

$$\delta = \varphi^T \theta \quad (3)$$

where, $\varphi^T = [-\delta, F]$ and $\theta^T = [b/k, 1/k]$. Once the contact model is expressed as an auto regressive process the identification can be performed.

3. Identification Algorithm

The state diagram of the identification algorithm proposed in this work is depicted in Fig. 2 and it consists of the following states:

(1) **Initialization.** This state computes the initial position of the robotic leg at the starting configuration: resting on the ground with zero velocity. The transition from this state to the following is done when the leg begins the stance phase.

(2) **Identify ground impedance.** In this state, the identification is performed by applying the RLS algorithm to the previously described ARMA model (see equation (3)). In the identification process, the ground penetration (δ) is used as the output and is computed as the difference between the foot's position (L) and the compression of the spring, while

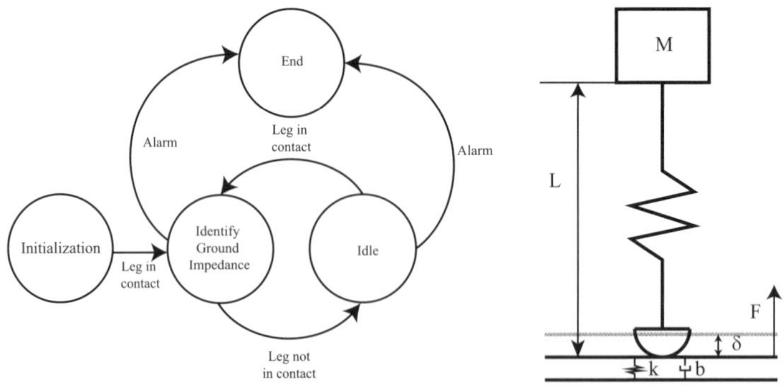

Fig. 2. State diagram of the identification algorithm (left) and parameters used by the identification algorithm (right)

the exerted force (F is used as input (see Fig. 2). The identification occurs from the moment when the leg comes in contact with the ground to the beginning of the next swing phase, which leads to the Idle state. Another possibility is to go directly to the final state (End) if an alarm is sent to the program (e.g. a command to terminate the experiment).

(3) **Idle.** In this state, the program merely waits until an alarm is set or when the leg comes in contact with the ground one more time. In the former case the next state is the final state and in the latter case the next state is the identification state.

(5) **End.** This is the final state of the algorithm, in this state the obtained data is saved to a file and all the shutdown procedure is performed in order to safely terminate the experiment.

4. Experimental Setup and Results

The experimental setup consists of: The HADE leg (see Fig. 1); an adjustable impedance box,[9] designed to behave as close as possible to the linear spring dashpot model (see equation (1)); and a box containing river sand (see Fig. 3). The experiments consist on commanding the leg to move vertically from an initial position (resting on the contact surface) in such manner that it moves upward until it loses contact with the ground. Next the leg is commanded to remain with no contact with the ground for 0.5 s and then move back to its initial position and stay in that configuration for another 0.5 s. This sequence of movements is repeated twice on each experiment.

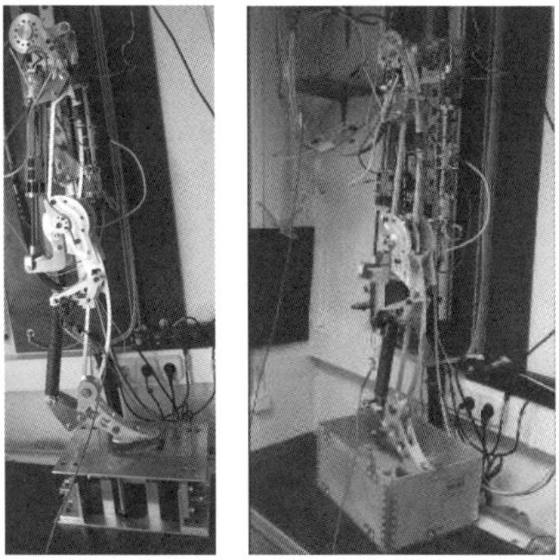

Fig. 3. Experimental setup. Leg stepping over the adjustable-impedance box (left) and the sand (right)

Figure 4 shows the mean value of the force measurements taken over 10 repetitions of the experiments for each material (adjustable impedance box (left) and river sand (right)) against time. In this figure it can be seen that the heel strike can be detected by monitoring the ankle's force measurement. Moreover, it can be observed that the response obtained for the adjustable impedance box is smoother than the response obtained with the river sand. This is explained because the river sand has a nonlinear behavior. On the other hand, as the river sand is softer than the box, the leg sinks, thus the resulting force is less than the force measured with the box.

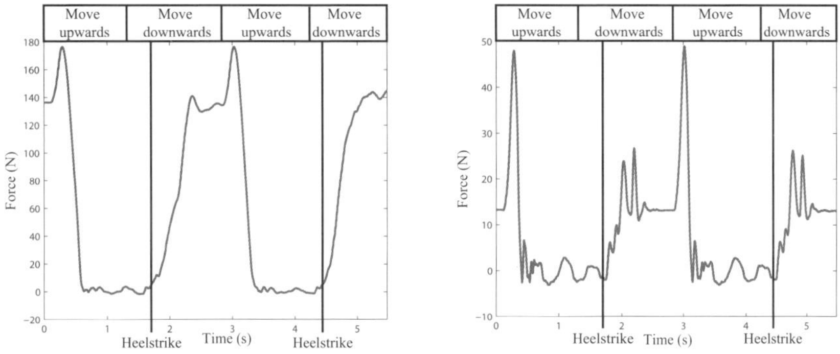

Fig. 4. Ankle force vs. time using the adjustable-impedance box (left) and the sand (right)

4.1. *Application of the identification algorithm*

To apply the proposed identification algorithm it is necessary to compute the ground penetration. The ground penetration is computed by subtracting the ankle vertical position Y_{ankle} to the foot's position Y_{foot} and this difference is what we used as the ground penetration δ (see Fig. 5).

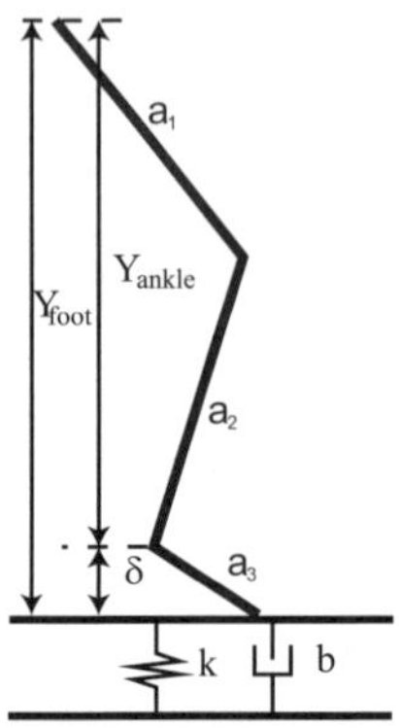

Fig. 5. Parameters used by the identification algorithm on the HADE leg

One important thing to note is that, as it was stated previously, the leg presents different amount of sinking (given different terrains). This is a limitation in this work, in other words, if the leg has different sinking it is necessary to have an additional source of information in order to be able to take it into account when computing the ground penetration. Because if the hip has a different vertical position at each step then the approximation of δ being only the ankle displacement is no longer valid, i.e. the vertical displacement of the hip needs to be included in the ground penetration.

4.2. *Identification results*

In this section we perform an offline evaluation of the identification algorithm to see if it can be later applied online. The results of applying the identification algorithm to the data obtained with the leg are presented in Fig. 6. The figure shows the value of the parameters θ_i for the adjustable impedance box (top-left) and the river sand (top-right); and the prediction error for the box (bottom-left) and the river sand (bottom-right).

The results shown in Fig. 6 are the average value of the 10 repetitions of the experiments for each material, and they were obtained only using the algorithm over the first step. As it can be seen in the figures, the parameter

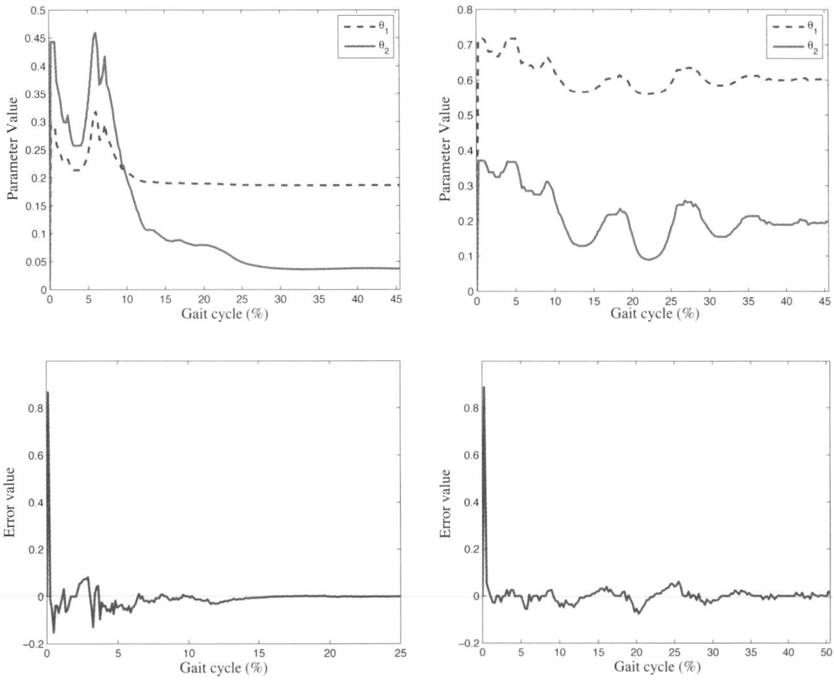

Fig. 6. Estimated parameters obtained with the box (top-left) and the sand (top-right), and estimated error obtained for the box (bottom-left) and the sand (bottom-right) against gait cycle percentage.

estimation for the box converges at around 25 % of the gait cycle while the sand converges at approximately 40 % of the gait cycle. Therefore, as a running gait is defined as a gait where the duty factor is less than 50% the algorithm can be used to estimate the ground impedance while using a running gait.

Another interesting thing to note is that in the case of the box, the convergence is faster. This fact is expected because its behavior is approximately linear. The algorithm is still able to converge when applied to a nonlinear material, as is the case of the river sand, in only one step.

5. Conclusions

In order to increase the adaptability of walking robots it is necessary to have information about the contact properties of the terrain. However, there is no reason why this information should be available a priori.

Motivated by this fact, here we proposed an event driven system identification algorithm to identify the impedance of the ground. The method is based on the RLS algorithm and it is developed taking advantage of the mechanical structure of the robotic leg.

A series of experiments have been performed on the HADE leg prototype to identify the parameters of two different terrains: an adjustable impedance box and a box containing river sand. The results show that in both cases, the algorithm converges and exhibits a low error. Although in the case of the box the convergence is faster, in both cases the identification is performed within a single step, in 25% of the gait cycle and 40% of the gait cycle respectively.

Acknowledgements

This work has been partially funded by the Spanish National Plan for Research, Development and Innovation through grant DPI2010-18702.

Mr. Juan Carlos Arevalo would like to thank the Spanish National Research Council, and Mr. Manuel Cestari would like to thank the Spanish Ministry of Economy and Competitiveness for funding their PhD research.

References

1. E. Garcia, J. C. Arevalo, G. Muoz and P. Gonzalez-de Santos, On the biomimetic design of agile-robot legs, *Sensors* **11** (2011).
2. J. C. Arevalo and E. Garcia, Impedance control for legged robots: An insight into the concepts involved, *Systems, Man, and Cybernetics, Part C: Applications and Reviews, IEEE Transactions on* **42**, 1400 (nov. 2012).
3. G. Simon and J. Schoukens, Robust broadband periodic excitation design, *Instrumentation and Measurement, IEEE Transactions on* **49**, 270 (2000).
4. J. Pratt, B. Krupp and C. Morse, Series elastic actuators for high fidelity force control, *Industrial Robot: An International Journal* **29**, p. 234241 (2002).
5. J. C. Arevalo, M. Cestari, D. Sanz-Merodio and E. Garcia, Impedance control for a bioinspired underactuated leg, in *15th International Conference for Climbing and Walking Robots, CLAWAR'12*, 2012. Baltmore.
6. G. Giraldi and I. Dharf, Literature survey of contact dynamics modelling, *Mechanism and Machine Theory* , 1213 (2002).
7. L. Ljung, System identification: theory for the user, *Preniice Hall Inf and System Sciencess Series, New Jersey* **7632** (1987).
8. F. Gustafsson and F. Gustafsson, *Adaptive filtering and change detection* (Wiley Online Library, 2000).
9. J. C. Arevalo, X. Carrillo, M. Cestari, D. Sanz-Merodio and E. Garcia, System identification applied to contact modeling: An experimental investigation, in *Robotics and Automation, 2013. Proceedings. ICRA'13. IEEE International Conference on*, 2013. (Accepted).

THE TURTLE, A LEGGED SUBMERGED INSPECTION VEHICLE

JOHN BILLINGSLEY

University of Southern Queensland, Toowoomba, Queensland 4350, Australia.

A collaboration between the Petroleum Institute of Abu Dhabi and the University of Southern Queensland is leading to the construction of an underwater inspection vehicle, the Turtle. The design has profited from many earlier projects and is in the form of a six-legged machine with paddles on the legs. This will enable it to 'swim' to a location and alight onto a stable surface for steady monitoring of vision. Although funds have been made available for its construction, there is still debate over actuation details. Alternatives are presented.

1. Introduction

The Turtle will be constructed under the auspices of the Petroleum Institute, Abu Dhabi, for the purpose of inspecting underwater pipelines. It has been decided that propulsion should be by means of leg-mounted paddles, rather than the more conventional propellers. The legs have a secondary purpose of allowing the robot to alight and manoeuvre on the sea-bed for steady long-term surveillance.

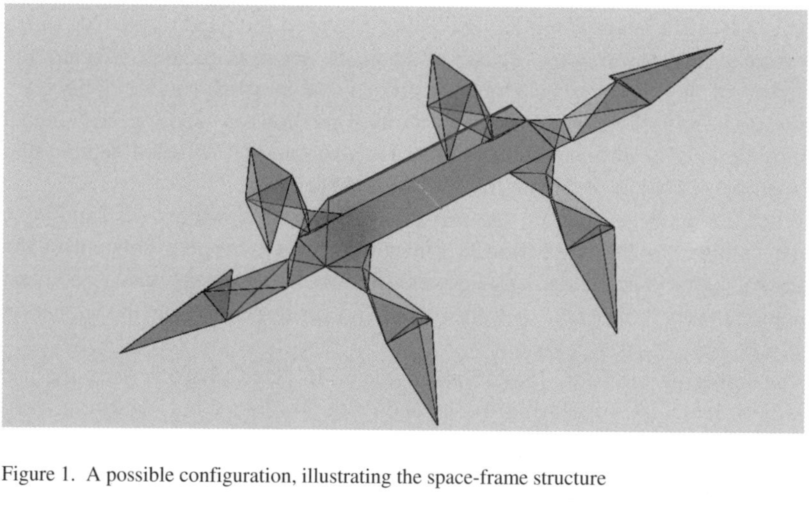

Figure 1. A possible configuration, illustrating the space-frame structure

Decisions have yet to be made on the means of actuation, to achieve a compromise between strength and endurance. Several earlier robots from the work of the collaboration have been actuated by means of pneumatic cylinders. These are of debatable merit in a vehicle such as the Turtle, but should not be summarily dismissed.

The 'space frame' concept of those earlier machines centred on the use of tetrahedral structures made up of light rods, rather than using linear limbs that had sufficient rigidity to resist bending. In the tetrahedral structure, adjoining elements pivot about two common vertices while an actuator of variable length is connected between a third vertex of each.

This principle has been carried forward into the new design, taking with it the requirement for a variable-length actuator.

2. Background and History

Early robots constructed by the Portsmouth Walking Robot Group had legs constructed from simple tubular rods, to reduce the robot mass. This led to some frailty if a side load was applied [1]. When Dr Cubero continued research on walking robots at the University of Southern Queensland, it was decided that a ‘space frame’ construction could give a considerable improvement in strength with a modest increase in mass. The resulting machine was named the STIC [2].

Each member consisted of a tetrahedron of rods, with adjoining members hinged about their common rod. Joints were articulated by a pneumatic cylinder that linked a third vertex of each member. Structurally the machine was very robust, but had shortcomings in performance that resulted from the limited force of the cylinders and the compliance of the air they contained [3].

Professor Billingsley had previous contact with a marine conservation project in Hoi Ha Wan, Hong Kong. This concerned the need to provide video surveillance of sea urchins, up to a depth of seventy metres. Again the provision of legs was considered to give a stable platform for long term observation. After considerable deliberation of methods to ‘depth proof’ such a system, the local team instead decided to use a submerged wheeled vehicle that could propel itself out of the depths and up the beach [4].

When Dr Cubero sparked the Petroleum Institute’s interest in funding a similar venture for the inspection of underwater pipe-work, the concept of the Turtle was born. The fundamental geometry of the space frame was agreed and simulations were animated. At this point the disagreement on the actuation arose.

Remembering the force limitations of the STIC, Dr Cubero is keen the use ball-screw actuators, preserving the geometry of a cylinder but offering a very much greater force to enable the robot to walk on dry land. However these would be ‘power hungry’ and would add considerable mass to the machine [5].

The opposing view is that a much lighter vehicle could achieve the same purpose, remaining submerged or floating at all times to limit the force requirements on the limbs. Options are discussed in the following sections.

3. Power options

The most straightforward option is the use of battery power, both to sustain the electronics and to power the propulsion.

Another option is the use of compressed air, in the form of a SCUBA-diving air bottle. These can be filled to 200 bar or more, that is 20 megapascals or 20 megaNewtons per square metre. For a twelve litre tank the energy can represent several hundred watt-hours.

Double-acting pneumatic cylinders have constant total volume, so the buoyancy of the vehicle would not be changed as the cylinders operated. Exhausted air can be allowed to bubble free, or can be trapped at the end of the mission to ensure a return to the surface. A chemical fuel could alternatively be used to generate pressurised gas.

4. Actuation options

If the pneumatic option is declined, many choices remain for electrical propulsion. With large variations of pressure with depth, any air voids are to be avoided. Thus conventional DC motors with commutation are not appropriate. Options that remain are brushless DC motors, stepper motors, AC induction motors or hysteresis motors.

Since the geometry is based on actuation by links that can vary in length, further variations that can be considered are linear motors, either induction or stepper, or mechanisms to convert rotation of a motor to linear movement. These can include ball-screw, rack and pinion or even drum-driven cables.

A further option is use rotary motion to drive the angle between two actuating blade-shaped levers, producing the same effect as the extension of a cylinder. Since the axis of the stress is well defined, these 'scissor blades' do not have to be massive.

5. Acknowledgments

This project is supported by the Petroleum Institute, Abu Dhabi. The author is grateful to Samuel Cubero for initiating the project and securing the funding. Many fruitful discussions will continue.

References

1. Luk, B.L.; Collie, A.A.; Billingsley, J.; , "Robug II: An intelligent wall climbing robot," *Robotics and Automation, 1991. Proceedings., 1991 IEEE International Conference on* , vol., no., pp.2342-2347 vol.3, 9-11 Apr 1991
2. Cubero, Samuel N., "Force, compliance and position control for a pneumatic quadruped robot." PhD dissertation, University of Southern Queensland library, 1997.
3. McLatchey, Graham, and John Billingsley. "Improving power to weight ratio of pneumatically powered legged robots." Advances in Climbing and Walking Robotics: Proceedings of the 10th International Conference (CLAWAR 2007). World Scientific, 2007.
4. Boh, T.; Billingsley, J.; Bradbeer, R.S.; Hodgson, P.; "Terramechanics based traction control of underwater wheeled robot," *OCEANS 2010 IEEE - Sydney* , vol., no., pp.1-3, 24-27 May 2010.
5. Cubero, Samuel N, "Design concepts for a hybrid swimming and walking vehicle," International Symposium on Robotics and Intelligent Sensors 2012.

AN ACTUATED CONTINUOUS SPRING LOADED INVERTED PENDULUM (SLIP) MODEL FOR THE ANALYSIS OF BOUNCING GAITS

DANIEL A. JACOBS, LINUS J. PARK and KENNETH J. WALDRON

Mechanical Engineering, Stanford University
Stanford, CA 94305, USA
** E-mail: dajacobs@stanford.edu*
www.stanford.edu

The Spring Loaded Inverted Pendulum (SLIP) model is a simple model for analyzing bouncing type legged gaits such as hopping and running. Reducing the full system dynamics allows for simplified analysis of major characteristics of locomotion for a wide variety of systems. However, the implementation of the SLIP model in the design and control of robotic locomotion can benefit from including more representative stance and flight dynamics. For example, in the SLIP model, the stance dynamics are fully conservative. However, physical systems will exhibit energy loss as a consequence of ground contact and will also do negative work through friction and actuator losses. To recover this lost work, the stance phase must be asymmetric and must include a thrusting portion. An actuated form of the SLIP model with fully continuous flight and stance dynamics is presented to analyze the stability space presented by the original SLIP model. The combined dynamic and control system stability is analyzed in the return map method and the simulated results show that heuristic control of the leg angle can stabilize the hopping gait only under certain circumstances.

Keywords: SLIP Model, Stability Analysis, Legged Locomotion

1. INTRODUCTION

The Spring Loaded Inverted Pendulum (SLIP) model shown is a useful model for representing the stance dynamics of legged bouncing gaits.[1]

Full and Koditschek[2] separated locomotion models into two groups, templates and anchors. They defined templates as reductive models that "advance intrinsic hypothesis" about the underlying dynamics and control of a task, where as anchors are the high order systems i.e. robots and animals that "admit the realization of the lower order template".

One major focus of current research is defining the parameter space where the SLIP model exhibits passive stability. At carefully defined kinematic and geometric parameters, it has been shown running gaits achieved by the SLIP model can be robust to perturbations.[3]

One of the major issues in transitioning from the SLIP model to an anchor system is sustaining the energy of the system in the presence of negative work during stance and energy loss to ground contact.

For a non-conservative system, the restitution phase must be able to generate positive work to replace what has been lost during the compression portion of stance. Using a SLIP type model, Ruina et al. showed that the energetic costs of transport are minimized with pseudo-elastic collisions, i.e. collisions that behave elastically even if the lost work is generated entirely by the leg.[5]

For the analysis of animal running and to aid in the design and control of robotic systems, our goal is to analyze running gaits in an anchor type formulation of the SLIP model. We present a simply actuated continuous SLIP-type model. The stance dynamics are represented with two continuous models, the Hunt-Crossley model[6] for the normal force of contact and the Mitiguy-Banerjee frictional model[7] for the tangential stick-slip forces.

2. The Actuated Spring Loaded Inverted Pendulum Model

The proposed model consists of three masses, a HAT particle(head arms torso), a shank body, and a foot particle. The shank center of mass is located at midpoint of the rest leg length, L_n. The inertia of the shank is calculated as a thin rod of length L_n. The HAT and foot masses are connected linearly by an actuator representing the leg as shown in Figure 1. The state vector of the system can be written as $x, \dot{x}, y, \dot{y}, L, \dot{L}, \theta, \dot{\theta}$, where x and y are the position of the HAT particle from the origin of the Newtonian Frame. The full model parameter set is shown in Table 1

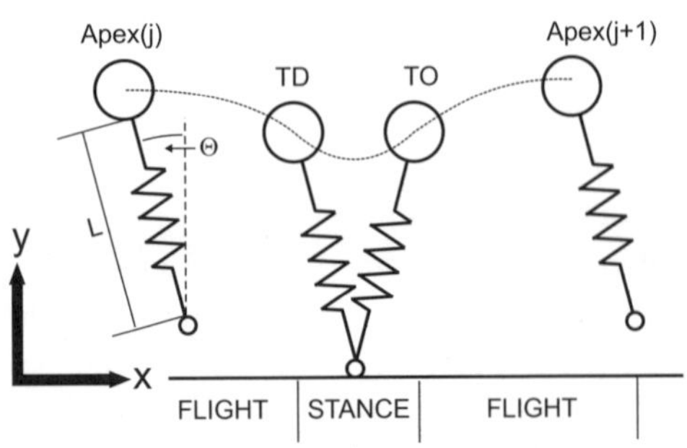

Fig. 1. The continuous spring loaded inverted pendulum model.

Parameter	Value	Units
HAT(Head,Arms,Torso) Mass	44.09	kg
Shank Mass	13.50	kg
Foot Mass	01.00	kg
L_n	.8562	m
k_{Leg}	24	$\frac{kN}{m}$
b_{Leg}	0.31	$\frac{kN*s}{m}$
c_{Thrust}	.8562	m
k	2.782 E5	$\frac{N}{m^{1.5}}$
λ	$\frac{3}{2}$(.45)k	$\frac{Ns}{m^{2.5}}$
n	1.5	(none)
μ_k	0.6	(none)
eps	1E-4	(none)

Similar to other formulations of the SLIP model, we assume that the orientation of the leg is controlled by a torque actuator between the Newtonian reference frame and the leg. For a legged robot, this assumption is equivalent to a torque actuator on a boom. During stance, the actuator is inactive.

During stance, the actuator in the system is modeled with three parameters: a spring constant k_{leg}, a damping constant b_{leg}, and a thrust constant c_{Thrust}. The thrust constant acts a multiplier for the elastic potential energy in the spring and maintains system energy over multiple hops. The equations for the actuator are given in (1,2).

$$F_{compression} = -k_{Leg}\left(L - L_n\right) - b_{Leg}\dot{L} \tag{1}$$

$$F_{restitution} = -k_{Leg}\left(L - L_n\right) c_{Thrust} \tag{2}$$

The physical parameters for the model are based off of the reported human geometric and mass parameters for males in De Leva[8] and the human leg stiffness estimates of Aramptzis et. al.[9]

The ground contact forces are modeled continuously using a normal force model and a tangential friction model. The normal forces are modeled using the formulation by Hunt-Crossley model.

The Hunt-Crossley model[6] is shown below:

$$F_c = \lambda(s^n)\dot{s} + k(s^n) \tag{3}$$

The Hunt-Crossley model has three parameters, a stiffness parameter, k, a damping parameter, λ, and a power parameter, n. The contact force generated by the model is a function of the penetration s and penetration rate $\dot{s}$.

The contact model parameters were generated based on the collision of a plantar soft tissue and concrete. The material properties of the plantar soft tissue are given by Ledoux et. al.[10] The combined stiffness of the interface and damping constant were then calculated using the technique proposed by Hunt and Crossley.

The frictional forces are modeled continuously using the Mitiguy-Banerjee model.[7] The Mititguy-Banerjee model is given as:

$$F_f = \mu_k |F_n| \frac{\mathbf{v}}{|\mathbf{v}| + \epsilon} \tag{4}$$

where μ_k is the kinetic coefficient of friction, F_n is the normal force of contact, $|\mathbf{v}|$ is the velocity of the contact point, and ϵ is a small number that controls the trade-off between accuracy and simulation time.

Our proposed model is capable of modeling the stick-slip behavior including slip reversal as well as complex rebound behavior such as chatter. Overall, these improvements impose more realistic limitations on the periodic solutions of the SLIP model template and support the transition from simulation to experimental vehicle.

3. Simulation and Stability Analysis

For a nonlinear system of equations, when a periodic solution exists, the Poincaré map or return map can be used to determine the stability of that solution. Although there is no rigid proof that a solution exists given a general nonlinear equation, for a legged robotic system it can be assumed safely that there are periodic solutions about a given point in the stride. For reference, the authors point the reader to the discussion in Chapter 5 of Westervelt et. al.[11]

We define the subspace intersected by the return map as the apex of the flight phase. At the top of flight, the leg is at the nominal length, oriented at touched down angle, and the vertical velocity is zero. Consequently, the dynamic state space is reduced from $x, \dot{x}, y, \dot{y}, L, \dot{L}, \theta, \dot{\theta}$ to $\dot{x}, y, \theta$. The thrust controller parameter is also included in the search for the fixed point solution. The inclusion of additional parameters in the return map was demonstrated on passive dynamic running robots by McGeer.[12] McGeer defined the parameter γ to represent the required slope necessary to maintain passive running motion in the presence of contact losses. Here, we use the thrust control parameter c_{Thrust} in a similar manner to control the energy added to the system.

The fixed periodic solutions of the routine will approximately satisfy the following eigenvalue equation:

$$\Delta V = \nabla \vec{S} \Delta V \tag{5}$$

Thus, the stability of the system can be calculated from the eigenvalues of the gradient matrix at the fixed points. If all of the eigenvalues have magnitude less than 1, then the perturbation at a fixed point decays over subsequent steps.

4. Heurisitic Control of the Actuated SLIP Model

For the set of parameters given in Table 1, the fixed point solutions of the actuated continuous slip model are shown in Figure 2

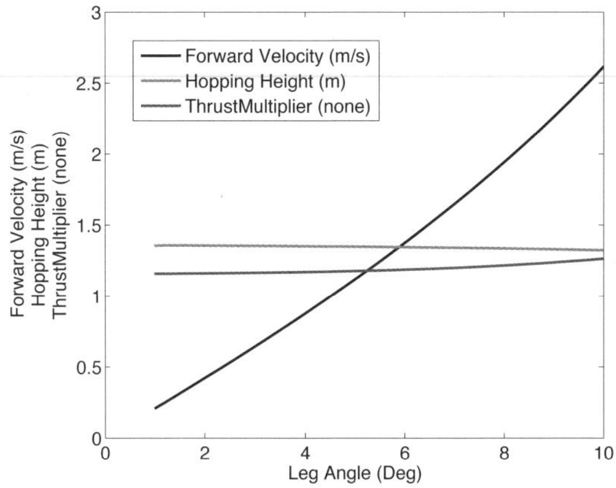

Fig. 2. Parameter Values of the fixed points solutions in the range of 1 to 10 degrees.

Examination of the fixed point solutions reveals similar behavior between the proposed system and human running. The hop height is relatively constant as a function of leg angle which shows that the increase in forward velocity is primarily a function of increasing leg angle. For the fixed points calculated, the thrust multiplier increases the spring potential energy by 15.8% to 26.4% to account for the damping losses in the leg and the ground.

Near ten degrees of leg swing, the hop height decreases and the thrust multiplier increases. This effect is the consequence of fixing the leg stiffness.

As the forward velocity increases, the asymmetry of the stance increases. As a result, more of the positive work in the restitution phase goes to maintaining the horizontal velocity and the hop height diminishes. Without increasing the leg stiffness, these trends of hop height and thrust continue until the system eventually trips.

Raibert's Leg Lab showed that heuristic control can be an effective method for controlling robots characteristics by the SLIP Model. To stabilize the unstable forward velocity eigenvalue, we add a simple heuristic control to the leg orientation control. If the velocity of the hip point is faster than the fixed point velocity during flight, the leg should be swung forward to increase the horizontal component of the braking force during stance compression and vice verse to regain velocity. The control law used to update the desired leg angle is as follows:

$$\theta_{des} = \theta_{fp} + k_h \left(\dot{x}_{flight} - \dot{x}_{fp} \right) \tag{6}$$

where the subscript fp refers to the fixed point values, $\dot{x}_{flight}$ is the horizontal velocity in flight, and k_h is a control constant. For the results shown, the value of k_h is 0.1.

In many analyses of the SLIP model, the physical system and the control method are evaluated separately. In our actuated model, the thrusting controller is evaluated as part of the dynamic system. Thus changing the parameters of the thrust controller affects both the location of the fixed points and their eigenvalues for certain manifolds of the map space.

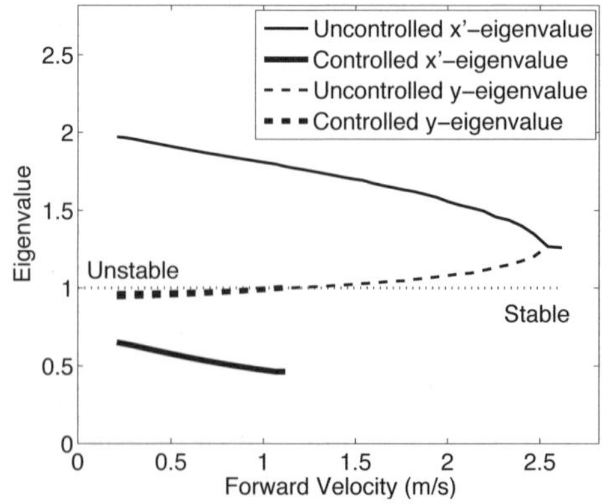

Fig. 3. Eigenvalues of the period one and period two systems under heuristic control.

To analyze the stability of the heuristic controller, we look at the stability of the period two solutions i.e. from Apex_j to Apex_{j+2}. The resulting fixed points are shown in Figure 3.

Figure 3 shows the eigenvalues for the forward velocity and hop height as a function of the fixed point value of the forward velocity. The forward velocity eigenvalue without heursitic control is unstable for the entire studied range. The horizontal dashed line shows that in the lower range, the uncontrolled hop height eigenvalue is less than one but near one.

The heuristic controller effects only the forward velocity eigenvalue and is only able to created a limit stability range for the system. The transition of the hop height eigenvalue from stable to unstable near 1.25 m/s shows the limits of heuristic control in the stabilization of this system. The described controller only functions as long as the hop height is stable without control.

5. Conclusion

In this paper, we proposed a actuated version of the SLIP model with fully continuous stance and flight dynamics for the analysis of bouncing gaits. We developed a dynamic simulation of a hopper with human-like parameters and report the fixed periodic solutions using the return map method. Our analysis of the eigenvalues of the fixed points shows that for the range of motion considered, the solutions were unstable. Using heuristic control, we show that bouncing motion can be stabilized under certain conditions and that the control can be shown to be asymptotically stable. We demonstrate that the stability of the combined dynamic and control system can be evaluated by considering apex intersections of two hops.

ACKNOWLEDGMENT

The authors would like to thank Samuel Hamner for sharing his motion capture data of running and muscle actuated simulations with us.

References

1. R. Blickhan, "The spring-mass model for running and hopping," *Journal of Biomechanics*, vol. 22, no. 11-12, pp. 12171227, 1989. [Online]. Available: http://www.sciencedirect.com/science/article/pii/0021929089902248
2. R. J. Full and D. E. Koditschek, "Templates and anchors: neuromechanical hypotheses of legged locomotion on land," *Journal of Experimental Biology*, vol. 202, no. 23, pp. 33253332, Dec. 1999. [Online]. Available: http://jeb.biologists.org/content/202/23/3325

3. A. Seyfarth, H. Geyer, M. Gnther, and R. Blickhan, "A movement criterion for running," *Journal of Biomechanics*, vol. 35, no. 5, pp. 649655, May 2002. [Online]. Available: http://www.sciencedirect.com/science/article/pii/S0021929001002457
4. M. A. Daley and A. A. Biewener, "Running over rough terrain reveals limb control for intrinsic stability," *Proceedings of the National Academy of Sciences*, vol. 103, no. 42, pp. 1568115686, Oct. 2006. [Online]. Available: http://www.pnas.org/content/103/42/15681.abstract
5. A. Ruina, J. E. Bertram, and M. Srinivasan, "A collisional model of the energetic cost of support work qualitatively explains leg sequencing in walking and galloping, pseudo-elastic leg behavior in running and the walk-to-run transition," *Journal of Theoretical Biology*, vol. 237, no. 2, pp. 170192, Nov. 2005. [Online]. Available: http://www.sciencedirect.com/science/article/pii/S0022519305001591
6. K. H. Hunt and F. R. E. Crossley, "Coefficient of restitution interpreted as damping in vibroimpact," *Journal of Applied Mechanics*, vol. 42, no. 2, pp. 440445, June 1975. [Online]. Available: http://dx.doi.org/10.1115/1.3423596
7. P. C. Mitiguy and A. K. Banerjee, "Efficient simulation of motions involving coulomb friction," *Journal of guidance, control, and dynamics*, vol. 22, no. 1, pp. 7886, 1999.
8. P. de Leva, "Adjustments to zatsiorsky-seluyanovs segment inertia parameters," *Journal of Biomechanics*, vol. 29, no. 9, pp. 12231230, Sept. 1996. [Online]. Available: http://www.jbiomech.com/article/0021-9290(95)00178-6/abstract
9. A. Arampatzis, G.-P. Brggemann, and V. Metzler, "The effect of speed on leg stiffness and joint kinetics in human running," *Journal of Biomechanics*, vol. 32, no. 12, pp. 13491353, Dec. 1999. [Online]. Available: http://www.sciencedirect.com/science/article/pii/S0021929099001335
10. W. R. Ledoux and J. J. Blevins, "The compressive material properties of the plantar soft tissue," *Journal of Biomechanics*, vol. 40, no. 13, pp. 29752981, 2007. [Online]. Available: http://www.sciencedirect.com/science/article/pii/S0021929007000747
11. E. Westervelt, J. Grizzle, C. Chevallereau, J. Choi, and B. Morris, *Feedback control of dynamic bipedal robot locomotion*. Boca Raton: CRC Press, 2007.
12. T. McGeer, "Passive bipedal running," *Proceedings of the Royal Society of London. B. Biological Sciences*, vol. 240, no. 1297, pp. 107134, May 1990. [Online]. Available: http://rspb.royalsocietypublishing.org/content/240/1297/107.abstract

DESIGN AND PROTOTYPING OF ACTIVE-SUSPENSION-BASED 4-LEGGED RUNNING ROBOT WITH ANTEROPOSTERIOR ASYMMETRY IN BODY

KYOKO NISHIMURA, KAZUYOSHI TSUTSUMI,
MASANORI MIZOGUCHI, SHINJI SAKIO AND KOJI SHIBUYA

Department of Mechanical and Systems Engineering, Ryukoku University
Seta-Oe-cho, Otsu-shi, Shiga-ken 520-2194, Japan

This paper describes our new 4-legged running robot. In our research group, we have been developing some mechatronic active-suspension-based 4-legged running robots step-by-step. This time, based on our experience, we have designed and prototyped a new robot with 10 pcs of DC motors, aiming at actualizing faster moving velocity. In this robot, the active-suspension mechanisms are employed not only for hopping and jumping but also for striking the ground to advance. From the result of analysis of the torque for necessary motion in each leg, the robotic body has been designed with anteroposterior asymmetry so that it has stronger rear legs compared with the four legs.

1. Introduction

Many kinds of 4-legged animals can move very fast, making full use of their well-developed musculoskeletal systems. Based on ingenious leg control mechanisms including the switching of several kinds of gait, they can achieve stable, efficient, and high-speed running even on rough terrain. It is quite fascinating in the creation of a moving robot to understand the motion mechanisms of those 4-legged animals. Research on them seems to be of great significance also in the sense that materialization of a robot mimicking the structure of a 4-legged animal can contribute to grasping what components are important for fast running in biological mechanisms.

Thus far, from those viewpoints, various 4-legged robots have been developed and studied [1]-[5]. Among them, the 4-legged robot built by M. Raibert, which was composed of spring elements and hydraulic actuators, is well known as a pioneering work. In our research group, we have been developing some active-suspension-based 4-legged running robots step-by-step [6][7]. This time, based on our experience, we have designed and prototyped a new robot with 10 pcs of DC motors, aiming at actualizing faster moving velocity. In this robot, mechatronic active-suspension mechanisms are employed not only for hopping and jumping but also for striking the ground to advance.

From the analysis of the torque for necessary motion in each leg, the robotic body has been designed with anteroposterior asymmetry so that it has stronger rear legs compared with the fore legs.

In the following sections, we precisely describe our new 4-legged running robot from the viewpoints of design and prototyping.

2. Design Policy

In the design, we tried to modularize robotic elements as thoroughly as possible in order to improve the maintenance characteristics. The four leg modules, in particular, have basically the same structure except that the number of actuators is different in the fore and rear legs as stated later. Each leg module consists of one submodule for hopping and one for leg-swinging; each submodule is composed of a compression spring, a rack-and-pinion, and a DC motor(s) as an actuator(s).

To achieve the active-suspension mechanism suitable for running, the choice of DC motors becomes important, as well as that of compression springs. As for determining the specs of a DC motor in a submodule for hopping, we use a simple 1/4 model with one mass and one compression spring as shown in Fig. 1. The left figure indicates the moment when the 1/4 body of the robot with mass m [kg] starts to jump at leg swing angle θ [rad]; the right one depicts when it gets to the highest point of h [m]. Here, k [N/m] is the spring constant. When the spring is contracted by distance x [m] from an equilibrium state, the relationship between potential energy of the spring and kinetic energy of the body jumping upward with a certain initial speed v_0 [m/s] is written as follows:

$$\frac{1}{2}kx^2 = \frac{1}{2}mv_0^{\ 2}. \tag{1}$$

If we set the horizontal component of robot speed to be v_x [m/s], the following relationship holds in terms of potential energy and kinetic energy:

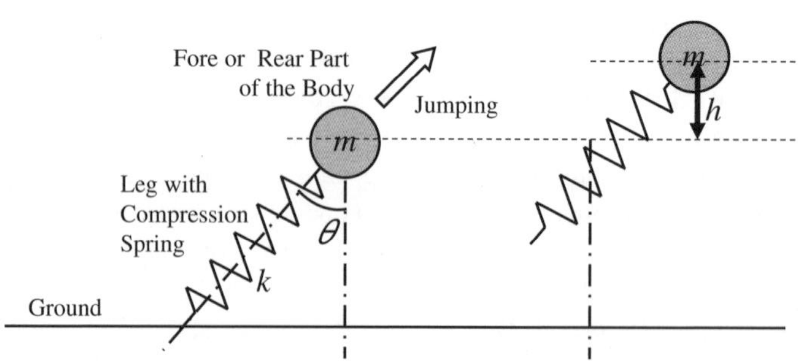

Fig. 1 1/4 model with one mass and one compression spring while jumping.

$$\frac{1}{2}mv_0^{\ 2} = \frac{1}{2}mv_x^{\ 2} + mgh\ . \tag{2}$$

Here, g [m/s^2] is gravitational acceleration. Considering the relationship $v_x = v_0 \sin\theta$, Eq. (2) can be modified to

$$v_0^{\ 2} = \frac{2gh}{\cos^2\theta}\ . \tag{3}$$

Applying Eq. (3) to Eq. (1) and rearranging it yields

$$x = \sqrt{\frac{2mgh}{k}}\frac{1}{\cos\theta}\ . \tag{4}$$

Thus, we can easily calculate how the actuator has to compress the spring in order to achieve the hopping height necessary in running motion. Based on this result, the rated torque and the rated rotational speed of the DC motor can further be determined.

As for a submodule for leg-swinging, we employ a 1/2 model with two masses and one leg with fixed length supposing bounding gait as illustrated in Fig. 2. Figure 2(a) indicates a moment when the fore leg lands on the ground during bounding gait. The torque necessary for rotating the fore leg joint is composed of that for bringing the propulsion force of the body (#F1) and that for not having the fore part of the body fall rearward due to its weight (#F2). Then, the torque for ensuring that the rear part of the body is going to fall downward (#F3) appears, but it generates an effect to assist #F2.

Figure 2(b), on the other hand, depicts a moment when the rear leg lands on the ground also during bounding gait. In a similar way to the previous case, the torque necessary for rotating the rear leg joint is composed of that for bringing the propulsion force of the whole body (#R1) and that for not making the rear part of the body fall backward due to its weight (#R2). Then, the torque for

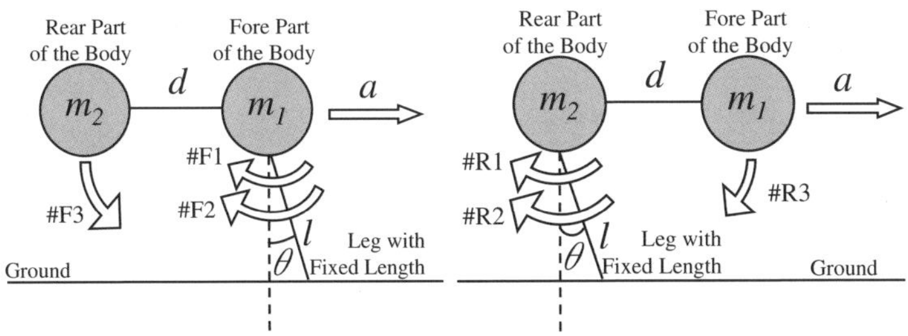

(a) When fore leg lands on the ground. (b) When rear leg lands on the ground.

Fig. 2 1/2 model with two masses and one leg with fixed length while running.

ensuring that the fore part of the body is going to fall downward (#R3) appears, and it acts to suppress #R2; therefore larger torque for rotating the rear leg joint is needed to compensate for #R3.

We define d [m] as length between fore and rear leg joints, l [m] as leg length, m_1 [kg] as mass of fore body, m_2 [kg] as mass of rear body and θ [rad] as swing angle. We think approximately of the models in Fig. 2(b) in order to consider the torque required for the rear leg joint. When the robot is going to accelerate in acceleration a [m/s^2], required force F [N] is

$$ma = F\cos\theta \Leftrightarrow F = \frac{ma}{\cos\theta}. \tag{5}$$

Torque to move forward T_1 is

$$T_1 = \frac{m_2 a}{\cos\theta} l. \tag{6}$$

Torque to support the rear part of the body T_2 is

$$T_2 = m_2 g l \sin\theta. \tag{7}$$

Torque to support the fore body T_3 is

$$T_3 = m_1 g d. \tag{8}$$

Thus, necessary torque for rear leg swing T_r is

$$T_r = T_1 + T_2 + T_3 = \frac{m_2 a}{\cos\theta} l + m_2 g l \sin\theta + m_1 g d. \tag{9}$$

In a similar way, necessary torque for the fore leg swing T_f is

$$T_f = \frac{m_1 a}{\cos\theta} l + m_1 g l \sin\theta - m_2 g d. \tag{10}$$

From the above, necessary torque for rotating the rear leg joints T_r should be larger than that for driving the fore ones T_f, though its degree depends on the moving speed of the robot. We use the same motors in the fore and rear legs of the robot to improve the maintenance characteristics. If we select larger torque motors to conform to the rear legs, the robot cannot run at goal-velocity because of the lack of rotation speed. On the other hand, if we select motors to conform to the fore legs, the torque is insufficient in the rear legs.

For these reasons, the robot is designed with anteroposterior asymmetry so that one DC motor is employed in a fore submodule for leg swinging and two

DC motors are used in a rear submodule. We had selected motors for swing and jump as the following tables illustrate.

Table 1. Motor specifications

Maxon Japan Co., Ltd. RE30(Part number 310007)			
Rated output [W]	60	Supply voltage [V]	24.0
Stalling torque [mNm]	1020	Maximum continuous torque [mNm]	85.0
No-load speed [rpm]	8810	Nominal speed[rpm]	8050

Table 2. Gear-head specification (For jumping motor)

Maxon Japan Co., Ltd. GP32BZ (Part number 358975)	
Reduction	3.7 : 1
Reduction absolute	63 : 17

Table 3. Gear-head specification (For swinging motor)

Maxon Japan Co., Ltd. GP32A (Part number 166158)	
Reduction	14 : 1
Reduction absolute	676 : 49

3. Prototyping

Figure 3 illustrates 3D-CAD drawings of the robot built in this study. Figure 4 shows a picture of the actual robot with body length of 555 mm, body width of 375 mm, length between fore and rear leg joints of 385 mm, leg length of 250 mm, and total weight of 14.5 kg. Subsequently, we reduced the robotic weight to 12.5 kg by changing material or lightening some parts. The distinguishing features of the robot can be summarized as follows;

- Modularization and sub-modularization of robot elements for improving maintenance characteristics;
- Many uses of robotic parts machined from aluminum and poli-acetal resin blocks for obtaining high rigidity;
- Simple construction suitably employing both prismatic joints and rotational joints;
- Employment of mechatronic active-suspension type of submodule with a compression spring, a rack-and-pinion, and a DC motor(s);
- Robot structure with anteroposterior asymmetry due to the employment of stronger rear legs.

Several kinds of sensors, including the one for judgment on landing, are installed on the robot. Based on information from them, we can achieve various sorts of gait from the robot by expanding and contracting each active-suspension type of submodule adequately.

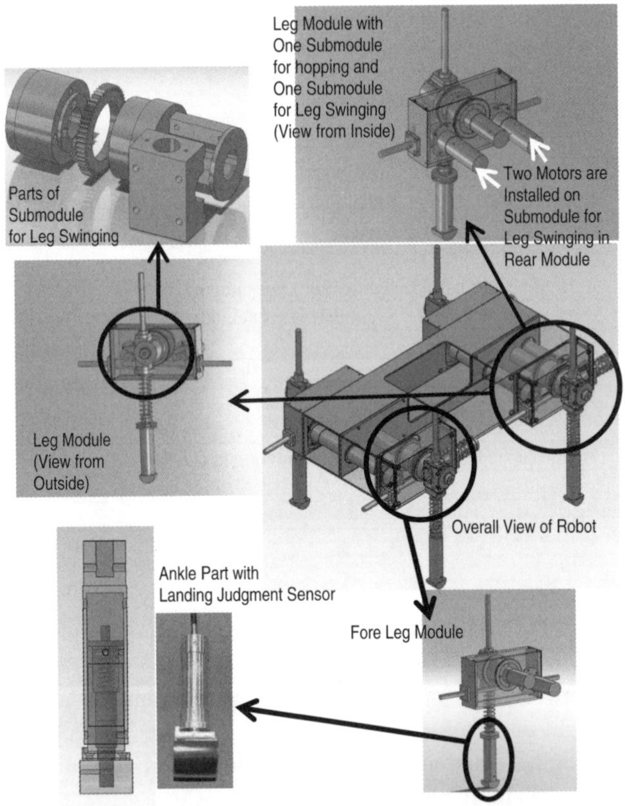

Fig.3 3D-CAD drawings of the robot and its robotic components

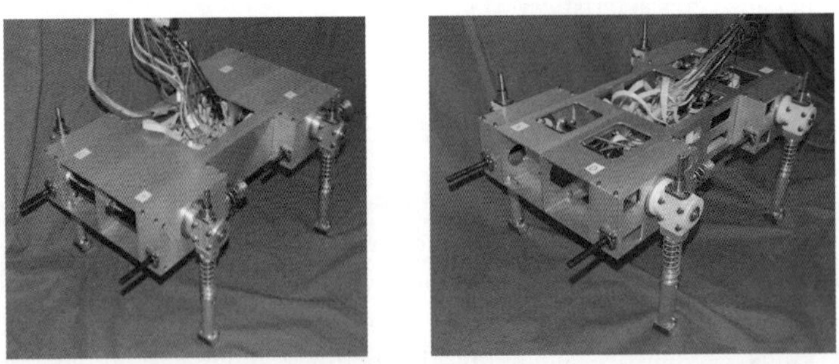

(a) First version before weight reduction (b) Present version after weight reduction

Fig.4 Picture of robot with modular construction but anteroposterior asymmetry in body

4. Test Running

We conducted some basic experiments for validation of robot motion, aiming at examining whether the robot acts properly under the control, and whether it does not have any problems in terms of strength. Since two motors are installed in a rear leg module as stated in the previous sections, the weight of the rear body is larger than that of the fore body; so the center of gravity exists on the rear side. Therefore, the amount of extraction and contraction in the springs necessary to have the robot jump upward uniformly must be different between the fore legs and rear legs. In this test using trot gait, only the amount of extraction and contraction of the springs for hopping are changed in stages, under the condition that the amount of all leg swings is common. Experimental conditions and results are shown respectively in Table 4 and Fig. 5. Figure 5 shows that the moving speed changes depending on the extraction and contraction amount in the rear legs. When the amount of extraction and contraction in the rear legs is set to be smaller than that in the fore legs, the robot sometimes made a false step. On the other hand, the amount of extraction and contraction in the rear legs is set to be much larger than that in the fore legs, the gait was disordered. These results in the test suggest that the extraction and contraction amount in the rear legs must be within a certain range of the proportion to that in the fore legs, and that the moving speed is maximized when the former is at a certain proportion to the latter. Through repeated tests, it was also confirmed that the robot had sufficient strength without any faults, such that gears did not come off and/or axles were not bent.

Table 4. Experiment condition

Anteroposterior asymmetry	amount of compression of fore leg spring [mm]	amount of compression of rear leg spring [mm]
Not consider	15	15
consider	15	22.9
	15	21.9
	15	20.9

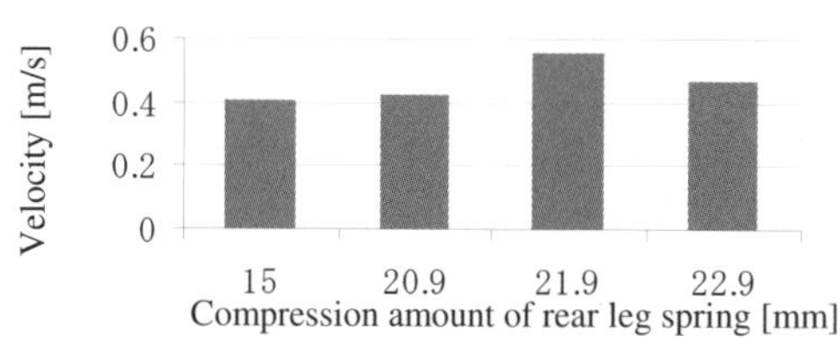

Fig.5 Result of test running
(The amount of expansion and contraction of the fore leg spring is constant)

5. Conclusions

In this study, we have designed and prototyped a mechatronic active-suspension-based 4-legged robot with anteroposterior asymmetry, which will produce a high potential to strike the ground strongly during acceleration. As a next step, we intend to conduct various experiments on gait such as walking, trotting, bounding, and galloping. Although, in such cases, control algorithms in consideration of the anteroposterior asymmetry in the robot body should be experimentally examined, a theoretical approach is also essential for determining adequate control quantity and timing. If we employ the 1/2 robotic model composed of two masses shown in Fig. 2, the robot in the air can be described as a free motion of a two-mass system on which only gravity acts. When fore legs or rear legs strike the ground, we will be able to describe the robot as a double pendulum system in which an actuator is installed in a joint between two links (a leg and a body). If the dynamical robot motion is divided into several cases such as fore-leg landing, in the air, and rear-leg landing, and those cases are appropriately switched, the total system while running can be precisely written by a set of those motion equations. Through combining theoretical and simulation studies with experimental studies using the robot that we designed and built in this study, in future work, we will do our best to uncover usefull mechanisms for faster operation of a 4-legged robot.

References

1. M. Raibert, Legged Robots That Balance, The MIT Press, 1986
2. K. Ito, "Rhythmic Pattern Generation during Locomotion," *J. of the Robotics Society of Japan*, Vol. 11, No. 3, pp.320-325, 1993
3. H. Kimura, "Dynamic Walk of the Quadruped Robot", *J. of the Robotics Society of Japan*, Vol. 11, No. 3, pp.372-377, 1993
4. T. Masuda, H. Kimura, T. Suehiro, K. Takase, "Emergence of Bound Gait Using Quadruped Robot "Rush"", *J. of Japan Society of Mechanical Engineers,* Vol.76, No. 761, pp.126-131, 2010
5. Y. Fukuoka, H. Kimura, "A Quadruped Robot Walking and Running with One Mechanism", *J. of Japan Society of Mechanical Engineers,* Vol.75, No. 750, pp.390-395, 2009
6. H. Okasyo, K. Hatta, T. Iinuma, K. Shibuya, K. Tsutsumi, "A Study on Improvement of Moving Speed for a 4-Legged Running Robot with Spring Elements", *Proc. of JSME Robotics and Mechatronics Conference 2006*, 2A1-C28, 2006
7. K. Hatta, H. Okasyo, K. Shibuya, K. Tsutsumi, "On the Relationship between Kicking and Moving Speed for a 4-Legged Running Robot", *Proc. of JSME Robotics and Mechatronics Conference 2007*, 1A1-F11, 2007

DESIGN OF LEAPING BEHAVIOR IN A PLANAR MODEL WITH THREE COMPLIANT AND ROLLING LEGS*

YA-CHENG CHOU, KE-JUNG HUANG, WEI-SHUN YU, AND PEI-CHUN LIN†

National Taiwan University, Mechanical Engineering
No.1 Sec. 4 Roosevelt Rd., ME, Eng. Bldg.
Taipei 106, Taiwan

We report on the development of a planar three-leg robot model for leaping behavior. The model is composed of a rigid body and three massless legs. The model for each leg is extracted from the virtual leg model of a rolling spring loaded inverted pendulum, which has two segments connected by a torsional spring. The quantitative equations of motions are formulated. By evaluating the dynamic behavior of the three-leg model, a 2-step leaping maneuver is developed. During the 1st-step leaping, the running legs thrust to induce a leap of the body. This leap serves two purposes: to provide sufficient time for the legs to reconfigure their poses; and to adjust the body pitch, preparing for the 2nd-step leap. After the robot lands on the ground with the desired body pose and leg configurations, all legs provide full thrust and initiate the 2nd-step leaping, which makes the body fly over the obstacle. The trajectory for leaping behavior is designed based on this dynamic model and is implemented in the RHex-style robot for experimental evaluation.

1. Introduction

Leaping is one of the unique actions performed by legged animals, which allows them to rapidly change their motion status. Animals use leaping in life-dependent activities such as hunting prey or escaping from predators. Animals also leap to overcome rough terrain and obstacles. For a robot, leaping behavior is also desirable. Land textures are often rough, whether in natural or artificial environments, and leaping can negotiate this rough terrain much faster than other types of gaits such as crawling.

Because inducing leaping/jumping motion on the robot usually requires actuators with considerable power to maneuver the body in a specific manner, it is difficult to generate this behavior on ordinary robots designed for walking or running. Therefore, jumping-specialized robots have been designed, while some walking robots were equipped with extra actuators and mechanisms for jumping. Examples include the miniature 7g jumping robot [1], the robot Grillo [2], the

---

* This work is supported by the National Science Council (NSC), Taiwan, under contract 100-2628-E-002 -021 -MY3.

† Corresponding information: peichunlin@ntu.edu.tw

frog robot Mowgli [3], the Mini-Whegs [4], and the wheeled robot SandFlea [5]. Few works relate to jumping-like behavior in the ordinary walking/running robots. On the simulation side, Wong et. al. simulated a quadruped performing a standing jump over an obstacle [6]. On the robot side, the quadruped LittleDog can jump to pass an obstacle by an optimization method [7]. The Hexapod RHex can jump in small steps by a pronking gait [8]. The robot PAW can jump a short step by matching front and hind leg motions [9]. However, the motions performed by the above empirical robots are not leaping behavior with large-scale ballistic flight. To the best of our knowledge, the quadruped BigDog is the only robot showing leaping behavior which is transient from running. However, the demonstration was revealed only in movie clips with no detailed documentation released [10].

Aiming to develop leaping behavior and its transition from ordinary running behavior in an empirical RHex-style robot, we report on the development of a planar reduced-order model, which serves as a guideline for determining how the legs of the robot should move. The main contribution of this work lies in the methodology of model construction, analysis of the model's behavior, and the systematic variation of the reduced-order parameters to search for a feasible domain of leaping behavior.

2. Model Formulation

Because leaping behavior mainly resides on the robot state in the sagittal plane, a planar model is desired. We also found that the intrinsic model for running (such as a spring-loaded inverted pendulum (SLIP) with only one virtual leg) is not entirely suitable for fast state changes. Thus, a three-leg model was designed, which is composed of a front leg, a middle leg, and a hind leg as shown in Figure 1(a). This model preserves the six-legged structure of insects or hexapod robots by collapsing the right and left legs into one. The body is modeled as a rigid body with mass m at center of mass (COM) and with mass moment of inertia I_b, and the legs are modeled as massless springs. In addition to focusing on the model development, experimental verification is set as the final step of this work. Thus, the dynamic behavior of the experimental platform (i.e., the RHex-style robot [11]) should be considered in the modeling work. The compliant half-circular leg of the RHex-style robot has two characteristics which cannot be captured by an ordinary linear spring—rolling contact and varying compliance which is determined by the contact point at every instant. Thus, instead of the linear virtual spring that is widely utilized in the traditional spring-loaded inverted pendulum (SLIP) model, we adopted a leg model with a rolling contact and torsional spring, as reported in the recently-developed Rolling SLIP (R-SLIP) [12]. The virtual leg in the R-SLIP model has a point mass and two segments connected by a torsional spring. The lower segment has a circular shape, providing rolling contact to the ground. During the rolling motion, because the length of moment arm changes, the equivalent linear stiffness

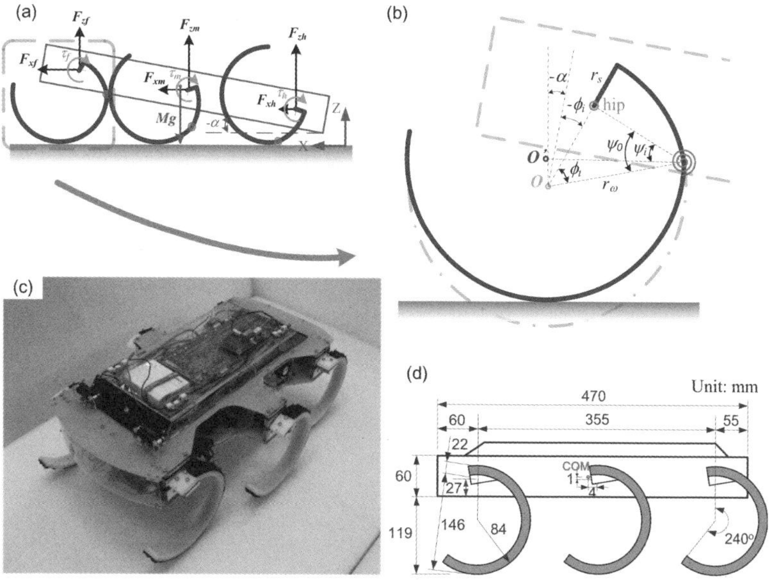

Figure 1. Illustration of the three-leg model (a)(b) and the experimental platform, the RHex-style robot (c)(d): (a) Overall illustration; (b) parameter definitions of the leg portion, where the robot body is sketched in dashed lines. The leg is composed of two segments (blue solid line and arc) connected by torsion springs (green spiral). The leg in its neutral configuration is plotted in an orange dash-dotted arc; (c) photo of the RHex-style robot; (d) dimensions of the robot.

between the robot's hip (or point mass) and ground-contact point varies accordingly. Thus, two characteristics of the half-circular leg described above can be adequately captured. Figure 1(b) shows leg parameters defined for model development: O is the center of the circular leg without leg compression; r_ω and r_s are leg geometrical parameters; ϕ_t and ψ_o are intrinsic leg parameters which present configuration of the torsion spring in its natural configuration; α is body pitch angle and $\phi_{i\ i=f,m,h}$ is orientation of the leg with respect to the robot body. The subscripts f, m, and h represent front, middle, and hind, respectively. When the leg contacts the ground and the torsion spring is compressed, the center of the circular leg moves to O' and the spring configuration can be captured by ψ_i. The legs are assumed to have pure rolling contact to the ground. With this assumption, the leg rotation angles, $\phi_{i\ i=f,m,h}$, and leg compressions, $\psi_{i\ i=f,m,h}$, are related to each other. As a result, the degrees of freedom (DOFs) of the model is reduced to three, and the three variables used to state the model formulation are orientation of the hind leg, ϕ_h, compression status of the hind leg, ψ_h, and body pitch angle, α.

We now give a brief description of the quantitative model formulation. There are nine external forces/moments acting on the body as shown in Figure 1(a), where each hip has one torque, $\tau_{i\ i=f,m,h}$, and two reaction forces, $F_{xi\ i=f,m,h}$

and $F_{zi\ i=f,m,h}$, in the horizontal (x) and vertical (z) axes. With these definitions, two force equations and one moment equation of motion (EOM) can be formulated. The acceleration of the body COM in EOMs can be derived by the configuration relating the hind leg to the COM. The torques in EOMs are motor torques, and these can be computed according to the DC motor model, which sets the motor torque as the function of the supplying voltage $V_{i\ i=f,m,h}$, motor speed (i.e., equal to hip speeds, $\phi_{i\ i=f,m,h}$), and motor internal characteristics. The reaction forces can then be computed according to static equilibrium. Finally, by importing motor torque, hip forces, and body acceleration into the EOMs, the equations of motion in state-space can be represented as:

$$\frac{d}{dt}\begin{bmatrix}\phi_h\\ \dot{\phi}_h\\ \psi_h\\ \dot{\psi}_h\\ \alpha\\ \dot{\alpha}\end{bmatrix}=\begin{bmatrix}\dot{\phi}_h\\ f_1(\phi_h,\dot{\phi}_h,\psi_h,\dot{\psi}_h,\alpha,\dot{\alpha},V_i)\\ \dot{\psi}_h\\ f_2(\phi_h,\dot{\phi}_h,\psi_h,\dot{\psi}_h,\alpha,\dot{\alpha},V_i)\\ \dot{\alpha}\\ f_3(\phi_h,\dot{\phi}_h,\psi_h,\dot{\psi}_h,\alpha,\dot{\alpha},V_i)\end{bmatrix}_{i=f,m,h} \tag{1}$$

With initial conditions and (1), the motion of the model in the stance phase can be simulated numerically until the model takes off for ballistic flight. Note that in the empirical evaluation, because ϕ_h, $\dot{\phi}_h$, and ψ_h are hard to measure and justify, these three states are further represented as a function of the COM displacement and velocity, where the mapping is one-to-one. Thus, the model can be regarded as having five varying initial conditions: α_0, $\dot{\alpha}_0$, Z_{c0}, $\dot{X}_{c0}$ and $\dot{Z}_{c0}$ as well as one fixed: ψ_h (i.e., simulation starts with the legs in a natural configuration). After the robot takes off, it moves according to the ballistic model.

3. Design of Leg Maneuver for Leaping Behavior

In general, if the legs of the robot (i) have sufficient DOFs for posture maneuver and (ii) have enough instantaneous power for fast leg-state changes, the robot may be able to transition from running to leaping in one step. However, the experimental platform for this work, the RHex-style robot, has only one active rotational DOF per leg. Though various behaviors can already be developed within this simple mechanical structure [11, 13, 14], the maneuverability of the body posture in a fast and dynamic manner is indeed limited. In addition, we also found that the motor power of the present robot is not enough to change phases of the tripods from 180 degrees (i.e., alternating tripod gait) to 0 degrees (i.e., leaping with all legs) in one stride. Therefore, a two-step leaping behavior is required to excite successful leaping in the RHex-style robot. The 1st-step leap serves two purposes: to adjust the body posture (i.e., pitch) and for practical implementation, to synchronize the phases of all legs in

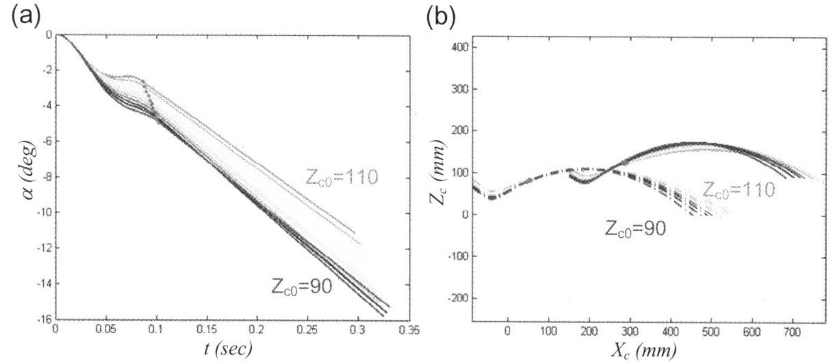

Figure 2. Simulation of various body states versus time with varying initial body height Z_{c0} (unit: mm) in the 1st-step leaping: (a) body pitch and (b) COM trajectory. Red dots represent take-off moment.

two tripods. In consequence, the body can adequately leap with full thrust from all legs in the 2nd-step leap.

Leaping is initiated and transient from tripod running, so the body state in running determines the initial conditions for leaping. Because the variation in pitch during running is small, the horizontal body posture (i.e., α_0 and $\dot{\alpha}_0$ at zero) can be treated as the initial condition for leaping, where the distances between the three hips and the ground are the same. At this configuration, if the three ground-contact legs (i.e., one tripod) are simultaneously driven with full power, the body state is varied as shown in Figure 2, where Z_{c0} is the initial body height. We specifically chose Z_{c0} as the active variable since it is the only variable which can be maneuvered easily in empirical operation among five initial conditions in the model (i.e., α_0, $\dot{\alpha}_0$, Z_{c0}, $\dot{X}_{c0}$ and $\dot{Z}_{c0}$). Figure 2(a) clearly reveals an important fact that, no matter at which body height Z_{c0} the robot starts to thrust, the body pitch decreases (i.e., head-up). This phenomenon is undesirable since the back side of the robot may hit the obstacle during leaping, and this posture is also not adequate to transition to running after leaping. Thus, in the current model formulation one-step leaping is not achievable. Because head-up of the model in 1st-step leaping is unavoidable, the 2nd-step leap is investigated with the initial conditions where the body is head-up. Figure 3 plots various states with time when the robot starts to thrust with full power at different initial body pitches, varying from -7° to -12°. Figure 3(a) reveals that the profiles of body pitch versus time can be dramatically different with different initial body pitches. In addition, final body pitch could be dramatically different from its initial head-up condition, which is desirable since this phenomenon allows the body pitch to be adjusted back to the desired level after leaping. As a result, to allow 2nd-step leaping to successfully fly over the obstacle, a suitable initial body pitch can be selected to make the final after-flight pitch more horizontal for proper landing. The figure reveals that -10° is around the right range.

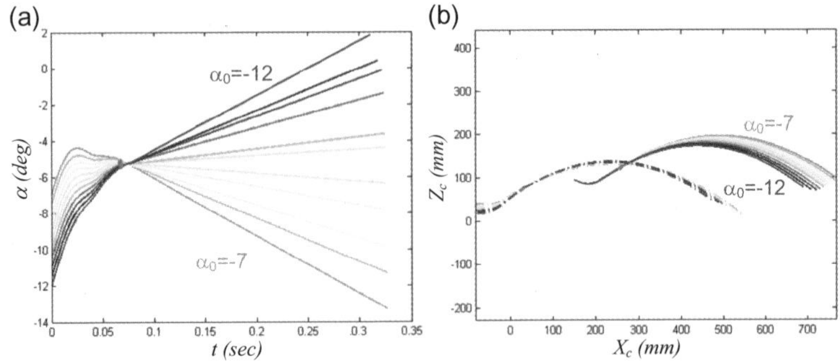

Figure 3. Simulation of various body states versus time with varying initial body pitch α (unit: deg) in the 2nd-step leaping: (a) body pitch and (b) COM trajectory. Red dots represent take-off moment.

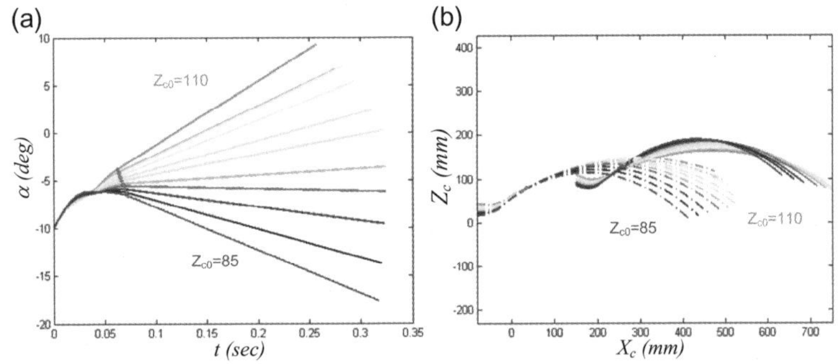

Figure 4. Simulation of various body states versus time with varying initial body heights Z_{c0} (unit: mm) in 2nd-step leaping: (a) body pitch and (b) COM trajectory. Red dots represent take-off moment.

Similar to the model in 1st-step leaping, the initial vertical height Z_{c0} is also the active parameter which can be varied in the 2nd-step leaping. With the body pitch set to -10°, Figure 4 shows variations of various states with time when the robot starts to thrust with full power at different initial robot heights. Figure 4(a) reveals that the initial height also has a significant effect on the overall body pitch trajectory. When the initial body height is higher (i.e., $Z_{c0}=110$), the body pitch is larger (i.e., head-down), which is undesirable since the robot may hit the ground. In contrast, if initial body height is lower (i.e., $Z_{c0}=85$), the body pitch remains negative, which may result in the robot's tail colliding with the obstacle. In addition, as shown in Figure 4(b), the flying distance of the robot's tail in this setting is too short. Therefore, a suitable initial body height should be selected—for example, around 0.095m—to yield horizontal flight.

The overall trajectory planning can now be constructed based on the above analysis. In 1st-step leaping, the legs land on the ground with selected body height and then provide full thrust, where the robot moves according to the trajectory shown in Figure 2(b). During the flight, the legs are synchronized and prepared to land with the desired body height and pitch. After touching the

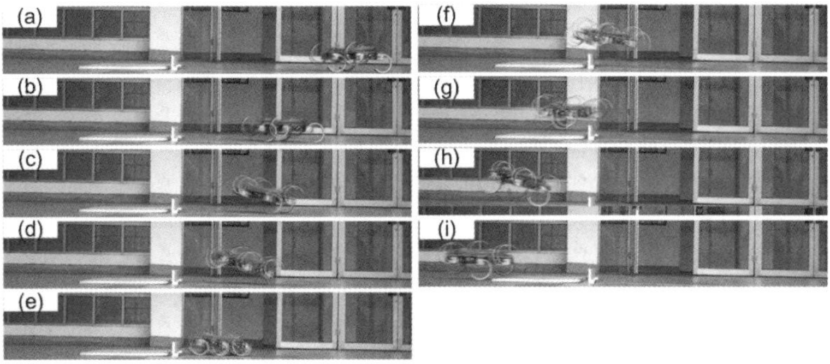

Figure 5. Sequence of images showing the robot leaping over the obstacle.

ground, all legs provide full thrust and initiate the 2nd-step leaping, which makes the robot move according to Figure 3(b) or 4(b).

4. Test of the Algorithm on the Empirical Robot

The RHex-style robot is utilized for evaluating leaping behavior. The photo and dimensions of the robot are shown in Figures 1(c) and 1(d). Further detailed specifications robot can be found in [15]. Figure 5 shows sequential images extracted from a video recording of the robot leaping over an obstacle. The height of the obstacle is 95mm. In the beginning, the robot runs toward the obstacle (a). Then the robot starts the leap (b), and takes off for the 1st flight (c). During this flight, two tripods are synchronized (d), preparing for landing with the correct body pitch and height from the ground (e). The robot initiates the 2nd-step leaping and its flight over the obstacle (f) (g). Next, the robot lands on the ground (h) and resumes running (i). Though more detailed investigation is required for evaluation, the video stills mostly confirm that, with the model-based trajectory planning method, the robot can indeed perform leaping behavior.

5. Conclusion

We report on the development of a planar three-legged model for leaping behavior. The model is composed of a rigid body and three massless legs. In order to catch the dynamic characteristics of the half-circle leg of the robot for experimental evaluation, the virtual leg in the R-SLIP model is adopted, which has two segments connected by a torsional spring. With assumption of rolling-contact, the model has 3 DOFs, which can be roughly mapped to the 3 planar states of the robot's COM, including vertical and horizontal motions as well as body pitch. The quantitative equations of motions are formulated and effects of model parameters are investigated. Through variations of the model parameters, a 2-step leaping maneuver is developed. At the 1st step of leaping, the legs land on the ground with specific body height and then provide full thrust. During the flight, the legs in the two tripods are synchronized to align in specific phases, preparing to land with the desired body height and body pitch. After the robot touches the ground, all legs provide full thrust and initiate the 2nd-step leaping, which makes the body fly over the obstacle. The planned trajectory is also implemented in the RHex-style robot. The video of our initial test shows that the robot can indeed perform leaping behavior.

References

[1] M. Kovac, M. Fuchs, A. Guignard, J. C. Zufferey, and D. Floreano, "A miniature 7g jumping robot," in *IEEE International Conference on Robotics and Automation*, 2008, pp. 373-378.

[2] U. Scarfogliero, C. Stefanini, and P. Dario, "The use of compliant joints and elastic energy storage in bio-inspired legged robots," *Mechanism and Machine Theory,* vol. 44, pp. 580-590, Mar 2009.

[3] R. Niiyama, A. Nagakubo, and Y. Kuniyoshi, "Mowgli: A Bipedal jumping and landing robot with an artificial musculoskeletal system," in *IEEE International Conference on Robotics and Automation*, 2007, pp. 2546-2551.

[4] B. G. A. Lambrecht, A. D. Horchler, and R. D. Quinn, "A small, insect-inspired robot that runs and jumps," in *IEEE International Conference on Robotics and Automation*, 2005, pp. 1240-1245.

[5] BostonDynamics. *SandFlea - Leaps small buildings in a single bound.* Available: http://www.bostondynamics.com/robot_sandflea.html

[6] H. C. Wong and D. E. Orin, "Dynamic control of a quadruped standing jump," in *IEEE International Conference on Robotics and Automation*, 1993, pp. 346-351.

[7] M. Zucker, N. Ratliff, M. Stolle, J. Chestnutt, J. A. Bagnell, C. G. Atkeson, *et al.*, "Optimization and learning for rough terrain legged locomotion," *International Journal of Robotics Research,* vol. 30, pp. 175-191, Feb 2011.

[8] D. McMordie and M. Buehler, "Towards pronking with a hexapod robot," in *4th Int. Conf. on Climbing and Walking Robots*, Karlsruhe,Germany, 2001, pp. 659-666.

[9] J. A. Smith, I. Sharf, and M. Trentini, "Bounding gait in a hybrid wheeled-leg robot," in *IEEE/RSJ International Conference on Intelligent Robots and Systems*, 2006, pp. 5750-5755.

[10] BostonDynamics. *BigDog - The most advanced rough-terrain robot on Earth* Available: http://www.bostondynamics.com/robot_bigdog.html

[11] U. Saranli, M. Buehler, and D. E. Koditschek, "RHex: A simple and highly mobile hexapod robot," *International Journal of Robotics Research,* vol. 20, pp. 616-631, Jul 2001.

[12] K. J. Huang and P. C. Lin, "Rolling SLIP: A model for running locomotion with rolling contact," in *IEEE/ASME International Conference on Advanced Intelligent Mechatronics*, 2012, pp. 21-26.

[13] E. Z. Moore, D. Campbell, F. Grimminger, and M. Buehler, "Reliable Stair Climbing in the Simple Hexapod 'RHex'," in *IEEE International Conference on Robotics and Automation (ICRA)*, 2002, pp. 2222-2227.

[14] D. Campbell and M. Buehler, "Stair descent in the simple hexapod 'RHex'," in *IEEE International Conference on Robotics and Automation (ICRA)*, 2003, pp. 1380-1385 vol.1.

[15] Y. C. Chou, W. S. Yu, K. J. Huang, and P. C. Lin, "Bio-inspired step-climbing in a hexapod robot," *Bioinspiration & Biomimetics,* vol. 7, p. 036008, Sep 2012.

WALKING DESPITE THE PASSIVE COMPLIANCE: TECHNIQUES FOR USING CONVENTIONAL PATTERN GENERATORS TO CONTROL INSTRINSICALLY COMPLIANT HUMANOID ROBOTS

P. KRYCZKA, K. HASHIMOTO, A. TAKANISHI

Graduate School of Science and Engineering, Faculty of Science and Engineering and Department of Modern Mechanical Engineering, Waseda Univesity,
2-2 Wakamatsu-cho, Shunjiku-ku, Tokyo, Japan
**E-mail: contact@takanishi.mech.waseda.ac.jp*

H. O. LIM

Faculty of Engineering, Kanagawa University,
3-27-1 Rokkakubashi, Kanagawa, Japan
E-mail: holim@kanagawa-u.ac.jp

P. KORMUSHEV, N. G. TSAGARAKIS, D. G. CALDWELL

Department of Advanced Robotics, Istituto Italiano di Tecnologia,
via Morego 30, 16163 Genova, Italy
E-mail: petar.kormushev@iit.it

There are several problems which arise when using a standard ZMP-based pattern generator to control an intrinsically compliant humanoid robot like COMAN. We present two techniques: pelvis forward trajectory smoothing and polynomial admittance gain modulation, which make it possible to use the conventional pattern generator to control such passively compliant robots. The former method modifies the reference of the pelvis trajectory to counteract its overshooting caused by the compliance of the legs. The latter method is meant to decrease the impact during initial contact and decrease the error between the foot position and original reference during the mid-stance caused by the use of admittance control. We explain details of both of the methods and show results from walking experiments with and without the controls, proving their effectiveness.

Keywords: passive compliance, biped locomotion, ZMP, admittance control

1. Introduction

Most of the walking humanoid robots capable of performing practical tasks use high-gain position control.[1,2] They are easier to control in a structured environment, but prone to model discrepancies or external disturbances. In order for humanoid robots to be more practical, they must be able to operate in unstructured environments. One way of improving the robot's robustness to external disturbances and environment model inaccuracies is to develop robots which are intrinsically compliant[3] or torque controlled.[4]

The intrinsic compliance can have numerous advantages, such as: mechanical separation of harmonic-drives from joints, that softens the impacts and thus increases the lifetime of harmonic-drives; overall compliance improving resistance to the ground unevenness or external disturbances; improvement of energy efficiency by storing and releasing the kinetic energy from the springs.[5] The compliance, however, also has disadvantages: the joint level position control is inaccurate; the leg joints deflect under the weight of the robot, thus making the walking control more difficult;[6] the overall system's resonant frequency is much lower than in high-gain position controlled robots and can affect walking stability. Moreover, most of the existing model-based pattern generators developed for humanoid robots assume ideal position control and thus cannot be directly used in compliant robots.

In this work we explain difficulties which arise when using the ZMP-based pattern generator (PG) to control passively-compliant robots. We also present techniques than can be used to control compliant robots with existing PGs. In the second section we introduce our research and discuss the consequences of passive compliance. In the third section we present the PG, developed control techniques and experimental results proving their effectiveness. We conclude the paper with summary of our work.

2. Problem formulation

In this work we focus on the description of the pattern modifications and real-time control necessary to execute the dynamically consistent gait pattern created when assuming a joint level position controllability on the compliant robot COMAN[3] Fig. 1. In this section we describe our research platform and the main consequences of the intrinsic joint-level compliance.

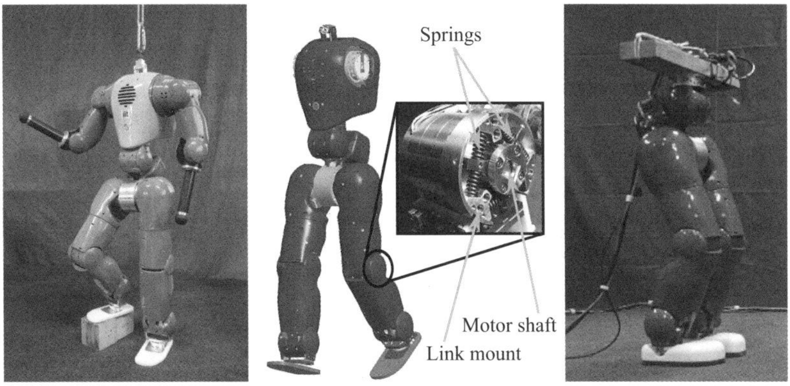

Fig. 1. Full body of our research platform COMAN. The figure in the middle shows knee's passive compliance mechanism. The picture on the right shows the version of the platform used in the experiments.

2.1. *COMAN structure*

The COmpliant huMANoid robot COMAN has an anthropomorphic bipedal structure equipped with series-elastic actuators. Every leg pitch joint (hip, knee, ankle) is actuated by a brushless DC motor connected to the joint by six linear springs as shown on Fig. 1. COMAN uses a distributed control system, that enables control of the motor position or joint torque. The robot with upper body is 0.93m high and weighs 28kg. The platform used in the experimental part of this work was without the upper body (see Fig. 1 right). The pelvis height is 0.5m and total weight of the lower body and waist is 18kg.

2.2. *Effects of the passive compliance on the robot's motion*

The passive compliance embedded in the COMAN's joints; while being advantageous in decreasing effects of sudden impacts on the structure and storing energy, is disadvantageous for joint level position control. When the position control is done on the motor level and the joint is subjected to high torque, the springs deflect and the joint diverges from the desired position. This makes the direct use of the trajectory references obtained from the pattern generator impossible. Whenever there is a sudden acceleration the springs deflect and the link does not reach the desired position. Furthermore, because of the compliance of the support leg the pelvis tends to sink, which causes discrepancies between the desired and real feet position. Using motor-position based control, this results in undesired early ground contact. Since the springs used in the structure have constant and relatively high stiffness, the structure is not able to damp the instantaneous impact occurring during the early ground contact, what causes a big disturbance and destabilizes the robot.

One possible solution to this problem is to detect early indications of instability (such as early ground contact) using automated real-time analysis of force sensor data from the feet.[7] Such information can be used to change the pattern dynamically during walking to improve stability. However, this would require real-time pattern regeneration capabilities, which is out of the scope of this paper and is our future work. In this paper, we investigate techniques for addressing this problem using only off-line pattern generation and feedback control.

3. Gait pattern generation and feedback control

As a source of the dynamically consistent gait patterns we use the ZMP-based preview controller with dynamic filter.[8] Due to the Multibody System (MBS) embedded in the dynamic filter, the resultant trajectories closely follow the reference ZMP that guarantee the system's stability, given the ideal experimental conditions (accurate robot model, joint level high-gain position control, flat ground etc.). Since the COMAN platform is equipped with mechanical compliance, the discrepancy between the real system and the MBS is too big to allow direct execution of the pre-calculated gait patterns on the robot. To cope with the difficulties described above and in 2.2 we implemented two techniques. The first one copes with the deformation of springs during high acceleration and deceleration of the pelvis, the second one copes with the impact caused by the early foot contact with the ground. In the following subsections we describe the techniques and present the experimental results confirming their performance. In all experiments we used the same gait parameters: step size 6cm, cycle time 1.75s, double support time 0.34s, foot clearance 4cm, steps number 21. A video of the conducted experiments can be found in Ref. 9.

3.1. *Pelvis sagittal trajectory smoothing*

Since the platform used in our experiments was not equipped with the upper body, the total Center of Mass (CoM) was located below the pelvis (37cm from the ground in free standing condition). Because of this, for CoM to follow a smooth reference generated by the preview controller the pelvis needs to move back and forth. This further results in high accelerations

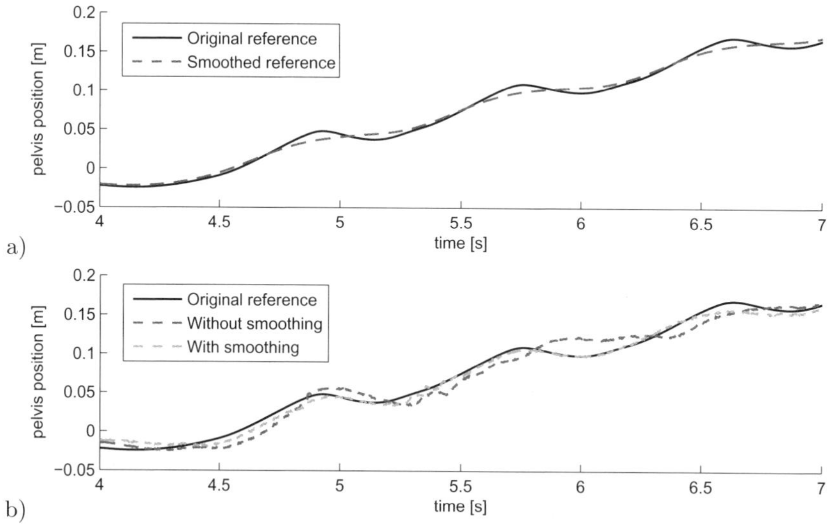

Fig. 2. The pelvis trajectory in the sagittal plane. Graph (a) shows the original trajectory and trajectory after smoothing and (b) shows the original trajectory and experimental results of walking with pelvis reference trajectory with and without smoothing.

and decelerations that cause excessive springs deflection and result in an unstable gait. To cope with the problem we smoothed the pelvis reference trajectory obtained from the preview controller, to reduce the acceleration and deceleration, and as a result follow the real reference closer than before. For smoothing we used the centered moving average filter with the averaging period 0.5s (the value was chosen experimentally). Thanks to use of the centered averaging filter we were avoid introducing lag into the reference. The fragment of trajectory before and after the smoothing is presented on Fig. 2a. The difference between the trajectory of the pelvis in walk with and without smoothing is presented on Fig. 2b. The trajectory was obtained using forward kinematics of joint angles data registered by joint encoders. We can see from the graph that the pelvis follows closer the original reference when executing the smoothed trajectory reference. The graph show only the first four steps. In the experiment without smoothing the robot tipped over after 7th step.

3.2. *Admittance control with polynomial gain modulation*

As described in chapter 2.2, because of the passive compliance, the pelvis sinks down during the single support phase. This results in the swing foot

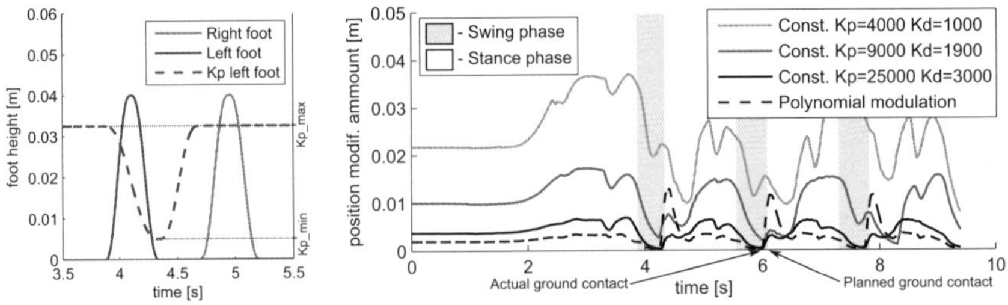

(a) (b)

Fig. 3. Graphs present (a) modulated admittance gain (Kp) reference for left foot during single gait cycle and (b) results from walking experiments comparing left foot height modification amount for the constant gain and the modulated gain. The modulated gains vary as follows: $Kp =$ 1000-50000 and $Kd =$ 750-2500.

being closer to the ground compared with the reference and thus causes early ground contact. This results in strong impact that disturbs the motion of the robot. To cope with the problem we implemented admittance control which adds active compliance to the leg by rising the foot according to the force feedback from the force-torque sensor located below the ankle joint.[10] The formula for calculation of the amount of the leg reference position modification is derived from the force balance of the parallel spring and damper mechanism and has the following form:

$$\Delta z[i] = \frac{F_z + \frac{K_d \Delta z[i-1]}{\Delta t}}{K_p + \frac{K_d}{\Delta t}}, \tag{1}$$

where, K_p and K_d are the spring and damping constants, respectively; the sampling time$\Delta\, t$ is 1ms and F_z is the measured vertical ground reaction force. Initially we tuned the proportional relation between the force and displacement. In order to compensate for the high impact we needed to set the spring constant of the admittance controller to a low value, this however resulted in even bigger sinking of the pelvis and respectively higher impact during the ground contact. To ensure both the soft ground contact and small sinking of the pelvis we developed the polynomial gain modulation. The gain of the admittance control starts from a very high value during the stance phase, then decreases to the minimal value during the swing phase to provide the soft contact and during the double support phase goes back to the high value to reduce the pelvis sinking (see Fig. 3a). We used the polynomial interpolation to extend the duration of the low gain period right before and after the initial contact. The experimental results presented on Fig. 3b show the foot vertical position modification amount resulting from admittance control. One can notice that the lower the Kp gain (spring constant) the higher the sinking of the robot during the stance phase, which not necessarily is followed by the big reaction during the ground contact. The results from experiments with polynomial gain modulation however show that the foot moves up by a significant distance right after the ground contact to decrease the impact and later moves down during the stance phase.

4. Summary

We presented two techniques that make it possible to successfully control the compliant humanoid robot COMAN with conventional pattern generator. We experimentally proved that the smoothing of the sagittal pelvis trajectory reference decreased overshooting of the real pelvis trajectory caused by spring deflection and helped in following the original reference. We also showed that the polynomial modulation of admittance gain helps in increasing the initial reaction to the ground contact and minimizes the amount the whole robot sinks during the stance phase. Application of both techniques results in the long stable gait, while omitting any of them, results in destabilization of the robot and eventual fall. Our future research will focus on the real-time pattern generation techniques, which would allow to regenerate the motion reference in case of big divergence from the original reference.

5. Acknowledgement

The research was performed during author's short stay at IIT. The author would like to thank Zhibin Li, Houman Dallali and Nicolas Perrin for their assistance during the stay.

This work was supported by MEXT/JSPS KAKENHI Grant Number (24360099). It is also partially supported by SolidWorks Japan K.K, DY-DEN Corporation and Cybernet Systems Co.,Ltd whom we thank for their financial and technical support.

References

1. K. Kaneko, F. Kanehiro, S. Kajita, H. Hirukawa, T. Kawasaki, M. Hirata, K. Akachi and T. Isozumi, Humanoid robot HRP-2, in *Proc. IEEE Int. Conf. Robotics and Automation ICRA*, 2004.
2. Y. Sakagami, R. Watanabe, C. Aoyama, S. Matsunaga, N. Higaki and K. Fujimura, The intelligent ASIMO: system overview and integration, in *Proc. IEEE/RSJ Int Intelligent Robots and Systems Conf*, 2002.

3. N. G. Tsagarakis, Z. Li, J. Saglia and D. G. Caldwell, The design of the lower body of the compliant humanoid robot cCub, in *Proc. IEEE Int Robotics and Automation (ICRA) Conf*, 2011.
4. M. Raibert, Dynamic legged robots for rough terrain, in *Proc. 10th IEEE-RAS Int Humanoid Robots (Humanoids) Conf*, 2010.
5. P. Kormushev, B. Ugurlu, S. Calinon, N. Tsagarakis and D. G. Caldwell, Bipedal walking energy minimization by reinforcement learning with evolving policy parameterization, in *Proc. IEEE/RSJ Intl Conf. on Intelligent Robots and Systems (IROS)*, (San Francisco, USA, 2011).
6. L. Colasanto, P. Kormushev, N. Tsagarakis and D. G. Caldwell, Optimization of a compact model for the compliant humanoid robot COMAN using reinforcement learning, *Journal of Cybernetics and Information Technologies* **12** (2012).
7. P. Kormushev, B. Ugurlu, L. Colasanto, N. G. Tsagarakis and D. G. Caldwell, The anatomy of a fall: Automated real-time analysis of raw force sensor data from bipedal walking robots and humans, in *Proc. IEEE/RSJ Intl Conf. on Intelligent Robots and Systems (IROS)*, 2012.
8. S. Kajita, F. Kanehiro, K. Kaneko, K. Fujiwara, K. Harada, K. Yokoi and H. Hirukawa, Biped walking pattern generation by using preview control of zero-moment point, in *Proc. IEEE Int. Conf. Robotics and Automation ICRA*, 2003.
9. Video accompanying the paper, available online: http://kormushev.com/goto/CLAWAR-2013
10. K. Hashimoto, Y. Sugahara, H. Sunazuka, C. Tanaka, A. Ohta, M. Kawase, H.-o. Lim and A. Takanishi, Biped landing pattern modification method with nonlinear compliance control, in *Robotics and Automation, 2006. ICRA 2006. Proceedings 2006 IEEE International Conference on*, 2006.

A BEHAVIOR-BASED LIBRARY FOR LOCOMOTION CONTROL OF KINEMATICALLY COMPLEX ROBOTS

MALTE LANGOSZ[1], LORENZ QUACK[2], ALEXANDER DETTMANN[2], SEBASTIAN BARTSCH[1], and FRANK KIRCHNER[1,2]

[1] *DFKI - Robotics Innovation Center Bremen, Robert-Hooke-Straße 5, 28359 Bremen, Germany*
{malte.langosz, sebastian.bartsch, frank.kirchner}@dfki.de

[2] *University of Bremen, FB3, Robotics Lab, Robert-Hooke-Straße 5, 28359 Bremen, Germany*
{lorenz.quack, alexander.dettmann}@uni-bremen.de

Controlling the locomotion of kinematically complex robots is a challenging task because different control approaches are needed to operate safely and efficiently in changing environments. This paper presents a graph-based behavior description which allows to dynamically replace behaviors on a robotic system. In the proposed approach, every behavior is represented as a directed graph that can be encoded into a data block which can be saved to or loaded from a behavior library. Since this is not a precompiled module like in other systems, the algorithm and parameters of a behavior can still be adapted online by modifying the data that represents the behavior. Thus, machine learning algorithms can optimize an existing behavior to an unknown situation, e.g., a new environment or a motor failure. With a first implementation, it is shown that the proposed behavior graphs are suited for controlling kinematically complex walking machines.

Keywords: Robot, Behavior, Control, Machine Learning

1. Introduction

Behavior-based systems are best suited for changing environments, where fast response and adaptivity are crucial[1]. Their distributed nature increases fault tolerance and promotes component reuse as well as distributed development[2,3]. Unfortunately, coordination effort of behaviors increases with system complexity. Especially, controlling the locomotion of kinematically complex robots is a challenging task. Different locomotion patterns, postures, or reflexes will be more or less efficient depending on the context, i.e., the external environment, the internal state, and the actual task. Actual systems react to context changes by tuning parameters of their existing behaviors[4–6], but work on online integration of new behaviors for completely new contexts is rather sparse.

If a robotic system gets into a new situation where no locomotion behavior of the system fits, it is desired that machine learning algorithms paired with simulation can be used to create new behaviors which can directly included into the running robot control. The goal of the behavior-based library, presented in this paper, is to create a flexible behavior description, which allows an easy creation of tools to design new and change existing behaviors. The concept of the behavior-based library has many similarities to the dataflow synchronous programming language LUSTRE[7] which is used for critical control software in aircraft, rail transportation, and nuclear power plants. The main differences are the modular concept of the behavior-based library and the time handling within the modules. Beside other collections of robotic libraries, like ROS[a] and Rock[b], the behavior-based library is designed to be accessible while a robotic system is operating, e.g., by accessing the behavior-based library from a Rock module one can dynamically change the functionality of that module. The behavior-based library is not designed to be an alternative to frameworks like ROS and Rock, instead it can be used additionally to such frameworks.

The behavior-based library is developed for kinematically complex robotic systems like the SpaceClimber[8–10] (Fig. 1), a walking robot with six legs, each having four degrees of freedom (DOF) and an additional body joint between the front and middle legs. The SpaceClimber has a lot of behaviors which enable it to traverse rough terrain. The robot control is based on cross-compiled C code which runs in MONSTER[11]. In the LIMES[c] project the system serves as test platform for the newly developed behavior-based library. Existing behavior modules shall be exchanged with behaviors from the behavior-based library while either maintaining the same robot behavior or adding additional functionality. Other systems of interest to apply the behavior-based library are introduced in.[12–14]

Section 2 describes the structure of the behavior-based library. Section 3 shows the exemplary realization of the SpaceClimber inverse kinematics and the locomotion pattern in the proposed graph-based approach. A final conclusion and an outlook are given in Sec 4.

---

[a]`http://www.ros.org`
[b]`http://rock-robotics.org`
[c]Learning Intelligent Motions for Kinematically Complex Legged Robots for Exploration in Space

Fig. 1. SpaceClimber: A target system for the behavior-based library.

2. Behavior Description

For the behavior-based library, a behavior module is defined as a block, that maps a defined number of input values to a defined number of output values (Fig. 2). To transfer a signal between different behaviors, a connection can be set up between an output and an input. A signal is defined as a tuple of a value and a weight. Inputs and outputs can have multiple connections, whereby, for each input a merge function defines how the input value is calculated. Possible merge functions are the weighted sum (SUM), the product of the signal values (PRODUCT), the value of the signal with the greatest weight (WTA), the maximum (MAX) and minimum value (MIN).

Additional meta information (textual descriptions) that can be attached to behavior modules, inputs, and outputs. The meta information for behavior modules can include information like name, description, and information that describes its suitability in certain scenarios like walking in a slope. This information is needed by higher decision layers or a human operator to select a suitable behavior for a given context. In addition, the search space for genetic algorithms which suppose to learn new behaviors for new contexts can be further limited, by defining dependencies and properties of behavior modules. The meta information attached to inputs and output can include names, type information, and expected input and output ranges. The latter can be used in conjunction with interval analysis[15] to check the behavior for consistency.

The mapping between the inputs and outputs is done by an algorithm which is either an atomic calculation or a network of behavior modules. Possible atomic calculations include PIPE which simply maps the input to the output, DIV which calculates the reciprocal of the input, SIN, COS,

TAN, ACOS, ATAN2 which calculate the respective trigonometric functions, POWER raises an input to the power of another, SIGMOID, and PHI. All of these take a single input and provide a single output with the exception of ATAN2 and POWER which take two inputs and PHI which takes three. PHI is a node that facilitates conditinal branching in our behavior representation. The idea, taken from code compiler's Static Single Assignment[16] (SSA) form, is to track every variable in the conditional branches independently. Afterwards the different versions of the variables are merged in the PHI (or Φ in SSA lingo) function into a single version depending on the condition of the branch. Therefore, the PHI function takes three inputs, the condition, the last variable version of the then-branch, and the last variable version of the else-branch. See Fig. 3 for a small example of how simple code containing a branching if-statement is turned into a behavior graph representation.

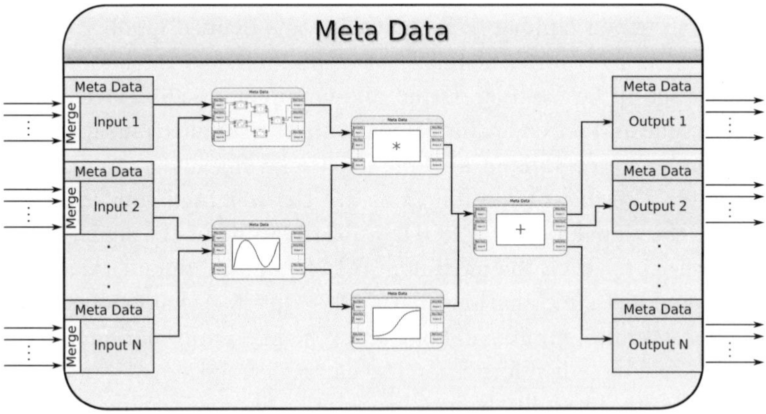

Fig. 2. Behavior module algorithm consisting of a network of atomic transfer functions.

If a behavior module includes a network of behavior modules it can be used on higher levels as a node of a network alongside nodes that include atomic functions or other behavior networks. In this way, a hierarchical decomposition of behaviors is possible without any special handling in the different layers.

Due to the graph representation of an algorithm, tools to work with and on behaviors can be designed without much effort. The behavior graphs can be saved in human readable files, like shown in Table 1. Functions from programming languages can be converted into behavior graphs. Machine

```
def simpleIf(x):
    ret = 0
    if x > 0:
        ret = 2*x
    else:
        ret = -x
    return ret
```

(a) branching code

```
def simpleIf_SSA(x):
    ret_0 = 0
    if x > 0:
        ret_1 = 2*x
    else:
        ret_2 = -x
    ret_3 = PHI(ret_1,ret_2)
    return ret_3
```

(b) branching code in SSA form

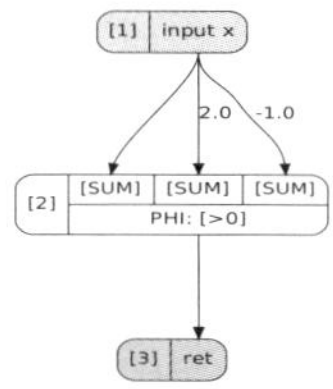

(c) behavior graph with PHI-node

Fig. 3. Simple example of conditional branching and the use of PHI-nodes in the behavior graph.

learning algorithms can easily change or optimize an algorithm by modifying the structure of the corresponding behavior graph. One example is the use of evolutionary computation to adapt the structure and the parameters of a graph. Similar work is successfully done in the field of neuroevolution or genetic algorithms.[17–19] Last but not least, the modular design of a behavior graph makes it easy to use interval arithmetics to validate that a behavior graph operates in a given output range and does not contain singularities like a division by zero.

Table 1. An example of a graph definition in YAML (www.yaml.org) format is shown. First, two nodes, one pipe and one output node, are defined followed by the declaration of the network's input nodes and the definition of an edge from the input node to the output node.


```
nodes:
  - id: 1
    type: "[PIPE]"
    inputs:
      idx: 0
      type: "[PRODUCT]"
      default: 1.0
    outputs:
      idx: 0
  - id: 6
    type: "[OUTPUT]"
    inputs:
      idx: 0
      type: "[SUM]"
      default: 1.0
    outputs:
      idx: 0

networkInputIds:
  - 1

edges:
  - fromNodeId: 1
    fromNodeOutputIdx: 0
    toNodeId: 6
    toNodeInputIdx: 0
    weight: 2.5
```


3. Proof of Concept

The non-reactive part of the locomotion control of SpaceClimber consists of a central pattern generator (CPG; Fig. 4(b)) that feeds a pulse into a pattern generator for each leg which in turn generate the feet's target cartesian coordinates. These are then translated into joint angles by the inverse kinematics module (Fig. 4(a)).

This part of the robot control software was chosen to perform a first proof of concept for the behavior-based library. The existing C code was reimplemented in Python[d]. This code was then translated by a script using Python's built-in parsing capabilities, i.e., the ast module, into nine behavior graphs. All behavior graphs are saved in the behavior library. By loading the top level graph (see Fig. 4(b)) the other eight graphs are loaded indirectly. The overall control network has no input and the outputs are transferred to the motors of a simulated SpaceClimber robot. For a single leg, the network consist of 264 nodes and 346 edges spread over eight graphs with the inverse kinematics alone contributing 45 nodes and 60 edges.

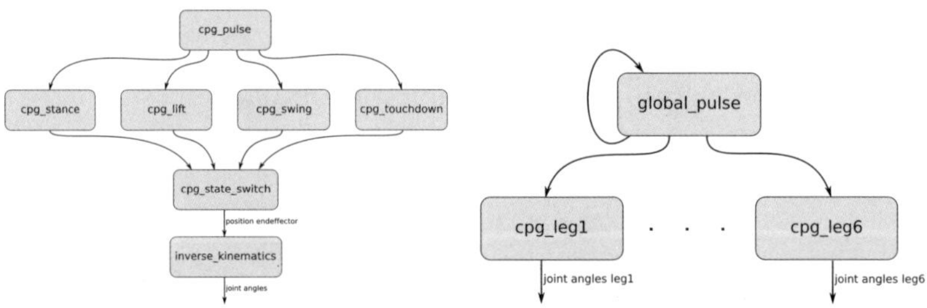

(a) Pattern generator for a single leg (b) CPG using Fig. 4(a) six times

Fig. 4. For a single leg (4(a)) all states are executed in parallel. They are subsequently merged in the cpg_state_switch by using the PHI-nodes discussed in Sec. 2. On a higher level (4(b)) the legs are coordinated by the CPG. Both figures depict hirarchical behavior graphs where each block uses other behavior modules. In a scenario of a changing environment, e.g., if the surface becomes icy, it would be possible to replace the `cpg_stance` module by a `cpg_stance_ice` module while the robot is operating.

As a first test of the applicability of the interval analysis the SpaceClimber's inverse kinematics for a single 4DOF leg was chosen as a target for analysis. The interval analysis determines the output intervals for a given a set of input intervals. We considered all possible inputs, i.e., the inputs are arbitrary real numbers in the interval $[-\infty, \infty]$. The results are shown in Table 2. The occurrence of NaN (Not a Number) in joint_2 indicates a potential problem in the code which turned out to be a division by zero error

[d] `http://www.python.org/`

that could only occur outside the robot's range of operation. This example demonstrates the benefits of interval analysis in this field, since a division by zero can lead to system crashes with unforeseeable consequences.

Table 2. Interval analysis for SpaceClimber's inverse kinematics.

Input intervals			Output intervals	
lean:	$[-\infty, \infty]$	interval analysis $\longrightarrow$ of inverse kinematics	joint_0:	$[-\infty, \infty]$
x:	$[-\infty, \infty]$		joint_1:	$[-4.71259, 3.05544]$
y:	$[-\infty, \infty]$		joint_2:	[−NaN, NaN]
z:	$[-\infty, \infty]$		joint_3:	$[-3.141, 3.14219]$

4. Conclusion and Outlook

The behavior-based library overcomes the gap between machine learning algorithms and the locomotion control of kinematically complex robots by allowing online adaptation of behaviors in the sense of adding, removing, or changing behaviors. The graph-based behavior representation gives the opportunity to analyze single parts of the network more in detail and only a small part of software has to be verified to ensure that no memory leaks, segmentation faults, or out of bounds errors crash the control software or transfer it into an undefined state. With a first implementation, it is shown that the behavior graph is able to represent the CPG-based control of a complex six-legged robot.

In future work, the execution of the behavior graphs will be implemented on the real SpaceClimber robot. Therefore, the robot's operating system MONSTER will be extended to execute the behavior graphs. Afterwards, an interface to parameter optimization algorithms will be developed and a genetic algorithm to evolve or modify the structure of a behavior will be designed. The later can be used to connect existing behavior modules as well.

Acknowledgments

The project LIMES is funded by the German Space Agency (DLR, Grant number: 50RA1218 / 50RA1219) with federal funds of the Federal Ministry of Economics and Technology (BMWi) in accordance with the parliamentary resolution of the German Parliament. LIMES is a cooperation between the DFKI Bremen Robotics Innovation Center and the University of Bremen.

References

1. B. Siciliano and O. Khatib, *Springer handbook of robotics* (Springer, 2008).
2. R. Brooks, A robust layered control system for a mobile robot *IEEE Journal of Robotics and Automation* (IEEE, 1986).
3. R. Arkin, *Behavior-based robotics* (MIT press, 1998).
4. J. Albiez, *Verhaltensnetzwerke zur adaptiven Steuerung biologisch motivierter Laufmaschinen* (GCA-Verlag, 2007).
5. B. Gaßmann, *Modellbasierte, sensorgestützte Navigation von Laufmaschinen im Gelände* (Univ.-Verlag Karlsruhe, 2007).
6. F. Michaud, Selecting behaviors using fuzzy logic, in *Proceedings of the Sixth IEEE International Conference on Fuzzy Systems*, 1997.
7. N. Halbwachs, P. Caspi, P. Raymond and D. Pilaud, The synchronous dataflow programming language LUSTRE, in *Proceedings of the IEEE*, 1991.
8. S. Bartsch, T. Birnschein, M. Römmermann, J. Hilljegerdes, D. Kühn and F. Kirchner, Development of the six-legged walking and climbing robot SpaceClimber *Journal of Field Robotics* (Wiley Subscription Services, 2012).
9. S. Bartsch, T. Birnschein, F. Cordes, D. Kuehn, P. Kampmann, J. Hilljegerdes, S. Planthaber, M. Roemmermann and F. Kirchner, SpaceClimber: Developement of a Six-Legged Climbing Robot for Space Exploration, in *Proceedings for the joint conference of ISR and ROBOTIK*, 2010.
10. M. Römmermann, D. Kuehn and F. Kirchner, Robot design for space missions using evolutionary computation, in *IEEE Congress on Evolutionary Computation*, 2009.
11. D. Spenneberg, M. Albrecht and T. Backhaus, M.O.N.S.T.E.R.: A New Behavior-Based Microkernel for Mobile Robots, in *Proceedings of the ECMR*, 2005.
12. M. Eich, F. Grimminger and F. Kirchner, A versatile stair-climbing robot for search and rescue applications, in *2008 IEEE International Workshop on Safety, Security and Rescue Robotics*, (IEEE, 2008).
13. D. Spenneberg and F. Kirchner, *The Bio-Inspired SCORPION Robot: Design, Control & Lessons Learned* (I-Tech Education and Publishing, 2007).
14. F. Spenneberg, D.; Kirchner, Omnidirectional walking in an eight legged robot., in *In Proceedings of the 2nd International Symposium on Robotics and Automation*, (VDE Verlag GmbH, 2000).
15. R. E. Moore, *Interval Analysis* (Prentice Hall, 1966).
16. R. A. Kelsey, A correspondence between continuation passing style and static single assignment form, in *Papers from the 1995 ACM SIGPLAN workshop on Intermediate representations*, (ACM, 1995).
17. L. Cardamone, D. Loiacono and P. L. Lanzi, *IEEE Trans. Comput. Intellig. and AI in Games* , 176 (2010).
18. Y. Kassahun, J. de Gea Fernández, M. Römmermann and F. Kirchner, On applying neuroevolutionary methods to complex robotic tasks, in *IEEE IROS Workshops on Exploring new horizons in Evolutionary Design of robots*, 2009.
19. M. Römmermann, M. Ahmed, L. Quack and Y. Kassahun, Modeling of leg soil interaction using genetic algorithms, in *Proceedings of International Conference of the International Society for Terrain-Vehicle Systems*, 2011.

EFFECTS OF NECK SWING MOTION ON THE BODY POSTURE OF A FOUR-LEGGED ROBOT

KOJI SHIBUYA

Dept. of Mechanical and Systems Engineering, Ryukoku University, 1-5, Yokotani Seta-Oe, Otsu, Shiga, 520-2194, Japan

KENJI MATSUHIRA

Dept. of Mechanical and Systems Engineering, Ryukoku University, 1-5, Yokotani Seta-Oe, Otsu, Shiga, 520-2194, Japan

This paper discusses how the neck swing motion of a four-legged robot affects its body posture, which is crucial for stabilized walking and running, through simulation. Using ODE (Open Dynamics Engine), we modeled a quadruped robot, based on a real robot that we have been developing, that has two joints in each front leg and three joints in each hind leg. First, we calculated the COG (center of gravity) at different neck joint angles. Then, we created a toe trajectory and chose the best set of the trajectory parameters by evaluating the pitch angle of the body. Finally, we investigated the effect of the neck swing motion on the body roll and pitch angles.

1. Introduction

To date, researchers have developed many legged systems [1]. We have focused on the four-legged animals, particularly horses, because of the horse energetics reported by Hoyt and Richard [2]. They concluded that horses use a gait pattern that best minimizes their energy consumption for locomotion per unit distance. If this result can be applied to a four-legged robot, we must choose the most suitable speed to each gait to reduce energy consumption. Another reason for our focus on horses is their neck swing motion. In the field of horse racing, some believe that this neck swing motion contributes to locomotion stability. Although we have not uncovered any academic literature supporting this hypothesis, we are of the opinion that this claim explains the stabilized body posture of racehorses.

We decided to build a four-legged robot to investigate these issues. In this paper, we focused on the relationship between neck swing motion and body posture. Before experimenting with an actual robot, we investigated this relationship through simulation. Therefore, the goal of this paper is to reveal the effects of the neck swing motion of a four-legged robot on its body posture through simulation.

Many quadruped robots have been constructed. For example, Raibert built a quadruped robot with a hopping motion [3]. At Boston Dynamics, he and his team also developed a four-legged robot with highly stable locomotion named BigDog [4]. This robot performed dynamic motions. Hirose et al. built a series of quadruped robots, some of which were applied to actual construction machineries [5]. Recently, Gonzalez et al. built a quadruped robot and discussed its control strategy for maintaining stability [6]. Basically, these robots adopted statically stable locomotion. However, we have not yet found four-legged robots utilizing neck swing motion to stabilize their body posture. In addition, the studies previously mentioned did not focus on the relationship between energetics, gait, and locomotion speed.

In our previous paper, we reported that the toe trajectory and timing of the neck swing motion significantly affect body posture and speed [7]. However, the relative length of the leg to the body was short compared with a real horse and the position of the robot's COG (Center Of Gravity) was low, resulting in stable locomotion. To investigate the effects of neck swing motion more clearly, we need to build a less stable robot with a higher COG position. Therefore, we built a new robot with longer legs.

In designing the robot, we referred to the link length ration of a real horse. We also created a new toe trajectory to prevent the robot from stumbling. We then performed a simulation using ODE (Open Dynamics Engine) [8] to determine the best locomotion parameters for stabilizing body posture in walking gait. Because walking gait is more stable than other gaits, such as gallop and trot, we chose the walking gait as the first step of this study. In this paper, we introduce the new robot and our simulation results.

2. New Robot and Simulation Model

Figure 1 shows a photograph of the new four-legged robot developed in our laboratory (left) and its simulation model in ODE (right). Each front leg has two joints, each hind leg has three joints, and the neck has one joint for a total of eleven joints. The ratio of lengths between each link is almost the same as those of real horses, resulting in a longer leg length and a higher COG position than those of our previous robot. The dimensions and masses of each link used in the simulation model (Fig. 1, right) are shown in Table 1.

It is impossible to completely mimic the ratios of the masses of all links because the robot was made of aluminum and the DC motors have densities that are larger than those of real horses. Because the purpose of mimicking a real horse is to raise its COG to a higher position in order to achieve a more unstable state, we focused on geometrical dimensions only. In ODE, the ground contact between the toe and the ground was modeled by a mass-spring-dashpot model [8].

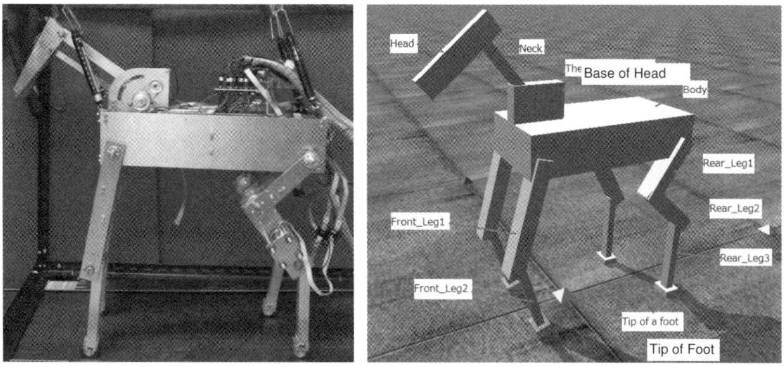

Figure 1. Left: Photograph of four-legged robot developed by our laboratory. Right: Simulation model based on the real robot. The dimensions and masses are almost the same.

Table 1. Dimensions and mass of each link and body used in simulation.

Link name	Dimensions [mm]	Mass [kg]
Body	256x520x120	7.0
Front_Leg1	56x45x335	0.9
Front_Leg2	30x25x230	0.28
Rear_Leg1	57x41x200	1.0
Rear_Leg2	47x42x215	1.4
Rear_Leg3	30x25x240	0.8
Tip of Foot	6x5x20	0.07
Base of Head	44x150x100	0.63
Neck	30x25x305	0.2
Head	36x240x36	0.1

3. Toe Trajectory

We created a new toe trajectory in the body frame for each foot of the new robot. Figure 2 shows the toe trajectory of both legs. The trajectory can be divided into two parts: the swing and the stance trajectories. In the swing phase of the hind leg, to prevent the toe from stumbling, the toe is lifted just after liftoff more quickly than with our previous trajectory. We determined the following five trajectory parameters: T [s] is the step period, A [m] is the half step length, and y_0 [m] and y_1 [m] are the vertical distances between the toe and the height of the shoulder or the hip joint axis at touchdown and in the swing phase, respectively. S [m] is the depth representing the distance from the ground to the lowest point of the trajectory. Considering the performance of the real robot, we set all the parameters as follows: T = 1.0 s, A = 0.1 m, y_0 = 0.5 m, and y_1 = 0.47 m. In simulation, S can be set at different values from 0.0001 to 0.02 m and we will examine the effects of the value of S on the body pitch angle in Section 4.2. Depth S cannot be measured in simulation and experiments using the real robot because we assumed that the ground is hard and the foot cannot "dig" into the ground in reality.

We calculated each joint angle from the trajectory data using inverse kinematics. Because the fore and hind legs have two and three joints and move in a plane, they were modeled as a two dimensional manipulator with two or three joints. The base joints (shoulder or hip) were fixed. To calculate the hind leg joint angles, we had to determine the direction of the hind leg foot because it has three joints. The direction was changed according to foot position. In the swing phase, the direction was reversed quickly just after liftoff. Figure 3 shows an example of a stick picture of the hind leg. The hip joint is fixed.

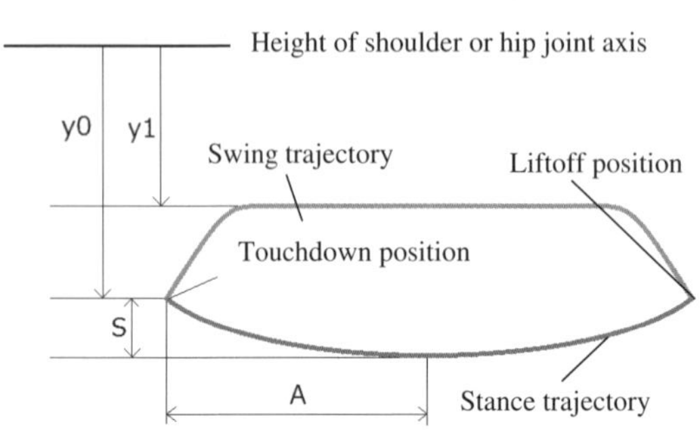

Figure 2. Schematic drawing of toe trajectory used in simulation.

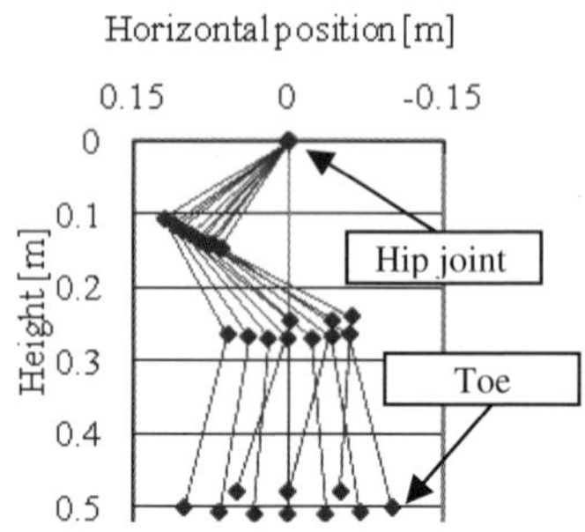

Figure 3 Stick picture of hind leg.

4. Simulation Results

4.1. *Effect of Head Mass and Neck Angle on COG*

We calculated the robot's COG to investigate the effect of head mass and neck angle. The masses added to the head were 0.0, 1.0, and 2.0 kg. The neck angle is the angle between the horizontal line and the neck link (Fig. 4), which was set at 0.0 deg to 90.0 deg.

Figure 4 shows the results. In this figure, a picture of the robot body is superimposed behind the positions of the COG. The solid and dashed lines represent the COG positions when the added head mass is 2.0 kg and 1.0 kg, respectively. For both lines, the lowest points and highest points represent the results when the neck angle is 0 and 90 deg. The rhombus plot represents the COG position when the added head mass is 0 kg. At the intersection point of the calculated values with the vertical axis (center of the body), the added head mass is 1.0 kg and the neck angle is approximately 40 deg. The position of the COG should be placed near the center of the body because it increases the degree of stability when the robot stands up with its four legs, which may increase the stability when walking and running. Therefore, the added head mass is 1.0 kg in this study.

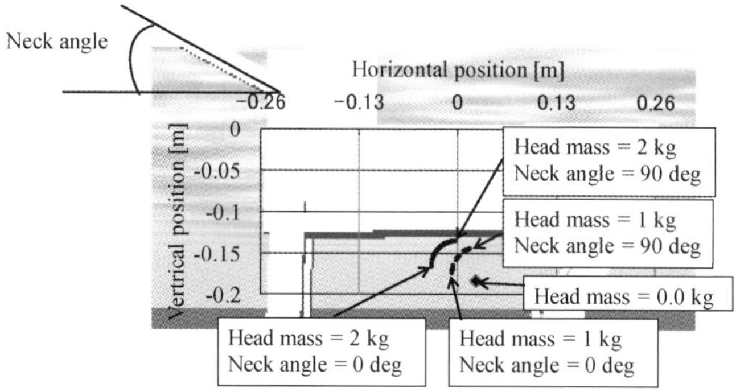

Figure 4. COG for various head masses and neck angles.

4.2. *Effects of Trajectory Parameters on Robot Pitch Angle*

We focused on the center position of the toe trajectory. We shifted it forward and backward (Fig. 5), and conducted a simulation in which the robot walked. We calculated the pitch angle of the body because the variation of the pitch angle seems to be important for racehorses. Of course, there are many animals that vary their body pitch angle in dynamic locomotion. However, racehorses seem to maintain their body posture, which is suitable for carrying loads, such as human riders. Therefore, we focused on the body pitch angle. In addition, although the roll body angle is also an important parameter, we ignored it in this section for the sake of simplicity.

In this simulation, depth S was 0.01 m and the neck angle was fixed to 40 deg. We examined various patterns of the center positions of the toe trajectories. Figure 6 shows an example of the pitch angle variation and Figure 7 shows the relationship between the shifting distances and the pitch angle variations. In the horizontal axis, minus indicates the backward direction. As illustrated in Figure 7, we found that the pitch angle variation is smallest when the front leg position is 0.01 m forward from the shoulder joint axis and the hind leg position is 0.03 m backward from the hip joint axis. The variation is approximately 4 deg.

We also compared the pitch angles in different duty factors and found that swing phase: stance phase = 3:7 is suitable because the variation is smaller than another duty factor (2:8).

We next investigated the effects of depth S. We examined various values in the front and hind leg depths and found that 0.02 m for the front leg and 0.01 m for the hind leg is most suitable in regards to the smallest pitch angle variation (Fig. 8).

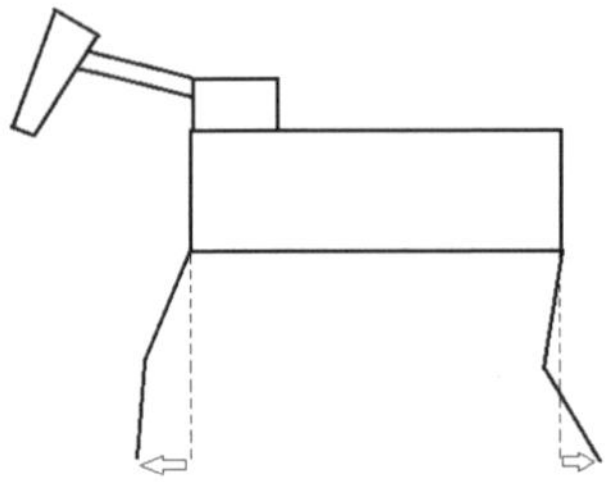

Figure 5. Shifting center of toe trajectory.

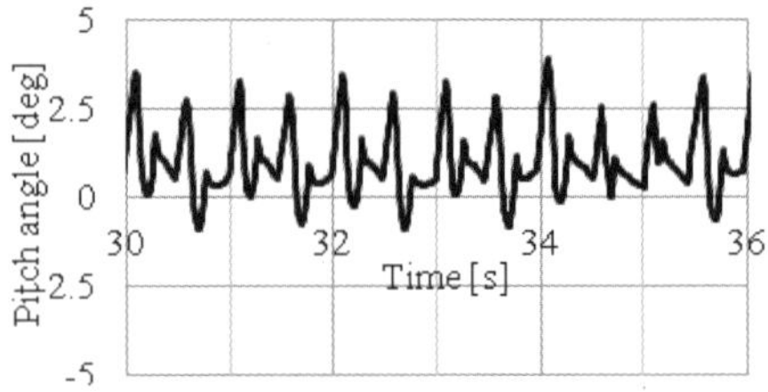

Figure 6. Pitch angle with the smallest variation of shifting center of toe trajectories in simulation. (Foreleg: 0.01 m, Hind leg: 0.03 m).

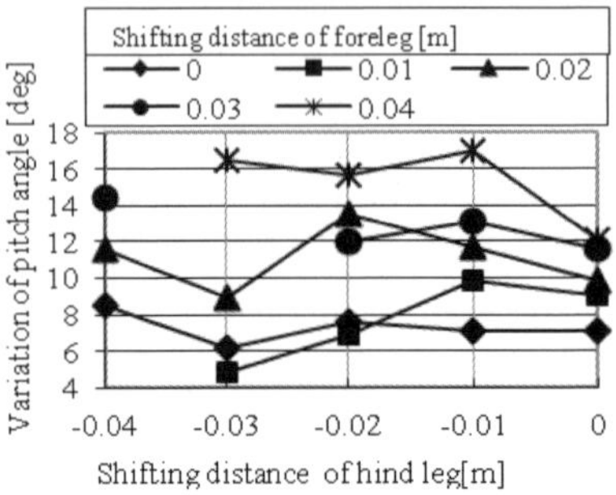

Figure 7. Relationship between variations of pitch angles and shifting distances.

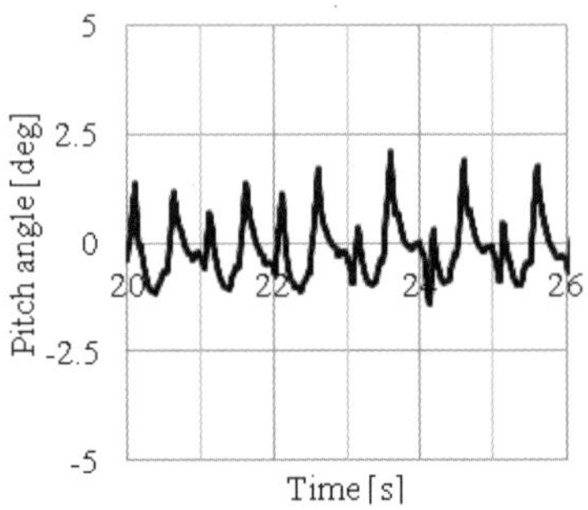

Figure 8. Pitch angle with the smallest variation of varying the depths in simulation. (Foreleg: 0.02 m, Hind leg: 0.001 m).

4.3. *Effects of Neck Swing Motion*

By using the trajectory parameters in 4.2, the robot walked for short periods. However, the robot could not walk for long periods when the neck joint was fixed. For example, when the neck joint is fixed at 60 deg, the robot walks for only 124 s before it falls down. In the simulation, the robot has inclination sensors and thus can change its leg motions (e.g., lengthening the leg length) according to the sensor values to lengthen the walking period. However, such feedback control should be minimized to reduce computational power. If other motions can increase body stability, we should apply these to the real robot. Therefore, we focused on the neck swing motion to compensate body stability. We investigated the effects of the neck swing motion on the robot's pitch and roll angles.

The time period of the swing motion is 0.5 s, which is half the step time of the toe. This means that the robot swings its neck two times per stride. When the toe of the front leg touches down, the head begins swinging backward and begins swinging forward when the other foreleg begins the swing phase (Fig. 9).

We simulated for 600 s, which is long enough to establish a steady walking state. First, we set the amplitude of the neck swing motion and found that the robot fell down within 100 s. We then, set it to 20 deg and examined the following eight angle patterns: 0-20, 10-30, 20-40, 30-50, 40-60, 50-70, 60-80, and 70-90 deg. In the 0-20-deg pattern, the robot fell down after 577 s. Also, in the 30-50-deg pattern, it fell down just after starting to walk. The other six patterns walked for 600 s.

Figures 10, 11, and 12 show examples of the results (10-30-deg, 40-60-deg, and 60-80-deg patterns). In Figure 10, the variation of the roll angle is relatively large and almost constant, while that of the pitch angle is small. The results for the other two patterns (0-20-deg and 20-40-deg) are almost the same as in Figure 10. In Figures 11 and 12, the variation of the roll angle decreases, while that of the pitch angle increases. The same tendency was observed in the 70-90-deg pattern.

If the maximum neck angle is below 40 deg, the roll angle is large. On the other hand, if the lowest neck angle is above 40 deg, the body posture pattern is different. The pitch angle increases, while the roll angle decreases. Although it

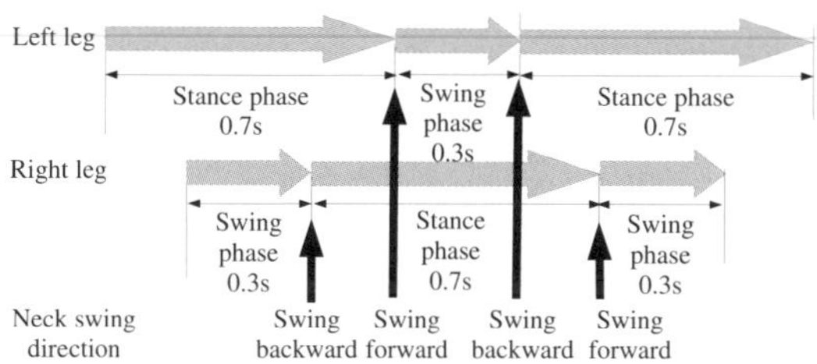

Figure 9. Swing timing chart.

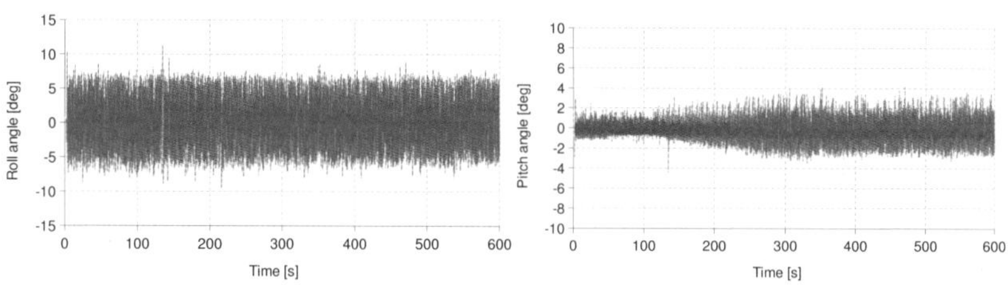

Figure 10. Example of roll and pitch angles (10-30-deg pattern).

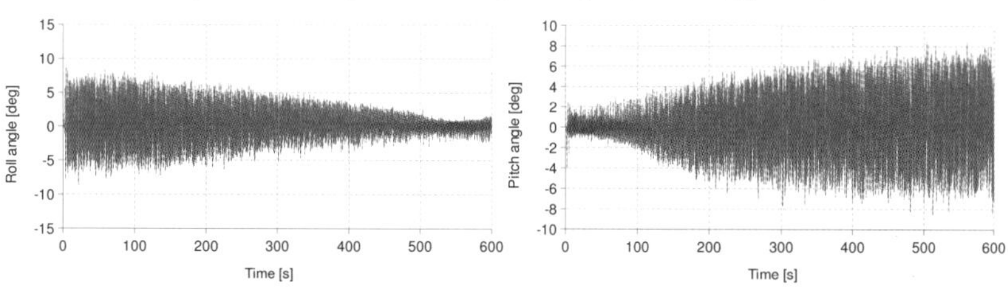

Figure 11. Example of roll and pitch angles (40-60-deg pattern).

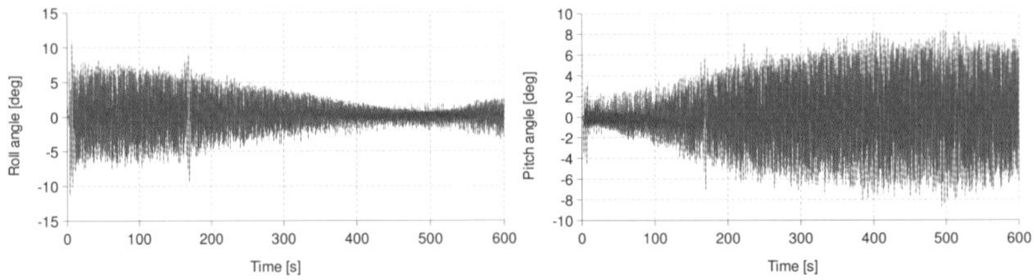

Figure 12. Example of roll and pitch angles (60-80-deg pattern).

is difficult to conclude which neck swing angle is best from the data, the neck swing motion affects body posture and we can utilize it to stabilize the robot.

5. Conclusion

In this paper, we introduced a new four-legged robot whose ratio between each link length is geometrically similar to that of horses. We then created a new toe trajectory resulting in low pitch angle variation. Using the trajectory, we investigated the effect of neck swing motion on the robot's posture. In the future, we want to conduct simulation for longer periods and investigate other dynamic gaits. Adding sideways motion may also be useful. We then will apply our results to the real robot. In addition, we intended to investigate the effects of neck swing motion on energy consumption.

References

1. D. C. Kar, Design of Statically Stable Walking Robot: A Review, *J. of Robotic Systems*, **20**(11), 671-686, (2003).
2. D. F. Hoyt and C. Richard, Gait and the Energetics of Locomotion in Horses, *Nature*, **292**, 16, (1981).
3. M. H. Raibert, Legged Robots That Balance, *The MIT Press*, (1986).
4. M. Raibert, K. Blankespoor, G. Nelson, R. Playter, and the BigDog Team, BigDog, The Rough-Terrain Quadruped Robot, *Proceedings of the 17th World Congress, The International Federation of Automatic Control* (*IFAC2008, Seoul, Korea*), July 6-11, (2008).
5. T. Doi, R. Hodoshima, Y. Fukuda, S. Hirose, T. Okamoto, and J. Mori, Development of Quadruped Walking Robot TITAN XI for Steep Slopes - Slope Map Generation and Map Information Application -, *J. of Robotics and Mechatronics*, **18**(3), 318-324, (2006).
6. P. Gonzalez de Santos, E. Garcia, and J. Estremera, Quadrupedal Locomotion: An Introduction to the Control of Four-legged Robots, *Springer*, (2006).
7. K. Shibuya and T. Fukui, Influences of Toe Trajectories and Head Motions on Gaits of a Four-legged Robot, *Proc. of 13th International Conference on Climbing and Walking Robots and the Support Technologies for Mobile Machines* (*CLAWAR2010, Nagoya, Japan*), 493-500, (2010).
8. http://ode.org/ode.html

ACTUATOR SIZING FOR HIGHLY-DYNAMIC QUADRUPED ROBOTS BASED ON SQUAT JUMPS AND RUNNING TROTS

H. KHAN*, C. SEMINI and D. G. CALDWELL

Department of Advanced Robotics, Istituto Italiano di Tecnologia (IIT)
Via Morego 30, 16163 Genova, Italy
**E-mail: hamza.khan@iit.it*
http://www.iit.it/en/advanced-robotics.html

V. BARASUOL

PPGEAS - Dept. of Automation and Systems
Federal University of Santa Catarina (UFSC)
Florianpolis, SC, Brazil 88040-970,
E-mail: victor@das.ufsc.br

It is challenging to design a quadruped robot that can perform highly dynamic tasks like jumping and running. Estimating appropriate range of joint torques and velocities is essential for the selection of the leg actuators. Jumping and running are considered as extreme tasks that push the actuators to their limits. In this paper we proposed a simple method that allows a quadruped robot designer to obtain approximate peak joint torques and joint velocities needed for a running trot at various forward velocities and squat jumps at different heights. A SLIP model is used for the mapping of CoM trajectory of a quadruped robot during a running tort. Experiments for a squat jump and running trot are performed with the highly dynamic quadruped robot HyQ for the validation of the proposed approaches. A case study is also discussed to demonstrate the usage of proposed tool.

Keywords: Quadruped Robot Design; Scaling; Actuator Sizing; Running Trot.

1. Introduction

Legged robots have become a much more prominent field of robotics in recent years, due to their potential versatility and capability of performing tasks that conventional vehicles are unable to do. Most of the recent legged robots, however, lack the versatility of performing both precise navigation over rough terrain and the strong, fast motions that are necessary for dynamic tasks such as jumping and running. Presently very few examples exist besides BigDog[1] by Boston Dynamics and HyQ[2] by IIT.

To design such kind of machines, firstly the designer has to define the tasks that the robot should accomplish then choose the actuator that satisfies the task requirements in terms of joint velocity and torque. It is easier for a designer to narrow down the desired dimensions and mass of the robot. But when it comes to the selection of actuators, it is not trivial to obtain appropriate joint velocity and torque limits without correctly modeling the whole robot in simulation and implementing stable locomotion controllers. The aim of this work is to build a scaling tool that helps the quadruped robot designers to select actuators that fulfill the desired requirements. We started our study using the squat jump as a characteristic motion for highly dynamic robots, obtaining peak values of joint velocity and torque in relation to robot mass, leg segment length and desired jump height.[3] This paper presents the extension of that study with a running trot to get peak torques and velocities in relation to forward velocity, robot mass and leg segment length. The validation of the proposed methodology is confirmed by performing a squat jump and running trot experiments with our highly dynamic quadruped robot HyQ.

2. State of the Art

In the early 1980's in the field of applied zoology, a number of experiments were conducted to study the dynamic similarities in quadruped mammals.[4–6] The hypotheses formulated by these studies layout the preliminary bio-inspired sizing guideline for the legged robot designers. Alexander[5] found that the legged animals of different sizes tend to move in dynamically similar fashion whenever their Froude numbers[a] are equal. In the light of dynamic similarity concept, it is also predicted that the geometrically similar animals of different sizes exhibit equal duty factors and equal relative stride lengths ($stridelength/h$), when they are traveling with equal Froude number.[6,8] To use these biologically inspired criteria for a legged robot design, a quadruped robot designer needs to solve an optimization problem by taking care of desired goals of machine.

Similar studies were conducted for the electric DC motor sizing of a bounding robot[9] and a hopping monopod robot.[10] But these studies were limited to quadruped robots with telescopic legs or a specific type of actuator. Another scaling study is done for Oncilla-robot (slightly bigger version of successor CheetahCub-robot)[11] but the study is only limited to

[a]Froude number is a dimensionless ratio used to study trends in animal gait patterns where the gravitational forces are important[7] and it is defined as $\frac{v^2}{gh}$, where v is forward velocity, g is the acceleration of free fall and h is the height of hip joint from the ground.

pantograph-based leg quadruped robots. The current work is for quadruped robots with articulated legs and it is not restricted to only one type of actuators.

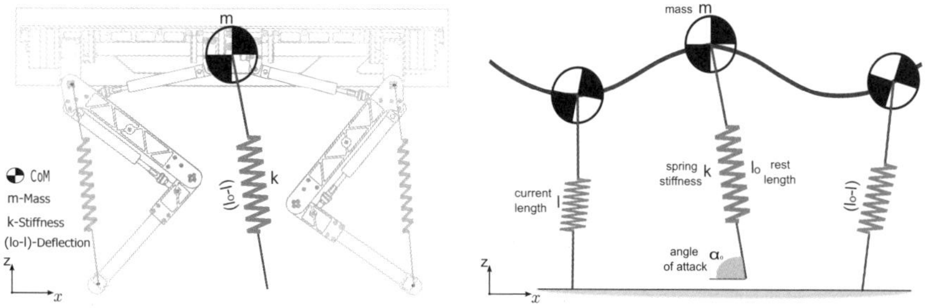

Fig. 1. **(a)** A quadruped robot with the equivalent virtual spring (connected to CoM) represents the virtual linear spring of a diagonal pair of legs **(b)** A CoM trajectory (purple solid) of the legged robot during a running trot represented by the SLIP model.

3. Simulation

Running Trot Simulation Desired joint torques are generated by placing a virtual linear spring between the hip and foot of each articulated robot leg[12] to control the motion of the leg during running. Each virtual spring of a diagonal leg pair is working synchronously in a trot and can be represented by one equivalent virtual spring, Fig. 1a. Blickhan and Full showed in one of their studies that the running motion for multilegged locomotion behaves like a bouncing monopod and they calculated the compression of a virtual monopod's leg and its stiffness from the animals' mechanical-energy fluctuation and ground-reaction force .[13] The center of mass (CoM) motion of any quadruped during a running trot can be approximated by a spring-loaded inverted pendulum (SLIP) model as it is observed in animals,[14] Fig. 1b. A point mass m attached to a mass-less linear spring can be described as a SLIP model where k is the spring stiffness, l is the current length and l_0 is the rest length. During stance phase, the spring force exerted on the ground is defined as $k(l_0 - l)$. During flight phase, the mass follows a parabolic trajectory under the law of gravitation. We define the spring rest length in relation to the leg segment length l_{ls} (Fig. 2b) of each articulated robot leg as follows: $l_0{=}K_l l_{ls}$, where K_l is the ratio between the leg segment length and the spring rest length. In our simulation K_l is equal to 1.41, which we assume to be a good value based on our experience with HyQ. From the SLIP model, we know the foot trajectory in Cartesian coordinates of the equivalent virtual spring leg during stance. Using the leg Jacobian, we can calculate the angular position, velocity and acceleration of hip and knee joints of the articulated robot leg. During the stance phase of a quadruped running trot only two feet touch the ground. So the ground reaction forces (GRF) for each articulated robot leg can be defined as $F = F_{vl}/2$ assuming evenly distributed load, where F_{vl} is the force for the equivalent virtual spring. The leg's GRF is then transformed into hip and knee joint torques with the Jacobian transpose.

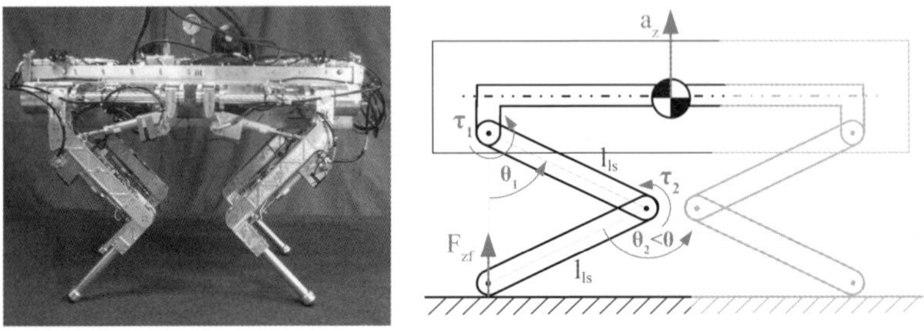

Fig. 2. HyQ: Hydraulic Quadruped robot. **(a)** picture and **(b)** sketch of side view of robot in squat posture, defining the centre of mass (CoM); acceleration vector a_z; joint angles $\theta_{1,2}$ and torques $\tau_{1,2}$ of the hip and knee joint, respectively; leg segment lengths l_{ls} and ground reaction force vector at the foot F_{zf}.

Squat jump Simulation As we published in[3] a squat jump is composed of several phases: first, a vertical *acceleration phase* from a squatting posture until lift-off; then, a parabolic *flight phase* with the legs moving to a suitable landing posture. The jump height is the crucial input to our simulation and is measured as the vertical distance that the CoM travels from the time the body lifts off the ground to the end of the upward motion. The maximum jump height of any object undergoing a parabolic flight phase is directly related to its lift-off velocity v_{lo} which can be obtained by equating the kinetic energy at lift-off with the potential energy at the maximum jump height. The estimated v_{lo} leads to the required constant vertical acceleration a_z and vertical force F_{zf} during the acceleration phase. The maximum vertical force is transformed into maximum knee joint torque with the Jacobian transpose. The above mentioned parameters are illustrated in Fig. 2b.

4. Experimental Platform: HyQ Robot

The experimental platform used in this study is the versatile, quadruped robot HyQ,[15] Fig. 2a, a hydraulically and electrically actuated machine that weighs $70kg$, is $1m$ long and has a leg segment length of $0.35m$. The robot's legs have three degrees of freedom each, two hydraulic joints in the sagittal plane (hip and knee flexion/extension) and one electric joint for hip abduction/adduction. Each joint has 120° range of motion and is controllable in torque and position. The maximum joint torque is 145Nm for the hydraulic and 152Nm for the electric joints.

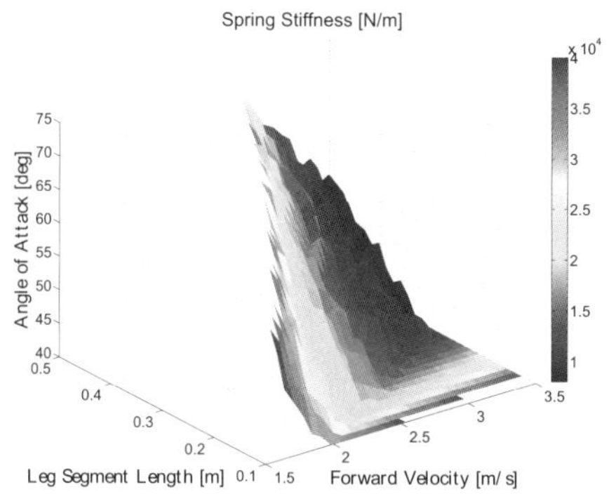

Fig. 3. Plot of stable SLIP parameters(Spring Stiffness and Angle of Attack) for different leg segment lengths and forward velocities at robot mass 73kg.

5. Results

Using the SLIP model we performed a number of trotting simulations for a range of forward velocities and leg segment lengths. First of all, suitable SLIP parameters (spring stiffness and angle of attack) had to be calculated for each input pair based on the steps-to-fall map.[16] The SLIP parameters with the lowest spring stiffness that resulted in stable hopping of 50 or more steps were then selected, shown in a Fig. 3. This way low impact peaks were obtained. A running trot experiment is performed on HyQ at forward velocity 1.6 m/s with 40% duty factor and it gives peak hip and knee joint torques equal to 24.3 N/m and 115.5 N/m respectively. 'HyQ Exp' arrows shown in Fig. 4 are the estimated peak hip and knee joint torques for a running trot at 1.6 m/s, equal to 21.9 Nm ($0.3Nm/kg * 73kg$) and 116.8 Nm ($1.6Nm/kg * 73kg$) respectively. Fig. 4 shows the peak joint torque scaled by the robot's body weight (BW) for different leg segment lengths and forward velocities. White areas indicate where the SLIP model failed to perform 50 steps for the given parameter range.

For a squat jump, Fig. 5a is a validation of proposed approach by comparing HyQ's experimental data with simulation.[3] The peak torques scaled by the BW are shown in Fig. 5b for different leg segment lengths and jump

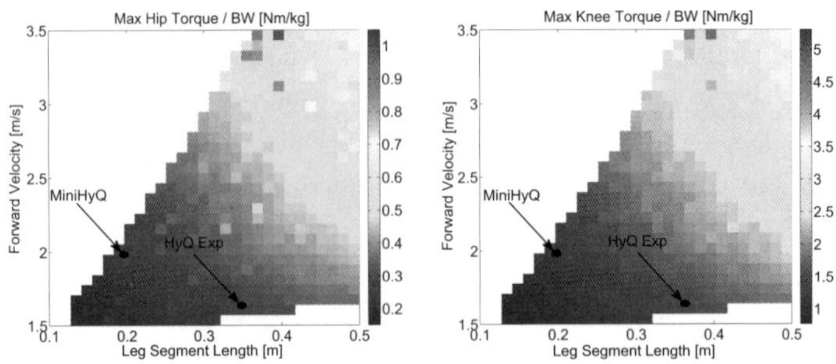

Fig. 4. **(a)** Plot of maximum hip joint torque scaled by BW for different leg segment lengths and forward velocities. **(b)** Plot of maximum knee joint torque. The arrows show the estimated result according to desired task specification for MiniHyQ (see Section 6).

heights. The three subplots of Fig. 5a show the data of the experiment (red solid line) and of the simulation (black dashed line) for the knee joint angle (top), knee joint torque (middle) and vertical ground reaction force (bottom). The acceleration phase starts at $0.1s$ and lasts till $0.4s$ when the torques go to zero. The robot touches down again at $0.78s$. The simulation calculates values only during the acceleration phase.

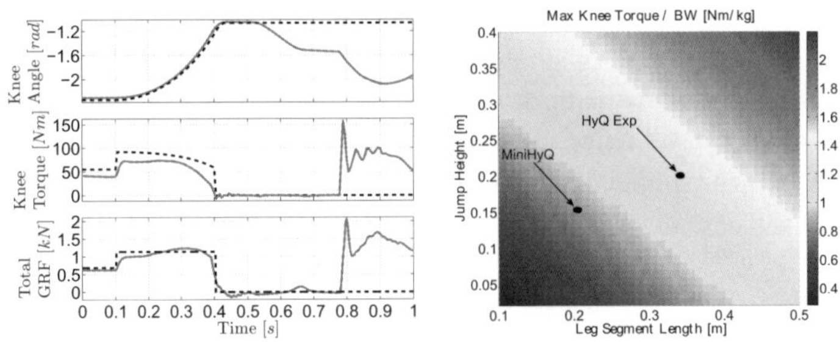

Fig. 5. **(a)** Plot of experimental data (red solid) and simulated data (black dashed) for a squat jump motion of 0.2m jump height. *Top:* knee joint angle θ_2; *middle:* knee torque τ_2; *bottom:* total ground reaction force $4F_{zf}$. **(b)** plot of maximum knee joint torque scaled by the BW for different leg segment lengths and jump heights. The arrows show HyQ experimental and MiniHyQ estimation results.

6. A Case Study: MiniHyQ

A case study is done for the development of a smaller version of a highly dynamic and versatile quadruped robot named *MiniHyQ*. MiniHyQ will be a torque-controlled versatile robot of small size that will be able to walk, move over rough terrain, jump and run. *Desired specifications for MiniHyQ:* robot mass 15 kg and leg segment length is 0.2 m. *Task performance limits:*

capable of running at 2 m/s and jump 0.15 m. *Estimated peak torques:* for a *running trot* peak hip and knee joint torques are equal to 5.25 Nm ($0.35Nm/kg * 15kg$) and 22.5 Nm ($1.5Nm/kg * 15kg$) respectively. Fig. 5b shows for a *squat jump* an estimated maximum knee joint torque of 13.5 Nm ($0.9Nm/kg * 15kg$) for MiniHyQ.

7. Discussion

In running trot simulation, it is assumed that the robot mass is evenly distributed and that the robot torso is always parallel to the ground. This assumption does not significantly influence the results even the biological studies have shown that all legged animals typically run with similar center of mass (CoM) motions relative to the (approximately) horizontal ground.[17] In this work ground stiffness is assumed constant and the equivalent virtual spring stiffness and angle of attack is chosen from the stability domain of steps-to-fall map of the SLIP model. Assumptions and limitation are discussed in detail for a squat jump simulation.[3]

8. Conclusion and Future Works

Above mentioned case study is an exemplary display of proposed tool for designing a versatile quadruped robot. This paper presented a study of actuator sizing for highly-dynamic quadruped robots and showed how one can estimate torque limits for tasks like a squat jump and running trot at various jump heights and forward velocities, respectively.

Future Works This study will be extended for the estimation of the hip abduction/adduction joint peak torques. Currently we are building a MiniHyQ (a smaller version of HyQ), whose actuators are sized and selected based on these estimations. These simple presented methods will also be validated by performing experiments on the MiniHyQ.

Acknowledgments

This research has been funded by the Fondazione Istituto Italiano di Tecnologia. The authors would like to thank CAPES for the scholarship granted to V. Barasuol (Grant Procs. 6463-11-8). The authors would like to thank also the colleagues that collaborated for the success of this project: Michele Focchi, Jake Goldsmith, Thiago Boaventura, Marco Frigerio, Ioannis Havoutis and our team of technicians.

References

1. M. Raibert, K. Blankespoor, G. Nelson, R. Playter, and the BigDog Team, "Bigdog, the rough-terrain quadruped robot," in *IFAC*, 2008.

2. C. Semini, N. G. Tsagarakis, E. Guglielmino, M. Focchi, F. Cannella, and D. G. Caldwell, "Design of HyQ - a hydraulically and electrically actuated quadruped robot," *IMechE Part I: J. of Systems and Control Engineering*, vol. 225, no. 6, pp. 831–849, 2011.
3. C. Semini, H. Khan, M. Frigerio, T. Boaventura, M. Focchi, J. Buchli, and D. G. Caldwell, "Design and scaling of versatile quadruped robots," in *CLAWAR Conference*, 2012.
4. R. M. Alexander, *Principles of Animal Locomotion.* Princeton University Press, 2003.
5. R. M. Alexander, "The gaits of bipedal and quadrupedal animals," *The International Journal of Robotics Research*, vol. 3, no. 2, pp. 49–59, 1984.
6. R. M. Alexander and A. S. Jayes, "A dynamic similarity hypothesis for the gaits of quadrupedal mammals," *Journal of Zoology*, vol. 201, no. 1, pp. 135–152, 1983.
7. W. J. Duncan, *Physical similarity and dimensional analysis.* Arnold (London), 1953.
8. A. A. Biewener, "Allometry of quadrupedal locomotion: The scaling of duty factor, bone curvature and limb orientation to body size," *J. of Experimental Biology*, vol. 105, pp. 147–171, 1983.
9. P. Chatzakos and V. Papadopoulos, "The influence of dc electric drives on sizing quadruped robots," in *IEEE ICRA Conference*, 2008.
10. P. Gregorio, M. Ahmadi, and M. Buehler, "Design, control, and energetics of an electrically actuated legged robot," *IEEE Transactions on Systems Man and Cybernetics*, vol. 27, pp. 626–634, 1997.
11. A. Sproewitz, T. Alexandre, M. D'Haene, R. Mockel, J. Degrave, M. Vespignani, S. Gay, A. Mostafa, B. Schrauwen, and A. J. Ijspeert, "Towards dynamically running quadruped robots: Performance, scaling, and comparison," in *AMAM Conference*, 2013.
12. J. Pratt, C. Chew, A. Torres, P. Dilworth, and G. Pratt, "Virtual model control: An intuitive approach for bipedal locomotion," *The International Journal of Robotics Research*, vol. 20, no. 2, pp. 129–143, 2001.
13. R. Blickhan and R. Full, "Similarity in multilegged locomotion: bouncing like a monopode," *Journal of Comparative Physiology*, vol. 173, pp. 509–517, 1993.
14. R. Blickhan, "The spring-mass model for running and hopping," *Journal of Biomechanics*, vol. 22, pp. 1217–1227, 1989.
15. C. Semini, *HyQ – Design and Development of a Hydraulically Actuated Quadruped Robot.* PhD thesis, Italian Institute of Technology and University of Genoa, 2010.
16. F. Peuker, A. Seyfarth, and S. Grimmer, "Inheritance of slip running stability to a single-legged and bipedal model with leg mass and damping," in *BioRob*, pp. 395 –400, june 2012.
17. M. Srinivasan and P. Holmes, "How well can spring-mass-like telescoping leg models fit multi-pedal sagittal-plane locomotion data?," *Journal of Theoretical Biology*, vol. 255, pp. 1–7, 2008.

PREVIEWED IMPEDANCE ADAPTATION TO COORDINATE UPPER-LIMB TRAJECTORY TRACKING AND POSTURAL BALANCE IN DISTURBED CONDITIONS

A. IBANEZ, P. BIDAUD and V. PADOIS

ISIR Institute for Intelligent Systems and Robotics CNRS UMR 7222,
Université Pierre et Marie Curie
Pyramide Tour 55, BC 173 — 4 Place Jussieu, F-75005 Paris, France
E-mail: {ibanez, bidaud, padois}@isir.upmc.fr

This paper proposes a preview control method for the whole-body motion of humanoid robots ensuring high performance in both interaction and postural balance tasks under large physical perturbations. By previewing the reduced coupled models of upper-limb interaction and postural balance dynamics, the proposed controller adapts simultaneously the impedance of the arms and the center of mass trajectory with respect to known external perturbations. Here, we show how the ZMP preview control formulation can be extended to account for disturbances resulting from the interacting arms dynamics of which control parameters are adapted online in order to maximize both interaction and balance performances over a preview horizon. The validity of this formulation is assessed through simulation considering a force applied at the humanoid hand level when it is walking.

Keywords: Humanoid, Biped walking, Legged robot, Adaptive impedance, Preview control, Zero-Moment Point, Postural balance, Manipulation, Interaction.

1. Introduction

Controlling the behavior of a humanoid robot placed in a dynamic environment in which it must perform manipulation tasks requires to ensure a robust control of postural balance and adaptation of handling tasks to make them compatible with the balance constraints.

In the case of human beings the coordination of motor activity can be interpreted[1] as a dialog in between two different parts of the central nervous system (CNS) involved in motor control: a predictive part relating high-level voluntary actions to carry out reaching a particular goal in a particular environment (which presupposes knowledge about expected action effects) and a reactive part performing local motor parametric adjustments imposed by the needed reactivity to the environment requests and the basic biological needs. The CNS generates Anticipatory Postural Adjustments (APA) that minimize the effects on balance of a simultaneous focal activity, while task-related motor activity adapts to constraints raised by postural balance. The whole motor response is organized with the aim to coordinate the APAs and the focal activity[2] in order to maximize the overall performance.

To obtain a similar control policy for humanoids, we propose a control scheme based on a multi-objective predictive/preview control of the impedance of the upper limbs and of the postural system.
Among existing contributions in this domain, control methods such as impedance controllers[3] allows to define the reactive behavior of a manipulator during interaction tasks of which parameters can be defined in an optimal way.[4] This resulting behavior of the upper-limbs induces perturbations[5] on the lower-bodies of the robot. The control of these two systems having coupled dynamics and independent objectives can be handled in an anticipative way using a distributed model predictive control (DMPC) method[6] dynamically tuning reactive control parameters.

The control method proposed in this paper relies on coupled reduced representations of two tasks dynamics, biped balance and upper-limb manipulation, expressed at the center of mass (CoM) and hand levels, respectively. For a given horizon of control parameters, the dynamics of the upper limbs can be previewed under an external force applied at the hand and an induced horizon of disturbance on the supporting platform is estimated, allowing for the evaluation over a finite horizon of a manipulation performance index and a ZMP-based balance stability criterion.
The novelty of the approach proposed in this work and the obtained results are presented in the next sections.

2. Formulation

The horizon of control parameters is defined as the horizon of CoM horizontal jerks $\mathbf{u}$ and cartesian control stiffness K and damping C at the hand level. At each control step, an optimization process is run over the control parameters to maximize the performance index of both balance and manipulation tasks over a finite horizon.

2.1. *Lower-limbs dynamics*

The lower-limbs motor activity mainly supports postural balance. Under this assumption and neglecting vertical and rotational effects we reduce the system to a punctual mass centered at the CoM of the robot of which horizontal dynamics $\hat{\mathbf{c}} = [\mathbf{c}\ \dot{\mathbf{c}}\ \ddot{\mathbf{c}}]^T$ are the ones of a 3D linear inverted pendulum (LIP3D) as seen in the works of Kajita *et al.*[7] An horizon of external force $\mathbf{F}^{e,c}$ applied on the CoM is supposed to be known. By rewriting the Zero-Moment Point (ZMP) equations for the LIP3D under $\mathbf{F}^{e,c}$ we can compute its position $\mathbf{p}$ in the ground plane as a function of $\hat{\mathbf{c}}$ and $\mathbf{F}^{e,c}$: under these hypotheses, the LIP3D model is subject to inertial effects, gravity and external actions, and the ZMP position $\mathbf{p}$ writes

$$\mathbf{p} = \mathbf{c} - \frac{Mz}{Mg - \mathbf{F}^{e,c}{}_{|z}}\ddot{\mathbf{c}} + \frac{z}{Mg - \mathbf{F}^{e,c}{}_{|z}}\mathbf{F}^{e,c}, \tag{1}$$

where z is the center of mass altitude, supposed in this model to be constant over time, $\mathbf{F}^{e,c}{}_{|z}$ is the vertical coordinate of the force applied on the CoM and M, g are the robot mass and gravity amplitude, respectively.

Balance is assumed to be maintained as long as the ZMP position is inside the convex hull of the contact points on the ground. The maximization of balance stability can be abridged as the minimization of the distance between the ZMP position $\mathbf{p}$ and a reference $\overline{\mathbf{p}}$ which is the center of the sustentation convex hull.

The CoM is controlled, similarly to the works of Kajita *et al.*,[7] through the computation of desired horizontal input jerk $\mathbf{u}$. Time is discretized over time intervals of constant lengths δt, and for a continuous function g of time g_k denotes $g(k\delta t)$. Given an horizon of $\mathbf{u}_k$ the approximated CoM dynamics can be previewed in the following way

$$\hat{\mathbf{c}}_k \equiv \begin{cases} \mathbf{c}_k = \mathbf{c}_{k-1} + \delta t\dot{\mathbf{c}}_{k-1} + \frac{\delta t^2}{2}\ddot{\mathbf{c}}_{k-1} + \frac{\delta t^3}{6}\mathbf{u}_{k-1} \\ \dot{\mathbf{c}}_k = \dot{\mathbf{c}}_{k-1} + \delta t\ddot{\mathbf{c}}_{k-1} + \frac{\delta t^2}{2}\mathbf{u}_{k-1} \\ \ddot{\mathbf{c}}_k = \ddot{\mathbf{c}}_{k-1} + \delta t\mathbf{u}_{k-1}. \end{cases} \tag{2}$$

A matrix-form of the preview of the ZMP position can be written, as seen in the works of Wieber,[8] using Eq. (1) and Eq. (2) expressed from current step k to final preview step $k+N$. Let us denote the horizons of previewed ZMP positions $\boldsymbol{\mathcal{P}}_k \equiv [\mathbf{p}_{k+1}\ \dots\ \mathbf{p}_{k+N}]^T$ and input jerks $\boldsymbol{\chi}_k^{\mathbf{c}} \equiv [\mathbf{u}_{k+1}\ \dots\ \mathbf{u}_{k+N}]^T$, which brings

$$\boldsymbol{\mathcal{P}}_k = \mathbf{P}_{\mathbf{x}}^{\mathbf{dis}}\hat{\mathbf{c}}_k + \mathbf{P}_{\mathbf{u}}^{\mathbf{dis}}\boldsymbol{\chi}_k^{\mathbf{c}} + \mathbf{D}_k, \tag{3}$$

where, writing $\alpha^{dis} \equiv \frac{1}{Mg-\mathbf{F}^{e,c}|_z}$,

$$\mathbf{P_x^{dis}} \equiv \begin{bmatrix} 1 & \delta t & \frac{\delta t^2}{2} - Mz\alpha^{dis}_{k+1} \\ \vdots & \vdots & \vdots \\ 1 & Ndt & N^2\frac{\delta t^2}{2} - Mz\alpha^{dis}_{k+N} \end{bmatrix}, \mathbf{P_u^{dis}} \equiv \begin{bmatrix} \frac{\delta t^3}{6} - \delta t Mz\alpha^{dis}_{k+1} & \cdots 0 \\ \vdots & \ddots 0 \\ (1+3N^2)\frac{\delta t^3}{6} - \delta t Mz\alpha^{dis}_{k+N} & \cdots * \end{bmatrix}$$

and finally

$$\mathbf{D}_k \equiv \left[z\mathbf{F}^{e,c}_{k+1}\alpha^{dis}_{k+1} \cdots z\mathbf{F}^{e,c}_{k+N}\alpha^{dis}_{k+N} \right]^T,$$

which can be interpreted as an horizon of disturbances on the stable ZMP reference $\overline{\mathbf{p}}$.

2.2. *Upper-limbs dynamics*

Upper-limbs dynamics are subject to control parameters and actions from the environment. Manipulation tasks requiring a certain level of precision (with respect to the interaction magnitude involved) are often achieved through controllers of high apparent stiffness, resulting in the transmission of actions through the upper-limbs towards the lower part of the system.

In the case of manipulation, we consider the external actions to be applied at the end-effector, ie the hand of the robot. A PD corrective controller is defined at this point. By comparing the operational space[9] dynamics of the effector under an external action $\mathbf{F}$ to the free ones, we identify a local impedant model of the hand dynamics $\hat{\mathbf{e}}$.

The whole-body control framework used in this work is a LQP-based[10] controller which computes optimal joint torques with respect to desired joint and cartesian accelerations and constraints, using joint-space dynamics subject to contact wrenches.

To capture the behavior of the upper-limbs, we reduce the humanoid system to a fixed-base manipulator (free-floating joints and contact forces are not considered). In the case of manipulation, given a desired acceleration $\mathbf{a}$

$$\mathbf{a} = K\left(\mathbf{e}^* - \mathbf{e}\right) + C\left(\dot{\mathbf{e}}^* - \dot{\mathbf{e}}\right) + \ddot{\mathbf{e}}^* \tag{4}$$

where K and C denote respectively cartesian control stiffness and damping, and $\overline{\mathbf{e}} = [\mathbf{e}^*\ \dot{\mathbf{e}}^*\ \ddot{\mathbf{e}}^*]$ are the reference dynamics of the effector, whole-body control consists in finding joint torques $\boldsymbol{\tau}$ that satisfy the free dynamics

$$\mathbf{H}\left(\mathbf{q}\right)\ddot{\mathbf{q}}^{\mathbf{a}} + \mathbf{N}\left(\mathbf{q}, \dot{\mathbf{q}}\right) + \mathbf{G}\left(\mathbf{q}\right) = \boldsymbol{\tau}, \tag{5}$$

where $\mathbf{H}$, $\mathbf{N}$ and $\mathbf{G}$ denote the joint-space mass, coriolis/centrifugal and gravity matrices, respectively. Dependency to $(\mathbf{q}, \dot{\mathbf{q}})$ is dropped in the rest of the paper. Joint-space dynamics under $\mathbf{F}$, however, write with $\mathbf{J}_e$ the Jacobian of the hand

$$\mathbf{H}\ddot{\mathbf{q}} + \mathbf{N} + \mathbf{G} = \boldsymbol{\tau} + \mathbf{J}_e^T\mathbf{F}. \tag{6}$$

Comparing Eq. (6) to Eq. (5) in the operational space under input joint torques $\boldsymbol{\tau}$ captures the local dynamics $\ddot{\mathbf{e}}$ of the effector subject to action $\mathbf{F}$

$$\mathbf{M}_e\ddot{\mathbf{e}} = \mathbf{F} - \mathbf{M}_e\dot{\mathbf{J}}_e\mathbf{J}_e^{\dagger}\left(\dot{\mathbf{e}} - \boldsymbol{\nu}\right) + \mathbf{K}_e\left(\mathbf{e}^* - \mathbf{e}\right) + \mathbf{C}_e\left(\dot{\mathbf{e}}^* - \dot{\mathbf{e}}\right) + \mathbf{M}_e\ddot{\mathbf{e}}^*, \tag{7}$$

where $\mathbf{M}_e = \left(\mathbf{J}_e\mathbf{H}^{-1}\mathbf{J}_e^T\right)^{-1}$, $\mathbf{K}_e = K\mathbf{M}_e$, $\mathbf{C}_e = C\mathbf{M}_e$ and

$$\boldsymbol{\nu} \quad \text{s.t.} \quad \boldsymbol{\nu} = \mathbf{J}_e\dot{\mathbf{q}}^{\mathbf{a}} \quad \text{and} \quad \dot{\mathbf{q}}^{\mathbf{a}} \quad \text{s.t.} \quad \mathbf{a} = \mathbf{J}_e\ddot{\mathbf{q}}^{\mathbf{a}} + \dot{\mathbf{J}}_e\dot{\mathbf{q}}^{\mathbf{a}}. \tag{8}$$

The model described in Eq. (7) exhibits terms related to inertial effects, external actions and impedant behavior. We assume the latter to be supported by the lower-limbs, considering that upper and lower models of the robot are serialized. That is,

$$\mathbf{F}^{e,c} = \mathbf{M}_e\dot{\mathbf{J}}_e\mathbf{J}_e^{\dagger}\left(\dot{\mathbf{e}} - \boldsymbol{\nu}\right) - \mathbf{K}_e\left(\mathbf{e}^* - \mathbf{e}\right) - \mathbf{C}_e\left(\dot{\mathbf{e}}^* - \dot{\mathbf{e}}\right) - \mathbf{M}_e\ddot{\mathbf{e}}^*, \tag{9}$$

where $\mathbf{J}_e^{\dagger}$ is the generalized inverse of the effector jacobian, written here as the dynamically consistent pseudoinverse minimizing kinetic energy.[9]

The following integration scheme is used

$$\begin{cases} \mathbf{e}_k = \mathbf{e}_{k-1} + \delta t\dot{\mathbf{e}}_{k-1} + \frac{\delta t^2}{2}\ddot{\mathbf{e}}_{k-1} \\ \dot{\mathbf{e}}_k = \dot{\mathbf{e}}_{k-1} + \delta t\ddot{\mathbf{e}}_{k-1} \\ \ddot{\mathbf{e}}_k = \mathbf{M}_e^{-1}\mathbf{F}_k + \mathbf{a}_k - \dot{\mathbf{J}}_e\mathbf{J}_e^{\dagger}\left(\dot{\mathbf{e}}_k - \boldsymbol{\nu}_k\right) \text{ from Eq. (7),} \end{cases} \tag{10}$$

where $\boldsymbol{\nu}_k$ is derived from Eq. (7) expressed at step $k-1$

$$\boldsymbol{\nu}_k = \dot{\mathbf{e}}_{k-1} + \left(\mathbf{M}_e\dot{\mathbf{J}}_e\mathbf{J}_e^{\dagger}\right)^{-1}\left[\mathbf{M}_e\left(\ddot{\mathbf{e}}_{k-1} - \mathbf{a}_{k-1}\right) - \mathbf{F}_{k-1}\right]. \tag{11}$$

Effector acceleration can hence be written, for $k > 0$

$$\ddot{\mathbf{e}}_k = \ddot{\mathbf{e}}_0 + \mathbf{M}_e^{-1}\left(\mathbf{F}_k - \mathbf{F}_0\right) + \left(\mathbf{a}_k - \mathbf{a}_0\right) - \dot{\mathbf{J}}_e\mathbf{J}_e^{\dagger}\left(\dot{\mathbf{e}}_k - \dot{\mathbf{e}}_0\right) \tag{12}$$

which allows to preview the effector dynamics over a finite horizon, knowing a future estimation of $\mathbf{F}_k$ applied on the hand.

3. Whole system preview control

The previewed performance of both balance and manipulation tasks under an external force $\mathbf{F}$ is optimized over a finite horizon. By serializing both upper and lower-limbs reduced models, their dynamics are coupled through the transmission of the external action, relatively to the control stiffness and damping K, C of the effector. The behavior of the resulting coupled model is previewed over a horizon of N time steps. Optimal control parameters are deduced from the preview and are set as active control parameters of both balance and manipulation reactive tasks, as shown in Fig. 1.

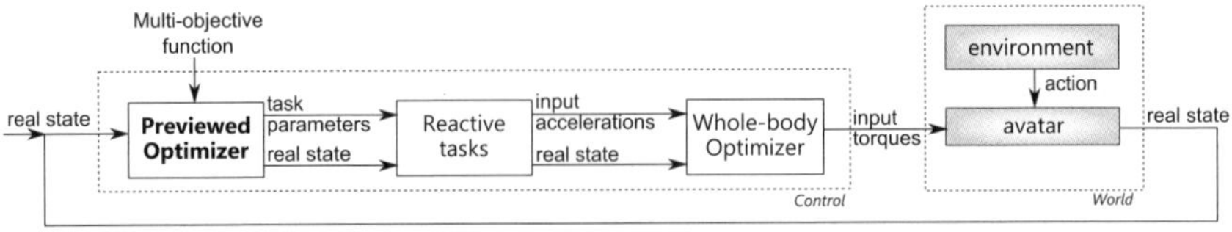

Fig. 1. Control architecture

3.1. *Optimization problem*

At each control step k, the manipulation and balance errors $\delta^{\mathbf{e}}$ and $\delta^{\mathbf{p}}$ are minimized for each preview step $i \in [k+1, k+N]$. If $\overline{\mathbf{e}}_i$ and $\overline{\mathbf{p}}_i$ denote respectively the objective of the manipulation and balance tasks, the following errors $\delta_i^{\mathbf{e}} \equiv \|\hat{\mathbf{e}}_i - \overline{\mathbf{e}}_i\|^2$ and $\delta_i^{\mathbf{p}} \equiv \|\mathbf{p}_i - \overline{\mathbf{p}}_i\|^2$ are to be minimized. To define the relative priority between these two objectives and capture the difference in magnitude of the two errors, a cost function of the form $\omega_{\mathbf{e}}\delta_i^{\mathbf{e}} + \omega_{\mathbf{p}}\delta_i^{\mathbf{p}}$ is minimized at each control step k, for each preview step $i \in [k+1, k+N]$. To enforce the stability of the minimization process we define a quadratic regularization term h and build the following cost function

$$f_k\left(\boldsymbol{\chi}_k\right) = \sum_{i=1}^{N} \left(\omega_{\mathbf{e}}\delta_{k+i}^{\mathbf{e}} + \omega_{\mathbf{p}}\delta_{k+i}^{\mathbf{p}} + h_{k+i}\right) \tag{13}$$

which is minimized with respect to control parameters $\boldsymbol{\chi}_k \in \mathbb{R}^{4N\times 1}$

$$\boldsymbol{\chi}_k = \left[K_{k+1}\ C_{k+1}\ \ldots\ K_{k+N}\ C_{k+N}\ \mathbf{u}_{k+1}\ \ldots\ \mathbf{u}_{k+N}\right]^T \equiv \left[\boldsymbol{\chi}_k^{\mathbf{e}}\ \boldsymbol{\chi}_k^{\mathbf{c}}\right]^T.$$

Regularization is achieved by minimizing the CoM input amplitude and the euclidean distance of manipulation control parameters K, C to a reference $\overline{K}, \overline{C}$, that is $h_i = \omega_{\mathbf{e}}^h\left(\|K_i - \overline{K}_i\|^2 + \|C_i - \overline{C}_i\|^2\right) + \omega_{\mathbf{p}}^h\|\mathbf{u}_i\|^2$.

3.2. *Resolution*

To reduce the dimensionality of the problem from $4N$ to $2N$, optimal input jerks $\boldsymbol{\chi}_k^{\mathbf{c}}$ in the sense of $\delta^{\mathbf{p}}$ are deduced from evaluated manipulation task parameters $\boldsymbol{\chi}_k^{\mathbf{e}}$. To do so, given a set of manipulation control parameters $\boldsymbol{\chi}_k^{\mathbf{e}}$, horizon $\boldsymbol{\chi}_k^{\mathbf{c}}$ are minimizing $\sum_{i=1}^{N}\omega_{\mathbf{p}}\delta_{k+i}^{\mathbf{p}} + \omega_{\mathbf{p}}^h\|\mathbf{u}_i\|^2$, and hence writes with Eq. (3)

$$\chi_k^{\mathbf{c}} = -\left(\mathbf{P}_{\mathbf{u}}^{\mathbf{dis}^T}\mathbf{P}_{\mathbf{u}}^{\mathbf{dis}} + \frac{\omega_{\mathbf{p}}^h}{\omega_{\mathbf{p}}} I_{2N\times 2N}\right)^{-1} \mathbf{P}_{\mathbf{u}}^{\mathbf{dis}^T}[\mathbf{P}_{\mathbf{x}}^{\mathbf{dis}}\hat{\mathbf{c}}_k - (\overline{\mathcal{P}}_k - \mathbf{D}_k)], \quad (14)$$

where $\overline{\mathcal{P}}_k \equiv \left[\overline{\mathbf{p}}_{k+1} \cdots \overline{\mathbf{p}}_{k+N}\right]^T$ denotes the horizon of ZMP references. Results in the following section where obtained with a Nelder-Mead simplex algorithm[11] run over $\chi^{\mathbf{e}}$ minimizing f as defined in Eq. (13).

4. Results

The validity of this control method is assessed through the simulation of a humanoid robot performing two concurrent tracking tasks: it maintains biped balance while following a pre-defined path with its right hand (*cf.* left hand side of Fig. 2) where is applied a known external effort in lateral and longitudinal directions as shown in Fig. 3.

Simulated experiments are performed using Arboris-Python,[12] an open-source dynamic simulator developed at ISIR with the Python programming language, involving a rather accurate model of the iCub robot[13] with 38 degrees of freedom and 4 contact points at each foot. Four controllers are compared:

(1) $C1$: ZMP Preview Control[7] with $\omega_{\mathbf{p}} = 1.e^6$ and $\omega_{\mathbf{p}}^h = 1$,
(2) $C2$: ZMP Preview Control accounting for external force at the hand $\mathbf{F}$ as directly applied on the CoM (*i.e.* $\mathbf{F}^{e,c} = \mathbf{F}$) with Eq. (14),
(3) $C3$: Optimized Control Parameters minimizing Eq. (13) with $\omega_{\mathbf{e}} = 1.e^7, \omega_{\mathbf{p}} = 1.e^6, \omega_{\mathbf{e}}^h = 10$ and $\omega_{\mathbf{p}}^h = 1$,
(4) $C4$: Identical to $C3$ with $\omega_{\mathbf{e}} = 1.e^6, \omega_{\mathbf{p}} = 1.e^7, \omega_{\mathbf{e}}^h = 10$ and $\omega_{\mathbf{p}}^h = 1$.

Controller $C3$ prioritizes the manipulation tracking performance over the ZMP task, and $C4$ is parameterized symmetrically. Cumulated normalized tracking errors for both tasks and the four controllers are presented on the right hand side of Fig. 2.

First, these results show the noticeable gain in balance provided by the introduction of an approximation (considered to be applied on the CoM) of the external effort. Second, the preponderance of the manipulation task in the controller $C3$ leads to a significant increase in the tracking performance of the hand though causing a slight prejudice to balance, while symmetric parameters lead to weaker gains but for both tracking tasks.

An intuitive trend in the stiffness strategy is computed online by controllers $C3$ and $C4$: as shown in Fig. 4 the stiffness parameter K tends to

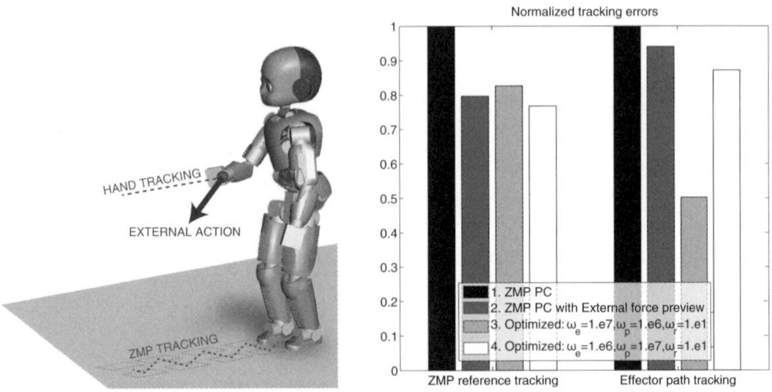

Fig. 2. Two disturbed tracking tasks: manipulation while walking and their normalized cumulated error for four controllers

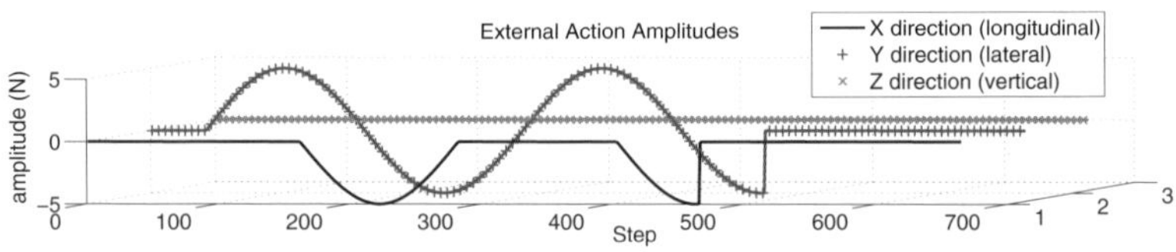

Fig. 3. External action applied on the hand in logitudinal and lateral directions

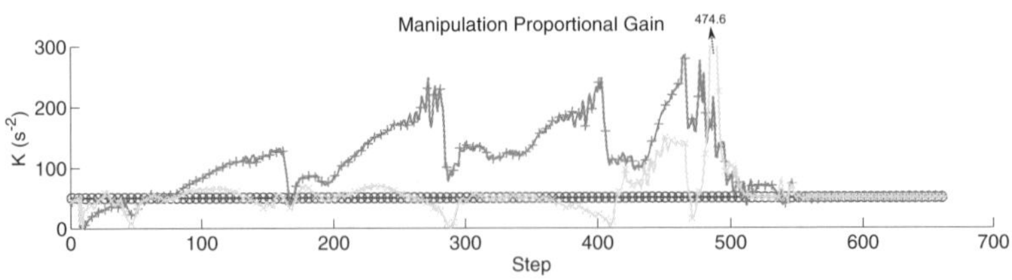

Fig. 4. Control strategy for the manipulation activity - black solid: $C1$, blue circle: $C2$, red plus: $C3$ and green cross: $C4$

be set higher than the reference $\overline{K} = 50s^{-2}$ by the controller $C3$ whereas $C4$ tends to loosen the manipulation task stiffness under $\overline{K}$ in favor of balance.

A detailed view of the ZMP lateral position is shown in Fig. 5. The decrease in ZMP tracking error resulting from the consideration of the external action is noticeable (*e.g.* at steps 250 and 450), and allows for an overall gain in balance. However, balance failure-prone short-term effects as visible

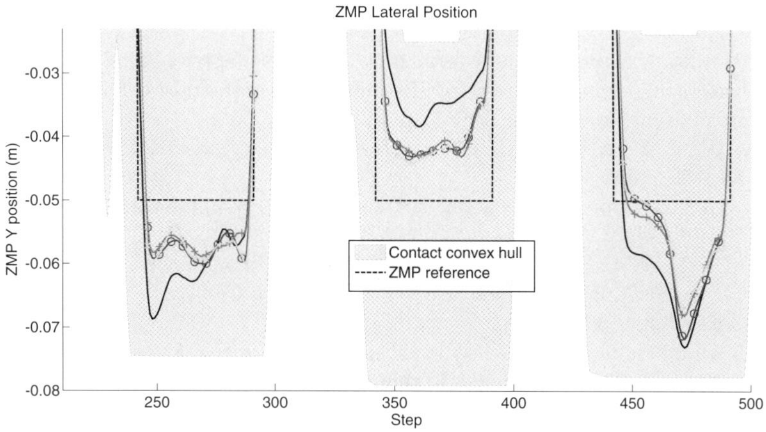

Fig. 5. Detailed evolution of the Zero-Moment Point in the lateral direction - black solid: $C1$, blue circle: $C2$, red plus: $C3$ and green cross: $C4$

at step 475 require a better estimation and adaptation of the transmission of the external action on the CoM that controllers $C3$ and $C4$ provide.

5. Conclusion

This work highlights the benefits of considering the coupling between concurrent tasks such as manipulation and balance. Focus has been put on identifying at a high level the influence of manipulation control parameters over the transmission of external actions towards the center of mass of a humanoid robot. This identification allows for an online adaptation of both simultaneous tasks in order to improve their performance based on a preview of external disturbances and their effects on selected criteria.

The resulting formulation produces a robust and parameterizable controller identifying in a predictive manner an optimal control strategy in favor of antagonist objectives.

We propose as future works a deeper study of the interdependence between locomotion and manipulation in a preview control framework.

Acknowledgment

This work was partially supported by the French Ministry of Higher Education and Research, by the European funds for regional development (FEDER3) for the ANIPEV project and by the RTE company through its chair "Robotics Systems for field intervention in constrained environments" hold by Vincent Padois.

References

1. J. Paillard, L'intégration sensori-motrice et idéomotrice, in *Traité de Psychologie Expérimentale*, eds. M. Richelle, J. Requin and M. Robert (Presses Universitaires de France, 1994) pp. 925–961.
2. A. Wing, J. R. Flanagan and J. Richardson, *Experimental Brain Research* **116**, 122 (1997).
3. N. Hogan, Impedance control: An approach to manipulation, in *American Control Conference*, 1984.
4. L. Love and W. Book, Environment estimation for enhanced impedance control, in *Proceedings of the IEEE International Conference on Robotics & Automation*, 1995.
5. A. Albu-Schaffer, C. Ott, U. Frese and G. Hirzinger, Cartesian impedance control of redundant robots: Recent results with the dlr-light-weight-arms, in *Proceedings of the IEEE International Conference on Robotics & Automation*, 2003.
6. A. N. Venkat, Distributed model predictive control : theory and applications, PhD thesis2006.
7. S. Kajita, F. Kanehiro, K. Kaneko, K. Kajiwara, K. Harada, K. Yokoi and H. Hirukawa, Biped walking pattern generation by using preview control of zero-moment point, in *Proceedings of the IEEE International Conference on Robotics & Automation*, 2003.
8. P.-B. Wieber, Trajectory free linear model predictive control for stable walking in the presence of strong perturbations, in *IEEE-RAS International Conference on Humanoid Robots*, 2006.
9. O. Khatib, A unified approach to motion and force control of robot manipulators: The operational space formulation, in *IEEE Journal on Robotics & Automation*, 1987.
10. J. Salini, V. Padois and P. Bidaud, Synthesis of complex humanoid whole-body behavior: a focus on sequencing and tasks transitions, in *Proceedings of the IEEE International Conference on Robotics & Automation*, 2011.
11. J. Nelder and R. Mead, *The Computer Journal* **7**, 308 (1965).
12. S. Barthelemy, J. Salini and A. Micaelli, Arboris-python.
13. G. Sandini, G. Metta and D. Vernon, The icub cognitive humanoid robot: An open-system research platform for enactive cognition, in *50 Years of Artificial Intelligence*, Lecture Notes in Computer Science (Springer, 2007) pp. 358–369.

A HIP JOINT STRUCTURE FOR BIPED ROBOT WITH REDUCED DOF'S OF MOTION

SATOSHI ITO*, AKIHIRO NAKAZAWA, TAKESHI ONOZAWA
SHINGO NISHIO and MINORU SASAKI

Department of Mechanical Engineering, Gifu University,
Yanagido 1-1, Gifu 501-1193, Japan
**E-mail: satoshi@gifu-u.ac.jp*
www.ics.human.gifu-u.ac.jp

In this paper, we propose a new hip joint structure for a biped robot that uses a small number of DoFs, reducing the number of actuators required, and thereby conserving weight. First, we will discuss the necessary number of DoFs to walk on a sloped surface without irregularities. Consequently, we designed a kneeless biped robot with six DoFs, two DoFs for each ankle, and the remaining two for the hip joint structure. The hip joint structure is required to produce two kinds of movements: the alternative swings between the left and right leg as well as the lateral weight shift required to lift up a foot. We invented a mechanism to realize these two movements using two actuators: a similar structure to the differential gear was applied to enable the leg to swing in the sagittal plane, and the parallel link structure was applied to produce lateral motions. In addition, we actually constructed a robot with the proposed mechanisms, and demonstrated the biped walking on a slope.

Keywords: biped robot, reduced DoF, hip joint structure, slope walk, CoP feedback

1. Introduction

Weight reduction is an important specification in mechanical design. This especially holds true for mobile machinery that changes its own spatial positions. When they are utilized as transportation vehicles, weight savings increase the total amount of baggage or passengers. More importantly, even if the machines do not transport anything, for instance, in the case of autonomous machines such as biped robots, weight savings directly reflects in the resulting energy efficiency. Some papers actually study the mechanical structure of biped robots.[1,2]

We studied biped robots and in particular, considered the way they walk in uncertain environments, such as with a variable ground gradient. One of the final goals in the development of biped robots is usually the ability to use robots or mechanical assemblies to accomplish human tasks, especially ones that necessitate spatial movement on uneven floors. Our main focus, in the field of biped robot study, is on the establishment of stability during upright standing or walking periods. The zero moment point (ZMP) criterion provides an effective method to generate a locomotion pattern, desired trajectories of joint angles or the center of mass (CoM) for the whole body in the motion planning stage, and a tracking control for trajectories designed to enable biped robots to progress using their legs without falling down even if disturbances are applied during their walk.

Our research group proposes a control method for the maintenance of balance using feedback from ground reaction forces, in particular, the center of pressure (CoP) of the ground reaction forces,[3] and applies it to weight shifting in the double support phase of movement[4] and the in-place stepping motion.[5] Our goal to experimentally confirm the validity of our control method is for a biped to walk on a slope with an unknown but constant gradient. However, we have not yet achieved this with the robots that we are developing. The main reason is the shortness of the joint torques: we designed the robot with many DoFs which require several servo motors. Most commercial servo motors have a low power-to-weight ratio, leading to the robot being excessively heavy in comparison with the torque of the motors.

For most servo-motors, radio control units used in hobby cars, planes, and helicopters generate a high output with respect to their weight. Some biped robots adopt these types of motors, or originally developed motors, and successfully produce certain agile human-like motions.[6–8] However, this kind of servo-system occasionally utilizes position control based on the pulse width of the digital output. Our control method, essentially torque control, is difficult to apply in its original form. In addition, angular sensors such as rotary encoders are not contained in the servo-motor, forcing us to install them on the robot afterwards. This is the reason that we are developing a biped robot ourselves. To reduce the weight of the robot, our present aim is to reduce the number of motors by designing a structure to couple some DoFs required for bipedal walking. This enables us to utilize high spec servo-motors that torque control is applicable to.

2. Design of a biped robot

2.1. *Purpose of our biped robot*

Our main objective is to demonstrate the effectiveness of our control method: accordingly, the goal of our bipedal robot is to walk not only on flat but also on sloping ground. In this paper, we achieved this goal using a new lighter hip structure with a reduced number of motors.

Bipedal robots move by repeating the support leg exchanges between the right and left legs.

For the hip joints, at least one DoF is required if the legs are swung symmetrically in the anterior-posterior direction within the sagittal plane. In the frontal plane, the robot's weight has to be moved in a lateral direction before lifting and swinging a leg. By constraining the legs such that they are parallel to each other, this can be accomplished with one DoF, although the ankle joints must then move in a coordinated manner. Mechanically integrating the sagittal and frontal plane movements at the hip joint structure reduces the required DoFs.

Knee joints are needed to lift the foot over obstacles or rough ground. In this paper, however, we limited ourselves to even sloping ground with no obstacles. Thus, a knee joint is unnecessary, and the hip joint can perform the lifting necessary for walking. This again reduces the DoF count.

To adapt to the slope in the single support phase, an ankle joint needs to have two DoFs for pitch and roll.

In summary, the bipedal robot used 6 DoFs with 2 in each ankle and 2 in the hip joint structure.

2.2. *DoF reduction*

2.3. *A new reduced structure of hip joint*

Based on the ideas presented above, we designed the actual structure of the robot as depicted in Figs. 1 and 2.

Fig. 1 details the sagittal swing motions. The differential gear appears on the left of the figure. A central pinion gear drives two side gears to swing the legs simultaneously in opposite directions. The differential gear mechanism itself limits the distance between the two hip joints in the frontal plane. This is the reason that we synchronized the function of the center gear shared with two coupled gears that were mechanically connected, for example, using a timing belt, as shown in the figure on the left. When driven, connected beveled gears produce rotation at right angles to the

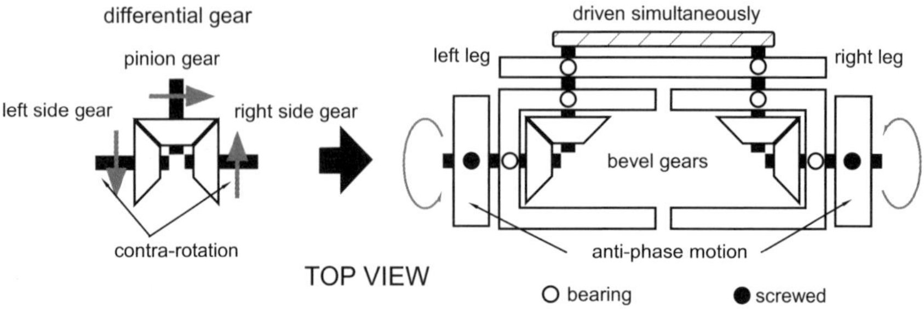

Fig. 1. Mechanism for alternative leg swing.

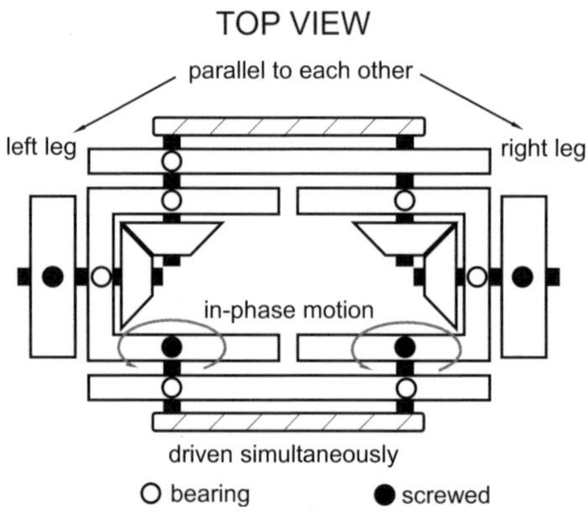

Fig. 2. Mechanisms to keep legs parallel.

drive axis, but in opposite directions, to produce leg swing motions in the sagittal plane.

For lateral motion, we introduced U-shaped bases, shown on the right in Fig. 1. Rotating the two U-shaped bases in phase as shown in Fig. 2 enables the realization of lateral movements. Another axis is set in the opposite direction to support the U-shaped base, sharing the rotation axis for the leg swing motion. A timing belt links the units across the front. When this belt is driven, the two U-shaped units slant together laterally. This is a kind of parallel link structure always keeping two legs parallel each other.

2.4. *Robot construction*

For easy handling of the robot during the experiment, we set a design goal of a small sized robot of around 30 cm in height and 3.5 kg in weight. This is

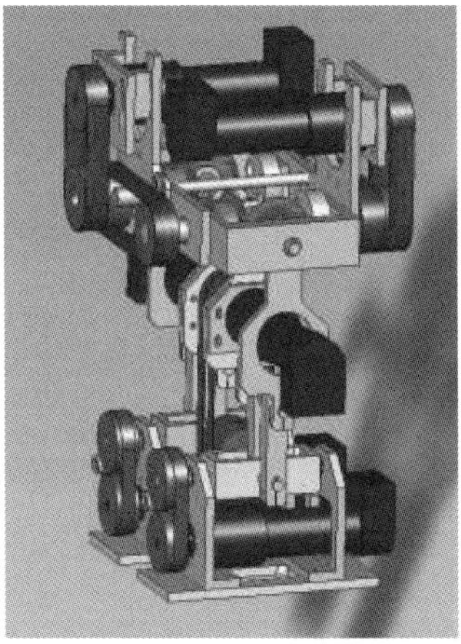

Fig. 3. Motor configuration.

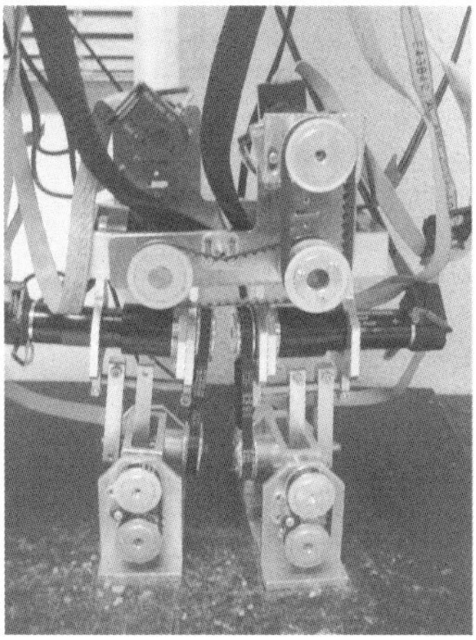

Fig. 4. Constructed biped robot.

sufficient for our purpose in this paper, i.e., to demonstrate the effectiveness of our control law.

Once the hip joint structure is realized, the motor configuration needs to be considered to avoid mechanical interference between links of the robot. Two motors were symmetrically installed on this hip joint structure for its actuation to balance the weight, as shown in Fig. 3. The timing belts transmit the driving force from the motor axis to the hip joint DoFs, and its tension can reduce the backlash between them.

The sagittal DoFs of the ankle joint are actuated by the motor placed in the middle of the leg, while the frontal ones are driven by the motor at the foot. To detect the position of the CoP, four load cells (KYOWA LMA-A-50N-P) were attached on the corner of each square sole.

The same motors, Maxon RE25 with 130g weight and 24.2mNm maximal continuous torque, were utilized for all DoF actuation. Figure 4 depicts the robot. It uses just six servomotors. It is 290 mm in height, 270 mm wide, and weighs 4.12 kg. Its feet are 160 mm long. The weight is greater than expected because of the large bevel gear.

3. Experiments

3.1. *Setups*

We used a personal computer (PC) as the controller, running a real-time operating system, ART-Linux. This computes the required torques for each of the six motors. It then outputs, using a D/A converter board (Interface

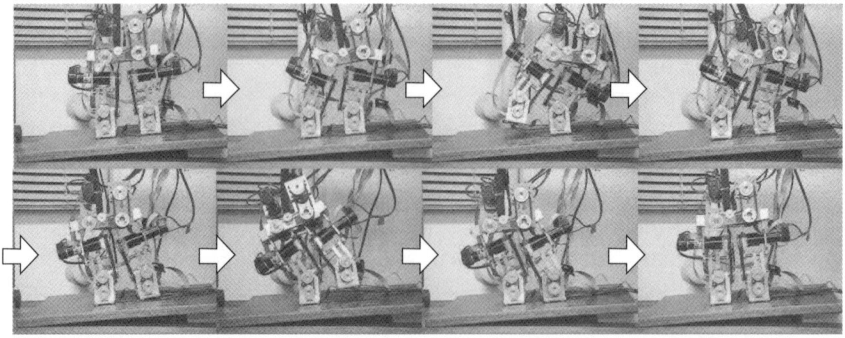

Fig. 5. Experiment on a leftward slope.

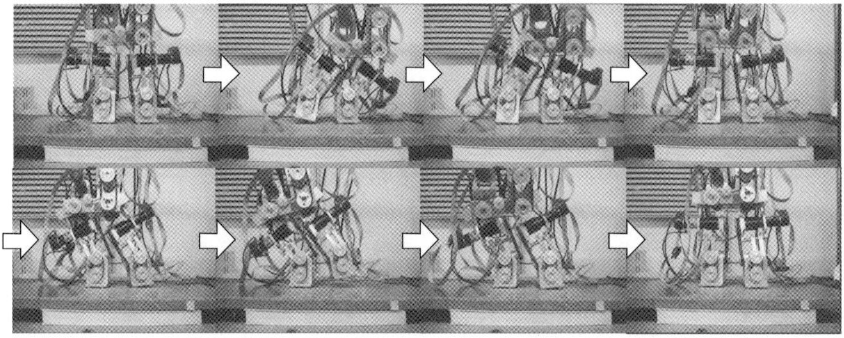

Fig. 6. Experiment on an upward slope.

PCI-3336), analog voltage corresponding to the magnitude of the torques, which is then applied to the motor drivers (TITECH Driver PC0121-1). These motor drivers supply the servo-motor with electric current from a 24V external power supply (ETA electric industry FHH24SX-U) to generate the required torque. The PC receives the joint angle information using the encoder counter board connected to the rotary encoder of the servo-motors. It also acquires, via the A/D converter board (Interface PCI-3153), information about the ground reaction force from load cells, the output of which is amplified in a multi-signal conditioner (KYOWA DPM-8K). The data detection and torque output process is executed every 1 ms.

3.2. *Control law*

The control method required to demonstrate the effectiveness of our solution is described in our previous paper.[5] It basically adopts CoP position

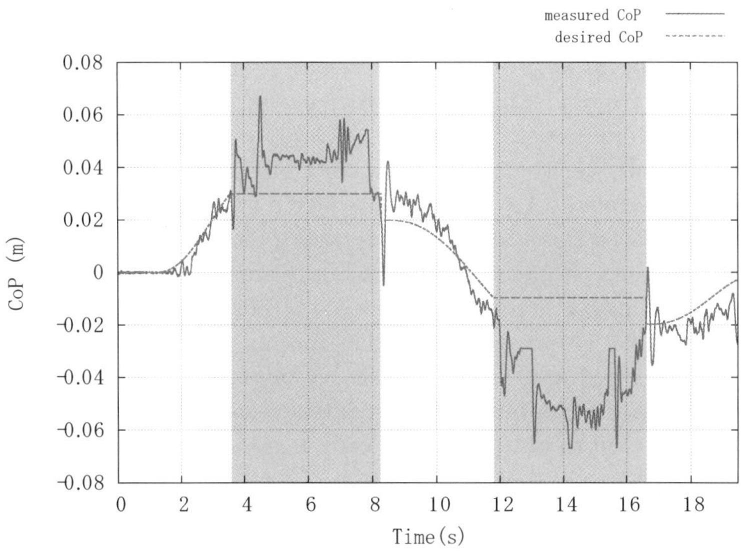

Fig. 7. Time course of CoP during the experiment.

feedback to maintain the balance of the biped robot.

$$\tau_\phi = -K_d\dot{\phi} + K_P(\phi_d - \phi) + K_f \int (P_d - P_{CoP})dt \tag{1}$$

where ϕ denotes the CoM of the whole robot, P_{CoP} is the position of CoP obtained from the load cells and P_d is a desired trajectory of CoP. K_d, K_p and K_f are the feedback gains. This ensures the local convergence of the CoP position if its desired trajectory varies slowly enough. Because the direct trajectory of the CoP is usually the same even if the slope of the ground is varied, this method allows biped robots to adapt to the slope without changing the control law.

3.3. *Results*

Bipedal walking was tried in two different conditions: a leftward slope and an upward slope. We used the robot described in the previous section. Figure 5 presents a sequence of photos showing actual walking on slopes for a leftward slope. An upward slope is shown in Fig. 6. Bipedal walking was achieved in both cases. The time course of the CoP position in the lateral direction is depicted with its desired trajectory in Fig. 7. The trajectory tracking of CoP was achieved in the double support phases (unshaded range).

4. Concluding remarks

To achieve bipedal walks on a sloping surface using the control law that we proposed, we required a biped robot with a control system that enables us to introduce torque control and measure joint angles and ground reaction forces. Thus far, we attempted to construct biped robots, but encountered a lack of motor power. We therefore invented a new, lighter hip joint structure with reduced DoFs that produces the leg alternative swing motion as well as the lateral sway motion. Adopting this new structure, we actually managed to construct a robot and achieved bipedal walks on a leftward and an upward slope. This demonstrates that this hip joint structure works well for an actual biped robot, and that our control law is effective for bipedal walk on a slope. For future walks, we intend to improve the speed of the bipedal walk, and to try walking on uneven ground with various gradient magnitudes along the route.

References

1. K. Yoneda, T. Tamaki, Y. Ota and R. Kurazume, *Journal of the Robotics Society of Japan (in Japanese)* **21**, 546 (2003).
2. I. M. C. Olaru, S. Krut and F. Pierrot, Novel mechanical design of biped robot sherpa using 2 dof cable differential modular joints, in *IEEE/RSJ International Conference on Intelligent Robots and Systems, 2009. IROS 2009.*, 2009.
3. S. Ito and H. Kawasaki, *Biological Cybernetics* **92**, 241 (2005).
4. S. Ito, H. Asano and H. Kawasaki, *A balance control in biped double support phase based on center of pressure of ground reaction forces*, in *Preprints of the 7th IFAC Symposium on Robot Control 2003 (SYROCO'03): , Wrocław, Poland*, 2003, pp. 189–194.
5. S. Ito and M. Sasaki, *Motion Control of Biped Lateral Stepping Based on Zero Moment Point Feedback for Adaptation to Slopes*, in *Biped Robots*, (Intech, 2011), pp. 15–34.
6. KONDO KAGAKU. home page, http://kondo-robot.com/ (2013).
7. Viston home page, http://www.vstone.co.jp/ (2013).
8. KYOSHO home page, http://www.kyosho.com/ (2013).

EXPLORING THE LOMBARD PARADOXON IN A BIPEDAL MUSCULOSKELETAL ROBOT

K. RADKHAH* and O. VON STRYK

Department of Computer Science, Technische Universität Darmstadt, 64289 Darmstadt, Germany
**E-mail: radkhah@sim.tu-darmstadt.de*
www.sim.tu-darmstadt.de

Towards advanced bipedal locomotion musculoskeletal system design has received much attention in recent years. It has been recognized that designing and developing new actuators with the properties of the human muscle-tendon complex is only one of the many tasks that have to be fulfilled in order to come close to the powerful human musculoskeletal system enabling the human to such versatile dynamic movements that no robot has been capable of replicating yet. But equally important is a technical implementation of the key characteristics of the human musculoskeletal leg system, segmentation and elastic leg behavior enabled by the mono- and biarticular muscles. So far, there has been an overwhelming consensus in biomechanics literature regarding the joint movements caused by biarticular muscles. In reality, however, they are responsible for an additional action during the second half of ground contact during fast dynamic motions in humans that has not yet been addressed by bipedal robot locomotion studies. Using BioBiped1, a bipedal compliant robot with human-inspired mono- and biarticular tendons, we demonstrate by means of a detailed multibody system dynamics simulation how this positive effect subserve energy-efficient dynamic 1D hopping motions and enables us to establish a novel bipedal locomotion model.

1. Introduction

Towards advanced bipedal locomotion musculoskeletal system design has received much attention in recent years. It has been recognized that designing and developing new actuators with the properties of the human muscle-tendon complex is only one of the many tasks that have to be fulfilled in order to come close to the powerful human musculoskeletal system enabling the human to such versatile dynamic movements that no robot has been capable of replicating yet.[1–3] But equally important is a technical implementation of the key characteristics of the human musculoskeletal leg system: (1) segmentation and (2) elastic leg behavior. These assumptions have motivated the studies by Hosoda, Klein, Niiyama and their co-authors.[4–6]

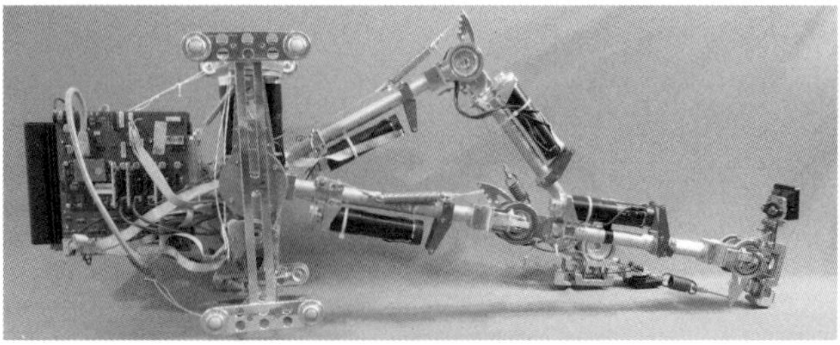

Figure 1. BioBiped1 robot: The first prototype of a planned series of robots with gradually increased bipedal locomotion capabilities.

Several different versions of human-like musculoskeletal leg prototypes were built with the intention to reproduce the motion capabilities of the human musculoskeletal system. Another recently launched project aims at developing a fast, efficient and robust bipedal robot inspired by the musculoskeletal system of the ostrich.[7] All these studies have in common the integration of muscle-tendon like structures and their functionalities, inspired either by humans or animals.

A similar approach is followed by the BioBiped project to explore whether a more human-like leg functionality can be achieved by harnessing the intrinsic dynamics of a properly designed mechanical musculoskeletal motion apparatus based on the human lower limb system.[8] The BioBiped1 robot, depicted in Fig. 1, features a highly compliant tendon-driven actuation system using active and passive human-like elastic tendons. The nine mono- and biarticular muscles shown in Fig. 2(a) are known to contribute to the dynamic locomotion behavior of humans by sharing the necessary work in a highly "intelligent" and very well organized way causing a chain of energy transfer. Of particular interest are the biarticular muscles as they affect simultaneously the movements of two joints. In biomechanics literature there is an overwhelming consensus regarding the joint movements caused by these muscles. For instance, the muscle GAS, abbreviated for *Gastrocnemius*, is reported to extend the ankle and flex the knee joint. However, in Ref. 9 we could recognize a paradoxical action of the GAS tendon by means of forward dynamics simulation of the BioBiped1 multibody system (MBS) dynamics model. It acted as synergist extending the knee joint at knee angles above a specific position. After extensive search, we found a few biomechanics studies confirming this additional action during second half of ground contact in human locomotion.[10] Apparently, it was observed by Lombard already in 1903 and labeled the "paradoxical" function of biar-

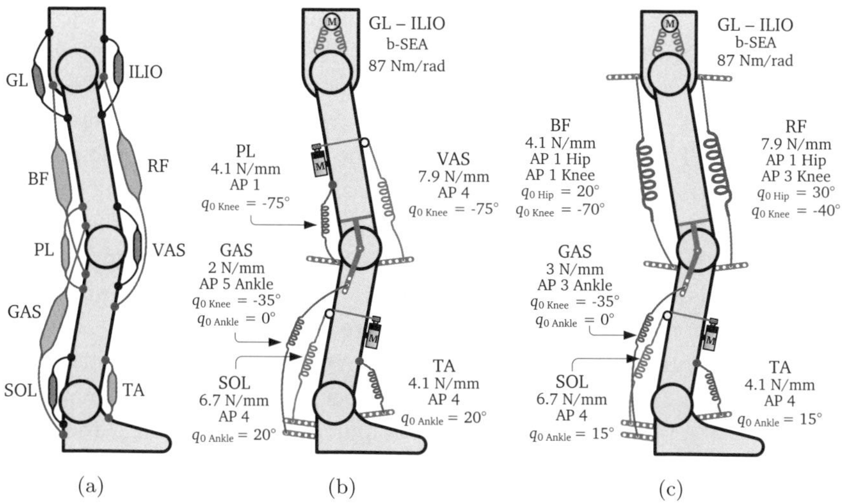

Figure 2. BioBiped1's musculoskeletal leg design (a) Essential human muscle groups during locomotion, tendons actuated by a motor in BioBiped1 are indicated by dark grey color; (b) locomotion model with the parameters used for the implemented structures (AP stands for the attachment point), studied in Section 3; (c) locomotion model suggested in Section 4, omitting the VAS motor.

ticular muscles. Interestingly, it has not yet been addressed by any bipedal robot locomotion studies.

In this paper we investigate the roles of the biarticular structures and elaborate on the reasons for this paradoxical behavior. Second we demonstrate the benefits rising from this understanding and analyze the impacts of this behavior on the interplay of the legs' active and passive tendons using forward dynamics simulation of the detailed BioBiped1 MBS dynamics model.[9,11] Finally, using these insights, we establish a novel bipedal locomotion model for energy-efficient dynamic 1D hopping motions.

2. Paradoxical Action of the Biarticular Structures

BioBiped1's legs are actuated by a combination of active and passive mono- and biarticular tendons mimicking the actions of the corresponding muscles that play an important role during human locomotion (cf. Fig. 2(a)). The monoarticular muscles comprise the antagonist-agonist pairs in each joint, *Gluteus Maximums* (GL) - *Iliacus* (IL) in the hip, *Vastus* (VAS) - *Popliteus* (PL) in the knee, *Soleus* (SOL) - *Tibialis anterior* (TA) in the ankle, and reported to strongly contribute to the task of power generation.[12] As for the technical realization in BioBiped1, the hip muscle pair is implemented by a bidirectional series elastic actuator (b-SEA), i.e. both flexion and extension of the hip joint are actively supported by a geared DC motor. For the knee and ankle joint it was decided to support only the extension movement by motor power. Thus, each knee and ankle joint is actuated by a combination of a unidirectional SEA (u-SEA) and its passive counterpart, as shown in Fig. 2(b). All remaining passive structures, including the biarticular structures, are integrated by a Dyneema tendon with built-in extension spring and can be detached or attached as desired. For more details regarding the actuator types and their implementations we refer to Ref. 9.

The biarticular muscles, of which the three most important are *Rectus Femoris* (RF), *Biceps Femoris* (BF) and GAS, in general are known to contribute to a proximodistal energy transport, i.e. from proximal to distal joints and in this way help to convert body segment rotations into desired translations of the body center of gravity.[13] While RF acts as combined knee extensor and hip flexor, BF, which is one of three muscles acting within the hamstrings muscle group, behaves exactly the other way. GAS extends the ankle and flexes the knee joint. However, this is a very common description of the above named muscles' actions and does not reveal their actually very complex, gait-dependent functionalities.

For instance, in human sprinting, the hamstrings muscle group has been found to be responsible for an additional action during the support phase.[14–16] Provided that the free end of the leg is guided, the hamstrings not only extends the hip, but also the knee joint.[10] The synchronous extension of hip and knee joint takes place for knee angles above -35 °[a]. Paradoxical muscle actions were also observed to be true for the GAS muscle during the last part of the ground phase during sprinting; there, GAS acted as synergist extending the knee joint at knee angles above -40 °.[17]

While it is an important insight that the roles of the biarticular muscles are manifold, there is only little information about the exact whole-body configuration and lever arms acting. Also, it would be interesting to investigate the nature of this phenomenon, whether it is gait-, phase- or configuration-dependent. Since detailed studies on human subjects are presumably required to fully understand the reasons for this paradoxical action, we will examine here this behavior by the laws of classical mechanics exemplary for the passive GAS tendon.

---

[a] A completely folded knee, resp. completely stretched knee, has an inner joint angle of -180 °, resp. 0 °.

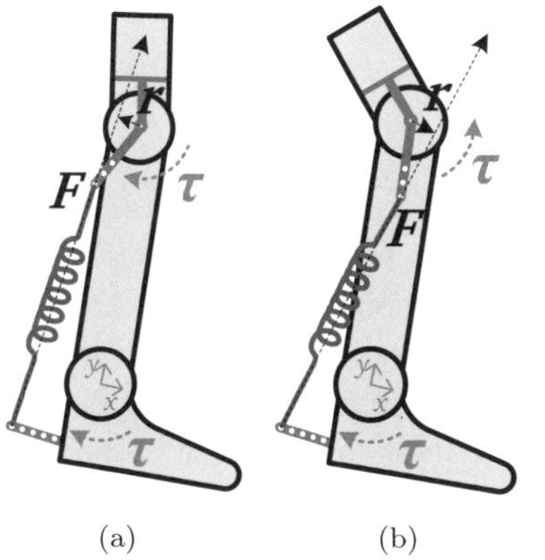

Figure 3. Paradoxical actions of the biarticular GAS structure can be strongly influenced by the manner the tendon is attached to the thigh: (a) GAS acts only as knee and ankle flexor; (b) GAS acts as ankle flexor and knee extensor.

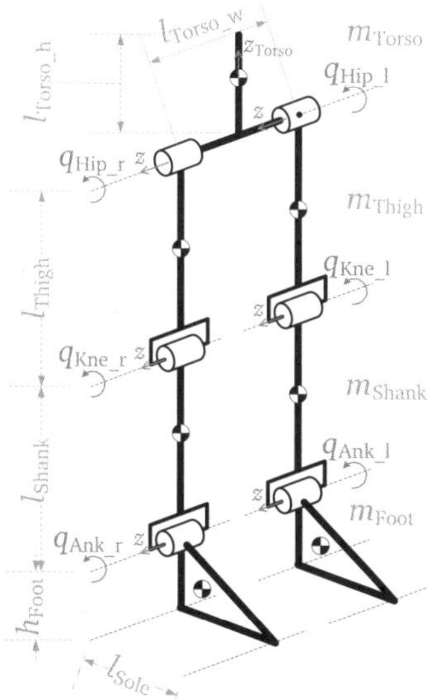

Dimensions and Masses
$l_{\text{Torso\_h}}$=269 mm; $l_{\text{Torso\_w}}$=120 mm;
$l_{\text{Thigh/Shank}}$=330 mm; l_{Foot}=122 mm;
h_{Foot}=67 mm; l_{Sole}=165 mm
m_{Torso}=5.332 kg; m_{Thigh}=0.843 kg;
m_{Shank}=0.804 kg; m_{Foot} =0.342 kg

Figure 4. Main kinematics and dynamics data of BioBiped1's rigid skeleton.

Let us draw your attention to Figs. 3(a) and 3(b). In order to change a flexion into an extension movement of a joint, the torques acting on that joint need to be reversed in their direction. According to the definition of torque as $\boldsymbol{\tau} = \boldsymbol{r} \times \boldsymbol{F}$ where $\boldsymbol{r}$ and $\boldsymbol{F}$ denote the lever arm and force vector, respectively, the torque reversion requires either a reversion of the force or lever arm. In Fig. 3(a) the GAS tendon still acts as both ankle and knee flexor. The tendon force applied to the thigh and the load experienced by the lever lie on the same side of the joint. In Fig. 3(b) we have depicted a possible construction for the attachment of GAS to the thigh such that the lever lies on the other side of the joint and herewith causes reversed torques to extend the knee joint. The hinge joint mounted on the thigh passes on the tendon forces to torques acting on the knee joint. Such construction can cause permanently an extension movement of the knee joint. Obviously, it is also possible to reverse the direction of the force vector. In general, a such permanent functionality may turn out to be quite beneficial as we will evaluate in the next section.

3. Evaluation of the Paradoxical Behavior of GAS by Open-Loop Controlled Trajectories for In-Place Hopping

In this section we will study the paradoxical behavior of the GAS tendon during in-place hopping motions in simulation.

3.1. *Simulation Model*

The simulation model of BioBiped1 consists of two levels containing the rigid joint-link structure, depicted in Fig. 4, and the underlying leg actuation design, displayed in Fig. 2(b). The robot is about 1.1 m tall in extended position and weighs around 9.3 kg. Its dynamic and kinematic parameters, such as link inertia and length, center of mass and mass of each link, were retrieved from the robot's CAD database. In order to detect collisions with the ground, each foot is assigned two designated point contacts, one each at the heel and the toe. The penalty-based contact model is described as a state machine that switches between normal force and stiction/friction. Details of the modular modeling approach and the foot-ground contact model by means of the self-developed libraries using MATLAB/Simulink and SimMechanics can be found in Ref. 11. The models of the actuators are discussed in detail in Ref. 9. The parameters of the contact model and geared DC motor are listed in Ref. 18. The parameters of the elastic transmissions, such as the stiffness, attachment points and rest angles of the implemented structures, are given in Fig. 2(b). In this study the model includes all monoarticular pairs in hip, knee and ankle joint and additionally the GAS tendon. In order to analyze the benefits of GAS as knee extensor, the tendon is attached to the thigh in the manner as shown in Figs. 2(b) and 3(b).

3.2. *Motion Generation*

Since hopping movements can be regarded as periodic oscillations between a flexed and extended leg configuration, we first select two desired leg configurations for the flexion and extension phase: $\boldsymbol{q}_{\mathrm{flex}} = (q_{\mathrm{Hip}}, q_{\mathrm{Kne}}, q_{\mathrm{Ank}}) = (26°, -63°, 13°)$, $\boldsymbol{q}_{\mathrm{ex}} = (q_{\mathrm{Hip}}, q_{\mathrm{Kne}}, q_{\mathrm{Ank}}) = (13°, -26°, -13°)$. The corresponding motor angles were then computed to compensate the gravitational forces in these configurations. For the generation of the trajectories we used the formula $y = A\sin(\omega\, t + \phi) + B$ with the amplitude A, angular frequency ω, phase ϕ and offset angle B. The desired fundamental frequency was set to $f_0 = 2\,\mathrm{Hz}$. As low gain parameters for each motor PD controller we chose $k_{\mathrm{p}} = 30$ and $k_{\mathrm{d}} = 8$.

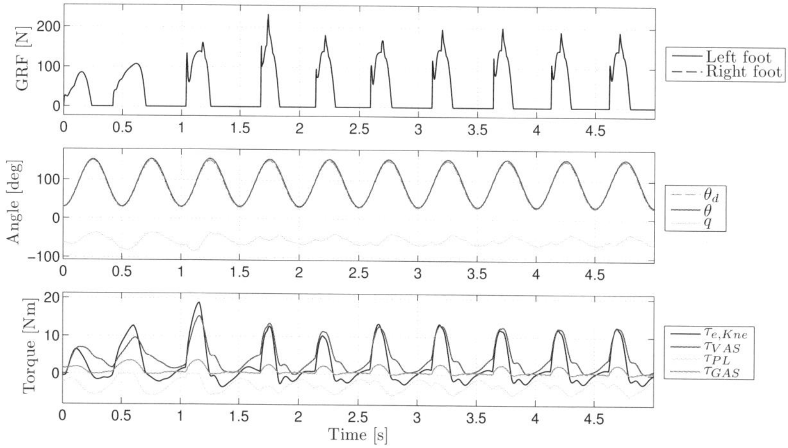

Figure 5. Simulation results with the locomotion model shown in Fig. 2(b): The top diagram displays the GRF; the middle diagram the desired and actual knee motor signal, i.e. θ_d and θ, and the actual knee joint angle q; the lower diagram shows the total knee joint torques, $\tau_{\mathrm{e,Kne}}$, and the single torques induced by all tendons coupling the knee (τ_{VAS} by the active VAS tendon, τ_{PL} by the passive PL tendon, τ_{GAS} by the passive biarticular GAS tendon).

3.3. *Forward Dynamics Simulation and Results*

The forward dynamics was computed in MATLAB/Simulink using the ode23 (Bogacki-Shampine) solver with variable step size, relative tolerance 10^{-3} and adaptive zero-crossing options. The outcome of this simulation were dynamic two-legged hopping motions with an average duty factor of 38.67 % and hopping height of 0.2218 m. The ground reaction forces (GRF) are displayed in the top diagram of Fig. 5. For these synchronous motions the GRF of both feet overlap. In the middle diagram the desired and actual knee motor signal, i.e. θ_d and θ, are displayed together with the actual knee joint angle q. The lower diagram is the most interesting, as it displays the total joint torques, $\tau_{\mathrm{e,Kne}}$, and the single torques induced by all tendons coupling the knee: τ_{VAS} by the active VAS tendon, τ_{PL} by the passive PL tendon, and τ_{GAS} by the passive biarticular GAS tendon. It can be recognized that GAS supports the actions of the knee motor by further extending the knee joint. In this leg actuation, however, this also results in higher PL torques, which in turn leads to higher VAS torques due to the interplay of the tendons. Therefore, with respect to an economical leg actuation design, it is thinkable to completely omit the knee motor for the extension, as for instance suggested in Ref. 19 for level-ground walking.

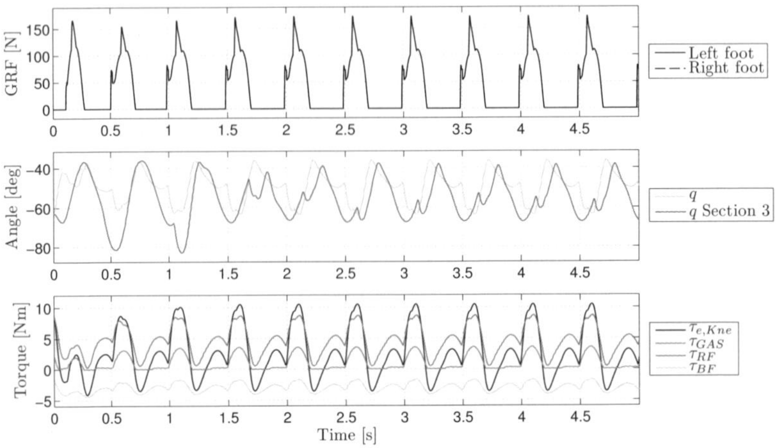

Figure 6. Simulation results with the locomotion model shown in Fig. 2(c): The top diagram displays the GRF; the middle diagram the actual knee joint angle q and for comparison that of the previous simulation; the lower diagram shows the total knee joint torques, $\tau_{\mathrm{e,Kne}}$, and the single torques induced by the passive biarticular tendons (τ_{GAS} by GAS, τ_{RF} by RF, τ_{BF} by BF).

4. Establishing a Novel Bipedal Locomotion Model

As one possible example, we suggest to include the RF and BF and to remove VAS and PL. The leg actuation design studied here is shown in Fig. 2(c). The same motor trajectories as in Section 3 are applied to this novel underactuated model, with the difference that the knee motions are now only influenced by passive biarticular tendons. As Fig. 6 indicates, this novel locomotion model is capable of highly dynamic two-legged hopping motions saving 62.59 % energy compared to the leg actuation discussed in Section 3 (see Table 1). With an average duty factor of 43.23 % and hopping height of 0.192 m, the motions are not as dynamic as those demonstrated in the previous section, but rather more regular when comparing the GRF patterns. The results suggest that dynamic hopping motions can be also performed without active knee extension benefiting from the paradoxical behavior of GAS. The only difference between the simulations of Section 3 and Section 4 concern the start of the simulation. The simulation model discussed in Section 3 was capable of starting directly from the ground, whereas for the robot model analyzed here we had to simplify the starting conditions by dropping the robot from 10 cm. By systematic optimization of the elastic transmission parameters in the ankle, an active lift-off from the ground is expected to be enabled even with this underactuated leg design.

Table 1. Energy consumption of the models discussed in Section 3 and 4.

Locomotion model	**Hopping height** [m]	**Hip** E [J]	**Knee** E [J]	**Ankle** E [J]	**Leg** $\mathbf{\Sigma E}$ [J]
With VAS (cf. Fig. 2(b))	0.2218	7.0526	304.3469	213.6929	525.0925
Without VAS (cf. Fig. 2(c))	0.192	8.5522	0	187.8801	196.4323

5. Conclusions

To our knowledge, this is the first paper in which the paradoxical action of biarticular muscles is discussed in the context of robot locomotion and exemplary analyzed for one muscle using the identified MBS dynamics simulation model of the BioBiped1 robot. The simulation results suggest to use the complex functions of these muscles beneficially to reduce the energy consumption. Also, the analyses raise the general question whether implemented human-like muscle-tendon structures have to act in conformity with biomechanical observations and suggestions. Rather, reported insights should be seen as valuable hints to enable the derivation of novel design guidelines that support dynamic robot locomotion.

Acknowledgements

This work was supported by the German Research Foundation (DFG) under grant no. STR 533/7-1. Special thanks go to Thomas Lens for valuable comments. The successful design and development of the BioBiped1 robot is due to a joint effort by the BioBiped project team members.

Bibliography

1. C.-M. Chew, J. Pratt, and G. Pratt, "Blind walking of a planar bipedal robot on sloped terrain," in *IEEE ICRA*, 1999, pp. 381 – 386.
2. J. W. Hurst, J. Chestnutt, and A. Rizzi, "An actuator with physically variable stiffness for highly dynamic legged locomotion," in *IEEE ICRA*, 2004, pp. 4662–4667.
3. R. V. Ham, B. Vanderborght, M. V. Damme, B. Verrelst, and D. Lefeber, "MACCEPA: the mechanically adjustable compliance and controllable equilibrium position actuator for 'controlled passive walking'," in *IEEE ICRA*, 2006, pp. 2195–2200.
4. K. Hosoda, T. Takuma, A. Nakamato, and S. Hayashi, "Biped robot design powered by antagonistic pneumatic actuators for multi-modal locomotion," *Robotics and Autonomous Systems*, vol. 56, pp. 46–53, 2007.

5. T. J. Klein, T. Phuam, and M. A. Lewis, "On the design of walking machines using biarticular actuators," in *CLAWAR*, 2008.
6. R. Niiyama, A. Nagakubo, and Y. Kuniyoshi, "Mowgli: A bipedal jumping and landing robot with an artificial musculoskeletal system," in *IEEE ICRA*, 2007, pp. 2546–2551.
7. S. Cotton, I. M. C. Olaru, M. Bellman, T. van der Ven, J. Godowski, and J. Pratt, "FastRunner: A fast, efficient and robust bipedal robot. concept and planar simulation." in *IEEE ICRA*, 2012, pp. 2358–2364.
8. K. Radkhah, C. Maufroy, M. Maus, D. Scholz, A. Seyfarth, and O. von Stryk, "Concept and design of the BioBiped1 robot for human-like walking and running," *International Journal of Humanoid Robotics*, vol. 8, no. 3, pp. 439–458, 2011.
9. K. Radkhah, T. Lens, and O. von Stryk, "Detailed dynamics modeling of BioBiped's monoarticular and biarticular tendon-driven actuation system," in *IEEE/RSJ IROS*, 2012, pp. 4243 – 4250.
10. K. Wiemann and G. Tidow, "Relative activity of hip and knee extensors in sprinting - implications for training," *New studies in Athletics 1*, vol. 10, pp. 29–49, 1995.
11. T. Lens, K. Radkhah, and O. von Stryk, "Simulation of dynamics and realistic contact forces for manipulators and legged robots with high joint elasticity," in *ICAR*, 2011, pp. 34–41.
12. L. Grègoire, H. E. Veeger, P. A. Huijing, and G. J. van Ingen Schenau, "Role of mono- and biarticular muscles in explosive movements," *International Journal of Sports Medicine*, vol. 5, pp. 301–305, 1984.
13. R. Jacobs, M. F. Bobbert, and G. J. van Ingen Schenau, "Mechanical output from individual muscles during explosive leg extensions: The role of biarticular muscles," *Journal of Biomechanics*, vol. 29, no. 4, pp. 513–523, April 1996.
14. K. Fischer, "Zur geführten Wirkung mehrgelenkiger Muskeln," *Zeitschrift für Anatomie und Entwicklungsgeschichte*, vol. 83, p. 7S2ff, 1927.
15. S. Molbech, "On the paradoxical effect of some two-joint muscles," *Acta Morphologica Neerlando-Scandinavica*, vol. 6, pp. 171 – 178, 1965.
16. J. G. Andrews, "A general method for determining the functional role of a muscle," *Journal of Biomechanical Engineering*, vol. 107, pp. 348 – 353, 1985.
17. E. B. Simonsen, L. Thomsen, and K. Klausen, "Activity of mono- and biarticular leg muscles during sprint running," *European Journal of Applied Physiology and Occupational Physiology*, vol. 54, no. 5, pp. 524–532, 1985.
18. K. Radkhah and O. von Stryk, "Model-based elastic tendon control for electrically actuated musculoskeletal bipeds," in *CLAWAR*, 2013.
19. K. Endo and H. Herr, "A model of muscle-tendon function in human walking," in *IEEE ICRA*, 2009, pp. 1909–1915.

AN AUTOMATED SYSTEM FOR SYSTEMATIC TESTING OF LOCOMOTION ON HETEROGENEOUS GRANULAR MEDIA*

FEIFEI QIAN, KEVIN DAFFON, TINGNAN ZHANG AND DANIEL I. GOLDMAN
School of Physics, Georgia Institute of Technology, 837 State St NW
Atlanta, GA 30332, USA

Particulate substrates like deserts or Martian terrain are often composed of collections of particles of different sizes and shapes. While much is known about how robots can effectively locomote on hard ground and increasingly on homogeneous granular ground, the principles of locomotion over heterogeneous granular substrates are relatively unexplored. In this study we test the locomotion performance of an open-loop controlled legged robot (Xplorerbot, 15 cm, 150 g) in a trackway filled with 3 mm diameter glass "fine grains", with two parallel lines of eight 25.4 mm diameter large glass "boulders" embedded within. We also develop an experimentally validated Discrete Element Method (DEM) simulation. In experiment and simulation, we observe three distinct modes of robot leg-ground interaction which influence locomotion performance. To systematically investigate how robot leg frequency, particle size and boulder distribution affect the interaction modes and robot speed and stability, we develop an automated system which can vary the properties of the heterogeneous granular substrate, as well as record robot locomotion performance. The system allows collection of ~200 runs/day facilitating systematic parameter exploration and comparison to simulation.

1. Introduction

Substrates that are rocky and loose are often found in environments that field robots must traverse; such terrains can contain granular media (GM) with particle sizes spanning multiple orders of magnitude (Figure 1b). When small robotic locomotors like PackBot or RHex (Figure 1a) travel across these "flowable" types of terrain [1], they exhibit characteristic failure modes (slips, unstable foot-holds, impassable barriers, or a limb/tread fluidization of a thin layer of smaller particles), which significantly affect robot stability, trafficability and power consumption. A major challenge in creating the next generation of mobile robots is expanding the scope of terramechanics [2, 3] from large tracked and treaded vehicles on homogeneous ground to arbitrarily shaped and actuated locomotors moving on and within complex heterogeneous terrestrial substrates. However, in typical heterogeneous environments, the force fluctuations introduced by heterogeneities during intrusion and drag can be large—

* This work is supported by a DARPA Young Faculty Award and the ARL MAST CTA.

comparable in size to the average force, making the applicability of continuum terramechanics [4, 5] unclear. Currently, most terrestrial vehicles (including mobile robots) are tested on substrates made of standardized homogenous media (e.g. Ottowa sand [6], lunar simulants [7]).

Figure 1 Natural heterogeneous terrain. (a) RHex robot traveling across heterogeneous gravel substrate (photo courtesy Alfred Rizzi, Boston Dynamics). (b) Sieved Mojave desert GM.

2. Experimental robot locomotion test and numerical modeling

To gain insight into robot locomotion on heterogeneous ground, in this paper we extend techniques from our previous studies [8] to create repeatable states of GM with controlled heterogeneity. We used a hexapedal locomotor: an open-loop controlled small legged robot (Xplorerbot, 15 cm, 150 g, Figure 2a, b) to perform laboratory experiments. We initially tested the robot in a model desert-like heterogeneous terrain: a bi-dispersed granular test bed filled with 3 mm diameter small particles (simplified fine grains) with larger 25.4 mm diameter glass particles (simplified "boulders") randomly embedded within (Figure 2c, d). The symmetrical geometry of the particles simplifies the leg-ground interaction, and makes it feasible to integrate experiment with our experimentally validated DEM simulations [9]. The kinematics of the robot were captured by two high speed cameras (AOS X-PRI) from both top and side views at a frame rate of 250 fps. High contrast dorsal markers were painted on the robot to obtain speed and trajectory information.

We extend our previous homogeneous granular media DEM simulation [9] for heterogeneous ground conditions. The granular bed in simulation ($\sim 2 \times 10^5$ particles) was 60 PD (particle diameters) in width, 15 PD in depth, and 180 PD in length, and has frictionless boundaries (Fig. 2d). In addition, we introduced heterogeneity to the sand bed by generating 10 ~ 20 randomly distributed glass boulders (25.4 mm diameter). To model the Xplorerbot, as in [9] we use a multi-body dynamic solver (MBDyn) and coupled it with our particle simulation (Figure 2b). By integrating the equations of motion using the force computed from DEM, we could reconstruct the locomotion of the robot on both homogeneous and heterogeneous GM.

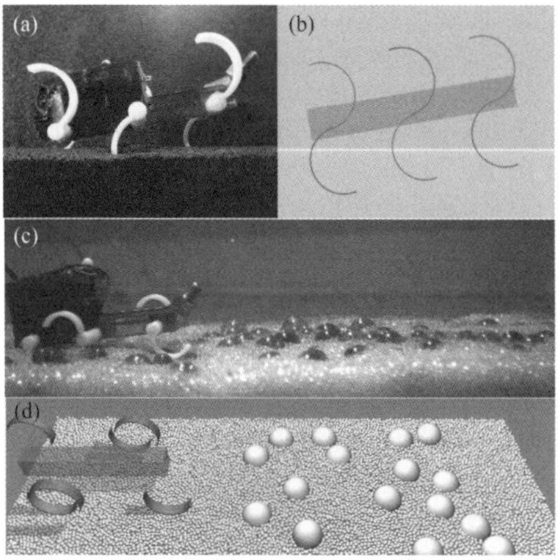

Figure 2 (a) A C-legged hexapedal robot (Xplorerbot, 15 cm, 150 g) standing on homogeneous fine grains (~1 mm poppy seeds). (b) The Xplorerbot in simulation constructed using MBDyn. (c) Xplorerbot traveling across model heterogeneous ground (bi-dispersed granular substrate as a randomized mixture of 3 mm fine grains with 25.4 mm larger glass boulders embedded withiin). (d) Xplorerbot traveling across model heterogeneous granular ground in DEM simulation.

3. Chaotic dynamics in robot trajectories

In both experiment and simulation on the random boulder field, we observed complex dynamics involving pitch, roll and yaw of the robot during transit. To simplify the problem and to make our system amenable to systematic study we next investigated locomotion on a boulder "lattice". We arranged eight 25.4 mm diameter glass boulders into a 4 x 2 lattice (Figure 3a, c), and we measured robot CoM trajectories on the horizontal plane from both experiment and simulation (Figure 3b, d). We found that the robot legs and body began to collide with the boulders after a few steps, and the direction of the robot altered after each contact. Starting from similar initial conditions, the robot CoM trajectories eventually diverged for different runs, much like that of electron beam scattering in a lattice (from a classical point of view) [10].

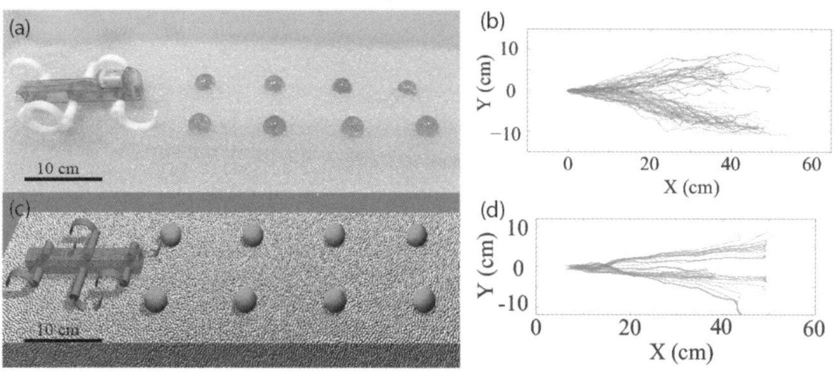

Figure 3 (a) Experiment setup of Xplorerbot running over 4 x 2 large boulder field; small particles are 3 mm diameter glass spheres. (b) Xplorerbot horizontal plane CoM trajectories measured in experiment. (c) Numerical simulation of Xplorerbot running over 4 x 2 large boulder field. (d) Xplorerbot horizontal plane CoM trajectories measured in DEM simulation.

Closer investigation in simulation suggested that the robot's CoM trajectory was sensitive to initial conditions. Fig. 4 shows an example of two simulation runs where the Xplorerbot's CoM initial position varied by 0.5 cm in both x and y directions (Figure 4a), while all other initial conditions were identical (e.g., the robot body axis was initialized to be parallel to the x-axis, the c-leg's initial phase was kept the same, and the boulders were distributed to the same locations and depths). For a short time, the two robot trajectories remained similar. After ~0.3 s, however, the middle left leg of the robot impacted boulder 1 at a different attack angle, resulting in two different leg-boulder contact modes. In the top trajectory, the leg forced the boulder to slide forward, and thus, the robot orientation was not significantly affected. In the bottom trajectory, the leg slipped off the boulder, generating a horizontal impulse that caused ~20 degree change in the yaw angle of the robot, and leading to a dramatically different trajectory (Figure 4b). This sensitivity to the initial condition indicates a signature of chaotic dynamics [11]. We intend to perform Lyapunov exponent analysis [12] in our future work to predict how nearby trajectories separate, and to explore the existence and type of attractors generated by these dynamics.

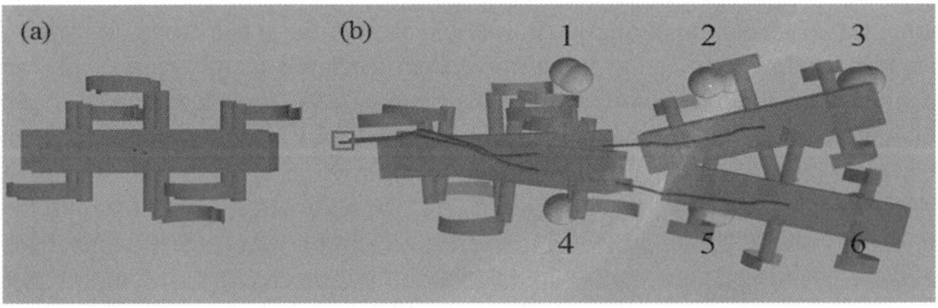

Figure 4 Two simulation runs with the CoM of the robot placed 0.5 cm apart initially. (a) Difference in the two initial locations. (b) Two trajectories. Red square indicates the robot initial position. Green filled circles indicate locations of 25.4 mm boulders. Gray background indicates 3 mm fine grains.

4. Leg-ground interaction modes

Based on our observations of locomotion sensitivity due to different limb-boulder interactions, we were inspired to characterize these interactions. We collected 67 locomotion runs on the boulder "lattice" ground, and 124 leg-ground interaction events were characterized. We observed three distinct modes (Figure 5) in both experiment and simulation:

a) A forced sliding mode, where the leg struck on the side of a slightly buried boulder, propelling the boulder forward or sideways (Figure 5a). The effect of this interaction on robot performance was small. We observed 53 cases when the robot exhibited this forced sliding mode (out of 59 cases where the robot leg struck on the side of the boulder), and the robot trajectories were not affected (yaw < 10 degree) in 51 cases of those (98.1%).

b) A slipping mode (Figure 5b), where the leg slid on the top of a deeply buried boulder, causing the robot to pitch/yaw/roll, while the boulder remained still or rotated against smaller grains. Robot stability was significantly affected in this mode. We observed 28 cases when the robot exhibited slipping mode (out of 31 cases where robot leg struck on top of deeply buried boulders), and the robot trajectory was significantly affected (yaw >10 degree) in 23 cases of those (82.1%).

c) A forced intrusion mode (Figure 5c), where the robot leg struck the top of the slightly buried boulder, forcing the boulder downward into the fine grains. By taking advantage of the mobility of obstacles towards leg intrusion direction, the robot reduced the impulse of the collision and maintained its stability in this mode. We observed 17 cases when the robot exhibited the forced intrusion mode (out of 36 cases where robot leg struck on top of slightly buried boulders), and the robot trajectory was affected significantly (yaw >10 degree) only in 1 case (0.06%).

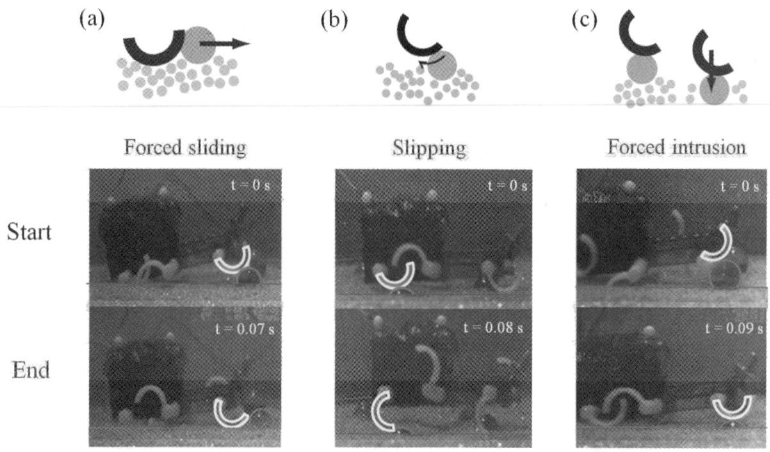

	Forced sliding mode	Slipping mode	Forced intrusion mode
Contact point on boulder	Side	Top	Top
Boulder depth before contact	Slightly buried	Deeply buried	Slightly buried
Effect on locomotion performance	Small	Large	Small

Figure 5 Three leg-ground interaction modes showing schematic (top row) and experimental data from side-view camera. (a) Forced sliding mode. (b) Slipping mode. (c) Forced intrusion mode.

5. Automated terrain creation and locomotion testing system

The complexity of the interactions that we observed even in the simple experimental setup compelled us to create an automated terrain creation and robot testing system (Figure 6). Such a system can allow comprehensive and systematic study of the effects of arbitrary heterogeneity and spatial distribution, relative sizes of grain/grain and grain/locomotor, and robot limb size and kinematics on interaction modes and performance. An automated system can also facilitate validation of the heterogeneous DEM simulation result.

The central structure of the system consists of an air fluidized bed trackway. Four vacuums (RIDGID, 16 gallon) are connected below the trackway, blowing air through a flow distributer (0.635 cm thick, 50 um pore size porous plastic) to evenly fluidize the ~1 mm diameter fine grains (poppy seeds) in the trackway, allowing control of the compaction and creation of repeatable homogeneous granular states of the fine grains. The trackway can also tilt to create inclined/declined granular environments.

To generate states of arbitrary heterogeneity, a 3-axis motor system (Copley, STA25, STB25, XTB38) is installed above the trackway, enabling the motor end-effector to move in three dimensions, driving a universal jamming gripper [13] (Figure 6c) to programmed locations, creating arbitrary distributions of multi-size granular particles.

The customized gripper assembly includes a balloon filled with granular material (a "universal jamming gripper" [13]), a support frame, and a HI-TEC servo (HSR-5980SG). The 3D-printed support frame connects the gripper to vacuum tubing through an air filter, enabling the granular material in the gripper balloon to achieve fluid-like or solid-like properties. The fluid-like property of

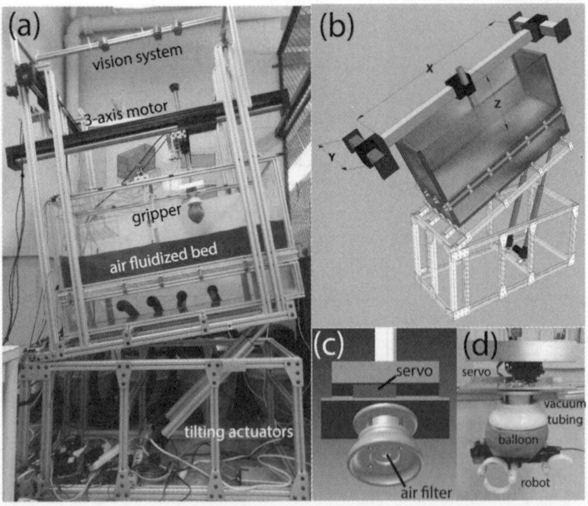

Figure 6. Automated terrain creation and locomotion testing system. (a) The automated system, including the vision system, the 3-axis motor, the universal jamming gripper, the air fluidized bed and the tilting actuators. (b) Mechanical drawing of the system: a 3-axis motor mounted on a tiltable trackway. (c) Mechanical design of the universal jamming gripper assembly, including the servo motor and the 3D-printed gripper support frame with air distribution structure. (d) The universal jamming gripper lifting the robot.

the granular media inside the balloon (when suction is off) allows the gripper to deform around the robot or boulders, while the solid property of the granular material (when suction is applied) enables the gripper to reach a jammed state, resulting in a rapid gripping of objects of complex shapes. A support frame provides attachment from the gripper balloon to the servo disk, enabling the gripper assembly to adjust the robot orientation after each locomotion test. The system allows distribution of boulders to designated locations before each locomotion test.

Kinematic information of the robot, including the x, y, z CoM position as well as the yaw, pitch, roll angle, is obtained and recorded by tracking with three top-view cameras (Naturalpoint, Flex13, 120 FPS) three IR-reflective markers attached to the robot. The cameras also monitor the location of the robot and the boulders before and after each test. This information is communicated to the motor system, so that the gripper can retrieve both the robot and boulders. All functions of the test bed are controlled by a single integrated LabVIEW program. The automated system can currently take more than 200 locomotion tests in one day, without human intervention. A set of sample trajectories is shown in Figure 7b.

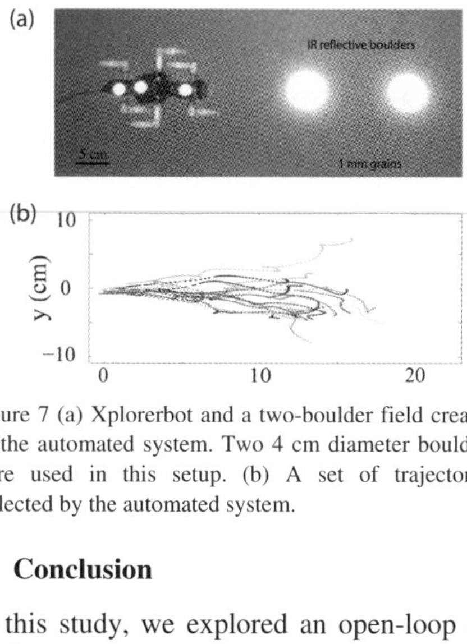

Figure 7 (a) Xplorerbot and a two-boulder field created by the automated system. Two 4 cm diameter boulders were used in this setup. (b) A set of trajectories collected by the automated system.

6. Conclusion

In this study, we explored an open-loop controlled legged robot's locomotion performance on heterogeneous granular ground, and characterized leg-ground interaction modes. We extended our previous DEM numerical simulation for heterogeneous ground application and revealed chaotic dynamics of the robot CoM trajectories caused by different types of leg-boulder interaction. We also designed and constructed an automated terrain-creation and locomotion testing system, which enabled the creation of repeatable heterogeneous granular substrate and systematic locomotion testing with minimal human intervention.

In the future, we will complement the above approaches with a continuum approach building on our "terradynamics" work in [14]. We will make modifications to existing continuum models so that they can be used to approximately model locomotor performance of robots in heterogeneous GM. We posit that for a certain range of particle sizes and interaction-types, we can modify the continuum equations through addition of stochastic terms to create a "fluctuating terradynamics" analogous to fluctuating hydrodynamics [15] that describes fluids near critical points [16, 17] and granular fluids [18].

Acknowledgments

We thank Andrei Savu for help with test bed construction, Duncan Hathaway for help with preliminary data collection, and Pierangelo Masarati for MBDyn support. We thank Paul Umbanhowar, Nick Gravish and Jeff Aguilar for helpful discussion.

References

1. C. Li, T Zhang, and Daniel I. Goldman, *Science*, **339**, 1408 (2013)
2. K. Terzaghi, R.B. Peck, and G. Mesri, *Soil Mechanics in Engineering Practice* (1996).
3. K. Iagnemma and S. Dubowsky, *Mobile Robots in Rough Terrain: Estimation, Motion Planning, and Control with application to Planetary Rovers* (2004).
4. M.G. Bekker, *Theory of Land Locomotion* (1956).
5. J.Y. Wong, *Terramechanics and off-road vehicles* (1989).
6. G. Meirion-Griffith and M. Spenko, *Journal of Terramechanics* **48**, 2 (2011).
7. G.H. Heiken, D.T. Vaniman, and B.M. French, *Lunar Sourcebook: A User's Guide to the Moon* (1991).
8. C. Li, et al., *Proc. Nat. Aca. Sci.* **106**, 9 (2009).
9. F. Qian, et al., *Proceedings of Robotics: Science and Systems* (2012).
10. M.A. McCord and M.J. Rooks. *SPIE Handbook of Microlithography, Micromachining and Microfabrication* (2000).
11. S.H. Strogatz, *Nonlinear Dynamics and Chaos* (2001).
12. E.N. Lorenz, *The Essence of Chaos* (1995).
13. E. Brown, et al., *Proc. Nat. Aca. Sci.* **107**, 44 (2010).
14. C. Li, et al., Science. **339**, 1408(2013).
15. L.D. Landau, *Fluid Mechanics*. (1959).
16. J.B. Swift, *Phys. Rev. A.* **15**, 319 (1977).
17. M. Wu, A. Guenter and D.S. Cannell, *Phys. Rev. Lett*, **75**, 1743 (1995).
18. D.I. Goldman, J.B. Swift and H.L. Swinney. *Phys. Rev. Lett.* **92**, 17 (2004).

INVERSE DYNAMICS FOR A QUADRUPED ROBOT LOCOMOTING ALONG SLIPPERY SURFACES

SAMUEL ZAPOLSKY[1], EVAN DRUMWRIGHT[1]
IOANNIS HAVOUTIS[2], JONAS BUCHLI[3], CLAUDIO SEMINI[2]

[1] *Positronics Lab,*
The George Washington University,
samzapo@gwmail.gwu.edu, drum@gwu.edu

[2] *Dep. of Advanced Robotics,*
Istituto Italiano di Tecnologia (IIT)
ioannis.havoutis@iit.it, claudio.semini@iit.it

[3] *Agile & Dexterous Robotics Lab,*
ETH Zurich
buchlij@ethz.ch

We describe a method for computing inverse dynamics forces suitable for robots locomoting along (or manipulating) sticky and slippery surfaces. Our method assumes rigid body dynamics, rigid body contact, and Coulomb friction, and is sufficiently fast for computation in real-time servos. We assess our method's performance on the HyQ quadrupedal robot in both simulation and *in situ*.

Keywords: Inverse dynamics; contact modeling; slipping locomotion.

1. Motivation

Strategies for controlling robots locomoting across terrain are generally predicated on sufficient traction. Locomotion across slippery surfaces (ice, grease, sand, *etc.*) can endanger the robot and its surroundings. We address this problem using an inverse dynamics (ID) controller that models variable surface friction in order to both minimize slipping and leverage high friction conditions (as available). This controller is able to utilize surface data—whether provided *a priori* or measured via sensory perception—to maximize robot locomotion performance on mixed sticky-slippery surfaces; the controller is even able to account for surfaces varying at the robot's individual feet. Our method employs quadratic programming (QP) models yet runs sufficiently quickly for use in hydraulic robot control loops and generates chatter-free torque profiles. We verify results using the HyQ quadrupedal robot platform.[1]

Notation

$\boldsymbol{m}$ the number of rigid bodies contacting the robot
$\boldsymbol{n}$ the number of contact points between the robot and the rigid bodies
$\boldsymbol{w}$ the number of generalized coordinates in the *system* (the robot and contacting bodies)
$\boldsymbol{fb}$ designation for the wrench or twist on the robot's floating base
$\boldsymbol{cb}$ designation for the m wrenches or twists corresponding to the contacting bodies
$\boldsymbol{c}_N$ the impulses to be applied to the contact points in the normal direction
$\boldsymbol{c}_D$ the impulses to be applied to the contact points in the tangential directions
$\mathbf{M}$ the generalized inertia matrix of the *system*
$\boldsymbol{v}$ the current generalized velocity of the *system*
$\boldsymbol{v}^*$ the generalized velocity after the contact model and external wrenches are applied
$\dot{\boldsymbol{q}}$ the current robot joint velocity
$\ddot{\boldsymbol{q}}$ the desired robot joint acceleration
$\boldsymbol{\tau}$ the vector of ID torques to be applied by the robot's actuators
$\boldsymbol{f}_{\mathbf{ext}}$ the vector of generalized external forces of the *system*
$\mathbf{N}$ the $\mathbb{R}^{w \times n}$ matrix of normal generalized contact wrenches of the *system*
$\mathbf{D}$ the $\mathbb{R}^{w \times r \cdot n}$ matrix of tangential generalized direction contact wrenches of the *system*; each contact uses r vectors in the tangent plane[5]
μ the coefficient of friction
$h \ll 1$ first-order approximation scalar

2. Inverse Dynamics Computation

We focus on the case where the desired robot acceleration is consistent with the contact constraints (*e.g.*, a robot is not commanded to accelerate its foot into a door). In computing ID, we assume that robots are modeled using rigid body dynamics with Coulomb contact friction. It is well known that—at the acceleration-level—rigid bodies in contact are inconsistent with the Coulomb friction model.[2] In order to admit solutions to arbitrary contact configurations, we will use the standard technique of moving to the velocity-level with *impulsive* forces (see, *e.g.*,[3]). We model rigid body contact using a convex contact model[4] with an approximate polygonal friction cone[5] in order to use fast pivoting QP algorithms. The process of determining the contact impulses that satisfy inverse dynamics and generate stable joint torques is a two-stage process, described by the convex QPs below.

$$\textbf{Stage I:} \min_{\boldsymbol{c}_N, \boldsymbol{c}_D} \frac{1}{2} \boldsymbol{v}^{*\mathsf{T}} \mathbf{M} \boldsymbol{v}^* \quad \text{(by principle of maximum dissipation)} \tag{1}$$

$$\mathbf{N}^{\mathsf{T}} \boldsymbol{v}^* \geq \mathbf{0} \quad \text{(non-interpenetration)} \tag{2}$$

$$\boldsymbol{c}_N \geq \mathbf{0}, \quad \boldsymbol{c}_D \geq \mathbf{0} \quad \text{(compressive force/polygon)} \tag{3}$$

$$\mu c_{N,i} \geq \mathbf{1}^{\mathsf{T}} \boldsymbol{c}_{D,i} \quad \text{(Coulomb friction)} \tag{4}$$

$$\boldsymbol{v}^*_{\text{robot.q}} = \dot{\boldsymbol{q}} + h\, \ddot{\boldsymbol{q}} \quad \text{(inverse dynamics)} \tag{5}$$

$$\mathbf{M}(\boldsymbol{v}^* - \boldsymbol{v}) = \mathbf{N}\boldsymbol{c}_N + \mathbf{D}\boldsymbol{c}_D + h\, (\boldsymbol{f}_{\text{ext}} + \begin{bmatrix} \mathbf{0} \\ \boldsymbol{\tau} \end{bmatrix}) \tag{6}$$

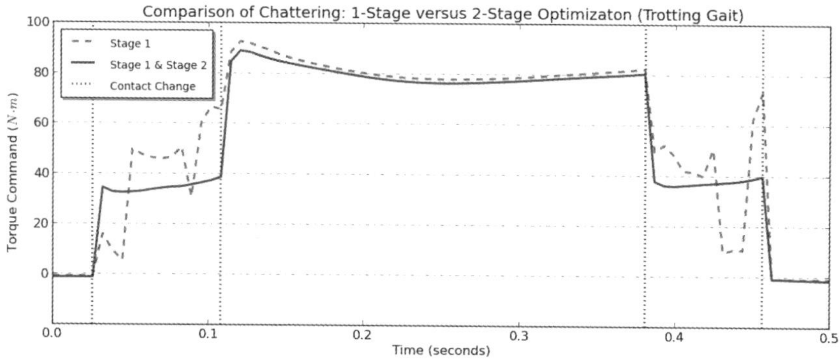

Fig. 1: *Comparison of torque profiles using only Stage I and both Stage I and Stage II. Proceeding chronologically, contact conditions are 1) three feet in contact; 2) four feet in contact; 3) two feet in contact; 4) four feet in contact; and 5) three feet in contact.*

$$\textbf{Stage II: } \min_{\mathbf{c}_N, \mathbf{c}_D} \frac{1}{2}\boldsymbol{\tau}^{\mathsf{T}}\boldsymbol{\tau} \quad \text{(chatter-free torque generation)} \tag{7}$$

$$\text{subject to } \textbf{Constraints } (2) - (6) \tag{8}$$

$$\text{and such that } \frac{1}{2}\boldsymbol{v}^{*\mathsf{T}}\mathbf{M}\boldsymbol{v}^{*} = \textbf{stage 1 objective} \quad \text{(energy)} \tag{9}$$

where $\boldsymbol{v}^* = \left[\boldsymbol{v}^{*}_{\text{robot.fb}}{}^{\mathsf{T}}\ \boldsymbol{v}^{*}_{\text{robot.q}}{}^{\mathsf{T}}\ \boldsymbol{v}^{*}_{\text{cb}}{}^{\mathsf{T}}\right]^{\mathsf{T}}$ and $\boldsymbol{f}_{\text{ext}} = \left[\boldsymbol{f}_{\text{fb}}{}^{\mathsf{T}}\ \boldsymbol{\tau}^{\mathsf{T}}\ \boldsymbol{f}_{\text{cb}}{}^{\mathsf{T}}\right]^{\mathsf{T}}$.

2.1. *Computing inverse dynamics when the desired robot acceleration is consistent with contact model (Stage I)*

Our method simultaneously computes the coupled contact and ID forces of an articulated rigid body system. The time complexity of solving the resulting convex QP is $O(n^3)$ in the number of contact points.[6] The QP for the contact model provably is always solvable, and the QP for ID is always solvable as well (predicated on consistency of desired robot acceleration with the contact model). The resulting QP is described by Eqns. 1–6.

2.2. *Simplifying the computation*

Assume that we first solve for the joint forces $\boldsymbol{f}_{ID}$ necessary to solve the ID problem under no contact constraints. The new velocity $\boldsymbol{v}^*$ is now defined as:

$$\boldsymbol{v}^* = \boldsymbol{v} + \mathbf{M}^{-1}(\mathbf{N}\boldsymbol{c}_N + \mathbf{D}\boldsymbol{c}_D + h\boldsymbol{f}_{ext} + \begin{bmatrix} \mathbf{0} \\ h(\boldsymbol{f}_{ID} + \boldsymbol{x}) \end{bmatrix}) \tag{10}$$

where we define $\boldsymbol{x}$ to be the actuator forces that are added to $\boldsymbol{f}_{ID}$ to counteract contact forces. To simplify our derivations, we will define the following vectors and matrices:

$$\mathbf{R} \triangleq \begin{bmatrix}\mathbf{N}\ \mathbf{D}\end{bmatrix} \tag{11}$$

$$\boldsymbol{z} \triangleq \begin{bmatrix}\boldsymbol{c}_N\ \boldsymbol{c}_D\end{bmatrix}^\intercal \tag{12}$$

$$\mathbf{M} \triangleq \begin{bmatrix}\mathbf{A} & \mathbf{B}\\ \mathbf{B}^\intercal & \mathbf{C}\end{bmatrix} \tag{13}$$

$$\mathbf{M}^{-1} \triangleq \begin{bmatrix}\mathbf{D} & \mathbf{E}\\ \mathbf{E}^\intercal & \mathbf{F}\end{bmatrix} \tag{14}$$

$$\boldsymbol{j} \triangleq \boldsymbol{v}_b + \begin{bmatrix}\mathbf{D}\ \mathbf{E}\end{bmatrix}(h\boldsymbol{f}_{ext} + \begin{bmatrix}\mathbf{0}\\ h\boldsymbol{f}_{ID}\end{bmatrix}) \tag{15}$$

$$\boldsymbol{k} \triangleq \boldsymbol{v}_q + \begin{bmatrix}\mathbf{E}^\intercal\ \mathbf{F}\end{bmatrix}(h\boldsymbol{f}_{ext} + \begin{bmatrix}\mathbf{0}\\ h\boldsymbol{f}_{ID}\end{bmatrix}) \tag{16}$$

The components of $\boldsymbol{v}^*$ are then defined as follows:

$$\boldsymbol{v}_b^* = \boldsymbol{j} + \begin{bmatrix}\mathbf{D}\ \mathbf{E}\end{bmatrix}(\mathbf{R}\boldsymbol{z} + \begin{bmatrix}\mathbf{0}\\ h\boldsymbol{x}\end{bmatrix}) \tag{17}$$

$$\boldsymbol{v}_q^* = \boldsymbol{k} + \begin{bmatrix}\mathbf{E}^\intercal\ \mathbf{F}\end{bmatrix}(\mathbf{R}\boldsymbol{z} + \begin{bmatrix}\mathbf{0}\\ h\boldsymbol{x}\end{bmatrix}) = \boldsymbol{v}_q + h\boldsymbol{a}_q \tag{18}$$

Using the latter equation, we can solve for $\boldsymbol{x}$:

$$\boldsymbol{x} = \frac{\mathbf{F}^{-1}(\boldsymbol{v}_q^* - \boldsymbol{k} - \begin{bmatrix}\mathbf{E}^\intercal\ \mathbf{F}\end{bmatrix}\mathbf{R}\boldsymbol{z})}{h} \tag{19}$$

Equation 19 indicates that once contact forces are computed, determining the actuator forces for ID requires solving only a linear equation. Substituting the solution for $\boldsymbol{x}$ into Eqn. 17 yields:

$$\boldsymbol{v}_b^* = \boldsymbol{j} + \begin{bmatrix}\mathbf{D}\ \mathbf{E}\end{bmatrix}\mathbf{R}\boldsymbol{z} + \mathbf{E}\mathbf{F}^{-1}(\boldsymbol{v}_q^* - \boldsymbol{k} - \begin{bmatrix}\mathbf{E}^\intercal\ \mathbf{F}\end{bmatrix}\mathbf{R}\boldsymbol{z}) \tag{20}$$

To simplify further derivation, we will define a new matrix and a new vector:

$$\mathbf{Z} \triangleq \begin{bmatrix}\mathbf{D}\ \mathbf{E}\end{bmatrix} - \mathbf{E}\mathbf{F}^{-1}\begin{bmatrix}\mathbf{E}^\intercal\ \mathbf{F}\end{bmatrix}\mathbf{R} \tag{21}$$

$$\boldsymbol{p} \triangleq \boldsymbol{j} + \mathbf{E}\mathbf{F}^{-1}(\boldsymbol{v}_q^* - \boldsymbol{k}) \tag{22}$$

Now, $\boldsymbol{v}_b^*$ can be defined simply, and solely in terms of $\boldsymbol{z}$, as:

$$\boldsymbol{v}_b^* = \mathbf{Z}\boldsymbol{z} + \boldsymbol{p} \tag{23}$$

We now represent the objective function (Eqn. 1) in block form as:

$$f(.) \triangleq \frac{1}{2}\begin{bmatrix}\boldsymbol{v}_b^*\\ \boldsymbol{v}_q^*\end{bmatrix}^\intercal \begin{bmatrix}\mathbf{A} & \mathbf{B}\\ \mathbf{B}^\intercal & \mathbf{C}\end{bmatrix}\begin{bmatrix}\boldsymbol{v}_b^*\\ \boldsymbol{v}_q^*\end{bmatrix} \tag{24}$$

which is identical to:

$$f(.) \triangleq \frac{1}{2}\boldsymbol{v}_b^{*\mathsf{T}}\mathbf{A}\boldsymbol{v}_b^* + \boldsymbol{v}_b^*\mathbf{B}^\mathsf{T}\boldsymbol{v}_q^* + \frac{1}{2}\boldsymbol{v}_q^{*\mathsf{T}}\mathbf{C}\boldsymbol{v}_q^* \tag{25}$$

As we will be attempting to minimize $f(.)$ with regard to $\boldsymbol{z}$, which the last term of the above equation does not depend on, we can ignore that term. Expanding remaining terms using Eqn. 17, the new objective function is:

$$f(.) \triangleq \frac{1}{2}\boldsymbol{z}^\mathsf{T}\mathbf{Z}^\mathsf{T}\mathbf{A}\mathbf{Z}\boldsymbol{z} + \boldsymbol{z}^\mathsf{T}\mathbf{Z}^\mathsf{T}\mathbf{A}\boldsymbol{p} + \boldsymbol{z}^\mathsf{T}\mathbf{Z}^\mathsf{T}\mathbf{B}\boldsymbol{v}_q^* \tag{26}$$

$$\triangleq \frac{1}{2}\boldsymbol{z}^\mathsf{T}\mathbf{Z}^\mathsf{T}\mathbf{A}\mathbf{Z}\boldsymbol{z} + \boldsymbol{z}^\mathsf{T}(\mathbf{Z}^\mathsf{T}\mathbf{A}\boldsymbol{p} + \mathbf{Z}^\mathsf{T}\mathbf{B}\boldsymbol{v}_q^*) \tag{27}$$

subject to the following constraints:

$$\mathbf{N}^\mathsf{T}\begin{bmatrix}\mathbf{Z}\boldsymbol{z} + \boldsymbol{p} \\ \boldsymbol{v}_q^*\end{bmatrix} \geq 0 \tag{28}$$

$$\boldsymbol{z}_i \geq 0 \qquad (\text{for } i = 1 \ldots n) \tag{29}$$

$$\mu \boldsymbol{z}_i \geq z_{n+i} + \ldots + z_{n+i+r-1} \qquad (\text{for } i = 1 \ldots n) \tag{30}$$

Symmetry and positive semi-definiteness of the QP follows from symmetry and positive definiteness of $\mathbf{A}$. Once the solution to this QP is determined, $\boldsymbol{x}$ can be recovered. The actuator forces determined via ID are then $\boldsymbol{f}_{ID} + \boldsymbol{x}$.

2.3. *Recomputing Inverse Dynamics to Stabilize Actuator Torques (Stage II)*

The optimization perfomed by Stage I is prone to selecting an ID solution in which only two legs receive the majority of joint torque; this artifact arises from the redundancy inherent in robots with three or more points of support. Such solutions may be acceptable in simulations but can produce control torque discontinuities harmful to physically situated robots. Stage II addresses this issue by selecting the joint torque with minimum ℓ_2-norm. We define the optimization variables of Stage II within the null-space of Eq. 1 (*i.e.*, optimizing in the null-space of system kinetic energy) to obviate inclusion of a quadratic constraint. Once Stage II has been solved, the desired actuator forces are recovered with matrix-vector arithmetic. The two-stage optimization results in joint torques that are safe for the robot.

3. Experiments

Simulation data was collected using SL.[8] We evaluated our optimization procedure using multiple experimental setups including a high friction surface (the control) and a comparatively low friction experimental surface.

We used a trotting gait[9] as the reference trajectory for locomotion trials, and we also employed a Figure-Eight task to avoid the issue of impact (our model assumes only "resting" contact or inelastic impact). We compared the performance of the robot between three control methods: a feedback-based control strategy, an existing (non slipping) ID controller,[7] and our ID method. These trials aimed to validate both the inclusion of Stage II into the optimization process (evidenced by Fig. 1) and performance increases on slippery and sticky surfaces (evidenced by Figs. 2 and 3, respectively).

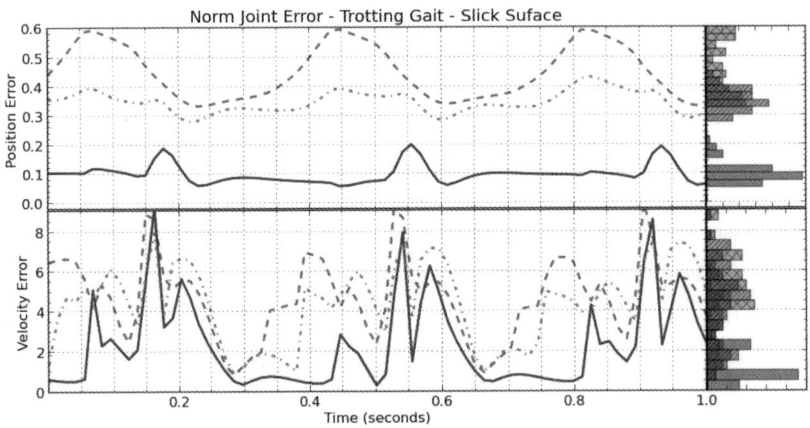

Fig. 2: *Norm of joint tracking error, comparing PID control* (green, double dash) *to our implementation of ID-based control* (blue, solid) *and an existing implementation*[7] *of non-slipping ID-based control* (red, dash-dot-dash) *on three steps of a walking trot (above) and one cycle of a Figure-Eight (below), both on surfaces with* $\mu = 0.1$. *A histogram of each plot's distribution is provided at right. Data was collected from the SL simulator.*

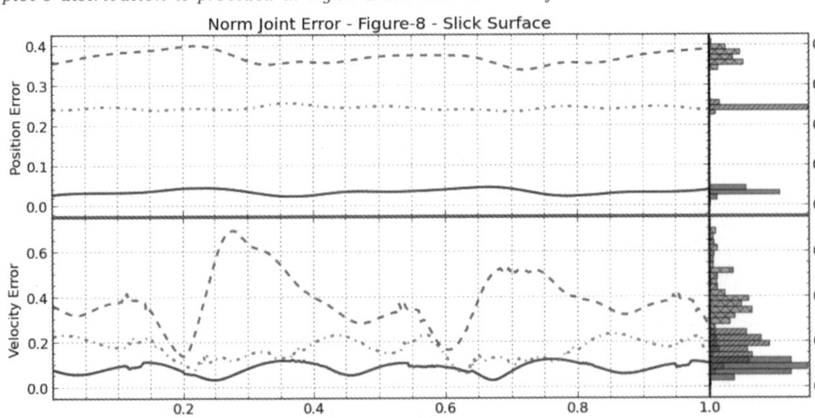

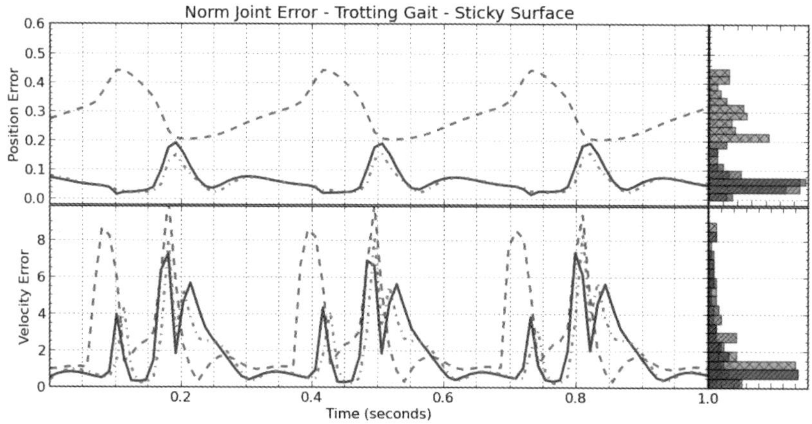

Fig. 3: *Norm of joint tracking error, comparing PID control* (green, double dash) *to our implementation of ID-based control* (blue, solid) *and an existing implementation*[7] *of non-slipping ID-based control* (red, dash-dot-dash) *on three steps of a walking trot (above) and one cycle of a Figure-Eight (below), both on surfaces with* $\mu = 1.0$. *A histogram of each plot's distribution is provided at right. Data was collected from the SL simulator.*

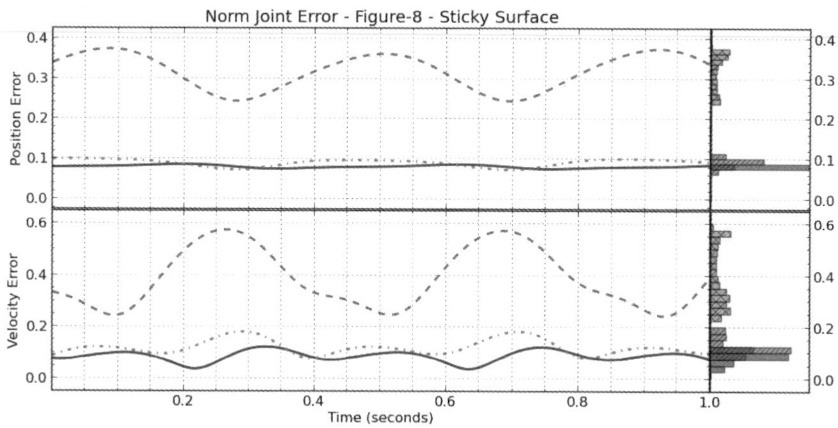

4. Results and Discussion

Mean position errors while performing a trot on simulated surface with $\mu = 0.1$ are reduced by 77.08% using our ID method over PID control. This performance differential exists even with the compliant (penalty-method-based) contact model used in SL (our ID model assumes rigid contact). Resulting tracking errors taken from the Figure-Eight reference trajectory on the slippery surface are even more compelling; position error was reduced 90.33% using our ID controller in place of PID control and 56.30% using our ID control in place of the existing (non slipping) ID control. Figure 3 shows that control accuracy on sticky surfaces is nearly identical between the two ID methods, as should be expected.

We also performed these experiments identically *in situ*, though these results were not as compelling. Neither ID controller was able to reduce position error substantially over PID control, though analysis indicates that the torques output by the three controllers are generally of the same direction and order of magnitude. Further analysis is likely to identify significant unmodeled phenomena, erroneous sensory data, or both.

Solving QPs and significant numerical linear algebra would seem to preclude the use of our approach in ID computation. Over sixty seconds of trotting, however, the two-stage optimization process completed in a mean time of 1ms and a worst-case time of 36ms. Thus, our method permits effective control rates of 250Hz or greater.

Acknowledgements

Zapolsky and Drumwright were funded by NSF award CMMI-110532.

References

1. C. Semini, N. G. Tsagarakis, E. Guglielmino, M. Focchi, F. Cannella and D. G. Caldwell, *Journal of Systems and Control Engineering* **225**, 831 (2011).
2. D. Stewart, *SIAM Rev.* **42**, 3(Mar 2000).
3. M. Anitescu and F. A. Potra, *Nonlin. Dyn.* **14**, 231 (1997).
4. E. Drumwright and D. A. Shell, Modeling contact friction and joint friction in dynamic robotic simulation using the principle of maximum dissipation, in *Springer Tracts in Advanced Robotics: Algorithmic Foundations of Robotics IX: Selected Contributions of the Ninth International Workshop on the Algorithmic Foundations of Robotics (WAFR)*, 2010.
5. D. Stewart and J. Trinkle, *Intl. J. Num. Meth. Engr.* **39**, 2673 (1996).
6. S. Boyd and L. Vandenberghe, *Convex Optimization* (Cambridge University Press, 2004).
7. M. Mistry, J. Buchli and S. Schaal, Inverse dynamics control of floating base systems using orthogonal decomposition, in *IEEE Intl. Conf. on Robotics and Automation (ICRA)*, 2010.
8. S. Schaal, *The SL simulation and real-time control software package*, tech. rep., Univ. Southern California (2009).
9. V. Barasuol, J. Buchli, C. Semini, M. Frigerio, E. R. De Pieri and D. G. Caldwell, A reactive controller framework for quadrupedal locomotion on challenging terrain, in *IEEE Intl. Conf. on Robotics and Automation (ICRA)*, 2013.

LAURON V: OPTIMIZED LEG CONFIGURATION FOR THE DESIGN OF A BIO-INSPIRED WALKING ROBOT

A. ROENNAU*, G. HEPPNER, L. PFOTZER, and R. DILLMANN
FZI Research Center for Information Technology,
Haid-und-Neu-Str. 10–14, D-76131 Karlsruhe, Germany
E-mail: {roennau, heppner, pfotzer, dillmann}@fzi.de, www.fzi.de

Walking robots are fascinating, but also complex mechatronic systems. Typical six-legged robots have at least 18 joints and therefore come along with a huge design space. This design challenge can be addressed by applying technical analyzes and optimization methods or following the design of biological role models. The design of the new six-legged walking robot LAURON V employs both approaches. An additional rotational joint was added to the bio-inspired leg design of the previous LAURON generation. In this work we present the combination methods of three optimization that were developed to optimize the leg configuration. The resulting leg mounting angles are compared to biological leg configuration of the stick insect.

Keywords: kinematic design, leg configuration, legged walking robot, hexapod

1. INTRODUCTION

Walking is the most natural form of locomotion, but compared to wheeled or tracked locomotion it is very complex. There are many different reasons to design and develop robotic systems that can walk: understand biological observations or prove new biological theories, design new lightweight, intelligent mechatronics or develop advanced walking robots for challenging tasks like the exploration or inspection of hazardous areas. Independent of the motivation there are two different approaches to develop new walking robots. On the one hand many scientists and engineers take a look at natural archetypes (typically stick insects, ants or cockroaches for hexapods) as inspiration for new designs and control methods. On the other hand it is possible to use engineering tools and analytical methods like the analysis of the kinematic workspace or load calculations to create new walking robots. But which approach leads to better walking robots?

There are many multi-legged walking robots like MESSOR,[1] BILL-Stick,[2] LAURON IVc[3] and Cheetah,[4] which have a bio-inspired kinematic design. For example, the walking robot BILL-Stick closely follows the kinematics of the stick insect and the EPFL Cheetah hind legs are 20% longer than its forelegs, which was inspired by the morphology of the cat. Other robots like RHex,[5] TITAN-X,[6] SILO-6[7] or the DLR Crawler[8] have a loose biological inspiration and follow technical optimization methods. The walking principle of RHex is clearly not natural due to its rotational character, nevertheless the idea was derived from observations of the cockroach.[5] In the work of Goerner the six-legged robot DLR Crawler was kinematically optimized with the help of workspace analyzes.[8] The goal of this optimization was to enlarge the workspace and therefore increase the step sizes and flexibility of the robot. SILO-6's leg configuration was optimized to reduce the required supporting foot forces.[7] The force is minimized by moving the mounting position of the middle legs outwards relative to the main body.

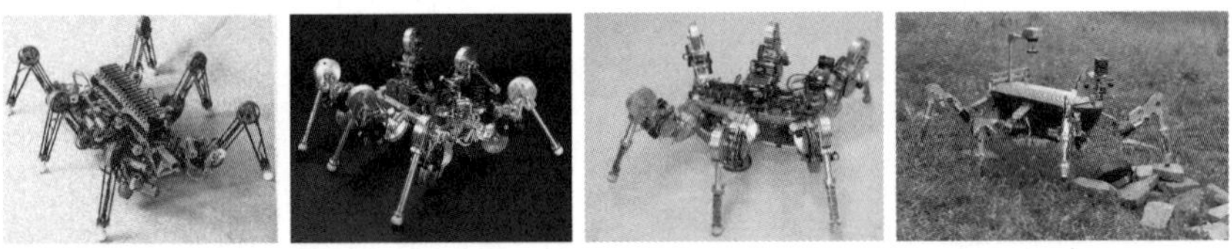

Fig. 1. LAURON Generations: from left to right - LAURON I, II, III, IVc

Fig. 2. (left) leg kinematics of stick insect,[9] (right) leg kinematics of LAURON V[10]

Typical leg configurations of current walking robots follow the principle of line or rotational symmetry. In many cases simple orthogonal mounting angles ($\Psi = 0°$) are used. This work will show, that the mounting angles can have significant influence on the performance of the walking robot.

The new LAURON V generation was developed to further improve the overall walking capabilities (step size, ground clearance, velocity, stability), increase the maneuverability by adding an additional joint to each leg and enable LAURON to manipulate objects with its front legs. We used the bio-inspired leg of LAURON V, which has a kinematic structure similar to the stick insect (see Fig. 2), and then applied optimization methods to determine the leg configuration.

After this short introduction, Section 2 and 3 will describe the pre-calculated data set and present the iterative optimization methods. The resulting leg configuration will be discussed and compared to the stick insect in Section 4. Finally, a short conclusion and outlook will complete this work.

2. APPROACH

All generations of the walking robot LAURON were inspired by the stick insect Carausius Morosus (see Fig. 1). First a fourth rotational joint δ was added to the leg design of the new LAURON V generation (see Fig. 2 and more details in[10]). The resulting larger workspace has some complex parts in the rear, but these are not relevant for the walking process and will not be discussed in this work (see left of Fig. 5). Next, the legs' mounting angles and positions relative to the main body (referred to as *leg configuration*) were determined. The two relevant mounting angles Ψ (roll) and η (yaw) have great influence on the max step sizes, stability, load distribution and flexibility of the robot (see angles in Fig. 3 and 4).

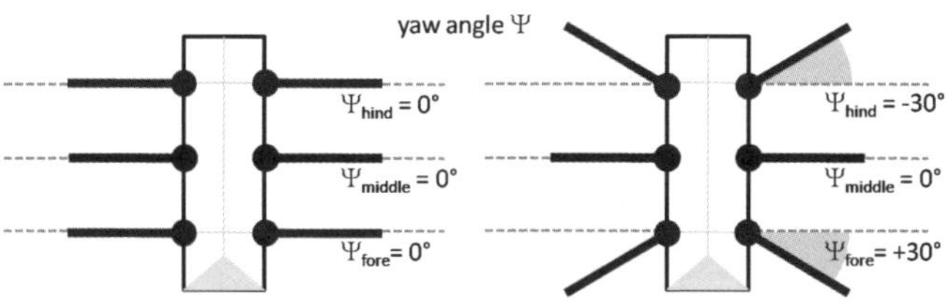

Fig. 3. Top View: examples for yaw mounting angles Ψ: (left) simple technical configuration with $\Psi = 0°$ for all legs, (right) bio-inspired configuration with a positive angle for the forelegs and negative Ψ angle for the hind legs

The leg configuration optimization methods are based on the workspace analysis following the criteria developed by Yoshikawa.[11] Yoshikawa's manipulability measure describes the movement of the foot tip in the Cartesian space that results from a deviation of the joint angles at the current position. This manipulability, which is calculated with the help of the Jacobian, can be illustrated as ellipsoid (see left of Fig. 5). Approximately 1 million joint angle configurations were pre-calculated[12] for the new LAURON V leg design and then normalized based on the volumes of the ellipsoids. Each joint angle configuration is represented by a 3D point with a value ranging from zero to one.The 3D point cloud is illustrated in the right part of Fig. 5 (red points = 1, blue points = 0). Three optimization methods were developed that use this 3D point cloud to improve the two mounting angles Ψ and η.

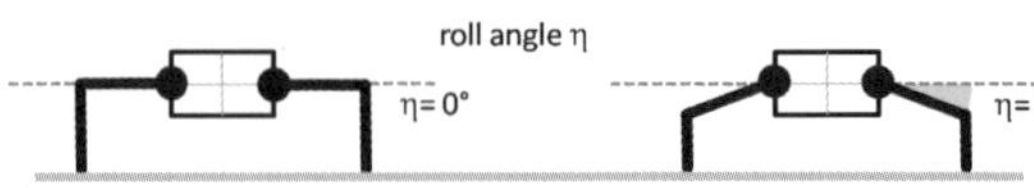

Fig. 4. Front View: examples for roll mounting angle η: (left) simple technical configuration with $\eta = 0°$, (right) bio-inspired configuration with a positive η angle

3. LEG CONFIGURATION OPTIMIZATION

The leg configuration is defined as the mounting orientation and position relative to the body frame of the robot. This work will focus on the mounting orientation, represented by the two angles Ψ (roll) and η (yaw). Instead of rotating the leg together with the Yoshikawa point cloud, all methods use intersections of this pre-calculated data set at different angles and heights. An example of these intersections is illustrated in the left of Fig. 6.

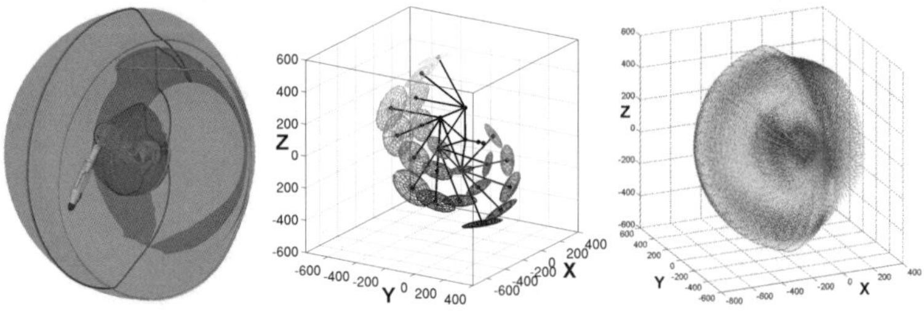

Fig. 5. (left) new workspace of leg with additional δ joint, (middle): Yoshikawa manipulability ellipsoids, (right): Yoshikawa point cloud: normalized ellipsoids as 3D points (manipulability value: red:= 1, blue:= 0)

Roll Angle: Large Manipulability Region

First the roll angle Ψ was analyzed. Due to the shape of the 3D point cloud and to the prior knowledge that this angle is positive for most insects, only positive roll angles were considered. In this first optimization step five configurations were pre-selected and analyzed: $\Psi = \{0°, 15°, 30°, 45°, 60°\}$. For each configuration 12 equidistant intersections with corresponding roll angle were taken from the 3D point cloud (see left of Fig. 6). The roll angles 30°, 45° and 60° showed the best manipulability (largest area with values

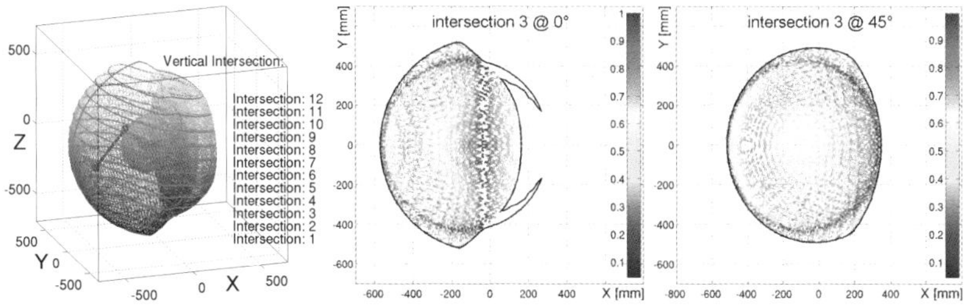

Fig. 6. (left): intersection at different heights and angles, (middle): intersection 3 with a roll angle of 0°, (right): intersection 3 with a roll angle of 45°

> 0.7). There are some flaws at 60° for high intersections and the area becomes rather small at 30° for low intersections. In the middle and right of Fig. 6 two intersections at the same height show the influence of the mounting angle Ψ. After this first step the best candidate roll angles were selected for further optimization: $\Psi = 30°$, $\Psi = 45°$, $\Psi = 60°$.

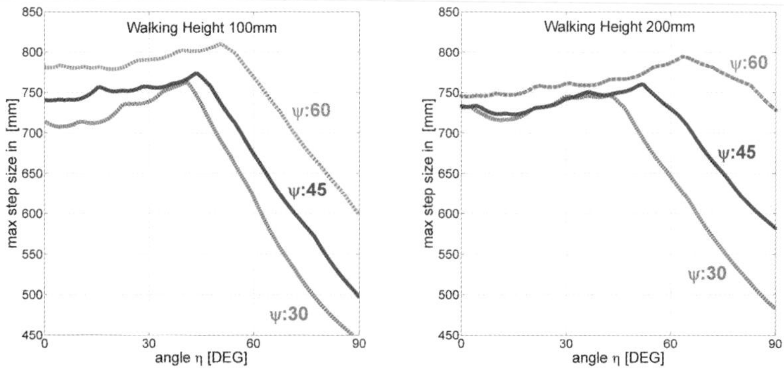

Fig. 7. Maximum step size for the walking heights 100mm (left) and 200mm (right)

Roll Angle: Flexible Standing Width

Next, the range of the standing width was analyzed. The max standing width improves stability and the range (max - min width) is an indicator for flexibility of the robot. It also defines the overall width of the robot.The values in the following Table show the max and min distance of the foot tip to the mounting point. Only positions within an area with good manipulability were considered. The roll angle $\Psi = 45°$ offers the biggest range, which increases flexibility, and has a good max width for stability.

Table 1. Min, Max Walking Width and Width Range [mm]

$\Psi = 30°$			$\Psi = 45°$			$\Psi = 60°$		
Min	Max	Range	Min	Max	Range	Min	Max	Range
262	564	302	237	558	321	280	543	263

Roll and Pitch Angle Optimization

Finally, the max step sizes were evaluated for the three candidate roll angles. In each intersection the longest straight trajectory, that can be fitted inside a region with good manipulability (> 0.7), was identified. The step sizes also depend on the walking height (clearance of the beta joint to the ground). For each candidate roll angle the yaw mounting angle η was varied in the range from 0 to 90°. The results of two analyzes are presented in Fig. 7 and show a effect that is visible in all results: the yaw mounting angle η has a positive effect on the step size up to a value of approx. 40°. The biggest step sizes can be achieved with the roll angle $\Psi = 60°$.

4. RESULTS AND DISCUSSION

A good manipulability value is a strong indicator that the foot tip position lies within the kinematic sweet spot of the robot. This allows the robot to perform fast movements and gives it a bigger kinematic flexibility. The results of all optimization methods showed that it is better to tilt (roll angle $\Psi > 0$) the mounting orientation. The first method showed that the roll angle can be used to enlarge the good manipulability region. The three candidate roll angles were selected with this approach. In all optimization steps the $\Psi = 45°$ configuration outperformed the 30° mounting orientation. The mounting angle $\Psi = 60°$ also performed well, but the width range and the size of the good manipulability region at high walking heights, was not as good as in the 45° configuration. Our last optimization method showed that yaw angles $\leq 40°$ have positive effects on the step width. The selected front yaw angle moves the workspace of the forelegs to the front and increases their overlap. This enables the robot to use its legs for manipulation tasks. The selected yaw angle of the hind legs moves the CoM to the rear and enlarges the support polygon in the main walking direction. The middle leg's yaw orientation enables the robot to better stand on its middle and hind legs, by enlarging the support polygon and allowing the mounting position of the middle legs to be moved to the rear. The final optimized leg configuration is illustrated in Fig. 8 and listed in the Table below.

Typical values for the leg configuration of the stick insect can be found in the work of H. Cruse[13] and are listed in Tab. 2 (right part). Comparing

Table 2. Leg configurations of LAURON V and the stick insect[13]

	LAURON V			stick insect		
angle	foreleg	middle leg	hind leg	foreleg	middle leg	hind leg
Ψ	+45°	+45°	+45°	+40°	+40°	+50°
η	+40°	+20°	-40°	+15°	+5°	-45°

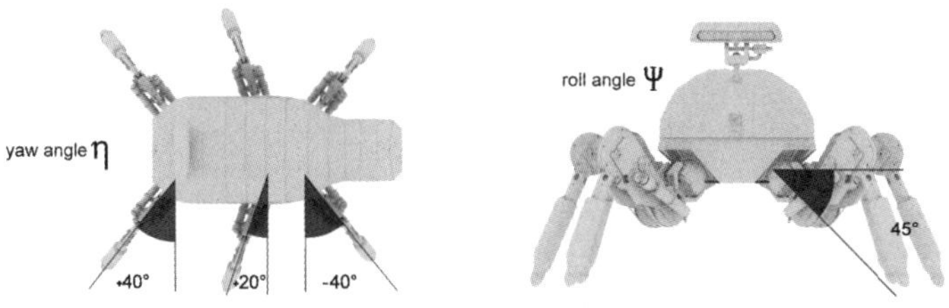

Fig. 8. Optimized Leg Configuration of LAURON V

the values to each other shows a strong correlation for the roll angle Ψ. The similarities of the technically optimizes yaw angle η are smaller, but still visible. These results demonstrate that the optimization methods iteratively lead to a bio-inspired joint angle configuration. Due to the fact, that in most cases the technical leg kinematics, torques and joint velocities do not scale well compared to the values of biological role models, it is a good idea to use optimization techniques to improve the kinematic design for a specific robotic leg. Nevertheless, in many cases this will lead to results similar to the values found in Nature. A combination of both design approaches offers a good chance to find a good, bio-inspired leg configurations, that can improve the walking robot's capabilities.

5. CONCLUSIONS AND FUTURE WORKS

Three kinematic design methods, that optimize the mounting angles of a walking robot, were presented in this work. All three methods base on the evaluation of multiple intersections of a pre-calculated 3D Yoshikawa point cloud. The resulting leg configuration has strong similarities to the configuration of the stick insect and has been used for the design of the walking robot LAURON V (see Fig. 9). Currently, extensive experiments are being conducted, showing the robustness and improved walking capabilities. The next steps will be the integration of a sensor head, grippers for the forelegs and the extensive evaluation of the advantages of the increased maneuverability resulting from the new additional leg joint.

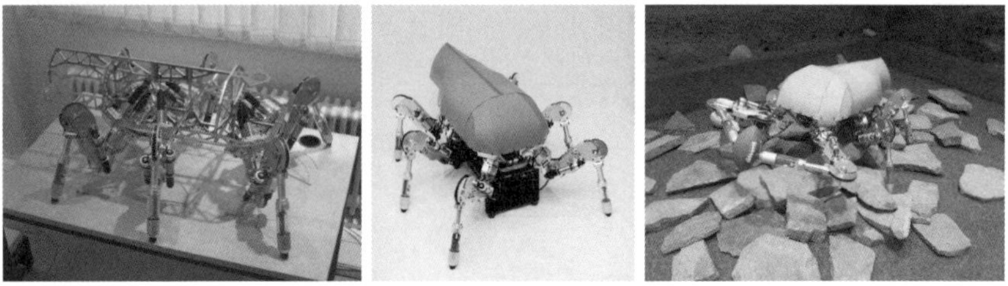

Fig. 9. New Six-Legged Walking Robot LAURON V with Optimized Leg Configuration

References

1. K. Walas and D. Belter, Messor-versatile walking robot for seach and rescue missions, in *Journal of Automation, Mobile Robotics & Intell. Systems*, 2011.
2. W. Lewinger, H. Reekie and B. Webb, A hexapod robot modeled on the stick insect, carausius morosus, in *IEEE ICAR*, 2011.
3. A. Roennau, G. Heppner, T. Kerscher and R. Dillmann, A behaviour-based free gait inspired by a walking stick insect, in *CLAWAR*, 2010.
4. S. Rutishauser and et al., Passive compliant quadruped robot using central pattern generators for locomotion control, in *IEEE EMBS BioRob*, 2008.
5. U. Saranli and et al., Rhex: A simple and highly mobile hexapod robot *Journal of Robotics Research* (SAGE Publications, 2001).
6. R. Hodoshima and et al., Development of track-changeable quadruped walking robot titan x, in *IEEE IROS*, 2010.
7. P. Gonzalez de Santos, J. Estremera and E. Garcia, Optimizing leg distribution around the body in walking robots, in *IEEE ICRA*, 2005.
8. M. Gorner and et al., The DLR-crawler: A testbed for actively compliant hexapod walking based on the fingers of DLR-Hand II, in *IEEE IROS*, 2008.
9. H.-J. Weidemann, F. F. Pfeiffer and J. Eltze, The six-legged tum walking robot, in *IEEE Intelligent Robots and Systems IROS'94*, 1994.
10. A. Roennau and et al., Design and kinematics of a biologically-inspired leg for a six-legged walking machine, in *IEEE RAS and EMBS BioRob*, 2010.
11. T. Yoshikawa, Manipulability of robotic mechanisms, in *The International Journal of Robotics Research*, 1985.
12. K. H. Koch, Lightweight design of a torso for the next generation of a six-legged robot, in *Masterthesis at FZI Karlsruhe, Germany*, Aug, 2010.
13. H. Cruse, The function of the legs in the free walking stick insect, carausius morosus, in *Journal of Comparative Physiology A*, 1976.

SHUFFLE TURN OF HUMANOID ROBOT SIMULATION BASED ON EMG MEASUREMENT

MASANAO KOEDA, TAKAYUKI SERIZAWA AND YUTA MATSUI
Osaka Electro-Communication University,
Faculty of Information Science and Arts,
Department of Computer Science
Kiyotaki 1130-70, Shijonawate, Osaka, 575-0063, JAPAN.

Recently, many researchers have been studying a method that involves the stepless slip motion of humanoid robots. However, it is not clear how it works or how it is controlled. In this research, we measured the muscle activity in the leg of a human subject performing a shuffle turning motion by using an electromyograph. The results indicated that the hip joint performed an important function in shuffle turning. Then, we verified this hypothesis using a life-size humanoid dynamic simulator that was constructed using ODE. By controlling the hip pitch joint angle and knee pitch angle using a proportional controller, the robot turned as much as a real human. Thus, our hypothesis was supported by this experiment.

1. Introduction

Biped robots conventionally perform walking and turning motions through repeated foot stepping. However, foot stepping is inefficient, time consuming, unstable, and generally unsuitable in narrow spaces under constrained postures, as illustrated in Figure 1. There are great expectations with the use of biped robots in a kitchen, assembly line, and energy plant inspection/destruction applications. Recently, many researchers have been studying a method that involves the stepless slip motion of humanoid robots to realize smooth, quick, and high stability movement [1, 2, 3], as shown in

Figure 1 Motivation

Figure 2. Miura et al. [4] proposed a control model for the slip turning motion of a humanoid robot. However, this model was not based on human motion and seems to be difficult to justify.

In this research, we investigated the muscle activity in the leg of a real human subject performing shuffling motion by using an EMG (electromyograph). Based on this measurement data, we determined the joint motion of the leg required for the movement. Finally, we verified this result using a dynamic humanoid simulator, which was implemented using an accurately sized and weighted human model.

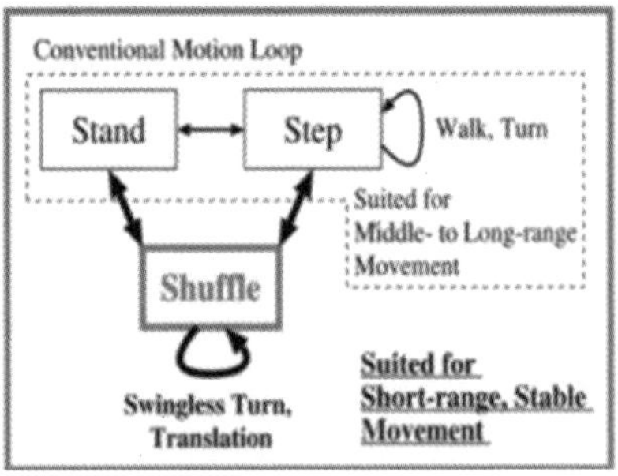

Figure 2 Conceptual diagram

Table 1 Measured leg muscles

CH	Muscle
1	Tensor fascia latae
2	Vastus laterails
3	Biceps femoris
4	Semitendinosus
5	Vastus medialis
6	Lateral head of gastrocnemius
7	Medial head of gastrocnemius
8	Tibialis anterior

2. Measurement of leg muscle activity during shuffle turning

2.1. *Experimental methodology*

EMG sensors were attached to leg muscles to determine their activity during a shuffling motion. The attached positions of these EMG sensors are listed in Table 1 and Figure 3. In addition, the whole body motion was also simultaneously measured using a motion capturing system. The subject of this experiment was 22-year-old male, with 70 kg and a height of 165 cm (Figure 4). This subject performed 45 and 90 deg shuffle turning movements in upright standing and knee bending postures.

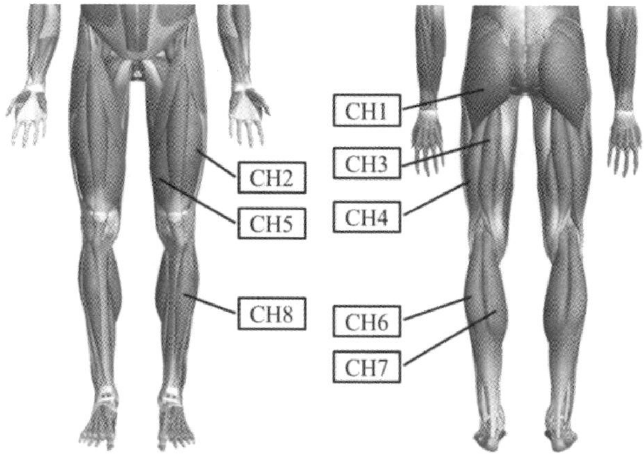

(a) Frontal view (b) Rear view

Figure 3 EMG attached muscles

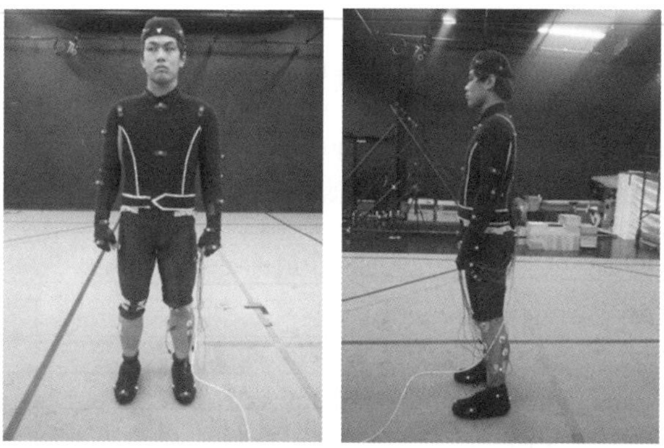

Figure 4 Overview of subject with EMG

P-EMGplus sensors (a wet-type sensor, with a maximum of 8 channels, 14-bit resolution, and 1-kHz sampling rate) manufactured by Oisaka Electronic Equipment Ltd. and a laptop PC (CF-W8, CPU: Intel Core2 Duo 1.2 GHz, RAM: 4 GB) manufactured by Panasonic were used for measuring the myoelectric potential of the leg. A VICON BLADE manufactured by Vicon was used for capturing the motion. This system is installed in Joint Institute for Advanced Multimedia Studio (JIAMS) of our university.

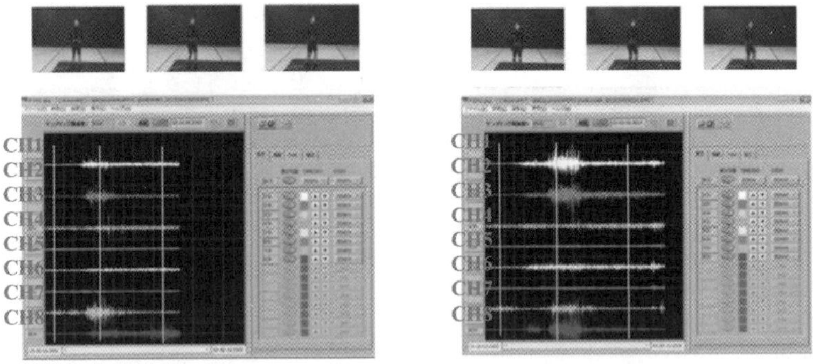

(a) Upright standing posture (b) Knee bending posture

Figure 5 EMG of left leg during shuffle turning in 45 deg counterclockwise direction

2.2. *Experiment and results*

The measured data shown by yellow lines on the EMG graph show the initial posture, start turning, and end turning moments. As you can see, CH1, CH2, CH7, and CH8 were active during turning (Figure 5). Taking into account these results, it is supposed that the hip joints are rotated in the pitch axis direction and knee joints are of little relevance to shuffle turning. The ankle joint should be controlled horizontally to maintain the body's balance.

3. Simulation

ODE (ver.0.11-1, double precision) was used to simulate the humanoid robot dynamics. The specifications of the computer were a CPU with an Intel Core2 Duo 1.2-GHz processor and 4GB of RAM running the Windows7 32-bit OS. The simulated robot had 12 DOFs in the lower body, as illustrated in Figure 6. The foot size, link length of the leg, and weight of the body parts were determined by referring to [5]

Table 2 Parameters of humanoid model in simulator

Name (ID No.)	Size
Body height (26)	1706[mm]
Knee height (160)	429.6[mm]
Crotch height (158)	759.7[mm]
Side neck height (31)	1440[mm]
Hip breadth (51)	340.3[mm]
Hip depth (167)	255.3[mm]
Thigh circumference (169)	536.4[mm]
Calf circumference (171)	366.8[mm]
Foot length (174)	249.4[mm]
Projected foot breadth (176)	96.3[mm]
Foot circumference (178)	247.8[mm]
Cervical height, sitting (76)	653.2[mm]
Head height (3)	237.1[mm]
Body weight (1)	64.9[kg]
Trunk weight (-)	44.9[kg]
Leg weight (-)	10.0[kg]

for a 24-29-year-olds male Japanese. The detailed parameters are listed in Table 2. In this simulator, the dynamic friction coefficient was set to 0.2, empirically. Based on the result of the EMG measurement, to conduct the shuffle turning, only the hip pitch joint and ankle pitch joints were proportionally controlled in the simulation in both the upright standing and knee bending postures (Figure 7).

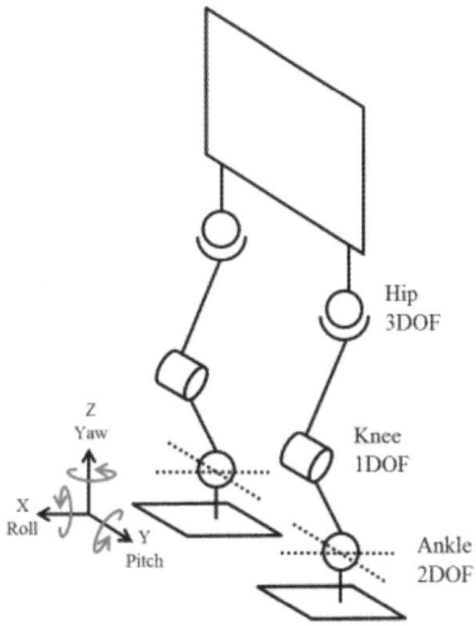

Figure 6 Simulation model

The initial posture for the upright standing turning included 0 deg angles for the hip pitch, knee, and ankle pitch joints. To simulate {45, 90} deg shuffle turns, the target angles for the hip pitch and ankle pitch joints were controlled in 0.2 deg steps in {100, 200} simulation loops, for total {increases, decreases} of {20, 40} deg, respectively, over approximately {3, 6} s. The final posture of the upright standing turning was the left hip at {20, 40} deg, right hip at {-20, -40} deg, knee at {0, 0} deg, left ankle at {-20, -40} deg, and right ankle at {20, 40} deg for the {45, 90} deg shuffle turns, respectively.

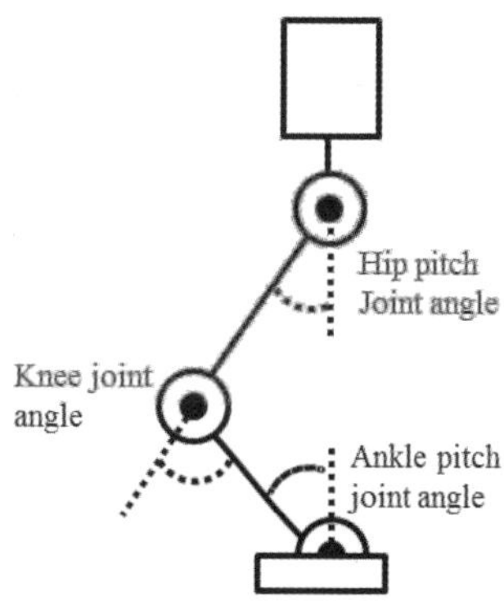

Figure 7 Joint angle

In the same way, the initial posture of the knee bending turning was 30 deg for the hip pitch joint, 60 deg for the knee joint, and 30 deg for the ankle pitch joint. The same angle control was used in this simulation. The final posture of the knee bending turning was the left hip at {50, 70} deg, right hip at {10, -10} deg, knee at {60, 60} deg, left ankle at {10, -10} deg, and right ankle at {50, 70} deg, respectively.

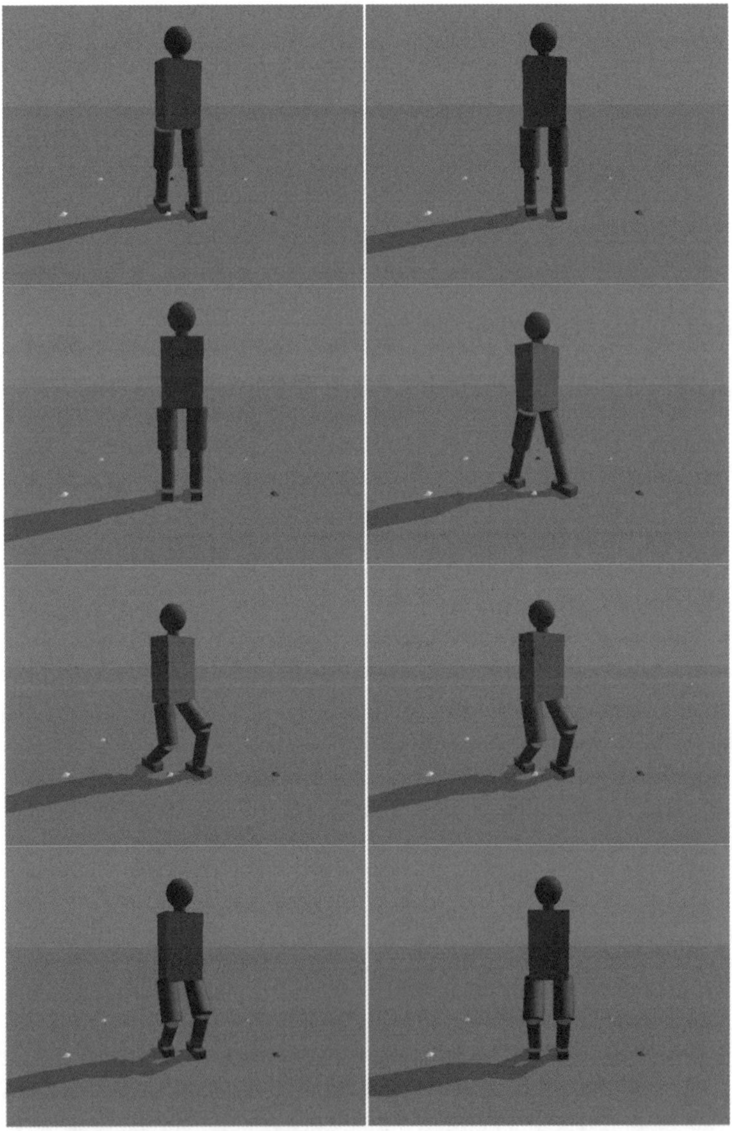

(a) Upright standing posture (b) Knee bending posture

Figure 8 45 deg shuffle turning in simulator

As a result, as shown in Figure 8, the simulation worked much like real human motion, and our control method was validated experimentally.

4. Conclusion

In this research, the EMG of the leg of a human during shuffling motion was measured, and it indicated that the hip joint motion plays an important function in shuffle turning. In addition, to verify this hypothesis, we built a life-size humanoid dynamic simulator using ODE, and simulated shuffle turning motions in two postures with very simple joint angle control. As a result, it was confirmed that this control method could turn the robot's body as much as the real motion of a human subject. In future work, we will specifically investigate whole body motion and examine the effect of the floor's friction on shuffle turning.

References

1. M. Koeda, et al., "Shuffle Turn and Translation of Humanoid Robots", In Proc. of 2011 IEEE Int. Conf. on Robotics and Automation, pp. 593-598 (2011).
2. K. Miura, et al., "Analysis on a Friction Based "Twirl" for Biped Robots", In Proc. of 2010 IEEE Int. Conf. on Robotics and Automation, pp. 4249-4255 (2010).
3. K. Hashimoto, et al., "Realization of Quick Turn of Biped Humanoid Robot by Using Slipping Motion with Both Feet", In Proc. of 2011 IEEE Int. Conf. on Robotics and Automation, pp. 2041-2046, pp. 2041-2046 (2011).
4. K. Miura, et al., "A Friction Based "Twirl" for Biped Robots", In Proc. of 8th IEEE-RAS Int. Conf. on Humanoid Robots, pp. 279-284 (2008).
5. Research Institute of Human Engineering for Quality Life, "Japanese body size data 1992-1994" (1997).

STUDIES OF TOTAL ADHESIVE FORCE OF MULTIPLE MAGNETIC WHEELS FOR A CLIMBING ROBOT

ARSIT BOONYAPRAPASORN, KANED THUNG-OD, RARDCHAWADEE SILAPUNT, THAVIDA MANEEWARN

Institute of Field Robotics (FIBO), King Mongkut University of Technology Thonburi 126 Pracha u-tid Road, Thungkru, Bangkok 10140, Thailand

E-mail:urarl.urarl9@gmail.com

Magnetic wheel climbing robots have been widely used in the ferro-magnetic structures inspection. Most research studies have investigated the effects of different parameters on the adhesive force of a single magnetic wheel. These investigations, however, are yet to examine multiple magnetic wheels assembled on the same shaft. Therefore, this study aimed to clarify the effects of the axial distance and the direction of the magnetization of each magnetic wheel on the total adhesive force. Finding suggested that for a certain range of the axial distance, the total adhesive force of double magnetic wheels was close to twice of the adhesive force of a single magnetic wheel when the direction of magnetization of double wheels are identical. Over the same range of the axial distance, the total adhesive force of double wheels approached twice the adhesive force of a single magnetic wheel as the axial distance increased. The interesting situation was when two magnetic wheels were in contact, the adhesive force of the same magnetization case was significantly higher than that of the opposite magnetization. The results of this study can be used for designing and controlling multiple magnetic wheels to achieve the desired adhesive force during the operation.

1. Introduction

Magnetic wheel climbing robots are utilized as inspection robots for various industries. Most components or structures that the robots have been designed to climb on are made of steel; for example, boilers and generators in electrical power plants [1]. In order to obtain the desired climbing ability for magnetic wheel climbing robots, the adhesive force of magnetic wheel is an important design aspect of these robots. Many previous works presented how to determine the adhesive force and analyzed the effects of parameters of wheels and surfaces to the adhesive forces; for example, properties of magnets and the thickness of the surface as shown in the previous works by Yukawa *et al.* [2], Fisher *et al.* [3], Rochat *et al.* [4], Hans *et al.* [5], Zhang *et al.* [6] and Zhang *et al.* [7].

In the situation when the design of magnetic wheels of wall climbing robots is required to have multiple wheels assembled on the same shaft, the effects of magnetic flux density from one magnetic wheel to other wheels nearby have been mentioned and discussed by [8], but the effect of direction of magnetization from each wheel have not been included yet. This work focuses on studying the effects of the axial distance between the magnetic wheels and the directions of magnetization to an adhesive force using an experimental approach.

This paper is organized as follows. First, the basic idea of magnetic force and the structure of a magnetic wheel are presented. Second, the experimental design is proposed. Then, the results is presented and discussed. Finally, the conclusion is stated.

2. Permanent Magnetic Wheel

2.1. *Structure of Permanent Magnetic Wheel*

In this study, the structure of a single permanent magnetic wheel consists of two steel plates, a plastic core, and thirty cylindrical N_dF_eB magnets as shown in Figure 1. Diameter and thickness of each magnet are 5mm and 10mm, respectively. Each steel plate has an outer diameter of 82 mm and the thickness of 3 mm. The diameter and thickness of the plastic core are 70 mm and 10 mm, respectively. The magnetic flux from the permanent magnets is transmitted to the ferromagnetic surface through the steel plate as a magnetic circuit.

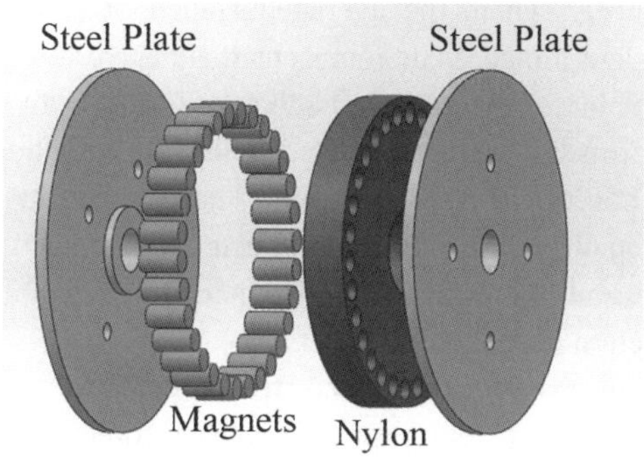

Figure 1. The structure of a single magnetic wheel.

2.2. *Magnetic force*

A magnetic wheel on the surface can be considered as a magnetic circuit. The adhesive force, F, by permanent magnets can be approximated as:

$$F = \frac{B^2 A_g}{2\mu_0} \tag{1}$$

where μ_0 is the permeability $\mu_0 = 4\pi$, B is the magnetic flux density in Tesla(T), and A_g is the pole area in m^2, [5],[6], [9], and [10]. Based on Eq. 1, if the pole area is fixed, the amount of an adhesive force is related to the amount of magnetic flux density as a quadratic function.

3. Experiment and Results

This study aimed to investigate the effect of the axial distance between two magnetic wheels along the shaft, L, and the direction of magnetization of the two magnetic wheels to the total adhesive force of two magnetic wheels assembled on the same shaft, F. The total adhesive force was determined by measuring the pulling force applied to separate the wheels from the steel plate with the thickness of 2 mm. Three experiments were done for three cases as follows: 1) a single magnetic wheel, 2) two magnetic wheels with an opposite direction of magnetization (the same magnetic polarity), and 3) two magnetic wheels with the same direction of magnetization (the opposite magnetic polarity). The first experiment was carried out repeatedly 10 times for each of the two single wheels. Then, the averaged valued of adhesive force of each single wheel was determined. The represented adhesive force of a single wheel was the averaged value of the averaged adhesive forces from each single wheel. The last two experiments were conducted under the variation of the distance: $L = 0,10,20,30,40,50,60,70,80,90,100$ mm. The experiments of case 2 and 3 were carried out repetitively for 10 times. Then, the averaged value of the force was used to represent the adhesive force at different combinations of axial distances and direction of magnetization.

Schematic diagrams of experiments for cases 1, 2, and 3 are shown in Figure 2. The experimental set up for the adhesive force measurement is shown in Figure 3. The experimental results are presented in Table 1. The graphs of total adhesive force and axial distance for case 2 and case 3 are presented in Figure 4.

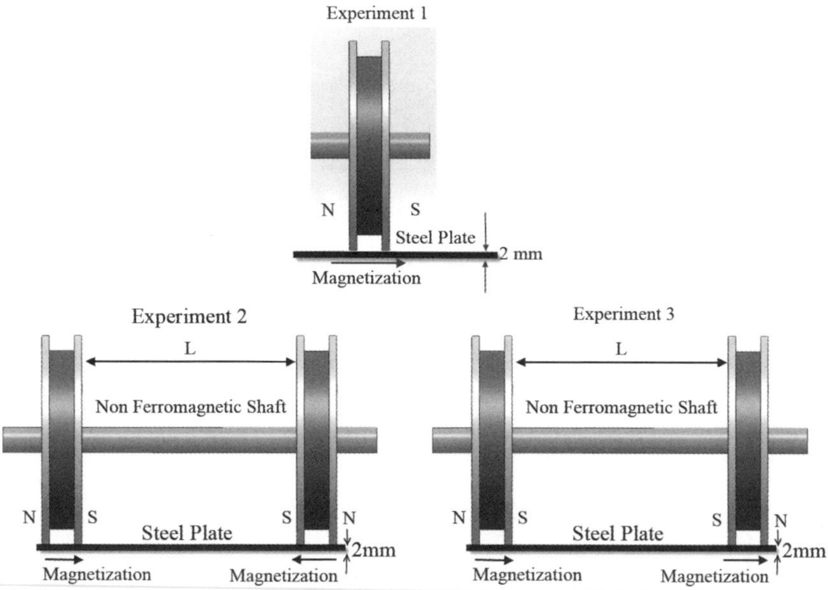

Figure 2. Schematic diagram of experiments of case 1: single magnetic wheel, case 2: two magnetic wheels with an opposite direction of magnetization, case 3: two magnetic wheels with the same direction of magnetization.

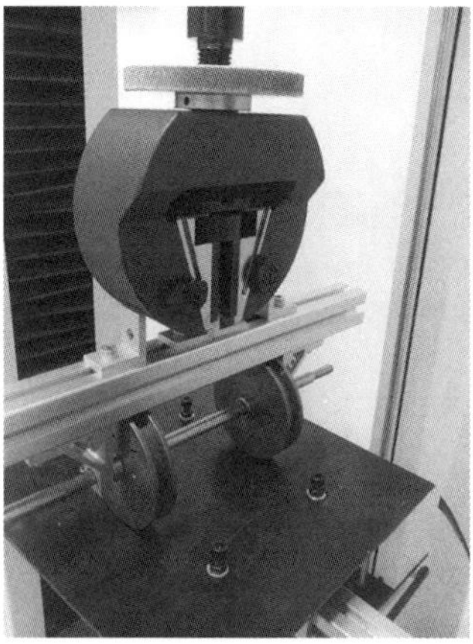

Figure 3. Real image of experiments set up of the total adhesive force measurement.

Table 1. The total adhesive force of each case at eleven different axial distances.

	Total Adhesive force (N)					
Case	L=0 mm	L=10 mm	L=20 mm	L=30 mm	L=40 mm	L=50 mm
1			83.3916			
2	158.85	166.12	165	162.50	166.78	159.67
3	97.67	105.5	138.33	147.13	156.17	151.67

	Total Adhesive force (N)				
Case	L=60 mm	L=70 mm	L=80 mm	L=90 mm	L=100 Mm
1			83.3916		
2	156.5	164.50	176.95	164.50	173.63
3	169.7	162.05	151.9	165.67	166.33

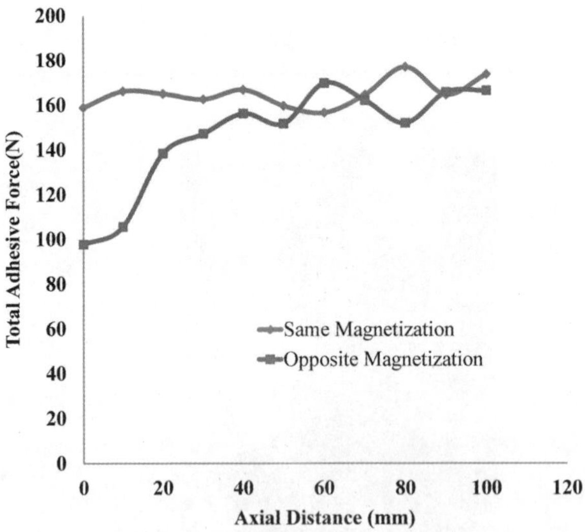

Figure 4. Plot of total adhesive force and axial distance of case 2 and case 3.

4. Discussion and Conclusions

Findings from the current experiment suggested the effects of the direction of the magnetization of the wheels. In the case of the same magnetization, the average total adhesive force was 165 N with the standard deviation of ±6.06 N. In the range of the axial distance between 0 and 70 mm, the average total magnetic force of double wheels corresponding to each axial distance was close to twice of the adhesive force of a single magnetic wheel, similar to [3] and close to the mean value. Most deviation was believed to be attributed from the bending of beam and plates that gradually formed as the axial distance increased beyond 70 mm. For the cases of the opposite magnetization, the average total adhesive force approached twice the adhesive force of a single magnetic wheel as the axial distance increased up to 70 mm. Similar to the previous case, deviation was more obvious when the axial distance was greater than 70 mm, These results suggested that the effect of magnetization for both configurations was strong below 70-mm axial distance. Undesirable beam and plate bendings started influencing the adhesive force measurement beyond that.

An interesting point is worth mentioned here. For both same and opposite magnetizations, at zero gap distance, the adhesive force of the same magnetization case was larger than that of the opposite magnetization. This confirmed the effect of the direction of magnetization on the total adhesive force of multiple wheels. The effects of other external variables, such as air gap and the thickness of the steel plate will be briefly discussed here. Whereas the increase of the gap and the decrease in the thickness of the surface led to the reduction of the adhesive force as proposed in [4] [6],[8],[10] and [11], these variables were not included in this study. The reason was because these variables could be treated as external variables. They varied based on the required mobility of magnetic wheels as well as the tasks characteristics of the surfaces and other environmental factors. The effects of these variables are generally considered in subsequent design phases. For example, in order to allow magnetic wheel climbing robots to travel on the ferromagnetic surface, the friction force of the wheel must be adequate, thus some materials such as rubber [4] can be used to cover the wheel to increase friction coefficient between surface and the magnetic wheel. However, this can cause air gap and consequently decrease the adhesive force of magnetic wheel [4]. Likewise, in the design of magnetic wheels for a thin surface, the surface thickness needs to be considered [6] and [8].

Implications from these findings should be applicable to the design of multiple magnetic wheels assembled on the same shaft to achieve the desired adhesive force for climbing robots. Based on the results of this experiment, the designers are able to gain the desired adhesive force provided by multiple magnetic wheels installed on the same shaft referring to the configuration of magnetization direction and the axial distance between the wheels.

References

1. Breitenmoser, F. Tach, C. Capari, R. Siegwart, and R. Moser, Proc. of 9th Int. Conf. on Autonomous Agents and. Multiagent Systems (AAMAS 2010), 1713 (2010).
2. T. Yukawa, M. Suzuki, Y. Satoh, H. Okano, IEEE Industrial Electronics, IECON 2006-32nd Annual Conference, (2006).
3. W. Fisher, F. Tache, and R. Seigwart, 6th International Conference on Field and Service Robotics-FRS 2007, (2007).
4. F. Rochat, P. Schoenneich, F. Mondana, H. Bleuler, and F. Tache, 13th International Conference on Climbing and Walking Robots and the Support Technologies for Mobile Machines, (2010).
5. S. C. Han, J. Kim, and H.C. Yi, H.C. , Journal of Precision Engineering and Manufacturing **Vol.10**, **No.4**, 143 (2009).
6. Y. Zhang, T. Dodd, K. Atallah, and I. Lyne, Proceeding of the 2010 IEEE International Conference on Mechatronics and Automation, (2010).
7. Y. Zhang, and T. Dodd, TAROS **Vol. 6856**, 414, Springer (2011)
8. W. fisher, F. Tache, R. Siegwart, 2007 IEEE/RSJ International Conference on Intelligent Robots and Systems, (2007)
9. X. Gao, D. Xu, Y. Wang, H. Pan, and W. Shen, Robotica **27**, 941, (2009).
10. P. Campbell, Permanent Magnet Materials and their Application, Cambridge University Press, New York U.S.A. (1994)
11. V. David, B. Marco, H. Ludek, and S. Petr, Journal of Magnetism and Magnetic Materials **321,** 3758 (2009)

QUANTITATIVE KINEMATIC PERFORMANCE COMPARISON OF RECONFIGURABLE LEG-WHEELED VEHICLES

ALIAKBAR ALAMDARI
Mechanical and Aerospace Engineering,
SUNY at Buffalo
Buffalo, NY, 14260
aalamdar@buffalo.edu

ROBIN HÉRIN
Mechanical and Aerospace Engineering,
SUNY at Buffalo
Buffalo, NY, 14260
robinher@buffalo.edu

VENKAT N. KROVI
Mechanical and Aerospace Engineering,
SUNY at Buffalo
Buffalo, NY, 14260
vkrovi@buffalo.edu

The Leg-Wheel-Vehicle paradigm offers remarkable and diverse opportunities for creation of mobile and maneuverable terrestrial locomotion systems. However, this capability needs to be unlocked by careful design and control of the individual articulations, the sub-chains and the systems as a whole. Viewing leg-wheel vehicles as another class of parallel- kinematic systems now facilitates the systematic application of the rich theory of Articulated Multi-body Systems to quantitatively evaluate design and control of such systems.

1. Introduction

Leg-Wheel-Vehicles (LWV) are a class of wheeled locomotion systems where the chassis is connected to a set of ground-contact wheels via actively- or passively-controlled legs, which can regulate wheel placement with respect to chassis during locomotion. These LWV systems exploit the reconfigurability and redundancy to realize significant benefits (improved stability, obstacle surmounting capability, and enhanced robustness) over both traditional wheeled and legged-systems. The combination of these benefits is extremely valuable in a variety of application settings from material handling on the shop floor to challenging uneven-terrain exploration.

However, LWVs are highly-constrained systems, subjected to both holonomic constraints (due to the multiple closed-loops) and non-holonomic constraints (due to wheel/ground contacts). Violation of these constraints, e.g. in terms of slipping and skidding at the ground-wheel contacts, results both in energy-dissipation and estimation-uncertainty. Hence considerable research has focused both on: (i) enhanced-suspension-design (kinematic and kinetostatic), to avoid constraint violation without either sacrificing payload capacity or increasing power consumption; and (ii) active-coordinated-control for enhancing mobility, stability and traction.

In this article we first present some background on such LWVs followed by a brief literature survey. We focus in on the computational/algorithmic implementation of screw-theoretic framework that aids the modeling, analysis, refinement and control of such systems. We conclude with a discussion of the promise and potential of the LWV paradigm by means of a quantitative comparative analysis of two LWV examples.

2. Background

Wheeled Mobile Platforms/Vehicles have been the preferred architecture for man-made terrestrial locomotion systems, by virtue of simplicity of mechanical construction and control, favorable payload-to-weight ratios, excellent load and tractive-force distribution; enhanced stability and energy-efficiency.

However, multiple disk-wheels cannot be attached arbitrarily to a single common platform/chassis without creating kinematic overconstraint (due to the lack of compatibility between the instantaneous motions of all moving parts). The enforcement of wheel ground contact constraints is solely via force-closure (and hence can be violated). This creates highly undesirable kinematic wheel-slip (skidding/slipping/scrubbing) resulting in reduced efficiency, poor odometry and unpredictable stick-slip behavior.

Kinematic overconstraint has traditionally been relieved by the addition of mechanical compliance (in the form of bushings and couplings) to mitigate the undesired constraint violation. A case can be made for the systematic and careful introduction of additional mechanical compliance – in the form of small articulated sub-chains with passive (springs/dampers) or semi-active (adjustable spring-dampers) or active (motorized) actuation. The resulting LWV systems form multiple closed kinematic loops with the ground that serve to constrain and redirect the effective forces and motions on the chassis.

Traditionally, wheeled mobile robots were considered to operate on planar surfaces, allowing the wheels to be modeled as thin disk-wheels (with holonomic rolling-without-slip constraint in the forward direction and non-holonomic no-side-slip constraint in the lateral direction). A number of authors have surveyed the various planar wheeled platform systems and their kinematic motion analysis in the plane [1,4]. Kinematic compatibility is established, evaluated and maintained in terms of matching of the Instantaneous Center of Rotations (ICR) of the disk-wheels and the chassis/platform. Campion et al. [6]

present a systematic, general and unifying approach for derivation of kinematic- and dynamic-models of planar wheeled vehicle, with arbitrary number of various types of planar articulated-leg-wheel chains attached to a common chassis.

In our own past work, we examine the potential for further generalizing and extending this work to create planar Composite Wheeled Vehicles that exploit active-or passive-articulated sub-chains for relaxing the rigid body constraints between the various axles by introducing further articulations [1, 3, 9, 14, 18]. The resulting articulations endow the composite vehicle with: (i) ability to accommodate changes in the relative configuration; (ii) redundant sensing for localizing the modules; and (iii) redundant actuation method for moving the common object to compensate for disturbances in the motions of the base.

Many new challenges are encountered when full three dimensional vehicle motion with varied wheel-ground contacts need to be considered. Various authors had noted that proper vehicle kinematic design, with appropriate internal reconfiguration degrees-of-freedom, is required to permit adaptation to uneven surfaces. Numerous articles examined passive accommodation of varied ground/wheel contact using an articulated axles/chassis e.g. Variable Length Axle [8, 16] and Passive Variable Camber [2, 7]. For rougher terrain applications, significantly more freedom becomes necessary between the chassis and ground contact, creating hybrid-articulated leg-wheel subsystem-designs. Numerous variants of such designs are possible depending upon the type, number, sequencing and nature of actuation (active/passive) of the joints. Examples range from the Mars Rover [11] and Shrimp [15] with rocker bogie suspensions, the WAAV [17] and Nomad [19] with articulated frames; to systems like the WorkPartner [12] and ALDURO [13] with powered legs and active/passive wheels.

3. Problem Formulation

A systematic design, analysis and control framework that builds upon individual component capability to examine system-level behavior is desirable. We are currently examining one such computational differential-geometry framework [9, 10] that builds upon the rich articulated multi-body literature and provides the tools to characterize, analyze, and validate seemingly disparate articulated-wheeled locomotion systems in a unified manner. The extension of the planar ICR theory to the instantaneous-screw-axes (ISA) theory proves critical for treatment of spatial cases but there is relatively limited work in this regard [5, 16]. Sreenivasan and Nanua [16] explored first- and second-order kinematic characteristics of wheeled vehicles on uneven terrain in order to determine vehicle mobility using screw-theory. Bruyninckx and Schutter [5] examined a description of the statics and velocity kinematics of serial, parallel and mobile robots based on the fundamental concepts of twists/ wrenches and reciprocity and proposed a unified treatment of serial, parallel and mobile robot kinematics.

Inspired by these efforts, in our work we examined a systematic and symbolic rapid computational formulation of kinematic models for the general class of LWVs, shown in Fig.1. Further, automating this process by using the symbolic toolbox in MATLAB, facilitates the rapid modeling and analysis of any given design of a LWVs [10]. Twist- and wrench-based approaches had previously been used to analyze motion and force capabilities of in-parallel articulated mechanical-systems (such as parallel manipulators or multi-fingered grasping) but never for articulated-systems with rolling wheel-ground contacts. Our contribution lay in extending the twist- and wrench-based modeling framework to such articulated wheeled robotic systems (and subsuming the extant specialized approaches for ordinary wheeled robots [5]). Automating this process by use of symbolic analysis methods facilitates the rapid analysis of various designs. The resulting modeling and analysis framework is well suited for both the design and control of such LWVs.

Twist of the contact point respect to the fixed frame:

$$
{}^{C}\left[{}^{F}V_{C}\right]=\left[{}^{C}\Gamma_{B}\right]{}^{B}\left[{}^{F}V_{B}\right]+\left[{}^{C}\Gamma_{1}\right]{}^{1}\left[{}^{B}V_{1}\right]+\ldots
+\left[{}^{C}\Gamma_{W}\right]{}^{W}\left[{}^{m}V_{W}\right]+{}^{C}\left[{}^{W}V_{C}\right]
$$

Non-holonomic Constraints at contact point:

$$
\left[S^{T}\right]_{4\times 6}\cdot{}^{C}\left[{}^{F}V_{C}\right]_{6\times 1}=[0\ \ 0\ \ 0\ \ 0]^{T}
$$

$$
\underbrace{S^{T}\left[{}^{C}\Gamma_{B}\right]}_{P_i}{}^{B}\left[{}^{F}V_{B}\right]=-B\dot{q}
$$

Assembling all constraints

$$
\begin{bmatrix} P_1 \\ P_2 \\ \cdot \\ \cdot \\ P_n \end{bmatrix}\times{}^{B}\left[{}^{F}V_{B}\right]=\begin{pmatrix} A_1 & & 0 \\ & \ddots & \\ 0 & & A_n \end{pmatrix}\cdot\dot{q}
$$

Figure 1: Computational implementation of twist approach [10]

4. Case Study: Comparative Analysis of Two Leg-Wheel-Vehicles

We now perform the comparative kinematic analysis of two highly maneuverable and reconfigurable LWVs. Fig.2 shows two examples of spatial leg-wheel sub-chains attached symmetrically to a triangular platform. Case 1 has five DOF for front leg-wheel and four DOF for rear leg-wheel, and case 2 has four DOF for front leg-wheel and three DOF for rear leg-wheel. One motor powers the wheel; one motor for steering of caster and one motor for changing the elevation of the link with respect to the platform and another one attached to the link where can change the geometry of the leg. The linear actuator in each leg plays an important role in supporting the weight of the chassis. We assign coordinate frames as shown in Figure 2 and set up the homogenous transformation between the various frames of references. Then, twist vectors expressed in local frames and contact twists expressed in contact frame are found, illustrated in Fig.1. After applying the non-holonomic constraints to the system and restricting the motion of the components of the system, we form the velocity level kinematic equations. As examples of parallel architecture systems, the final effective kinematics relations for our leg-wheel systems may be written as: $P_x \dot{\mathbf{x}} = A_q \dot{\mathbf{q}}$.

Investigating the singularities of Leg-Wheel-Vehicles pose substantially more problems, compared to a serial manipulator. Gosselin and Angeles [25] provide a general classification based on the input–output velocity map for a generic mechanism. Inverse kinematic singularity occurs when the determinant of A_q vanishes, namely; $\det(A_q) = 0$ and forward kinematic singularity occurs when the rank of P_x is less than 6, namely; $\text{rank}(P_x) \leq 5$. If $\text{rank}(P_x) \leq 5$ is satisfied, the moving platform gains 1 or more degrees of freedom. In other words, at a forward kinematic singular configuration, the manipulator cannot resist forces or moments in some directions.

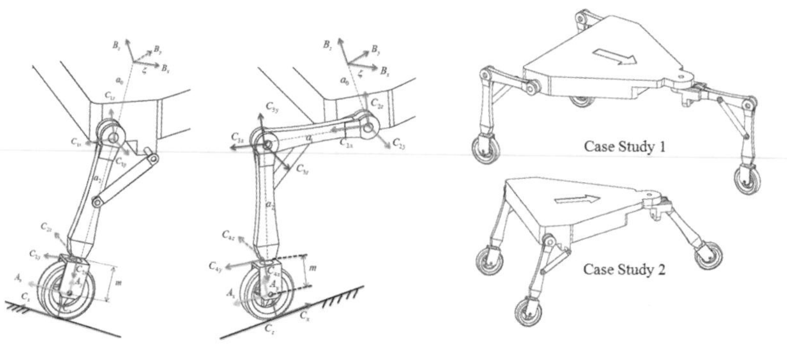

Figure 2: Two models for Reconfigurable Vehicles

Various performance indices such as manipulability [23] isotropy [24] may now be extended for such LWV. The performance indices are usually formed based on the evaluation of the determinant, norms, singular values and eigenvalues of the Jacobian matrix. Merlet reviewed the merits and weaknesses of these indices [20]. These indices have physical interpretation and are useful for design, control and optimization [21]. Care must be taken to ensure dimensional homogeneity of the Jacobian matrices, otherwise the ensuing results will depend on the physical units of parameters chosen [22, 23].

5. Results

We will assume that the legs are arranged symmetrically about the body. By virtue of the redundancy within the locomotion system, we note that the legs can assume a variety of configurations without altering the pose of the chassis. The choice of configuration however affects the mobility and maneuverability of the overall system.

Hence we conducted a numerical study to examine this as follows. We place an inertial frame of reference at the chassis frame of reference such that x, y, z, Roll, Pitch and Yaw take on the nominal values of $[0,0,0,0,0,0]^T$. Then relative to the inertial frame, we assume that axles of all the 3 leg-wheel systems are symmetrically distributed on a circle of radius (R) located at a height (H) below the XY plane of the chassis frame of reference.

We perform inverse kinematics to how determine the configuration angles of each leg, which then allows us to compute the manipulability of the system. In Figure 3 we compute and depict the translational manipulability of Leg Wheel Vehicle 1. The ultimate goal of this exercise was to aid the determination of the best pose (or sets of poses) for the internal configuration of the redundant LWV to maximize its manipulability.

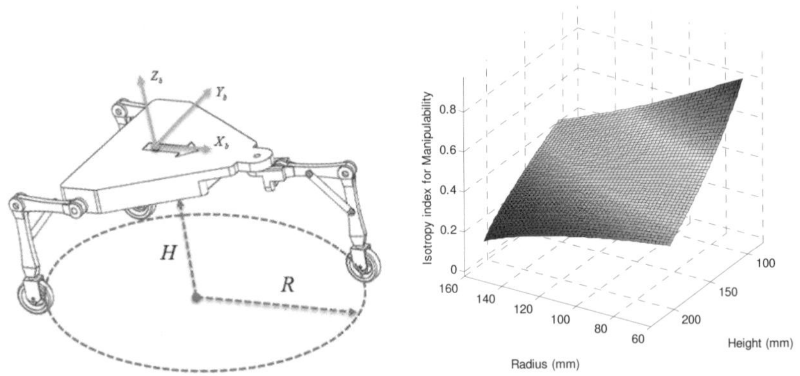

Figure 3: (A) Parameters (R & H); (B) Translational Manipulability of Leg Wheel Vehicle 1

6. Discussion

The Leg-Wheel-Vehicle paradigm offers remarkable and diverse opportunities for creation of very mobile and maneuverable terrestrial locomotion systems. However, the capabilities of Leg-Wheel locomotion systems to manipulate the chassis/payload (to improve obstacle surmounting capabilities and reduce actuation requirements) needs to be carefully unlocked by both design and control.

The nature and number of both the added wheels, together with the intermediate articulations, has a significant influence on the mobility, maneuverability, controllability, stability and efficiency of the wheeled vehicle. Viewing such LWVs as yet another class of parallel-kinematic chains (with multiple articulated leg-wheel branches attached to a common chassis) allows the systematic application of the rich theory of articulated multibody systems to design, analyze, simulate and control of the ensuing systems.

References

1. Abou-Samah, M., Tang, C., Bhatt, R., Krovi, V.: A kinematically compatible framework for cooperative payload transport by nonholonomic mobile manipulators. Autonomous Robots 21(3), 227–242 (2006)
2. Auchter, J., Moore, C., Ghosal, A.: A Novel Kinematic Model for Rough Terrain Robots, Lecture Notes in Electrical Engineering, vol. 14, chap. 16, pp. 215–234. Springer Netherlands (2009)
3. Bhatt, R.M., Tang, C.P., Krovi, V.N.: Formation optimization for a fleet of wheeled mobile robots a geometric approach. Robotics and Autonomous Systems 57(1), 102–120 (2009)
4. Borenstein, J., Everett, H.R., Feng, L.: Navigating Mobile Robots: Sensors and Techniques (1996)
5. Bruyninckx, H., Schutter, J.D.: Unified kinetostatics for serial, parallel and mobile robots (1998)
6. Campion, G., Bastin, G., Dandrea-Novel, B.: Structural properties and classification of kinematic and dynamic models of wheeled mobile robots. Robotics and Automation, IEEE Transactions on 12(1), 47–62 (1996)
7. Chakraborty, N., Ghosal, A.: Kinematics of wheeled mobile robots on uneven terrain. Mechanism and Machine Theory 39(12), 1273–1287 (2004)
8. Choi, B.J., Sreenivasan, S.V.: Gross motion characteristics of articulated mobile robots with pure rolling capability on smooth uneven surfaces. Robotics and Automation, IEEE Transactions on 15(2), 340–343 (1999).
9. Fu, Q., Krovi, V.: Articulated wheeled robots: Exploiting reconfigurability and redundancy. In: ASME 2008 Dynamic Systems and Control Conference (DSCC2008) (2008)
10. Fu, Q., Zhou, X., Krovi, V.: The reconfigurable omnidirectional articulated mobile robot (ROAMER). In: Proceedings of 12th International Symposium on Experimental Robotics. Delhi, India (Dec. 18-21, 2010)
11. Hacot, H.: Analysis and traction control of a rocker-bogie planetary rover. M.S. thesis, MIT (1998)
12. Halme, A., Leppanen, I., Suomela, J., Ylonen, S., Kettunen, I.: Workpartner: Interactive human-like service robot for outdoor applications. The International Journal of Robotics Research 22(7-8), 627–640 (2003)
13. Hiller, M., Germann, D.: Manoeuvrability of the legged and wheeled vehicle alduro in uneven terrain with consideration of nonholonomic constraints. In: Proceedings of 2002 International Symposium on Mechatronics (ISOM 2002)
14. Jun, S.K., White, G.D., Krovi, V.N.: Kinetostatic design considerations for an articulated leg-wheel locomotion subsystem. Journal of Dynamic Systems, Measurement, and Control 128(1), 112–121 (2006).

15. Siegwart, R., Lamon, P., Estier, T., Lauria, M., Piguet, R.: Innovative design for wheeled locomotion in rough terrain. Robotics and Autonomous Systems 40(23), 151–162 (2002)
16. Sreenivasan, S.V., Nanua, P.: Kinematic geometry of wheeled vehicle systems. Journal of Mechanical Design 121(1), 50–56 (1999)
17. Sreenivasan, S.V., Waldron, K.J.: Displacement analysis of an actively articulated wheeled vehicle configuration with extensions to motion planning on uneven terrain. Journal of Mechanical Design 118(2), 312–317 (1996)
18. Tang, C.P., Bhatt, R., Abou-Samah, M., Krovi, V.: Screw-theoretic analysis framework for cooperative payload transport by mobile manipulator collectives. Mechatronics, IEEE/ASME Transactions on 11(2), 169 –178 (2006). DOI 10.1109/TMECH.2006.871092
19. Wettergreen, D., Bualat, M., Christian, D., Schwehr, K., Thomas, H., Tucker, D., Zbinden, E.: Operating Nomad during the Atacama Desert Trek, chap. 14, pp. 82–89. Springer London (1998)
20. Merlet, J. P. (2006). Jacobian, manipulability, condition number, and accuracy of parallel robots, J. Mech. Des., Trans. ASME 128(1): 199–206.
21. Kucuk, S. & Bingul, Z. (2006). Comparative study of performance indices for fundamental robot manipulators, Rob. Autom. Syst. 54(7): 567–573.
22. Schwartz, E., Manseur, R. & Doty, K. (2002). Noncommensurate systems in robotics, Int. J. Rob. Autom. 17(2): 86–92.
23. Cardou, P., Bouchard, S. & Gosselin, C. (2010). Kinematic-sensitivity indices for dimensionally nonhomogeneous jacobian matrices, IEEE Trans. Rob. 26(1): 166–173.
24. Gosselin, C. & Angeles, J. (1991). Global performance index for the kinematic optimization of robotic manipulators, J. Mech. Transm. Autom. Des. 113(3): 220–226.

ENHANCING PERSONAL ELECTRIC VEHICLES USING RECONFIGURABLE DESIGN

ROBIN S. HÉRIN, YIN-CHI CHEN, ALIAKBAR ALAMDARI and VENKAT N. KROVI

Mechanical and Aerospace Department, State University of New York at Buffalo, Buffalo, NY 14260, USA
E-mail: vkrovi@buffalo.edu
http://mechatronics.eng.buffalo.edu/

Personal Electric Vehicles can leverage reconfigurability to allow better matching of their performance to the application environments. We are currently examining creation of a Computer-Aided Engineering environment that facilitates systematic and quantitative design refinement of such reconfigurable Personal Electric Vehicles.

1. Motivation

The Personal Electric Vehicle (PEV) paradigm has generated significant interest from multiple perspectives. They offer an innovative consumer product that fits into a niche market place for urban transport. These offer the prospect of improved efficiency by taking advantage of significant weight reduction, and use of new technologies (in-wheel electric motors, batteries). However, technological limitations preclude a simplistic drop-in replacement for the conventional combustion engine and drive train architecture. These challenges are even more pronounced for man-portable PEVs that could be used in sidewalks, malls, and on occasion on the local city streets.[1] On the sidewalks/malls, requirements include low-speed travel (5-10 mph), high maneuverability with a small footprint whereas, on the city streets, higher travel speeds (30-40 mph), improved stability and aerodynamics become necessary. One approach to accommodating both sets of requirements is via a reconfigurable design that can adjust selected parameters and allow for its performance modulation.

From a perspective of simplicity, we focus our attention on a battery-powered three-wheeled electric vehicle, inspired by the conceptual design by Alan Fratoni. A three-wheeled vehicle design allows for a smaller footprint, lower weight, drag and manufacturing costs compared to a four-wheeled vehicle. At the same time, the tripod support guarantees static stability, without the need for dynamic balance seen in two-wheeled vehicles. We seek to quantitatively evaluate three-wheeled designs, including its capability for reconfiguration in order to match its performance to desired application. The performance criteria can be specified using many factors. In this work, we will consider static stability, dynamic stability and aerodynamic drag. However, economic viability of PEVs for everyday use,[2] or analyzing its acceptance for human transportation[3] are beyond the scope of our work.

2. Background

The architectural layout of the wheels with respect to the chassis has permitted researchers to distinguish between unicycle-based design,[4] in-line & in-parallel bi-wheel designs,[15] tricycle designs[12,16] or the more conventional four-wheel designs.[13,14] Alternatively, PEVs have been categorized into stand-up scooters, seated cycles, mobility scooters by Ulrich.[7] Similarly, the notion of reconfigurability within the chassis is also not new. Wu & Lin developed a model of a reconfigurable two-wheeled vehicle (folding bicycle) named Union.[5] Another example is the HIRIKO[6] (the MIT City Car) which can switch between a non-operating collapsed position and the operating mode.

However, interest in systematically studying three-wheeled architecture, including vehicle dynamics,[8] passenger-vehicle handling and stability[9] is quite recent. In the case of tilting three-wheeled vehicles, the number of tilting wheels offers one possibility for classification.[10] Alternatively, another possible classification is to distinguish the type of tilting system.[12,16] Steering Tilt Control (STC) provides the dynamics behavior of a motorcycle at high speed but is unstable at low speed or even at standstill. Direct Tilt Control (DTC) guarantees stability at every speed, but tends to perform poorly at high speed. Some vehicles were tested to synchronize those two modes but with limited success.[11]

3. Case study

We envisage a flexible design of our three-wheeled vehicle with two-operating modes and one non-operating mode as shown in Figure 1. In the first operating mode (shown in Fig. 1b), the driver leans forward prone on her/his belly as in a tilting motorcycle; the second operating mode (as shown in Fig. 1c) is with the driver seated and the non-operating mode (in Fig. 1d) is the compact collapsed position, like an umbrella stroller.

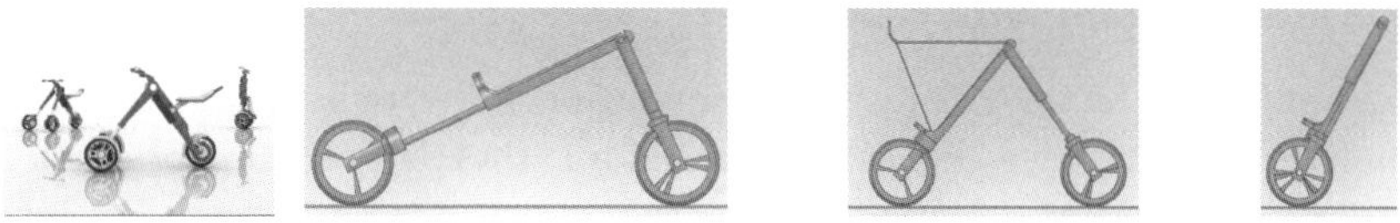

Fig. 1. (a) Personal Electric Vehicle by Alan Fratoni (b) 1st position (c) 2nd position (d) 3rd position

In order to realize the dynamic behavior of a motorcycle, some further constraints need to be placed on the design. The two front wheels are almost parallel and in our model the front legs remain in the same plane, even when leaning in turns. A tilting mechanism needs to be incorporated, either as a passive spring damper mechanism or via active control, to prevent the vehicle from rolling over in turns. A compliant mechanical linkage within the chassis now allows for tilting into turns like a motorcycle but now adds additional degrees of freedom (that need to be controlled). In this paper, we focus on assessing the link between the vehicle performances and the reconfigurable architecture using three different aspects. Fully exploiting

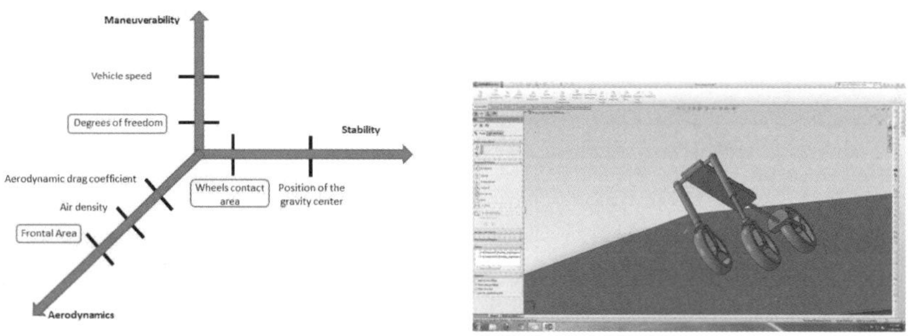

Fig. 2. (a) Quantitative evaluation of performance measure (b) Project in its virtual environment

the new capabilities endowed by additional interval of these reconfigurable degrees of freedom requires careful design, analysis and ultimately control and is best done within a quantitative framework. Specifically, we are pursuing these design-refinement/virtual prototyping efforts within a Computer Aided Engineering environment that leverages Solidworks, SimMechanics and the Matlab Optimization toolbox.

4. Modeling and simulation

Solidworks can be used as dynamic simulation software by using its Motion Analysis add-in. This add-in feature allows the performance of motion-studies by actuating selected joints within the system. Although the measurement and analysis capability provided by the add-in is very limited, Solidworks nevertheless remains an invaluable tool to generate 3-D parametric models of components and assemblies and for proof of concept simulations. We then examine the process of exporting the created model from Solidworks to a couple of alternate analysis modules (SimMechanics and ADAMS) to aid a more in-depth analysis (as shown in Fig. 3). Figures 4

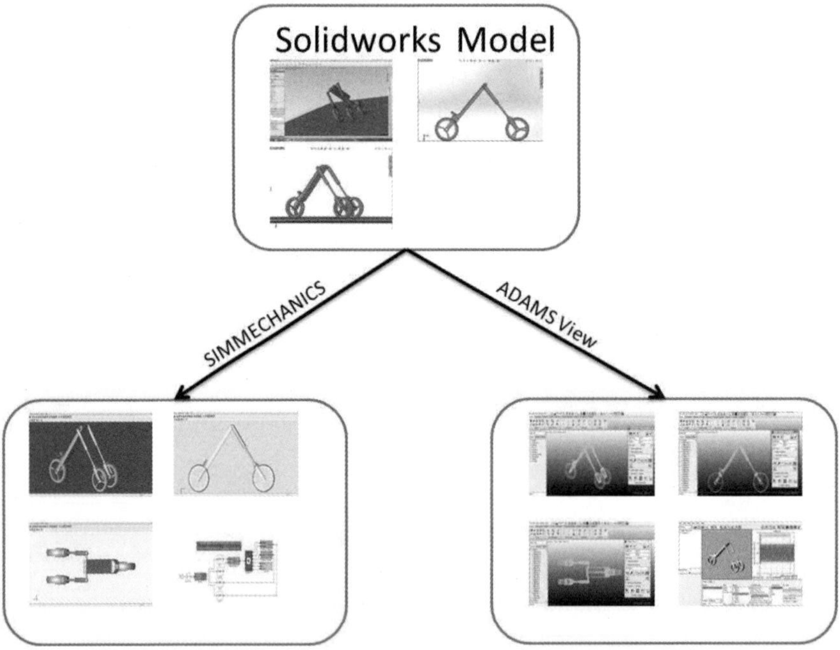

Fig. 3. Solidworks model serving as the starting point for creation of SimMechanics and ADAMS models.

and 5 illustrate the results based on our preliminary simulation and analysis studies of three wheeled vehicles. A more comprehensive numerical validation of such reconfigurable systems is a part of our future work.

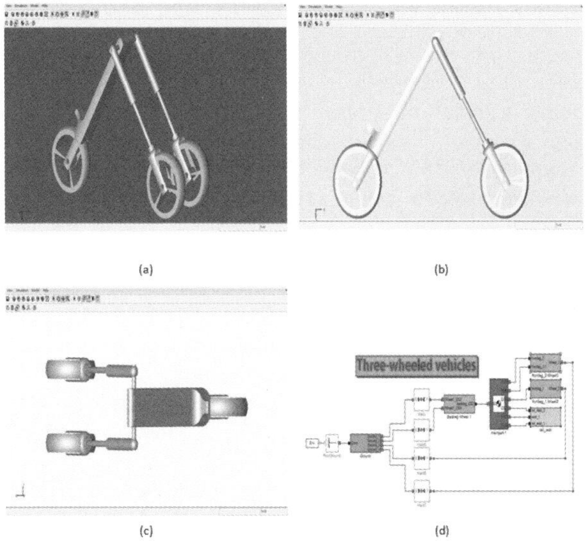

Fig. 4. SimMechanics Model: (a) Isometric view (b) Side view (c) Top view and (d) Block diagram view.

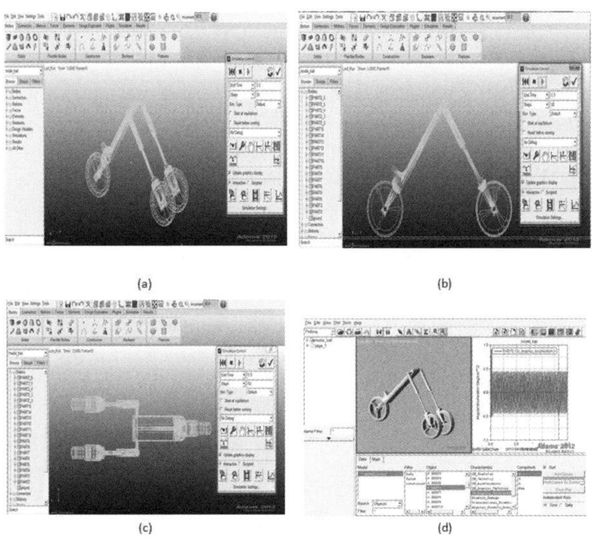

Fig. 5. Adams Model: (a) Isometric view (b) Side view (c) Top view and (d) Simulation results.

References

1. T. K. Hwang, L. C. Chu. *Enhancing Accessibility of Mass Transportation in Metropolitan Taipei by Formulating a Portable Personal Electric Vehicle* The International Association of Societies of Design Research, Seoul, 2009.
2. Nathaniel S. Pearre, Willett Kempton, Randall L. Guensler, Vetri V. Elango. *Electric vehicles: How much range is required for a day's driving?* Transportation Research Part C 19 (2011) 1171-1184.
3. Michael A. Tamor, Chris Gearhart, Ciro Soto. *A statistical approach to estimating acceptance of electric vehicles and electrification of personal transportation.* Transportation Research Part C 26 (2013) 125-134.
4. Gheorghe Deliu, Mariana Deliu. *Monowheel Dynamics.* International Conference on Economic Engineering and Manufacturing systems, Braşov, 2009.
5. Weiche Wu, Minhan Lin. *http://www.wuweiche.com/2008/05/union.html.*
6. William J. Mitchell, Chris E. Borroni-Bird and Lawrence D. Burns. ***Reinventing The Automobile*** *Personal Urban Mobility for the 21st Century.* Massachusetts Institute of Technology, 2010.
7. Karl T. Ulrich. *Estimating the technology frontier for personal electric vehicles*, Transportation Research Part C 13 (2005) 448-462, Philadelphia, 2005.
8. Jeffrey C. Huston, Brian J. Graves. *Three Wheeled Vehicle Dynamics.* SAE Technical Paper 820139, 1982.
9. Paul G. Van Valkenburgh and Richard H. Klein. *Three-Wheel Passenger Vehicle Stability and Handling.* SAE Technical Paper 820140, 1982.
10. Robert Q. Riley. *Alternative Cars in the 21st Century : a new personal transportation paradigm* 2004, ISBN : 0-7680-0874-3.
11. Sang-Gyun So. *Development of Dual Mode Automatic Tilt Control Systems for Ultra-Narrow Commuter Vehicle.* Ph.D Dissertation, University of California at Davis, 1996.
12. Johan J. H. Berote. *Dynamics and Control of a Tilting Three Wheeled Vehicle.* Ph.D Dissertation, University of Bath, 2010.
13. Milliken, William F, and Milliken, Douglas L. *Race Car Vehicle Dynamics* SAE 1995.
14. Rajesh Rajamani. *Vehicle Dynamics and Control* 2006.
15. Vittore Cossalter. *Motorcycle Dynamics* 2nd English Edition, 2006.
16. Nicola Amati, Andrea Festini, Luigi Pelizza, Andrea Tonoli. *Dynamic modelling and experimental validation of three wheeled tilting vehicles, Vehicle System Dynamics* International Journal of Vehicle Mechanics and Mobility, 2011, 49:6, 889-914.

HAND-MANOEUVRED WHEELCHAIR USING WHEELS FITTED WITH FEET FOR ENHANCED MOBILITY

T. OKADA[1], H. WADA[2], N. MIMURA[1], T. SHIMIZU[3] and K. NAGATA[1]
[1]*Niigata University*, [2]*DOUBLE R&D Co. Ltd.*, [3]*Ibaraki University*
950-2181, Niigata Japan *http://okada.eng.niigata-u.ac.jp*

This paper describes the design of a manually powered wheelchair which has eight feet fitted to each driving wheel for improving the mobility of popular wheelchairs. The prototype skeleton of the wheelchair fabricated in this study is same as that of the standard wheelchair available in the market. The wheels on most wheelchairs are designed to propel the wheelchairs over a flat terrain and they have trouble traversing a rough terrain, specifically in an environment with bumps, stairs, slippery rocks and tree roots. To cope with these difficulties, these play an important role instead of the wheels. The design for the prototype is faced the twin challenges of keeping the total weight low, and not invading the dynamic hand space of handrim-powered locomotion while retaining the mobility of the standard wheelchair. The experimental results demonstrate that a wheelchair with feet improves the mobility performance when encountering difficulties.

Keywords: Wheelchair, Feet, Mobility, Climbing stairs, Assistive Technology

1. Introduction

Wheelchairs transport people who have difficulty in walking resulting from weakness in the legs or disability. Therefore, wheelchairs are necessary instruments as an extension of self, and are needed to continue daily life. In addition, they enable users to enjoy sports such as tennis, basketball and soccer. For users' individual satisfaction, wheelchairs are available in various shapes, sizes, frames, driving methods, designs and applications.[1] Wheelchairs generally have four wheels, although there is a simple three-wheel design and a six wheel design using a rocker-bogie mechanism. The sizes of the front and rear wheels of many wheelchairs are different most having smaller front wheels. Of course, products of a reverse type are available in the market for aged persons whose mass centre and hand manoeuvring range have shifted towards the front of the seat. The small wheels are mostly free similar to casters and the large wheels are powered by hands or motors. The traveling direction is changed by driving theright and left wheels independently. Otherwise, the axial direction of the small wheels is controlled for steering.[2]

To manoeuvre in a narrow space, omnidirectional mobility is a necessity. Basically, three or four wheels having free rollers of a barrel form are activated cooperatively to make the platform move omnidirectionally. A wheelchair[3] can be given enhanced step-climbing capability by using two omnidirectional wheels at the front for passive rolling and two standard wheels at the rear for active driving. The four wheels of the Permoveh,[4] newly developed electric wheelchair have many active rollers on the wheel periphery orthogonal to the shaft of the driving wheels.

Most manual wheelchairs have handrims fitted to the right and left driving wheels. However, levers can take the place of handrims if the need arises. In particular, a wheelchair[5] generates all required motions including braking by manipulating only one lever. The wheelchair[2] derives driving power from pedals by employing a riding motion similar to that of a bicycle. This is intended to be utilised as a rehabilitation aid or walking or as therapy for regenerating motor nerves.[2]

These wheelchairs stand on rolling wheels at all times. However, the wheels can be modified to offer discontinuous landing by splitting circular treads into several sections or by fitting gripping claws for climbing up stairs.[6–8] Hybrid systems which combine wheels with legs have also been used. The wheels of the robot PEOPLER[9] is equipped with legs whose rotation is independent of the wheels, as it obtains torque from motors located in the main body.

This paper describes the development of a wheelchair for improving the mobility and performance of wheelchairs on an irregular terrain. We apply the structural concept used for developing the hybrid robot PEOPLER-IV.[9] However, the prototype in this paper has two free wheels at the front and moves without using electric power. We show the simplicity of attaching feet to the wheels and validate its use to cope with soft, rigid and slippery ground including obstacles such as bumps, bricks, debris and stairs. The feet said outdoor traveling even when encountering obstacles.

2. Reason for attaching feet to wheels

The characteristics of wheeled and legged machines are constrastive. Wheeled machines are ssuited to fast movement by driving the wheels at high speed over flat road ground. On the other hand, legged machines are suited to slow movement over a rough terrain by swinging their legs reciprocally. Therefore, wheeled machines find it difficult to cope with obstacles, while legged machines show the ability to handle this problem.

A wheeled machine may block the user from reaching his goal when encountered with obstacles, and the wheels are likely to sink below the tire level on soft land such as sandy or muddy areas. Also, it is difficult for a wheeled machine to escape from a slippery terrain. On the contrary, a legged machine can overcome difficulties by climbing up or over obstacles and enlarging its tread area to prevent sinking from the ground level (footwear called *kanjiki* in Japanese is an example). In addition, the legs can strike the ground to prevent slipping, similar to a spiked shoe. These characteristics of wheeled and legged machines led us to develop wheels having feet fitted to the wheel rims so that the wheelchair can behave as both a wheeled and legged machine by switching on demand.

3. Technical problems in designing wheels

Several technical problems were considered in designing wheelchairs fitted with feet because it was difficult to route the wire to supply electric power to the feet through the wheels. These problems are summarised as follows:

1) Attachment of feet to a main wheel driven by hand.
2) Axial parallelism of all the feet.
3) Driving mechanism of feet for simultaneous motion.
4) Feet attachment which does not disturb handrim operation.
5) Low level of hand power for changing feet angle.
6) Self-locking of feet angle while the wheel rotates.
7) Ease of feet angle calibration.
8) Number of feet without having to manipulate the spoke layout.

4. Initial conditions for wheelchair design

To save time and cost, we modified a wheelchair, a popular general transport model for the physically disabled as follows:

1) The body frame was taken from the AR-101 40 H51 wheelchair manufactured by Matsunaga Manufacturing Co. Ltd..
2) To transmit the feet-driving torque, we used chains with a half-inch pitch which are commercially available as bicycle parts. Sprockets were used as tensioners for not only guiding the chains but also adjusting the angular displacement.
3) The wheel diameter was 22 in, and there were eight feet per wheel. Each aluminum feet was 10 cm long (=37% of the wheel radius) and 5 mm thick.
4) There were 36 spokes as opposed to the standard 24 spokes.
5) Worm and worm-wheel gears were used to take advantage of self-locking function to fix the feet angles simply.

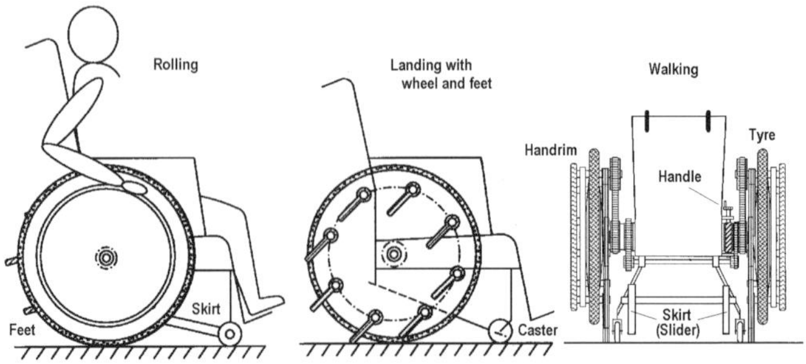

Fig. 1. Illustration of wheelchair with feet on the wheels (side skeleton and back).

6) Easy attach and detach of feet owing to the use of serration machining.
7) Two circular rings on both sides of each rim supported the feet axes.

5. Conceptual image of design

The wheelchair wheels at the back and small wheels at the front. This is beneficial for supporting a load at the back area and steering instantly by using the small wheels as free wheels. The minimum number of feet is two or three depending on length, and there is no limit to the maximum number. Of course, too many feet become redundant and behave as if they form the profile of a wheel tread. Therefore, the number might be determined by trial and error, and we considered that eight feet were reasonable with respect to the wheel diameter.

To maintain a balanced stance on stairs, it is necessary to avoid the front edge of a step from encountering into the lower space between the front and rear wheels. For this function, we used a long slippery board called *skirt* as a supplemental supporter for landing. To maintain the hand-motion space for driving the handrim, the feet were not allowed to invade the space. In addition, the feet attachment points need to be housed in a narrow interior space so that the handrim remains within a user's normal hand reach for driving. These considerations made us to show the images of the wheelchair (see Fig.1).

Movement by the feet is illustrated in Fig.2. The backwards and forwards feet inclinations are shown in Figs. 2a and 2b, respectively. The wheelchair swings vertically as if it were a cart having octagonal-asterisk rollers. Figure 2c shows the conceptual design of feet covers for safety under a conditions where the feet face downwards without facing upwards.

Two types of drivetrain transfer angular displacements from one axis to the other. Ropes, wires, chains and belts connect mechanical parts such as pulleys, drums and sprockets located on the axes. The angular velocity of

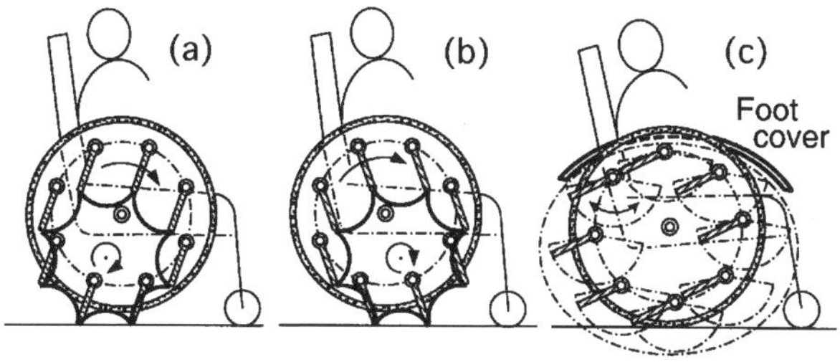

Fig. 2. Side sketches of movement by feet or wheel.

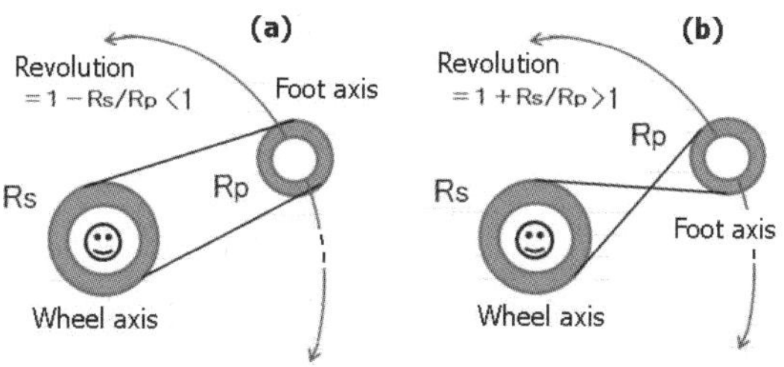

Fig. 3. Two drivetrains connecting foot and wheel axes for torque transition.

the feet depends on the sizes of the mechanical parts. Suppose that a rope in Fig.3 connects a star axis (wheel) and a planetary axis (feet), and that R_s and R_p represent the respective radii of the two axes. Then, one revolution of a wheel rim gives the angular speed of the feet expressed as 1-R_s/R_p and 1+R_s/R_p depending on types (a) and (b) in the figure, respectively. Clearly, the revolution equals 1 when R_s=R_p in type (a). This keeps the direction of the feet same without the influence of rim rotation. We designed the wheelchair using this feature to keep the mechanism simple.

Figure 4 shows detailed images of the design of wheels with feet, handrims and feet-driving mechanisms using two chains. The number of spokes is reduced to 32 by removing 4 of the original 36 so as to space the eight feet axes at the same intervals. The axes are allocated at the positions referred to by the spoke numbers 3-4, (8), 12-13, (17), 21-22, (26), 30-31, (36). We believe that the removal of the spokes does not disturb wheel configuration because two circular plates prevent the rim from deforming and weakening. An air valve is positioned between spoke numbers 18 and 19. The black small and big circles show the fixture points of the guide sprockets and handrims, respectively.

Figure 5 shows a design sketch of a one-hand handle torque transition route to the feet via two worm and worm-wheel gears. The handling shaft is combined with the worm gear on the right and left sides. However, the gears

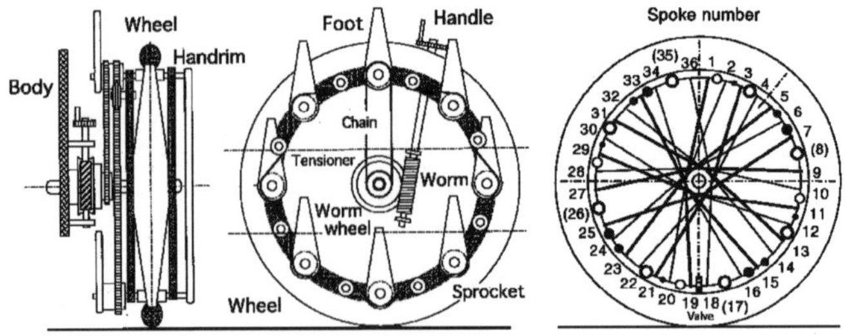

Fig. 4. Arrangement of feet near the rim.

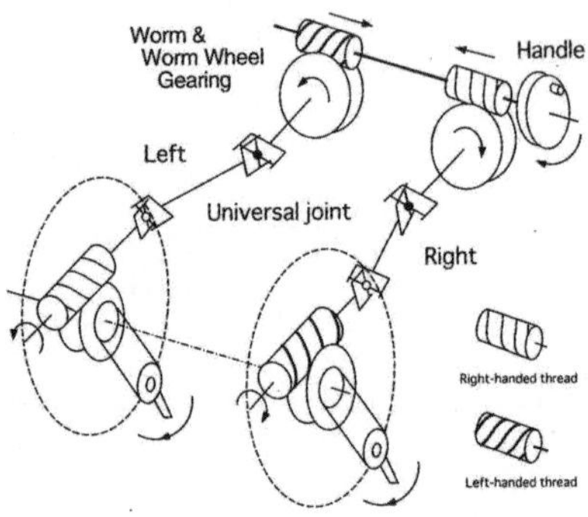

Fig. 5. Design of a driving mechanism for synchronous motion of feet on both sides.

have screw threads which are oriented in opposite directions in order to cancel the thrust force so as to simplify the design of their support flanges. Universal joints make the torque transition route flexible. Two worm and worm-wheel gears per side make it possible to change the feet angle with a small hand force.

6. Fabrication and assembly

6.1. *Parallel support of feet axes*

We constructed four circular planar parts with a thickness of 5 mm. Then we combined them to form a circular ring by joining two ends using patches and bolts (see Fig.6). Each ring is located close to either side of each wheel rim and connected using 35 mm long aluminum rods 12 mm in diameter. This assembly provides stable support because we intended that patched area does not meet in angular phase. For this purpose, the design of the two parts is dissimilar in angular phase. The rods also fix the rings to the rim with bolts similar to the fixture of the handrim to the rim.

Fig. 6. Planar rings attached to rim sides for supporting feet axes.

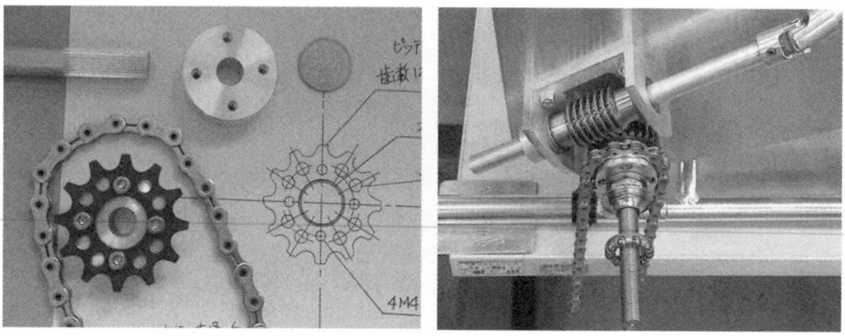

Fig. 7. Mechanical parts for transferring torque from handle to feet axes (left). The worm and worm-wheel gears connected to the spindle concentrically but independently of wheel rotation by bearings (right).

6.2. *Feet driving with synchronism in angular direction*

The chains are geared on 12-tooth sprockets which are fixed to nuts. These nuts slip onto the feet axes having a length of 8.5 cm and diameter 12 cm to the serration mechanism of 30 triangular threads. The right hand side in Fig.7 shows these connections. The image on the right shows a close-up of the worm and worm-wheel gear mechanism floating on 12 mm wheel spindle via bearings. The inner rod of the worm gear becomes a shaft driven by the hand force via universal joints. The reduction ratio of the gears is 1/50 which is sufficient to generate the self-locking function of feet angles.

6.3. *Skirts preventing loss of balance*

We found that skirts were suitable for supporting the load of a wheelchair partially and for sliding on the edges of stairs or over the top of uneven terrains. However, they were too long to be mounted in a low position. Therefore, we cut them in half and fixed the halves to the front and rear bridges connected to the right and left wheelchair skeletons. Figure 8 shows

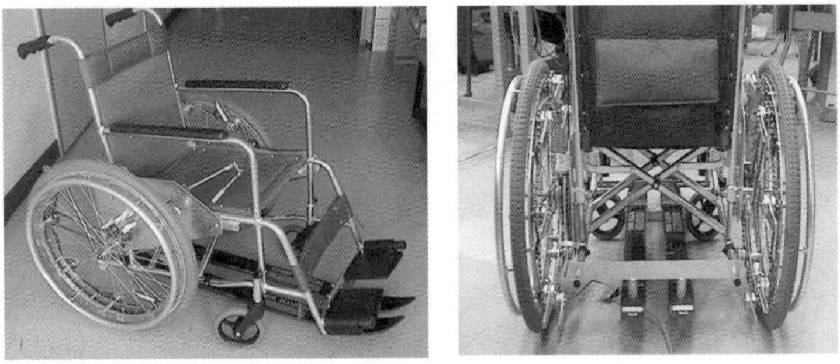

Fig. 8. Wheelchair prototype equipped with skirts in the lower section.

images of the assembled wheelchair. The boards are inclined upwards at 5 deg at their rears. It is desirable to increase the width of the rwo skirts or use one wide board such as a snowboard. Our wheelchair has two skirts mounted in parallel as wide apart as possible without any contact with the swing of the free wheels. The free wheels could be exchanged with omnidirectional wheels to widen the available space. We confirmed that the wheel degrades the smoothness of travel which does not agree with our basic design concept of maintaining the existing mobility.

7. Experiments

When the front wheels are smaller than the rear wheels, it is logical to climb upwards in the backwards direction. In addition, users are secure and safe when facing forwards while moving up the stairs. This implies that the right and left feet need to be controlled to have the same direction angles as those of the two wheels on both sides.

We confirmed that moving backwards and forwards by using the wheels was not difficult. The operation of the handrim was not disturbed by the feet because they stayed to one side facing backwards and downwards. We confirmed that the feet operated effectively in supporting the load with the wheels off the ground while the wheelchair was moving upwards and downwards. The wheelchair could climb up and over obstacles such as bumps and thresholds of doorways. Figure 9 shows an example of the wheelchair climbing over a 15 cm stack of lumber. Also, we confirm that the skirt contributed to maintaining a standing stance and preventing loss of balance.

Figure 10 shows sequential frame-grabs taken from a movie showing the wheelchair climbing an 8 cm wooden bump. In this experiment, the user noticed that the wheelchair was approaching a bump and rotated the handle to prepare the feet for standing. Then, the user drove the handrim to move backwards without bothering with the handle. Once the wheelchair

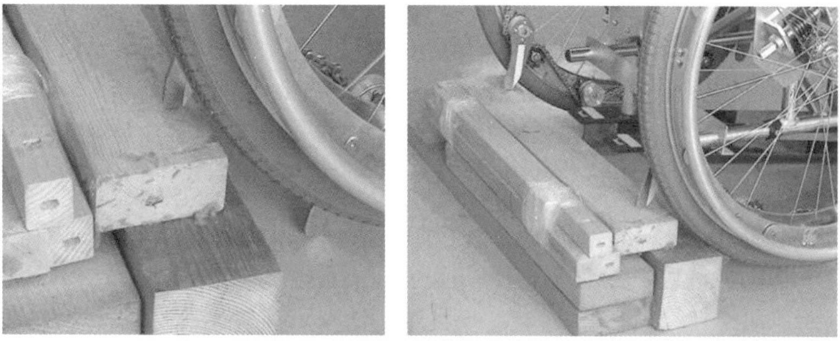

Fig. 9. Magnified images of landing by feet in climbing up steps.

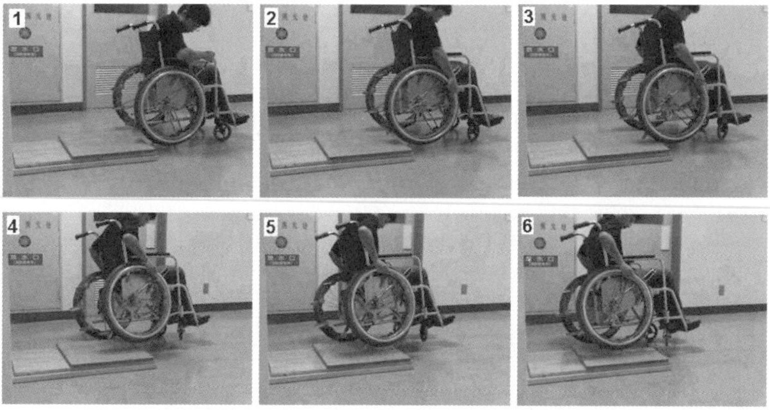

Fig. 10. Sequential snapshots of climbing motion on a board of height 8 cm.

was standing on its feet, it moved as if the wheel had been replaced by octagonal-edged wheels. Obviously, the climbable height of the bump is limited by the user's driving power.

8. Conclusion

To improve the mobility and performance of a wheelchair, we applied the results of a study on a four-wheeled skid-steering vehicle. The prototype wheelchair is powered by human hand and two of the four wheels are free wheels. By modifying only the wheels, it was easy to develop a prototype without degrading the mobility compared with a conventional wheelchair. The attachment of feet to the wheels made it possible for the wheelchair to move as a legged wheelchair which can climb up and over bumps and cope with stairs, in addition to moving on slippery, muddy or sandy roads. The results of this study are summarised as follows:

- Error of feet angles is less than ±7 deg.

- The wheelchair lands stably at four points on a rugged terrain of ± 3 cm.
- Angular change in feet requires a hand power of less than 1kg·m.
- Angular error in right and left feet is less than ± 6 deg.
- Maximum height of steps which can be climbed up and over is 15 cm.
- The wheelchair can stand on stairs with an inclination angle of less than 10 deg.

The wheelchair could be useful as a robust carrier which can push obstacles even on uneven terrain covered with debris and rubble.

We wish to thank the JST program under Contract No. AS2311365B for supporting this study. This study was also supported partially by a grant from KAKENHI(C24560287).

References

1. Rory A. Cooper : "Wheelchair selection and configuration", Demos Medical Publishing, Inc, 1998.
2. Yasunobu Handa : "Wheelchair of a front wheel driving type", Japanese Patent No.4195238, 2008.
3. Masayoshi Wada : "A 4WD omnidirectional wheelchair with enhanced step climbing capacity", J. of Robotics and Mechatronics, Vol.20, no.6, pp.846-853, 2008.
4. http ://www.youtube.com/watch?v=thKHERie-Ek
5. Akira Korosue : "Wheelchair driven by a hand lever operation", Japanese Patent No.1985-58868, 1985.
6. Taro Iwamoto and Koji Shibuya : "Concept and mechanism of robot vehicle using variable configuration segmented wheels", J. of JSME, Vol.C71, no.701, pp.171-177, 2005(in Japanese).
7. Kathryn A. Daltorio, Terence E. Wei, et al. : "MiniStep O-Whegs^{TM} climbs steep surfaces using insect-inspired attachment mechanisms", IJRR, Vol.28, no.2, pp.285-302, 2009.
8. Kan Taguchi : "Enhanced wheel system for step climbing", Advanced Robotics, Vol.9, no.2, pp.137-147, 1995.
9. Tokuji Okada, Satoru Tezuka and Hiroshi Wada : "Smooth walk of a skid-steering vehicle consisting of feet jointed to each wheel rim", Proc. of 15th CLAWAR, Baltimore, pp.585-594, 2012.

DESIGN AND LOCOMOTION MODES OF A SMALL WHEEL-LEGGED ROBOT

IOAN DOROFTEI

Robotics Laboratory, Mechanical Engineering, Mechatronics and Robotics Department, "Gheorghe Asachi" Technical University of Iasi, B-dul D. Mangeron, 61-63 700050-Iasi, Romania

ION ION

Manipulators and Robots Laboratory, Technology of Manufacturing Department, POLITEHNICA University of Bucharest, Splaiul Independentei, 313, 060042-Bucharest, Romania

Legged robots have superior terrain adaptability comparing to traditional wheeled vehicles. They also offer attractive capabilities in terms of agility and obstacle avoidance. On the other hand, traditional wheeled platforms provide sufficient robustness, mechanical simplicity and energetic performance. They are fast, powerful in terms of load to weight ratio, stable, and easy to control. Hybrid locomotion systems were developed to exploit the terrain adaptability of legs in rough terrain and simpler control as well as high speed associated with wheels. In this paper the design and the locomotion modes of a small wheel-legged robot are presented.

1. Introduction

Walking vehicles have superior terrain adaptability, comparing to other mobile robots. They also offer attractive capabilities in terms of agility and obstacle avoidance. Their use is convenient on soft ground or in unstructured environments, where the performance of wheels and tracks are considerable reduced. On the other hand, wheeled robots provide sufficient robustness, mechanical simplicity and energetic performance. They are fast, powerful in terms of load to weight ratio, stable, and easy to control. Later on, hybrid locomotion systems were developed to exploit the terrain adaptability of legs in rough terrain and simpler control as well as high speed associated with wheels.

Many hybrid locomotion systems with different architectures have been developed since today. In Japan [1] has been developed a vehicle with five locomotion devices, each device consisting of a wheel and a leg. Another hybrid robot, called SAPPHYR, with two free rear wheels and two traction legs in the front, was designed in France [2]. WHEELEG [3], developed in Italy, was a robot with two pneumatically actuated front legs, each one with three degrees of freedom (d.o.f.), and two independently actuated rear wheels. A hybrid wheelchair with two d.o.f. planar legs, and four wheels, has been developed at University of Pennsylvania [4], destined for use on uneven terrain,. In Finland, Halme *et al.* [5] designed a hybrid locomotion robot, called WorkPartner, using four legs equipped with wheels. Each wheel was used to serve as a foot during walking mode and as a wheel in wheeled mode. Roller-Walker, a hybrid mobile robot with a special foot mechanism in each leg was developed in Japan [6]. In Finland, has been developed a hybrid system, called Hybtor [7]. This robot has four wheeled legs, each consisting of a three d.o.f. mammal-type leg and a rubber wheel as a foot. The rolking mode of locomotion has been introduced. The advantages of this locomotion mode, comparing to normal walking, are better speed and stability. At Tohoku University, in Japan, has been developed a robot with four legs (two in the front and two rear legs) and two wheels in the middle, to develop a leg-wheel type robot for the exploration [8], [9]. In 2004, in Thailand, a leg-wheel hybrid robot, with two active two d.o.f. legs in the front and two passive rear wheels, has been developed [10, 11].

In this paper the design of a small wheel-legged robot and its locomotion modes will be presented. This report is the result of a research conducted at the Robotics Laboratory, Mechanical Engineering, Mechatronics and Robotics, "Gh. Asachi" Technical University of Iasi, Romania.

2. Robot Architecture

The robot developed by our laboratory [12] consists of a chassis, two legs with two d.o.f. each and two passive wheels as feet, in the front, and two rear actuated/passive wheels, respectively. The rear wheels can be passive or actuated, thanks to the two electromagnetic clutches that connect these wheels and the shafts of their driving motors.

The robots using the same architecture (two legs and two wheels), developed before [3, 10, 11], are able to move using only hybrid locomotion mode. Our hybrid locomotion robot (see Figure 1) can move using three locomotion modes: wheeled locomotion mode, using the rear actuated wheels and the two passive wheels in the feet; the first hybrid locomotion mode, using the active front legs and the passive rear wheels; the second hybrid locomotion mode, using the legs and active rear wheels.

Another advantage of this configuration (see Figure 2.b), comparing to other previous designs kinematics (Figure 2.a) [10, 11], is an improved stability, when only one leg is in support phase.

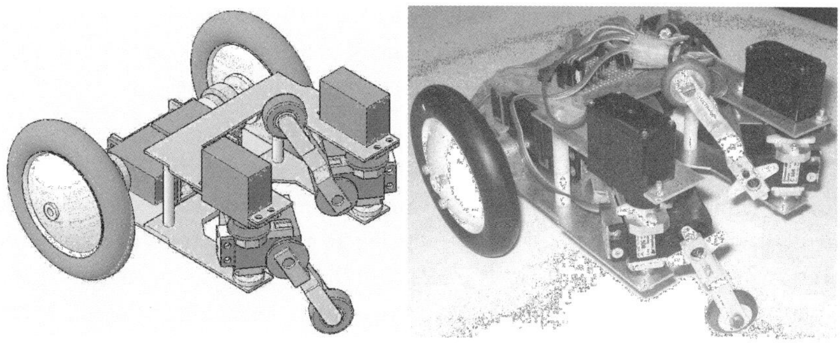

(a) (b)

Figure 1. Hybrid locomotion robot: a) CAD model; b) real prototype

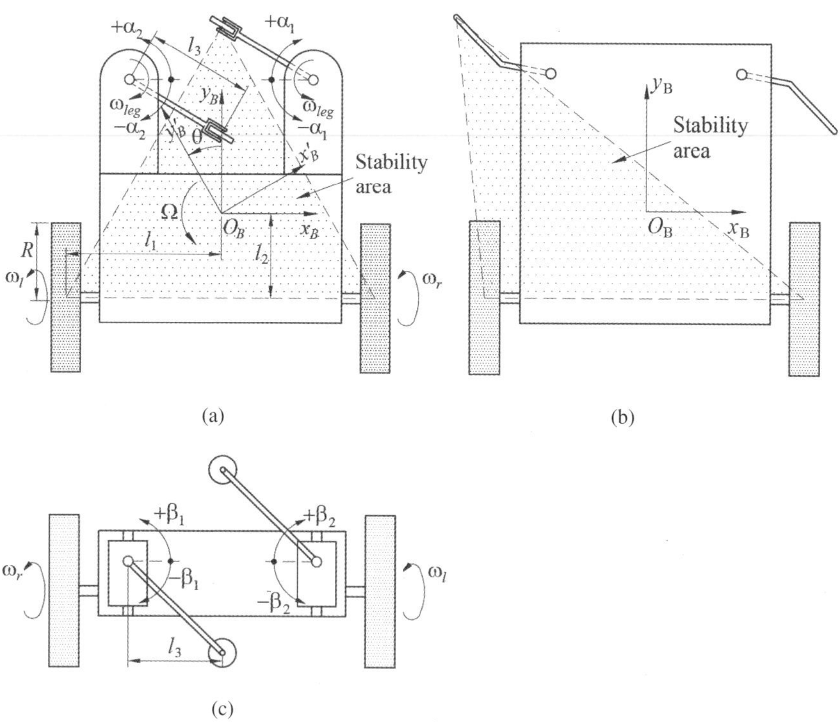

Figure 2. Robot stability area and kinematics: a) other solutions; b) our robot – top view; c) our robot – front view

The prototype developed by us is a small scale hybrid locomotion platform, with 230 x 186 x 125 [mm] as overall dimensions *(H* x *l* x *h*). At a real scale, it may be a robotized chair for people with locomotion disabilities.

3. Locomotion Modes

Three locomotion modes can be implemented on this robot: wheeled locomotion mode, using the big actuated wheels (acting in the front, for this locomotion mode) and the two passive wheels in the feet; the first hybrid locomotion mode, using the active front legs and passive rear wheels; the second hybrid locomotion mode, using the legs and active rear wheels.

3.1. *Wheeled Locomotion Mode*

The two actuated wheels (big ones) act in the front of the robot and the small wheels, as feet of the passive legs, are passive.

One of the small passive wheels, or both of them (depending on the instantaneously trajectory of the robot), will be in contact with the ground, only to keep the stability of the robot. We will have two small passive wheels on the ground when the robot is moving forward/backward on a straight or curved trajectory (see Figure 3), and a single passive wheel in contact with the ground for turning on the spot (see Figure 4).

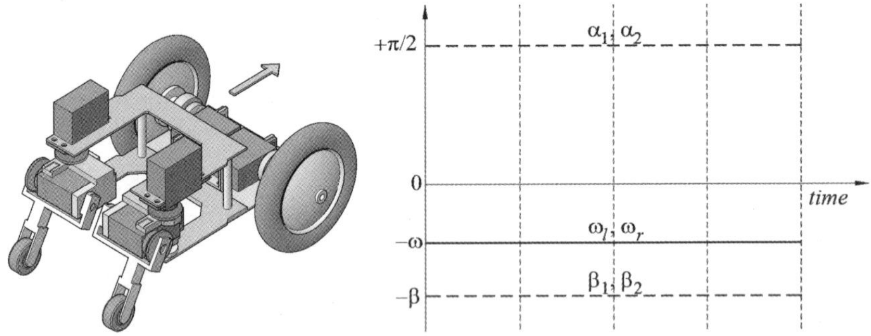

(a) (b)

Figure 3. Wheeled locomotion mode on a straight trajectory: a) robot configuration; b) diagram of motors parameters (angular position of the two legs servos; angular velocities of rear wheels)

As we may see in Figure 3.b, in this sequence, the two legs are orientated in the front of the robot, with the feet (small passive wheels) on the ground. Rear wheels are active and will usually rotate as well as the robot will move backward. Is preferred this direction because the robot could climb obstacles, thanks to the big diameters of the actuated wheels. If the robot moves forward, the small passive wheels will not be able to climb. Anyway, the robot can move forward on a flat terrain or for some maneuvers.

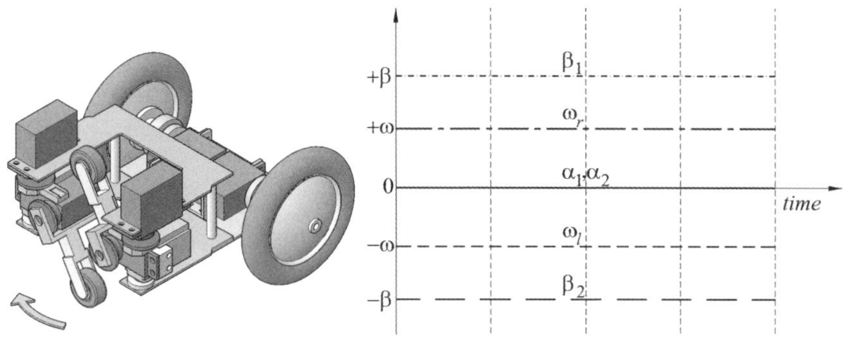

(a) (b)

Figure 4. Turning on the spot: a) robot configuration; b) diagram of motors parameters (angular position of the two legs servos; angular velocities of rear wheels)

This sequence is using to change the robot trajectory direction. In this case, one leg is touching the ground (both the legs are in the neutral position, according to the vertical axes) and the rear wheels have opposite rotating directions.

3.2. *Hybrid Locomotion Modes*

There are two hybrid locomotion modes that could be implemented on this mall platform

- *First hybrid locomotion mode*: the two legs are actuated and the rear wheels are free (decoupled from their motors shafts,);
- *Second hybrid locomotion mode*: the legs and the rear wheels are simultaneously actuated (in order to increase the power of the robot).

The single difference between these two hybrid locomotion modes consists in the actuation (or not) of the rear wheels. The second solution has the advantage of improving the traction force of the vehicle, when it moves in an unstructured terrain. Some intermediary sequences of the robot configuration during one cycle of the hybrid locomotion mode are shown in Figure 5.

Because the trajectories of the legs are crossing in the support phases, few precise rules should be established for the case when the robot is using hybrid locomotion mode:

- The legs could not be simultaneous in transfer phase (in that case the robot will fall down);
- The legs could not be simultaneously in support phase otherwise they will cross each other and they can be destroyed;
- When a leg is in support phase the other one should be in transfer phase, moving in opposite direction.

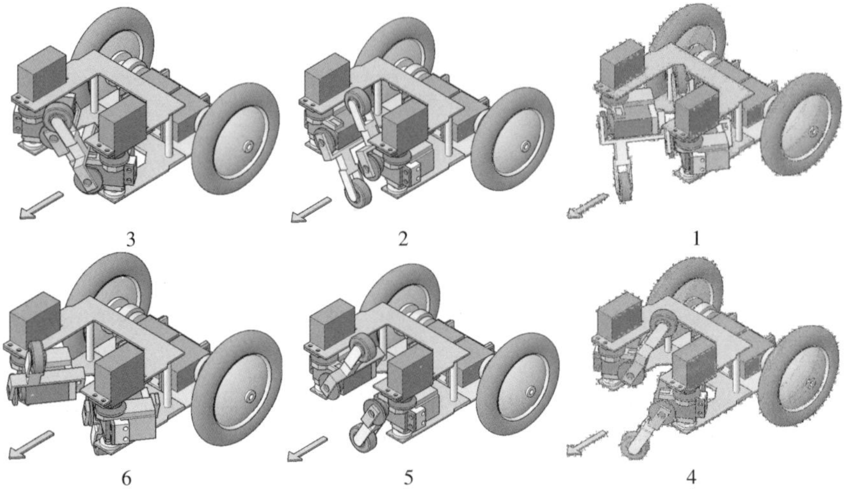

(a)

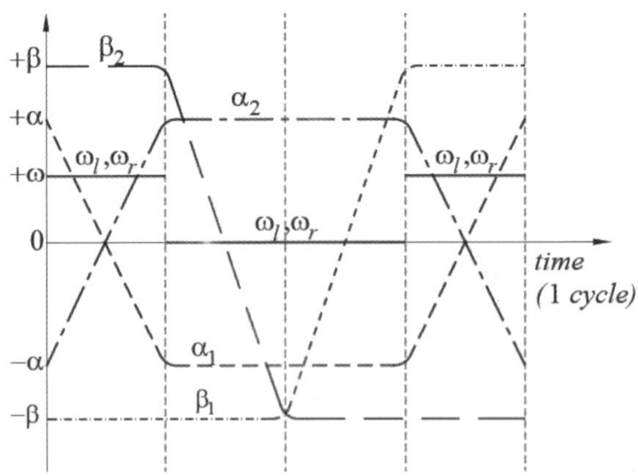

(b)

Figure 5. Hybrid locomotion mode: a) locomotion sequences; b) diagram of motors parameters (angular position of the two legs servos; angular velocities of rear wheels)

In Figure 5.a, six sequences of the hybrid locomotion modes are demonstrated, starting from the first robot configuration (noted with 1) till the last configuration (noted with 6) of a locomotion cycle (two steps).

Even if the diagram shown in Figure 5.b is identical for both hybrid locomotion modes, in the first hybrid locomotion mode, when the rear wheels are passive, their angular velocities, ω_l and ω_r, are generated by the displacement of the support leg.

In order to avoid legs/wheels slippage, for the second hybrid locomotion mode next relation should be respected:

$$\omega_{med} \cdot R = \omega_{leg} \cdot l_3 \tag{1}$$

where

$$\omega_{med} = \frac{\omega_r + \omega_l}{2} \tag{2}$$

l_3 is the horizontal projection of the leg length, ω_l and ω_r are the angular speeds of the left and right rear wheels, ω_{leg} is the angular speed of the support leg, R is the radius of rear wheels (see Figure 2.a). In order to simplify the control algorithm, the robot has been designed as well as $l_3 = R$.

4. Conclusion

The testing gave a qualitative view of the system's mobility performance. All the locomotion modes have been tested, using: forward, backward, turning right on the spot and turning left on the spot sequences, curved trajectory. All the tests have been done in the laboratory. We should also test the robot on a soft ground. The accuracy of direction and movement of the mobile robot is improved thanks to the small passive wheels used as feet. These wheels avoid the legs slippage. An open loop control was enough to test the mobility of the vehicle and no sensors have been used till now. A closed loop control should be implemented in the future, using encoders for the wheels and legs, touch sensors for the feet and range sensors for the robot.

References

1. Y. Ichikawa, N. Ozaki and K. Sadakane, A hybrid locomotion vehicle for nuclear power plants. IEEE Transactions Systems, Man and Cybernetics 13:(6), pp. 1089–1093 (1983).

2. M. Guihard, P. Gorce and J.G. Fontaine, Sapphyr: Legs to pull a wheel structure. Proc IEEE Int Conf on Robotics and Automation, Nagoya, Japan, pp. 1303–1308 (1995).
3. G. Muscato and G. Nunnari, Legs or wheels? Wheeleg - a hybrid solution. Proceedings of the 1st International Conference on Climbing and Walking Robots - CLAWAR-99 (editors G.S. Virk, M. Randall, and D. Howard), Portsmouth, UK. Professional Engineering Publishing, pp. 173–180 (1999).
4. V. Krovi and V. Kumar, Modeling and control of a hybrid locomotion system, ASME J Mech Des 121:(3), 448–455 (1999).
5. A. Halme, L. Leppanen and S. Salmi, Development of workpartner robot—design of actuating and motion control system. Proceedings of the 2nd International Conference on Climbing and Walking Robots - CLAWAR-99 (editors G.S. Virk, M. Randall, and D. Howard), Portsmouth, UK, Professional Engineering Publishing, pp. 657–665 (1999).
6. S. Hirose, and H. Takeuchi, Study on roller-walk (basic characteristics and its control), in Proc IEEE International Conference on Robotics and Automation, pp. 3265–3270 (1996).
7. A. Halme, L. Leppanen, S. Salmi and S. Ylonen, Hybrid locomotion of a wheel-legged machine. Proceedings of the Third International Conference on Climbing and Walking Robots - CLAWAR-2000 (editors M. Armada and P. Gonzalez de Santos), Madrid, Spain, Professional Engineering Publishing, pp. 167–173 (2000).
8. Y.J. Dai, E. Nakano, T. Takahashi and H. Ookubo, H., Motion control of leg-wheel robot for an unexplored outdoor environment. Proceedings of the 1996 IEEE/RSJ International Conference on Intelligent Robots and Systems, vol. 2, Nov. 4-8, pp. 402-409 (1996).
9. N. Eiji and N. Sei, Leg-wheel robot: a futuristic mobile platform for forestry industry. IEEE/Tsukuba International Workshop: Can Robots Contribute to Preventing Environmental Deterioration?, Nov. 8-9, pp. 109-112 (1993).
10. K. Suwannasit and S. Laksanacharoen, A BIO-Inspired Hybrid Leg-Wheel robot. TENCON 2004, (2004 IEEE Region 10 Conference), Proceedings Analog and Digital Techniques in Electrical Engineering, 21-24 Nov., Chiang Mai, Thailand (2004).
11. K. Suwannasit and S. Laksanacharoen, A Hybrid Leg-Wheel Robot. The 18th Conference of Mechanical Engineering Network of Thailand, KhonKan, Thailand, October 18-20 (2004).
12. I. Doroftei, C. Marta, C.O. Hamat, L. Suciu, G. Prisacaru, A Hybrid Wheel-Leg Mobile Robot, Annals of DAAAM for 2008, ISSN 1726-9679 (2008).

COMPLIANCE CONTACT CONTROL OF MECANUM WHEELED MOBILE ROBOT FOR IMPROVING SAFETY

Naotaka NISHIGAMI, Naoyuki TAKESUE*

Tokyo Metropolitan University,
6-6 Asahigaoka, Hino, Tokyo, 191-0065, Japan
**E-mail: ntakesue@sd.tmu.ac.jp*

Rikiya MAKINO, Kouhei KIKUCHI, Kousyun FUJIWARA

Toyota Motor Corporation,
Toyota-cho, Toyota, Aichi, 471-8571, Japan

Hideo FUJIMOTO

Nagoya Institute of Technology,
Gokiso-cho, Showa-ku, Nagoya, Aichi, 466-8555, Japan

Recently, robots sharing spaces with humans have been spreading. Therefore, since the situations in which the robots physically contact with humans and/or environments will increase inevitably, the kinetic control in consideration of the circumference is needed. In general, the non-contact sensor is used on mobile robots. However, since such a sensor has a dead angle and it may break down, another way improving the safety is required. Although a force or tactile sensor may be another solution, a lot of sensors are needed in order to compensate a dead angle. Therefore, in this paper, we propose a method to improve the safety by estimating the external force applied on the wheeled mobile robot and by using compliance control based on virtual conveyor. The proposed method doesn't need an additional cost of sensor, and has an advantage because of having no dead angle about driving directions of the robot.

Keywords: Compliance Control, Force Estimation, Force Sensorless, Safety, Mecanum Wheel, Mobile Robot

1. Introduction

Recently, robots sharing spaces with humans have been spreading. Therefore, the situations in which the robots physically contact with humans and/or environments will increase inevitably. If the robot is operated without detecting the collision, there are risks of injuring humans and of damaging the robot itself or environments. As a solution of such problems, the kinetic control in consideration of the circumference is needed.

In general, the non-contact sensor such as a laser range finder which detects an obstacle's approach is used on mobile robots.[1] However, since such a sensor has a dead angle and it may break down, another way improving the safety is required. Although a force or tactile sensor is one of possible solutions,[2,3] such sensor is expensive and a lot of sensors are needed in order to compensate a dead angle. As a result, it is difficult to put the additional sensor for detecting collision from point of view of cost and space.

Therefore, in this paper, we propose a method to improve the safety by estimating the external force applied on the wheeled mobile robot and by using compliance control based on virtual conveyor. The proposed method doesn't need an additional cost of sensor, and has an advantage because of having no dead angle about driving directions of the robot. It can be a complementary protective measure.[4]

2. Mecanum Wheeled Mobile Robot

Figure 1 shows the mecanum wheeled mobile robot used in this study. Figure 2 shows the wiring/circuit diagram of the mecanum wheeled mobile robot. It has two 12V batteries, four motor drivers, four DC motors with gears and rotary encoders on wheel axes. It also has D/A converter, DIO and counter modules that are connected to notebook PC via USB cables. The robot can omnidirectionally move by four mecanum wheels.[5]

The wheel angles are obtained from the rotary encoders. The motor output torques are commanded from PC via D/A and the motor drivers.

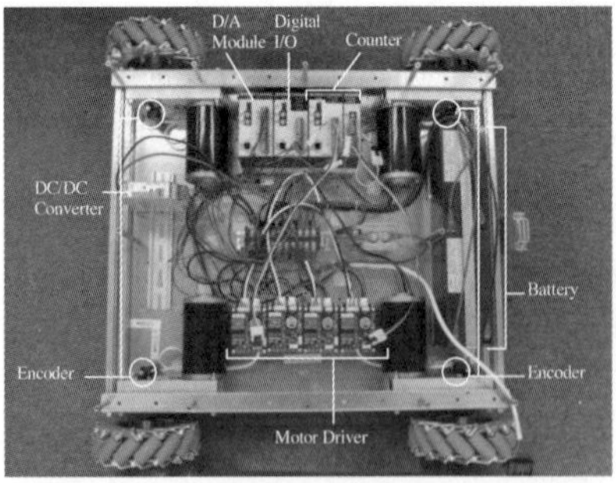

Fig. 1. Mecanum Wheeled Mobile Robot

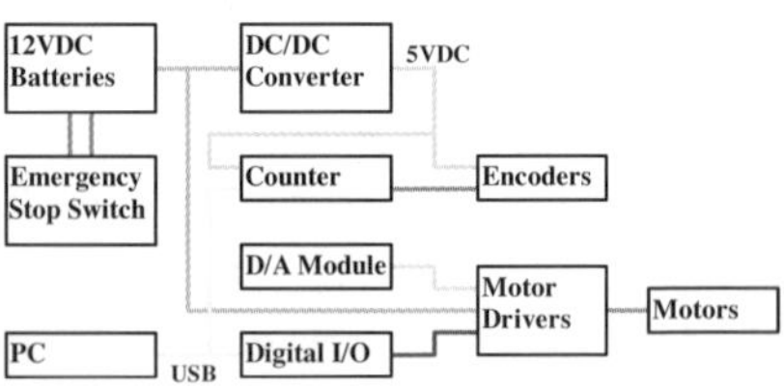

Fig. 2. Wiring/Circuit Diagram

3. Kinematics of Robot

In this section, the kinematics of the wheeled mobile robot is considered. Two coordinate systems are defined as shown in Figs. 3 and 4.

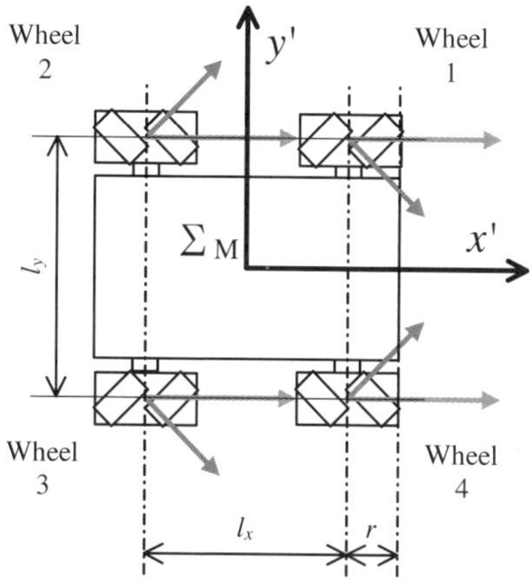

Fig. 3. Robot Coordinate System

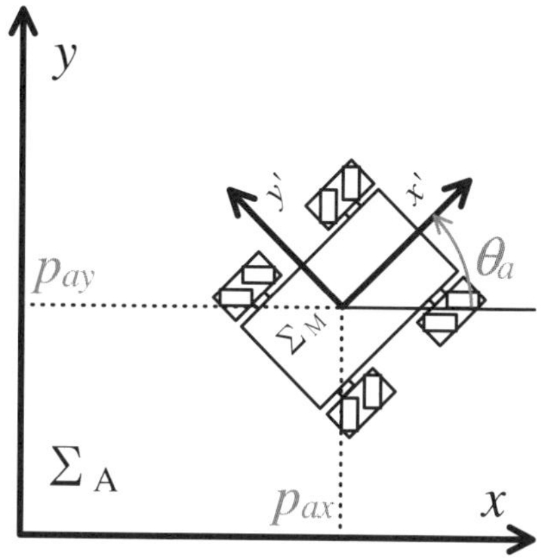

Fig. 4. Absolute Coordinate System

Figure 3 shows the robot coordinate system Σ_M which sets x'-axis as the forward direction and y'-axis as the left direction. l_x indicates the distance between the front and rear wheels, l_y depicts the distance between the left and right wheels and r is the radius of wheel. Four wheels are numbered as shown in Fig. 3. Figure 4 shows the absolute coordinate system Σ_A which sets x, y-axes on the ground surface.

In the robot coordinate, the relation between the vector of the wheel angular velocities $\boldsymbol{\omega}_w = [\omega_1, \omega_2, \omega_3, \omega_4]^T$ and the vector of the robot's velocity $\boldsymbol{v}_m = [v_{mx}, v_{my}, \omega_m]^T$ can be written as follows:

$$\boldsymbol{v}_m = J\boldsymbol{\omega}_w = \begin{bmatrix} a & a & a & a \\ -a & a & -a & a \\ -b & -b & b & b \end{bmatrix} \begin{bmatrix} \omega_1 \\ \omega_2 \\ \omega_3 \\ \omega_4 \end{bmatrix} \tag{1}$$

where J is the Jacobian matrix, $a = r/4$, and $b = r/(2(l_x + l_y))$.

Moreover, in the absolute coordinate, the vector of the robot's velocity $\boldsymbol{v}_a = [v_{ax}, v_{ay}, \omega_a]^T$ can be written as follows:

$$\boldsymbol{v}_a = R_z(\theta_a)\boldsymbol{v}_m = R_z(\theta_a)J\boldsymbol{\omega}_w \tag{2}$$

$$= \begin{bmatrix} \cos\theta_a & -\sin\theta_a & 0 \\ \sin\theta_a & \cos\theta_a & 0 \\ 0 & 0 & 1 \end{bmatrix} \begin{bmatrix} a & a & a & a \\ -a & a & -a & a \\ -b & -b & b & b \end{bmatrix} \begin{bmatrix} \omega_1 \\ \omega_2 \\ \omega_3 \\ \omega_4 \end{bmatrix} \tag{3}$$

On the other hand, in order to move the robot as the desired velocity, the desired angular velocity of wheels $\boldsymbol{\omega}_{wd}$ can be described as follows:

$$\boldsymbol{v}_{md} = R_z(\theta_a)^{-1}\boldsymbol{v}_{ad} \tag{4}$$

$$\boldsymbol{\omega}_{wd} = J^+\boldsymbol{v}_{md} = \begin{bmatrix} c & -c & -d \\ c & c & -d \\ c & -c & d \\ c & c & d \end{bmatrix} \begin{bmatrix} v_{mdx} \\ v_{mdy} \\ \omega_{md} \end{bmatrix} \tag{5}$$

where $\boldsymbol{v}_{md}$ is the robot's desired velocity in the robot coordinate system, $\boldsymbol{v}_{ad}$ is that in the robot absolute system and J^+ is the pseudo inverse Jacobian whose $c = 1/r$ and $d = (l_x + l_y)/r$.

In addition, the statics can be written as follows:

$$\boldsymbol{\tau}_w = J^T\boldsymbol{f}_m \tag{6}$$

where $\boldsymbol{\tau}_w = [\tau_1, \tau_2, \tau_3, \tau_4]^T$ is the vector of torques on wheels, J^T is the transpose of the Jacobian and $\boldsymbol{f}_m = [f_{mx}, f_{my}, m_{mz}]^T$ is the vector of forces on the robot body.

Therefore, if the disturbance torques on the wheels can be observed, the external forces applied on the robot body will be estimated as follows:

$$\hat{\boldsymbol{f}}_m = J^{+T}\hat{\boldsymbol{\tau}}_e \tag{7}$$

$$\hat{\boldsymbol{f}}_m = R_z(\theta_a)^{-1}\hat{\boldsymbol{f}}_a \tag{8}$$

where $\hat{\boldsymbol{f}}_m$ is the vector of estimated external forces on the robot coordinate system, $\hat{\boldsymbol{f}}_a$ is the vector of estimated external forces on the absolute coordinate system, J^{+T} is the transpose of the pseudo inverse Jacobian and $\hat{\boldsymbol{\tau}}_e$ is the vector of estimated external torques on wheels.

In this study, the frictions in the wheel drive system (motors and gears) are identified via experiments and the external torques (disturbance torques) are estimated.

4. External Torque Estimation

In order to estimate the external torque, the friction characteristics of the wheel drive system are identified as the preliminary experiment. The equation of motion of each wheel drive system can be described as follows:

$$I_i\,\dot{\omega}_i + F_{Di}\omega_i + F_{Ei}\,\mathrm{sgn}(\omega_i) = \tau_i + \tau_{ei} \qquad (i = 1 \cdots 4) \tag{9}$$

where I_i is the moment of inertia, ω_i is the angular velocity, F_{Di} is the viscous friction coefficient, F_{Ei} is the Coulomb friction, $\mathrm{sgn}(\omega_i)$ is the sign of ω_i, τ_i is the output torque and τ_{ei} is the external torque. If the angular acceleration and the external torque of each wheel is nearly zero, the friction

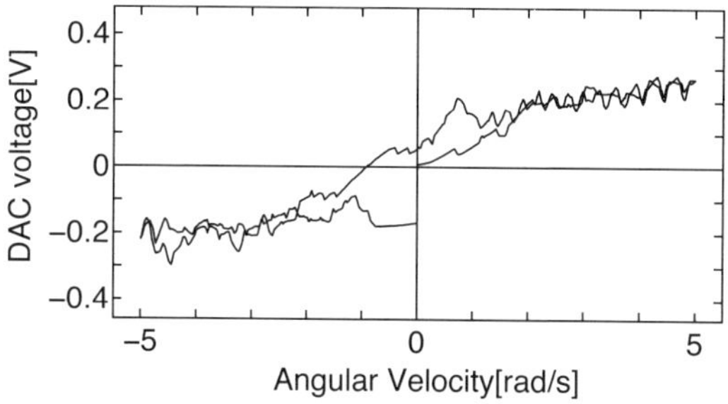

Fig. 5. Relation of Angular Velocity and Commanded Voltage

torque (viscous and Coulomb frictions) can be identified by the output torque of each motor.

In friction characteristics identification experiment, the each wheel is rotated by angular velocity control in the air in which the external torque can be regarded as zero. An experimental result of wheel 1 is shown in Fig. 5.

Figure 5 shows the relation between the angular velocity ω_1 and the commanded voltage of D/A (i.e. the output torque τ_1). Since the wheel was rotated in the air, this figure shows friction characteristics of the wheel drive system. F_i, the friction characteristic of each wheel drive system, can be written as follows:

$$F_i = F_{Di}\,\omega_i + F_{Ei}\,\mathrm{sgn}(\omega_i) \tag{10}$$

In Fig. 5, the y-intercept F_{Ei} represents the Coulomb friction and the slope F_{Di} depicts the viscous friction coefficient. Table 1 shows the identified parameter of each wheel drive system.

Table 1. Identified Parameters

Parameter	Wheel 1	Wheel 2	Wheel 3	Wheel 4
F_{Di} [V/(rad/s)]	0.045	0.036	0.054	0.04
F_{Ei} [V]	0.025	0.02	0.03	0.02

5. Compliance Control of Mecanum Wheeled Mobile Robot

In this section, the compliance control of the wheeled mobile robot is considered.

First, in this study, the compliance motion is defined as follows:

- When the external force is detected, the robot stops immediately by lowering an output.

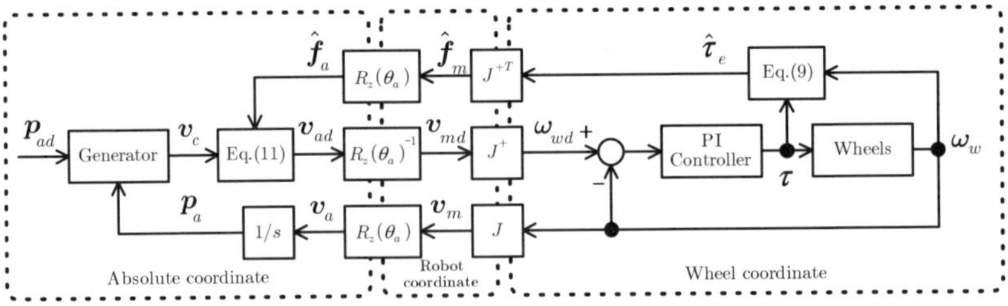

Fig. 6. Control System

- When the external force is detected continuously, the robot continues stopping.
- When the larger external force is detected, the robot should have evacuation motion.
- The robot returns to the original motion smoothly after being released from external force.

To achieve these items, the virtual conveyor is introduced for the compliance control. The virtual conveyor is a control method based on admittance control so that the robot moves on the virtual conveyor. The compliance control based on virtual conveyor realizes smooth acceleration and deceleration.

The dynamical model for the compliance control is described as follows:

$$M_v \dot{\boldsymbol{v}}_{ad} + F_v(\boldsymbol{v}_{ad} - \boldsymbol{v}_c) = \hat{\boldsymbol{f}}_a \tag{11}$$

where M_v is a virtual mass, F_v is a virtual friction including viscous and Coulomb friction, $\boldsymbol{v}_{ad}$ is a desired velocity vector of the robot, $\hat{\boldsymbol{f}}_a$ is an estimated external force vector and $\boldsymbol{v}_c$ is a velocity vector of virtual conveyor. Based on the above model, the desired velocity of the robot, $\boldsymbol{v}_{ad}$, is derived from the estimated force.[6]

Finally, the whole of control system can be illustrated in Fig. 6.

6. Experiments

In this section, the compliance control experiment based on the virtual conveyor method is carried out. Figure 7 shows the target trajectory of the experiment. First, the robot moves 1 meter in the positive direction of y-axis of the absolute coordinate. Next, it goes 1 meter in the negative direction of x-axis after a rotation of 90 [deg]. And, it goes back along its trajectory. Then, a man stands on the way and contacts with the robot for three times.

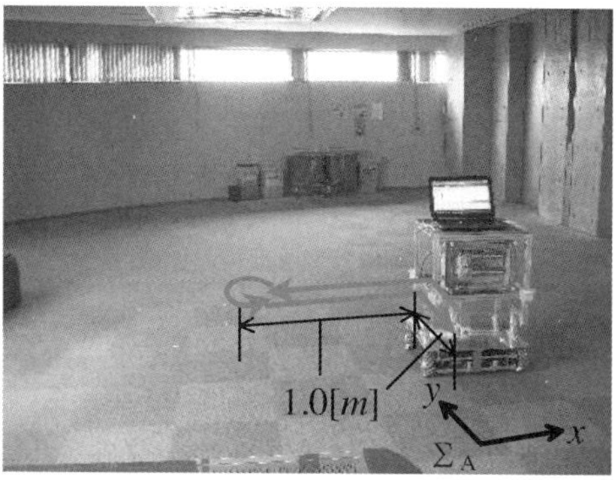

Fig. 7. Target Trajectory of Robot

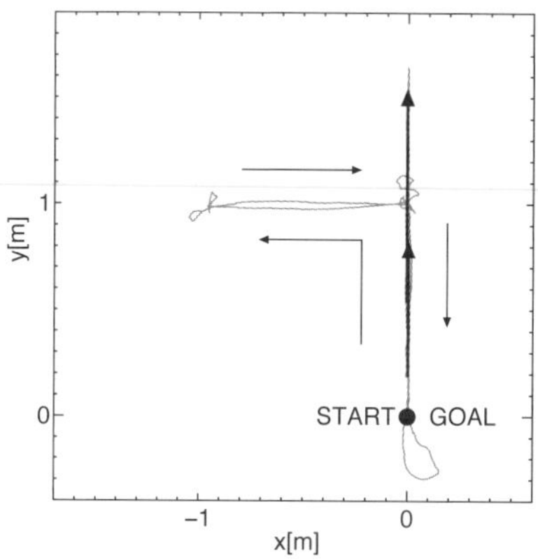

Fig. 8. Trajectory and Estimated External Force (vectors)

The experimental results are shown in Figs. 8 to 10.

Figure 8 shows the trajectory of robot and estimated external force by contact with a man. The gray line indicates the actual trajectory of robot (odometry). The black arrows represent the external force applied to the robot. As seen from the figure, when the robot detected larger external force, it operated evacuation behavior.

Figure 9 shows the output voltage of wheel 1 over time. Figure 10 shows the position on absolute coordinate system and velocity of wheel 1 over time. The dash and dot lines represent the x and y-position of the robot, respectively. The black solid line shows the tangential velocity of wheel 1. As seen from the figures, when a man contacted with the robot, the output voltage, position and velocity were changed to stop and evacuate.

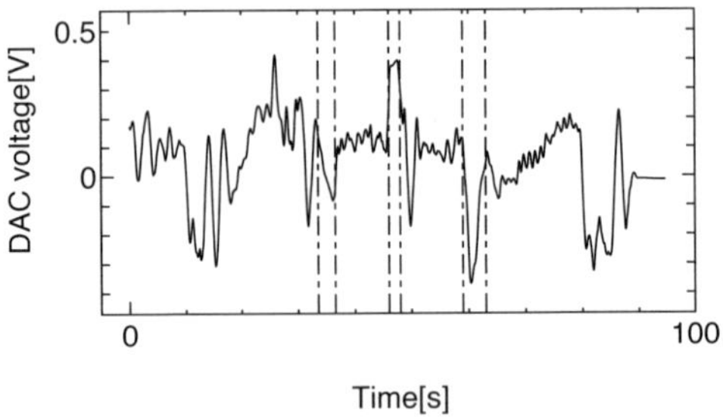

Fig. 9. Output Voltage of Wheel 1

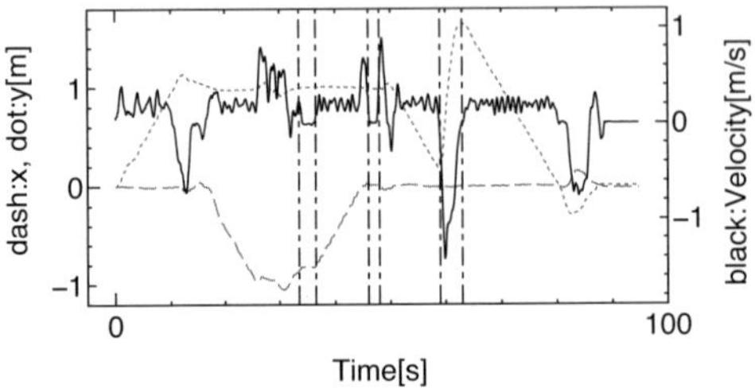

Fig. 10. Position and Velocity Displacement

7. Conclusions

In this paper, in order to improve the safety of mobile robot which shares an environment with humans, we have proposed that the compliance control using estimated external forces is applied to the wheeled mobile robot. The experiments were conducted and the effectiveness of the proposed method was shown. As future works, we will examine the behavior according to ground conditions.

References

1. C. S. Park, et. al.: Characterization of the Hokuyo UBG-04LX-F01 2D Laser Rangefinder, Proc. of IEEE RO-MAN, pp.385-390, 2010.
2. T. Tsuji: Impedance Control of Mobile Robot with Shell-shaped Force Sensor, Proc. of 2009 IEEE Int. Conf. on Mechatronics, pp.1-6, 2009.
3. S. Katsura, K. Ohnishi: Human Cooperative Wheelchair for Haptic Interaction Based on Dual Compliance Control, IEEE Trans. on Industrial Electronics, Vol.51, No.1, pp.221-228, 2004.
4. ISO 12100, Safety of machinery—General principles for design—Risk assessment and risk reduction, 2010.
5. P. Viboonchaicheep, et. al.: Position Rectification Control for Mecanum Wheeled Omni-directional Vehicles, Proc. of IEEE IECON, pp.854-859, 2003.
6. R. Kikuuwe, N. Takesue, A. Sano, H. Mochiyama and H. Fujimoto: Admittance and Impedance Representations of Friction Based on Implicit Euler Integration, IEEE Trans. on Robotics, Vol.22, No.6, pp.1176-1188, 2006.

NEURAL CONTROL OF A THREE-LEGGED RECONFIGURABLE ROBOT WITH OMNIDIRECTIONAL WHEELS

P. MANOONPONG* and F. WÖRGÖTTER

Bernstein Center for Computational Neuroscience,
The Third Institute of Physics, University of Göttingen,
Göttingen 37077, Germany
**E-mail: poramate@physik3.gwdg.de*

P. LAKSANACHAROEN

Mechanical and Aerospace Engineering Department,
King Mongkut's University of Technology North Bangkok,
Bangkok 10800, Thailand
E-mail: stl@kmutnb.ac.th

In this article, we present neural control of a three-legged reconfigurable robot with omnidirectional wheels. It is systematically synthesized based on a modular structure such that the neuromodules are small and their structure-function relationship can be understood. The resulting network consists of four main modules. A so-called minimal recurrent control (MRC) module is for sensory signal processing and state memorization. It directly drives the motion of two front wheels while a rear wheel is indirectly controlled through a velocity regulating network (VRN) module. In parallel, a simple neural oscillator network module serves as a central pattern generator (CPG) producing basic rhythmic signals for sidestepping where stepping directions are controlled by a phase switching network (PSN) module. The combination of these neuromodules generates various locomotion patterns. Applying sensor inputs to the neural controller enables the robot to avoid obstacles as well as a corner. The presented neuromodules are developed and firstly tested using a physical simulation environment, and then finally transferred to the real robot.

Keywords: Neural networks, Mobile robot control, Autonomous robots, Obstacle avoidance.

1. Introduction

During the last years, we have developed a physical three-legged reconfigurable robot with omnidirectional wheels.[1] It combines the concept of using legs, wheels, and rolling sphere for multi-locomotion modes. Due to its closed-spherical shape, it can roll passively where this rolling motion can minimize the friction and lead to energy efficiency.[2] For autonomous exploration, it will transform into two inter-connected hemispheres with extending its three legs for locomotion using wheels. If the wheels are broken, it could use the legs for further locomotion. To the best of our knowledge, this type of robot, which combines legs, wheels, and a rolling sphere for multi-modal locomotion, so far has not been developed by others. In general, there are several leg-wheel hybrid robots but without rolling sphere[3,4] while there are spherical rolling robots but without legs and wheels.[2,5]

Continuing the development of our robot system, this article presents neural control of the robot for the generation of active locomotion using wheels and legs as well as controlling a reactive obstacle avoidance behavior. Neural control exhibits dynamical features (e.g., periodic and hysteresis behavior) which are here exploited for the locomotion generation and robot behavior control. This neural network control also has a modular structure consisting of four modules. Due to its modularity, the controller is robust to changes of structures; i.e., modules can be completely removed leading to graceful degradation of the agent's functionality while as a whole the system can still function partially. We believe that our neural modules can be important components for locomotion generation in other complex robotic systems or they can serve as useful modules for other module-based neural control applications.

2. Three-Legged Reconfigurable Robot with Omnidirectional Wheels

The robot consists of three legs, omnidirectional wheels, and a body with two (hemi)spherical shells (Fig. 1). It has a total of eight DC servo motors. There are two motors at each leg where each of them moves the leg up and down and drives the wheel. There are two motors at the hinge (middle) joint for a transformation process. The process will allow one side hemisphere to open up another side of hemisphere at a time.

The robot is generally designed based on the concept of a spherical form where its three identical legs with the wheels are kept inside its shells (body) in order to perform passive rolling motion. This form (called dormant mode, Fig. 1(a)) provides the compact shape of the robot. For active locomotion, it will transform into two hemispherical shells where the wheeled legs are projected out of the shells (called transformed mode, Fig. 1(b)). To assure the proper positioning that the robot is allowed to split into two hemispheres, we use the data read from an accelerometer at the onset of an expansion sequence.[1] The robot also has two infrared (IR) sensors installed at its front

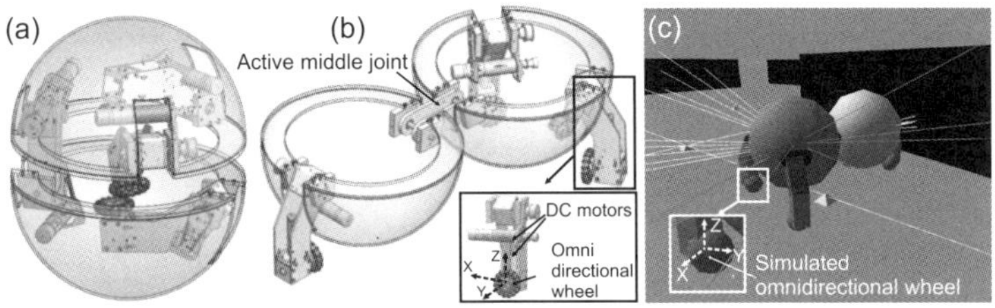

Fig. 1. (a), (b) The physical robot in dormant and transformed modes, respectively. (c) Simulated robot. To simulate the omnidirectional wheels of the robot, we set friction coefficients for two orthogonal directions (x- and y-axes) of each wheel independently. As a consequence, the wheel shows the unique property of rolling freely in the direction of its axis (i.e., freely rotating around x-axis), while operating as a normal wheel in the direction perpendicular to its axis (i.e., actively rotating around y-axis).

to detect obstacles. We use the physics simulation environment called Yet Another Robot Simulator (YARS)[6,7] to simulate the robot (Fig. 1(c)). The simulated robot is qualitatively consistent with the real one in the aspect of geometry, dimensions, mass distribution, motor torque, and sensors while frictional coefficient between the robot and ground is roughly estimated. Note that we simulate spherical shells for body parts instead of hemispherical shells since YARS has not yet provided a hemispherical geometry. However, the mass distribution and the total weight of the simulated and real robots are qualitatively consistent.

3. Neural Control for Locomotion and Obstacle Avoidance

Neural control is based on a modular structure (Fig. 2). It consists of four main modules: a minimal recurrent control network (MRC), a velocity regulating network (VRN), a neural oscillator network (abbreviated CPG, see below), and a phase switching network (PSN). The neural modules have been developed and applied to four-, six- and eight-legged robots as well as two wheeled robots in part.[6–8] Here, we for the first time combine them for wheeled and legged locomotion and a reactive obstacle avoidance behavior of a three-legged reconfigurable robot with omnidirectional wheels. We only discuss their functions used for the application here since the details of the neural modules together with the setup of their weights have already been presented in previous studies.[6–8] All neurons of the controller (i.e., hidden and motor neurons (Fig. 2)) are modelled as non-spiking neurons. Their activity develops according to $a_i(t+1) = \sum_{j=1}^{n} w_{ij}\, o_j(t) + b_i$; $i = 1, \ldots, n$ where n denotes the number of units, b_i represents a fixed internal bias term of neuron i, a_i their activity, w_{ij} the synaptic strength of the connection

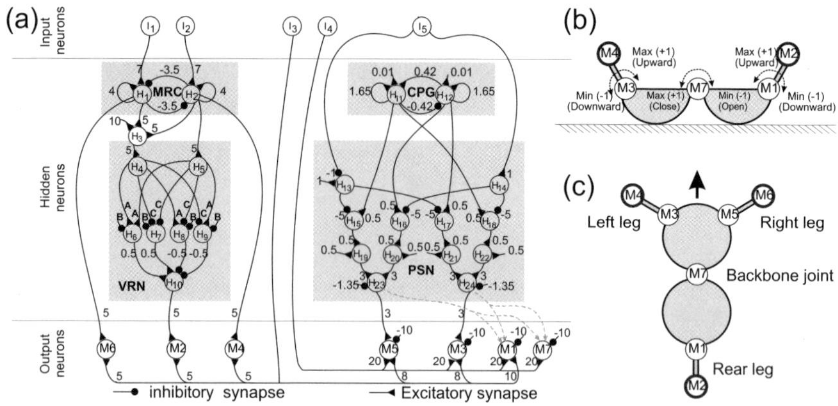

Fig. 2. (a) Neural control of the three-legged reconfigurable robot with omnidirectional wheels. There are three different neuron groups: input, hidden, and output. Input neurons I receive sensory signals controlling robot behavior ($I_{1,2,3,4,5}$). Hidden neurons H are divided into four subgroups or modules (MRC, VRN, CPG, and PSN) having different functionalities (see text for details). Output neurons are described as motor neurons ($M_{1,...,7}$) where $M_{1,3,5}$ are for controlling leg joints, $M_{2,4,6}$ are for controlling wheels, and M_7 is for controlling the body joint. All connection strengths together with bias terms are indicated by the small numbers except some parameters of the VRN given by $A = 1.7246$, $B = -2.48285$, $C = -1.7246$. Note that dashed arrows indicate additional synapses which can be added to obtain more locomotion behaviors. (b) The movements of the leg joints and the body joint. (c) The location of the motor neurons on the robot.

from neuron j to neuron i. The neuron output o_i is given by a hyperbolic tangent (tanh) transfer function $o_i = \tanh(a_i) = \frac{2}{1+e^{-2a_i}} - 1$. Input neurons $I_{1,2,3,4,5}$ are here configured as linear buffers ($a_i = o_i$).

The MRC module has been originally evolved through the evolutionary algorithm ENS[3] for generating obstacle avoidance behavior of a miniature Khepera robot with two wheels.[6] Due to its recurrent connections, the MRC exhibits hysteresis effects which allow an agent to keep on doing a task till the task is completed even if the stimulus has decayed or is removed. The hysteresis phenomena have already been discussed as models for short-term memory.[9] Without such memory, the agent might switch between tasks reactively without completing any of them and, thus fails to complete tasks. Here, we apply the MRC to directly drive the two front wheels (M_4, M_6) of our robot and exploit its hysteresis effects for controlling obstacle avoidance behavior and filtering sensory noise. We use two IR sensor signals $IR_{1,2}$ for obstacle detection at the front of the robot. They are transmitted to the inputs $I_{1,2}$ projecting to the MRC. I_1 corresponds to the left IR sensor signal and I_2 to that of the right one. Applying the output signals of H_1 and H_2 directly to their target motor neurons M_6, M_4 and indirectly to the motor neuron M_2 (Fig. 2) via the VRN module enables the robot to autonomously change its motion. For instance, the robot changes from moving forward to turning left when there are obstacles on the right, and vice versa. This way, it can avoid obstacles and escape from corners as well as deadlock situations.

The VRN module[7] basically changes the rotational direction of M_2 with respect to turning motion. The network approximately works as a multiplication operator. It was constructed as a feedforward network with two input, four hidden, and one output neurons. It was trained by using the backpropagation algorithm. The neuron H_3 is added to combine both outputs of the MRC. The output of H_3 projects to the input neuron H_4 of the VRN. Another input neuron H_5 of the VRN receives its input from the neuron H_2 of the MRC. By doing so, the VRN drives the rear wheel M_2 with low ≈ -1 activation to rotate clockwise leading to a right turn when there are obstacles on the left (i.e., the output of H_1 of the MRC is $\approx +1$ while the output of H_2 of the MRC is ≈ -1) and vice versa. In case no obstacle is detected (i.e., the outputs of $H_{1,2}$ are ≈ -1), the rear wheel will be inactive (zero activation) leading to forward motion.

The CPG module[7] generates basic rhythmic signals which control leg movements (M_3, M_5) in directly through the PSN module. The leg movements result in sidestepping. The CPG is realized by using the dynamics of a simple 2-neuron network with full connectivity and biases. Its weights were adjusted such that the network generates periodic attractors.[7] The network with the resulting weights produces rhythmic outputs that differ in phase by $\pi/2$ with a frequency of approximately 0.8 Hz. According to the robot configuration, making the front legs moves out of phase to each other by $\pi/2$, we obtain sidestepping.

The PSN module[8] is used to steer the sidestepping directions (i.e., lateral motions to the left and right). The network basically switches the phase of the two rhythmic signals originally coming from the CPG to lead or lag behind each other by $\pi/2$ in phase when the input I_5 is changed from 0 to 1 and vice versa. By applying this network property, the movements of the left and right legs will be reversed corresponding to the modification of I_5. Consequently, the robot will change its sidestepping directions from the right to the left and vice versa. The PSN is a hand-designed feedforward network consisting of four hierarchical layers with 12 neurons. The synaptic weights and bias terms of the network were determined in a way that they do not change the periodic form of its input signals and keep the amplitude of the signals as high as possible.

4. Experiments and Results

The first experiment presents locomotion behaviors of the robot using legs and wheels in the physics simulator (YARS). In this case, all input neurons ($I_{1,2,3,4,5}$, Fig. 2) were set manually to clearly observe the robot locomotion behaviors. The inputs $I_{1,2}$ were set to -1 and the input I_4 was set to 0 (i.e., the transformed mode with forward motion). We let the robot move over flat terrain and continuously changed inputs $I_{3,5}$ to investigate its basic locomotion. By simply controlling the input I_3, the robot changes its locomotion from using wheels to legs and vice versa. During legged locomotion, changing the input parameter I_5 from zero to one leads to sidestepping to the left and changing I_5 back to zero leads to sidestepping to the right. The input parameters and motor signals during the experiment are shown in Fig. 3. The video clip of this experiment showing forward motion using wheels and sidestepping using legs can be found at http://www.manoonpong.com/CLAWAR2013/S1.wmv.

The second experiment shows the performance of the controller implemented on the real robot. Figure 4(a) shows transformation behavior where the robot transformed from the dormant mode (Fig. 1(a)) to the transformed mode (Fig. 1(b)). In this case, the input I_4 was set from 1 to 0 while the other inputs were set to 0. Note that the expansion is automatically stopped as soon as the legs reach a desired position determined by the potentiometers of the leg joints. After this transformation, the robot will be able to locomote using wheels or legs. Figure 4(b) shows wheeled locomotion with a reactive obstacle avoidance behavior. For this scenario, the inputs $I_{1,2}$ received IR sensory signals while the inputs $I_{3,4,5}$ were set to 0. The robot moved toward a corner and then autonomously turned left since its IR sensor detected the right wall. Due to the hysteresis effects of the MRC, the effects allow the robot to keep on turning for longer than the stimulus itself (i.e., IR signal). This way, the robot performed a large turning angle; thereby easily avoiding the corner. In contrast, using finite state control or classical Braitenberg control[10] without state memory the robot needs to turn several times in order to avoid obstacles or it sometimes gets stuck. Figure 4(c) exemplifies legged locomotion (i.e., sidestepping to the right). The robot performed rhythmic leg movements where the inputs were set as $I_{1,2} = -1$, $I_3 = 1$, $I_{4,5} = 0$. Due to mechanical problems of this first prototype robot (i.e., backlash and slip of the leg driving mechanisms), its legs cannot follow the motor commands all the times as expected. As a result, sidestepping using its legs cannot be effectively performed.

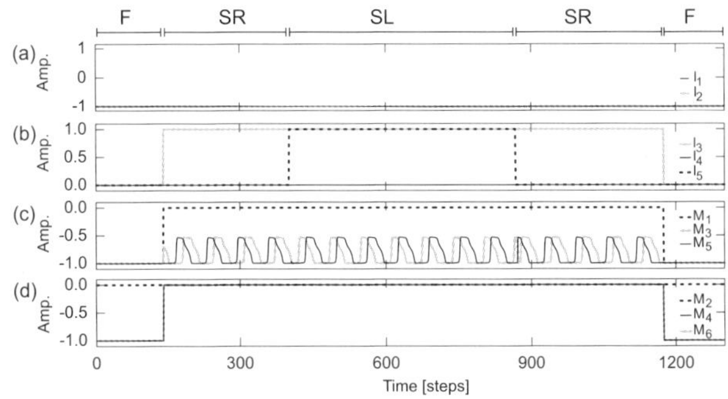

Fig. 3. (a) Inputs $I_{1,2}$ were set to -1 at all times resulting in only forward motion. (b) Inputs $I_{3,4,5}$. I_3 was used to switch between wheeled ($I_3 = 0$) and legged ($I_3 = 1$) locomotion. I_4 was here set to 0 in order to keep the robot in the transformed mode. Setting I_4 to 1 leads to the dormant mode. I_5 is used to steer the sidestepping directions. Setting I_5 to 0 leads to the sidestepping to the right SR and setting it to 1 leads to the left SL. (c) Motor signals at the leg joints ($M_{1,3,5}$). The motors $M_{3,5}$ show the periodic signals when the legged locomotion mode was activated. One can observe that when the robot steps sideways to its right the periodic signal of M_5 leads the one of M_3 by $\pi/2$ in phase and vice versa when the robot steps sideways to its left. (d) Motor signals at the wheels ($M_{2,4,6}$). Low ≈ -1 activation drives the wheels such that the robot moves forward F while zero activation means the wheels roll freely. Backward motion using the wheels can be achieved by setting $I_{1,2}$ to 1 and $I_{3,4,5}$ to 0.

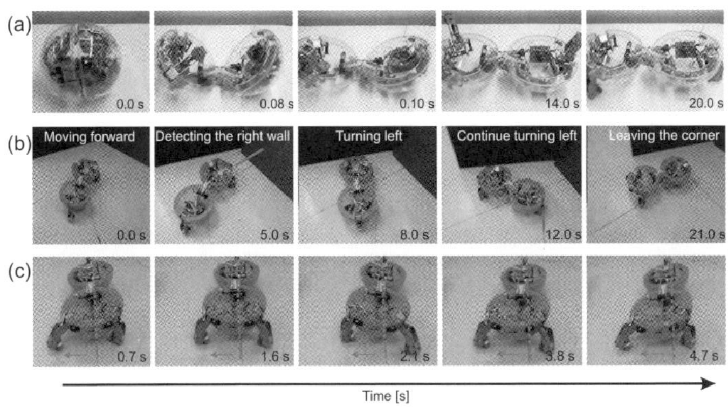

Fig. 4. (a) Transformation from the dormant mode to the transformed mode. (b) Obstacle avoidance behavior using wheels. (c) Sidestepping to the right using legs. The video clip of the tests including passive rolling motion can be see at http://www.manoonpong.com/CLAWAR2013/S2.wmv.

5. Conclusion

We presented neural control of a leg-wheel robot having two (hemi) spherical body shells and three legs with omnidirectional wheels. The controller was designed as a modular structure composed of four modules (MRC, VRN, CPG, and PSN). The MRC and CPG modules were developed by realizing dynamical properties of recurrent neural networks while the VRN and PSN modules were developed as feedforward networks. This neural controller generates active locomotion behaviors using wheels (i.e., forward, backward, turn left/right) and legs (i.e., sidestepping) as well as a reactive obstacle avoidance behavior. The behaviors were activated through the five inputs of the controller. Besides the active locomotion behaviors, the robot can perform passive rolling using its closed-spherical body.

Our next step will be the improvement of the leg driving mechanisms to obtain a better legged locomotion. In addition to this, we will use proprioceptive sensors (i.e., rotational sensors of wheels and joint angle sensors of leg joints) for damage detection and apply neural learning[11] to find useful motor responses after damage.

Acknowledgments: This research was supported by Emmy Noether grant MA4464/3-1, BCCNII grant 01GQ1005A (project D1), and the Office of the Higher Education Commission, Thailand. We thank Natthaphon Bunathuek for his help in real robot experiments.

References

1. N. Chadil, M. Phadoognsidhi, K. Suwannasit, P. Manoonpong and P. Laksanacharoen, A reconfigurable spherical robot, in *Proc. IEEE Int. on Robotics and Automation (ICRA)*, 2011.
2. Y. Kim, S. Ahn and Y. Lee, Kisbot: new spherical robot with arms, in *Proc. of Int. on Robotics, control and manufacturing technology (WSEAS)*, 2010.
3. M. Eich, F. Grimminger, S. Bosse, D. Spenneberg and F. Kirchner, Asguard: A hybrid legged wheel security and sar-robot using bio-inspired locomotion for rough terrain, in *Proc. of the IARP/EURON Workshop*, 2008.
4. S. Nakajima and E. Nakano, *J. Robot Mechatronics* **20**, 912 (2008).
5. R. Armour and J. Vincent, *J. Bionic Eng.* **3**, 195 (2006).
6. M. Hülse, S. Wischmann and F. Pasemann, *Connect. Sci.* **16**, 249 (2004).
7. P. Manoonpong, *Neural preprocessing and control of reactive walking machines: Towards versatile artificial perceptionaction systems* (Springer, 2007).
8. P. Manoonpong, F. Pasemann and F. Wörgötter, *Robot. Auton. Syst.* **56**, 265 (2008).
9. E. Harth, T. Csermely, B. Beek and R. Lindsay, *J. Theor. Biol.* **26**, 93 (1970).
10. V. Braitenberg, *Vehicles: Experiments in synthetic psychology* (MIT, 1984).
11. P. Manoonpong, C. Kolodziejski, F. Wörgötter and J. Morimoto, *Adv. Complex Syst.* (2013).

THE PASSIVE DYNAMICS OF WALKING AND BRACHIATING ROBOTS: RESULTS ON THE TOPOLOGY AND STABILITY OF PASSIVE GAITS

N. ROSA Jr.* and K. M. LYNCH

Mechanical Engineering, Northwestern University,
Evanston, IL 60208, USA
**E-mail: nr at u.northwestern.edu*

Simple walking models, like the compass-gait model, have yielded useful insight into the basic mechanics of walking. A similar model serves as a template for brachiation. With the ability of a two-link robot to walk and swing, we explore the multi-locomotion capability of a generalized two-link model with potential footholds at any location. We focus on the connected components of passive gaits in a five-dimensional state-time space. Our main results are: (1) a walking gait and a brachiating gait cannot be in the same connected component and (2) the stability of a gait depends on whether impacts are state-based (e.g., footfall in a biped walker) or time-based (e.g., time between clamping brachiator hands to a wall). For the same connected component of gaits, the different impact types result in different bifurcations.

Keywords: brachiation; walking; compass-gait; simplest walking model

1. Introduction

We study the passive hybrid dynamics of planar, single-joint, two-link walking and brachiating robot models. These simplified models of walking and brachiating have been studied extensively in the past.[1–4] These locomotors are modeled as a hybrid system, where two continuous swing motions are separated by an impact with a surface causing a discontinuous jump in velocity.

We generalize the two-link walking models[1,3] and brachiator models[2,5] to a single model that can both walk and brachiate. This allows us to study robots like our wall-climbing Gibbot robot,[4] which can achieve a "foothold" anywhere on a vertical plane. Since the "feet" of the generalized model can achieve a foothold at any location on a vertical plane, we can study two impact strategies: switching the stance foot when (1) it "collides" with a fixed virtual slope (state-based impacts) or (2) a set period of time has elapsed (time-based impacts). Figure 1 shows a simple two-link model

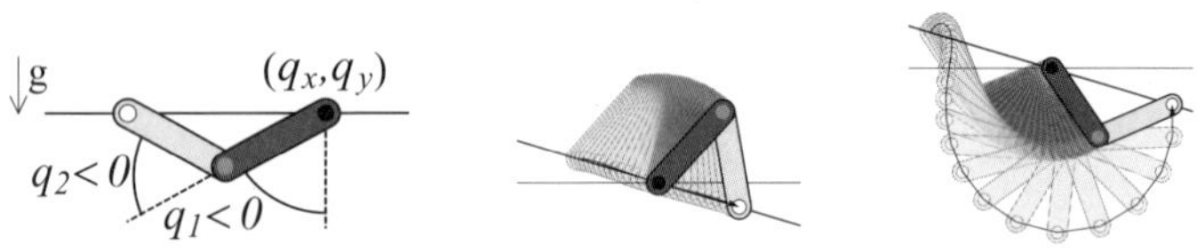

(a) the model (b) a walker (c) a brachiator

Fig. 1. An example of (a) the two-link model (b) walking and (c) brachiating on the same downhill slope. The swing leg is the lightly shaded link.

walking and brachiating on the same downhill slope.

In the past, simple two-link models have successfully guided the design of several walking and brachiating robots.[5–8] Yet, with the exception of work by Fukuda et al.,[9] walking and brachiating robots are treated separately because of their distinct environments. Given our generalized model, could walking and brachiating gaits actually be considered examples of the same type of gait? Furthermore, the same gait can be achieved by either state-based or time-based impact. But is there a difference in stability depending on the type of impact?

We address these two questions by studying the connected components and stability of gaits in our generalized model. We define a gait as a period-one fixed point of the hybrid dynamics in a five-dimensional state-time space. For a period-one fixed point, the state corresponds to the robot's post-impact state at the beginning of a new swing and time corresponds to the period of time between impacts. Gaits in the same connected component have a path connecting them.

Definition 1. A *path* between two gaits a and b is a continuous function f from the unit interval $[0, 1]$ to the set of gaits in $\mathbb{R}^5$ such that $f(0) = a$ and $f(1) = b$.

If a path exists between two gaits, then it is possible to continuously deform one gait into the other. We give an example in Section 3.

Similar to the terminology used in the walking community, we refer to the link that is in contact with the surface as the stance leg and the other link as the swing leg. After an impact the roles of the two links switch. For motions of the swing leg relative to the stance leg where the net displacement does not exceed one revolution during a step, the net displacement can only have two possible values. These values map the swing leg trajectory to a net motion where the links cross once or not at all. We define walking and brachiating gaits based on this notion of links crossing.

Definition 2. The links *cross* when the net angular displacement of the swing leg relative to the stance leg does not equal the difference between the final and initial angles of the swing leg relative to the stance leg. For a *walking* gait, the links cross. For a *brachiating* gait, the links do not cross.

We return to this definition in Section 3. Our contributions are

(1) **Walking and brachiating fixed points are not in the same connected component of gaits.** We show that walking and brachiating gaits are two disjoint sets. We explore a connected component in each set through numerical simulation and show that both gaits can passively locomote above or below a fixed slope. For example, we show that it is possible to continuously deform a walking gait that walks on ground to a gait that passively "walks" on an inclined ceiling.
(2) **The impact strategy affects the stability of a gait.** We find that state-based switching leads to more stable gaits than time-based switching. When gaits become unstable, state-based switching leads to period-doubling bifurcations (stable period-$2n$ gaits, $n \geq 1$), and time-based switching leads to Neimark-Sacker bifurcations (n-periodic or quasiperiodic gaits, $n > 1$).

We define the hybrid dynamics of the system in Section 2. In Sections 3 and Section 4 we show (1) and (2), respectively. We conclude in Section 5.

2. The Hybrid Dynamics

The physical parameters of each link of the robot are identical to each other, which allows us to define a gait in half a swing of the robot. The configuration vector (Figure 1(a)) of the robot is $q' = [q_x, q_y, q_1, q_2]^T$, where (q_x, q_y) are the x-y coordinates of the stance leg in a world frame, q_1 is the angle of the stance leg from the vertical, and q_2 is the angle of the swing leg relative to the stance leg. The pivot point (q_x, q_y) remains fixed throughout the swing motion. For convenience, during the swing we use a reduced configuration vector $q = [q_1, q_2]^T$ with state vector $x = [q^T, \dot{q}^T]^T$. We define the flow of the continuous dynamics of the double pendulum as

$$x(x_0, t) = x_0 + \int_0^t F(x(s))\, ds,$$

where x_0 is the initial state of the robot at time $s = 0$, t is the impact time, and $F(x(s)) = [\dot{q}(s)^T, \ddot{q}(s)^T]^T$.

When the robot impacts, it undergoes an instantaneous, plastic impact. The configuration of the robot does not change, but its velocity does. The impact map H relates the pre-impact state to the post-impact state of the robot at the time of impact and is defined as

$$H(x) = \begin{bmatrix} A & 0 \\ 0 & P(q) \end{bmatrix} x + \begin{bmatrix} b^T \; 0 \; 0 \end{bmatrix}^T ,$$

where $A = \left[\begin{smallmatrix} 1 & 1 \\ 0 & -1 \end{smallmatrix}\right] \in \mathbb{R}^{2x2}$ and $b = \left[\begin{smallmatrix} \pi \\ 0 \end{smallmatrix}\right] \in \mathbb{R}^2$ flip the coordinate system, and $P(q) \in \mathbb{R}^{2x2}$ relates the pre-impact velocity $\dot{q}^-$ to the post-impact velocity $\dot{q}^+$ such that $\dot{q}^+ = P(q^-)\dot{q}^-$. A detailed derivation of the equations and parameters used for our two-link model can be found in a previous paper.[4]

In sum, the hybrid system is written as a discrete map $G\colon \mathbb{R}^4 \times \mathbb{R} \to \mathbb{R}^4$

$$x_1 = H(x(x_0, t)) = G(x_0, t),$$

where x_0 is the initial state, $x(x_0, t)$ is the pre-impact state, and x_1 is the final state after the impact at time t. A "gait" is a period-one fixed point (x_0^*, t^*) satisfying $x_0^* = G(x_0^*, t^*)$. The fixed point is stable if the maximum eigenvalue of $\frac{\partial G}{\partial x_0}(x_0^*, t^*)$ is inside the unit circle.

3. Connected Components in the State-Time Space

In our search for gaits, we allow the state and impact time to vary. In this five-dimensional state-time space, we define the set of all gaits as

$$X = \left\{(x_0, t) \in \mathbb{R}^5 \mid G(x_0, t) - x_0 = 0\right\} ,$$

where x_0, t, and G are defined in Section 2. Let $S = (X, \mathcal{O} \cap X)$ be a subspace topology of the standard topology $T = (\mathbb{R}^5, \mathcal{O})$, where $\mathcal{O}$ is the collection of open sets in $\mathbb{R}^5$. By the implicit function theorem, if the Jacobian matrix

$$J(x_0^*, t^*) = \begin{bmatrix} \frac{\partial G}{\partial x_0} & \frac{\partial G}{\partial t} \end{bmatrix} (x_0^*, t^*) - \begin{bmatrix} \frac{\partial x_0}{\partial x_0} & \frac{\partial x_0}{\partial t} \end{bmatrix}$$

has maximal rank at a fixed point (x_0^*, t^*), then nearby solutions form a 1-D curve passing through (x_0^*, t^*) in the state-time space. If $J(x_0^*, t^*)$ does not have maximal rank, then locally the nature of the connected component requires further analysis. It is at these fixed points where multiple 1-D curves can intersect. We trace the connected components of X using an Euler-Newton numerical continuation method.[10] There are many connected components in the 5-D state-time space. Figure 2 shows parts of two connected components projected onto a 2-D subspace of the state-time space. The connected component of walking gaits in Figure 2(a) is comprised of three 1-D curves glued together at two fixed points $([\pi, \pi, 0, 0]^T, 1.59)$ and

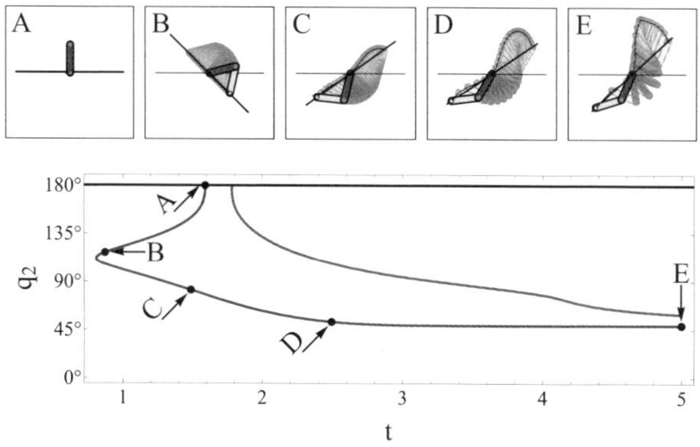

(a) part of a walking connected component

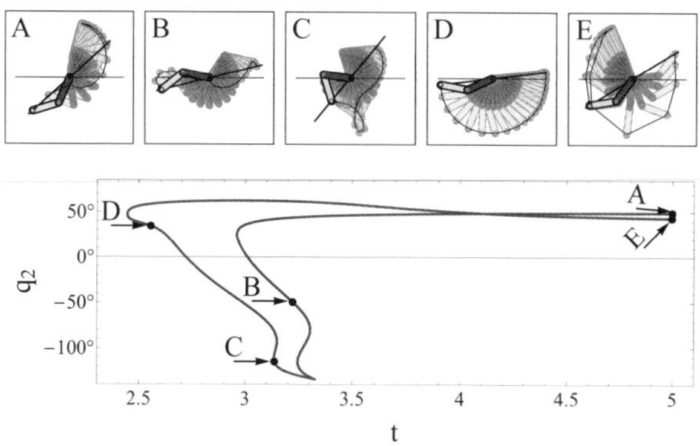

(b) part of a brachiating connected component

Fig. 2. Parts of two connected components projected onto the q_2-t plane. Every point is a gait in the state-time space. The cartoon animations are the swing motions at points A-E. The walking connected component has an infinite number of solutions[3] branching off the curve at $q_2 = 180°$, while the brachiating curve does not have branching solutions.

$([\pi, \pi, 0, 0]^T, 1.78)$, where $J(x_0^*, t^*)$ is not maximal rank. The black, straight-line curve at $q_2 = 180°$ consists of the state $[\pi, \pi, 0, 0]^T$ for all switching times t^*. The state is an equilibrium point of the double pendulum dynamics. For the connected component of brachiating gaits in Figure 2(b), $J(x_0^*, t^*)$ evaluated at every fixed point along the curve has maximal rank. Because of this, the curve has a unique arc-length parameterization.[11] There are other 1-D solution curves of brachiating gaits (see, e.g., Figure 4), but these curves are not connected to each other.

In the Introduction, we defined walking and brachiating based on whether the links cross or not. For a trajectory to be a fixed point, the

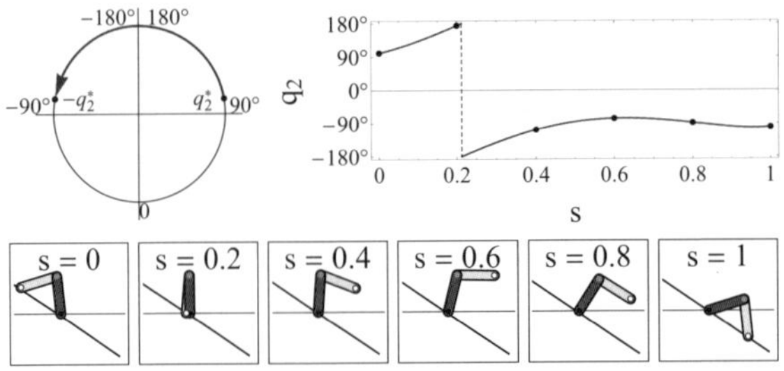

(a) walking

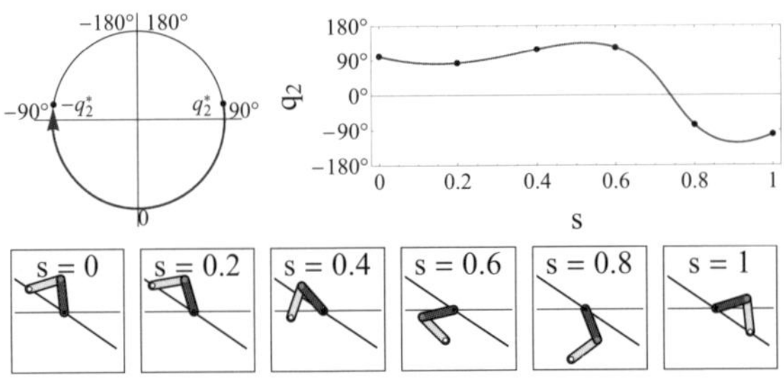

(b) brachiating

Fig. 3. The trajectory of the swing leg of two gaits with the same initial configuration $q \approx (-174.1°, 100.6°)$ are shown. The arrow on the circle corresponds to the net displacement of the swing leg relative to the stance leg, the plot is the swing leg's trajectory scaled to each gait's switching time, and the animations are snapshots of the robot's trajectory over time. The net displacement of the walking gait (a) crosses 180° (i.e., the links cross), while the brachiating gait (b) crosses zero (i.e., the links did not cross).

angle of the swing leg must equal $q_2(t) = -q_2(0) = -q_2^*$ prior to impact. This puts a constraint on the net angular displacement of the swing leg

$$\Delta q_2 = \int_0^t \dot{q}_2(s)ds = 2\pi k - 2q_2^*,$$

where k is an integer. For $|\Delta q_2| \leq 2\pi$, the net angular displacement has the intuitive meaning of the links crossing or not crossing (Figure 3). If $q_2^* \in [-\pi, \pi]$,[a] then $k \in \{-1, 0, 1\}$, where $k = 0$ is a brachiating gait and $|k| = 1$ is a walking gait. We have now partitioned the set of fixed points into two sets. Let B and W be the set of fixed points with values of $|k|$ equal to 0 and 1, respectively. By construction, we have that $B \cup W = X$. If there is a continuous path between fixed points in B and W, then there must exist a fixed point on the path that is in both sets. Such a fixed

[a]Removing $-\pi$ or π excludes brachiating either counterclockwise or clockwise at $q_2^* = \pi$.

point cannot exist, because this would mean that the same state has two different trajectories—one where the links cross and another where they do not. This cannot be true as solutions to the differential equation of the double pendulum are unique. Hence, $B \cap W = \emptyset$ and no path can exist between gaits in these sets.

While a continuous path cannot exist between a walking gait and a brachiating gait, connected components in both B and W connect gaits that locomote above and below the surface. Figure 2(a) shows a path in W that continuously deforms a walking gait starting at the gait labeled B to what is often referred to as "over-hand"[9] brachiation (gait C). Topologically, this gait is equivalent to a gait that "walks" below the surface.

4. Stability and Bifurcations

The switching time t plays an interesting role in our model. If we treat it as a free parameter, then we can impact whenever we want. If instead the robot impacts when it returns back to its initial slope σ, then $t = t(x_0)$ is dependent on the state. Figure 4 shows the stability of a 1-D curve of brachiating gaits under the two switching strategies as we move along the curve (this curve does not intersect the curve in Figure 2(b), which does not have stable time-based gaits). In the example of Figure 4, the set of stable time-based switching gaits is a subset of the set of stable state-based switching gaits. In our experience, all gaits that are stable under time-based switching are also stable under state-based switching, but the converse is not true. While both switching strategies give rise to fold bifurcations, we have also

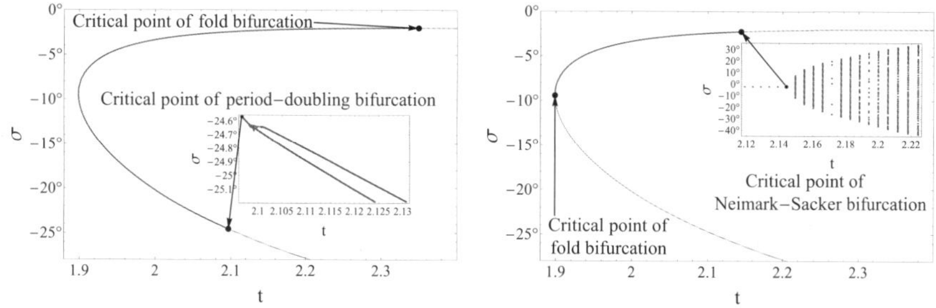

(a) state-based switching

(b) time-based switching

Fig. 4. Stability of gaits on a brachiating solution curve under (a) state-based and (b) time-based impacts. The impacts can occur at a fixed slope σ or switching time t. The blue segments of the curve are stable fixed points and the red are unstable. The types of bifurcation are also highlighted. The insets show the higher period gaits that result.

observed period-doubling bifurcations in the case of state-based switching and Neimark-Sacker bifurcations[12] in the case of time-based switching.

5. Conclusion

We have presented results on the connected components and stability of gaits for a two-link robot capable of brachiating and walking using a generalized two-link model. We have shown that walking and brachiating are distinct gaits, but they are not differentiated by moving above or below a slope. We have also shown that the impact strategy affects the stability of a gait, where state-based impacts are more stable than time-based impacts.

For future work, we plan to extend our model to powered gaits in a state-control space that includes control parameters for an actuator at the joint. Do connected components remain disconnected in this higher-dimensional state-control space or are they part of the same connected component that appears disconnected when projected onto the 5-D state-time space of passive gaits studied in this paper?

References

1. A. Goswami, B. Thuilot and B. Espiau, *The International Journal of Robotics Research* **17**, 1282 (1998).
2. M. W. Gomes and A. L. Ruina, *J. Theoretical Biology* **237**, 265 (2005).
3. M. Garcia, A. Chatterjee, A. Ruina and M. Coleman, *J. of Biomechanical Engineering* **120**, 281 (1998).
4. N. Rosa, A. Barber, R. D. Gregg and K. M. Lynch, Stable open-loop brachiation on a vertical wall, in *IEEE Int. Conf. Robotics & Automation*, 2012.
5. J. Nakanishi, T. Fukuda and D. Koditschek, *IEEE Trans. Robotics & Automation* **16**, 109 (2000).
6. P. A. Bhounsule, J. Cortell and A. Ruina, Design and control of ranger: an energy-efficient, dynamics walking robot, in *Climbing and Walking Robots CLAWAR*, 2012.
7. K. Sreenath, H.-W. Park, I. Poulakakis and J. Grizzle, *The International Journal of Robotics Research* **30**, 1170 (2010).
8. J. Nakanishi and S. Vijayakumar, Exploiting passive dynamics with variable stiffness actuation in robot brachiation, in *Robotics: Science and Systems*, 2012.
9. T. Fukuda, Y. Hasegawa, K. Sekiyama and T. Aoyama, *Multi-Locomotion Robotic Systems - New Concepts of Bio-inspired Robotics*, Springer Tracts in Advanced Robotics, Vol. 81 (Springer, 2012).
10. E. Allgower and K. Georg, *Numerical Continuation Methods, An Introduction* (Springer-Verlag New York, Inc., New York, NY, 1990).
11. A. Pressley, *Elementary Differential Geometry* (Springer Verlag, 2010).
12. Y. Kuznetsov, *Elements of Applied Bifurcation Theory* (Springer-Verlag New York, Inc., New York, NY, 2004).

SECTION–8

MANIPULATION AND GRIPPING

MODELING AND ANALYSIS OF ROBOTIC GRASPING USING SOFT FINGERTIPS

AKHTAR KHURSHID
Department of Mechatronics Engineering, Air University, Sector E-9
Islamabad 44000, Pakistan

ABDUL GHAFOOR
School of Mechanical and Manufacturing Engineering, National University of Sciences and Technology Sector H-12, Islamabad 44000, Pakistan

M AFZAAL MALIK
Department of Mechatronics Engineering, Air University, Sector E-9
Islamabad 44000, Pakistan

YASAR AYAZ
School of Mechanical and Manufacturing Engineering, National University of Sciences and Technology Sector H-12, Islamabad 44000, Pakistan

The ability to create stable, encompassing grasps with subsets of fingers is greatly increased by using soft fingertips that deform during contact and apply a larger space of frictional forces and moments than their rigid counterparts. This is true not only for human grasping, but also for robotic hands using fingertips made of soft materials. Soft fingers contribute to dexterous grasping on account of the area contact and high friction involved. Devices such as robotic hands and mechanical grippers, having more than two fingers, are widely used in practical applications. About 90% of the grips involved in industrial applications can be realized with a three finger hand. This paper presents a novel approach in modeling of soft contacts between three soft fingertips and a cylindrical object using viscoelastic material and analyses its characteristics employing BondGraph Methods. The fingers are made viscoelastic by using springs and dampers. The paper shows how the weight of the object moving downward is controlled by the friction between the fingers and the object during the application of contact forces by varying the damping and the stiffness in the soft finger.

1. Introduction

A manipulator is almost of no use without an end-effector since this is the only means by which manipulation is ultimately accomplished. Thus end-effectors and their design are as old as manipulators themselves. End-effectors are used mainly for grasping and manipulation of an object. Other usages of manipulators include sensing about an object or performing a manufacturing operation. For the first two uses, the task is accomplished by a gripper or hand.

Mechanical grippers either depend solely on direct contact with the object or friction to accomplish their tasks. Most of these are two-fingered [1], but they are made for specific tasks. We have worked for a multipurpose gripper so a three fingers gripper is targeted for analyses.

Typical hand grips can be grouped, more or less exhaustively, into six grip classes. Cylindrical hollow grip, tip grip, hook grip, three finger grip, hand palm grip and tong grip. If the consideration is restricted to human activities necessary for industrial work, a direct relationship between the hand with the necessary tools and the number of fingers involved in the specific work may be observed. It has been presented in "Robot Grippers" by Gareth J. Monkman, Stefan Hesse, Ralf Steinmann and Henrik Schunk [1a]. Consequently it is shown that about 90% of the grips involved in industrial applications can be realized with the three fingers hand. Furthermore, all fingers do not possess the same strength. The middle finger is the strongest one and the little finger the weakest. Grasping operations are always an integral part of more complicated handling strategies even in cases when they are performed automatically.

Substantial research literature has been produced in the study of grasping [2-5], but relatively low effort has been spent in the study of in-grasp fine manipulation. During grasping, slippage of the grasped object is controlled by overcoming its weight with the friction between the grasping fingers and the object. This friction further depends upon the applied force. For securing the object from damaging, the contact fingers are made soft by introducing springs and dampers at contacts.

Human finger tips are fleshy, soft and deformable. They locally mold to the shape of a touched or grasped object due to their viscoelastic behavior, and for these reasons, are capable of extremely dexterous manipulation tasks. Viscoelastic materials have an interesting mix of material properties that exhibit viscous behavior (like the gradual deformation of molasses) as well as elasticity (like a rubber band that stretches instantaneously and quickly returns to its original state once a load is removed). The clearest way to visualize the behavior of a material containing both elastic and viscous components is to think of a spring (exerting forces to return to its unstressed state) in series with a dashpot (a damper that resists sudden motion, similar to the pneumatic cylinder that prevents a storm door from slamming shut).

Most robot fingers are crude and therefore rather limited in capability. This realization has led to the investigation of robotic manipulation with soft, human like fingers, for example, Sun and Howe [6], Trembley and Cutkosky[7], Howe and Cutkosky [8], Russel and Parkinsan[9], and Shinofa and Goldenburg[10] report on experiments in which either foam-backed or fluid-filled fingers successfully enhanced dexterous capability. However in these works the material used for soft grasping is chosen first and then its performance in soft grasping is investigated. An alternate approach could be to determine the required properties of the soft material for stable grasping first and then designing a gripper using material chosen according to these properties. In this paper we try to identify such required properties through bondgraph modeling and experimentation in virtual environment.

In this work, the robot fingers are made soft by introducing springs and dampers at contact. Thus by varying the damping and stiffness, control of the grasped object is achieved and in-grasp fine manipulation can be done [11-13]. Bondgraph method is used to represent the system dynamics as shown in Figures 1 and 2. The effects due to softness of the finger tips while manipulating an object and due to the friction at the finger contacts, and their internal damping and stiffness are modeled in this paper by using bond-graphs.

Bondgraph method is an attempt to explore the modeling intricacies encountered in the system, using an alternative but powerful modeling technique. Khurshid and Malik [14-16] have used bond graph techniques for modeling and simulation of different dynamic systems. Bond-graphs represent the dynamics of the system pictorially and are extensively used for the modeling of physical system dynamics in multiple energy domains, as discussed by Khurshid and Malik [17]. A bond-graph model is based on the interaction of power between the elements of the system. The cause-effect relationships help in deriving the system state equations. Further, the model yields insight into various aspects of the control of the system [18, 19]. The traditional graphical techniques include block diagrams, signal flow graphs, and circuit diagrams. While the block diagrams and signal flow graphs preserve only the computational structure of a system, at least for analog computations, circuit diagrams reflect only the topological structure. Moreover, signal flow graphs and circuit diagrams are restricted to use in electrical systems only and hence lack the much desired unified approach for systems analysis. For these reasons, Paynter [20], worked for another graphical representation of dynamic systems, which could show simultaneously the topological and computational structure, and which would be general enough to be applicable to all kinds of physical systems so that a unified modeling approach could be evolved for dynamic systems. This led him to bond graph modeling based on state-space formulation.

In this paper the effects of the damping of soft finger, the stiffness of the springs in the soft finger and the friction between the soft contact surfaces with the grasped object are modeled by using bond graph methods. In section 2 the soft contact model is considered for bond graph modelling, simulation and analysis. In section 3 the bond graph model is developed and presented. In section 4 the simulated results are shown and in section 5 results are discussed. Conclusion is presented in section 6.

2. Soft Contact Model

Figure 1 shows a human hand using three fingers grasping a cylindrical object. Here we construct a model for a robotic gripper to grasp an object in a similar manner and then workout the properties of softness required to accomplish a stable grasp. It is shown in Figure 2 as schematic diagram.

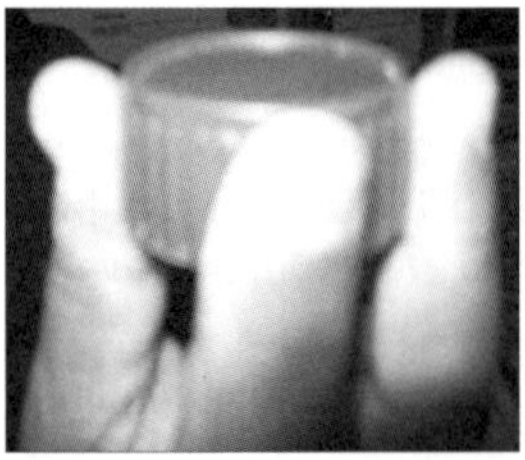

Figure 1: Three fingers grasping an object.

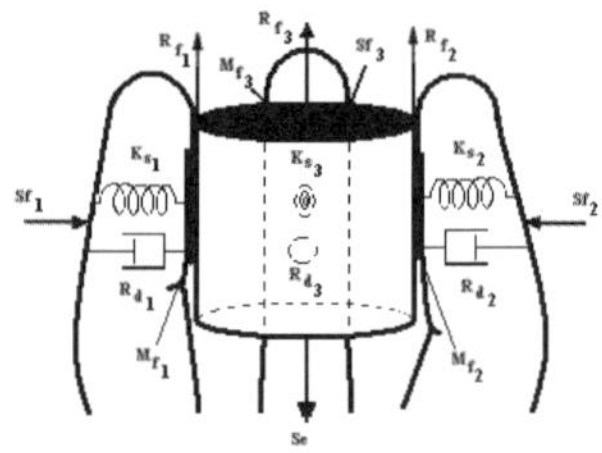

Figure 2: Schematic diagram

The schematic diagram of the three fingers grasping the object in Figure 1 is shown in Figure 2. It consists of three fingers which are used to manipulate the object as done by human fingers. The three fingers are made soft by introducing linear mass, spring, and damper effects in these. Forces Sf_1, Sf_2 and Sf_3 are applied to all the three fingers respectively for the grasping of the object. The weight of the object causes it to slip from the grasping of fingers, whereas the friction between the fingers contact surfaces with the object balance it. Friction is represented as damping by R_{f1}, R_{f2} and R_{f3} at the finger's contact surfaces with the object and the dampers are part of the fingers having damping R_{d1}, R_{d2} and R_{d3}. The stiffness of the springs used in the fingers is K_{s1}, K_{s2} and K_{s3}. The mass of the outer surface layers of the fingers in contact with the object is M_{f1}, M_{f2} and M_{f3}. The mass of object is M_o and its weight is taken as Se. We have assumed that the soft material shows linear viscoelasticity. Parameters R_f, K_s, and R_d are time-invariant and positive.

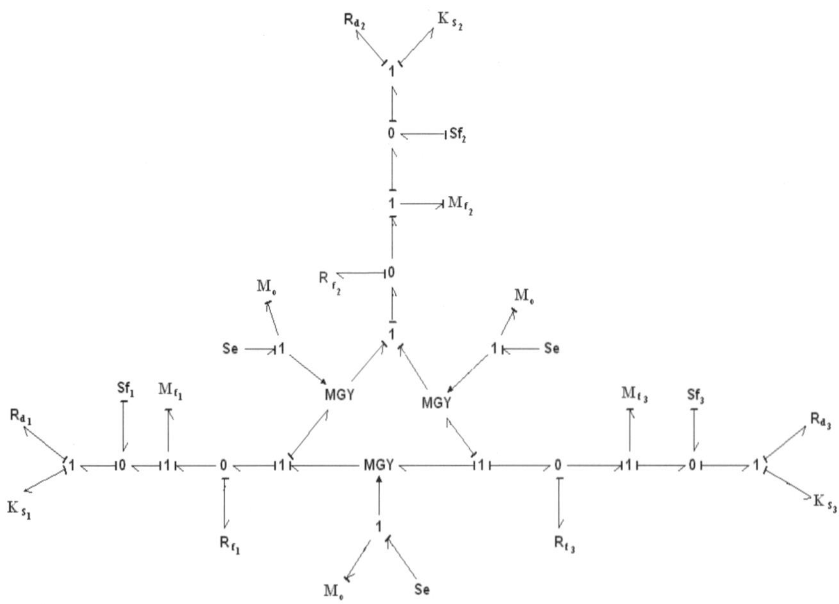

Figure 3: The bond graph model of three fingers grasping an object as shown in figures 1 and 2.

3. BondGraph Model

In this paper the effects of the damping of soft finger, the stiffness of the springs in the soft finger and the friction between the soft contact surfaces with the grasped object are modeled by using bond graph methods.

The physical system shown in Figure 1 and 2 is converted into bond graph shown in Figure 3. Following is the description of this bondgraph from its construction point of view.

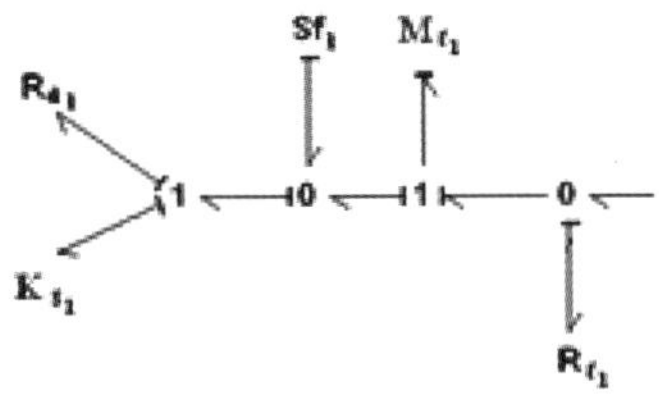

Fig 4: Bondgraph of one finger

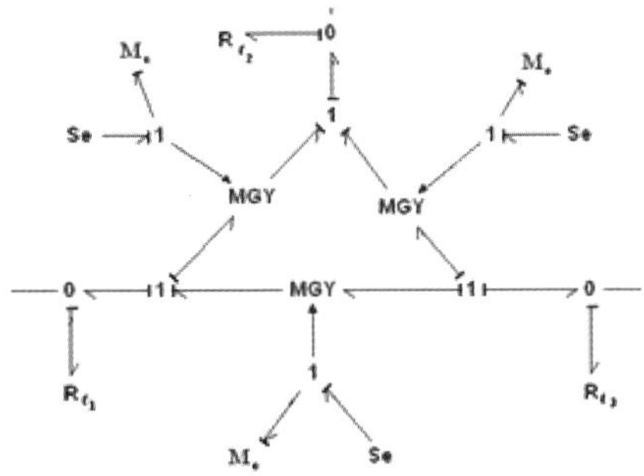

Fig 5: Bondgraph of 3-D rigid body motion

Figure 4 is showing the BondGraph of one soft finger. Similarly we have BondGraph for three soft fingers in complete BondGraph. Figure 5 is showing the BondGraph of three dimensional rigid body motion taken from book system dynamics [19]. The symbols shown are explained below.

A bond graph model is a precise mathematical description of the physical system in the sense that it leads to the state space description of the system.

S_f = applied force on each soft finger (N)

R_d = damping in each soft finger (Ns/m),

R_f = friction at each soft finger contact (Ns/m)

M_f = mass of each finger (kg)

M_o = mass of the object (kg),

S_e = weight of the object (N),

K_s = springs stiffness in each soft finger (N/m)

TABLE 1. SIMULATION RESULTS

Sr #	$K_{s_1} = K_{s_2} = K_{s_3}$ [N/m]	$R_{d_1} = R_{d_2} = R_{d_3}$ [Ns/m]	$R_{f_1} = R_{f_2} = R_{f_3}$ [Ns/m]	Peak Velocity [μm/s]			Settling Time for Steady State Velocity [s]			Peak Displacement [mm]			Steady State Value of Displacement [mm]		
				v_x	v_y	v_z	v_x	v_y	v_z	x	y	z	x	y	z
1	10	5	5	-1.2	-1.7	-1.4	0.4	0.5	0.4	-0.036	-0.12	-0.05	-0.025	-0.043	-0.032
2	10	10	10	-2.6	-2.84	-2.7	0.5	0.8	0.5	-0.06	-0.14	-0.08	-0.025	-0.043	-0.032
3	10	15	15	-4.0	-4.2	-4.1	0.3	1.0	0.6	-0.072	-0.14	-0.089	-0.028	-0.037	-0.032
4	10	20	20	-5.5	-5.5	-5.5	0.4	0.8	0.7	-0.08	-0.137	-0.09	-0.030	-0.037	-0.032
5	20	5	5	-1.3	-1.7	-1.4	0.6	0.8	0.6	-0.037	-0.126	-0.091	-0.032	-0.036	-0.033
6	20	10	10	-2.7	-2.9	-2.8	0.6	0.9	0.6	-0.06	-0.14	-0.08	-0.032	-0.036	-0.033
7	20	20	20	-5.4	-5.5	-5.5	0.4	0.8	0.6	-0.08	-0.137	-0.09	-0.032	-0.036	-0.033

4. Simulation

For simulation of the state space equations of the physical system, we have used 20-sim computer software [21]. Some of the results of grasping the object by soft contact fingers based on BondGraph modeling and simulation are shown in Figures 6 below. However, the results of total simulated experiments are consolidated in tabulated form as shown in Table 1.

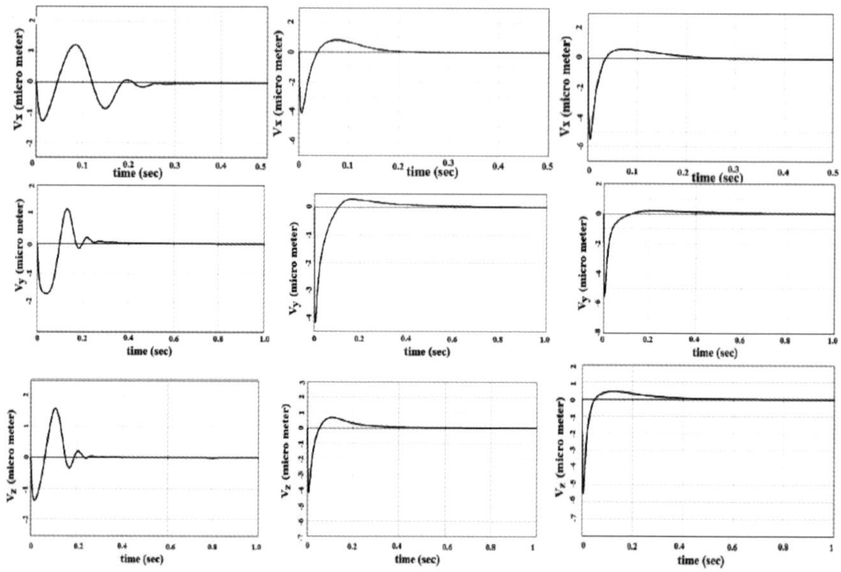

Figure 6: Velocities v_x, v_y and v_z of object during grasping at different damping and stiffness values of soft fingers

5. Results and Discussion

The objective of this work is to design and develop a robotic gripper which has soft fingers like human fingers. Soft fingers have ability to provide area contact which helps in dexterous grasping and fine manipulation of the gripping object. This work is a step towards this final goal. We have carried out a detailed parametric study of the dynamic system and have observed the effects of changing material properties on the dynamics of the soft contact grasping system. It helps in building the intuition about soft finger model. The object mass is 100g and fixed for all experiments.

Changes are made in stiffness and damping. Then repeating the simulated experiments by varying these values, obtaining its effect on the friction and stability of the system the final results are achieved. The final results are shown in Table 1. R_d the damping and R_f the friction in each soft finger are increased from 5Ns/m to 20Ns/m. K_s the spring stiffness in each soft finger is increased from 10 to 20 N/m [22]. Then by applying 1N force the velocity of the object and its displacement is observed. Velocity curves show the velocities against time in x,y and z directions. In figure 6, top three curves are displayed to show the results of velocity v_x vs time tabulated at serial number 1, 3 and 7 of the table 1. The object slips during grasping with peak velocity $v_x = 1.2\ \mu m/s$ and just within 400 mille seconds settle down to steady state zero value tabulated at Sr # 1 in table 1. The peak displacement is 0.036 mm and settles down to steady state value of 0.025mm. Similarly the object slips during grasping with peak velocity $v_x = 4.0\ \mu m/s$ and just within 300 mille seconds settle down to steady state zero value tabulated at Sr # 3 in table 1. The peak displacement is 0.072 mm and settles down to steady state value of 0.028mm. Finally the peak velocity $v_x = 5.4\ \mu m/s$ is observed when the stiffness of fingers is doubled from 10 to 20 N/m and it settles down to zero steady state value after 400 mille seconds tabulated at Sr # 7 in table 1. The peak displacement is 0.08 mm and settles down to steady state value of 0.032mm.

Results for v_y *and* v_z can be observed from the table 1. The curves in second row of figure 6 are showing v_y and last row for v_z. Due to lack of space all the curves cannot be shown in this paper thus the simulated results are shown in tabulated form.

6. Conclusion

A novel approach to model and simulate an effective soft contact grasping system is presented in this research paper. The parametric study is made to evolve suitable values of material properties for an effective grasping. The bond graph modeling technique has been applied to obtain the precise mathematical model of the three soft contact robotic fingers. The three fingers are made soft by introducing linear mass, spring, and damper effects in them. It would have been much more tedious to get these results using traditional methods.

From the simulated results presented in Table 1, it is concluded that by varying the friction and damping between the contact surfaces the object velocity during grasping and its displacement can be controlled. They also control the peak values of velocity and displacement t and the stability of the object as well. The stiffness of the spring does not have much effect on the dynamics of the object in comparison to the damping and friction. It is also concluded that the damping of the soft finger and the friction between the soft contact surfaces effect considerably in manipulation of the object. Therefore, these parameters can be used in doing in-grasp fine manipulation of the object.

Acknowledgments

The authors are indebted to Air University and National University of Sciences and Technology, Islamabad, Pakistan, for having made this research work possible.

References

1. E.N.Ohwovoriole "Kinematics and Friction in Grasping by Robotic Hands" 398/ Vol. 109, Sep 1987, ASME Transactions.
1. **[a]** Robot Grippers" by Gareth J. Monkman, Stefan Hesse, Ralf Steinmann and Henrik Schunk, Lakshminarayana, K.,"Mechanics of Form Closure",1978, ASME 78-DET- 32.
2. Trinkle, J.C, Abel, J.M. and Paul, R. P., 1988, "An Investigation of Enveloping Grasping in the Plane", International Journal of Robotics Research, vol. 3 no. pp. 33-55.
3. Trinkle, J.C., 'On the Stability and Instantaneous Velocity of Grasped Frictionless Objects', IEEE J. Robotics and Automation, vol. 8, no. 5, 1992, pp. 560-572.
4. Zefran, M., Kumar, V., and Croke, C., 1999, Metrics and Connections for Rigid Body Kinematics. International Journal of Robotic Research., vol. 18, No. 2, pp. 243-258.
5. Zefran, M. and Kumar, V., 2002, A Geometrical Approach to the Study of the Cartesian Stiffness Matrix. ASME J Mech Des., vol. 124, pp. 30-38.
6. J.S. Son, E.A. Monteverde, and R.D. Howe, "A Tactile Sensor for Localizing Transient Events in Manipulation," Proceedings of the 1994 IEEE International. Conference on Robotics and Automation pp. 471-476, San Diego, May 1994.
7. M. Tremblay and M.R. Cutkosky, "Estimating friction using incipient slip sensing during a manipulation task," Proceedings of the 1993 IEEE International Conference on Robotics and Automation , pp. 429-434, Atlanta, Georgia, May 1993.
8. R.D. Howe and M.R. Cutkosky, "Sensing skin acceleration for texture and slip perception," Proceedings of the 1989 IEEE International Conference on Robotics and Automation, pp. 145-150, Scottsdale, Arizona, May 1989.
9. R.A. Russell, S. Parkinson, "Sensing Surface Shape by Touch," Proceedings of the 1993 IEEE International Conference on Robotics and Automation, pp. 423-428, Atlanta, Georgia, May 1993.
10. K.B. Shimoga and A.A. Goldenberg, "Soft Materials for Robotic Fingers," Proceedings of the 1992 IEEE International Conference on Robotics and Automation pp. 1300-1305, Nice, France, May 1992.
11. Ghafoor, A, Dai, J.S. and Duffy, J. Fine Motion Control Based on Constraint Criteria Under Pre-Loading Configurations, Journal of Robotic Systems, 17(4), 2000, pp. 171-185.
12. Ghafoor, A., Dai, J.S. and Duffy, J., Stiffness Modelling of a Soft-Finger Contact in Robotic Grasping, Transactions of ASME: Journal of Mechanical Design, 126(4): 646-656, 2004.
13. Dai, J.S. and Kerr, D.R., Six-Component Contact Force Measurement Device Based on the Stewart Platform, Journal of Mechanical Engineering Science, Proc. of IMechE, 214(5): 687-697.
14. Khurshid and Malik, "Modeling and Simulation of an automotive system by using Bond Graphs" 10th International Symposium on Advanced Materials ISAM 2007 Islamabad, Pakistan.
15. Khurshid and Malik, "Bond Graph Modeling and Simulation of Impact Dynamics of a Car Crash" 5th International Bhurban Conference On Applied Sciences And Technology 5th IBCAST-2007, Islamabad, Pakistan.
16. Khurshid and Malik, "Modeling and Simulation of a Quarter Car Suspension System using Bond Graphs" 9th International Symposium on Advanced Materials ISAM 2005, Islamabad, Pakistan.
17. Khurshid and Malik, "Bond Graph Modeling and Simulation of Mechatronic Systems" International Multi-topic Conference 2003, INMIC 2003, In association with IEEE, Islamabad, Pakistan.
18. A. Mukherjee, R. Karmakar, Modeling and simulation of engineering systems through bond graphs, Narosa Publishing House, New Delhi, 2000.
19. D. C. Karnopp, D. L. Margolis, and R. C. Rosenberg, System Dynamics: Modeling and simulation of mechatronic systems, third edition, Wiley Inter-science, 2000.
20. Paynter, Analysis and Design of Engineering Systems, MIT press, Camdridgs, MA, 1961.
21. 20-sim Control Laboratory, University of Twente Controllab Products B.V. Drienerlolaan 5 EL-CE, 7522 NB Enschede the Netherlands. 2003.
22. Khurshid, Ghafoor and Malik, "Modeling and Analysis of Soft Contact in Robotic Grasping Using BondGraph Methods" Advanced Materials Research

FORCE SENSING AND CONTROL DURING MOVEMENT AND OBJECT MANIPULATION IN *MERO* WALKING ROBOTS

ION ION[1], ADRIAN CURAJ[2], AURELIAN VASILE[3], DUMITRU IULIA[2] STAMATESCU GRIGORE[2]

[1] *Manufacturing Technology Department,* [2] *Automatic Control and Industrial Informatics Department,* [3]*Mechanism and Robot Theory Department,*

POLITEHNICA University of Bucharest Romania

Walking robots are built to safely transport and assure a stable position of the technological equipment on-board the robot's platform on unarranged terrain and may have a lower or higher autonomy degree. To assure the stable position of the walking robots, they are equipped with transducers and sensors, necessary to determine the force distribution in the shifting mechanisms. For safe handling, position errors must be sensed and controlled through the movement of the finger grip attached to the technological installation. The paper analyzes two sensing and motion control devices in robotic systems and the complex sliding phenomenon that aims both robotic technological equipment (moving an object manipulated during the technological process) and walking robots (moving them in terms under conditions).

1. Introduction

Due to the mechanical properties of soil, walking robots have a inherent potential advantage on wheeled vehicles when moving on land leading to reduced stiffness and tackling irregular surfaces. There are two general ways in which steps can be controlled in order for the robot to adapt to the natural ground: walking controlled by video or proximity sensors and walking blindly.

By using video sensors (cameras) or proximity transducers, the robot can predict the shape of the land surface whilst moving. In this way we can define an internal model of the environment geometry and thus can plan trajectory extremities, legs and stepping mode. [1], [6], [3].

The second way of robot moving on natural ground (irregular) is known as blindly walking. The robot changes its step length and direction of movement steps based on where the positions of points in the leg ends meet the land area

Walking robot motion control using walking blindly moving can be decomposed into two complementary processes, namely active control and outfit control.

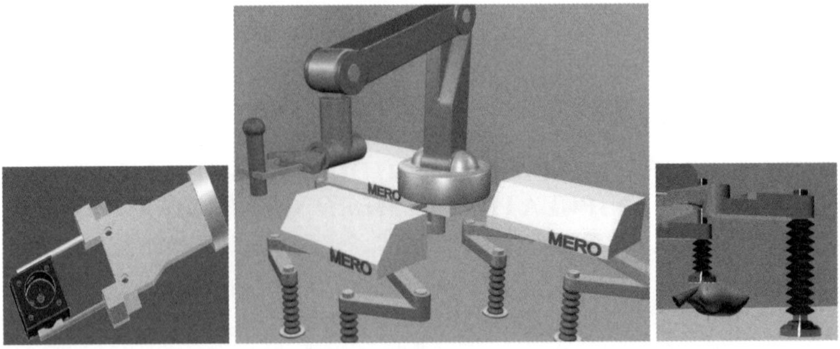

Figure.1 The modular walking robot MERO- support for the "DISROB" sensing and motion control in robotic systems (possible applications)

Active control uses body bias to maintain a desired orientation so that the vertical projection of the centre of gravity of the robot to find inside the support polygon, to achieve static conditions.

Active control is provided via contact sensors and transducers that measure forces at the contact areas of feet with, the ground.

Active control extends platform joints, to create the conditions for a robot motion control by force steps through uneven terrain displacement.

Contact sensors ensure and protect the load. [1], [2], [3]

To control the walking robot shift in structured or less structured environments we need the following specific functions:

- the environmental perception and shaping using a multisensorial system for acquiring data;
- data collecting and defining the field configuration;
- movement planning;
- analysis of the scenes;
- control over the objects handling.

Data acquisition system of "MERO" modular walking robot figure 1 (will be made up of the following parts): [2], [5], [4]..

- the position sensor measuring the pair's variables, namely the relative positions of the kinematic elements adjacent to the driving pairs;
- the feet's tactile transducers signaling the feet's contact with the support area;
- the force cell measuring the reaction forces between ground and feet;
- the verticality sensors measuring the platform's deviations from the horizontal positioning unless any further conditions;
- the measuring systems of the platform's height as compared to the ground's surface;

- the measuring devices in the transfer phase of the feet's height as against the ground's level.

These sensors convert the mechanical variation of a quantity (force, torque, linear displacement, angular) the variability of electrical parameters (resistance, voltage, current) through electrorezistive transducers (TER).

The modular walking robot "MERO" (Figure 1 new model –in constrution) will be equipped with six identical sensors. The information provided by the sensors is acquired and processed by the management and control system of the robot.

2. System model description DISROB

The device force sensing and motion control DISROB is attached Fig.1 onto the transport systems of walking robots and technological equipment in the gripper. It refers to the complex phenomenon of slip (tangential component), which through the two different types of sensors (sliding on two axes and power) controlled through an integrated microcontroller system (power regulator) allows adapting them to displacement on unimproved terrain.

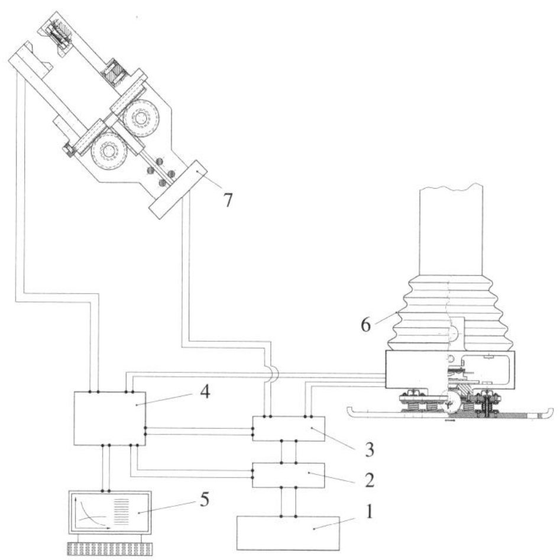

Figure.2 "DISROB" sensing and motion control in robotic systems

The DISROB (Fig.2) sensing and motion control by force is a hardware and software system (how to up-grade) that has two different types of sensors (sliding on two axes and strength).The intelligence given by a micro -integrated controller provides a force control loop to keep it from slipping. It controls allowable values and is attached to the feet o robot (6) and for technological equipment in the gripper fingers position (7) and has the composition and distribution mains pressurized fluid (1), proportional pressure regulator (2), electronic module setting reference point (3), power regulator (4), PC controller (5). Fig 3a.b

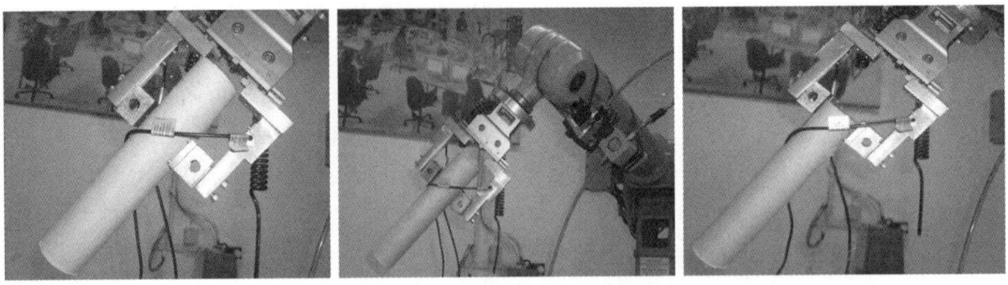

Figure.3a Force sensing system and sliding for gripper

At position 6 contact sensors are components / slip and transducers that measure forces at the contact areas of the feet to the ground, providing security and stability of the walking robot movement through active control.

Element-gripper (finger) positioning must be. developed for controlling gripping force to overcome the weight and inertial forces manipulated object and maintain it in the finger grip and handling during fixation.

On the other hand, the gripping elements should not produce excessive force to damage the manipulated object.

The sensing and motion control allow identification and measurement of contact forces with the ground through a transducer element whose elastic deformation is caused by a ball that sends robot step foot contact with the ground under the weight of the burden robot foot and additional load for transported technological installations. Contact forces are determined using strain gauges of the measurement module whose deformation generates an analog signal processed with the power regulator.

Measuring relative sliding is through an optical sensor without physical contact with the track that provides a measure displacement resolution 1000dpi/inch extent at resolution of 0.024 mm.

The sensor is based on Laser Stream technology, which measures changes in position by acquiring consecutive images (frames interference) and mathematically determining the direction and magnitude of movement.

The force sensing and control module attached to a robot's gripper is composed of the following parts:

- two motion sensors
- two force sensors
- two pressure direction change electro valves
- two numerically controlled pressure regulators

The two optical movement sensors are attached to the gripper's fingers. They are based on the Laser Stream technology and provide 10000 dpi resolutions – about 0.0025 mm. These sensors output on a SPI port the detected relative movements Δx and Δy to a central unit.

The force sensors are also located on the gripper's fingers and are composed by a tensometric stamp and a 12bit operational amplifier that provide 0.02 FSR (Full Scale Range) resolutions. Information from these sensors is read through AD (analog to digital) converters by a central microcontroller based unit.

The pressure direction change electro valves are controlled form the main unit through digital outputs.

The main unit includes 12bit DA (digital to analog) converters used to control the pressure regulators.

The robot's gripper, when grasping, moving or letting go of an object, through the main microcontroller-based unit has the appropriate information about the object's state relative to the gripper's fingers. In the grasping phase the gripper's fingers grab the object until the force sensors indicate a minimum force level applied on the object. The next step in that the gripper tries to pick-up the object and moves it. The motion sensors tell the main unit if the object is slipping, how fast and in what direction.

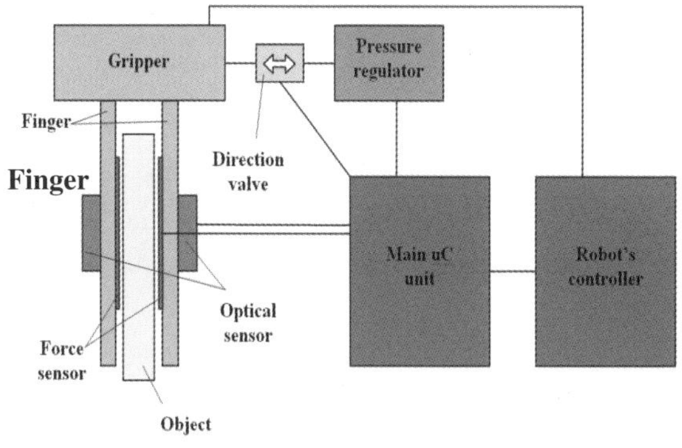

Fig.3b Force sensing system and sliding for gripper (scheme)

The main unit introduces this data into an incremental PID regulating algorithm that outputs the values for the DA converter of the pressure regulators thus tightening the grasp on the object in order to have a slipping-free movement. In the letting-go phase, the gripper lowers the force exerted by the fingers on the object through the pressure direction change valves and tries to move away. If the optical motion sensors do not detect a motion of the fingers relative to the object equal with the motion of the gripper itself, it means that the gripper is still attached to the object and commands through the PID algorithm a further decrease in force of the gripper's fingers.

Force transducers (Fig.4) located at the ends of the modular walking robot foot and incremental hold sensor allows the robot to control the orientation of the tilt adjustment in both the sagittal and frontal plane.(Fig 5)

In order to verify the results obtained by solving analytical elastic element of the transducer was analyzed by finite element method.

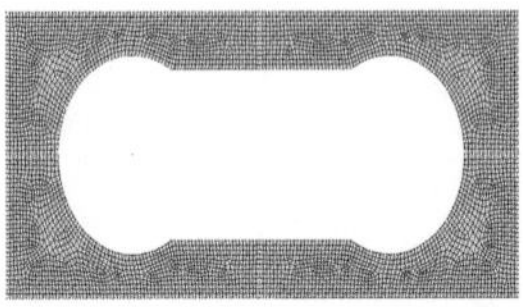

Fig.4 Force transducer

Fig.5 Force sensing system and sliding for leg

Force transducer that can be used in the construction of the robot were designed and built. It has four full bridge transducers connected electrorezistive (type Wheatstone). Elastic element, which are mounted transducers Fig.3 electrorezistive, is determined as a static framework and plan was sized for permissible stress of the material σ = 270 MPa and maximum load F = 1000 N. sensor.

2.1. Simulation

In order to better understand and test the two types of force sensing and control devices, one for the leg of the MERO robot and one for the gripper attached to the technical installation onboard the robot, a few computer simulation were carried out.

The virtual simulation environment was created using DirectX and OpenGL technologies in a Windows based software developed by the authors. This environment includes the different types of terrains that a robot could encounter in real world condition like straight and smooth sections, rough sections with hurdles, gaps and obstacles, steps, angled planes and many others, basically any kind of real world terrain type that the authors considered and entered into the software's terrain library.

For the simulation of the force sensing device attached to the leg of the robot, the RRP- model of the robot was also constructed virtually and introduced in the virtual environment. The ground contact, slip possibility and applied force provided by the force sensing device were generated by a software function and fed into the movement algorithm of the MERO robot. Different terrain types and different slipping scenarios were generated in order to observe the influence of the leg's slip possibility and to get a starting point in introducing the parameters that the force sensing and control device provide, into the movement algorithm.

Tests and simulations were also done on the gripper attached module. In this case, an incremental PID pressure regulating algorithm is used to provide enough force to the gripper in order to restrict slipping of the handled object in lifting or moving operations. A PID regulator model for the gripper was included in the software that in conjunction with a slipping and force sensing generator function that provides test data to the virtual model helps in gathering test data for the real model's optimization.

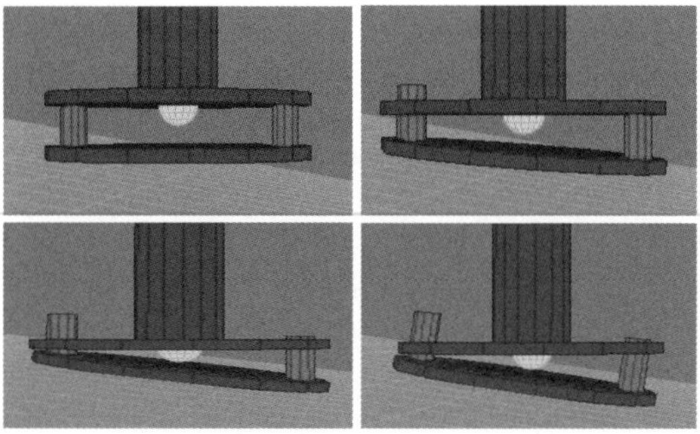

Fig.6 Computer simulation for applications with MERO robot leg

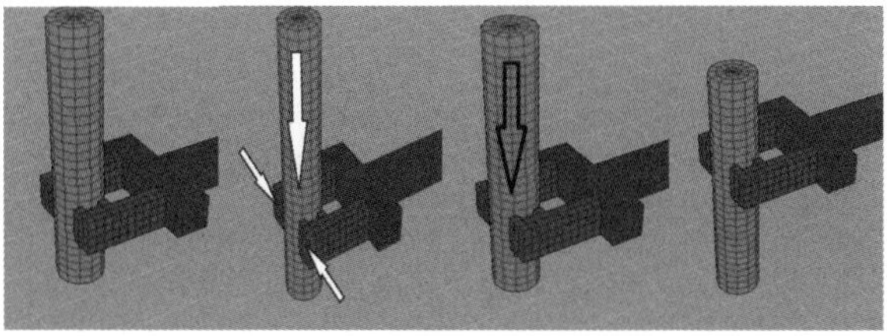

Fig.7 Computer simulation applications with technological equipment attached gripper robot MERO

The application described here was developed with the purpose of easing the simulation and providing a better understanding of different strategies for a "contact" signal was received from one of the feet's touch sensors Fig.6 and Fig.7 sensors attached gripper robot MERO

Results and conclusion

The MERO type tansducers used in walking robots offers both force control and robot protection.

Solutions found by the authors has enabled sensing and control device DISROB, which senses touch motion, force measurement and slip to transfer objects through grasping and fixation devices attached to robots and robots stepped force and slip that in a control slip by force increase positioning accuracy and safety requirements of the application,

Acknowledgments

This research was partially supported by a grand of the Romanian National Authority for Scientific Research, CNDI –UEFISCDI project number PN-II-IN-CI-2012-.1-0242

References

1. Ion I., I Simionescu,., A Curaj, A. Vasile MERO Modular Walking Robots, Solution for Displacing Technological Equipments on Irregular Terrains *Proceedings of the 13th World Congress in Mechanism and Machine Science(2011)*,, Guanajuato, México, 19-25 June, 2011
2. Ion I., Curaj A.and Vladareanu L, "Design and Motion Synthesis of Modular Walking Robot MERO" *Journal of Automation Mobile Robotics & Intelligent Systems* ISSN 1897-8649 Vol2 No4 pag.25-30(2008)
3. Kumar, V. and Waldron, K.J., (1988) Force Distribution in Walking Vehicle,in *Proc. 20th ASME Mechanism Conference*, Orlando, Florida,
4. K.J., Waldron Force and Motion Management in Legged Locomotion, *Proceedings of the 24th IEEE Conference on Decision and Control,* Fort Lauderdale, (1991) pp.12-17
5. Mahdi A., Stephen N. Foot Force Criterion for Robot Stability *Proceedings of the 13th International Conference on Climbing and Walking robots and the Support Technologies for Mobile Machines CLAWAR 2012,* 23-26 July– 2012, Baltimore USA,
6. Roennau A., G Heppner., Pfotzer L., R Dillmann., (2012)," Foot Design Evaluation for a six Leged Walking Robot *Proceedings of the 13th International Conference on Climbing and Walking robots and the Support Technologies for Mobile Machines* CLAWAR 2012, 23-26 July–2 012, 511-518 Baltimore USA,
7. Mikhail S.Jones, Jonatahan H., Effects of Leg Configuration on Running and Walking Robots Proceedings of the 13th International Conference on Climbing and Walking robots and the Support Technologies for Mobile Machines CLAWAR 2012, 23-26 July– 2012, 519-526 Baltimore USA,

OBJECT GRASPING WITH DUAL ROBOT MANIPULATOR USING GENETIC ALGORITHMS

M. A. H. HASSAN and M. O. TOKHI
Dept. of Automatic Control and Systems Engineering
University of Sheffield, Sheffield, UK
cop11mah@sheffield.ac.uk

An approach for object grasping with dual robot manipulator is presented. Dynamic and kinematic models of the manipulator are a developed based on point to point trajectory planning. A genetic algorithm is proposed to optimise the trajectory parameters using quadrinomial and quintic polynomials approach to minimise travelling time and space. The approach describes the segments that connect initial, intermediate and final point in the joint space. An objective function derived for object grasping is introduced in driving the genetic algorithm.

Keywords—flexible robot manipulator; object grasping; genetic algorithm

1. Introduction

Robotic manipulators have been studied for more than a decade in an attempt to replicate the human hand capabilities. Most of the research carried out has emphasised fingertip grasping, which is considered as traditional approach for robotic grippers and hand. A new concept of whole arm manipulation [1] describes the ability of the entire manipulator's body to be used to grasp the object. This concept has been proved to deliver more force for grasping as equivalent expressions of "power grasp" [2] or "enveloping grasp" [3, 4]. The whole arm grasping technique has been able to accomplish several tasks such as pushing, searching, enclosure and exclusion [5]. The enclosure grasp of objects is a significant application for whole arm manipulation and offers a robust and stable grasp for larger and heavier objects. The multiple points of contact between object grasped and the surfaces of the manipulator links enable increased load capacity. The ability to use the entire body of the manipulator for grasping also allows objects of various dimensions to be grasped [6].

Optimisation of trajectory planning using genetic algorithms (GAs) has been proposed in [7, 8]. GA is population-based, stochastic, and global search method. The global search ability in GA provides the possibility to find global solutions of redundancy compared to classical techniques [9]. A path planning method based on a GA has been proposed in [10] while adopting the direct kinematics and the inverse dynamics to minimise the path length, the ripple in the time evolution and the energy requirements, without any collision with the obstacle in the workspace. The point to point trajectory planning proposed by [11] focused on minimising vibration of a flexible redundant manipulator (FRM) in joint space using GA. The authors have used quadrinomial and quintic polynomials to describe the segment, which connects the initial, intermediate and final points in the joint space.

This paper presents a strategy for object grasping with 3-link robot manipulators using GA. The paper is structured as follows: Dynamic and kinematic models of 3-links flexible robot manipulator are introduced in Section 2. Section 3 presents the kinematics of the 3-link robot manipulator. Quadrinomial and quintic polynomial are used to describe the motion planning for the system, which connects the initial, intermediate and final points in the joint space. Genetic algorithm optimisation is also described in Section 3. Finally, dual three flexible links is used in simulations, and two case studies are conducted and discussed in Section 4.

2. Dynamic and Kinematic Models

2.1. *Dynamic Model of 3-link Robot Manipulator*

The dynamic model for an n-joint, revolute, direct drive robot manipulator is described by the following expression [12].

$$M(q)\ddot{q} + N(q,\dot{q}) + D(q,\dot{q}) = \tau \tag{1}$$

where $M(q) \in Rnxn$ represents the inertia effects, $N(q,\dot{q}) \in Rn$ represents the remaining dynamic terms, such as the centrifugal-coriolis effects, gravitational effects, and frictional effects, $D(q,\dot{q}) \in Rn$ represents the contact forces placed on the robot manipulator by the environment, $\tau(t) \in Rn$ represents the input torque vector. The subsequent development is based on the assumptions that $q(t)$ and $\dot{q}(t)$ are measurable, $M(q), N(q,\dot{q})$, and $D(q,\dot{q})$ are unknown, second order differentiable, functions of $q(t)$ and $\dot{q}(t)$.

2.2. *Kinematic Model of 3-link Robot Manipulator*

Consider a planar three-link robot arm with revolute joints depicted in figure 1. The mass, inertia and length of the three-link are denoted by m_i, l_i and τ_i respectively. The three joint angles are labeled as q_1, q_2 and q_3. The coordinate of the end effector is defined by position and orientation, with the reference point of the end effector (x, y) and the orientation angle Φ. Hence, the coordinate (x, y, Φ) completely specifies the position and orientation of the end effector.

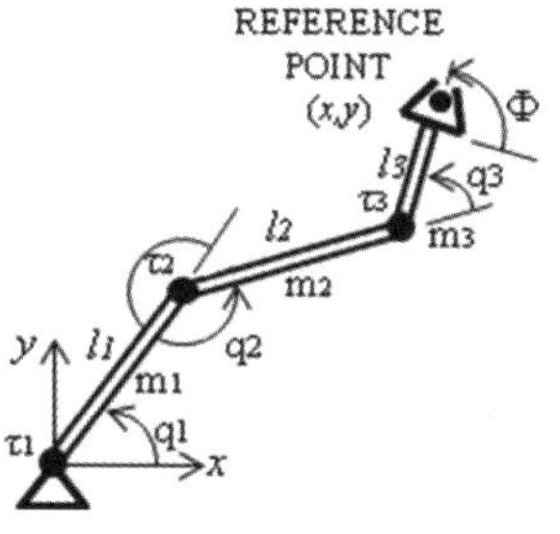

Figure 1: The three-link robot manipulator

2.2.1. *Direct Kinematics*

From basic trigonometry, the position and orientation of the end effector can be written in terms of the joint coordinates as follows:

$$x = l_1 \cos q_1 + l_2 \cos(q_1 + q_2) + l_3 \cos(q_1 + q_2 + q_3) \tag{2}$$

$$y = l_1 \sin q_1 + l_2 \sin(q_1 + q_2) + l_3 \sin(q_1 + q_2 + q_3) \tag{3}$$

$$\Phi = (q_1 + q_2 + q_3) \tag{4}$$

2.2.2. *Inverse Kinematics*

The inverse kinematics is the computation of joint coordinates for a given set of the end effector coordinates, and involves solving a set of nonlinear and complex equations. The equation can be derived from direct kinematics by assuming that Cartesian coordinates, x, y and Φ are given. The analytical expressions for the joint angles q_1, q_2 and q_3 in terms of the Cartesian coordinates can be derived by substituting equation (4) into equations (2) and (3) and solving the equation to form a single nonlinear equation in term of q_1 is obtained as:

$$(-2l_1 x')\cos q_1 + (-2l_1 y')\sin q_1 + (x'^2 + y'^2 + l_1^2 + l_2^2) = 0 \tag{5}$$

After numerical computation, two solutions are obtained for q_1 as:

$$q_1 = \gamma + \sigma \cos^{-1}\left(\frac{(x'^2 + y'^2 + l_1^2 + l_2^2)}{2l_1\sqrt{x'^2 + y'^2}}\right),\ \gamma = a\tan 2\left(\frac{-y'}{\sqrt{x'^2 + y'^2}}, \frac{-x'}{\sqrt{x'^2 + y'^2}}\right),\ \sigma = \pm 1 \tag{6}$$

The two solutions of q_1 can be represented as $\sigma = \pm 1$ is the left arm position while $\sigma = -1$ represents right arm position.

q_2 can be solved using *atan2* function using solution in (6) as follow:

$$q_2 = \text{atan2}\left(\frac{y' - l_1 \sin q_1}{l_2}, \frac{x' - l_1 \cos q_1}{}\right) - q_1 \tag{7}$$

Thus, for each solution of q_1, there is one unique solution for q_2. Finally, q_3 can be easily determined from equation (4):

$$q_3 = \Phi - (q_1 + q_2) \tag{8}$$

3. Motion Planning and GA optimization

3.1. *Motion Planning Strategy*

The trajectory of the 3-link manipulator depends on the position, velocity and acceleration for each degree of freedom. The point-to-point trajectory illustrated in Figure 2 is connected by several segments with continuous acceleration at the intermediate via point. The intermediate points can be given as particular points that should be passed through. Assume that there are m_p intermediate via points between the initial and final points.

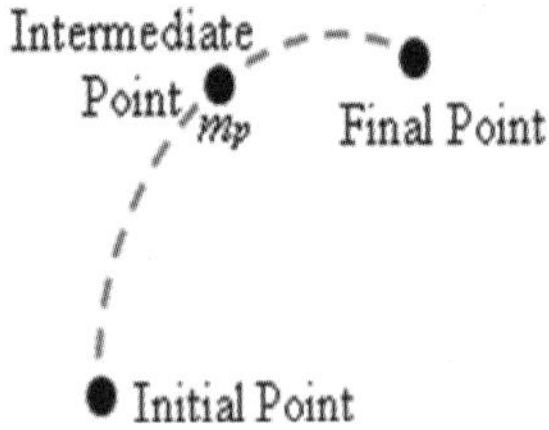

Figure 2: Intermediate points on the point-to-point trajectory

Between the initial points to m_p intermediate via points, a quadrinomial is used to describe these segments as [6]:

$$q_{i,i+1}(t) = a_{i0} + a_{i1}(t_1) + a_{i2}(t_1^2) + a_{i3}(t_1^3) + a_{i4}(t_1^4), (i = 0, \ldots, m_p - 1) \tag{9}$$

where $(a_{i0}, \ldots, a_{i4})$ are constants and the constraint and t_1 is the execution time from point i to point $i+1$. The intermediate point $i+1$'s acceleration can be obtained as:

$$\ddot{q}_{i+1} = 2a_{i2} + 6a_{i3}t_i + 12a_{i4}t_i^2 \tag{10}$$

The segment between the m_p intermediate points and the final point can be described by quintic polynomial as:

$$q_{i,i+1}(t)=b_{i0}+b_{i1}(t_1)+b_{i2}(t_1^2)+b_{i3}(t_1^3)+b_{i4}(t_1^4)+b_{i5}(t_1^5),(i=m_p) \quad (11)$$

where $(b_{i0},..., b_{i5})$ are the unknown constants.

All of the above parameters can be determined by using the following optimisation method.

3.2. *Trajectory Optimization Using GAs*

For 3-link robot manipulator using point to point trajectory planning, there are nine parameters to be optimised, and these are the following [13]:

$$[q_1,q_2,q_3,q_g,\dot{q}_1,\dot{q}_2,\dot{q}_3,t_1,t_2] \quad (12)$$

where q_i and $\dot{q}_i$ are intermediate joint angle and velocity for *i*th joint respectively, q_g is the global angle of the final configuration of the end-effector, and equals the sum of joint angles of the manipulator [14], t_1 is execution time from initial to intermediate via point, and t_2 is execution time from intermediate to final point.

Figure 3 shows the flow chart of GA operation. The initial population is generated at random and carried out among the population. The evaluation of the population in GA replaces the worst with new elements. Every generation will go through different operators adopted in GA as reproduction, crossover

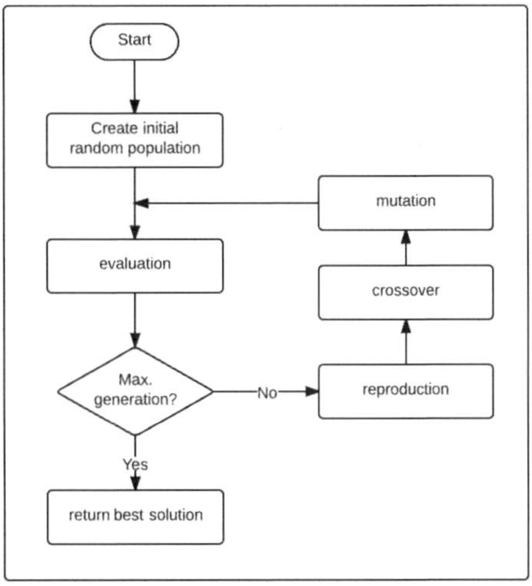

Figure 3: Flow chart of GA operation

and mutation. Each index is computed individually and is integrated with the fitness function as follows:

$$f_f = \sum_{j=1}^{b} \sum_{j=1}^{a} f_i, f_i^j = \begin{cases} 0 & |\tau_i^j| < \tau_{i\max} \\ |\tau_i^j| - \tau_{i\max} & otherwise \end{cases} \tag{13}$$

where f_i^j is the cost function, a is number of robot links, and b is number of joint positions from initial to final configuration.

For object grasp, the object function f_{ob} has been combined with free space fitness function to form overall fitness function f, as follows:

$$f = f_f / f_{ob} \tag{14}$$

The objective function is devised so as to avoid intersection between the link and the object while grasping [15]. The objective function is represented as follows:

$$f_{ob} = \begin{cases} 1 & \sum_{j=1}^{b} \sum_{i=1}^{a} (link_{ij} \cap object) = 0 \\ 0 & otherwise \end{cases} \tag{15}$$

4. Simulation results and discussion

This section discusses the results of the 3-link robot arm simulation. The parameters of the system are given in Table 1.

Table 1: Three Links Robot Arm Parameters

	Left Arm			Right Arm		
	Mass, m_i kg	Length, l_i m	Torque, τ_i Nm	Mass, m_i kg	Length, l_i m	Torque, τ_i Nm
Link 1	1.00	1.00	50	1.00	1.00	50
Link 2	1.00	1.00	30	1.00	1.00	30
Link 3	0.50	0.50	10	0.50	0.50	10

The initial end effector position of the 3-link robot arm for left arm is $(x_{L0} = -2.5m, y_{L0} = -0.5m, q_{Lg} = -90°)$ and for right arm is $(x_{R0} = 2.5m, y_{R0} = 0.5m, q_{Rg} = 90°)$. The joint velocities and accelerations of the initial and final configuration are assumed zero. The range of robot joints' rotation is between (π to $-\pi$) to avoid the links overlap with one another. The final position of the end effector was set to $(x_{Lf} = -0.2m, y_{Lf} = 1.9m)$ for left arm and $(x_{Rf} = 0.2m, y_{Rf} = 1.9m)$ for right arm. The object to be grasped has circular shape with radius 0.90 m.

The GA runs for a maximum generation value of 50, population of 40 elements, crossover probability of 0.8 per chromosome and a mutation probability of 0.05 per locus. At the initial generation of population, GA generates the chromosome elements with specified range of $q_i = [-\pi, \pi]$ rad, $q_g = [-\pi, \pi]$ rad, $\dot{q}_i = [-0.4, 0.4]$ rad/s and $t_i = [0,5]$ sec.

The simulation without GA required the parameters to be set heuristically within the range of $q_i, q_g, \dot{q}_i$ and t_i. Those parameters were set as follows: Left arm $[5\pi/6, -\pi/12, -\pi/2, \pi/18, -0.1, -0.2, 0.2, 3, 3]$ and right arm $[\pi/6, \pi/8, \pi/2, 10\pi/11, 0.2, 0.1, 0.05, 3, 3]$. The GA was automatically tuning the parameters in the range of $q_i, q_g, \dot{q}_i$ and t_i, and generated best solution for shorter travelling time and trajectory length for both arms.

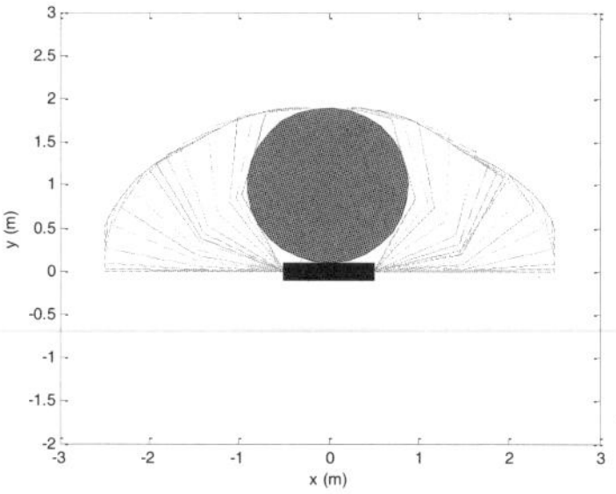

Figure 4: Object grasping with dual flexible manipulator

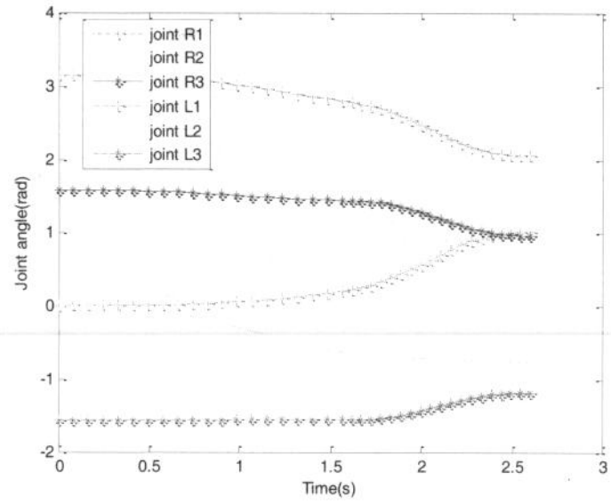

Figure 5: Joint angle with GA optimisation

Figure 4 shows the object grasping using dual 3-links flexible robot manipulator. In both cases, the end-effectors of robot arms are accomplished to travel to the final position to grasp the object. The joints angle in Figures 5 is representing the grasping position for each link as shown in Figure 5. Table 2 shows the travel time and trajectory length for simulations with and without GA.

Table 2: Optimisation results

Result value	Left Arm		Right Arm	
	Without GA	With GA	Without GA	With GA
Total travelling time (sec)	6.00	2.60	6.00	2.60
Total Cartesian trajectory length (m)	2.97	2.87	2.96	2.88

5. Conclusion

An object grasping mechanism with dual robot manipulator using genetic algorithm has been presented. An optimisation of point to point trajectory planning on kinematics redundancy parameters that enables end-effector position tracking information as well as object grasping has been developed. The optimisation showed that the object grasping using GA was efficient, and reduced the travelling time and space. The GA has been shown to be able to achieve multi-objective optimisation effectively on kinematics redundancy with specified objective functions.

References

1. K. Salisbury, "Whole Arm Manipulation," Proc. 4th Int. Symposium Robotics Research, 1987,pp. 183-189.
2. M. J. Sheridan, S. C. Ahalt and D. E. Orin, "Fuzzy Control for Robotic Power Grasp," *Robotics Society of Japan – Advanced Robotics*, vol. 9, no. 5, 1995, pp. 535-546.
3. J. C. Trinkle, J. M. Abel and R. P. Paul, "An Investigation of Frictionless Enveloping Grasping in the plane," Int. J. Robot. Res., vol. 7, no. 3, 1988, pp.33-51.
4. C. Xiong, H. Ding, Y. Xiong, "Fundamentals of Robotic Grasping and Fixturing," vo. 3, 2007, pp. 1-8.
5. K. Salisbury. W. Townsend, B. Ebrman, D. Dipietro, "Preliminary Design of Whole Arm Manipulation System (WAMS)," Proc. Of IEEE Int. Conf. On Robotics and Automation, IEEE Comput. Soc. Press, 1988, pp. 254-260.
6. M. D. Hanes, S. C. Ahalt, K. Mirza, and D. E. Orin, "Power grasp force distribution control using artificial neural networks," *Journal of Robotic Systems,* vol. 9, pp. 635-661, Jul 1992.
7. N. Kubota, T. Arakawa, T. Fukuda, K. Shimojima, and Ieee, "Trajectory generation for redundant manipulator using virus evolutionary genetic algorithm," in *1997 IEEE International Conference on Robotics and Automation (ICRA97) - Teaming to Make an Impact*, Albuquerque, Nm, 1997.
8. S. D. Sun, A. S. Morris, and A. M. S. Zalzala, "Trajectory planning of multiple coordinating robots using genetic algorithms," *Robotica,* vol. 14, Mar-Apr 1996.
9. D. B. Fogel, "An introduction to simulated evolutionary optimisation," *Ieee Transactions on Neural Networks,* vol. 5, Jan 1994.
10. E. J. S. Pires, J. A. T. Machado, and I. Ieee, "A GA perspective of the energy requirements for manipulators maneuvering in a workspace with obstacles," *Proceedings of the 2000 Congress on Evolutionary Computation, Vols 1 and 2,* 2000.
11. S. G. Yue, D. Henrich, W. L. Xu, and S. K. Tso, "Point-to-Point trajectory planning of flexible redundant robot manipulators using genetic algorithms," *Robotica,* vol. 20, May-Jun 2002.
12. F. L. Lewis, D. M. Dawson and C. T. Abdallah, *Robot Manipulator Control: Theory and Practice*, Marcel Dekker Publishing Co., New York, NY, 2003.
13. I. K. Bahaa, I. M. Ali, T. O. Ali, "Motion planning for a robot arm by using Genetic Algorithm" JJMIE, vol. 2, pp. 131-136, Sep 2008.
14. T. W. Lung "Robot Analysis, the Mechanism of Serial and Parallel Manipulators" John Wiley &Sons, Inc., 1999.
15. D. Xin, C. Hua, G. W. Kang, "Neural Network And Genetic Algorithm Based Global Path Planning In A Static Environment", Journal Of Zhejiang University Science, Vol. 6A No. 6, 2005, 549-554.

A NOVEL HYBRID SPIRAL-DYNAMICS RANDOM-CHEMOTAXIS OPTIMIZATION ALGORITHM WITH APPLICATION TO MODELLING OF A FLEXIBLE ROBOT MANIPULATOR

A. N. K. NASIR and M. O. TOKHI

Department of Automatic Control & Systems Engineering, University of Sheffield, UK

This paper presents a hybrid spiral-dynamics algorithm with random-chemotaxis; a synergy between chemotactic strategy of bacteria and spiral dynamics algorithm. Bacterial foraging algorithm has a good local search technique through random tumble and swim action. However it has slow convergence speed and high processing time. On the contrary, spiral dynamics algorithm is relatively simple and has very good global search technique based on its spiral model. This results in fast convergence speed and short computation time. However, the insufficient local search strategy may possibly trap the algorithm to local optima thus reduced accuracy. Incorporating bacteria chemotactic strategy into spiral dynamics algorithm can effectively solve the problems. Moreover, the strengths of both original algorithms can be preserved hence improving their performances. In this work, the proposed algorithm is used as an optimization tool to estimate parameters of high order dynamic model of a flexible manipulator system. The results show that the proposed algorithm has faster convergence speed and higher accuracy in comparison to its predecessor algorithms. On the other hand, time-domain and frequency-domain responses show that the algorithm can predict better model for the flexible manipulator system.

1. Introduction

Metaheuristic algorithms have been considered by researchers as important tools to solve real world problems due to their capability of producing reliable solutions and optimum results. Most of the algorithms have been developed based on adaptations that are either biological-inspired (living organism) or nature-inspired (natural-phenomena). Bacterial foraging algorithm (BFA) and Spiral dynamics algorithm (SDA) are two examples of metaheuristic algorithm that can be classed under biological-inspired and nature-inspired respectively.

BFA is the adaptation of foraging strategy of Escherichia Coli bacteria to survive in their life [1]. The algorithm consists of 3 basic phases namely chemotaxis, reproduction and elimination-dispersal. In the chemotaxis phase, each bacterium tumbles one step randomly around its current location looking for higher nutrient location. The bacterium swims few more steps toward similar direction of that tumble if the position after each motion has higher nutrient than the previous location. This technique is called chemotactic strategy and is continuously repeated until the end of the bacterium life. Chemotaxis is the most important phase in which it contains a very crucial parameter, namely bacteria step size, C, to move from one location to another. The parameter can predominantly determine the algorithm performance. A large step size may be defined to expedite bacteria movement or to accelerate the algorithm convergence speed. However, it may cause the algorithm to get trapped into local optima if the global optimum point is located in remote area hence decrease the accuracy. Therefore, in this case, a small step size should be defined to increase the accuracy of the solution but it may lead to slower convergence since more steps are required to reach the global optimum point. Reproduction is another phase after the chemotaxis where half of the bacteria in their population with higher fitness are reproduced to increase the health of the overall bacteria population. It may increase the possibility as well expedite the time for the bacteria to reach the highest nutrient location within the search area. From an optimization algorithm point of view, this strategy can help the algorithm to converge faster since it brings half of the bacteria in the population to a closer location to the optimum solution. In order to maintain the total number of bacteria in the population, half of the bacteria with lower fitness are eliminated. Before moving to the next cycles, the whole bacteria population is distributed randomly within the search space. This is to give the opportunity for population to be placed nearer to the optimum location hence improving the performance. Application of BFA can be mostly found in science and engineering, economic, medical and from literature, it has been proven that the solution of this algorithm is reliable and optimum.

SDA has recently gained attention since it is a relatively simple algorithm and it may give competitive performance in comparison to other popular algorithms in solving optimization problems [2]. The development of the algorithm is generally based on the natural spiral phenomena such as hurricanes and tornados, spiral galaxy, and etc. The most important part of SDA is the spiral model in which it contains two crucial parameters that may significantly determine its performance and they are spiral radius, r and angular displacement, θ. The advantage of this algorithm is that it has diversification and intensification at the initial and final phases of the spiral motion respectively. These two properties are very important to ensure that the algorithm converges to the optimum solution. Moreover, the spiral model produces dynamic step size in its spiral trajectory, which can give smooth convergence and it gives higher opportunity for the search point to reach remote location. The diversification enables the algorithm to explore the whole search area with large step size while the intensification allows the exploitation around particular point. In general, the movements of the whole search points for SDA are directed towards a global best position However, the movement of individual search point which can represent a very good strategy for local search may not exist. Therefore, there is no balance between the exploration and exploitation strategy hence causing limitation in the performance.

Since the introduction of BFA and SDA, a lot of developments have been made to improve their performance mostly using hybrid approaches. This paper proposed a synergy between chemotactic strategy of bacteria and spiral dynamics algorithm, which increases the overall performance of its predecessor algorithms. Moreover, the application of the proposed algorithm as a tool to optimally estimate dynamic parametric model of a flexible manipulator system is also presented. Section 2 of this paper presents the details of the hybrid algorithm. The implementation and comparative results of the algorithm to model hub angle and end-point acceleration of a flexible manipulator system are presented in Section 3 while Section 4 concludes the paper.

2. Hybrid Spiral-Dynamics Random-Chemotaxis Algorithm

Hybrid spiral-dynamics random-chemotaxis (HSDBC-R) algorithm is the incorporation of chemotaxis phase of BFA into SDA where it preserves the strentghs and eliminates the drawbacks and thus improves the performance over both predecessor algorithms. The SDA has very good strategy for global search but is limited in local search and vice versa. Incorporating chemotactic strategy into SDA balances the exploration and exploitation parts of the algorithms. The details of HSDC-R parameters are presented in Table 1 while its flowchart is depicted in Figure 1. Notice that, after the bacteria move one-step ahead using spiral model, they perform randomly tumble and swim actions without referring to the global best position to represent local search strategy.

Moreover, an additional step is incorporated into the chemotaxis phase through the introduction of fbest and xbest. This is to ensure the best solution and best fitness are always preserved when the bacteria move from one location to another when performing tumble and swim actions. In addition, it eliminates the oscillation problem in the algorithm convergence.

Table 1. Parameter description for HSDBC-R

Parameter	Description	Parameter	Description
θ	Angular displacement	m	Number of bacteria
r	Spiral radius	Sn	Square matrix
kmax	Maximum iteration	fbest	Best fitness
Ns	Number of swim	xbest	Best location

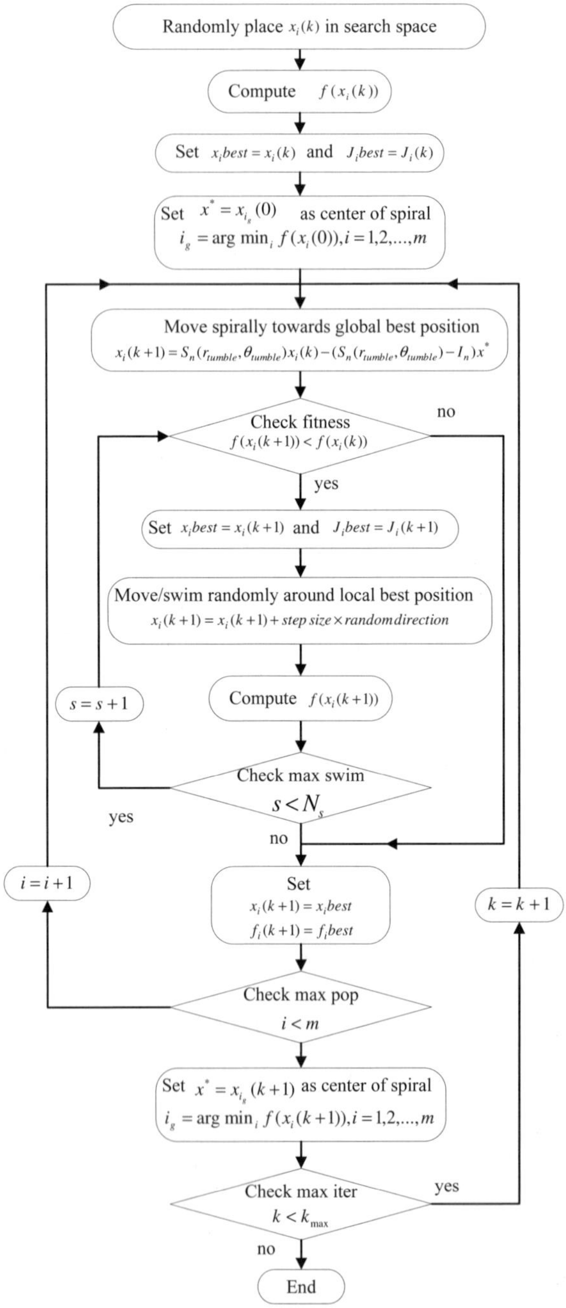

Figure 1. Flowchart of HSDBC-R.

3. Modelling of Flexible Manipulator System and Results

Flexible manipulator system is commonly used as a platform to test designed controllers. A single-link flexible manipulator is used in this study [3], [4]. In order to accurately design a controller, dynamic model for the system must be reliable and precise. Due to natural vibrational behavior in the flexible body and nonlinear characteristic, modelling of a flexible manipulator system is very challenging and difficult. Flexible manipulator consists of multiple outputs namely hub angle, hub velocity and end-point acceleration. From control point of view, modeling of hub angle and end-point acceleration are of interest since they are commonly used to test input tracking and vibration suppression capability. This paper is dealing with the modelling for end-point acceleration, which has higher dimensions in the dynamic model due to the presence of both rigid and flexible dynamics.

In this work, an eight-order ARX model is chosen as a candidate structure for end-point acceleration. This selection is based on a good performance shown by the model in a study conducted previously [5]. The mathematical formulation for the ARX model can be represented as:

$$\hat{y}(t)=\frac{b_0z^{-1}+b_1z^{-2}+b_2z^{-3}+b_3z^{-4}+b_4z^{-5}+b_5z^{-6}+b_6z^{-7}+b_7z^{-8}}{1+a_1z^{-1}+a_2z^{-2}+a_3z^{-3}+a_4z^{-4}+a_5z^{-5}+a_6z^{-6}+a_7z^{-7}+a_8z^{-8}} \tag{1}$$

Notice that, 8 zeros and 8 poles values are to be identified making it as a 16 dimensions problem. In the algorithms under study, 50 search points or bacteria were defined while their initial locations were randomly placed on a feasible search region. The total iteration number for SDA and HSDBC-R was defined as 1000 iterations whereas for BFA it was 3200 iterations since it depends on total number of chemotaxis, Nc, reproduction, Nre and elimination-dispersal, Ned. The optimization parameters for BFA were defined as C = 0.008, Ns = 4, Nc = 200, Nre = 4, and Ned = 4 while for SDA $r=0.99$ and $\theta=0.381$. The HSDBC-R was at its optimum performance when the parameters were defined as C = 0.008, Ns = 3, $r=0.97$ and $\theta=0.48$. The adaptive formulation for spiral radius in HSDBC-R was defined as [6]:

$$r=\frac{(0.97-1)}{1+\dfrac{0.001}{|f(x_i(k))|}}+1 \tag{2}$$

The numerical results for the end-point acceleration model based on SDA, BFA and HSDBC-R are presented in Table 2. Notice that, the best fitness value of 25.03, the fastest convergence over at 356 iterations, the smallest error range in both modelling and validation parts were presented by HSDBC-R.

Table 2. Numerical result for end-point acceleration

Algorithm	Best fitness	Convergence step at 35	Range of error (modelling)	Range of error (validation)
BFA	27.91	1440/3200	[-0.0936, 0.1054]	[-0.0987, 0.1076]
SDA	32.41	645/1000	[-0.1100, 0.0970]	[-0.0983, 0.1274]
HSDBC-R	25.03	356/1000	[-0.0770, 0.0842]	[-0.0802, 0.0977]

The results in terms of time-domain and frequency-domain responses are depicted in Figure 2. Time-domain responses show all the optimization algorithms can closely follow the actual response captured from hardware. Moreover, the power spectral density graph confirms the accuracy of the estimated models where for 1000 Hz sampling frequency, end-point acceleration must display three resonance frequencies approximately at 11.72 Hz, 39.06 Hz and 66.41 Hz in its dynamic model [7].

4. Conclusion

A novel hybrid spiral-dynamics random-chemotaxis that synergizes chemotactic strategy of bacteria and spiral dynamics algorithm has been developed. Combination of some parts from BFA and the whole part of SDA complements to each other and balances the global search and local search strategy thus preserves the strength of both algorithms. In addition to that, an additional step has been incorporated into the chemotaxis phase to sustain the best fitness and best location in each motion hence avoiding the oscillation problem in the convergence. Application of the proposed algorithms to modelling end-point acceleration for the flexible manipulator system shows that it has outperformed the original algorithms both in terms of convergence speed and accuracy. Time domain and frequency domain responses show that it has estimated better model for end-point acceleration.

Acknowledgments

Ahmad Nor Kasruddin Nasir is on study leave from Faculty of Electrical & Electronics Engineering (FKEE), Universiti Malaysia Pahang (UMP) and is sponsored by Ministry of Higher Education (MOHE), Malaysia.

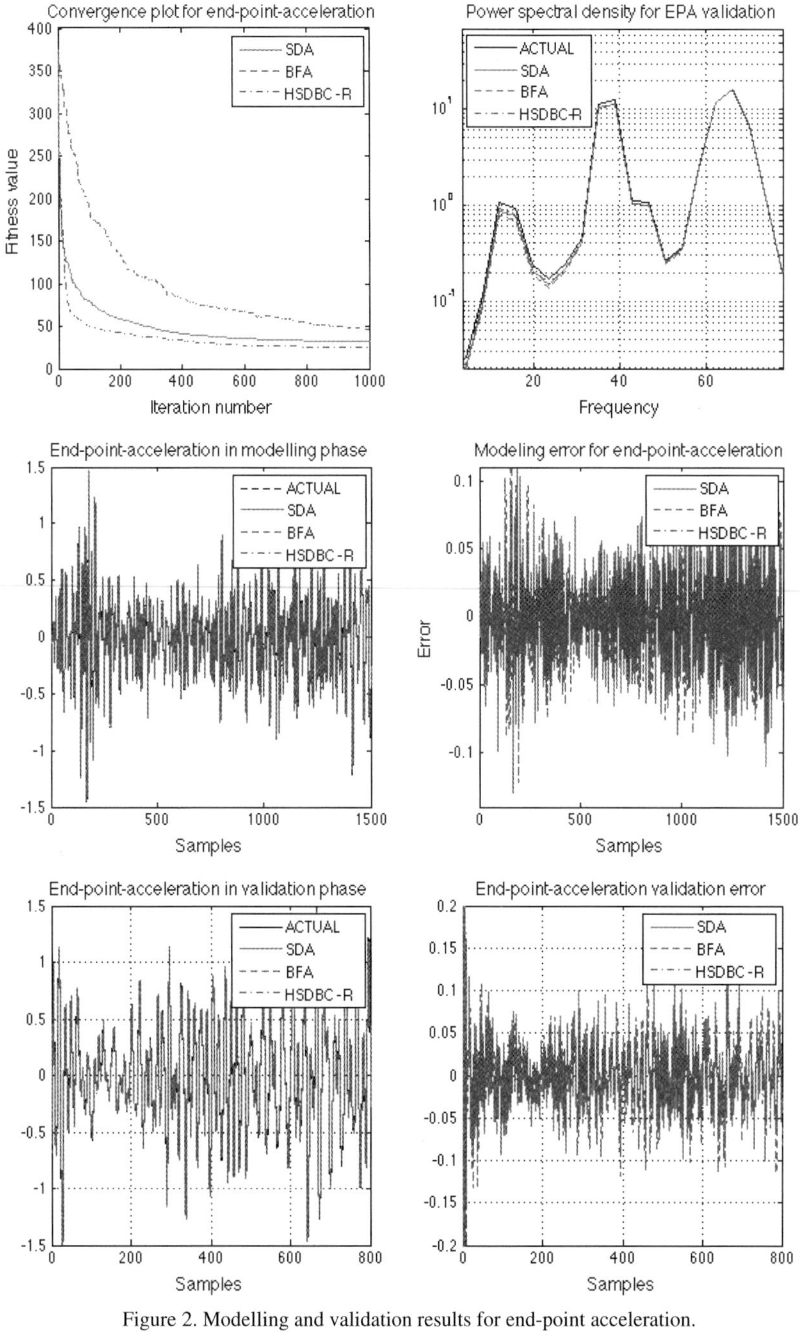

Figure 2. Modelling and validation results for end-point acceleration.

References

1. K. M. Passino, "Biomimicry of bacterial foraging for distributed optimization and control," IEEE Control System Magazine, June 2002, pp. 52 – 67.
2. K. Tamura, and K. Yasuda, "Primary study of spiral dynamics inspired optimization," IEEJ Transactions on Electrical and Electronic Engineering, **6**(S1), 2011, pp. 98 – 100.
3. A.K.M. Azad, "Analysis and design of control mechanism for flexible manipulator systems." PhD thesis, Dept of Automatic Control & Systems Eng., The University of Sheffield, UK, 1994.
4. Poerwanto, H, "Dynamics simulation and control of flexible manipulator systems", PhD thesis, Dept of Automatic Control & Systems Eng., The University of Sheffield, UK, 1998.
5. A.N.K. Nasir and M.O. Tokhi, "Novel hybrid bacteria foraging and spiral dynamics algorithms with application to modelling of flexible system", Proceeding of 5th International Conference on Modelling, Simulation and Applied Optimization (ICMSAO' 2013), 28 – 30 April 2013, Hammamet, Tunisia.
6. A.N.K. Nasir, M.O. Tokhi, N.M. Abd Ghani, R.M.T. Raja Ismail, "Novel Adaptive Spiral Dynamics Algorithms for Global Optimization", Proceeding of 11th IEEE Conference on Cybernatic Intelligence Systems 2012, Limerick Ireland, 23 – 24 August 2012, pp. 99 – 104.
7. Supriyono, H. "Novel bacterial foraging optimization algorithms with application to modelling and control of flexible manipulator systems", PhD thesis, Dept of Automatic Control & Systems Eng., The University of Sheffield, UK, 2012.

SECTION–9

MODELLING AND SIMULATION OF CLAWAR

MODELLING OF MOVEMENT OF THE THREE-LINK ROBOT WITH OPERATED FRICTION FORCES ON THE HORIZONTAL SURFACE

S.F. JATSUN, L.YU. VOLKOVA, G.S. NAUMOV and A.S. YATSUN

Department of Theoretical Mechanics and Mechatronics, South-West State University, 305040, 50 let Oktyabry, 94, Kursk, Russia

In this article the three-link robot moving on a horizontal surface is considered, the mathematical model of object is developed, results of movement numerical modeling of the device are presented.

1. Introduction

Nowadays different scientists provide in the field of the mechanisms, which imitate behavior of real biological prototypes. Copying of the main principles of the biological objects behavior allows to create the devices possessing unique properties which find concrete practical application. Snake-like robots belong to such devices. Further under the snake-like robot we mean solid bodies connected among themselves by joints which are equipped with rotary drives, it allows to links to make relative movement [1-9]. As a result of links interaction with a supporting surface the mechanical system makes some movement in a certain direction.

2. Description of the three-link robot

Let's consider the crawling robot consisting of links 1, 2 and 3 which general view is given in figure 1. Links 1 and 2, 2 and 3 are connected among themselves by rotary drives 4 and 5. The robot interacts with surface by contact elements 6 – 9.

The described robot differs from already known ones that extreme points of the device's links are equipped with contact elements 6 - 9, friction forces in which with a surface are operated sizes that is reached by use of the special drives changing properties of contact surfaces. It leads that in certain timepoints friction forces can fix points of the robot links on a surface and can be equal to zero. It means that there are two essentially various phases of movement of object. In the first phase when the friction forces aren't equal zero the robot moves on a rough surface, the position of the object's center of mass is change. In the second phase when the friction forces are equal zero the robot moves on a smooth surface, such type of movement is used for change of the device configuration.

The first phase of the robot movement is the most difficult for modeling therefore in this article the special attention to research of the device behavior in this phase is paid.

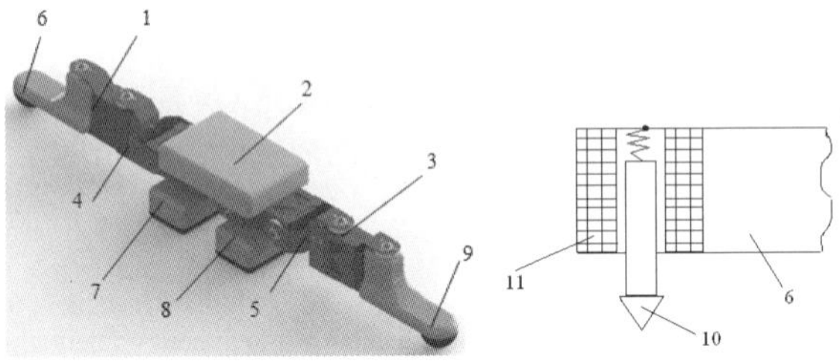

a b

Figure 1. a - general view of the robot, b - control module of the friction forces: 1 – 3 – links, 4, 5 – drives, 6 – 9 – contact elements, 10 – edge element; 11 – electromagnetic drive.

3. Mathematical model of the robot

For the description of the multilink robot movement we will enter absolute motionless system of coordinates Oxy and relative systems of coordinates $O_ix_iy_i$ which are rigidly connected with points O_1, O_2, O_3. Angles φ_i define rotate of coordinate systems $O_ix_iy_i$ concerning Oxy (figure 2).

We will consider that all robot links i=1÷3 are absolutely rigid bodies and represent cores lengths l_i which masses m_i are concentrated in the centers of their symmetry C_i.

The robot moves on a rough horizontal surface under the action of the torques M_{12} (M_{21}) and M_{32} (M_{23}) formed by drives 4 and 5 and providing relative movements of links 1 and 2, 2 and 3, and at the expense of friction forces F_1, F_2, F_3 and F_4 arising in points O_1, O_2, O_3, O_4 and directed opposite vectors of velocities in these points. Each of friction force F_k, k=1–4 is described by the following law:

$$F_k = \begin{cases} F_k^0, \ 0 \le t < t_1^k, \\ 0, \ t_1^k \le t < t_2^k. \end{cases} \quad (1)$$

where t_1^k – time of shutdown of force F_k, t_2^k – time of inclusion of force of F_k,.

The links positions is described by a vector of the generalized coordinates

$$\overline{q} = (q_1 \quad q_2 \quad q_3 \quad q_4 \quad q_5)^T, \quad (2)$$

at which $q_1=x_{C1}$, $q_2=y_{C1}$ – coordinates of the center mass of the link 1 in the system Oxy, $q_3=\varphi_1$, $q_4=\varphi_2$ and $q_5=\varphi_3$ – angles of links rotation.

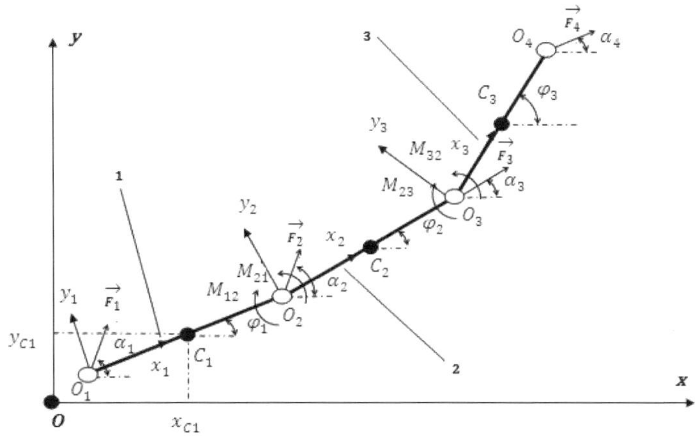

Figure 2. Settlement scheme of the robot.

The system of the differential equations of the device movement is formed with use of the Lagrange equations of the second kind and in a matrix form looks like:

$$A(q)\ddot{q} + B(q)\dot{q}^2 = F\,. \tag{3}$$

where $A(q)$, $B(q)$ – matrixes of coefficients, F – matrix of generalized forces:

$$A(q) = \begin{pmatrix} a_{11} & a_{12} & a_{13} & a_{14} & a_{15} \\ a_{21} & a_{22} & a_{23} & a_{24} & a_{25} \\ a_{31} & a_{32} & a_{33} & a_{34} & a_{35} \\ a_{41} & a_{42} & a_{43} & a_{44} & a_{45} \\ a_{51} & a_{52} & a_{53} & a_{54} & a_{55} \end{pmatrix},\ B(q) = \begin{pmatrix} b_{11} & b_{12} & b_{13} & b_{14} & b_{15} \\ b_{21} & b_{22} & b_{23} & b_{24} & b_{25} \\ b_{31} & b_{32} & b_{33} & b_{34} & b_{35} \\ b_{41} & b_{42} & b_{43} & b_{44} & b_{45} \\ b_{51} & b_{52} & b_{53} & b_{54} & b_{55} \end{pmatrix} \tag{4}$$

$$F = \begin{pmatrix} F_{11} & F_{21} & F_{31} & F_{41} & F_{51} \end{pmatrix}^T. \tag{5}$$

Coefficients of matrix $A(q)$ are determined by formulas:

$$a_{11} = m_1 + m_2 + m_3,\ a_{12} = 0,\ a_{13} = -(l_1/2)\sin\varphi_1(m_2 + m_3),$$
$$a_{14} = -(l_2/2)\sin\varphi_2(m_2 + 2m_3),\ a_{15} = -m_3(l_3/2)\sin\varphi_3,\ a_{21} = 0,$$
$$a_{22} = m_1 + m_2 + m_3,\ a_{23} = (l_1/2)\cos\varphi_1(m_2 + m_3),\ a_{24} = (l_2/2)\cos\varphi_2(m_2 + 2m_3),$$
$$a_{25} = m_3(l_3/2)\cos\varphi_3,\quad a_{31} = -(m_2 + m_3)(l_1/2)\sin\varphi_1,$$
$$a_{32} = (m_2 + m_3)(l_1/2)\cos\varphi,\ a_{33} = J_1 + 0.25m_2l_1^2 + 0.25m_3l_1^2,$$

$$a_{34} = (0.5m_2 + m_3)(l_1 l_2/2)\cos(\varphi_1 - \varphi_2),\ a_{35} = m_3(l_1 l_3/4)\cos(\varphi_1 - \varphi_3),$$
$$a_{41} = -(m_2 + 2m_3)(l_2/2)\sin\varphi_2,\ a_{42} = (m_2 + 2m_3)(l_2/2)\cos\varphi_2,$$
$$a_{43} = (0.5m_2 + m_3)(l_1 l_2/2)\cos(\varphi_1 - \varphi_2),\ a_{44} = (J_2 + 0.25m_2 l_2^2 + m_3 l_2^2),$$
$$a_{45} = m_3(l_2 l_3/2)\cos(\varphi_2 - \varphi_3),\ a_{51} = -m_3(l_3/2)\sin\varphi_3,\ a_{52} = m_3(l_3/2)\cos\varphi_3,$$
$$a_{53} = m_3(l_1 l_3/4)\cos(\varphi_1 - \varphi_3),\ a_{54} = (l_2 l_3/2)m_3\cos(\varphi_2 - \varphi_3),$$
$$a_{55} = J_3 + 0.25m_3 l_3^2$$

Coefficients of matrix $B(q)$ are equal:

$$b_{11} = 0,\ b_{12} = 0,\ b_{13} = -(m_2 + m_3)(l_1/2)\cos\varphi_1,\ b_{14} = -(m_3 + 2m_3)(l_2/2)\sin\varphi_2,$$
$$b_{15} = -m_3(l_3/2)\cos\varphi_3,\ b_{21} = 0,\ b_{22} = 0,\ b_{23} = -(m_2 + m_3)(l_1/2)\sin\varphi_1,$$
$$b_{24} = -(m_2 + 2m_3)(l_2/2)\sin\varphi_2,\ b_{25} = -m_3(l_3/2)\sin\varphi_3,\ b_{31} = 0,\ b_{32} = 0,\ b_{33} = 0,$$
$$b_{34} = (0.5m_2 + m_3)(l_1 l_2/2)\sin(\varphi_1 - \varphi_2),\ b_{35} = m_3(l_1 l_3/4)\sin(\varphi_1 - \varphi_3),\ b_{41} = 0,$$
$$b_{42} = 0,\ b_{43} = (0.5m_2 + m_3)(l_1 l_2/2)\sin(\varphi_2 - \varphi_1),\ b_{44} = 0,$$
$$b_{45} = m_3(l_2 l_3/2)\sin(\varphi_2 - \varphi_3),\ b_{51} = 0,\ b_{52} = 0,\ b_{53} = m_3(l_1 l_3/4)\sin(\varphi_3 - \varphi_1),$$
$$b_{54} = m_3(l_2 l_3/2)\sin(\varphi_3 - \varphi_2),\ b_{55} = 0.$$

Coefficients of matrix F are presented in the following look:

$$F_{11} = F_1\cos\alpha_1 + F_2\cos\alpha_2 + F_3\cos\alpha_3 + F_4\cos\alpha_4,$$
$$F_{21} = F_1\sin\alpha_1 + F_2\sin\alpha_2 + F_3\sin\alpha_3 + F_4\sin\alpha_4,$$
$$F_{31} = M_{12} + F_2(l_1/2)\sin(\alpha_2 - \varphi_1) + F_3(l_1/2)\sin(\alpha_3 - \varphi_1) + F_4(l_1/2)\sin(\alpha_4 - \varphi_1),$$
$$F_{41} = M_{21} + M_{23} + F_3 l_2\sin(\varphi_2 - \alpha_3) + F_4 l_2\sin(\varphi_2 - \alpha_4),$$
$$F_{51} = M_{32} + F_4 l_3\sin(\varphi_3 - \alpha_4).$$

4. Modeling of the robot movement

As object of modeling the three-link robot is considered, masses and lengths of which links are equal: m_i=1 kg, l_i=1 m. Modeling is carried out under zero initial conditions. For definition of linear and angular movements, velocities and accelerations of multilink system we will use specially developed algorithm with an integration method of Verlet.

The purpose of modeling is research influence of the torques M_{12} (M_{21}) and M_{32} (M_{23}) on the robot links movements in the phase when friction forces are equal to zero.

Let's consider robot movement by torque M_{32} which is described by a law:

$$M_{32} = \begin{cases} M_{32}^{0}, & 0 < t < t^*, \\ 0, & t > t^*, \end{cases} \quad \text{at } t^* = 2\text{ s},\ M_{32}^{0} = 1\text{ Nm}.$$

Movement of the robot stops when the angle of the link 3 are equal $\varphi_3=\pi$. The positions of links of the mechanism on the plane to some timepoints are provided on figure 3.

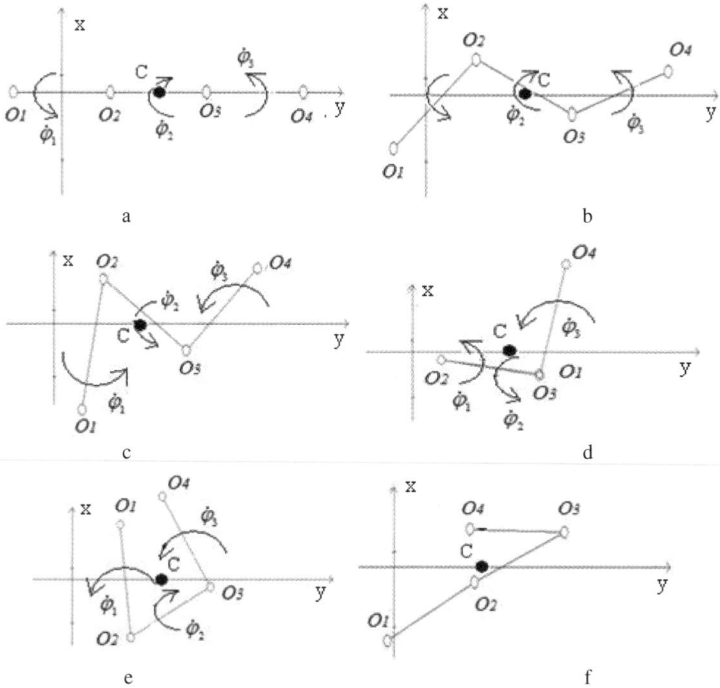

Figure 3. The positions of links on the plane in timepoints: a) t=0 s (initial position); b) t=0,25 s; c) t=0,5 s; d) t=0,75 s; e) t=0,9 s; f) t=1,05 s (final position) at action of torque M_{32}.

According to these dependences it is visible that links of system make complex movement, links 1 and 3 always rotate counterclockwise, and the link 2 rotates in the beginning towards to links 1 and 3 and then changes the rotation directions for the opposite. The center of mass of the system - a point C - thus remains motionless.

In case the system moves under the action of the torque M_{21} describing the law:

$$M_{21} = \begin{cases} M_{21}^{0}, & 0 < t < t^*, \\ 0, & t > t^*, \end{cases} \qquad \text{at } t^* = 2\text{ s}, M_{21}^{0}=1\text{ Nm},$$

before condition $\varphi_2=\pi$ performance, the positions of the robot links to the fixed timepoints are presented in figure 4.

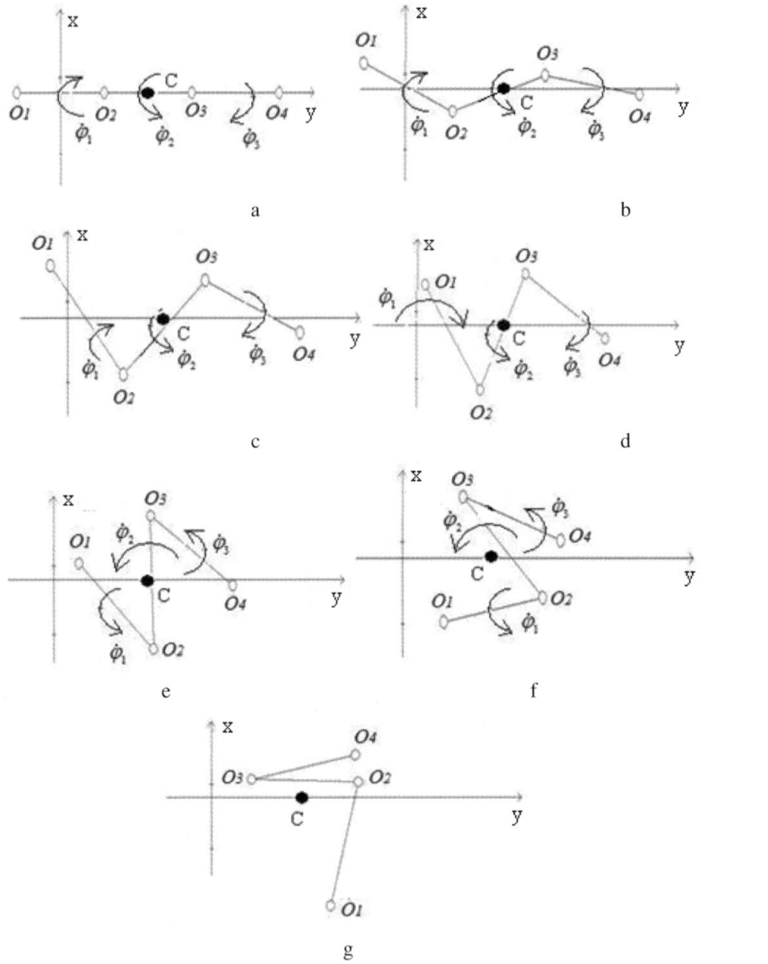

Figure 4. The positions of links on the plane in timepoints: a) t=0 s (initial position); b) t=0,25 s; c) t=0,5 s; d) t=0,75 s; e) t=1 s; f) t=1,25 s; g) t=1.5 s (final position) at action of torque M_{21}.

According to dependences it is visible that the device link 2 rotates counterclockwise, links 1 and 3 thus will be satisfied in the beginning rotate clockwise, and then change the direction of rotation and all links of the device rotate counterclockwise, position of the center of mass of system also remains invariable.

Let's consider object movement under the action of the torques M_{12} and M_{32} having equal modules, both opposite directions and changing on the following laws:

$$M_{12} = \begin{cases} M_{12}^{0}, \ 0 < t < t^{*}, \\ 0, \ t > t^{*}, \end{cases} \quad M_{32} = \begin{cases} M_{32}^{0}, \ 0 < t < t^{*}, \\ 0, \ t > t^{*}, \end{cases}$$

at $t^{*} = 2$ s, $M_{12}^{0} = -1$ Nm, $M_{32}^{0} = 1$ Nm.

The robot moves until the condition $\varphi_3 = \pi/2$, $\varphi_1 = -\pi/2$ will be satisfied. In this case symmetric movement of links 1 and 3 concerning the center of mass of system is observed, the link 1 rotates clockwise, and a link 3 counterclockwise, the link 2 moves progressively (figure 5). At the following phase there is a fixing of points O_2 and O_3 and rotation of links 1 and 3 to the opposite side.

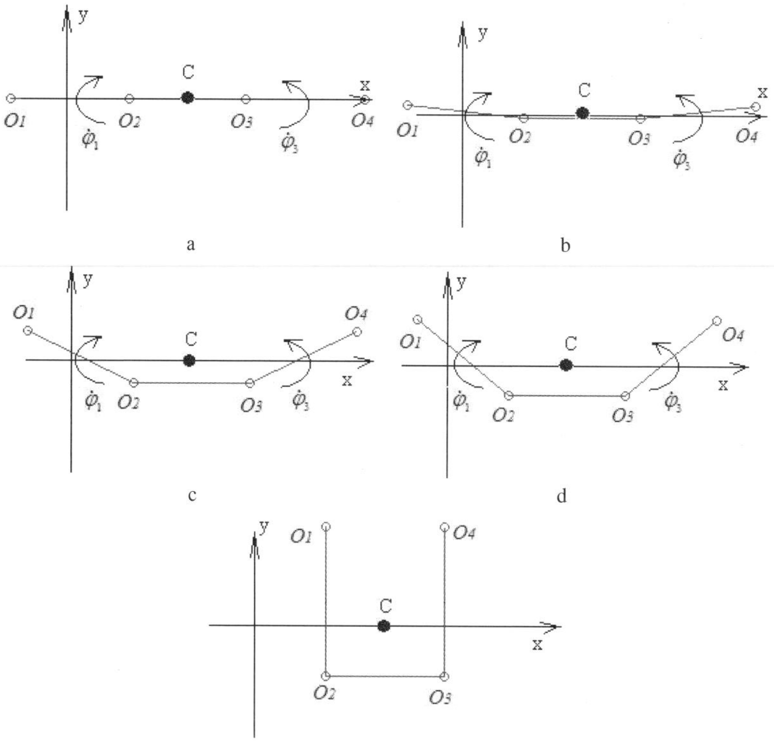

Figure 5. The positions of links on the plane in timepoints: a) t=0 s (initial position); b) t=0,2 s; c) t=0,4 s; d) t=0,6 s; e) t=0,8 s (final position) at action of torques M_{12} and M_{32}.

5. Conclusions

In article the three-link robot is considered, forces of interaction of which contact elements with a surface are operated sizes. The mathematical model of object describing phases of movement of the device at absence and existence of friction of contact elements of the robot construction with a surface that corresponds to movement of the device on rough and smooth surfaces is presented. The algorithm of movement of the device is developed, results of numerical modeling of the robot movement on a horizontal surface without friction forces are received and analysed. Such results could be useful for design and investigation of real snake-like robot prototype construction and for optimization control algorithms.

References

1. Lavrentiev M. A., Lavrentiev M. M. *About One Principle of Creation of Traction Force of Movement* // Journal of Applied Mechanics and Technical Physics. 1962. No. 4. P. 3-9.

2. Hirose S., Monshima /1. *Design and Control of a Mobile Robot with an Articulated Body* // International Journal of Robotic Research. 1990. Vol. 9, No. 2. P. 99-115.

3. Hirose S. *Biologically Inspired Robots: Snake-like Locomotors and Manipulators* // Oxford: Oxford Univ. Press, 1993. 220 p.

4. Li N., Furuta K., Chernousko F. L. *Motion Generation of the Capsubot Using Internal Force and Static Friction* // Proc. 45th IEEE Conf. on Decision and Control. San Diego. USA. 2006. P. 6575-6580.

5. Burdick J. W., Radford J., Chirikjan G. S. A *Sidewinding Locomotion Gait for Hyper-Redundant Robots* // Proc. 1993 IEEE Intern. Conf. on Robotics and Automation. Atlanta. 1993. Vol. 3. P. 101–106.

6. Chernousko F.L. *Optimum Control of Movement of Multilink System in the Environment with a Resistance* // PMM. 2012. Vol. 76. No. 3. P. 355-373.

7. Chernousko F.L., Shunderuk M.M. *Influence of Friction Forces on Dynamics of the Two-Link Mobile Robot* // PMM. 2010. Vol. 74. No. 1. P. 22-36.

8. Chernousko F.L. *Optimum Movement of Multilink System in the Environment With a Resistance* // Tr. Inst. Mat. Mekh. UrO RAN. 2011. Vol. 17. No 2. . P. 240–255.

9. Bolotnik N.N., Figurina T.Yu., Chernousko F.L. *Optimum Control of Rectilinear Movement of Two Bodies System in the Resisting Environment* // PMM. 2012. Vol. 76. No. 1. P. 3-22.

TRANSITION ANALYSIS OF A BIPED POLE-CLIMBING ROBOT — CLIMBOT

HAIFEI ZHU[1], YISHENG GUAN[2], WENQIANG WU[1], XUEFENG ZHOU[3] and HONG ZHANG[4]

[1]*School of Mechanical and Automotive Engineering,*
[3]*School of Electronic and Information Engineering,*
South China University of Technology,
Guangzhou, Guangdong, 510640, China

[2]*School of Mechanical and Electrical Engineering,*
Guangdong University of Technology,
Guangzhou, Guangdong, 510006, China
E-mail: ysguan@scut.edu.cn

[4]*Department of Computing Science, University of Alberta,*
Edmonton, AB, Canada, T6G 2E8
E-mail: hzhang@ualberta.ca

Transition is a desired and critical capability for robots climbing in trusses. In this paper, the issue of transition with a 5-DoFs biped pole-climbing robot between two staggered squared poles in arbitrary relative configuration is addressed and investigated. Two important propositions are first presented and proved. An algorithm based on the analysis is then presented to figure out the optional grasping regions on the current and the target grasping poles. Finally, simulations are conducted to verify and demonstrate the effectiveness of the transition analysis and the proposed algorithm.

Keywords: Climbing Robot; Transition Analysis; Motion Planning.

1. Introduction

To carry out high-rise tasks in place of humans, a series of pole-climbing robots have been developed and investigated in the past decades, such as UT-PCR,[1] CPR,[2] Shady3D,[3] 3DCLIMBER,[4] Climbot,[5] Treebot,[6] and so on. Among them, those with an arm-like body in the middle and grippers at the two ends are considered to be dominant, for they inherit the flexibility and dexterity of the climbing patterns of some animals like inchworms and slothes. Robots with this locomotion are referred to Biped Pole-Climbing Robots (hereafter BiPCRs), typical examples of which are Shady3D, Climbot, 3DCLIMBER, and Treebot.

Being able to transit among poles is one of the distinguished advantages of BiPCRs. However, most of research attention has been paid to robotic system development and simple climbing motion on a pole, less work focusing on the transition issue. To the best of our knowledge, only transition process of BiPCRs was qualitatively described in Refs. 3,7, lacking of specific analysis of how to plan the transiting motion with a BiPCR. As an actual and critical demand, a pole-climbing robot needs to figure out whether the transition is possible, where to start the transition and which are the optional reachable grasping points on poles. In other words, the feasibility and implementation of transition among poles should be addressed.

This paper focuses on the transition analysis of BiPCRs like the 5-DoFs biped climbing robot Climbot. Although a BiPCR with 5 DoFs cannot reach arbitrary configuration with its swinging gripper, it is still capable of performing some transitions between two staggered poles. We propose and prove two propositions which are important to analyze the transition problem with 5-DoFs BiPCRs. In addition, we will present an algorithm to solve the problem mentioned in the previous paragraph.

2. Problem Statement

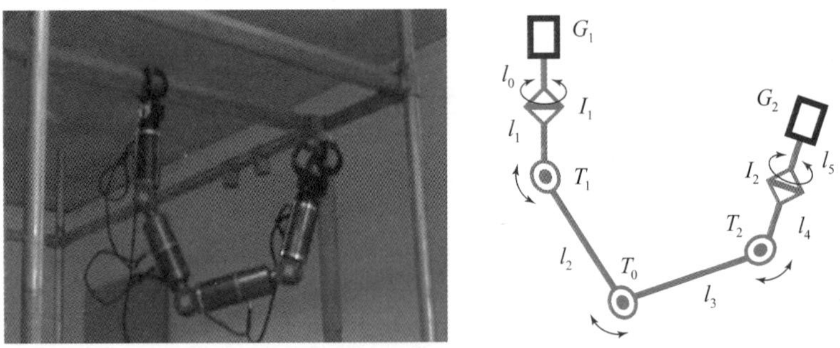

Fig. 1. Climbot transiting among poles and its kinematic model

Climbot is a biped climbing robot with five degrees of freedom we built with modular method, as shown in Fig. 1. At each end of Climbot is a gripper consisting of two fingers with perpendicular V-shaped grooves. Climbot is thus capable of grasping cylindrical and squared poles of various sizes. We focus on the transition scene similar to that in Fig. 1, that is, transition between two squared poles with arbitrary relative configuration. Considering grasping orientation constrains, grasps may be achieved from four perpendicular directions depending on the squared pole (see Fig. 2). As a result,

16 possible transition cases should be checked and confirmed in total. For each possibility, the optional grasping regions on two poles should be found.

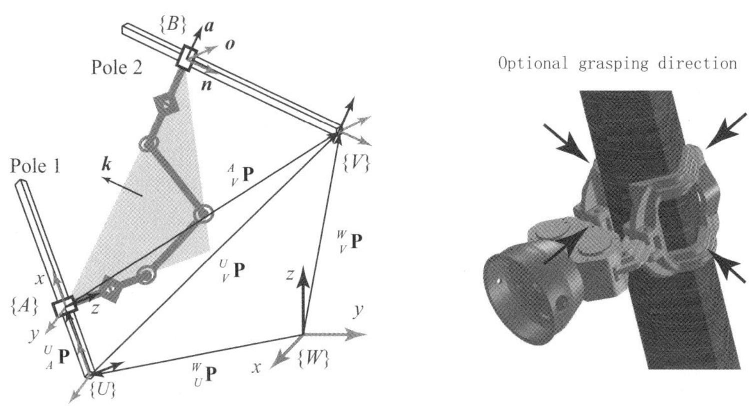

Fig. 2. The transition model of Climbot

For simplify of the analysis and without loss of generality, we take one of the 16 possible transitions for example. The transition may be depicted by a geometric model shown in Fig. 2. Let $\{W\}$ denote the world coordination frame, and the origins of frames $\{U\}$ and $\{V\}$ be the reference points of two poles, and the origins of frames $\{A\}$ and $\{B\}$ the current and the target grasping points, respectively. We have the following problem statement.

Problem : *Given two squared poles* ${}^{W}_{A}\boldsymbol{P} = {}^{W}_{U}\boldsymbol{P} + t_1{}^{W}\boldsymbol{d}_1$ *and* ${}^{W}_{B}\boldsymbol{P} = {}^{W}_{V}\boldsymbol{P} + t_2{}^{W}\boldsymbol{d}_2$ *(where* ${}^{W}\boldsymbol{d}_1$ *and* ${}^{W}\boldsymbol{d}_2$ *are the unit vectors of the pole axes,* t_1 *and* t_2 *are two scales), the grasping orientations* ${}^{W}_{U}\mathbf{R}$ *and* ${}^{W}_{V}\mathbf{R}$*, and Climbot's inverse kinematics* ${}^{B}_{A}\mathbf{T}$*, to find the optional grasping regions on* ${}^{W}_{A}\boldsymbol{P}$ *and* ${}^{W}_{B}\boldsymbol{P}$ *to perform a transition between them.*

3. Analysis

Clearly, if the robot may reach frames $\{A\}$ and $\{B\}$ with its two grippers in desired orientation, then the transition between them is possible. In other words, the transition feasibility is equivalent to the existence of solutions of the robotic inverse kinematics ${}^{B}_{A}\mathbf{T}$. Because of this, we need to describe *Pole 2* with respect to frame $\{A\}$. From Fig. 2, we have the homogeneous transformation ${}^{U}_{A}\mathbf{T} = \begin{bmatrix} \mathbf{I}_{3\times 3} & {}^{U}_{A}\boldsymbol{P} \\ \mathbf{0} & 1 \end{bmatrix}$ and ${}^{U}_{A}\boldsymbol{P} = \begin{bmatrix} t_1 \; 0 \; 0 \end{bmatrix}^{\mathrm{T}}$. Thus,

$$ {}^{A}_{V}\boldsymbol{P} = {}^{A}_{U}\mathbf{T}{}^{U}_{V}\boldsymbol{P} = ({}^{U}_{A}\mathbf{T})^{-1}{}^{U}_{V}\boldsymbol{P} = {}^{U}_{V}\boldsymbol{P} - {}^{U}_{A}\boldsymbol{P}. \tag{1} $$

Similarly, we can express $\{V\}$ with respect to $\{U\}$ as:

$$\begin{bmatrix} {}^U_V\boldsymbol{P} \\ 1 \end{bmatrix} = {}^U_W\mathbf{T} \begin{bmatrix} {}^W_V\boldsymbol{P} \\ 1 \end{bmatrix} = \begin{bmatrix} {}^W_U\mathbf{R}^{\mathrm{T}} & -{}^W_U\mathbf{R}^{\mathrm{T}}{}^W_U\boldsymbol{P} \\ \mathbf{0} & 1 \end{bmatrix} \begin{bmatrix} {}^W_V\boldsymbol{P} \\ 1 \end{bmatrix} = \begin{bmatrix} {}^W_U\mathbf{R}^{\mathrm{T}}({}^W_V\boldsymbol{P} - {}^W_U\boldsymbol{P}) \\ 1 \end{bmatrix}. \quad (2)$$

Substituting Eq. (2) into Eq. (1), we may obtain the reference origin of *Pole 2* with respect to $\{A\}$ as,

$${}^A_V\boldsymbol{P} = {}^W_U\mathbf{R}^{\mathrm{T}}({}^W_V\boldsymbol{P} - {}^W_U\boldsymbol{P}) - {}^U_A\boldsymbol{P}. \quad (3)$$

The direction unit vector of *Pole 2* can be easily transformed to $\{A\}$ as ${}^A\boldsymbol{d}_2 = {}^U_A\mathbf{T}^{-1}{}^W_U\mathbf{T}^{-1}{}^W\boldsymbol{d}_2$. Thus, we have ${}^A_B\boldsymbol{P} = {}^A_V\boldsymbol{P} + t_2{}^A\boldsymbol{d}_2$.

In the following subsections, we will further analyze the transition problem under consideration of orientation constrains and position constrains.

3.1. *Orientation Constrains*

Considering orientation constrains, we propose and prove two propositions which are critical for analyzing the pole-transition problem with climbing robots like Climbot. Note that variable symbols without superscript are used with respect to frame $\{A\}$ hereafter.

Proposition 3.1. *Given the current base grasping at $\{A\}$ and the target grasping orientation ${}^A_B\mathbf{R} = \begin{bmatrix} \boldsymbol{n} & \boldsymbol{o} & \boldsymbol{a} \end{bmatrix}$, there exists only one possible configuration for Climbot to transit between the two staggered poles.*

Due to the special structure configuration of Climbot, its links are always in a plane ("robot plane" for short, shown in gray in Fig. 2). Therefore, the following orientation constrains must be satisfied,

$$\boldsymbol{k} \cdot \boldsymbol{a} = 0, \quad \boldsymbol{n} \cdot \boldsymbol{a} = 0, \quad \|\boldsymbol{a}\| = 1, \quad (4)$$

where $\boldsymbol{k} = \begin{bmatrix} -\sin\alpha & \cos\alpha & 0 \end{bmatrix}^{\mathrm{T}}$ represents the norm vector of the robot plane, $\tan\alpha = a_y/a_x$. Substituting the equation of *Pole 2* into (4) yields

$$t_2 = \frac{p_{Vx}a_y - p_{Vy}a_x}{n_y a_x - n_x a_y}. \quad (5)$$

Hence we may obtain a unique possible grasping position as ${}^A_B\boldsymbol{P} = {}^A_V\boldsymbol{P} + t_2{}^A\boldsymbol{d}_2$. Proposition 3.1 is proved.

Proposition 3.2. *For Climbot to transit between two staggered poles with the base and target grasping orientations ${}^W_U\mathbf{R}$ and ${}^W_V\mathbf{R}$ respectively, the optional grasping region on Pole 2 is linear with the possible base grasping region on Pole 1.*

Recall Eq. (3), and set $\boldsymbol{\lambda} = {}^W_U\mathbf{R}^{\mathrm{T}}({}^W_V\boldsymbol{P} - {}^W_U\boldsymbol{P})$, ${}^A_V\boldsymbol{P}$ may be written as,

$$ {}^A_V\boldsymbol{P} = [p_{Vx} \;\; p_{Vy} \;\; p_{Vz}]^T = [-t_1 + \lambda_x \;\; \lambda_y \;\; \lambda_z]^T. \quad (6) $$

Note that, since ${}^W_U\mathbf{R}$, ${}^W_V\boldsymbol{P}$ and ${}^W_U\boldsymbol{P}$ are known invariants, $\boldsymbol{\lambda}$ is actually a constant vector. Substituting Eq. (6) into Eq. (5) results in

$$ t_2 = \frac{(-t_1 + \lambda_x)a_y - \lambda_y a_x}{n_y a_x - n_x a_y} = \sigma t_1 + \delta, \quad (7) $$

where $\sigma = -a_y/(n_y a_x - n_x a_y)$, $\delta = (\lambda_x a_y - \lambda_y a_x)/(n_y a_x - n_x a_y)$, both are constants. Therefore, t_2 and t_1 have a linear relationship. Once t_1 is determined, t_2 may be calculated by Eq. (7). The proposition is proved.

3.2. *Position Constrains*

To further determine whether the transition is possible, except for the above orientation constrains, we need to ensure that ${}^A_B\boldsymbol{P}$ is inside the reachable regions of the robot. To solve the inverse kinematics, the following inequality must be met,

$$ \left| c_3 = \frac{(p'_x c_1 + p'_y s_1)^2 + (p'_z - l_{01})^2 - (l_2^2 + l_3^2)}{2l_2 l_3} \right| \leq 1, \quad (8) $$

where $c_1 \triangleq \cos\theta_1$, $s_1 \triangleq \sin\theta_1$, $l_{01} \triangleq l_0 + l_1$ (similar expressions are used for shorthand hereafter), $\tan\theta_1 = a_y/a_x$, $\boldsymbol{P'} = \boldsymbol{P}_V + t_2\boldsymbol{n} - l_{45}\boldsymbol{a}$, is the position vector of the wrist joint (say, T_2 in Fig. 1) of Climbot.

Setting $\omega_1 = p'_x c_1 + p'_y s_1$, $\omega_2 = p'_z - l_{01}$, and considering the rotation limitation of joint T_0 and $l_2 = l_3$ for Climbot, Eq. (8) may be simplified as,

$$ 2l_2 \cos\frac{\theta_{lim}}{2} \leq \omega_1^2 + \omega_2^2 \leq 4l_2^2, \quad (9) $$

where θ_{lim} is the rotation range of joint T_0. Utilizing t_1(Eq. (7)), we have

$$ \omega_1^2 + \omega_2^2 = (A^2 + C^2)t_1^2 + 2(AB + CD)t_1 + B^2 + D^2, \quad (10) $$

where $A = \sigma(n_x c_1 + n_y s_1) - c_1$, $B = (\delta n_x - a_x l_{45} + \lambda_x)c_1 + (\delta n_y - a_y l_{45} + \lambda_y)s_1$, $C = \sigma n_z$, $D = \delta n_z - a_z l_{45} + \lambda_z - l_{01}$.

Let $f(t_1) = \omega_1^2 + \omega_2^2 - 4l_2^2$, $g(t_1) = \omega_1^2 + \omega_2^2 - 2l_2 \cos\frac{\theta_{lim}}{2}$, Eq. (9) may be transformed to two quadratic functions, as

$$ f(t_1) = Et_1^2 + Ft_1 + G \leq 0 \quad (11) $$

$$ g(t_1) = Et_1^2 + Ft_1 + H \geq 0 \quad (12) $$

where $E = (A^2 + C^2)$, $F = 2(AB + CD)$, $G = B^2 + D^2 - 4l_2^2$, $H = B^2 + D^2 - 2l_2 \cos\frac{\theta_{lim}}{2}$.

In fact, Eq. (11) describes the accessibility of the target grasping point, and Eq. (12) describes the rotation limitation depending on different robots. Hence, the discriminants of $f(t_1)$ and $g(t_1)$ corresponding to transition situations are listed in Tab. 1.

Table 1. Discriminants corresponding to the transition situations.

Discriminants	Characterization[a]	Optional transition regions
$\Delta_1 < 0$	Out of workspace	No
$\Delta_1 = 0,\ \Delta_2 > 0$	Out of rotation limitation	No
$\Delta_1 = 0,\ \Delta_2 \leq 0$	On the boundary of workspace	Unique
$\Delta_1 > 0,\ \Delta_2 \leq 0$	Some inside ws. and rot. lim.[b]	A continuous segment
$\Delta_1 > 0,\ \Delta_2 > 0$	Some inside ws. but part out of rot. lim.	Two segments

Note: [a] Refer to the target grasping points satisfied orientation constrains.
[b] Terms "ws." stands for workspace, while "rot. lim." for rotation limitation of T_0.

Solving Eq. (11) and (12), we may figure out the potential grasping regions on *Pole 1* and *Pole 2* for Climbot to perform transitions. However, since a pole usually has limited length, a test about whether the grasping points are on the poles should be carried out. Besides, if the rotation limitation of joints T_1 and T_2 is taken into account, another test by solving the inverse kinematics should be carried out. The value ranges of t_1 and t_2 corresponding to invalid movement should be removed.

3.3. *The Algorithm*

The procedure of transition analysis is described in Algorithm 3.1. If it is feasible for Climbot to perform transition, the optional grasping regions described by t_1 and t_2 will be returned. Note that the rotation axis $\mathbf{n}$ in line 21 is the first column vector of ${}^{W}_{U}\mathbf{R}$, $\mathbf{n}$ in line 23 is similar.

4. Simulations

In order to verify the above method for the transition analysis, simulations are conducted with MATLAB in this section. One of the examples is shown in Fig. 3, where transition between two staggered squared poles with arbitrary relative configuration is performed. In this example, there exist 6 possible transitions from different sides of the poles, with the optional grasping regions on *Pole 1* (the green segment). In planning a global climbing path on trusses, grasping points for transition need to be determined within these two regions. The procedure takes about $25 \sim 35$ ms in total.

Algorithm 3.1 Finding the optional grasping regions for Climbot to transit

Input: ${}^W_U\boldsymbol{P}$ and ${}^W_V\boldsymbol{P}$: reference points of the two poles;
${}^W_U\mathbf{R}$ and ${}^W_V\mathbf{R}$: grasping orientations on the two poles.

Output: t_1 and t_2: optional grasping regions on the two poles.

for $i = 1$ to 4 **do**
  for $j = 1$ to 4 **do**
    calculate intermediate variables: σ, δ, $A \sim H$;
    $\Delta_1 \Leftarrow F^2 - 4EG$ and $\Delta_2 \Leftarrow F^2 - 4EH$;
    if $\Delta_1 < 0$ or ($\Delta_1 = 0$ and $\Delta_2 > 0$) **then**
      $t_1 \Leftarrow \emptyset$, $t_2 \Leftarrow \emptyset$;
      continue;
    else if $\Delta_1 = 0$ **then**
      $t_1 \Leftarrow -\frac{F}{2E}$;
    else if $\Delta_1 > 0$ and $\Delta_2 \leq 0$ **then**
      $t_1 \Leftarrow [\frac{-F-\sqrt{\Delta_1}}{2E}, -\frac{F+\sqrt{\Delta_1}}{2E}]$;
    else if $\Delta_1 > 0$ and $\Delta_2 > 0$ **then**
      $t_1 \Leftarrow [\frac{-F-\sqrt{\Delta_1}}{2E}, \frac{-F-\sqrt{\Delta_2}}{2E}]$ *and* $[\frac{-F+\sqrt{\Delta_2}}{2E}, \frac{-F+\sqrt{\Delta_1}}{2E}]$;
    end if
    check rotation limitations of other joints and modify t_1;
    restrict t_1 according to the lengths of the poles;
    if $t_1 \neq \emptyset$ **then**
      $t_2 = \sigma t_1 + \delta$;
      save t_1 and t_2;
    end if
    ${}^W_V\mathbf{R} \Leftarrow R_n(90°){}^W_V\mathbf{R}$;
  end for
  ${}^W_U\mathbf{R} \Leftarrow R_n(90°){}^W_U\mathbf{R}$;
end for

5. Conclusions and Future Work

In this paper, the issue of transition between poles with 5 DoFs BiPCRs is addressed and investigated. Two propositions have been presented and proved for pole-transiting with orientation and position constraints, and an algorithm has been proposed to figure out the optional grasping regions on both poles. Simulations with Climbot transiting between two staggered squared poles have veryfied the validity and the effectiveness of the analysis and the proposed algorithm. In the near future, the algorithm will be utilized in global path planning for autonomous climbing with Climbot.

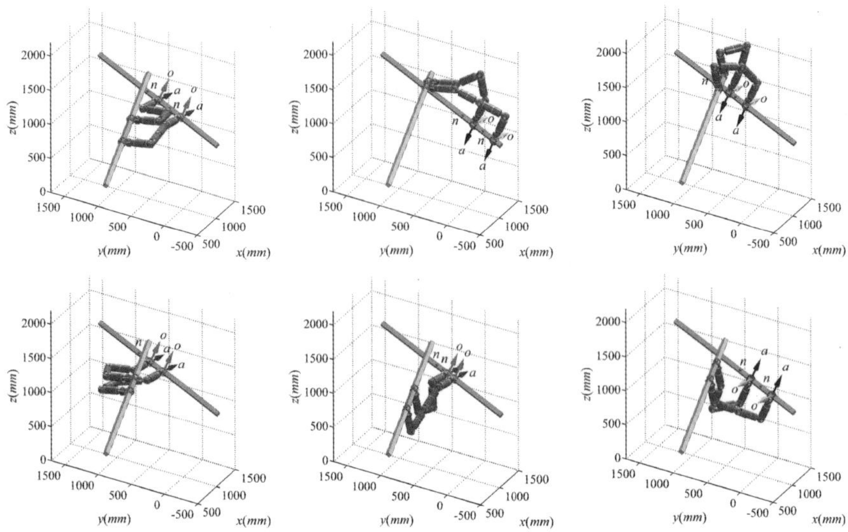

Fig. 3. Simulation of transiting between two staggered squared poles

Acknowledgments

This research is supported in part by the National Natural Science Foundation of China (Grant No.50975089), the National High Technology Research and Development (863) Program of China (Grant No.2009AA04Z204), and the China Postdoctoral Science Foundation (Grant No.2012M521600).

References

1. A. Baghani, M. Ahmadabadi and A. Harati, Kinematics modeling of a wheel-based pole climbing robot (ut-pcr), in *Proc. IEEE Int. Conf. on Robotics and Automation*, 2005.
2. R. Aracil, R. Saltarn and O. Reinoso, *IEEE Robotics and Automation Magazine* **2006**, 16 (2006).
3. Y. Yoon and D. Rus, Shady3d: A robot that climbs 3d trusses, in *Proc. IEEE Int. Conf. on Robotics and Automation*, 2007.
4. M. Tavakoli, A. Marjovi, L. Marques and A. de Almeida, 3dclimber: A climbing robot for inspection of 3d human made structures, in *Proc. IEEE/RSJ Int. Conf. on Intelligent Robots and Systems*, 2008.
5. Y. Guan, L. Jiang and H. Zhu, *et al*, Climbot: A modular bio-inspired biped climbing robot, in *IEEE Int. Conf. on Intelligent Robots & Systems*, 2011.
6. T. L. Lam and Y. Xu, A flexible tree climbing robot: Treebot c design and implementation, in *Proc. IEEE Int. Conf. on Robotics and Automation*, 2011.
7. W. Chung and Y. Xu, Path planning algorithm for space manipulator with a minimum energy demand, in *Proc. IEEE Int. Conf. on Robotics and Biomimetics*, 2012.

DESIGN OF SMA BASED ACTUATORS USED IN ROBOTICS

MUHAMMET ÖZTÜRK
Aeronautical and Astronautical Engineering, Istanbul Technical University, Maslak, Istanbul 34469, Turkey

BAHADIR KOÇER
Mechatronics, Yildiz Technical University, Beşiktaş, Istanbul 34369, Turkey

RECEP UYGUN
Mechanical Engineering, Hitit University, Çorum 19030, Turkey

This work introduces to Shape Memory Alloy (SMA) using in Robotics as actuators. We investigated a quadruped platform actuated by SMAs which act with variable elasticity. SMAs ought to prove necessary forces for all gaits. The main purpose of this study; determine the necessary for all actuation region and specify the essential actuation.

1. Introduction

The engineering science works on the variable crafts like airplane, spacecraft, car, robot etc. But how much the scientists improve the models, control algorithms and other scientific knowledge if you don't have the material that has the features you need. If you don't have it means that you have a distinct.

Shape Memory Alloys are very useful in comparing to classical actuators. When you use SMAs weight becomes lighter, you have more power comparing to electric motors, and you don't have noise like hydraulic systems. However, SMAs have some disadvantages like control that their behavior is nonlinear. SMAs have been used in many different robotic systems as actuators. The use of SMAs in applications involving actuation has several advantages such as large deformation, excellent power-to-mass ratio, maintainability, reliability, and clean and silent actuation. The disadvantages are slow response, low energy efficiency due to conversion of heat to mechanical energy, and also motion control difficulties due to hysteresis, nonlinearities, parameter uncertainties, and unmodeled dynamics [1].

The use of SMAs as a power source of a device or mechanism has several advantages over traditional actuators including, silent and smooth operation, and direct actuation, simple operation as a resistive heating device, compact size and excellent power to weight ratio [2]. SMA actuators have several advantages such as excellent power-to-mass ratio, maintainability, reliability, and clean and silent actuation. The disadvantages are low energy efficiency due to conversion of heat to mechanical energy, inaccurate motion control due to hysteresis, nonlinearities, parameter uncertainties, difficulty in measuring variables such as temperature, and slow response due to the thermal process involved in the working principle [3]

2. Shape Memory Alloy Features

There are different alloys that are used nowadays but especially in robotics the most useful of SMAs is Nickel-Titanium Alloy. Nickel-Titanium Alloy (Nitinol) is generally a mixture of %50 Nickel and %50 Titanium (of course the rate can change arbitrarily).

When we use Nitinols in robotic we must know the features of Nitinols. SMA materials have two condition named Austenite and Martensite. Austenite is the high temperature situation of SMA material and the Martensite is the low temperature situation of SMA material. Those situations are important that materials structure changes and so deformations change rapidly.

In Austenite phase SMA material have cubic Crystal Structures. In Martensite phase SMA materials have Monoclinic Crystal Structure (Figure 3).

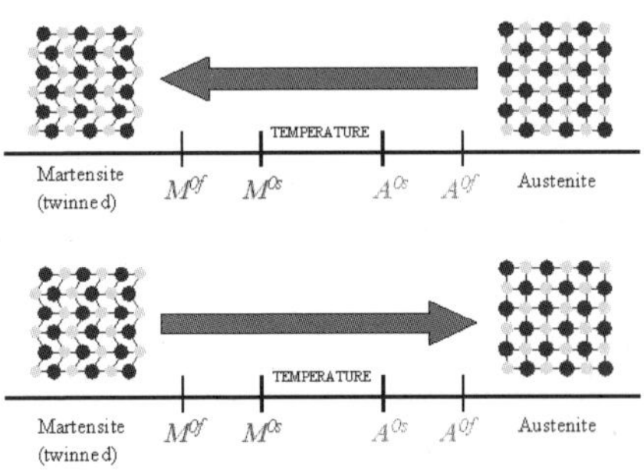

Figure 1. SMA structures with respect to temperature [11]

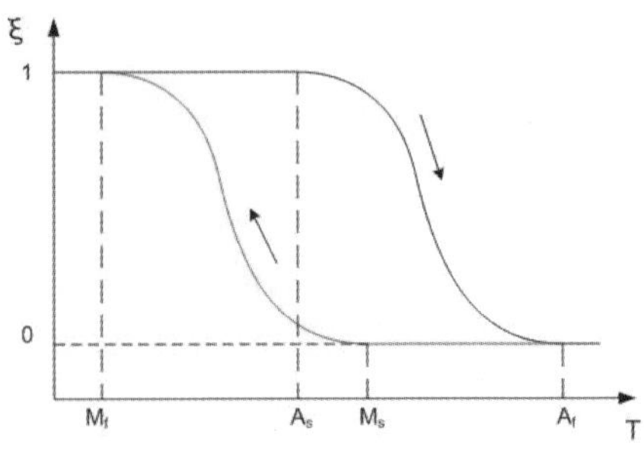

Figure 2. Temperature-stress comparison [10]

SMA materials exist in the austenite phase at zero-stress and high temperatures but when we decrease the temperature, the material undergoes a self-accommodating crystal transformation into the martensite phase. The transformation temperatures at zero-stress state of an SMA material are M_f (martensite finish), M_s (martensite start), A_s (austenite start), A_f(austenite finish). Upon cooling, the transformation to the martensite phase starts at M_s and completes at M_f whereas, on heating the transformation to the austenite phase starts at A_s and completes at A_f. This material has multiple variants and twins present, all with crystallographically equivalent but with differing orientations. When the stress level is increased multiple martensitic twins start converting into stress-preferred twins; this process is called "detwinning"[4].

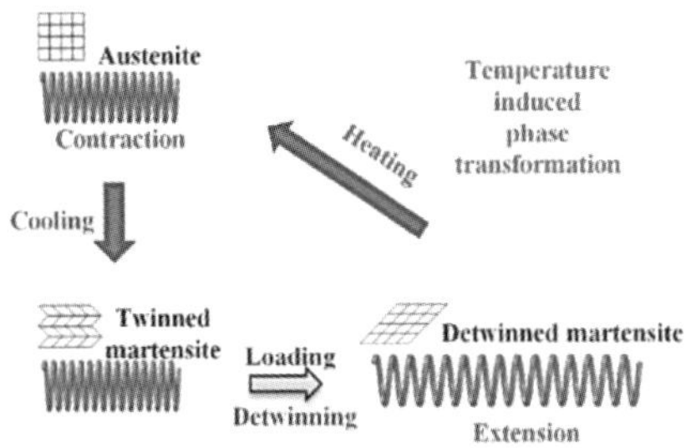

Figure 3. Temperature Induced Phase Transformation [5]

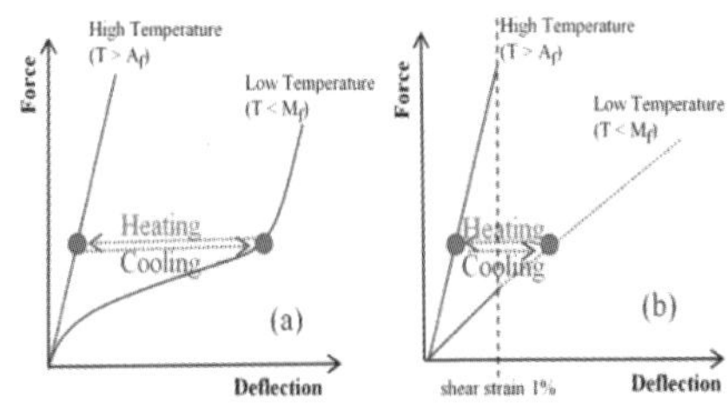

Figure 4. Force-deflection curve for (a) actual behavior and (b) curve based on conventional coil spring design [5]

3. Modeling SMA Coil Spring Actuator

We have wanted to use a system like human muscle system. Human muscle system is more advantageous that the muscles are near to the skeleton system and so doesn't need to a huge area and have a big torque/force rate. Besides it has a huge power/weight ratio.

The Titanium and Nickel Alloy of MMF (Metal Muscle Fiber) that we used in our work starts to contract when reach to $70°C$. When it cooled, it relaxes and prolongs its length to the normal state. Besides, in normally SMA actuators can only be operated by heat energy but MMF SMA can be applied electric energy as well. Instead of heating the material, a suitable amount of electric voltage is applied to the terminals of wire to obtain the same operation. This makes easy MMF SMA to control in many engineering applications. SMA possesses several advantages over public actuators. It shows a large ratio of length change. The light weight is another important advantage of SMA. By simple calculation of the parameters, one can see that a 200 mm MMF150 wire with the weight of 22.4 mg can contract by 9 mm while carrying the load of 1.8 kg. In other words, it implies that the wire may carry an 80000 times heavier load than its weight while contracting by 9 mm [6].

When we start to model we must know how much force will be needed. Afterwards we must calculate the other parameters like deflection. In generally some shattering powers like bending moment, axial force and shear force are being neglected so we can have a simple formulation.

$$F=\frac{G*d^4}{8D^3n}*\delta \quad (1)$$ [5]

In formulation G is shear modulus, d is wire diameter, D is spring diameter, n is number of active coils [5]. This formulation is fine but if you need a very sensitive dust you need the bending moment, axial force and shear force.

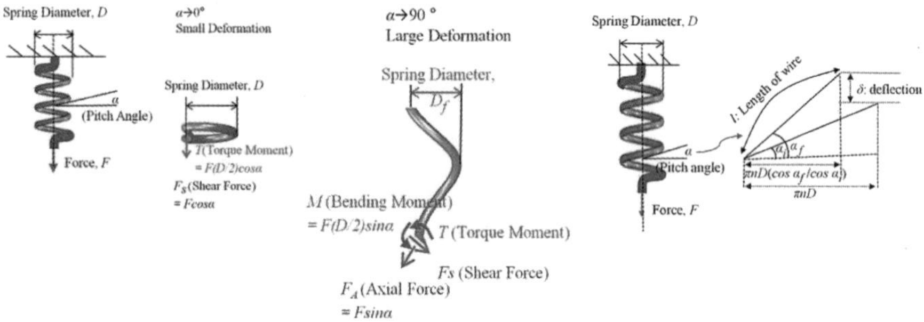

Figure 5. Axially loaded SMA coil spring features [5]

Complementary function isn't enough to control the SMA actuators for every situation. If we want to define the event totally we can use some other models like two-state model.

$$F=\frac{G*d^4}{8D^3n}*\left(\frac{\cos^3\alpha_i}{\cos^2\alpha_f\left(\cos^2\alpha_f+\frac{\sin^2\alpha_f}{1+v}\right)}\right)*\delta \quad (2)$$ [5]

4. SMA Based Walking Robot Structure

We designed the robot like in Figure 6. Our robots weight 47 gram and the 4% mass of his legs. The body is 100mm-40mm and the upper legs are 37mm and the lower legs are 33mm. We put the SMAs between upper and front foots.

The most important thing in the designing of a robot is to design the leg system so firstly we prepared the leg structure. Like you see in the structure the SMA spring was in the cylbase that is connected to foot.

The SMA spring will be in the Cyl so the spring can't get in any other shape and the rod will have pressure on the spring. Besides we will be able to control the temperature of the SMA because of the closed area.

When you want you can have a heat system in cyl to control the temperature of the spring.

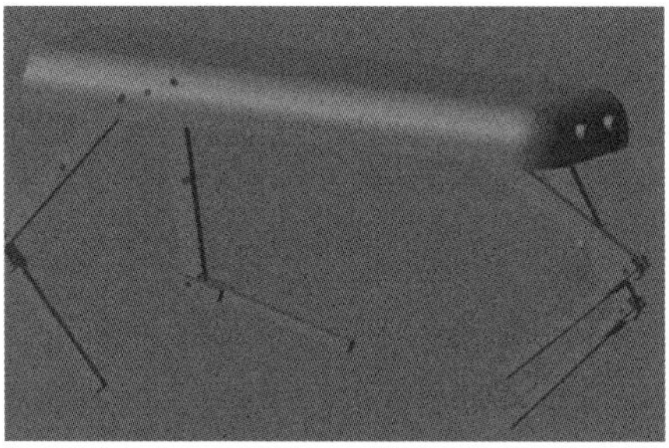

Figure 6. General design

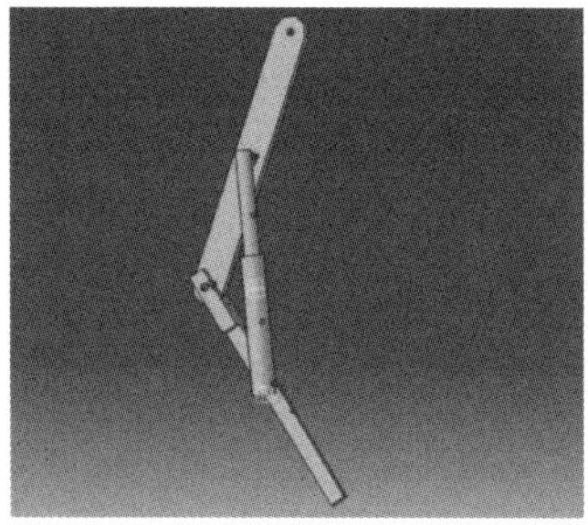

Figure 7. Leg system design

5. Dynamic Analysis

In our system it is important that how much power the SMA actuator must be satisfying. So we have calculated the maximum and minimum power. When we build the structure on the system we have used used only standart earth gravity that to stay on its feet how much power we will need. When we have an analysis we have seen that our SMA actuator must satisfy the $2.835*10^{-18}$ N. When we have wanted a velocity the wish will change.

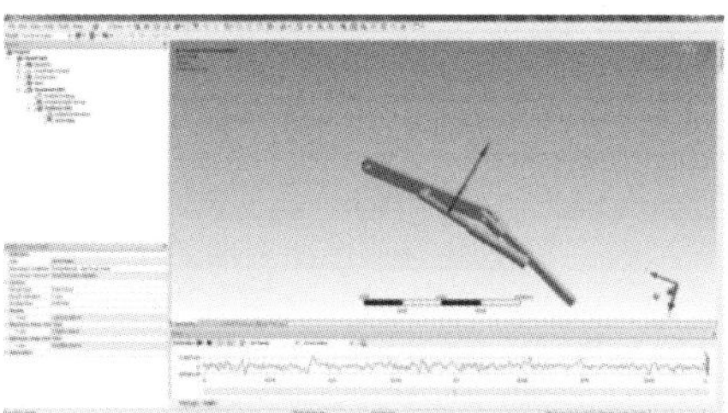

Figure 8. Leg system analysis

Table 1: Different Velocity and Necessity Forces

Velocity (rad/s)	Force (N)
1	$3{,}2441 * 10^{-18}$
2	$3{,}3068 * 10^{-18}$
10	$9{,}1507 * 10^{-17}$

But this system is only to hold a leg system on its way. Especially when we put a load on the leg like the body what would do the system. We have seen on the Catia that the system will be 0.045 kg but in normal the weights doesn't be like that.

The SMA actuator based quadruped walking robots becomes lights and generally used the stainless steel to be lighter. A simple SMA based robot can have the weight of 3kgf with the size of (Height: 150 mm, Width: 320 mm, Length: 300mm) and with the applied 12 SMA actuators.[7]

6. Control

General control structure is designed for stable gaits. So we will need some needs to build a model that: A real manufacture of the multi body structure, a convenient calculation time, convenient for simple integration of several controllers and easily understandable and adjustable [8]. The control system of a leg has been shown below.

In this toolbox we can have a force on the leg and the with the PID controller we have controlled the force in the wanted distance and system has reached to stability in a second (Figure 12). It shows that the system is working in good way.

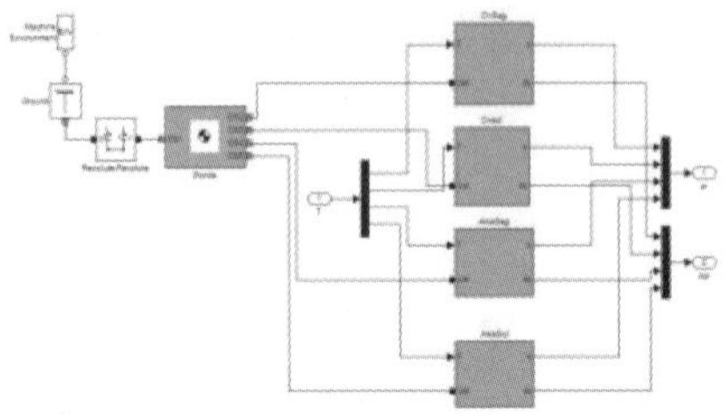

Figure 9. Leg control system

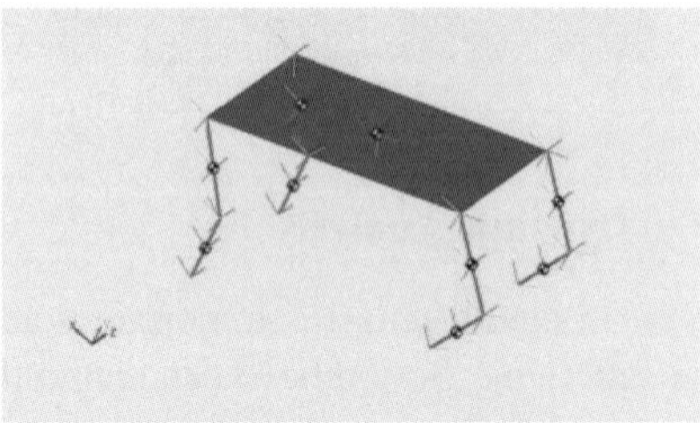

Figure 10. Leg control system animation

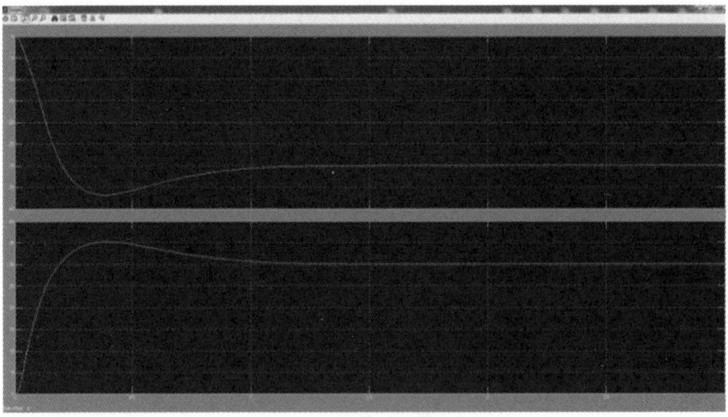

Figure 11. The stability time

7. SMA Actuator Design

In the designing of SMA Actuator we must choose an SMA (especially nitinol) and we must know properties of it.The SMA spring wire that we have chosen has the nominal values like the ambient temperature (20 C).

To calculate the system we can use the formula (1). We will put the SMA spring in the Cyl (Figure 8) so diameter must be 2 mm and when we accept d=0.2 mm

$$9{,}1507*10^{-17}\ \mathrm{N} = \frac{10.8*10^{9} N/m*(2*10^{-4})^{4} m^{4}}{8(0.002)^{3} m^{3}*3} * \delta \qquad \delta = 1.016744 * 10^{-18} m$$

Table 2. Physical Properties of SMA Wire [12]

Physical Properties	Metric	English
Density	6.45 g/cc	0.233 lb/in³
Mechanical Properties	**Metric**	**English**
Tensile Strength, Ultimate	754 - 960 MPa	109000 - 139000 psi
Tensile Strength, Yield	100 MPa	14500 psi
Elongation at Break	15.5 %	15.5 %
Modulus of Elasticity	28.0 GPa	4060 ksi
Poissons Ratio	0.300	0.300
Shear Modulus	10.8 GPa	1570 ksi
Electrical Properties	**Metric**	**English**
Electrical Resistivity	0.0000760 ohm-cm	0.0000760 ohm-cm
Magnetic Susceptibility	0.00000250	0.00000250
Thermal Properties	**Metric**	**English**
Specific Heat Capacity	0.320 J/g-°C	0.0765 BTU/lb-°F
Thermal Conductivity	10.0 W/m-K	69.4 BTU-in/hr-ft²-°F
Melting Point	1240 - 1310 °C	2260 - 2390 °F
Solidus	1240 °C	2260 °F
Liquidus	1310 °C	2390 °F
Component Elements Properties	**Metric**	**English**
Nickel, Ni	55.0 %	55.0 %
Titanium, Ti	45.0 %	45.0 %

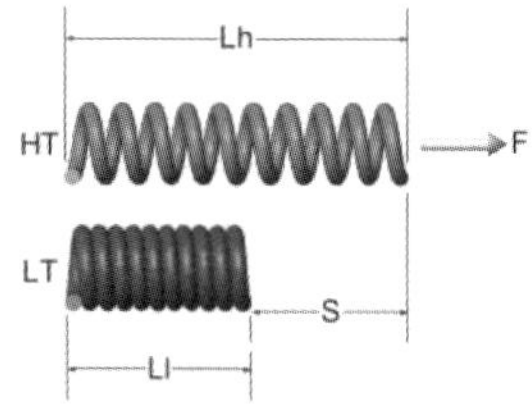

Figure 12. SMA compression spring actuator [9]

After all the works we have to know that how much electric we will need. Because of their metallurgical composition shape memory alloys possess an inherently high resistivity (especially Ni-Ti), which means they can be heated and subsequently activated by passing a direct current through them. The resistance can be evaluated as a function of their length and cross-sectional area a follows [9].

$$R = \frac{\rho * L}{A} \quad (3)$$

Where ρ equals the SMA resistivity, L is the free length of the element, and A its cross-sectional area.

From Ohm's Law we can calculate the voltage required for activation as:

$$V = I * R \quad (4)$$

The power requirement can be evaluated as:

$$W = I * V = I^2 * R \quad (5)$$

The current value can be work on ambient conditions so if we will have a model we must care on thermal activity and air conditions [9]. So we must have test on the SMA actuator.

8. Conclusion

In this paper, we described about SMAs usefulness and about their features. SMA Coil Spring actuators have been discussed. We have seen that the SMA actuators are very useful but some distinct must be known to effectively use in any system. The SMA actuators have been worked to be used like muscle in human body and wanted to have a quadruped robot model with SMA actuator. The dynamic analysis has been done by Ansys and the control working has been done with Simmechanics.

References

[1] EhsanTarkeshEsfahani and Mohammad H. Elahinia, "Stable Walking Pattern for an SMA-Actuated Biped," *IEEE/ASME Transection on Mechatronics*, vol.12, No. 5, October 2007

[2] JunghyukKo, Martin B. Jun, Gabriele Gilardi, Edmund Haslam, Edward J. Park, "Fuzzy PWM-PID control of cocontracting antagonistic shape memory alloy muscle pairs in an artificial finger," *Mechatronic Systems Engineering, School of Engineering Science, Simon Fraser University*, Canada, pp.1190-1202, July 2011

[3] Roberto Romano and Eduardo AounTannuri "Modeling, control and experimental validation of a novel actuator based on shape memory alloys," *Mechatronics 19,* pp.1169-1177, 22 March 2009

[4] Koray K. Safak and George G. Adams "Modeling and simulation of an artificial muscle and its application to biomimetic robot posture control," *Robotics and Autonomous Systems 41*, pp. 225–243, 14 June 2002

[5] Sung-Min An, JunghyunRyu, Maenghyo Cho and Kyu-Jin Cho "Engineering design framework for a shape memory alloy coil spring actuator using a static two-state model", *Smart Materials And Structures 21,* 16pp., 24 April 2012

[6] Thanhtam Ho and Sangyoon Lee" Design and Implementation of an SMA-Actuated Jumping Robot", *The 2010 IEEE/RSJ International Conference on Intelligent Robots and Systems,* Taipei, Taiwan, 18-22 October 2010

[6] Thanhtam Ho and Sangyoon Lee" Design and Implementation of an SMA-Actuated Jumping Robot", *The 2010 IEEE/RSJ International Conference on Intelligent Robots and Systems,* Taipei, Taiwan, 18-22 October 2010

[7] Hyung-Min Son, Jun-Bum Gu, Se-HoonPark, Yun-Jung Lee and Tae-Hyun Nam "Design of Ne Quadruped Robot with SMA actuators for Dynamic Walking", *SICE-ICASE International Joint Conference 2006,* Bexco, Busan, Korea, 18-21 October 2006

[8] DominikNäf "Quadruped walking/running simulation", *Swiss Federal Institute of Technology Zurich,* Spring 2011, Germany

[9] Jose R. Santıago Anadon "Large Force Shape Memory Alloy Linear Actuator", *University of Florida,* 2002

[10] http://en.wikipedia.org/wiki/Shape-memory_alloy

[11] http://smart.tamu.edu/overview/smaintro/simple/definition.html

[12] http://www.matweb.com/search/datasheet_print.aspx?matguid=44afc7d3c6eb4829bc2df27884fd2d6c

FINITE ELEMENT MODELLING OF THE ADHESION SYSTEM OF A VORTEX BASED CLIMBING ROBOT

FILIPPO BONACCORSO, SALVATORE D'URSO, DOMENICO LONGO, GIOVANNI MUSCATO

Dipartimento di Ingegneria Elettrica Elettronica e Informatica, Università degli Studi di Catania, Viale A. Doria 6, 95125 Catania Italy
e-mail: filippo.bonaccorso@dieei.unict.it, dlongo@unict.it, gmuscato@dieei.unict.it

Vortex adhesion is one of the most promising techniques for the design of lightweight climbing robot. However the results presented up to now are mainly simple demonstrations of prototypes and do not give precise criteria for the design. In this paper we present our last results toward a better understanding of vortex technology and to the development of a simulation strategy based on finite element modelling. The vortex-based suction cup has been modelled and simulated by using finite element and the results are compared with experimental trials.

1. Introduction

The first problem to solve in the design of a climbing robot is the choice of the adhesion technique to adopt. Many different solutions have been proposed in the past, each one with advantages and drawbacks with respect to the specific environment [1].

One of the most promising techniques that have been experimented by several researchers, is the pneumatic adhesion obtained by using a vortex generated by a rotating fan. Although the first experiment concerning this method were presented more than ten years ago, there are not yet many methodological results available concerning how to design a vortex based suction system.

Main results are simple experimental trials and it is not yet clear how to dimension the fan generating the vortex, how to choose the shape of the impeller, which rotating speed to adopt, etc. However it has been experimentally observed that small vortex-based climbing robots have a very good weight/force of suction ratio, allowing to build autonomous systems capable to perform the ground –wall transition. On the other side increasing the size of the systems and consequently the weight of the robot does not produce a sufficient rise of the adhesion force. As a consequence only small robots based on vortex with a small payload have been proposed. Such small robots can be autonomous, without the need of a power cable, since can be powered by on-board batteries whose weight can be supported by the robot. The main applications for these small systems are inspection tasks. For other more complex tasks as painting, or cleaning, when a greater payload is required and the adoption of a cable carrying the power supply is not a big problem, pneumatic suction was more successful.

The purpose of this paper is to adopt finite element modelling as a simulation strategy to model a vortex suction system in order to predict the suction force.
In the next sections a brief review of vortex adhesion and on the previous application of vortex for climbing robots will be reported; after that the experimental device and the designed robot adopted for our test will be briefly described. Then the modelling strategy based on finite element will be reported. The results of the simulation with a comparison to the experimental measurements and some concluding remarks will conclude the paper.

2. Vortex adhesion

Vortex adhesion is based on the generation of an air vortex inside a suction cup. This is generated by using a spinning rotor connected with some turbo vanes inside a cup. In this way a rotating column of air is created similar to those found in a tornado, with an interior pressure lower than the outside air pressure. This pressure difference causes an attraction force inside the cup as it is described in Figure 1.

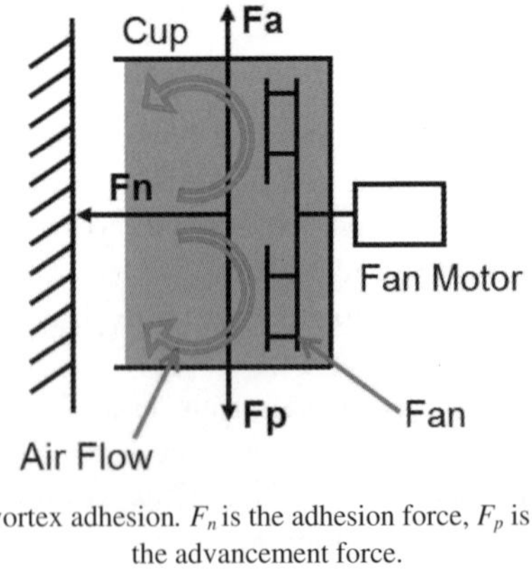

Figure 1. The principle of vortex adhesion. F_n is the adhesion force, F_p is the gravitational load, F_a is the advancement force.

The adoption of vortex as suction principle for climbing robot was firstly reported in some patents [2], [3] and in the report [4]. This idea was adopted by researcher at Duke University also to build several prototypes of robots that won two CLAWAR climbing robots competitions [5]. A company was also established for commercializing climbing robots developed with this principle [6].

In the PhD thesis [7] vortex adhesion was tested as adhesion method for a biologically inspired robot. Many different types of impellers were built in order to understand which of their features were the most important. However the reported experimental trials are rather limited and showed an adhesion force not useful for the planned application. According to [7] the force could be predicted by the formula:

$$F = \frac{\pi\rho\Omega^2 R^4}{2} \tag{1}$$

Where: ρ is the air density, Ω is the angular velocity and R is the impeller radius.

However experimental results have shown that this formula overestimates the adhesion force. In particular for a 10.16cm impeller at 4000RPM the formula predicted an adhesion force of 2.29N, while experimental results have shown a force between 0.9N and 1.1N [7].

Also in [8] and [9] the adoption of vortex technology was initially tested for the City-Climber robot prototype. However simulations and experimental trials resulted in an adhesion force not sufficient for the particular application.
In [10] and [11] neural networks were proposed to identify a model of vortex suction, while in [12] the relation between the pressure and the speed was interpolated with a polynomial and then the adhesion force computed.

Finally in [13] vortex adhesion was adopted for the PARASWIFT robot, a climbing robot that is able to climb a vertical wall and deploy a paraglider for a remote controlled return to ground.

3. Experimental data

The ALICIA VTX robot, shown in Figure 2, is a climbing robot prototype previously built in our laboratory. This robot is capable to perform a ground to wall transition [10] and was adopted also to demonstrate the cooperation between UGV and climbing robots. An experimental test bed was also set-up in order to acquire both the pressure inside different positions of the cup and the adhesion force. Details are reported in [11] and [12].

Figure 2. The Alicia VTX climbing robot prototype and its exploded view.

4. Finite element modelling

Since theoretical models are not reliable and experimental trials are time consuming and expensive, we decided to perform some simulations by using finite element method.

Figure 3. The Impeller and the suction cup.

The main interest was dedicated to model the air flux incoming within the suction cone and exiting from the fan, to investigate how the adhesion force varies with respect to the radius and to the speed of the impeller (Figure 3).
Figure 4 shows a radial section of the cup. In yellow it is reported the suction area while in light blue the bladed area of the fan. The arrows show the air motion in proximity of the fan. The most interesting part is the light blue area. This is the area where the air is moved thus producing the pressure variation in the different regions of the cup and consequently allowing the adhesion.

The main data adopted for the simulation are:

- Cup diameter 19.2 cm
- Cup height 3.1cm
- Impeller external diameter 12.6cm
- Impeller internal diameter 3cm
- Gap between impeller and surface 0.5cm
- Impeller thickness 2mm
- Number of fun blades 18

Initially the impeller was considered with rectangular blades and then it was simulated as the real experimental one with trapezoidal blades.

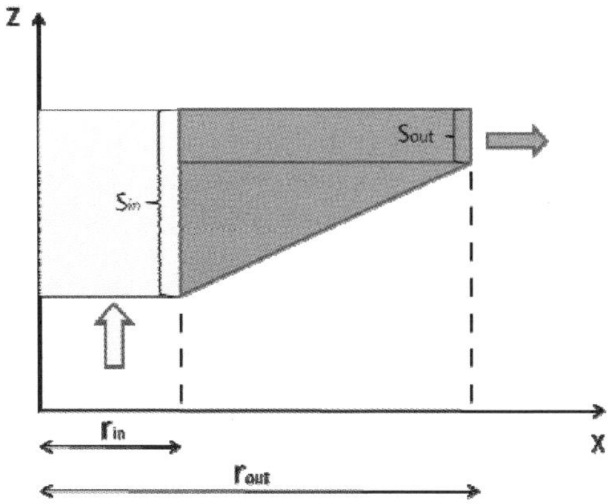

Figure 4. Section of the suction cup.

The ANSYS^{TM} package was adopted to build the 3D mesh of the cup and of the impeller, as it is shown in Figure 5, and to perform the finite element modelling and simulation. All the needed parameters for the simulation as boundary conditions, number of iterations, solver type, have been selected in order to obtain a good solution.

In particular the mesh was built with 7216 nodes and 3554 elements. The impeller was simulated as rotating at a constant angular speed within the cup. A turbulence model was selected for the fluid, while for all the remaining parameters the default settings were adopted with RMS residual type as convergence criteria.

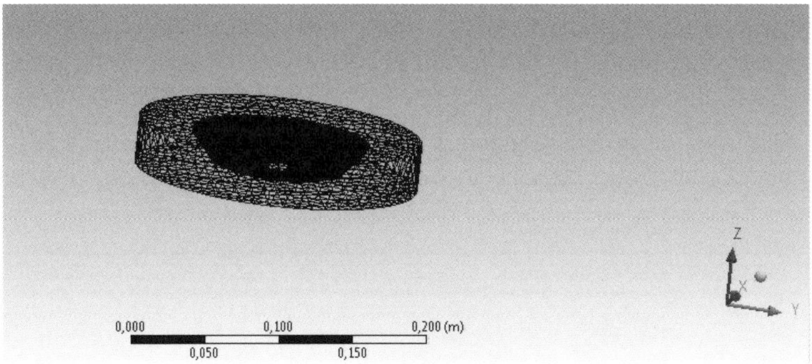

Figure 5. Mesh of the structure.

5. Results

In Fig. 6 an example of the obtained pressure distribution inside the cup for a specific speed is reported. Integrating the pressure inside the cup it was possible to obtain the adhesion force at the different speeds and this was compared with the experimental data as shown in Figure 7. As it can be observed, although there are some differences, the obtained values are very similar in a wide range of speeds. Further corrections could be introduced taking into account the unavoidable simplifications that are introduced in the simulations.

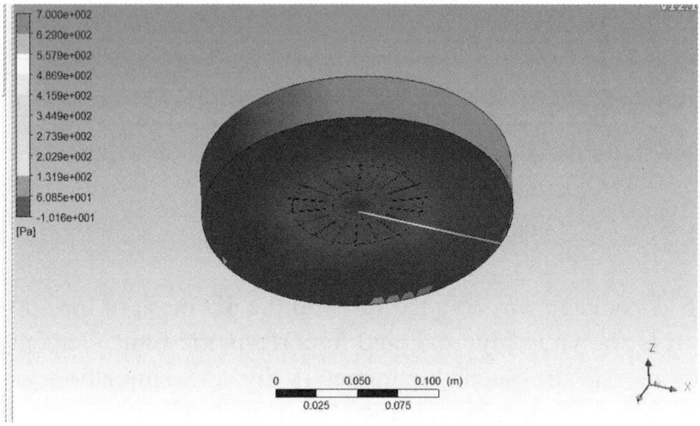

Figure 6. Pressure distribution for a speed of 683rad/s.

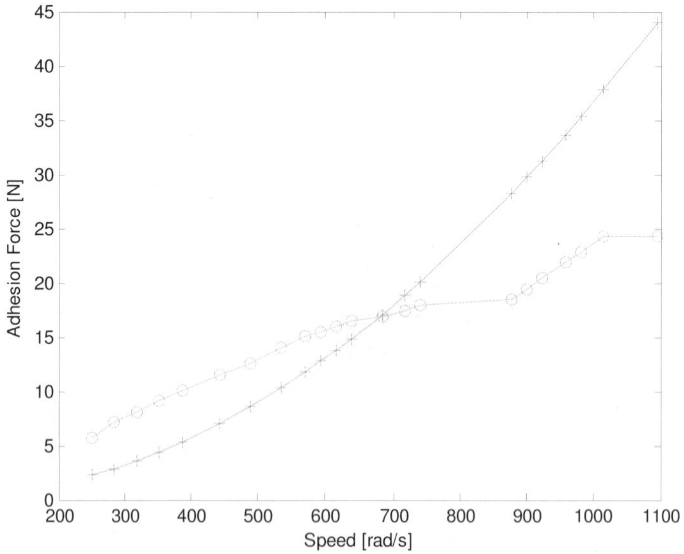

Figure 7. Comparison between experimental (Red) and simulated (Blue) data for different speeds.

Finally in Figure 8 the simulations for two different impeller diameters are reported.

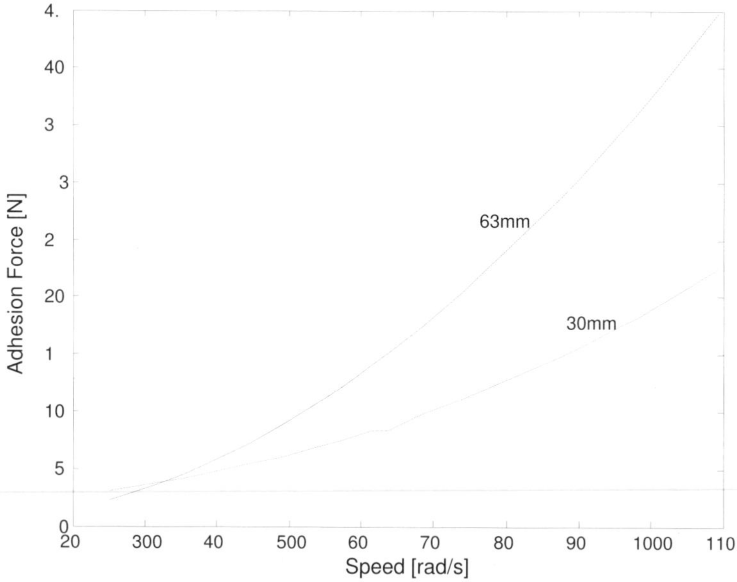

Figure 8. Force variation for two different impeller radius (Blue=63mm; Red=30mm).

6. Conclusion

This paper has summarized the results of a finite element model of a vortex based adhesion system for a climbing robot. The obtained results are in agreement with the experimental data. Further work is under investigation in order to improve this model and to obtain the relationships between speed and radius of the impeller for different shapes of the fan.

We believe that these relationships could represent useful design formulas allowing also to better understand when a vortex based adhesion system could be adopted.

References

1. D. Longo, G. Muscato, "Adhesion techniques for climbing robots: State of the art and experimental considerations", *11th International Conference on Climbing and Walking Robots and the Support Technologies for Mobile Machines*, CLAWAR'08, 08-10 September 2008, Coimbra, Portugal.
2. L. Illingworth and D. Reinfeld, "Vortex attractor for planar and nonplanar surfaces," *U.S. Patent* 6,619,922, Sep. 16, 2003.

3. D. Reinfeld, and L. Illingworth, (2002). VORTEX GENERATOR – patent:EP1224396, http://www.freepatentsonline.com/EP1224396A1.html.
4. J. Janet, D. Reinfeld and B. Wiedeman (2004), "Vortex Regenerative Air Movement: Attraction and Attachment on Vertical and Inverted Surfaces: A Simple Method for Static and Mobile Robots for Climbing on Walls and Ceilings", *U.S. Army Soldier and Biological Chemical Command*, Soldier Systems Center, Natick, Massachusetts. NATICK/TR-04/013L, May 2004. DTIC# ABV299170.
5. http://secure.pratt.duke.edu/pratt_press/web.php?sid=186
6. http://www.clarifyingtech.com/public/robots/robots_public.html
7. T. Wei , "A robot designed for walking and climbing based on abstracted cockroach locomotion mechanisms", *PhD Thesis*, Department of Mechanical and Aerospace Engineering, Case Western Reserve University, January 2006
8. J. Z. Xiao, A. Sadegh, M. Elliot, A. Calle, A. Persad, H. M. Chiu,"Design of Mobile Robots with Wall Climbing Capability", *Proceedings of the 2005 IEEE/ASME International Conference on Advanced Intelligent Mechatronics*, pp438~443, July 2005.
9. M. Elliott, J.Z. Xiao, W. Morris, A. Calle, "City-Climbers at Work", Video *Proceeding of the 2007 IEEE International Conference on Robotics and Automation*, pp2764-2765, Roma, Italy, April 2007.
10. F. Bonaccorso, C. Bruno, D. Longo G. Muscato, "Structure And Model Identification Of A Vortex-based Suction Cup", *Proceedings of the11th International Conference on Climbing and Walking Robots and the Support Technologies for Mobile Machines* CLAWAR'08, 8-10 September 2008, Coimbra, Portugal.
11. F. Bonaccorso, D. Longo, G. Muscato, "Modeling of an innovative actuator for climbing robot adhesion"; *Proceedings of the 12th International Conference on Climbing and Walking Robots and the Support Technologies for Mobile Machines CLAWAR'09*, Istanbul, Turkey, 09-11 September 2009, pp.891-900.
12. F. Bonaccorso, G. Muscato, A. Pagano, A. Fichera, "A new approach to the modelling of vortex-based suction cups for climbing robot adhesion", *Proceedings of the 15th International Conference on Climbing and Walking Robots and the Support Technologies for Mobile Machines* CLAWAR 2012, Baltimore (USA), 23-26 July 2012.
13. L. Geissmann et al., "PARASWIFT – A Hybrid Climbing And Base Jumping Robot For Entertainment", *Proceedings of the 14th International Conference on Climbing and Walking Robots and the Support Technologies for Mobile Machines* CLAWAR 2011, Paris, France, 6 – 8 September 2011, pp 317 - 324.

DEVELOPMENT OF A QUADRUPED ROBOT MODEL IN SIMMECHANICS™

M. F. SILVA*

INESC TEC - INESC Technology and Science (formerly INESC Porto) and ISEP/IPP - School of Engineering, Polytechnic Institute of Porto Porto, 4200-072 Porto, Portugal
**E-mail: mss@isep.ipp.pt*
http://ave.dee.isep.ipp.pt/~mss/

R. S. BARBOSA and T. S. CASTRO

ISEP/IPP - School of Engineering, Polytechnic Institute of Porto Porto, 4200-072 Porto, Portugal
E-mail: rsb@isep.ipp.pt, tomassoutocastro@gmail.com

The presented work allowed the creation of a quadruped robot model (with flexible body, its legs and its hip and knee joints) in the virtual simulation program Simmechanics™. The final objective is to use this simulation model to optimize quadruped robot locomotion. The model developed accepts different gaits by direct introduction of the angular positions of the knee and hip joints. Feet-ground interaction was also modelled using a theoretic model described in the literature.

Keywords: Legged Quadruped Robots, Modelling, MATLAB, SimMechanics.

1. Introduction

Increased research in robotics has been remarkable, but the costs and time invested in the development of prototypes are very high which encourages the search for new ways of working that enable the development and testing of machines more quickly, economically and efficiently.

There are many tools on the market that allow to design, simulate and test mechanical systems, such as Adams,[1] Webots,[2] VisSim,[3] Simpack.[4] All these programs allow the analysis and simulation of dynamic systems, with real-time view of the equipment, thus helping to detect problems that may occur and facilitating the design stage of mechanical parts.

The programs for dynamic systems simulation are used in various industries, with particular interest in robotics due to the possibilities for modelling and simulating different types of systems with varying degrees of complexity. In addition to simplifying the design, these programs also assist in testing the equipments. In some cases it is possible to test the robots in 3-D virtual environment. Although its use does not completely eliminate the need to build a working prototype, it decreases the probability of error due to various tests that can be done before building a real model.

Bearing these ideas in mind, the goal of this paper is to present the development of a quadruped robot model, in SimMechanics™, with the objective of creating a tool for the optimization of quadruped legged robot parameters. The remainder of this paper is organized as follows. Section two briefly presents the SimMechanics™ environment and section three the robot model adopted. Sections four and five present the implementation of the robot model in the software, and some simulation results, respectively. Finally, section six outlines the main conclusions of this study.

2. SimMechanics Simulation Application

Simmechanics™ is a three dimensional modelling tool that allows the simulation of multibody models.[5] Its interface works based on blocks representing bodies (links), joints, motion constraints and forces and torques. The various elements are joined together using lines representing the signals transmitted from one block to another. The blocks required for modelling a robot are of the type *Body*, *Revolute Joint*, *Joint Actuator*, *Body Actuator*, *Ground*, *Machine Environment*, *Body Sensor*, and *Joint Sensor*.[5,6]

Although Simmechanics™ simplifies the process of modelling / simulation of the desired system, its implementation can become complex. In the implemented model (Figura 1(a)), the legs are connected to the robot body through joints named "hips". Each leg consists of two parts: (*i*) the top one ("upper leg") is connected to the robot body by the "hip", and (*ii*) at the other end to the "lower leg" through a "knee" joint.

3. Model of the Quadruped Robot

3.1. *Kinematics and Trajectory Planning*

It is considered a walking system (Figure 1(b)) with $n = 4$ legs, equally distributed along both sides of the robot body, having each one two rotational joints (*i.e.*, $j = \{1, 2\} \equiv \{\text{hip, knee}\}$).

It was chosen a forward orientation of the joint that connects the upper to the lower part of the leg. The angles of the joints are defined as measured relative to each other (Figure 1(c)): the angle of the hip is measured from the horizontal body position, and the knee angle is measured relative to an

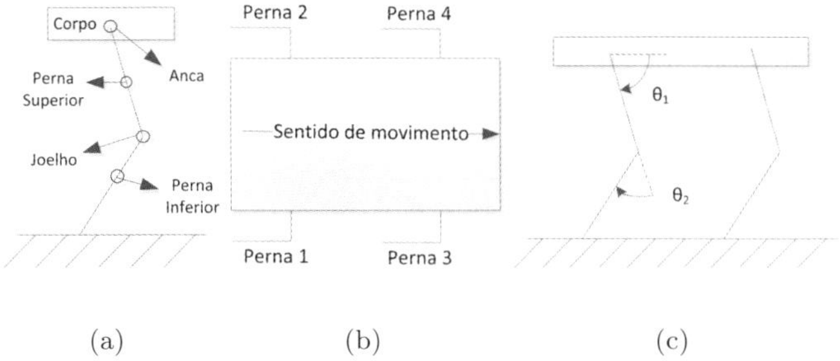

Fig. 1. Scheme of the constitution of the robot leg (a), numbering of the robot legs (b) and disposition of the joints angles (c).

imaginary line extending from the upper leg. This angular relationship is used by Simulink™ and is quite common in robotics, because it facilitates the visualization of the position of the members.[7]

In this work is intended to get some flexibility regarding the gait. For that reason, actuators were placed in the knee and hip joints. These actuators receive the angular positions that allow the definition of the robot gait, and therefore variables have been defined to characterize the desired motion of the robot (Table 1). The variables shown in Table 1 are defined as matrices with 2 columns: the first column represents sample time, and in the second column are the desired values for the joints positions.

	Leg 1	Leg 2	Leg 3	Leg 4
Hip angular position	*hip_1*	*hip_2*	*hip_3*	*hip_4*
Knee angular position	*knee_1*	*knee_2*	*knee_3*	*knee_4*

Although this robot allows the introduction of different trajectories on each leg, in this case it was chosen to define the movement of leg 4 equal to the first leg and third leg equal to leg 2.

3.2. *Robot Dynamic Model*

The inverse dynamics is used to calculate the torques/forces needed for the leg to describe a certain movement. For the program to calculate the forces necessary for the intended motion, the inertia matrices of the various body components have to be defined.

Based on previous studies, it was concluded that to increase the efficiency and stability of locomotion, the robot body should present compliance. Therefore, a compliant robot body model was implemented.

3.3. *Foot-Ground Contact Model*

The model of contact with the ground considers a contact point on the lower end of the leg. When this comes into contact with a given point on the ground, a tri-dimensional force is exerted on the ground, and a force with equal intensity and opposite direction in the robot foot. In the work described are only considered two components of the contact force: normal force (exerted on y direction, and responsible for ensuring that the robot is always above ground) and tangential force (exerted towards x, imposes the robot motion).

The two main simulation methods of contact with the ground are rigid or compliant contacts. The method used in this paper calculates the contact forces through an approximate model of the ground based on models obtained by the study of soil mechanics. It was chosen to use a model of contact that uses a spring-damper type system, described by equations (1) and (2).[7] In equation (2) the exponent m allows the user to more precisely adjust the characteristics of the soil.

$$F_x = -K_x(x - x_0) - B_x\dot{x} \tag{1}$$

$$F_y = -K_y y - B_y(-y)^m \dot{y} \tag{2}$$

4. Implementation of the Robot Model

4.1. *Modelling of the Robot Body*

The body of the robot is composed of four identical blocks, interconnected by a spring and damper arranged in parallel to simulate a backbone, making the robot walk softer and cushioning the more abrupt contacts. The body is modelled using four *Body Blocks*, connected by a universal type joint actuated by a "*Joint Spring and Damper*" block type (Figure 2(a)).

In the body blocks must be introduced its mass, inertia matrix and various CS that will allow to connect the legs to these blocks and also the joints that connect the four blocks constituting the body of the robot. Each block has a mass of 9 kg and a size of 0.3 m by 0.2 m with 0.15 m high. The mass of the blocks is calculated based on its volume and the density of the used material (2.8 kg/m^3 for the legs and 1.0 kg/m^3 for the body).

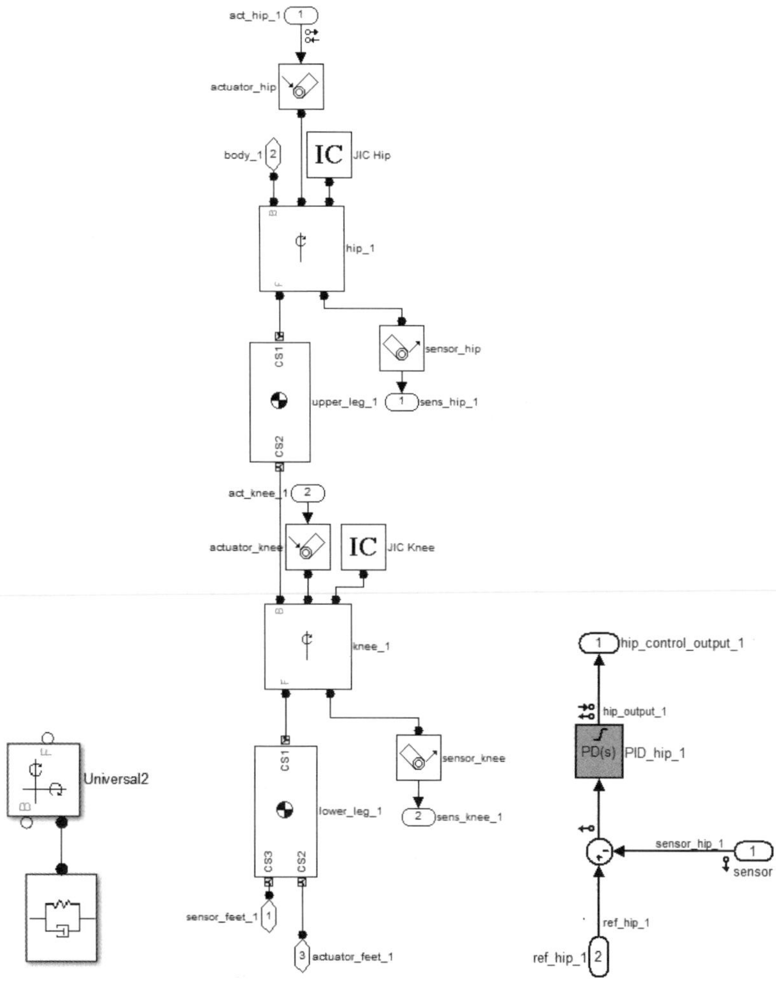

(a) (b) (c)

Fig. 2. Spring-damper system of the body (a), model of a robot's leg (b), and PD controller (c).

It is necessary to join the different blocks and define the CS, so that the body takes the form previously established. These CS are used as reference to connect a joint or body actuator / sensor.

In the robot presented are used relative referrals. A *"Ground"* block is connected in one of the body first block CS using a *"6 Degrees Of Freedom"* joint type; this assembly enables the robot to move in three-dimensional space. The position of the remaining blocks of the body is defined based on the option *"adjoining"* of CS. This option "tells" the CS that its position is referenced to the CS which is connected to the adjacent block.

4.2. *Modelling of the Robot Legs and its Control System*

The two parts that constitute the robot's legs are represented in this model by rigid cylinders. The legs have two rotary joints with one degree of freedom each. The upper joint simulates the hip rotation and the lower one the knee. Figure 2(b) presents the Simmechanics^{TM} model used to model the robot's legs.

The block *body_1* is connected on one side to the CS of the body block 1 and on the other side (Figure 2(b)) to a rotational type joint (*hip_1*), that models the robot hip. The upper leg segment is depicted in Figure 2(b) with the name *"upper_leg_1"*. The CS of the upper leg segment, which connects the leg to the body, is defined based on the body CS. Both the CG and the CS2 of the upper leg segment are defined based on CS1. The configuration of the knee rotational joint equals the hip one, being this block identified in Figure 2(b) with the name *"knee_1"*.

The lower leg segment is very similar to the upper one in the way its CG is set. In addition to the CS being used to connect the lower leg segment to the upper one, there must also be defined two CS at its lower end. These CS (CS2 and CS3) will allow the soil model to "read" the position and speed of the foot and also actuate it with the ground reaction forces.

The actuation of the members is carried out directly on the joints through joint actuators. Controllers are connected to these actuators enabling the control of the joints movement. There are also sensors attached to these joints of the legs. These sensors gather the value of the actual position of the joint enabling controllers to calculate the error between the desired position and the actual position and thereby adjusting the control signal. The controller chosen to control the joints of the hips and knees of the robot is of the proportional and derivative (PD) type. Its implementation was performed using the PID block, part of the Simulink^{TM} tool (Figure 2(c)).

5. Final Robot Assembly and Model Tests

The total robot body weight is 36 kg. It was determined that both the robot upper and lower leg segments would have a length of 0.2 m and a radius of

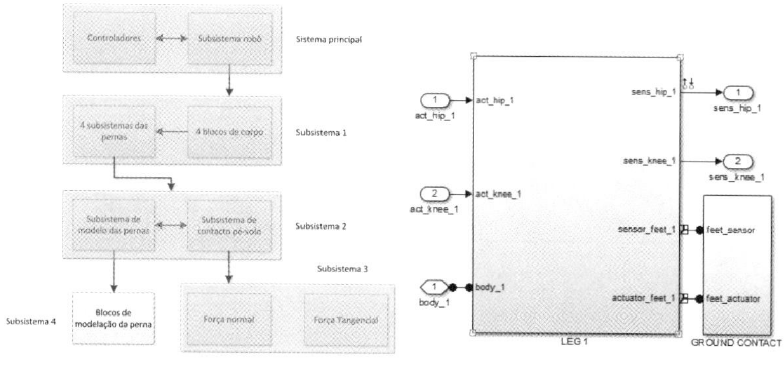

(a) (b)

Fig. 3. Organization scheme of the robot model (a) and subsystem 2 (b).

0.02 m, leading to a final weight of 1.4 kg per leg. Thus, the total weight of the robot is 41.6296 kg.

Figure 3(a) presents a schematic organization of the model to facilitate its understanding. Within the *Subsystem 1* are the four blocks that make up the body as well as the universal joints that unite them. Each body block is connected to a subsystem identified in Figure 3(a) as *Subsystem 2* where are defined the subsystems of contact with the ground and the subsystem with the leg modelling blocks (Figure 3(b)).

After tuning the controllers of the robot joints, it was proceeded to the analysis of its movement, as shown in Figure 4, being possible to observe the robot taking several steps.

6. Conclusions

This paper described the creation of a quadruped robot model in Simmechanics^{TM}, characterized by a flexible body, and the modelling of its feet-ground interaction. The model developed accepts different gaits by direct introduction of the angular positions of the knee and hip joints. It is the authors intention to further improve this model in order to allow its use to testing distinct gaits and the locomotion in 3-D and in irregular terrains.

Acknowledgments

This work is funded (or part-funded) by the ERDF - European Regional Development Fund through the COMPETE Programme (operational pro-

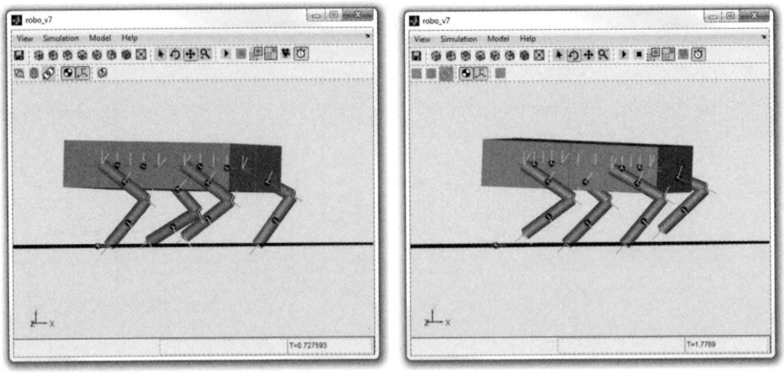

(a) (b)

Fig. 4. Robot walking with legs 2 and 3 in the transfer phase (a) and with legs 1 and 4 in the transfer phase (b).

gramme for competitiveness) and by National Funds through the FCT - Fundação para a Ciência e a Tecnologia (Portuguese Foundation for Science and Technology) within project FCOMP - 01-0124-FEDER-022701.

References

1. M. Software, Adams - the multibody dynamics simulation solution `http://www.mscsoftware.com/Products/CAE-Tools/Adams.aspx`, Last accessed on February, 2013.
2. Cyberbotics, Webots: the mobile robotics simulation software `http://www.cyberbotics.com`, Last accessed on February, 2013.
3. V. S. Incorporated, Vissim - a graphical language for simulation and model-based embedded development `http://www.vissim.com/products/vissim.html`, Last accessed on February, 2013.
4. S. AG, Simpack - multi-body simulation software `http://www.simpack.com/`, Last accessed on February, 2013.
5. Mathworks, Simmechanics - model and simulate multibody mechanical systems `http://www.mathworks.com/products/simmechanics/`, Last accessed on February, 2013.
6. Mathworks, Simmechanics™ link users guide `http://www.mathworks.cn/help/pdf_doc/physmod/smlink/smlink_ug.pdf`, Last accessed on February, 2013.
7. M. F. Silva, J. A. T. Machado and A. M. Lopes, *ROBOTICA* **23**, 595 (2005).

SECTION–10

PLANNING AND CONTROL

MODEL-BASED ELASTIC TENDON CONTROL FOR ELECTRICALLY ACTUATED MUSCULOSKELETAL BIPEDS

K. RADKHAH* and O. VON STRYK

Department of Computer Science, Technische Universität Darmstadt,
64289 Darmstadt, Germany
**E-mail: radkhah@sim.tu-darmstadt.de*
www.sim.tu-darmstadt.de

Human-inspired musculoskeletal design of bipedal robots offers great potential towards enhanced dynamic and energy-efficient locomotion but imposes also major challenges on their control. In this paper we present an analytical model-based controller that takes into account the system's complex musculoskeletal actuation dynamics in order to fully exploit the intrinsic dynamics. The effectiveness of the proposed approach is evaluated for hopping-in-place motions on the simulation model of the BioBiped1 robot, a human-inspired musculoskeletal biped featuring a highly compliant tendon-driven actuation system.

1. Introduction

This paper addresses the challenging question of motion generation for a musculoskeletal biped with three-segmented legs driven by a number of active and passive, mono- and biarticular structures. Despite recent increase in popularity (see Refs. 1–4), such designs are so far quite unique and open up a number of questions regarding the degree of design complexity and suitable control concepts. For pneumatically actuated systems basic feedforward controllers for the operation of the valves have been employed based on ground contact sensing to realize walking, running and jumping motions.[5] The timely operation of the valves based on experimental tuning was particularly important since the structures can get in the way of each other. Sinusoidal excitation of electrical motors may be successful for rather simple mechanical systems with few passive parameters and with linear or even direct drive transmissions.[6] An important issue to be considered during a such motion generation process is choosing the step frequency correctly, which is quite difficult for a nonlinear complex musculoskeletal leg design. Second, not all motions can be represented by basic sinusoids. Further, particularly in case of nonlinear actuation dynamics, identifying the corresponding motor actions without model knowledge is not feasible.

Therefore, for these kind of complex musculoskeletal systems we suggest a model-based approach for trajectory planning and execution. Using a similar concept, introduced in Ref. 7 for flexible manipulators, an earlier version of this method had been successfully applied to a bipedal robot with a very basic linear elastic actuation system for dimensioning the required motor-gearboxes.[8] In this paper we describe and discuss the approach in a detailed manner exemplary for the complex nonlinear elastic actuation system of the bipedal robot BioBiped1, a novel highly elastic actuation system comprising human-inspired active and passive mono- and biarticular muscle-tendon structures.[9] As proof of concept, the method is applied in simulation to the identified robot model to generate hopping-in-place motions. The resulting motions are compared to motions obtained by motor excitation, a quite popular approach often employed for underactuated robots. It is demonstrated that the model-based approach is capable of realizing the desired motions without increasing the power consumption. Furthermore, comparability and evaluation of different leg actuation designs for diverse gaits and performance criteria are facilitated by this method.

2. BioBiped's Musculoskeletal Actuation Dynamics

The BioBiped1 robot, shown in Fig. 1(c) and built within the BioBiped project, is characterized by highly compliant active and passive actuation comprised of nine mono- and biarticular human-like muscle-tendon structures.[9] These structures represent the muscle groups that are essential during human locomotion (cf. Fig. 1(a)). We differentiate between bidirectional and unidirectional electrically driven series elastic actuators, referred to as b-SEA and u-SEA, respectively.[10] As shown in Fig. 1(b), the hip joint is actuated by a b-SEA representing the antagonist-agonist muscle pair *Gluteus Maximus* (GL) - *Iliopsoas* (ILIO). The knee and ankle joints are each actuated by a combination of a u-SEA and its passive counterpart: in the knee *Vastus lateralis* (VAS) - *Popliteus* (PL) and in the ankle *Soleus* (SOL) - *Tibilias anterior* (TA). While the extensor function is actively supported, flexion in the opposing direction is invoked by a passive monoarticular structure. In addition, each leg has three passive biarticular structures, each connecting two joints and transferring power from the proximal to the distal joints: *Rectus femoris* (RF), *Biceps Femoris* (BF), and *Gastrocnemius*

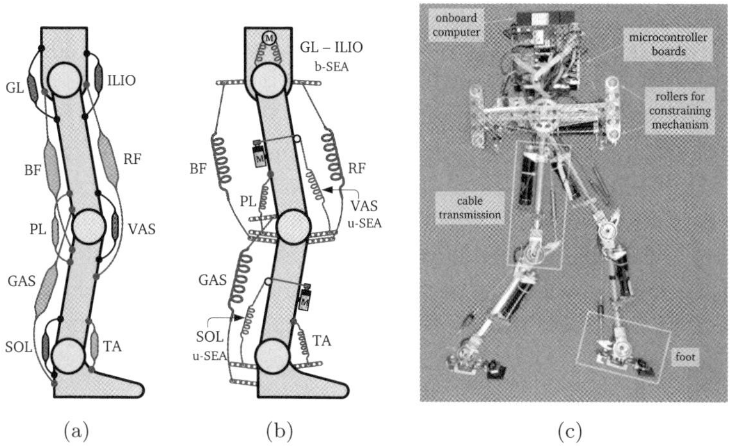

Figure 1. Technical realization of BioBiped1's actuation system: (a) Essential human muscle groups during locomotion, structures contributing mostly to the power generation are indicated by dark grey color; (b) technical realization of the mono- and biarticular elastic structures in the legs of BioBiped1;[10] (c) real BioBiped1 platform.

(GAS). The technical realization of all structures is described in Ref. 9. All passive structures can be attached and detached as desired.

To summarize, each joint is affected by exactly one actuated structure comprised of a DC motor and the coupled spring-tendon, but additionally also by a number of passive structures. Note that the biarticular structures affect two joints simultaneously, e.g. $\tau_{\mathrm{RF,H}}$ is the torque induced by RF on the hip joint, whereas $\tau_{\mathrm{RF,K}}$ indicates the torque caused by RF on the knee joint. For brevity, the output torque function for each joint is listed below:

$$\tau_{\mathrm{e,H}}(q_{\mathrm{H}}, \theta_{\mathrm{H}}, q_{\mathrm{K}}) = \tau_{\mathrm{GLILIO}}(q_{\mathrm{H}}, \theta_{\mathrm{H}}) + \tau_{\mathrm{RF,H}}(q_{\mathrm{H}}, q_{\mathrm{K}}) + \tau_{\mathrm{BF,H}}(q_{\mathrm{H}}, q_{\mathrm{K}}) \quad (1)$$

$$\begin{aligned}\tau_{\mathrm{e,K}}(q_{\mathrm{K}}, \theta_{\mathrm{K}}, q_{\mathrm{A}}) &= \tau_{\mathrm{VAS}}(q_{\mathrm{K}}, \theta_{\mathrm{K}}) + \tau_{\mathrm{PL}}(q_{\mathrm{K}}) + \tau_{\mathrm{GAS,K}}(q_{\mathrm{K}}, q_{\mathrm{A}}) \\ &\quad + \tau_{\mathrm{RF,K}}(q_{\mathrm{K}}, q_{\mathrm{H}}) + \tau_{\mathrm{BF,K}}(q_{\mathrm{K}}, q_{\mathrm{H}})\end{aligned} \quad (2)$$

$$\tau_{\mathrm{e,A}}(q_{\mathrm{A}}, \theta_{\mathrm{A}}, q_{\mathrm{K}}) = \tau_{\mathrm{SOL}}(q_{\mathrm{A}}, \theta_{\mathrm{A}}) + \tau_{\mathrm{TA}}(q_{\mathrm{A}}) + \tau_{\mathrm{GAS,A}}(q_{\mathrm{A}}, q_{\mathrm{K}}) \quad (3)$$

with q_{H}, q_{K} and q_{A} denoting the hip, knee and ankle joint angle, respectively. θ_{H}, θ_{K} and θ_{A} stand for the hip, knee and ankle motor position, respectively. The single torque terms refer to the torques induced by the corresponding structure, e.g. τ_{GLILIO} is the torque exerted by the b-SEA on the hip joint. The knee joint possesses the most complex dynamics behavior, as its function includes torques applied by five out of nine structures in total. In general, the output functions are highly complex and nonlinear.

Not only does each joint torque function include several torque terms, but each torque term depends also on various parameters, such as spring stiffness, rest length and force application points to attach a tendon to the joint, where applicable. Another feature of the musculoskeletal actuation system is the high number of integrated redundant functionalities. For instance, the knee joint flexion is not only due to the passive action of PL, but might also result from the action of GAS or BF. Such redundant functionalities can be found for all joints, both for flexion and extension. This complex actuation dynamics raises the question of suitable controllers and motivates the choice of model-based trajectory planning and execution which facilitates the comparison of various leg actuation designs.

3. Method

In this section we propose a new method for model-based motion generation and control combining feedforward and feedback control for musculoskeletal robots. Based on the detailed mathematical models of the different actuator types presented in Ref. 10, we derive step-by-step the required control signals for given reference motions. An overview of the method is given in Fig. 2.

Step 1: In the first step we derive the actuated joint torques for given reference trajectories $\boldsymbol{q}_{\mathrm{d}}$. This problem is ill-posed, as the constraint forces, that are required for solving for the actuated joint torques, depend on the applied actuation torques. Therefore, a low-gain PD controller determines the required joint torques, denoted as $\hat{\boldsymbol{\tau}}_{\mathrm{e,J}}$, to move the rigid robot model along the desired motion trajectories during the forward dynamics simulations. Instead of a PD controller, it is also possible to carry out an orthogonal decomposition to project the robot dynamics onto a reduced dimensional space.[11]

Step 2: Using the numerical results obtained from the forward dynamics computation of the rigid joint-link structure, $\hat{\boldsymbol{\tau}}_{\mathrm{e,J}}$ and $\hat{\boldsymbol{q}}$, the motor angles and torques required for the elastic robot, $\boldsymbol{\theta}_{\mathrm{d}}$ and $\boldsymbol{\tau}_{\mathrm{m,d}}$, are computed analytically based on the corresponding models of the b-SEA, u-SEA and passive mono- and biarticular structure. The subscript $_d$ indicates the desired and computed variables. For brevity, only the important equations will be listed here, and not the full derivations.

The mechanical motor dynamics is composed of a frictional term, an inertial term and the torques induced by the elastic transmission on the motor, denoted as $\boldsymbol{\tau}_{\mathrm{e,M}}$:

$$\boldsymbol{\tau}_{\mathrm{m,d}} = \boldsymbol{I}_{\mathrm{m}}\, \ddot{\boldsymbol{\theta}}_{\mathrm{d}} + \boldsymbol{D}_{\mathrm{vg}}\, \dot{\boldsymbol{\theta}}_{\mathrm{d}} + \boldsymbol{\tau}_{\mathrm{e,M}} \tag{4}$$

with $\boldsymbol{D}_{\mathrm{vg}} \equiv \boldsymbol{E}(\boldsymbol{d}_{\mathrm{r}} + \boldsymbol{d}_{\mathrm{g}})$ where $\boldsymbol{E}$ is the identity matrix and $\boldsymbol{d}_{\mathrm{r}}$ and $\boldsymbol{d}_{\mathrm{g}}$ stand for the rotor and gearbox damping vectors, respectively. $\boldsymbol{I}_{\mathrm{m}}$ is the diagonal inertia matrix, computed by $\boldsymbol{I}_{\mathrm{m}} = \boldsymbol{E}\,(\boldsymbol{I}_{\mathrm{r}} + \boldsymbol{I}_{\mathrm{g}})$, where $\boldsymbol{I}_{\mathrm{r}}$ and $\boldsymbol{I}_{\mathrm{g}}$ refer to the rotor and gearbox inertia vectors, respectively.

Using $\hat{\boldsymbol{\tau}}_{\mathrm{e,J}}$ and $\hat{\boldsymbol{q}}$, we can solve Eqs. (1) - (3) for the required motor positions $\boldsymbol{\theta}_{\mathrm{d}}$. The only terms that depend on the motor position are those caused by a b-SEA or u-SEA: τ_{GLILIO}, τ_{VAS} and τ_{SOL}. In addition to the actuators, a number of passive structures contribute to the joint dynamics. Their contributions can be determined based on the simulated joint angles $\hat{\boldsymbol{q}}$. Rewriting Eqs. (1) - (3) with respect to the active terms we obtain:

$$\boldsymbol{\tau}_{\mathrm{GLILIO}}(\hat{\boldsymbol{q}}_{\mathrm{H}}, \boldsymbol{\theta}_{\mathrm{H}}) = \hat{\boldsymbol{\tau}}_{\mathrm{e,H}} - \boldsymbol{\tau}_{\mathrm{RF,H}}(\hat{\boldsymbol{q}}_{\mathrm{H}}, \hat{\boldsymbol{q}}_{\mathrm{K}}) - \boldsymbol{\tau}_{\mathrm{BF,H}}(\hat{\boldsymbol{q}}_{\mathrm{H}}, \hat{\boldsymbol{q}}_{\mathrm{K}}) \tag{5}$$

$$\begin{aligned}\boldsymbol{\tau}_{\mathrm{VAS}}(\hat{\boldsymbol{q}}_{\mathrm{K}}, \boldsymbol{\theta}_{\mathrm{K}}) = {} & \hat{\boldsymbol{\tau}}_{\mathrm{e,K}} - \boldsymbol{\tau}_{\mathrm{PL}}(\hat{\boldsymbol{q}}_{K}) - \boldsymbol{\tau}_{\mathrm{RF,K}}(\hat{\boldsymbol{q}}_{\mathrm{K}}, \hat{\boldsymbol{q}}_{\mathrm{H}}) \\ & - \boldsymbol{\tau}_{\mathrm{BF,K}}(\hat{\boldsymbol{q}}_{\mathrm{K}}, \hat{\boldsymbol{q}}_{\mathrm{H}}) - \boldsymbol{\tau}_{\mathrm{GAS,K}}(\hat{\boldsymbol{q}}_{\mathrm{K}}, \hat{\boldsymbol{q}}_{\mathrm{A}})\end{aligned} \tag{6}$$

$$\boldsymbol{\tau}_{\mathrm{SOL}}(\hat{\boldsymbol{q}}_{\mathrm{A}}, \boldsymbol{\theta}_{\mathrm{A}}) = \hat{\boldsymbol{\tau}}_{\mathrm{e,A}} - \boldsymbol{\tau}_{\mathrm{TA}}(\hat{\boldsymbol{q}}_{\mathrm{A}}) - \boldsymbol{\tau}_{\mathrm{GAS,A}}(\hat{\boldsymbol{q}}_{\mathrm{A}}, \hat{\boldsymbol{q}}_{\mathrm{K}}) \tag{7}$$

The right-hand sides of Eqs. (5) - (7) now contain only the known total joint torques and the torques generated by the passive, mono- and biarticular structures. The left-hand sides represent the torques excerted by the active structures and include the unknown motor positions as linear terms. The torques transmitted by these actuator types can be computed by means of the mathematical models given in Ref. 10. Combining the solutions of the left-hand sides for $\boldsymbol{\theta}_{\mathrm{H}}$, $\boldsymbol{\theta}_{\mathrm{K}}$ and $\boldsymbol{\theta}_{\mathrm{A}}$ with the right-hand sides of Eqs. (5) - (7) yields the desired motor position trajectories. Let us exemplary formulate the equation for the hip motor position where $\boldsymbol{\tau}_{\mathrm{GLILIO}} = k_{\mathrm{e}}\,(\boldsymbol{\theta}_{\mathrm{H}} - \hat{\boldsymbol{q}}_{\mathrm{H}})$:

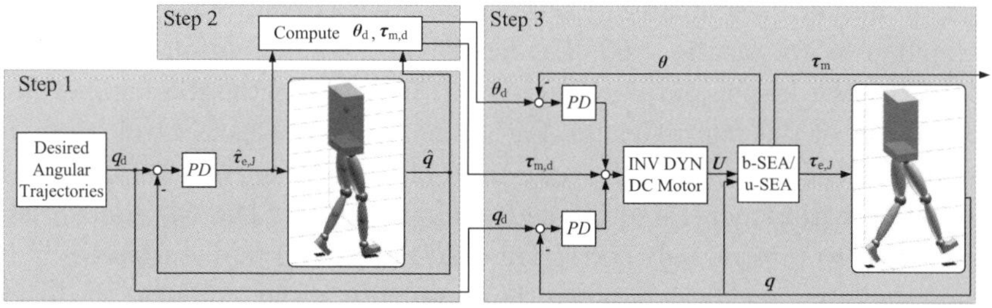

Figure 2. Controller scheme to derive the required motor control signals for given joint reference trajectories (adapted from Ref. 8).

$$\boldsymbol{\theta}_{\mathrm{H}} = \hat{\boldsymbol{q}}_{\mathrm{H}} + \frac{1}{k_{\mathrm{e}}}\,\hat{\boldsymbol{\tau}}_{\mathrm{e,H}} - \frac{1}{k_{\mathrm{e}}}\,\boldsymbol{\tau}_{\mathrm{RF,H}}(\hat{\boldsymbol{q}}_{\mathrm{H}}, \hat{\boldsymbol{q}}_{\mathrm{K}}) - \frac{1}{k_{\mathrm{e}}}\,\boldsymbol{\tau}_{\mathrm{BF,H}}(\hat{\boldsymbol{q}}_{\mathrm{H}}, \hat{\boldsymbol{q}}_{\mathrm{K}}) \tag{8}$$

k_{e} stands for the constant torsional stiffness of the elastic spring. In order to determine the full motor torques using Eq. (4), the transmission torques induced on the motor side are needed as well. For this specific actuator, transmission torques on motor and joint side are equal, i.e. $\tau_{\mathrm{e,M\,H}} \equiv \tau_{\mathrm{GLILIO}}$ holds for variables reflected through the transmission. Due to the linearity of the transmission the solution for $\boldsymbol{\theta}_{\mathrm{H}}$ (cf. Eq. 8) is rather straightforward, in contrast to the nonlinear transmission of the u-SEA.

The u-SEA of the knee, as well as of the ankle joint, has a more complex output torque function involving a cross product computation:

$$\tau_{\mathrm{VAS}}(\hat{q}_{\mathrm{K}}, \theta_{\mathrm{K}}) = \left({}^{K}\boldsymbol{p}_{\mathrm{KAJ}} \times {}^{K}\boldsymbol{F}(\hat{q}_{\mathrm{K}}, \theta_{\mathrm{K}})\right) \bullet e_Z \tag{9}$$

$$= \left({}^{K}\boldsymbol{p}_{\mathrm{KAJ}} \times \left(k_{\mathrm{e}}\,\Delta l(\hat{q}_{\mathrm{K}}, \theta_{\mathrm{K}}) \frac{{}^{K}\boldsymbol{l}(\hat{q}_{\mathrm{K}})}{\|{}^{K}\boldsymbol{l}(\hat{q}_{\mathrm{K}})\|}\right)\right) \bullet e_Z \tag{10}$$

with ${}^{K}\boldsymbol{p}_{\mathrm{KAJ}} \in \mathbf{R}^3$ denoting the vector to the moving tendon fixation point. ${}^{K}\boldsymbol{F} \in \mathbf{R}^3$ is the tendon force vector with respect to the knee coordinate system. A schematic drawing of this actuation concept as well as definitions for the spring elongation Δl and tendon length vector ${}^{K}\boldsymbol{l} \in \mathbf{R}^3$ are presented in Ref. 10. Note also that the transmission system exerts different torques on the motor side:

$$\tau_{\mathrm{e,MK}}(\hat{q}_{\mathrm{K}}, \theta_{\mathrm{K}}) = -r_{\mathrm{M}}\, \left\|{}^{K}\boldsymbol{F}(\hat{q}_{\mathrm{K}}, \theta_{\mathrm{K}})\right\| \tag{11}$$

with r_{M} denoting the radius of the gearhead output shaft.

For simplicity and consistent treatment, we neglect damping terms in the computation of the inverse models. However, damping is incorporated in the dynamics equations of the motor, transmission and joint for the forwards dynamics simulation of the complete model.[10]

Step 3: In this last step, the forward dynamics of the elastic robot, i.e. the complete multibody system (MBS) dynamics model including its complex actuation, is simulated. Each actuator is PD controlled to track the computed motor positions $\boldsymbol{\theta}_{\mathrm{d}}$. The feedforward compensation term, represented by the precomputed torques $\boldsymbol{\tau}_{\mathrm{m,d}}$, constitutes the most important component of this controller (cf. Fig. 2), as it reduces the control effort. It unburdens the control loop because most of the necessary power converter excitation can be generated in the feedforward path. The PD control law is required to provide only corrections and to respond to disturbances.[12] It is also possible to choose between the two modes, either complete feedback control with high gain parameters tuned to track the reference signals very precisely or sole feedforward control exploiting the eigendynamics.

4. Simulation Study and Results

In this section we compare the model-based approach, outlined in the previous section, with open-loop feedforward excitation of the motors. In order to realize in-place hopping motions, we first create sinusoidal patterns with the formula $y = A \sin(\omega t + \phi) + B$ based on desired leg configurations for the flexion and extension phase at the fundamental frequency $f_0 = 2\,\text{Hz}$ ($\boldsymbol{q}_{\text{flex}} = (q_\text{H}, q_\text{K}, q_\text{A}) = (26°, -63°, 13°)$, $\boldsymbol{q}_{\text{ex}} = (q_\text{H}, q_\text{K}, q_\text{A}) = (13°, -26°, -13°)$). Using the motor positions required to compensate for the gravitational forces occurring during these configurations, the amplitude A, angular frequency ω, phase ϕ and offset angle B are computed to subsequently generate the motor trajectories. As, at this stage, postural stability is still neglected, the robot's upper body movements are constrained to 1D motions.[9] With the simulation parameters given in Table 1 the forward dynamics of the BioBiped1 robot model is simulated for ten cycles in MATLAB/Simulink using the ode23 (Bogacki-Shampine) solver with variable step size, relative tolerance 10^{-3} and adaptive zero-crossing options. The robot is about 1.1 m tall in extended position and weighs around 10 kg. For the segment masses and link lengths we refer to Ref. 13.

Utilizing $\hat{\boldsymbol{\tau}}_{\text{e,J}}$ and $\hat{\boldsymbol{q}}$ obtained from the above simulation run, we determine the required motor positions $\boldsymbol{\theta}_\text{d}$ and torques $\boldsymbol{\tau}_{\text{m,d}}$ to apply the model-based approach and proceed as described in *Step 2* and *Step 3*. In Fig. 3 snapshots of this model-based realization of the hopping motions are illustrated. The corresponding ground reaction forces (GRF) occurring in each cycle are given below. The average duty factor and hopping height amount to 25.57 % and 0.26 m, respectively. The realized motions match very well the motions obtained by the open-loop feedforward excitation method. For the standard deviations see Table 2. Considering a motion range of about 30 ° in the hip, 40 ° in the knee and 33 ° in the ankle joint, the obtained values for σ_q can be regarded as insignificant. These deviations result from the deviations existing between the actual motor positions of the open-loop excitation and model-based method which are due to various reasons. Since the signals from *Step 1* are further processed to determine the motor positions and torques which include implicitly the fourfold derivation of $\hat{\boldsymbol{q}}$, it is essential to apply filtering to the signals. Filtering is also applied to the computed signals in *Step 2* to ensure that irrelevant peaks are not unnecessarily tracked by the method. This signal filtering may lead to changed configurations of $\boldsymbol{q}$ and $\boldsymbol{\theta}$ which can cause tendon slackening. Besides it should be noted, as previously mentioned, that the computation of the hip transmission torque does not consider any friction causing slight deviations, as

Table 1. Simulation parameters

Geared DC motor in hip, knee, ankle	
Motor torque constant	$k_t = 2.6 \cdot 10^{(-2)}$ Nm/A
Motor rotor inertia	$I_r = 3.3 \cdot 10^{(-6)}$ kg m^2
Motor speed constant	$k_v = 2.6 \cdot 10^{(-2)}$ Vs/rad
Motor armature resistance	$R_a = 0.611$ Ohm
Gearbox ratio	$n_g = 66$
Gearbox inertia	$I_g = 7 \cdot 10^{(-8)}$ kg m^2
Rotor and gearbox viscous damping	$d_{vg} = 1.5 \cdot 10^{(-4)}$ Nms/rad

Contact model parameters[14]	
Vertical collision force constant	$k_c = 8 \cdot 10^3$ N/m
Collision damping coefficient	$\lambda_c = 10^4$ Ns/m^2
Sliding friction coefficient	$\mu_{fk} = 0.6$
Sliding friction to stiction transition velocity limit	$v_{stic} = 0.001$ m/s
Maximum stiction force coefficient	$\mu_{fs} = 0.8$
Horizontal ground interaction stiffness	$k_{fs} = 10^4$ N/m
Stiction damper	$d_{fs} = 40$ Ns/m

Elastic transmissions			
Structure	**Stiffness** [N/mm]	**Attachment** [number]	**Rest angle** [deg]
VAS	15.5	4	-70
PL	6.7	1	-70
SOL	13.3	4	15
TA	4.1	1	15

Motor controller gains	
P-gain hip b-SEA $k_{p,H} = 200$	D-gain hip b-SEA $k_{d,H} = 50$
P-gain knee VAS $k_{p,K} = 30$	D-gain knee VAS $k_{d,K} = 8$
P-gain ankle SOL $k_{p,A} = 30$	D-gain ankle SOL $k_{d,A} = 4$

friction is included in the forward model. Hence, it is even more noteworthy that the model-based approach is capable of tracking the original motions incorporating dynamic ground contacts so well, without any ground contact sensing. The most important result is that the motions can be tracked without increasing the power consumption (see the values for σ_u and σ_i). Also, it should be mentioned that the actually generated motor torques $\boldsymbol{\tau}_m$ agree very well with the computed feedforward torques $\boldsymbol{\tau}_{m,d}$. This feedforward compensation term can highly reduce the efforts of the feedback control and in this way turn the motion control more robust in the presence of external disturbances.

Table 2. Standard deviations of the model-based generated from the open-loop excitation trajectories.

Joint	**Hip**	**Knee**	**Ankle**
$\sigma_q[°]$	0.8252	1.7741	3.0559
$\sigma_\theta[°]$	0.2789	4.2434	3.2883
σ_u[V]	0.22277	3.1192	1.7518
σ_i[A]	0.3485	0.8337	0.5538

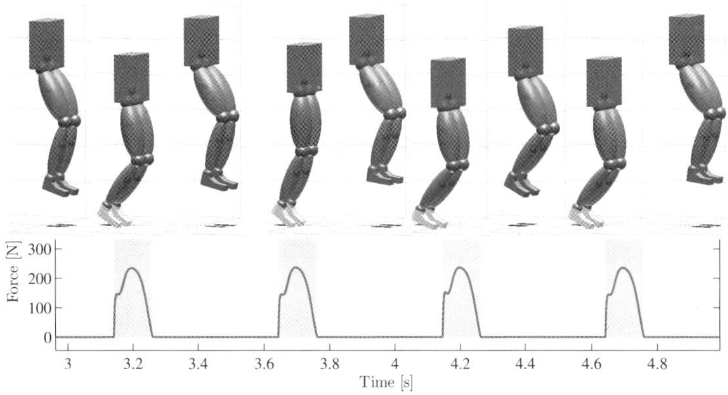

Figure 3. Snapshots of the simulated BioBiped1 robot tracking the in-place hopping motions obtained from the open-loop excitation of the motors based on the model-based approach. In the lower diagram the corresponding patterns of the GRF of both legs, which overlap due to the symmetric movements, are depicted.

5. Conclusions

In this paper we presented a model-based elastic tendon controller for musculoskeletal robots using as reference motions joint angle trajectories. The standard deviations showed that the method performs as well as open-loop feedforward oscillatory excitation providing additionally a number of benefits.

By this method, the complete musculoskeletal system including all tendon structures and the corresponding dimension space can be studied. For instance, it is possible to vary the spring stiffness or the attachment points for a defined leg actuation design and recompute the changed actuation requirements demanded of the motors. Being well suited for systems with highly redundant actuation, the method allows to precompute each structure's contribution to the overall joint dynamics and to adapt the requirements demanded of the motors to achieve, for example, energy-efficient locomotion. It can be analyzed how varied actuation requirements for the same reference motions result in different locomotion performance characterized by selected criteria such as the duty factor, energy consumption or hopping height. In this context, the method has clear advantages over basic sinusoidal motor excitations. Such concept is tremendously important to advance musculoskeletal designs and their use.

An important application of this method is also the dimensioning of the motor-gearboxes prior to a robot's construction.[8] Future work includes the implementation of the method on the hardware platform. Further, optimization of selected performance criteria will be conducted incorporating the outcomes of the approach as initial trajectories. In this context, it is also interesting to exploit the benefits of machine learning methods.[15]

Acknowledgements

This work was supported by the German Research Foundation (DFG) under grant no. STR 533/7-1. Special thanks go to Thomas Lens for helpful discussions. The BioBiped1 robot was developed within the BioBiped project.

Bibliography

1. K. Hosoda *et al.*, "Pneumatic-driven jumping robot with anthropomorphic muscular skeleton structure," *Autonomous Robots*, vol. 28, no. 3, pp. 307–316, 2010.
2. T. J. Klein, T. Phuam, and M. A. Lewis, "On the design of walking machines using biarticular actuators," in *CLAWAR*, 2008.
3. R. Niiyama, "Design of a musculoskeletal Athlete robot: A biomechanical approach," in *CLAWAR*, 2009, pp. 173–180.
4. Y. Asano *et al.*, "Lower thigh design of detailed musculoskeletal humanoid "Kenshiro"," in *IROS*, 2012, pp. 4367–4372.
5. K. Hosoda *et al.*, "Biped robot design powered by antagonistic pneumatic actuators for multi-modal locomotion," *Robotics and Autonomous Systems*, vol. 56, pp. 46–53, 2007.
6. F. Iida, J. Rummel, and A. Seyfarth, "Bipedal walking and running with compliant legs," in *ICRA*, 2007, pp. 3970–3975.
7. A. D. Luca, "Feedforward/feedback laws for the control of flexible robots," in *ICRA*, 2000, pp. 233 –239.
8. K. Radkhah and O. von Stryk, "Actuation requirements for hopping and running of the musculoskeletal robot BioBiped1," in *IROS*, 2011, pp. 4811–4818.
9. K. Radkhah *et al.*, "Concept and design of the BioBiped1 robot for human-like walking and running," *IJHR*, vol. 8, no. 3, pp. 439–458, 2011.
10. K. Radkhah, T. Lens, and O. von Stryk, "Detailed dynamics modeling of BioBiped's monoarticular and biarticular tendon-driven actuation system," in *IROS*, 2012, pp. 4243 – 4250.
11. M. Mistry, J. Buchli, and S. Schaal, "Inverse dynamics control of floating base systems using orthogonal decomposition," in *ICRA*, 2010, pp. 3406–3410.
12. G. Ellis, *Control System Design Guide.* Elsevier Academic Press, 2004.
13. K. Radkhah and O. von Stryk, "Exploring the Lombard paradoxon in a bipedal musculoskeletal robot," in *CLAWAR*, 2013.
14. T. Lens, K. Radkhah, and O. von Stryk, "Simulation of dynamics and realistic contact forces for manipulators and legged robots with high joint elasticity," in *ICAR*, 2011, pp. 34–41.
15. D. Mitrovic, S. Klanke, and S. Vijayakumar, "Adaptive optimal feedback control with learned internal dynamics models," in *From Motor Learning to Interaction Learning in Robots*, Springer, 2010, vol. 264, pp. 65–84.

VELOCITY CONTROL OF SERIAL ELASTIC ACTUATOR BASED ON EKF ESTIMATOR AND NEURAL NETWORK

Yichao MAO, Rong XIONG*, Qiuguo ZHU, and Jian CHU

State Key Laboratory of Industrial Control Technology, Zhejiang University, Hangzhou, 310027, China

** Corresponding author, E-mail: rxiong@iipc.zju.edu.cn*

Abstract: Serial Elastic Actuator (SEA) is a kind of advanced actuator which can efficiently adapt environment disturbance because it has better compliance against external impact than traditional actuator. Previous research work of SEA control are mainly focused on torque control. However, in applications such as manipulator motion planning, the position following accuracy is considered a more critical performance metric, which leads to the requirement of delicate velocity control of SEA. Different with previous PID control method, this paper proposes a velocity controller based on neural network and extended kalman filter (EKF). Neual network is utilized to learn a feed-forward dynamic model of the controller under the condition with a certain payload and pose of SEA, which can compensate nonlinear disturbances due to gravity and the elasticity of spring. And the proposed neural network model is adjusted adaptively using observation data of the actuator payload and configuration acquired by a EKF estimator. Experimental results demonstrate the validity and robustness of the proposed method.

Keywords: Series elastic actuator, velocity control, neural network, EKF estimator.

1. Introduction

Series Elastic Actuator (SEA) is first introduced by Gill A. Pratt etc. in 1990th.[1] Different to the structure of motor-reducer-load used in the traditional actuator, a spring is added in SEA to link the reducer and the load of the actuator. Thus, SEA can achieve toque control more precisely with the capability of measuring the torque applied on the load directly through the deflection of the spring. Such a torque control mode performs high compliance to the external impact. The MIT leg Laboratory applied SEA in their bipedal robot "Spring Flamingo" and "M2",[2,3] and implemented the walking control based on SEA torque control. In recent years, with the increasing requirement of safe interaction or collaboration between human and machine in the domain of service robot and industrial robot,

SEA also have been applied to different kinds of manipulator to enhance their essential safety.[4–6]

Correspond to the work requirement of manipulator, not only the output toque of SEA should been controlled accurately, but also the output position and velocity should been controlled as accurate as possible. Here the velocity control becomes an essential element that has strong influence on the performance of the whole system. Depend on the precise control of the torque applied on the load, the velocity controller can be designed based on the toque control, i.e, the SEA velocity controller is usually constructed with the torque inner loop. The PID controller and the LOG controller based on the optimal control theory have been successfully employed.[7] However, those SEA velocity controllers have two problems:(1)the control period of the velocity controller based on toque control is much more longer than the velocity control period of traditional motor, which makes it difficult to cope with the nonlinear disturbance due to gravity, the elasticity of spring, as well as the mechanic damping. These nonlinear disturbance will result in the deterioration of robustness of the whole system. (2) in practice, the payload of SEA will change when the manipulator execute grasping or other tasks and the pose variation of manipulator will bring out the change of the initial pose of SEA, while current velocity controllers of SEA lack the adaptability to the deviation of the joint payload and pose, which will affect the performance of the control system.

To overcome these two problems in SEA velocity control, this paper proposes an adaptive velocity control algorithm using a feed-forward model learned offline and an extended Kalman filter (EKF) estimator online. The main contribution of the algorithm includes: (1) use a feed-forward model learned by neural network under a certain condition of payload and pose of SEA to compensate various nonlinear factors that influence the control performance;(2) estimate the states of payload and pose of SEA with EKF to increase the adaptability to the state variation. The experimental results validates the efficiency and robustness of the method.

In the following of this paper, Section II gives the framework of the whole structure of the controller, Section III introduces the method of the feed-forward model learning algorithm based on neural network, Section IV describes the method of the adaptive controller based on EKF estimator, Section V testifies the efficiency of the controller designed in this paper.

2. The structure of the velocity controller

The structure of the velocity controller based on EKF estimator and neural network feed-forward model is shown in Fig. 1, where the statement of the notations is listed in Table 1.

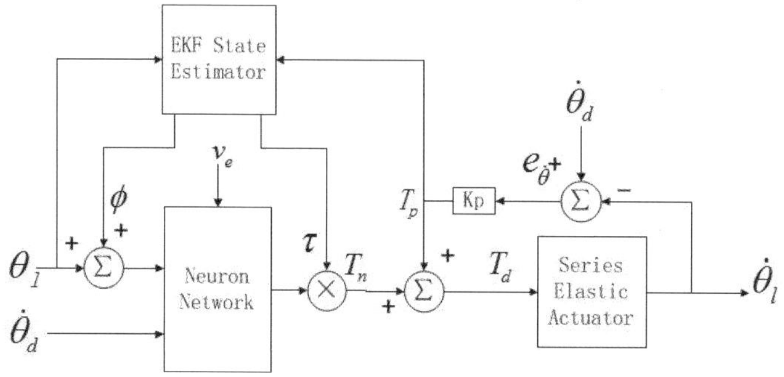

Fig. 1. The overview of the structure of velocity controller.

Table 1. Statement of the Notations.

Notation	Statement
θ_l	Current angle of the load
$dot\theta_l$	Actual velocity of the load
$dot\theta_d$	reference velocity of the load
$e_{\dot{\theta}}$	Velocity tracing error
T_d	Ultimate output torque of controller
T_p	Output torque of proportional controller
K_p	Proportional parameter of proportional controller
T_n	Output torque of feed-forward model
ϕ	Angle of pose
τ	Gravity torque of the load
v_e	Criteria function of the neural network

The output of the controller is:

$$T_d = T_n + T_p \tag{1}$$

It is consisted of two parts:

(1) Feedback control of the following error of SEA, where proportional controller is adopted. The output of this part is:

$$T_p = K_p e_{\dot{\theta}} \tag{2}$$

This part is constructed to guarantee the load running in reference velocity. Simultaneously, it acts as a role of a deviation estimator in the feed-forward model.

(2) Feed-forward model, which is provided by a neural network with EKF estimator. The output of the feed-forward model at k^{th} cycle can be described as :

$$T_n = \tau(k) Neuron(\theta + \phi(k), \dot{\theta}) \tag{3}$$

where $Neuron(\theta, \dot{\theta})$ is the normalized neural network model function. It employs the output of proportional controller as learning error. After the parameters of neural network being optimized, it functions as a feed-forward model under specific condition. With system states estimation through an EKF estimator, the feed-forward model can be closer to the actual control model.

3. Feed-forward model learning based on neural network

In this paper, reference compensation technique(RCT) is used for the neural network.[8] The current pose angle θ_l and the reference velocity $\dot{\theta}_d$ are the input of the neural network while the feed-forward torque T_n is the output. Each neuron is a nonlinear function:

$$f_{neuron}(x) = \frac{1}{1 + e^{-x}} \tag{4}$$

The neural network training error can be defined as:

$$v_e = K_p e_{\dot{\theta}} \tag{5}$$

which is same as the output of the proportional controller. So the objective function to be optimized can be set as:

$$F = \frac{1}{2} v_e^2 \tag{6}$$

From Eq. (1), it can be derived that, when F is optimized, the neural network output T_n will converge to the controller output T_d.

Assume that the inverse of the SEA dynamic model $f(\theta, \dot{\theta})$ is described as:

$$f(\theta, \dot{\theta}) = T_d + \varepsilon \tag{7}$$

where ε is the error, which depends on the performance of the controller. If and only if the controller sustains the SEA rotate at the reference velocity, then ε equals 0, which means the controller output T_d is consistent to the inverse dynamic model of SEA $f(\theta, \dot{\theta})$.

Although the initial neural network output T_n is widely different with $f(\theta, \dot{\theta})$ that sustains the SEA rotate at a certain angular velocity and determined angle because of the poor initial parameters of the neural network, it will becomes closer and closer to the real model $f(\theta, \dot{\theta})$ by the method of iterative learning. First, the compensation of the proportional controller will control the load to run in a neighborhood range of the reference velocity. Thus T_d is approximate to $f(\theta, \dot{\theta})$. Second, as we mentioned above, when F is optimized, T_n will converge to T_d. So T_n will becomes closer to $f(\theta, \dot{\theta})$. Consequently, the performance of the controller improves, which reduces the error of the load velocity and makes the actual control torque T_d becomes more approximate to $f(\theta, \dot{\theta})$. After the iterative learning, a precise feed-forward model will be obtained. The relationship between T_d, T_n and $f(\theta, \dot{\theta})$ is showed in Fig. 2.

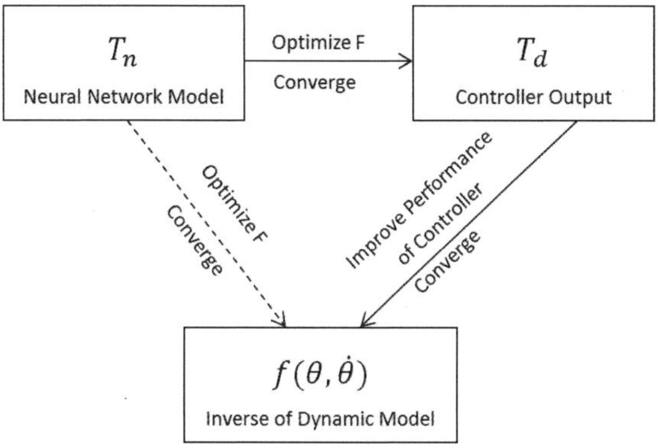

Fig. 2. The relationship between T_d, T_n and $f(\theta, \dot{\theta})$.

4. Adaptive controller based on EKF state estimator

The feed-forward model based on neural network learning could improve the controller performance. The equation of feed-forward model Eq. (3) means that the normalized function $Neuron(\theta, \dot{\theta})$ in the neural network learning process determines the shape of the feed-forward function. But it is learned with a determined payload and a determined pose of SEA. They are not consistent with the real dynamic model. To make the controller more flexible with changing states of the payload and pose, this section will use a EKF estimator to estimate the system states, to make the feed-forward model converge to the real dynamic model. In practice, the yaw angle of pose has little influence on the velocity control and the pitch angle of pose is influenced by the change of the payload, hence only the payload τ and the roll angle of pose ϕ are set as system states to be estimated.

For a single SEA without vision and gyro to measure the external environment, it lacks enough information to construct an accurate motion model. So it is reasonable to regard the payload and the angle of pose as constant, and to approximate the real model by introducing the estimation of the covariance. According to the analysis, the motion model can be established as:

$$\begin{bmatrix} \tau(k) \\ \phi(k) \end{bmatrix} = \begin{bmatrix} 1 & 0 \\ 0 & 1 \end{bmatrix} \begin{bmatrix} \tau(k-1) \\ \phi(k-1) \end{bmatrix} + N(0, R_t) \tag{8}$$

where the R_t is the covariance of the motion model, and $N(.)$ is the normal distribution.

And the measurement can be established as:

$$T_d(k) = T_n(k) + N(0, Q_t) \tag{9}$$

where the Q_t is the covariance of the measurement. The basic thought of this is to regard $T_d(k)$ as the real measurement, while $T_n(k)$ as desired measurement based on the estimation of the current state, the error between them is the output of PID controller. With the similar thought of the iterative learning of the neural network, the system states can be adjusted in a recursive mode to guarantee the feed forward model converge to the real inverse dynamic model of SEA $f(\theta, \dot{\theta})$

The initial states are:

$$\tau(0) = \tau_0$$
$$\phi(0) = 0$$

5. Experimental results

5.1. *Experimental platform*

The experiments are conducted on a self-designed SEA platform shown in Fig. 3. The elastic coefficient is $k_s = 1.48Nm$, and the limit of the deflected angle is $\theta_l - \theta_m = \pm 4^\circ$. All of the experiments set the demanded velocity of $\frac{\pi}{3}$rad/s, and change the payload by adding or reducing the number of block fixed on the arm of the SEA. Every block is weighted 500g. The feed-forward model of 2 blocks is first learned by the neural network.

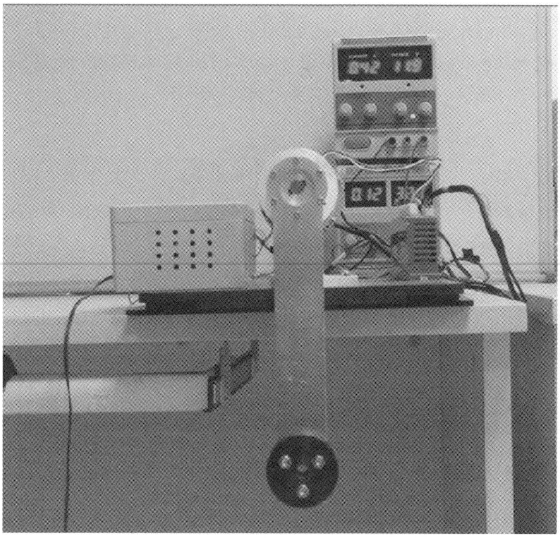

Fig. 3. Experimental platform.

5.2. *Comparison between PID controller and feed-forward controller based on neural network*

The first experiment is to compare the performance of the regular PID controller and the neural network feed-forward controller when the states of the system are constant. When the manipulator loads 2 blocks and keep the platform horizontal, the comparison of the two controllers is showed in Fig. 4.

From the profile of the output velocity, it is seen that disparity between the output and the reference velocity exists for PID controller to track the changing gravity torque, and changes periodically according to the position

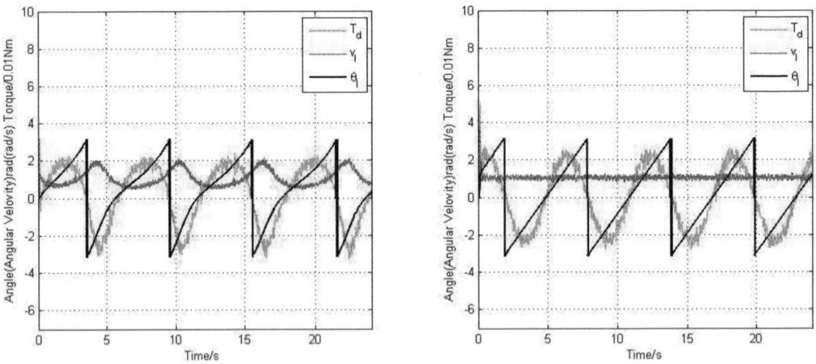

Fig. 4. Comparison between PID controller and neural network feed-forward controller. (Left: PID controller, right: neural network feed-forward controller.)

changing of the load. In contrast, the feed forward controller with neural network has high performance and guarantees the load velocity in a small neighborhood around the reference value.

The feed-forward profile of the neural network, the output profile of the proportional controller and the total output profile of the controller is showed in Fig. 5. From the profiles we can see that the feed-forward profile of the neural network is overlapping the control output, so the proportional controller just need to respond to the noises of the environment to sustain the rotate of the SEA load. The feed-forward improve the efficiency and performance of the controller.

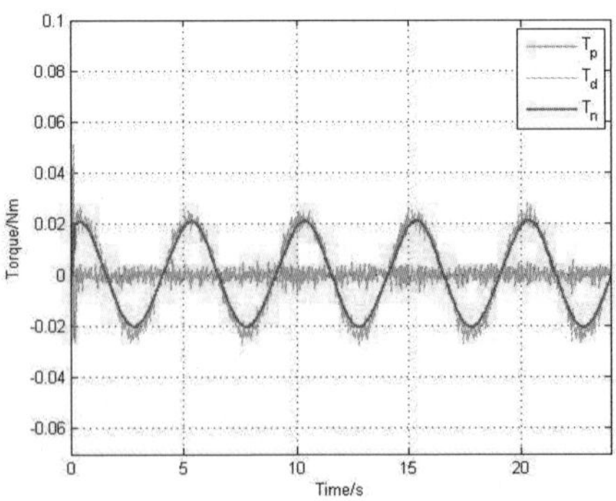

Fig. 5. Feed-forward output of the neural network

5.3. *Comparison between controllers with and without EKF estimator*

This section will test the adaptive controller based on EKF estimator and neural network strictly. We reduce the number of blocks from 2 to 1 and rotate the platform 90 degree. Under such condition, the performance of the standard neural network feed-forward controller and the adaptive controller based on EKF estimator and neural network is showed in Fig. 6. The feed-forward profile and the total output profile of the each controller is showed in Fig. 7:

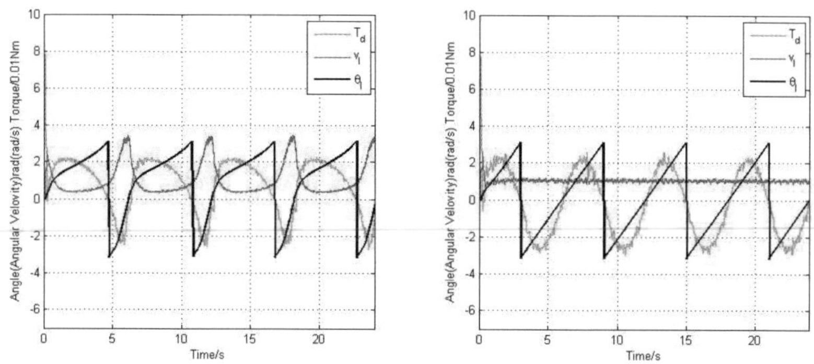

Fig. 6. Left: Neural network controller Right: The neural network with EKF estimator Comparison of the performance under condition of changed load and pose.

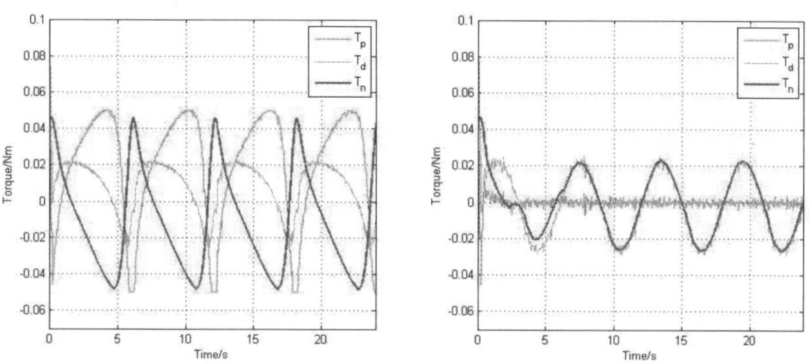

Fig. 7. Left: Neural network controller Right: The neural network with EKF estimator Comparison of the feed-forward output under condition of changed load and pose.

The control profile shows that the adaptive controller based on EKF estimator has well adaptability to severe pose deviation and payload change. The system states estimation at different control cycles is showed in Fig. 8:

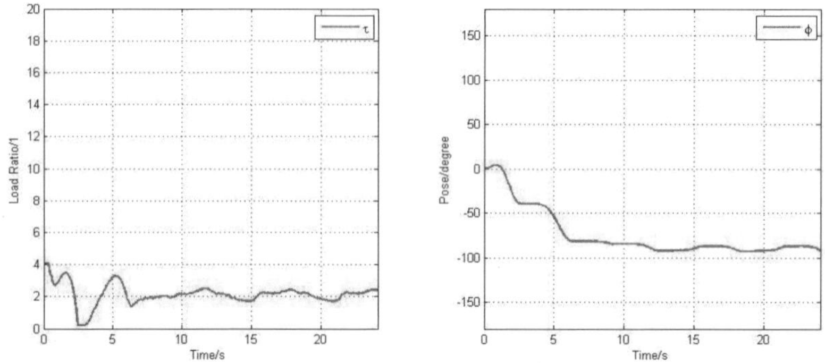

Fig. 8. The estimation of states at different cycles Left: The estimation of load Right: The estimation of pose

We can see that the system states converge rapidly to the neighborhood of the real states by the estimation of EKF estimator. It testifies that the EKF estimator could correctly estimate the states of the system as well as adjust the feed-forward profile.

6. Conclusion

This paper proposes a method to control the SEA velocity based on torque control by applying the neural network to obtain the feed-forward model and EKF estimator to improve the adaptability to the external environmental change. Through the result figures, the efficiency and robustness of the algorithm are validated.

The next work is to study how to apply the algorithm to multiple joints system. And based on the SEA technics, our institute is involved in manufacturing new type of leap robot with single actuator. It is believed that performance of biped robot will be improved by introducing the SEA technics. Additionally, although the algorithm proposed in this paper can adapt the change of the payload and angle of pose, the estimation will deviate when the robot is suffered irregular external force, because the real model is not the SEA dynamic model yet, but the model mixed with the dynamic model and the external force. This is the problem we should deal with next.

7. Acknowledgements

This work was supported by the National Nature Science Foundation of China (Grant No.NSFC: 61075078), the Natural Science Foundation of Zhejiang Province (Grant No.LQ12F03009) and the Fundamental Research Funds for the Central Universities(Grant No.2012FZA5014).

References

1. G. A. Pratt, M. M. Williamson. Series Elastic Actuator, *IEEE* International Conference on Intelligent Robots and Systems, Pittsburgh, PA, USA, Vol.1, 1995: 399-406.
2. J. E. Pratt. Exploiting Inherent Robustness and Natural Dynamics in the Control of Bipedal Walking Robots. PhD thesis, Massachusetts Institute of Technology, 2000.
3. D. W. Robinson, J. E. Pratt, D. J. Paluska, *et.al.* Series Elastic Actuator Development for a Biomimetic Walking Robot, *IEEE/ASME* International Conference on Advanced Intelligent Mechatronics, Atlanta, USA, 19-23 September, 1999: 561-568.
4. URL: http://yobotics.com
5. M. A. Diftler, J. S. Mehling, M. E. Abdallah, *et.al.* Robonaut 2-The first Humanoid Robot in Space, *IEEE* International Conference on Robotics and Automation, Shanghai, China, 9-13 May, 2011: 2178-2183.
6. URL: http://mekabot.com
7. Marco Hutter, C. David Remy, *et.al.* Scarl*ETH*: Design and Control of a Planar Running Robot, *IEEE/RSJ* International Conference on Intelligent Robots and Systems, San Francisco, CA, USA, 25-30 September, 2011: 562-567.
8. S. Jung and T. C. Hsia, Neural network inverse control techniques for PD controlled robot manipulator, ROBOTICA, Vol.19, No.3, 2002: 305C314.

LOW-DIMENSIONAL USER CONTROL OF AUTONOMOUSLY PLANNED WHOLE-BODY HUMANOID LOCOMOTION MOTION TOWARDS BRAIN-COMPUTER INTERFACE APPLICATIONS

KARIM BOUYARMANE[1*], JORIS VAILLANT[2], and JUN MORIMOTO[1]

[1] *ATR Computational Neuroscience Laboratories,*
2-2-2 Hikaridai, Kyoto 619-0288, Japan
[*] *Corresponding author, e-mail: karim.bouyarmane at atr.jp*

[2] *Montpellier 2 University, CNRS LIRMM,*
161 rue Ada, 34095 Montpellier, France
E-mail: joris.vaillant at lirmm.fr

The goal of the presented work is to use a low-dimensional control interface to control the high-DOF whole-body locomotion motion of a humanoid robot, for instance using a 2D keyboard or joystick interface (up-down, left-right) to control a 30 DOF acyclic locomotion task (reaching a goal position by climbing on a stair). This work is motivated by the use of non-invasive BCI, which offers only such low-dimensional control signals, for low-level motion control of humanoids or exoskeletons in the assistive robotics domain of applications. The methodology is the following: given the target complex locomotion task, the humanoid autonomously plans the high-DOF motion and then executes it allowing the user to control on-line some low-dimensional features of the motion, namely way-points of the moving end-links. The approach is based on the two-stage contact-before-motion planning paradigm, which autonomously plans a sequence of transition contacts in its first stage, then executes the collision-free dynamics-consistent motion in the second stage, keeping dynamics balance of the motion. Example of this control approach is demonstrated in dynamic simulation.

Keywords: Whole-Body Motion Planning; Humanoid Locomotion; Motion Autonomy.

1. Motivation of the Work

Control of humanoid robots through brain-computer interface (BCI) is an active field of research that raises challenging planning and control problems and that can develop into promising assistive applications for the elderly or reduced-mobility people. The humanoid design allows the user to intuitively comprehend the kinematics and dynamics of the controlled device, and allows the device to be readily used in a human-targeted living environment as such, i.e. without adaptation of the living environment to the device.

Previous work on control of such humanoid systems do not however take full profit of the dexterity of the humanoid design. In Refs. 1 and 2, the humanoid is seen as a continuously walking machine that is "steered" with the BCI the same way a mobile manipulator would be, for example as it is actually the case in Ref. 3. The specificity of a legged device over a wheeled one thus do not emerge within the range of applications spanned in these works. In particular, dealing with slightly differently structured environment, such as one with stairs, would not be straightforward. We do acknowledge that since these works rely on state-of-the-art walking pattern generators, they will naturally benefit from the advances in such pattern generation methods that can, for example, autonomously deal with variable stair heights or non-flat terrain.[4–7] They can also benefit from the hierarchical architectures in which they are embedded[3,8] that would allow them to be wired to alternative behaviors when facing stairs for instance. We nevertheless investigate a different approach, based on our previous work of complex whole-body motion planning and control of humanoid robots. We believe that through a more generic motion approach, the device can prove more efficient in coping with the many unpredictable situations that often occur in real life and would not have been accounted for in the said hierarchical architectures.

As we are aiming for non-invasive brain-signal-acquisition methods such as using an EEG cap for example, one can only expect to have at his disposal low-dimensional low-resolution signals to be mapped to the high-DOF kinematics of the humanoid. State-of-the-art non-invasive BCI control achieved precise control of a two-dimensional cursor position on a screen by spinal cord injury patients.[9] In Ref. 10, subjects rapidly learned to accurately control the position and speed of a one-dimensional cursor on a screen using various motor imagery (imagining repeating the word "move", or tongue/shoulder/hand imagery). One of the latest studies carried out at our institute shows that two-DOF finger movement can be reconstructed in real time from the brain activities extracted by Magnetoencephalogram (MEG).[11] From these studies, we can expect that the development of a method to extract two-DOF control command from the brain activities by using non-invasive and not-too-expensive measurement device in real time will become realistic. We take these results as a premise for the controlled feature selection in the high-DOF motion of the robot. We thus present in this work a control interface for two 1-dimensional (2x1D) features of the movement (step way-point target, step way-point threshold) than can be later interfaced with motor imagery BCI, proposing an alternative to the currently investigated visual-stimulation-based EEG such as SSVEP[2,3,8] or P300.[1]

2. Problem Formulation

We formulate our problem as follows: Given an initial and goal configuration for a humanoid robot, in arbitrary contact state with the environment, use a two-dimensional control interface to execute the motion of the humanoid from the initial to the goal configuration.

Let N be the number of degrees of freedom of the humanoid robot ($N \gg 2$, e.g. $N = 30$). The humanoid robot motion control requires an N-DOF command signal in the joint space, the problem can thus be reformulated as mapping the 2-DOF user interface command signal to the N-DOF whole-body motion of the humanoid robot, as schematically represented in Fig. 1.

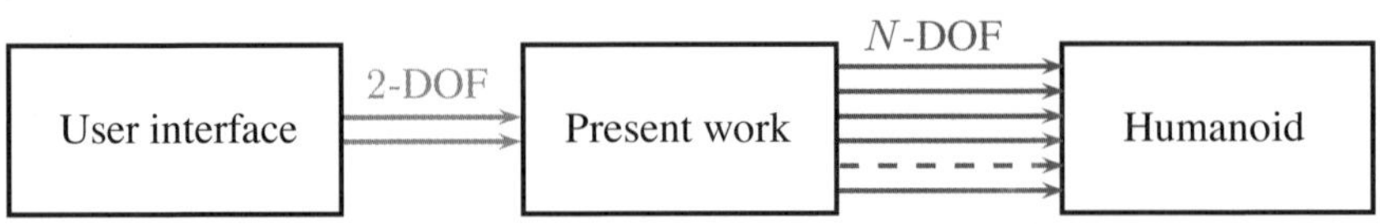

Fig. 1. Schematic representation of the problem formulation.

3. Methodology

The proposed approach is based on off-line autonomous planning of the N-DOF whole-body motion followed by on-line local modification of the planned motion, using the 2-DOF command to help the execution or correct the shortcomings of the autonomously planned motion, as schematically depicted in Fig. 2.

The motion autonomy framework consists of two components: an off-line contact transitions planner and an on-line controller that follows the prescribed contact transitions.

The initial and goal configurations of the humanoid robot do not necessarily share the same contact state with the environment (as it would be the case in Refs. 12–15), thus defining a locomotion problem, for which cyclic walking based on footsptep planning [16–18] does not necessarily provide a solution, since we allow the use of various links/bodies for support. We thus resort to our previous acyclic multi-contact planning framework.[19]

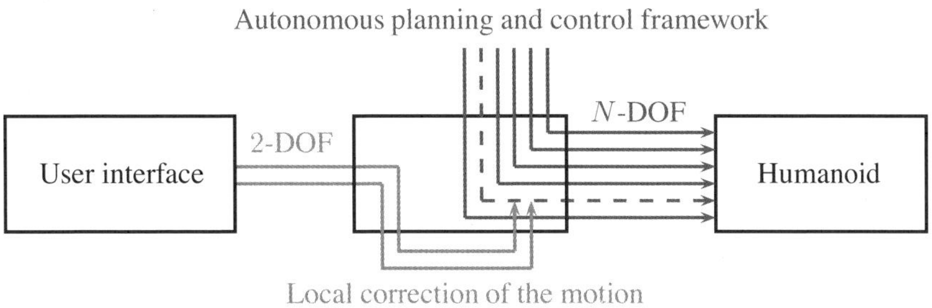

Fig. 2. Schematic representation of the methodology.

In this respect, the off-line planner only considers these initial and goal contact states, as inferred from the initial and goal configurations, and plans a feasible sequence of contact transitions between the two states, using a best-first search algorithm in the contact-state space.[20] Each contact transition is associated with a corresponding whole-body configuration generated by a physics-constrained inverse-kinematics solver.[21]

Given this planned sequence of contact states and associated configurations, the on-line controller[22] executes a whole body motion of the humanoid going through the sequence of steps by solving at control frequency a quadratic program (QP) in the generalized accelerations, contact forces, and actuation torques of the robot, minimizing a weighted sum of squared task acceleration errors, under whole-body dynamics equation of motion, actuation, joint and torque limit, and friction cone linear constraints. The tasks are defined on the center of mass of the robot, on the moving end-link that changes its contact state in the current step, and on the whole-body configuration, according to a finite-state machine (FSM) that goes through three major states while executing the steps: 1) moving the whole-body while staying in the same contact state, 2) lift-off of a contact, and 3) touch-down of contact. The transition between the lift-off and touch-down phase when moving a contact link is triggered when the link goes through a heuristics-defined way-point (Full details on the online-generation of the whole-body motion is found in Ref. 22).

In the present work we add as a main contribution to this on-line controller an autonomous collision avoidance linear constraint in the QP formulated as a velocity-damper based on the Faverjon and Tournassoud method.[23] It acts as a repulsive field that keeps the link of the robot away from the obstacle, without preventing it from reaching its goal contact location, that acts in turn as an attracting field. This formulation might however get the moving link trapped in undesired equilibrium situations (that can be pictured as local minima of the resultant field), in which the link stays immobile under the combined actions of the repulsive and attractive fields.

This is where the user can use the two-dimensional command at his disposal in order to untrap the link, by moving the location of the way-point through which the link was heuristically prescribed to go. We thus choose to use the two-dimensional command signal to control the motion of the moving end-link during the step through intuitive control of the way-point. Since we do not want the moving link to stop (halt) at the way-point, in order to have smooth motion, we define the way-point only as far-away target to reach, and the transition is triggered when the link crosses a threshold plan while aiming for the far-away target, see Fig 3. One variable of the 2-DOF signal will be used to control the height of the way-point, while the other variable is used to control the height of the threshold plan, see Fig. 4 and details in Appendix A.

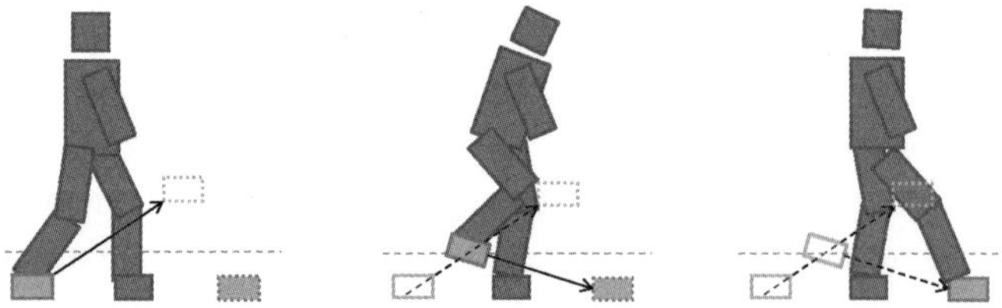

Fig. 3. The way-point and threshold strategy.

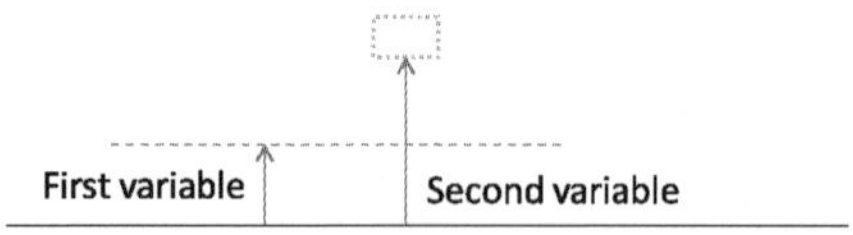

Fig. 4. The controlled variables with user interface.

4. Sample Results and Discussion

The video that can be downloaded at `www.cns.atr.jp/~karim/clawar2013.wmv` or `www.cns.atr.jp/~karim/clawar2013.mp4` demonstrates an example of use of the framework. In this example, the specified goal location of the robot is up on the stair. The framework autonomously plans the contact locations and intermediate static postures in the off-line stage to climb the stair (Fig. 5), then the on-line motion control scheme starts.

Fig. 5. The autonomously planned sequence of steps.

Whenever the FSM enters the contact lift-off state, the ambient color in the screen changes to a blue ambient color to indicate that the user can now control some features of the motion being executed. A black sphere and a blue wire frame plane are displayed to represent respectively the tracked way-point and the threshold plane (Fig. 6). When the contact body starts to move, it targets the position of the black sphere, until crossing the blue plane. At that point the FSM goes to the contact touch-down state, the ambient color of the screen is changed back to normal, and the user has no longer control on the motion.

Fig. 6. Left: in user-control mode. The blue wire frame plan represents the threshold, the black sphere represents the way-point. Middle and right: two example transitions from autonomous mode to user-control mode.

In the video we demonstrate some extreme cases. By moving the blue plane high enough we are actually able to simply control the position of the contact body that tracks the black sphere. This allows to test the said extreme behaviors such as bringing the foot upwards until reaching hip and knee joint limits (Fig. 7, right), or positioning the black sphere inside an obstacle (Fig. 7, left). As required, in the first case the robot does not loose balance and the knee and hip joint do not violates the joint limits, and in the second case the collision avoidance constraints prevents the body from "penetrating in" (colliding with) the stair.

Four computer keyboard keys are used to move the black sphere up and down and the blue plane up and down. This control interface can be replaced

Fig. 7. Left: when the back sphere is positioned by the user inside an obstacle, the obstacle-collision-avaoidance constraint prevents the collision. Right: extreme motions imposed by the user, joint limits and equilibrium constraints keep the robot in safe configuration

by other devices, such as joystick, or, as discussed in the motivations of the work, a motor imagery BCI.

Finally, we plan in our future work to port this method from simulation to real robot. Among others, issues regarding the integration of real-time feedback of robot sensors should be dealt with in this perspective.

Acknowledgements

This work is supported with a JSPS Postodoctoral Fellowship for Foreign Researchers, ID No. P12707. A part of this study is the result of "Brain Machine Interface Development" carried out under the Strategic Research Program for Brain Sciences by the Ministry of Education, Culture, Sports, Science and Technology (MEXT) of Japan. The authors would like to thank Abderrahmane Kheddar and François Keith for valuable use of the AMELIF dynamics simulation framework.

Appendix A. Low-Dimensional User Control Variable Derivation

Figure A1 illustrates the following derivations. Let P_o denote the origin position of the contact body at the start of the current step, P_g be its aimed goal position at the end of the step (as planned off-line), $\mathbf{N}$ the normal to the contact surface at the goal position. The way-point W is defined as

$$W = P_o + \alpha\,(P_g - P_o) + h_1\,\mathbf{N}\,, \tag{A.1}$$

where α is a fixed coefficient between 0 and 1, typically $\alpha = 0.5$, and h_1 is the height of the way-point ($h_1 \in \mathbb{R}$).

The threshold is defined as a plan $(T, \mathbf{N})$ of origin point T and normal $\mathbf{N}$, T being defined as

$$T = P_g + h_2\,\mathbf{N}\,, \tag{A.2}$$

where h_2 is the height of the threshold ($h_2 \in \mathbb{R}$).

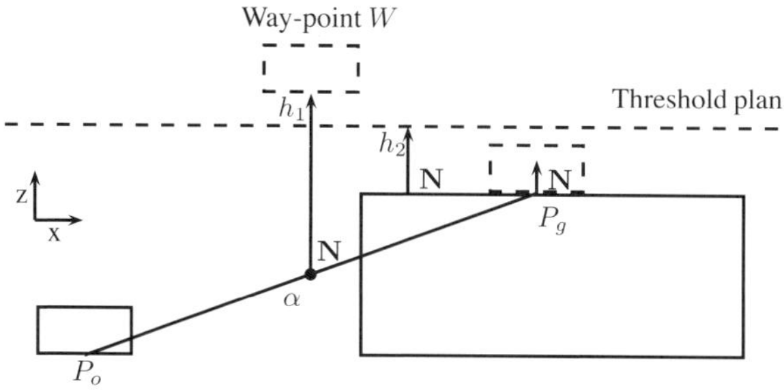

Fig. A1. User-controlled variables

The low-dimensional on-line controlled variables that we retained in this study are thus $(h_1, h_2) \in \mathbb{R}^2$.

References

1. C. J. Bell, P. Shenoy, R. Chalodhorn and R. P. N. Rao, *Journal of Neural Engineering* **5**, 214 (2008).
2. P. Gergondet, S. Druon, A. Kheddar, C. Hintermuller, C. Guger and M. Slater, Using Brain-Computer Interface to Steer a Humanoid Robot, in *Proc. of IEEE International Conference on Robotics and Biomimetics*, 2011.
3. M. Bryan, J. Green, M. Chung, L. Chang, R. Scherery, J. Smith and R. P. N. Rao, An Adaptive Brain-Computer Interface for Humanoid Robot Control, in *Proc. of IEEE-RAS International Conference on Humanoid Robots*, 2011.
4. M. Morisawa, F. Kanehiro, K. Kaneko, S. Kajita and K. Yokoi, Reactive biped walking control for a collision of a swinging foot on uneven terrain, in *Proceedings of the IEEE-RAS International Conference on Humanoid Robots*, 2011.
5. A. Herdt, H. Diedam, P.-B. Wieber, D. Dimitrov, K. Mombaur and M. Diehl, *Advanced Robotics* **24**, 719 (2010).
6. A. Takanishi and I. Kato, Development of a biped walking robot adapting to a horizontally uneven surface, in *Proceedings of the IEEE-RSJ International Conference on Intelligent Robots and Systems*, 1994.
7. K. Hashimoto, Y. Sugahara, M. Kawase, A. Ohta, C. Tanaka, A. Hayashi, N. Endo, T. Sawato, H. ok Lim and A. Takanishi, Landing pattern modification method with predictive attitude and compliance control to deal with uneven terrain, in *Proceedings of the IEEE-RSJ International Conference on Intelligent Robots and Systems*, 2006.
8. M. Chung, W. Cheung, R. Scherer and R. P. N. Rao, A Hierarcgical Architecture for Adaptive Brain-Computer Interfacing, in *Proceedings of the International Joint Conferences on Artificial Intelligence*, 2011.
9. J. R. Wolpaw and D. J. McFarland, *Proceedings of the National Academy of Sciences of the United States of Americ* **101**, 17849 (2004).
10. K. J. Miller, G. Schalk, E. E. Fetza, M. den Nijs, J. G. Ojemanne and R. P.

Rao, *Proceedings of the National Academy of Sciences of the United States of Americ* (2010).
11. A. Toda, H. Imamizu, M. Kawato and M. A. Sato, *NeuroImage* **54**, 892 (2011).
12. J. Kuffner, S. Kagami, K. Nishiwaki, M. Inaba and H. Inoue, *Autonomous Robots* **12**, 105 (2002).
13. K. Yamane, J. Kuffner and J. K. Hodgins, *ACM Transactions on Graphics (Proc. SIGGRAPH 2004)* **23**(August 2004).
14. E. Yoshida, O. Kanoun, C. Esteves-Jaramillo and J. P. Laumond, Task-driven Support Polygon Reshaping for Humanoids, in *Proceedings of the IEEE-RAS International Conference on Humanoid Robots*, 2006.
15. E. Yoshida, J.-P. Laumond, C. Esteves, O. Kanoun, T. Sakaguchi and K. Yokoi, Whole-Body Locomotion, Manipulation and Reaching for Humanoids, in *Motion In Games*, 2008.
16. J. Kuffner, K. Nishiwaki, S. Kagami, M. Inaba and H. Inoue, Footstep Planning Among Obstacles for Biped Robots, in *IEEE/RSJ International Conference on Intelligent Robots and Systems*, 2001.
17. J. Chestnutt, J. Kuffner, K. Nishiwaki and S. Kagami, Planning Biped Navigation Strategies in Complex Environments, in *Proceedings of the IEEE-RAS International Conference on Humanoid Robots*, 2003.
18. J. Chestnutt, M. Lau, J. Kuffner, G. Cheung, J. Hodgins and T. Kanade, Footstep Planning for the ASIMO Humanoid Robot, in *Proceedings of the IEEE International Conference on Robotics and Automation*, 2005.
19. K. Bouyarmane and A. Kheddar, *Advanced Robotics* **26** (2012).
20. K. Bouyarmane and A. Kheddar, Multi-contact stances planning for multiple agents, in *IEEE International Conference on Robotics and Automation*, 2011.
21. K. Bouyarmane and A. Kheddar, Static multi-contact inverse problem for multiple humanoid robots and manipulated objects, in *IEEE-RAS International Conference on Humanoid Robots*, 2010.
22. K. Bouyarmane and A. Kheddar, Using a multi-objective controller to synthesize simulated humanoid robot motion with changing contact configurations, in *IEEE/RSJ Int. Conf. on Intelligent Robots and Systems*, 2011.
23. B. Faverjon and P. Tournassoud, Planning of manipulators with a high number of degrees of freedom, in *IEEE International Conference on Robotics and Automation*, 1987.

STATE TO STATE MOTION PLANNING FOR UNDERACTUATED SYSTEMS USING A MODIFIED RAPIDLY EXPLORING RANDOM TREE ALGORITHM

R. SHVARTSMAN*, Y. TAN and D. OETOMO

Melbourne School of Engineering,
The University of Melbourne, Australia
**E-mail: rinas@pgrad.unimelb.edu.au*

The interest in underactuated mechanisms has seen an increase in the robotics community in the application of agile dynamic motions such as human-like walking. Rapidly-exploring Random Trees (RRT) algorithms serve as a successful and practical approach for path planning in high-dimensional systems, but their applicability for systems with underactuation has been less successful. In this paper, we propose two modifications to the RRT algorithm which facilitate the search of a start-to-target actuation command policy for underactuated systems. The first modification addresses the distance cost function which defines proximity of states. Specifically, we estimate the reachability property from one state to another with a simple geometrical heuristic. The second modification was geared to bias the distribution of the random samples in the state space towards the desired target states. We present the results on a torque-limited simple pendulum as well as on an underactuated model of a double pendulum.

Keywords: Motion Planning; Underactuated Systems; RG-RRT; Non-Euclidian Distance Metric; Nested Search.

1. Introduction

The study of underactuated systems has gained increased attention in the recent years, especially due to the increased interest in the study of a bipedal robot performing dynamic gait. When the gait cycle of a bipedal system includes single-stance toe-roll or point-feet locomotion,[1] unactuated rotational degrees of freedom (DoF) are added to the system in the interface between the foot and the ground. Despite having unactuated degrees of freedom, underactuated systems can be controllable, and in many applications it is possible to find an appropriate set of command signals for the actuators so that, via the dynamic coupling between the actuated and the unactuated DoF, the system is manipulated from one state to another, according to the desired specifications of the problem.[2] However, the task of finding an appropriate set of such command signals is non-trivial, and is in many cases considered as the "bottle neck" of approaches to the design of stable motions for underactuated systems. The reason for this difficulty is that - unlike a fully actuated system (which can easily track a given desired trajectory) - an underactuated system cannot possess independent accelerations for each of its DoF. Hence, the motion control of underactuated systems usually involves some form of motion planning based on the integration of the system's equations of motions (EoM).

The need to find a feasible trajectory between two states of a system can occur in many applications, for example, when planning a limit cycle for a periodic gait of an underactuated biped. In the design of a bipedal gait, the states may refer to those immediately after and before an impact, representing the start and end states, respectively. The impact of the foot with the walking surface then maps the target state back to the start state, making the trajectory periodic.[3,4] It is therefore necessary in this case to focus on the design of a motion trajectory that arrives at a given target state from a given start state through any feasible trajectory, even if it is not unique or optimal in any sense. The aim is not to track a particular trajectory, but rather, to find a set of actuation commands resulting in a trajectory which eventually approaches the target. It is desired, however, to exploit the knowledge of the dynamics of the system to make the search more efficient.

There exist numerous approaches for achieving this goal. Traditional approaches include: the *shooting methods*, which parameterize the input and optimizes the parameters such that the target is reached via the integration of the EoM,[3,4] the *energy shaping* methods,[5] and the *optimality-criterion-based* approaches, such as dynamic programming,[6] among others. All of these methods have their own limitations, especially when used for underactuated multi-dimensional systems. Relatively new approaches include the *Rapidly-exploring Random Trees* (*RRT-based*) search algorithms[7] which attempt to generate many feasible sub-trajectories in random directions in state space as well as towards the target state. RRT-based approaches are becoming well-used as they overcome many of the limitations possessed by the traditional approaches. This paper aims to build on some of the existing modifications recently proposed for the RRT algorithm by suggesting a modified distance heuristic and a nested search approach which gradually reduces the search area.

2. Background

The method of RRT aims to grow a search tree in state space by iteratively expanding one of the nodes in the tree towards a randomly sampled state. In order to bias the search towards a target state, the target is occasionally chosen as the sample instead of a random state. The expansion procedure involves two main stages: (1) the distance between all nodes in the tree and the sample is calculated (or estimated) to find the nearest node, which is then chosen to be expanded (2) an appropriate command torque is then chosen to be applied for a short time duration, either from a discrete set of possible torques or by analytically calculating the most appropriate torque. A forward integration of the equations of motion is then performed and the resulting final state is added as a node in the tree if the path leading to it does not violate any constraints.

The *Reachability Guided* RRT (RG-RRT) method,[8] builds on the RRT algorithm: it expands the tree towards random samples, but it also rejects some of the samples using a simple but very useful rejection criterion. Every node in the tree holds information of the next set of states towards which it can be expanded in one step, namely, the *reachable states*. The union of all these states is called "the reachable set". The rejection criterion for a given sample is if the sample is closer to the tree itself (i.e. any of the nodes comprising it) rather than to the reachable set. This encourages the tree to reach out more rapidly and avoid forming a dense tangle. The logic behind this criterion is that when a node is close to a sample, this does not necessarily imply that expanding the node would grow the tree towards the sample when kinodynamic constraint is present, such as underactuation, which is of particular interest in this paper.

Conventionally, the distance function used to define how close two states are is Euclidian in most cases found in literature. This is sometimes disadvantageous, as the Euclidian distance between two states is not necessary a good indication for how easy it is to reach one state from the other in an underactuated system.[9] Clearly, the chance that a certain node of the search tree will develop towards a sample is not determined solely by the Euclidian distance between the node and the sample, but it is also affected by the *directionality* of the tree at the neighborhood of the node. This observation is illustrated in Fig. 1, where two nodes are shown in the neighborhood of a target state. The closest to the target belongs to node B (in terms of the Euclidian distance d_1), but the reachable states of node A has more chance of reaching the target state with less computational effort. We would prefer to have more nodes like A in the vicinity of the target state. For this purpose, we introduce a "directionality" term d_2 into the distance function in order to encourage the selection of nodes like A instead of nodes like B.

The method proposed in this paper is based on a geometrical analysis of the states as opposed to other variations to the distance function suggested in the literature, which, in the absence of kinematic constraints, rely mostly on performing an analytical estimation of the optimal 'cost-to-go' function.[10,11] We believe that our approach is both easier to implement (when no optimality criteria are specified) and more time-effective in calculating the distance between two states.

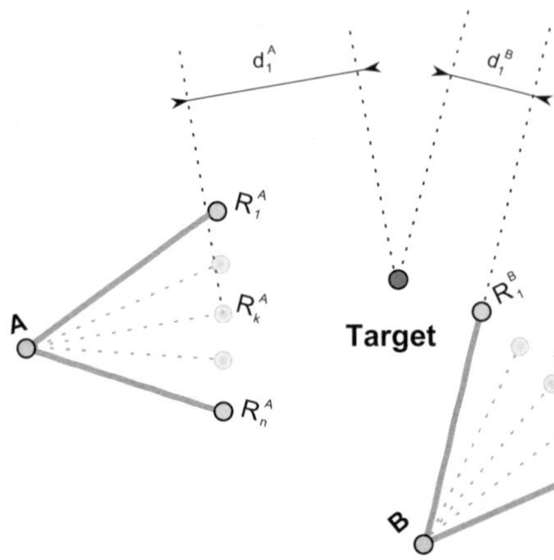

Fig. 1. Two nodes, A and B, in a tree and their reachable states. It is clear in this example that it is easier to develop node A to reach the target than node B. The measure of Euclidian distance, however, does not indicate the level of difficulty to reach the target from each node.

3. Proposed Method

3.1. *Directed Distance Function*

Suppose that a node N in the tree has $n > 1$ reachable states and S is a sample state. In order for our algorithm to prefer node A over B in the example illustrated at Fig. 1, the standard Euclidian distance function was replaced with the following *modified distance function*:

$$D(N, S) = d_1 + d_2 \tag{1}$$

where: $d_1 = \sqrt{\langle \overrightarrow{NS}, \overrightarrow{NS} \rangle}$ is the standard Euclidian distance between the N and S states; and d_2, which is the *directionality* term, is defined as the scaled angular shift from N to S:

$$d_2 = \lambda \cdot \begin{cases} \frac{L_{\text{mean}}^2}{W} \min_k \left[1 - \cos(\theta_k)\right] & d_1 < \gamma L_{\max} \\ 2\frac{L_{\text{mean}}^2}{W} & \text{otherwise} \end{cases} \tag{2}$$

where γ and λ are scalar design parameters; $L_{\max}$ is the distance between N and its furthest reachable state; L_{mean} is the mean distance between N and all of its reachable states; θ_k is the angle between the two vectors rooted at N and pointing towards S and the kth reachable state of N $(1 \leq k \leq n)$, R_k , respectively:

$$\cos(\theta_k) = \frac{\left\langle \overrightarrow{NS}, \overrightarrow{NR_k} \right\rangle}{\sqrt{\left\langle \overrightarrow{NS}, \overrightarrow{NS} \right\rangle \left\langle \overrightarrow{NR_k}, \overrightarrow{NR_k} \right\rangle}} \tag{3}$$

and W is the "opening length" of the node defined as: $W = \sqrt{\left\langle \overrightarrow{R_1R_n}, \overrightarrow{R_1R_n} \right\rangle}$.

A disadvantage of this distance function is that it takes substantially more computation time than the simple Euclidian distance function. Note, however, that d_2 is replaced by a conservative upper bound when the sample is relatively far from the node and hence the dynamics in the vicinity of N and its reachable states is irrelevant. Note also that W, $L_{\max}$ and L_{mean} are only calculated once when the node is created. In addition, by comparing the coverage area of non-targeted trees over grid-discretized state-space,[10] it was observed that using the modified distance function $d_1 + d_2$ makes the tree cover fewer cells in the grid than using the Euclidian distance function d_1. Hence, it is more beneficial to use the modified distance function only when the chosen sample, S, happens to be the target state. In this manner, computation time is significantly decreased with the tree being well dispersed in state-space, while around the vicinity of the target, it obtains good node density directed towards the target.

3.2. *Nested Search*

A successful search is generally defined as a one which returns a node within distance ε to the target, where ε is the tolerance allowed. If ε is very small, then a tree with long trajectories between its nodes and their parents is not well suited for this task. On the other hand, very fine trees fail to expand quickly enough to reach various states in the state-space. In order to overcome this dilemma of choosing a suitable time interval between nodes, Δt, and a maximal number of iterations before stopping the algorithm, a nested search was proposed in order to gradually converge to the target without wasting too much time on very far branches of the tree. The additional motivation stems from the exponential-time nature of tree-searches: it is much less time-consuming to have c searches of I iterations each, than to have a single search with cI iterations.

The idea behind a nested search is that after a substantial number of nodes reached the neighborhood of the target, the search area (i.e. the portion of the state-space inside which random samples are generated) can be shrunk and only the closest set of nodes was considered for the subsequent iterations. Moreover, Δt, and possibly the torque resolution between sibling nodes, Δu, may be reduced in magnitude as well.

The main drawback in performing a nested search is the loss of completeness guarantee that holds for the non-nested RRT searches because it might occur that all the nodes in the set used to seed the next stage of the search will never lead to the target. However, this can be resolved by going back to the previous stage and expanding the tree more before reducing the search area. Later on, we show that this undesired scenario is less probable when using the modified distance function presented in Eq. (1).

The proposed nested-search algorithm is summarized in Algorithm 1 below. The idea is to define a series of P-dimensional spherical volumes in the state-space, B_i, of radius r_i centered about the target, where we define

$$r_i = \frac{\sqrt{\left\langle \overrightarrow{StartTarget}, \overrightarrow{StartTarget} \right\rangle}}{2^i} \quad , \quad (i \geq 0) \tag{4}$$

where P is the number of states in the system.

The stack named "$Tree1$" holds all the nodes created in the whole search and the information about their parents for later retrieval of the path. The RG-RRT search[8] (line 06, see Algorithm 1) is performed at each stage on a temporary stack called "$Tree2$". At the end of each refinement stage s ($1 \leq s \leq NumOfStages$), we insert all the nodes which are within B_s (i.e., with distance $d_1 \leq r_s$ from the target) in the stack named "$Tree3$" (lines 12-14), which is then used to initialize $Tree2$ in stage $s+1$ (line 15). At the end of each stage of the nested search, the search area, the integration interval and the torque resolution are decreased (provided $Tree3 \neq \emptyset$) in order to make the next stage search more focused around the target (Line 17).

Algorithm 1: Directionality-Guided-Nested-RG-RRT $(Start, Target)$

01: $\{SearchArea;\ \Delta t;\ \Delta u\}$ = InitializeSearchParameters();
02: $\{Tree1;\ Tree2;\ Tree3\} = \{Start;\ Start;\ Null\}$;
03: $InitDist$ = EuclidianDistance($Start$, $Target$)
04: **For:** $s = 1$ **to** $NumOfStages$
05: **For:** $i = 1$ **to** $NumOfIterationsPerStage$
06: $\{Node^*\ ,\ k^*\ \} \longleftarrow$ ChooseNodeForExpansionRG-RRT($Tree2$);
07: $N \longleftarrow$ CreateNewNode($State \longleftarrow R_{k^*}^{Node^*}$, $Parent \longleftarrow Node^*$);
08: $\{R_1^N;\ ...\ ;\ R_n^N\ \} \longleftarrow$ CreatePrimaryReachableStates(N);
09: $\{E_{11}^N;\ ...\ ;\ E_{nn}^N\} \longleftarrow$ CreateSecondaryReachableStates($\{R_1^N\ ;\ ...\ ;\ R_n^N\ \}$);
10: $\{\ Tree1\ ;\ Tree2\ \} \longleftarrow \{Tree1 \cup N;\ Tree2 \cup N\}$;
11: **If** $(\|N - Target\| < \varepsilon)$ **GoTo** Line 17 ; % solution found
12: **ForAll** $N_j \in Tree1$
13: **If** $\left(\|N_j - Target\| < \frac{InitDist}{2^s}\right)$ % see Eq. (4) using $i = s$
14: $Tree3 \longleftarrow Tree3 \cup N_j$;
15: $Tree2 \longleftarrow Tree3$;
16: $Tree3 \longleftarrow Null$;
17: $\{SearchArea;\ \Delta t;\ \Delta u\}$ = ResizeSearchParameters
($\{SearchArea;\ \Delta t;\ \Delta u;\ Tree1\}$);
18: **Return** PrintPathFromStartToBest(Tree);

Line 05 of the algorithm makes use of RG-RRT algorithm to generate an admissible sample and the distance function discussed previously to choose the most appropriate node for development at each iteration. In line 07, a database for the new node N is created and it includes the state of the node, the time it occurs, the torque used to produce it from its parent and a pointer to the parent node. Similarly, n primary reachable states associated with N, R_1^N, ..., R_n^N, are created (line 08). In order to use the RG-RRT method using Eq. (1) as the distance function, each primary reachable state R_k^N has to have its n secondary reachable states: E_{k1}^N, ..., E_{kn}^N. All the n^2 secondary reachable states associated with node N are created in line 09.

4. Simulation Set-up

The algorithm was applied on two systems [a]:
A torque limited simple pendulum with a point mass of $m_1 = 1[Kg]$ at a distance $L_1 = 1[m]$ from the hinge. The task was to move the pendulum state $(\alpha, \dot{\alpha})$ from horizontal rest state $\left(\frac{\pi}{2}, 0\right)$ to a chosen target state $(-1, 3)$. The magnitude of the torque was upper bounded at $10[Nm]$.
An underactuated double pendulum with no actuation at the base

[a] A separate reference frame was used for each system.

hinge, with a point mass m_1 at a distance L_1 from the base hinge and a point mass $m_2 = 2[Kg]$ at a distance $L_2 = 2[m]$ from the actuated joint. The task was to move the double pendulum to state $(\alpha_1, \dot{\alpha}_1, \alpha_2, \dot{\alpha}_2)$ from horizontal rest state $(0, 0, 0, 0)$ to the vertical downwards rest target state $(-\frac{\pi}{2}, 0, 0, 0)$. No torque upper bound was set for the actuated joint.
For both systems, each node had $n = 3$ primary reachable states (which themselves had $n = 3$ secondary reachable states each) that were produced by three different torques (for the simple pendulum, these were $U = \{+10, 0, -10\}\,[Nm]$, and for the double pendulum these were $u_{\text{parent}} + U$) applied for the interval Δt. Initially, Δt was set to $0.1[sec]$ and it was halved at every following stage. The torque resolution was not changed across the different stages. The search interval was halved at every stage in each dimension, decreasing the search area by a factor of 4 for the simple pendulum and 16 for the double pendulum. No obstacles or kinematic constraints were prescribed.
The algorithm described above was implemented with two cost functions for each of the two systems: the *modified distance function* implemented as in Eq. (1) and the *conventional Euclidian distance function*, implemented here as a comparison of performance. Note that in the implementation of the Euclidian distance function, no secondary reachable states were created. 10,000 iterations were performed at each stage of the search for the simple pendulum while 100,000 iterations were run for each stage for the double pendulum.

5. Results and Discussion

5.1. *Torque Limited Simple Pendulum*

The resulting distribution of nodes for three consecutive stages is presented in Table 1. It is shown that the density of the nodes near the target is greater when using the modified distance function, (i.e. $d_1 + d_2$), and hence the odds of converging to the target with high precision are greater. This trend was observed for different levels of the nested search (shown in Table 1 for $s = 1, 2, 3$). Note that distance to the target reduces exponentially with i (see Eq. (4)). The resulting tree is shown in Fig. 2. The overall motion trajectories from the start to the target states are drawn in solid line in Fig. 2 (a) and (b) for the two cost functions being compared. Closer views around the target states of the tree are given in Fig. 2 (c) and (d), and Fig. 3(a)-(d). The effect of the nested search can be seen in Fig. 2 (c) and (d) where node density is higher in the later stages of the search, marked by

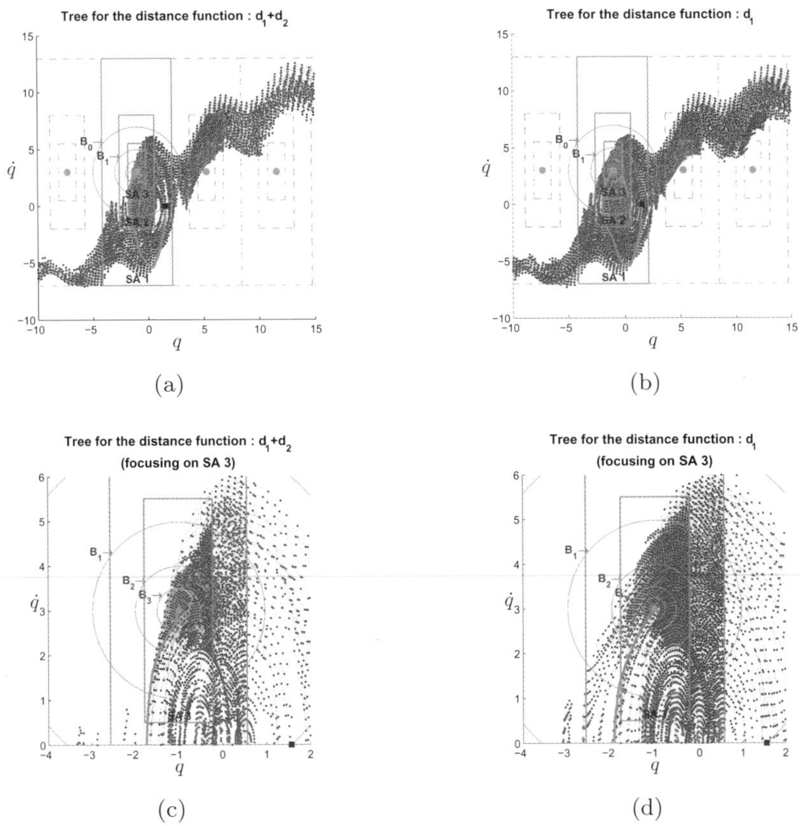

Fig. 2. A comparison between the resulting trees for the torque-limited simple pendulum with $NumOfStages = 3$. Plots (a), and its magnified view (c), refer to the modified distance function Eq. (1), whereas plots (b), and its magnified view (d), refer to the conventional Euclidian distance function. The rectangles represent the three nested search areas (SA) used for the three stages and the circles represent the different B_i. The start states and target states (which is 2π-periodic) are marked with bold square and circles, respectively. The best path found is shown in solid line.

rectangular boxes. It can also be seen that the distribution of nodes is more concentrated around the target state for the modified distance function (Fig. 2(c) and Fig. 3(a)) than that for the conventional Euclidian distance function (Fig. 2(d) and Fig. 3(b)). A closer look at the immediate vicinity of the target (Fig. 3 (c) and (d)) reveals not only the higher number of nodes near the target state as the result of implementing the proposed modified distance function but also a more accurate arrival at the target point.

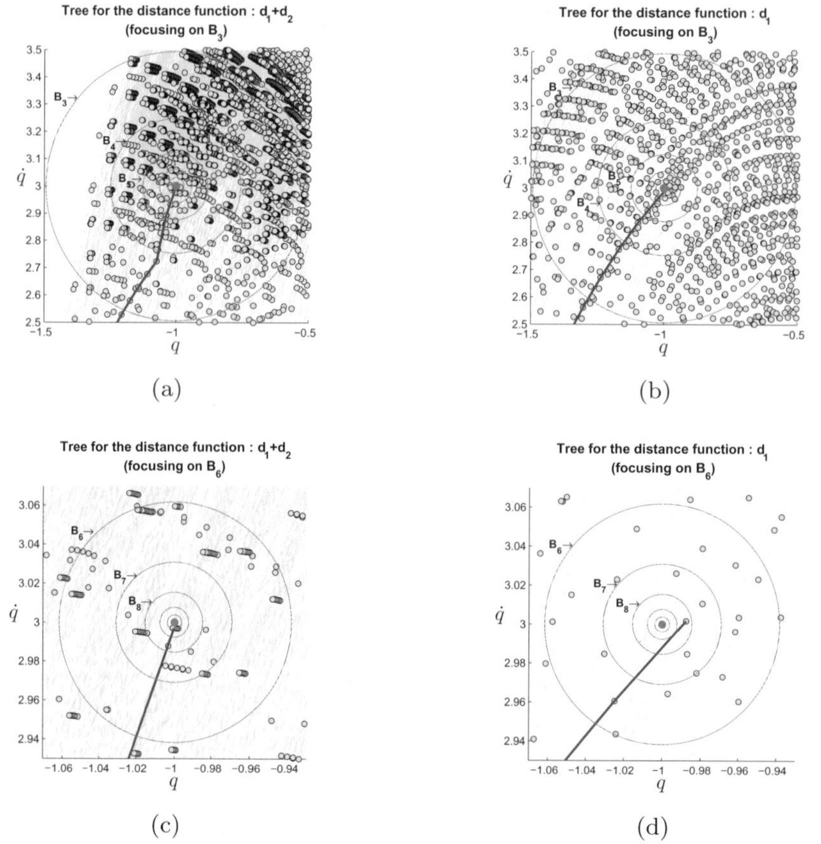

Fig. 3. The magnified views of Fig. 2 in the vicinity of the target. The accurate distribution of nodes in the various B_i for the two cost functions at each stage of the search is presented in Table 1.

5.2. *Underactuated Double Pendulum*

The resulting nodes distribution for the double pendulum system is shown in Table 2 across 4 stages of the nested search. Note that at every stage, *Tree*1 for both cost functions being compared have the same number of nodes. Results similar to that of the simple pendulum were confirmed in this underactuated scenario, where the density of the nodes is higher near the target for the proposed modified distance function. Furthermore, it is expected that directionality plays an even more important role in underactuated cases, due to the limitation in the possible directions of actuation (hence direction of local tree growth). Due to the higher dimension of the

Table 1. A comparison of the number of nodes in the tree (for the two distance functions) within the various B_i for the **simple pendulum model** searched with $NumOfStages = 3$. Numbers in bold indicate the size of $Tree3$ used to initiate the search of the next stage (if performed).

stage	$s=1$ ($\Delta t = 0.1$)		$s=2$ ($\Delta t = 0.05$)		$s=3$ ($\Delta t = 0.025$)	
distance function	d_1	d_1+d_2	d_1	d_1+d_2	d_1	d_1+d_2
$i>0$	**910**	**4,429**	4,264	11,073	11,950	20,355
$i>1$	198	3,983	**1,029**	**9,228**	4,270	16,715
$i>2$	49	615	232	3,316	**971**	**9,377**
$i>3$	11	142	58	511	236	2,471
$i>4$	5	46	17	110	59	484
$i>5$	1	15	7	28	20	136
$i>6$	—	3	1	11	5	32
$i>7$	—	1	—	1	1	8
$i>8$	—	—	—	—	—	7
$i>9$	—	—	—	—	—	7
$i>10$	—	—	—	—	—	—
best Euclidian distance [m]	0.0416	0.0127	0.0309	0.0127	0.0126	0.0030
elapsed time [sec]	50	45	156	169	414	450

Table 2. A comparison of the number of nodes in the tree (for the different distance functions) within the various B_i for the **double pendulum model** searched with $NumOfStages = 3$. Numbers in bold indicate the size of $Tree3$ used to initiate the search of the next stage (if performed).

stage	$s=1$ ($\Delta t = 0.1$)		$s=2$ ($\Delta t = 0.05$)		$s=3$ ($\Delta t = 0.025$)		$s=4$ ($\Delta t = 0.0125$)	
distance function	d_1	d_1+d_2	d_1	d_1+d_2	d_1	d_1+d_2	d_1	d_1+d_2
$i>0$	**21**	**6,665**	311	32,077	14,917	88,510	71,620	168,976
$i>1$	—	115	**26**	**2,144**	468	48,120	14,488	106,520
$i>2$	—	4	—	1,482	**11**	**41,165**	207	76,604
$i>3$	—	—	—	35	—	979	**0**	**1005**
$i>4$	—	—	—	—	—	—	—	—
best Euclidian distance [m]	0.4325	0.1538	0.2311	0.0883	0.1571	0.0879	0.1571	0.0879
elapsed time [sec]	7,987	9,055	14,011	17,560	20,645	25,368	27,674	38,829

system, it is not possible to present the plots of the tree clearly. It can be noted from Table 2 that the number of nodes at every stage of nested search is of order(s) of magnitude higher for the proposed modified distance function than that for the conventional Euclidean distance function. Note also that for $s = 3$ for the double pendulum, $Tree3$ is an empty set for the Euclidian distance function, meaning that no further refinement stage could be attempted without further developing the previous stages, while the tree from the modified distance function is still capable of further refinement.

6. Conclusion

This paper presented a computationally efficient method to estimate the reachability of states by biasing the choice of nodes towards a target state utilizing the knowledge of the system dynamics. Additionally, a nested search technique was proposed in order to focus the search more rapidly around the target. It was demonstrated that the modified distance function which comprises both a Euclidian distance term and a directionality term was more successful in directing the search towards the target. By comparing two trees, one grown with the Euclidian distance function and the other with the modified one, it was shown that there were more nodes in the vicinity of the target for the latter case. Moreover, the directionality term, in contrast to the Euclidian term, helps develop the tree nodes in accordance to the knowledge of the system dynamics, which is highly advantageous in the case of underactuated systems. Future work would concentrate on the development of efficient gait limit cycles for bipedal dynamic walking.

References

1. M. Spong and G. Bhatia, Further results on control of the compass gait biped, in *Procs IEEE/RSJ Intl' Conf. Intel. Rob. & Syst.*, 2003.
2. M. Spong, *Control Problems in Robotics and Automation*, 135 (1998).
3. M. Srinivasan and A. Ruina, *Nature* **439**, 72 (2005).
4. E. R. Westervelt, J. W. Grizzle, C. Chevallereau, J. H. Choi and B. Morris, Chapter 6, in *Feedback Control of Dynamic Bipedal Robot Locomotion*, eds. S. S. Ge and F. Lewis (CRC Press, U.S., 2007) pp. 137–190.
5. M. W. Spong, *Control Systems, IEEE* **15**, 49 (1995).
6. S. LaValle, *Control Problems in Robotics* , 19 (2003).
7. S. M. Lavalle and J. J. Kuffner, Randomized kinodynamic planning, in *Procs. IEEE Intl. Conf. Rob. Autom.*, 1999.
8. A. Shkolnik, M. Walter and R. Tedrake, Reachability-guided sampling for planning under differential constraints, in *Procs. IEEE Intl. Conf. Rob. Autom.*, 2009.
9. P. Cheng and S. M. LaValle, Reducing metric sensitivity in randomized trajectory design, in *Procs. IEEE/RSJ Intl. Conf. Intel. Rob. & Syst.*, 2001.
10. E. Glassman and R. Tedrake, A quadratic regulator-based heuristic for rapidly exploring state space, in *Procs. IEEE Intl. Conf. Rob. Autom.*, 2010.
11. R. Tedrake, LQR-trees: Feedback motion planning on sparse randomized trees, in *Proc. Rob. Sci. & Syst.*, (Seattle, USA, 2009).

BIPED WALKING OVER ROUGH TERRAIN BY ADAPTIVE GROUND REFERENCE MAP

NING WU[†], CHEE-MENG CHEW, AUN NEOW POO

Mechanical Engineering, National University of Singapore,
9 Engineering Drive 1, EA-03-06, Singapore, 117575

In this paper, we propose a constructive and robust control design for online stabilization of biped robot dynamic motion, especially walking on rough or uneven terrains. We proposed an online moving ground reference map based on both present and future information, which is obtained via minimizing the error of the footstep, and the error of the position and velocity of the Center of Mass (CoM) of the robot. By applying this reference map, robust CoM trajectories can be generated based on preview control. The dynamic simulation results show that the proposed approach can significantly improve walking stability and minimize the error in tracking the pre-defined trajectory even on uneven terrains. The technique is very general and can be applied to a wide variety of humanoid robots.

1. Introduction

In the field of robotics, research into bipedal locomotion of humanoid robots has been very active in recent years. By virtue of their mechanical structure, one of the significant advantages of legged robots over other types of robots is their ability to navigate various terrains usually accessed by human beings. Walking on flat terrains has been well studied [1-3], but walking on rough terrains remains a challenge. In general, there are two main groups of approaches for achieving stable bipedal robot locomotion over uneven terrains. In the first group, the focus is on developing a motion plan based on knowledge of the profile of the terrain, which can achieve stable robot locomotion [4, 5]. If the robot's control systems can then ensure the accuracy of tracking of the predefined trajectories, stable walking will be achieved. In the other group, the assumption made is that there is insufficient knowledge of the terrain to develop a motion plan to achieve stable walking. The focus of this group is then on developing in the robot strong disturbance rejecting capabilities so that the robot

[†] Work corresponding author. Email: wuning@nus.edu.sg.

can maintain stable walking without falling even when faced with disturbances and unexpected unevenness in the terrain.

In previous studies, several robots have realized stable walking over uneven terrains by maintaining the zero-moment point (ZMP) within the support polygon [4-6]. However, robustness is poor and the approaches fail in the presence of significant unknown disturbances. Some researchers have attempted to improve the disturbance rejection ability by adjusting the ZMP reference online [6-8] and some improvements have resulted. However, there is still a need to develop an effective strategy which can adapt to large disturbances. An effective strategy would be one which can adapt to unknown disturbances such as unevenness in the terrain and include preview action to overcome large disturbances, such as stairs and steep slopes, when these can be known beforehand.

The approach proposed in this paper, which is based on the preview controller first proposed by Kajita, et, al. [4], is shown to significantly improve the robot's ability to reject disturbances. In this approach, the support foot reference on the ground is modified by regulating the speed of the CoM to make it more suitable for the current robot state. For example, if the actual velocity of the CoM is higher than the pre-planned velocity when the foot lands on the ground, the subsequent step support foot reference will be placed further. A CoM trajectory is generated to track a moving ground reference map based on an optimized preview control. The trajectories of all the joints are then generated through inverse kinematics based on the positions of the CoM and the pre-planned foot positions.

The paper is organized as follows. Section 2 introduces the proposed approach using a moving ground reference map based on the linear inverted pendulum model and an optimized preview control. Section 3 presents the experiments and a discussion of the results obtained. The conclusion is presented in Section 4.

2. Methodology

2.1. *Background*

In this section, we review preview control based on the ZMP equation on a slope [4, 5]. Here the mass of the robot is assumed to be lumped at the CoM and the height of the CoM with respect to the slope is assumed to be maintained constant as the robot moves forward along the slope. The inclination of the slope is $\theta°$ as shown in Fig. 1. In Fig. 1, X'Y'Z' is the reference frame attached to the slope with its X'-axis going up the slope in the steepest direction, the Y'-axis is horizontal, and the Z'-axis orthogonal to both the X'- and the Y'-axes. From dynamic analysis, we have

$$x'_{\mathrm{p}} = x' - \ddot{x}' \cdot \frac{z_{\mathrm{c}}}{g\cos\theta} + z_{\mathrm{c}} \cdot \tan\theta \tag{1}$$

$$y'_{\mathrm{p}} = y' - \ddot{y}' \cdot \frac{z_{\mathrm{c}}}{g\cos\theta} \tag{2}$$

where x' and y' are the displacements of the CoM and x'_{p} and y'_{p} are the position of the ZMP along the X'- and the Y'-axis respectively. z_{c} is a constant representing the height of CoM in the same frame.

In the design of the control system for the robot, the objective is to have the output track the command signals without any steady-state errors. Tomizuka and Rosenthal [9] developed a digital *preview controller*, which includes the integral of the ZMP tracking error and state feedback and also makes use of information on future reference inputs [4]. It is shown that the use of future information is very effective in achieving good tracking performance, resulting in stable walking following closely the desired or reference walking pattern.

Unfortunately, in order to create the ZMP reference needed to generate a suitable and stable walking pattern, accurate information on the terrain is required. However, except for a specially controlled environment, accurate information on the terrain usually not available. Some disturbance will always be present, for example undulation or unevenness in the terrain which can prevent the foot from reaching the expected location. Such disturbances will create errors between the support foot position and the ZMP reference as shown simulation example in Fig. 2. For the simulation shown, the robot's preplanned path is for a flat ground but unexpectedly steps onto a flat plate, of height 10mm, placed in its path when it is about 3s into its locomotion. As can be seen, this caused a significant error to occur between the foot position and the ZMP reference which becomes larger and larger with each step taken by the robot.

2.2. *Moving Ground Reference Map*

In order to maintain stable walking avoid falling in spite of unknown disturbances, an online adjustable moving ground reference map is proposed here. This map takes advantage of available information on both future desired stepping positions and the current robot state (CoM position, velocity and foot position).

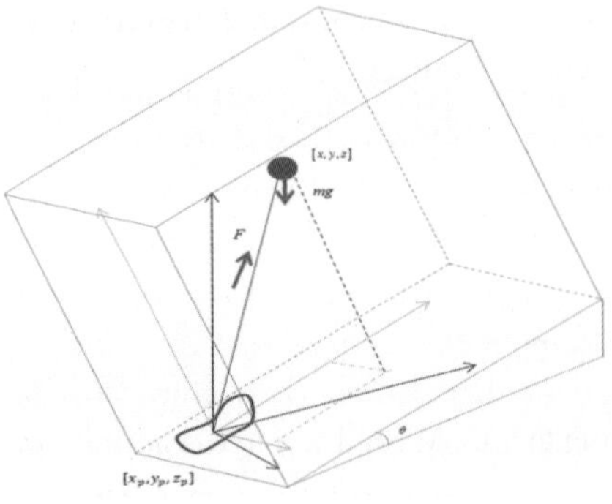

Figure 1. Linear Inverted Pendulum Model on the slope reference.

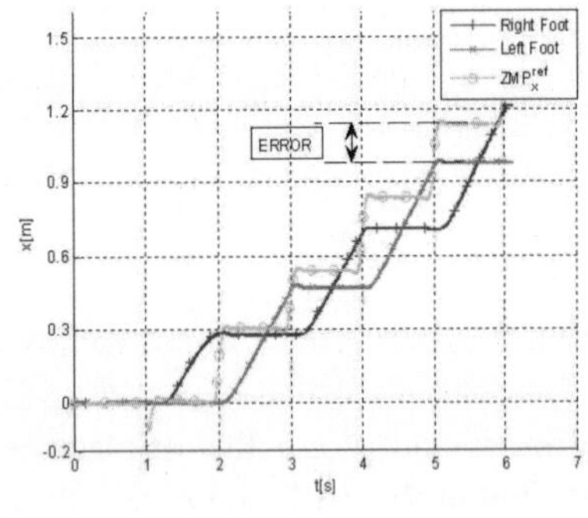

Figure 2. The support feet cannot follow ZMP when disturbance happened.

The moving ground reference map contains the most suitable ZMP ground reference points based on the current state of the robot and the allowable stepping locations within the next N_l finite steps.

The relationship between the initial state $(x_0^{(n)}, \dot{x}_0^{(n)})$ and the final state $(x_T^{(n)}, \dot{x}_T^{(n)})$ for the nth step [10] are

$$\begin{aligned} x_T^{(n)} &= C_T x_0^{(n)} + \frac{S_T \dot{x}_0^{(n)}}{w} \\ \dot{x}_T^{(n)} &= w S_T x_0^{(n)} + C_T \dot{x}_0^{(n)} \end{aligned} \tag{3}$$

where x and $\dot{x}$ are the position and the velocity respectively of the CoM, $w = \sqrt{g/z_c}$, z_c is the height of CoM, g is the acceleration due to gravity and $S_T = \sinh(wt)$, $C_T = \cosh(wt)$.

The new ZMP reference for the $(n+1)^{\text{th}}$ step is calculated at the nth step by minimizing the errors in the position and the velocity of the CoM, and the reference tracking error. This is done by treating it as a Quadratic Programming problem. We have

$$\min \frac{1}{2} P \left\| e_{fT}^{(n+1)} \right\|^2 + \frac{1}{2} Q \left\| e_{xT}^{(n+1)} \right\|^2 + \frac{1}{2} R \left\| e_{\dot{x}T}^{(n+1)} \right\|^2 \tag{4}$$

where $e_{fT}^{(n)} = x_{p(ref)}^{(n)} - x_{fT}^{(n)}$ is the error between the pre-defined ZMP reference $x_{p(ref)}^{(n)}$ and the position of the foot $x_{fT}^{(n)}$, $e_{xT}^{(n)} = x_{T_d}^{(n)} - x_T^{(n)}$ is the error between the desired position $x_{T_d}^{(n)}$ and the actual position $x_T^{(n)}$ of the CoM, and $e_{\dot{x}T}^{(n)} = \dot{x}_{T_d}^{(n)} - x_T^{(n)}$ is the velocity error of the CoM all defined for the instance at the end of the nth step. P, Q, and R are weight parameters. By substituting Eq. 3

into Eq. 4, differentiating and letting the differential equals to zero, the optimized location of the position of the foot at the end of the nth step, or the next foot position is obtained as

$$\hat{x}_{fT}^{(n+1)} = \frac{\left[C_T\left(x_{T_d}^{(n+1)} - \frac{S_T}{w}\dot{x}_T^{(n)}\right) + S_T w\left(\dot{x}_{T_d}^{(n+1)} - C_T\dot{x}_T^{(n)}\right) + \left(x_{p_{(ref)}}^{(n+1)} - x_{fT}^{(n)}\right)\right.}{D_T}$$

where $D_T = C_T^2 + S_T^2 w^2 + 1$, and $\hat{x}_{fT}^{(n)} \in \left[x_{f_{min}}^{(n)}, x_{f_{max}}^{(n)}\right]$.

Therefore, we have the moving ground ZMP reference:

$$\hat{x}_p^{ref} = \begin{bmatrix} \mathbf{v}_1 \\ \mathbf{v}_0 \\ \mathbf{v}_0 \\ \vdots \\ \mathbf{v}_0 \end{bmatrix} x_{fT}^{(n)} + \begin{bmatrix} \mathbf{v}_0 \\ \mathbf{v}_1 \\ \mathbf{v}_0 \\ \vdots \\ \mathbf{v}_0 \end{bmatrix} \hat{x}_{fT}^{(n+1)} + \cdots + \begin{bmatrix} \mathbf{v}_0 \\ \vdots \\ \mathbf{v}_0 \\ \mathbf{v}_1 \\ \mathbf{v}_0 \end{bmatrix} x_{fT}^{(n+N_L-1)} + \begin{bmatrix} \mathbf{v}_0 \\ \vdots \\ \mathbf{v}_0 \\ \mathbf{v}_0 \\ \mathbf{v}_1 \end{bmatrix} x_{fT}^{(n+N_L)} \tag{5}$$

with each term on the right of Eq. 5 representing one foot position and with preview of N_L is the number of future ZMP steps in preview control. The rows of Eq. 5 represents sampling instances with the column vectors $\mathbf{v}_1$ and $\mathbf{v}_0$ containing m number of 1's or 0's respectively with m being the number of sampling periods required for completing one foot step. The vector $\hat{x}_p^{ref}$ is a column vector comprising $m \times (N_L + 1)$ values of the foot positions for the $m \times (N_L + 1)$ sampling periods required to move from the start of the nth step to the end of the $(n+N_L)^{\text{th}}$ step. From Eq. 5, the first m values of $\hat{x}_p^{ref}$ will be $x_{fT}^{(n)}$ for the nth step, followed by m values of $\hat{x}_{fT}^{(n+1)}$ for the next $(n+1)^{\text{th}}$ step, and so on. According to this moving ground reference map, a more robust and stable walking pattern generator can be obtained.

2.3. *Overall Algorithm of Proposed Approach*

Fig. 3 shows the general structure of the walking pattern generation process. The pre-defined ZMP map x_p^{ref} is used to generate the initial CoM trajectory. Subsequently at every step, a new optimal ZMP reference trajectory is computed using Eq. 5 to minimize the errors in the CoM position and velocity, and in the the ZMP. Converting Eq. 1 and Eq. 2 into state space format, and discretizing the system, we have

$$\begin{aligned} \mathbf{x}(k+1) &= A\mathbf{x}(k) + Bu(k) \\ x_p(k) &= C\mathbf{x}(k) \end{aligned} \tag{5}$$

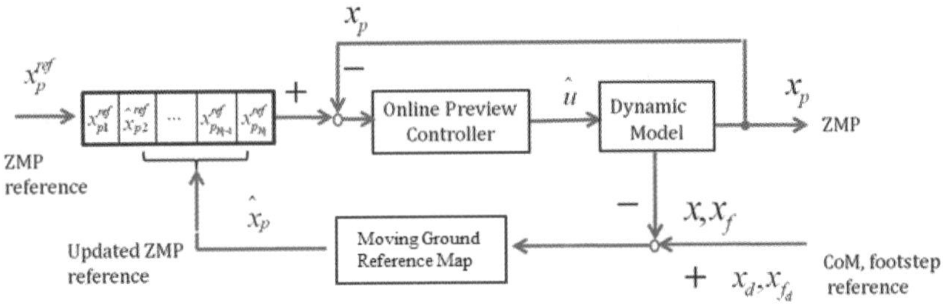

Figure 3. Pipeline Visualization of the Walking Pattern Generation.

where $\mathbf{x}(k) = [x(k) \quad v(k) \quad x_p(k)]^T$ and x(k) and v(k) are the position and the velocity, respectively, of the CoM. u(k) is the control effort and $x_p(k)$, which is the ZMP position in the x-direction, is the output to be controlled. The matrices A, B, and C can be found in reference [4]. In order to track the moving ZMP reference map $\hat{x}_p^{ref}$, we introduce a controller $\hat{u}(k)$ based on the preview control approach to generate the CoM trajectory with

$$\hat{u}(k) = -G_i \sum_{i=0}^{k} e(i) - G_x \mathbf{x}(k) - \sum_{j=1}^{N_L} G_d(j) \hat{x}_p^{ref}(k+j) \tag{6}$$

where N_L is the number of future ZMP steps in preview control, $e(i) = x_p^{ref}(i) - x_p(i)$ is the error of the ZMP position and G_i, G_x and G_d are gain factors [9]. Therefore, a closed loop control scheme is achieved by using the online moving ground future reference.

3. Simulation Results

In order to evaluate the proposed approach, simulations in Webots were conducted. A humanoid robot model is developed as the simulation subject. The robot stands 1.7 m, weighs 86.59 kg (distributed mass), and has 6 degrees of freedom in each of its two limbs. In the simulation, it is equipped with GPS, gyro, accelerometer in the pelvis (CoM) and force sensors in the foot.

For the simulation, the robot is made to walk along a horizontal terrain and then up a slope. The slope is inclined at 8 degrees but the information provided to the robot for this is 5 degrees to simulate inaccurate terrain information obtained from the sensors. Although in practice the sensor error will unlikely to be so large, a large sensor error was used to evaluate the efficiency of the proposed walking algorithm.

Without any adaption of the pre-planned ZMP trajectory, simulations showed that the robot is not able to achieve stable locomotion up the slope without falling. However, with the application of the adaptive moving ground reference map, stable walking up the slope was achieved and the resulting ZMP and CoM trajectories obtained are shown in Fig. 4a for the x-direction. In the figure, ZMP_d, shown by the blue dash line, is the pre-planned "desired" ZMP trajectory based on the 5-degree slope. ZMP_g is the re-computed trajectory generated online using the moving ground reference map. ZMP_{real} is the trajectory of the ZMP of the robot based on its actual CoM position, the latter being calculated using data from the inertia sensors (gyro and accelerator). It can be seen from Fig. 4b, a zoomed in version of Fig. 4a, that the generated ZMP_g (pink dash line) deviates from the pre-defined ZMP_d even from the second step. This indicates that using the proposed method, the generated ZMP_g had adapted quite quickly to the actual terrain profile. The real ZMP_{real} (green solid line) calculated based on the real CoM (blue solid line) is close to the generated ZMP. This simulation results also show that with preview control, the generated ZMP has been tracked accurately with a smooth CoM trajectory achieved. The corresponding trajectories in the y-direction are shown in Fig. 5.

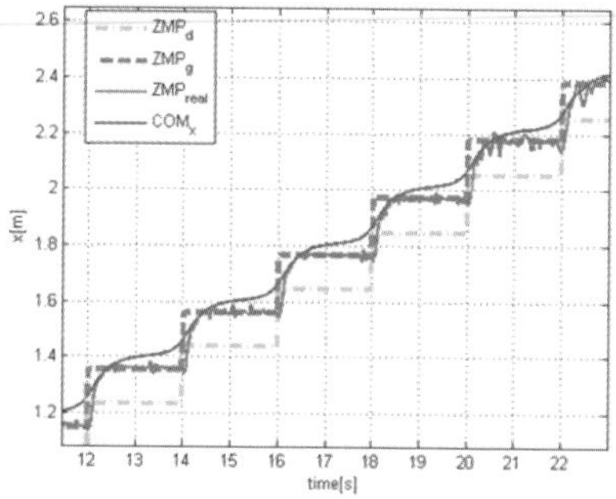

Figure 4a. CoM and ZMP trajectories (x direction).

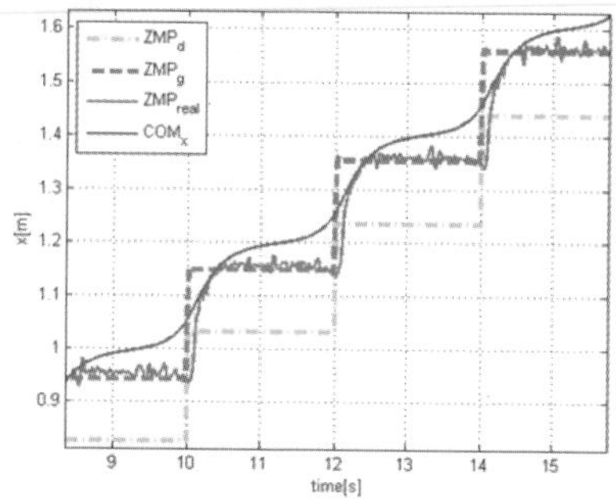

Figure 4b: Zoom in of Fig. 4a.

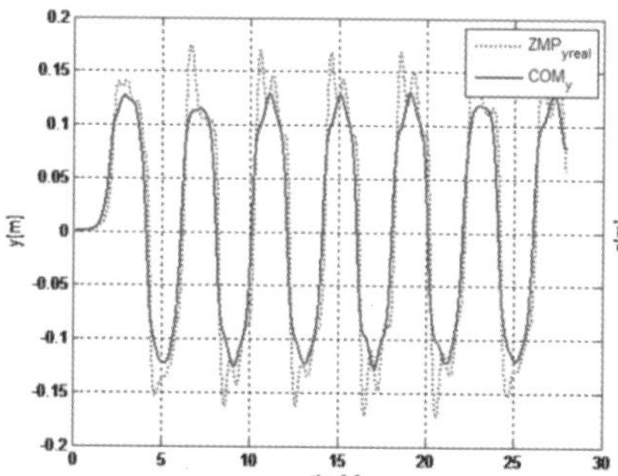

Figure 5. CoM and ZMP trajectories (y direction).

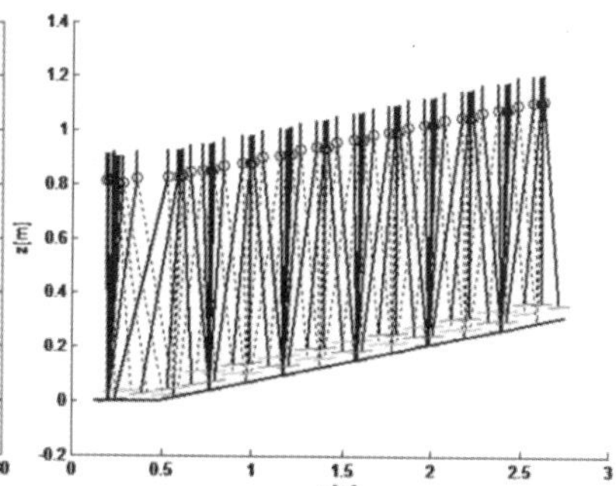

Figure 6. Stick diagram of biped walking.

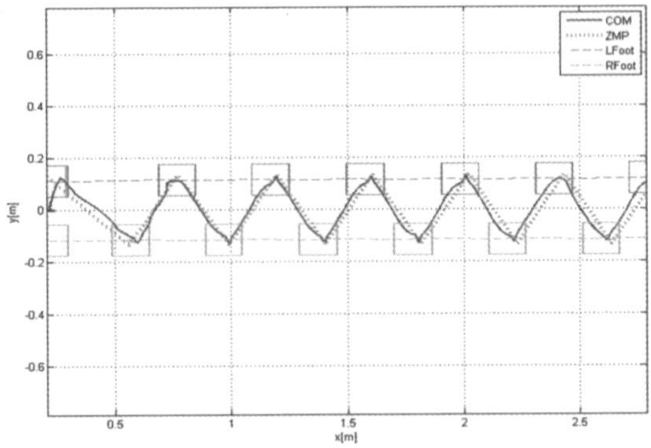

Figure 7. Ground projections of CoM and ZMP on the uneven terrain.

The stick diagram of the biped walking on the slope is shown in Fig. 6. It can be seen that, with the proposed approach, the biped can adapt to the unknown slope properly.

Fig. 7 is the ground projection of the real CoM and the real ZMP trajectories on the uneven terrain. As the robot moves, it can be observed that the generated ZMP, which is followed closely by the real ZMP to moves beyond the CoM creating added momentum to facilitate walking up the slope. Hence, using the proposed adaptive ground reference map, the CoM trajectory has been modified to cater to external disturbances, such as inaccurate terraing information, and stable locomotion is effectively achieved.

4. Conclusions

In this paper, an approach using an adaptive ground reference map to achieve optimized bipedal locomotion over rough and uneven terrains is presented. This ZMP reference map is continuously adjusted in real time based on the robot's actual dynamics during locomotion to maintain stable walking in the face of external disturbances. Preview control is also incorporated to achieve better ZMP tracking. Simulation results show the effectiveness of the proposed approach with the robot achieving stable and smooth locomotion in spite of external disturbances and inaccurate information of the terrain.

References

1. Sakagami, Y., R. Watanabe, C. Aoyama, S. Matsunaga, N. Higaki and K. FujiMura. *The intelligent ASIMO: system overview and integration.* in *Intelligent Robots and Systems, 2002. IEEE/RSJ International Conference on.* 2002.

2. Kaneko, K., F. Kanehiro, S. Kajita, H. Hirukawa, T. Kawasaki, M. Hirata, K. Akachi and T. Kazuhiko. *Humanoid robot HRP-2*. in *Robotics and Automation, 2004. Proceedings. ICRA '04. 2004 IEEE International Conference on*. 2004.
3. Chew, C.M. and G.A. Pratt. *Dynamic bipedal walking assisted by learning*. Robotica, 2002. **20**(5): p. 477-491.
4. Kajita, S., F. Kanehiro, K. KANEKO, K. Fujiwara, K. Harada, K. Yokoi and H. Hirukawa. *Biped walking pattern generation by using preview control of zero-moment point*. 2003. Taipei.
5. Huang, W., C.M. Chew, Y. Zheng and G.S. Hong. *Pattern generation for bipedal walking on slopes and stairs*. in *Humanoid Robots, 2008. Humanoids 2008. 8th IEEE-RAS International Conference on*. 2008.
6. Kajita, S., M. Morisawa, K. Harada, K. KANEKO, F. Kanehiro, K. Fujiwara and H. Hirukawa. *Biped Walking Pattern Generator allowing Auxiliary ZMP Control*. in *Intelligent Robots and Systems, 2006 IEEE/RSJ International Conference on*. 2006.
7. Nishiwaki, K. and S. Kagami. *Strategies for adjusting the ZMP reference trajectory for maintaining balance in humanoid walking*. in *Robotics and Automation (ICRA), 2010 IEEE International Conference on*. 2010.
8. Diedam, H., D. Dimitrov, P.B. Wieber, K. Mombaur and M. Diehl. *Online walking gait generation with adaptive foot positioning through Linear Model Predictive control*. in *Intelligent Robots and Systems, 2008. IROS 2008. IEEE/RSJ International Conference on*. 2008.
9. Tomizuka, M. and D.E. Rosenthal. *On the Optimal Digital State Vector Feedback Controller With Integral and Preview Actions*. Journal of Dynamic Systems, Measurement, and Control, 1979/06/00/. **101**(2): p. 7.
10. Kajita, S., F. Kanehiro, K. Kaneko, K. Fujiwara, K. Yokoi and H. Hirukawa. *Biped walking pattern generation by a simple three-dimensional inverted pendulum model*. Advanced Robotics, 2003. **17**(2): p. 131-147.

SECTION–11

POSITIONING, LOCALIZATION AND PERCEPTION

METHOD FOR ESTIMATING LOCATION AND YAW ANGLE OF A SIX-LEGGED ROBOT FOR OMNI-DIRECTIONAL WALKING CONTROL*

HIROAKI UCHIDA†, KOUHEI TAKAHASHI, MASAYA SUZUKI

Department of Mechanical Engineering, Kisarazu National College of Technology, 2-11-1 Kiyomidai-higashi, Kisarazu, Chiba 292-0041, Japan

The present paper discusses a method for estimating the location (x,y) and yaw of the body of a six-legged robot using a laser range finder. In the present study, two outside fixed points are set before walking. During walking, the location and yaw angle of the body are calculated by referring to these two fixed points. The proposed estimation method is verified experimentally by straight walking of a six-legged robot.

1. Introduction

Multi-leg robots are expected to be used in extreme environments that are dangerous for humans, such as mine fields, nuclear power plants, and disaster areas. We have examined six-degree-of-freedom control methods for the body of a six-legged robot and proposed a three-degree-of-freedom posture control method for controlling the height of the body, the pitch angle, and the roll angle using virtual impedance. The effectiveness was verified through 3D simulations [1] and walking experiments. Moreover, an omni-directional control method to control the location of the body (x, y) and the yaw angle was verified through 3D simulations [2]. However, in applying the proposed omni-directional control method, we need to correctly detect the location of the body and the yaw angle in global coordinates. In the present study, we discuss a method for estimating the location and yaw angle of a six-legged robot using a laser range finder (LRF). There were a lot of reports measuring the external environment using LRF equipped on a mobile robot [3]-[7]. These reports were almost dealing with the wheel-type robots, and, we find a few reports dealing with a multi-legged robot [8]. The multi-legged robot is aimed for work under the environment where the people such as nuclear power plants are hard to work, and it is necessary to control the location of the body (x, y) and the yaw angle exactly. In this study, we examine the method to detect the location of the robot and a yaw angle as the first step to establish method to control the location of the robot and a yaw angle.

2. Laser Range Finder

Table 1 shows the specifications of the LRF used in the present study. This LRF can measure the distance between the sensor and the obstacle in the walking area. In the present study, the LRF is attached to the body.

Table 1. Specifications of the LRF

Optical Source	Semiconductor laser : λ=785nm Laser power : 0.8mW
Measurement distance	20～5600mm
Accuracy	0.06～1m : ±10mm , 1～4m : 1% of distance
Resolution	1mm
Scan angle	240 degree
Angle resolution	Approximately 0.36 degree （360°/1024）
Scanning time	500 ms/scan

3. Location Estimation Method

3.1. *Setting the Outside Fixed Points*

The location of the body (x,y) and the yaw angle are estimated using the LRF for the case of a moving six-legged robot. The purpose of the present study is to estimate the location and yaw angle of the body in the walking area. As such, the location and yaw angle in global coordinates are calculated by setting two fixed points in the walking area. Figure 1 shows the model measuring the external environment using LRF on the body of a six-legged robot. Here, y is the walking direction, x is the vertical direction to y, and in the LRF, an angle from the x-axis is $\theta_i(i=1,\cdots,127)$. The distance between the LRF and a measurement object is $r_{\theta i}(i=1,\cdots,127)$. Figure 2 shows a detecting method of fixed points A and B. In Fig.2, the horizontal axis shows the detecting angle by the LRF, and the vertical axis shows the distance. In Fig.2, we suppose that a rate of change $diff(\Delta_{\theta i})$ becomes maximum around point θ_i, θ_i is set as a fixed point. The rate of change $diff(\Delta_{\theta i})$ is calculated as follows.

$$diff(\Delta_{\theta i}) = (\Delta_{\theta i} - \Delta_{\theta(i-1)})^2 \tag{1}$$

where,

$$\Delta_{\theta i} = \theta_i - \theta_{i-1}$$

In this study, one fixed point is detected in each range $0 < \theta_i < \pi/2$, and $\pi/2 < \theta_i < \pi$ of LRF. The fixed point is the maximum value of the rate of change $diff(\Delta_{\theta i})$ calculated by Eq.(1).

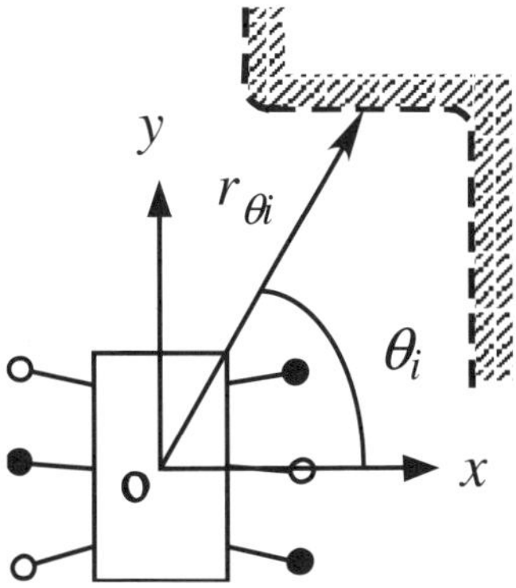

Figure 1 Coordinate and measurement model using the LRF

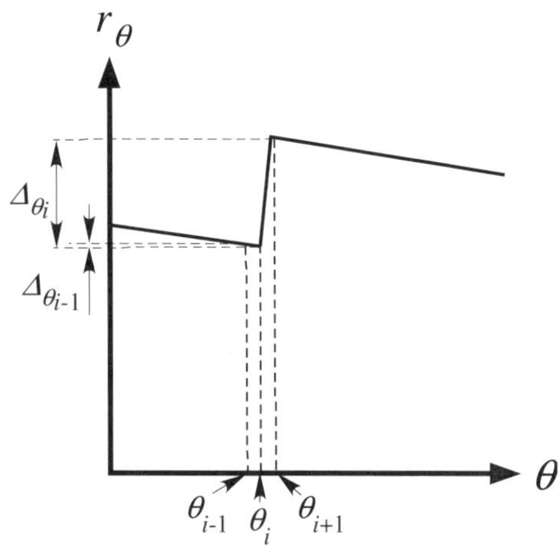

Figure 2 Detecting method of external fixed point

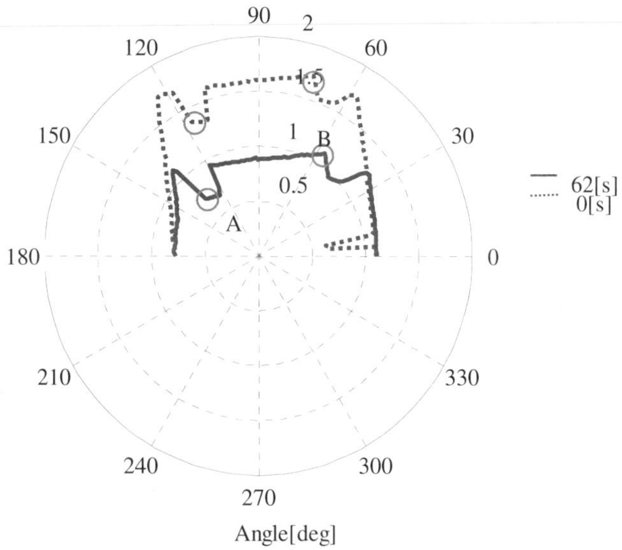

Figure 3 Example of data obtained using the LRF

Figure 3 shows the relationship between the distance from the body to the external obstacle and the angle in the walking area. In Fig. 3, the solid line shows the data at 62 s, and the dotted line shows the data at 0 s. In this case, the six-legged robot walks ahead approximately 0.6 m. In Fig. 3, the two fixed points are denoted by A and B.

3.2. *Calculation of Location and Yaw Angle using Two Fixed Points*

This section describes the method used to calculate the location and yaw angle of the body in global coordinates using the fixed points described in Section 3.1. Using the two fixed points, the location and yaw angle of the body are calculated from the variation of the position of the LRF resulting from the robot motion. Figure 4 shows the measurement model. In Fig. 4, φ_1 and φ_2 are the angles between A and the *x*-coordinate, and B and *x*-coordinate, respectively, at the start position. Moreover, φ_3 and φ_4 are these respective angles at the measurement position. l_1 and l_2 show the distance between the LRF at the start point and the fixed points A and B, l_3 and l_4 show the distance between the LRF point in the walking and the fixed points A and B, respectively. As a result, the yaw angle θ_z, *x*-position, and *y*-position are calculated by the following equations:

$$\theta_z = -\{(\varphi_3 - \varphi_1) - (\varphi_4 - \varphi_2)\}/2 \tag{2}$$

$$x = \{l_1 \cos \varphi_1 - l_3 \cos(\varphi_3 + \theta_z) + l_2 \cos \varphi_2 - l_4 \cos(\varphi_4 + \theta_z)\}/2 \tag{3}$$

$$y = \{l_1 \sin \varphi_1 - l_3 \sin(\varphi_3 + \theta_z) + l_2 \sin \varphi_2 - l_4 \sin(\varphi_4 + \theta_z)\}/2 \tag{4}$$

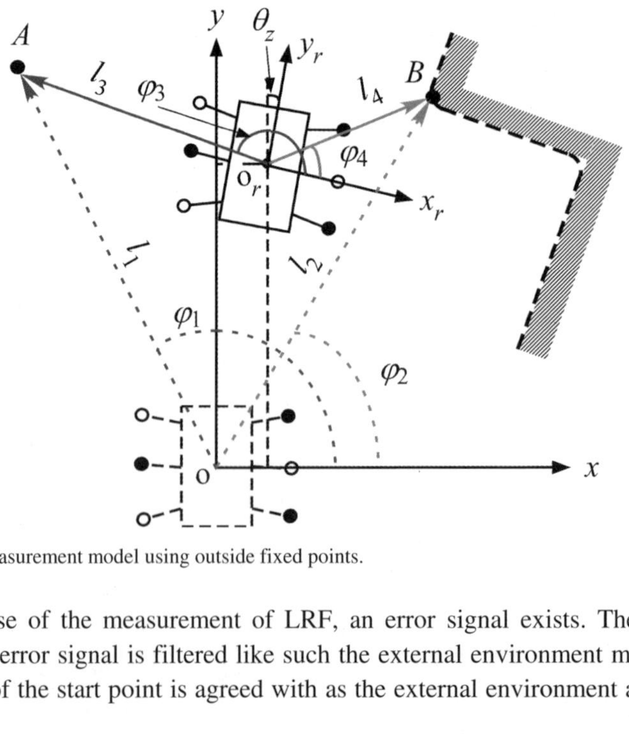

Figure 4 Measurement model using outside fixed points.

In the case of the measurement of LRF, an error signal exists. Then, in this study, an error signal is filtered like such the external environment measured at the time of the start point is agreed with as the external environment at the time of the walking. The external environment is measured several times before the walking and the fixed points A and B are set in a state without an error. At the walking time, the fixed points A and B are obtained by two methods, one method is the measurement values by the LRF, the other one is estimated values using the location (x, y) and the yaw angle calculated by Eqs.(2)-(4). The measurement values of the fixed points are compared with the estimated values every time. If the measurement value is different from the estimated value over the threshold, the location and the yaw angle are set by the estimated values.

4. Walking Experiments

4.1. *Six-legged Robot*

Figure 5 shows a six-legged robot using this research, and its dimensions are shown in Table 2. One leg of the six-legged robot consists of three joints that are a rotating part (θ_{1i}), a thigh part (θ_{2i}) and a shank part (θ_{3i}). Here, i is leg number $(i = 1, \cdots, 6)$.

Table 2 Dimension of robot

Height	500mm	Weight of robot	24.9kg
Length	666mm	Length of thigh	168mm
Width	710mm	Length of shank	312mm
Moment of Inertia I_p		0.6804kgm^2	
Moment of Inertia I_r		0.2697kgm^2	

Figure 5 Six-legged robot

4.2. *Straight Walking by Three Support Legs*

The six-legged robot walks using three support legs and three swinging legs. The orbit of the six-legged robot is a straight walking of 6 cm per step, where one cycle of walking is 5 s in duration and the total number of steps is 10. In this case, the y direction is the walking direction. The six-legged robot used in the present study has the three rotating joints in the rotating part, the thigh part, and the shank part in one leg. The reference signals of each joint angle are calculated based on inverse kinematics from the toe orbit. Each leg joint is controlled by the PD control to follow the reference signal. The proportional and differential gains are determined by trial and error.

4.3. *Measurement Results and Discussion*

Figure 5 shows the results of fixed point A and B measured by the LRF and estimated values. In Fig.5, the red solid line and the dashed and the dotted line show the measurement values (A_x, A_y) of fixed point A, the blue dashed line and the dotted line show the measurement values (B_x, B_y) of fixed point B, the black solid line and the dashed and the dotted line show the estimated values $(\hat{A}_x, \hat{A}_y)$ of fixed point A, the green dashed and the dotted line the estimated values $(\hat{B}_x, \hat{B}_y)$ of fixed point B. In the *y*-direction, it is seen that the estimated values are almost agree with the measurement values of both A and B. However, in the *x*-direction, there are error between the estimated values and the measurement values after 30 s.

Figure 6 shows the results of the walking experiment. Figures 6(a) and 6(b) show the results for the *x*- and *y*-directions, and Fig. 6(c) shows the results for the yaw angle of the body. The solid line indicates the values calculated using Eqs. (2) through (4), and the dotted line indicates the reference orbit designed in Section 4.1. Figure 3 shows that the tracking errors for the *x*- and *y*-directions and the yaw angle of the body accumulate because of the only local feedback (FB) control method of each leg joint. Moreover, the position of the robot body vibrates in the *x*-*y* plane.

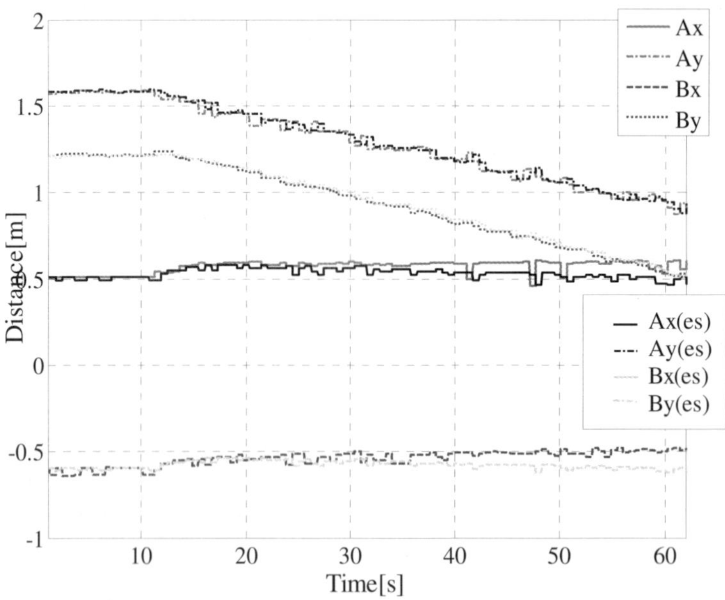

Figure 5 Measurement values and estimated values of straight walking.

In this study, we discuss the detecting method of the location (*x*, *y*) of the robot and the yaw angle in order to realize the trajectory following control proposed in Reference [2]. In the detecting method of this study, there are error between the measurement values and the estimated values, then, it isn't useful for applying the trajectory following control. In Reference [2], the mathematical model in which the outputs are the location (*x*, *y*) of the robot and the yaw angle was derived. Next step, with this mathematics model, we'll calculate the estimated values of the location (*x*, *y*) and the yaw angle by constituting Kalman filter and will examine the estimating method to filter the measurement error in the actual value.

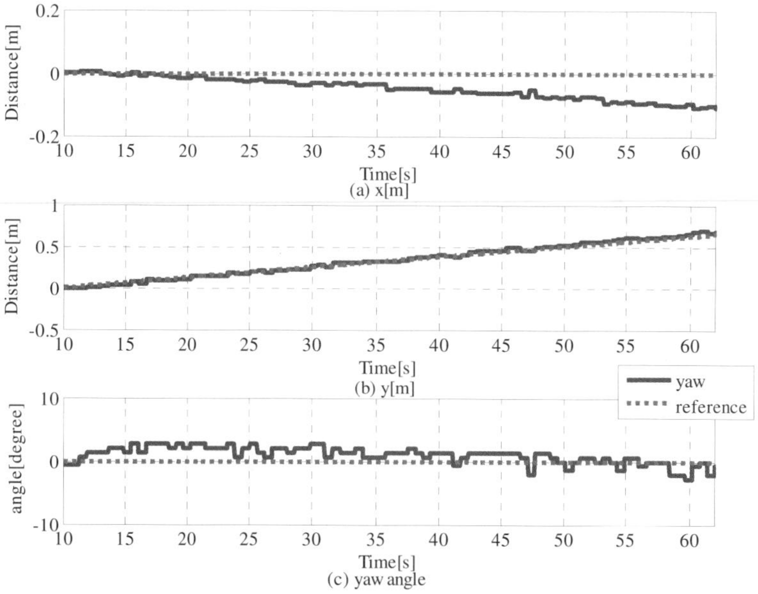

Figure 6 Results of straight walking.

5. Conclusion

The present study describes a method for measuring the location (*x*,*y*) and yaw angle of the body of a six-legged robot using an LRF. The measurement method and calculation algorithm were verified by straight walking of the six-legged robot.

For future work, the estimated method of the location (*x*, *y*) of the robot and the yaw angle will be examined using Kalman filter. Moreover, the trajectory following control method of the location and the yaw angle will be verified by 3D simulations and the walking experiments of a six-legged robot.

References

1. H. Uchida and K. Nonami, Attitude Control of a Six-legged Robot in Consideration of Actuator Dynamics by Optimal Servo Control System, *Climbing & Walking Robots Towards New Application, I-Tech,* ***pp*** 299-312 (2007).
2. H. Uchida, N. Shiina and H. Kondoh Trajectory Following Control for Omni-Directional Walking of a Six-legged Robot, *Proc. 15th Int. Conf. Climbing and Walking Robots and the Support Tech. for Mobile Machines,* ***pp*** 697-704 (2012).
3. Jr-Hung Guo, Kuo-Lan Su, Chung-Chieh Wang and Chia-Ju Wu, Laser Range Finder Applying in Motion Control System of Mobile Robots, *Proc. of 4th International Conference on Innovative Computing, Information and Control (ICICIC),* ***pp*** 536 – 539 (2009).
4. Takasuka, N., Tanaka, T., Kaneko, S., Tada, T. and Suzuki, S., Self-localization of a mobile robot on irregular ground using a laser range finder, *Proc. of World Automation Congress (WAC),* ***pp*** 1-6(2010).
5. Nishimoto, K., Sagami, A. and Kaneyama, K., Three dimensional recognition of environments for a mobile robot using laser range finder, *Proc. of SICE, 2007 Annual Conference,* ***pp*** 401-405(2007).
6. Jin Baek Kim and Byung Kook Kim, Calibrated localization with 2-D laser range finder for indoor mobile robots, *Proc. of International Conference on Control Automation and Systems (ICCAS),* ***pp*** 551 - 556 (2010).
7. Hee Jin Sohn and Byung Kook Kim, A Robust Localization Algorithm for Mobile Robots with Laser Range Finders, *Proceedings of the 2005 IEEE International Conference on Robotics and Automation(ICRA),* ***pp*** 3545 – 3550(2005).
8. Plagemann, C., Mischke, S., Prentice,S., Kersting,K., Roy, N. and Burgard, W., Learning predictive terrain models for legged robot locomotion, *Proc. of IEEE/RSJ International Conference on Intelligent Robots and Systems (IROS), pp* 3545 – 3552(2008).

COOPERATIVE MULTI-ROBOT LOCALIZATION WITH RADIO-BASED SENSORS

L. PFOTZER*, S. BOHN, J. OBERLAENDER, G. HEPPNER,
A. ROENNAU and R. DILLMANN

FZI Research Center for Information Technology,
Interactive Diagnosis- and Servicesystems (IDS),
Haid-und-Neu-Str. 10-14, 76131 Karlsruhe, Germany
**E-mail: pfotzer@fzi.de, www.fzi.de/ids*

This article describes a cooperative localization method for reconfigurable robots and multi-robot teams. In our approach, we use radio-based communication modules to determine the distances between multiple robots. With the help of cheap and simple sensors and suitable models of sensor measurements and robot motion, a particle filter can be applied to estimate the relative robot poses. In addition to tracking relative robot poses, our approach is also capable of solving the kidnapped robot problem for multiple cooperative robots.

Keywords: Cooperative localization; Radio-based Distance Estimation; Particle Filter; Reconfigurable Robots.

1. Introduction

In the field of cooperative robotics, knowledge about the position and orientation of all involved robots is required in order to perform cooperative tasks. Determining such multiple robot poses is often described as multi-robot localization.[1] Imagine mobile robots which are distributed in the environment and the task of joining them together to form a single robotic system. These reconfigurable robots are commonly very small and have limitations concerning power, size and weight. Because of these restrictions and of course the costs, it is inefficient to equip each robot with sensors like laser scanners or time of flight (TOF) cameras. Without the use of sophisticated sensors, determining an accurate pose of each robot is challenging, especially outdoors.

The term of multi-robot localization covers a large amount of different approaches trying to solve the position tracking, global localization and the kidnapped robot problem.[2,3] Zhang et al.,[4] for example, use Monte Carlo localization with self-adaptive samples (SAMCL) extended by a position mapping (PM) algorithm to handle multi-robot localization. They assume that every robot has a map of the environment. Charrow et al.[5] presented a method for localizing a fixed target using several mobile robots based on a particle filter as described in Hoffmann et al.[6] The main focus of their work is to overcome the limited sensor information by taking advantage of robots' mobility to rapidly improve the estimate of the target position. Inspired by these approaches, we realized a cooperative relative localization of multiple mobile robots.

The localization method introduced in our paper aims to cooperatively solve the problem of multi-robot localization without the need of large, heavy and expensive sensors. Furthermore, our approach allows reliable localization in various kinds of environments ranging from indoor areas as well as unstructured outdoor fields. This cooperative localization method is applicable for example in a team of search and rescue robots for civil protection or with reconfigurable robots performing inspection and maintenance tasks.

2. Approach

2.1. *Hardware*

There exist many sensors like laser scanners, TOF cameras, GPS, etc., which are suitable to realize a precise localization for single robots in dedicated environments. However, using reconfigurable robots both indoors and outdoors induces several constraints on selecting proper sensors. Firstly, the sensor has to be cheap, small and lightweight because these robots are often small and cannot carry heavy payloads. Secondly, the sensor should be resistant against sunlight and independent from visibility influences. Especially outdoors, laser scanners or cameras often have problems with these influences. On the other side sensor systems like GPS are suitable outdoors but often have difficulties indoors or in urban areas with high buildings.

2.1.1. *Radio-based Distance Sensor*

We decided to use radio-based sensors because they fulfil most of the defined constraints. Particularly, we investigated two different types of commercially available radio modules: Digi's XBee ZB radios and Anaren Integrated Radio (AIR) modules, carrying a CC 1101 from Texas Instruments, which are both very lightweight, small and cost efficient. The XBee ZB communication modules are based on the IEEE 802.15.4-2003 standard and allow baud rates of up to 250 kbit/s at a frequency of 2.4 GHz. On the other hand the AIR modules use the 868 MHz band which allows longer transmission ranges than XBee ZB modules. Both radio sensors have the ability to measure the received signal strength indicator (RSSI) value.

2.1.2. *Reconfigurable Robot System KAIRO II*

The modular inspection robot KAIRO II[7] is applied as evaluation platform for our new localization approach. KAIRO II is a heterogeneous chain-typed robot, which consists of two different types of modules. Drive modules equipped with wheels can be alternately connected through active joint modules. As shown in Fig. 1, we use a set-up of three robots, each consisting of two drive modules and one respectively two joint modules. Every single robot is equipped with a radio sensor and is able to drive independently.

Fig. 1. Three self-contained robots build from KAIRO II modules. Two robots consist of two drive and two joint modules each, while the third robot has only one joint module.

2.2. *Methodology*

To realize cooperative localization, the previously introduced radio sensors are used to determine the distances between multiple robots, depending on the RSSI value. As we do not have information about the robot orientation and the direction of the measured distances, the distribution of the robot poses is multi-modal, non-linear and not normally distributed. Hence, we utilized a particle filter to estimate the robot poses, because in contrast to Kalman filters, it can handle arbitrary distributions and non-linearities.[2] The overall localization precision is increased by specific relative movements of the robots.

2.2.1. *Measurement Model*

First of all, an error model for the radio-based sensors has been established. To this end, we used a test bed and performed different measurements to determine the accuracy of the radio sensors. A SICK LMS 100 laser range sensor was used to measure the real distance while simultaneously gathering the RSSI values between two radio sensor modules. Fig. 2 shows the measured values for the XBee module. We got similar results for the AIR module. The measurements were then fit into the logarithmic distance model

$$d\,(\mathrm{RSSI}) = a \cdot \log\left(b \cdot \mathrm{RSSI}\right) + c \qquad (1)$$

by estimating the parameters a, b and c. This logarithmic function is also illustrated in Fig. 2. From these recorded data we could also identify a linear error model $\sigma\,(d) = m \cdot d + y$, with an ascending variance σ^2 according to our measured sensor error. In combination with the distance model from Eq. 1, we estimated the measuring error of our radio sensors as a normal distribution $\mathcal{N}\left(d\,(\mathrm{RSSI}), \sigma^2\,(d\,(\mathrm{RSSI}))\right)$.

In the next step, a measurement model for the particle filter could be deduced from the previously introduced sensor error model. The previously defined normal distribution is used to rate the measurements. Thus, the weight of each particle depending on the amount of total radio connections $N = \frac{n \cdot (n-1)}{2}$ between n robots can be calculated:

$$\prod_{i}^{N} \mathcal{N}\left(m_i, d_i, \sigma\,(d_i)\right) \qquad (2)$$

whereas m_i is the corresponding particle distance, d_i the measured distance and i the radio connection between two robots. The weight is modelled as a Gaussian $\mathcal{N}\,(x, \mu, \sigma) = e^{\frac{-(x-\mu)^2}{2\sigma^2}}$ with the maximum of 1 at μ and the variance σ.

2.2.2. *Robot Motion Model*

The prediction step of the particle filter, also needs a realistic robot motion model. Due to the special kinematics of our previously introduced KAIRO II robots, which are similar to an articulated steering,[8] we developed a motion model to get an adequate prediction of the robot movement. This approach considers the rigid axis of the drive modules and takes the instantaneous center of rotation *icr* into account, which is shown in Fig. 3.

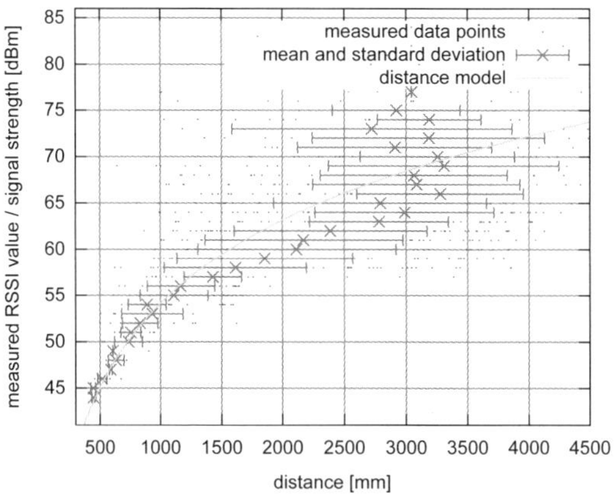

Fig. 2. RSSI values and mean filtered RSSI values with standard deviation measured at different distances with the XBee radio sensors. The distance model has been fitted as a logarithmic function.

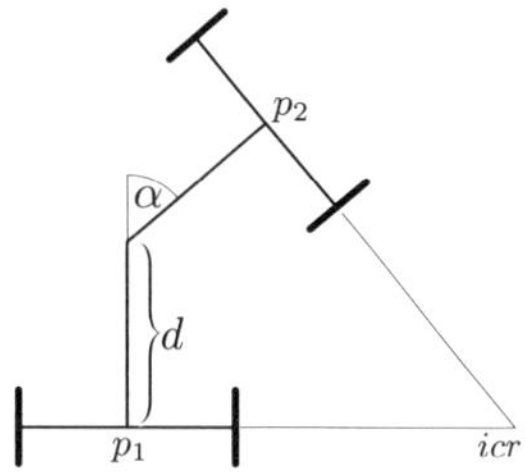

Fig. 3. Simplified representation of an articulated steering[8] used as motion model for every subsystem of KAIRO II with two drive modules.

The relative pose between the drive module axis with a half joint module length d is defined as $p_\alpha = (d + d\cos\alpha, d\sin\alpha, \alpha)$. The joint angle α can be assumed as normally distributed: $\hat{\alpha} \sim \mathcal{N}(\mu_\alpha, \sigma_\alpha^2)$. Then, we create independent samples for the relative poses $\hat{p_i} = (\hat{x_i}, \hat{y_i}, \hat{\theta}_i)^T$ of the front ($i = 2$) and rear ($i = 1$) axis, using the differential drive motion model as proposed by Eliazar et al.[9] The assumption of rigid axes leads to pure translation components for the relative poses:

$$\hat{q}_i = (\hat{x}_i, \hat{y}_i)^T \ , \quad \text{thus } \hat{p}_i = \begin{pmatrix} \hat{q}_i \\ \hat{\theta}_i \end{pmatrix} \tag{3}$$

The icr is calculated by intersecting two lines running through the origin of the drive module axis and orthogonally to the translation vectors $\hat{q}_1$ and $\hat{q}_2$. Thus, the curve radius r can be determined. To model the slip of the wheels, we interpolate linearly between the two poses $\hat{q}_1$ and $\hat{q}_2$:

$$\hat{q} = t \cdot \hat{q}_1 + (1 - t) \cdot \hat{q}_2, \quad \text{with } t \sim \mathcal{U}\,(0, 1) \tag{4}$$

sampled from a symmetric triangular distribution $\mathcal{U}$. The angle β is then determined by applying the translation vector $\hat{q} = (\hat{x}, \hat{y})^T$ along the arc around icr with radius r: $\beta = \frac{\hat{q}}{r}$. The resulting translation $\hat{q}'$ can be calculated with the following equation:

$$\begin{pmatrix} \hat{q}' \\ 1 \end{pmatrix} = T_c^{-1} R_\beta T_c \begin{pmatrix} \hat{q} \\ 1 \end{pmatrix}, \quad \text{with } T_c = \begin{pmatrix} 1 & 0 & c_x \\ 0 & 1 & c_y \\ 0 & 0 & 1 \end{pmatrix} \tag{5}$$

Finally, this leads to pose samples $\hat{p}' = (\hat{q}', \beta)^T$, which are necessary for the pose estimation step of the particle filter.

Both measurement model and robot motion model are necessary for the particle filter. Moreover, a customized model generates the initial distribution (see Fig. 6 (a)), which helps to keep the amount of particles small. This preprocessing step is based on a circle-intersection method.

3. Experiments and Results

Evaluating the efficiency and the limits of our implemented approach, we defined different scenarios for simulation as well as for testing with real robots. Based on these scenarios we run several tests.

3.1. *Simulation*

A simulation of both robot motion and measured distances was implemented to gather a large amount of data for determining computation time, accuracy and possible sampling rate of the particle filter.

In a first scenario, we randomly choose the starting positions and the steering angle of three robots. In every simulation step the robots were driven a distance between 5 mm and 15 mm. We run automated test series recording run time and localization error while varying the amount of particles between 1000 and 15000 in steps of 500. As expected by the application of a particle filter, the computation time increases linearly with the amount of particles. With a realistic value of 5000 particles, we reach a frequency of 30 Hz for running the filter steps, which will be sufficient for a real world application. Also, the localization error with previously unknown poses converges very fast, as illustrated in the first 1000 steps of Fig. 4.

Furthermore, we investigated the tracking capabilities over a long period as shown in Fig. 4. The results indicate that our mean localization error, which was caused by the previously unknown robot poses, stays nearly constant at about 500 mm. It is observable that the localization error is decreasing slowly while tracking the robot poses.

In a third test scenario, we initialized the robots again with random positions and started the tracking. Suddenly, we transferred one robot to a new pose about 3 m away from the old position to simulate the kidnapped robot problem. The results illustrated in Fig. 5 indicate that the particle filter reacts by re-sampling the particles to reduce the localization error after about 5000 steps.

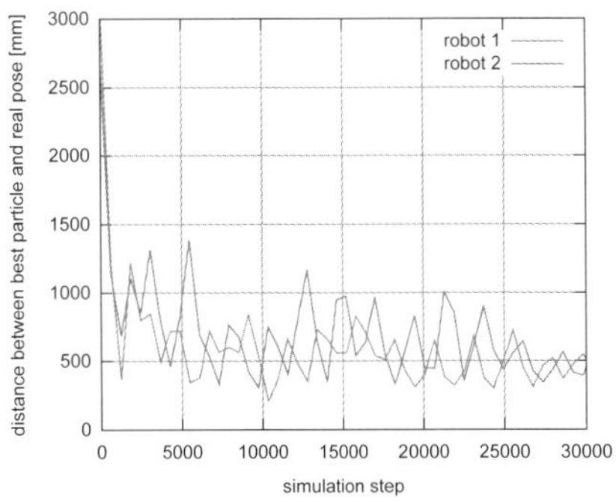

Fig. 4. Starting with initially unknown poses the localization error decrease very fast in the first 1000 simulation steps. The further course shows the tracking of multiple robots with a slightly decreasing mean error.

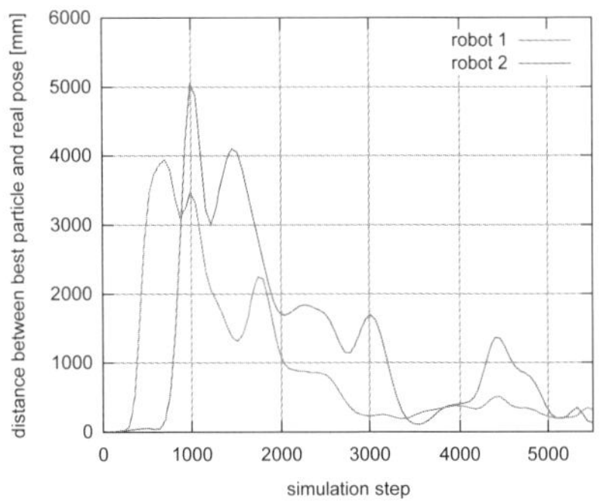

Fig. 5. Progress of the localization error while running a simulation of the kidnaped robot problem. The pose of robot 1 was suddenly changed after 200 steps.

3.2. *Evaluation with* KAIRO II

For a real world evaluation of our implemented approach, we used three subsystems of the modular robot KAIRO II as described in section 2.1.2. These robots were driven manually by human operators, while the localization program was running on an external computer. At the same time the measured distance values of the XBee modules were covered on every robot and together with the odometry values, transmitted to the localization software. In this experiment, we distributed several robots in the environment at randomly chosen initial positions. Fig. 6 (a) shows the visualization of the initially distributed particles and Fig. 6 (b) the estimated poses after 800 filter steps and a overall robot movement of about two meters.

(a) Initial distribution of particles. (b) Distribution after 800 steps.

Fig. 6. Visualization of the particles used by the filter to estimate the robot poses. Relative poses of three robots are illustrated, with the first robot located at the origin.

4. Conclusion and Future Work

Summarizing, we notice that our localization approach is applicable to approximate relative poses for reconfigurable robots and multi-robot systems like KAIRO II. Our results showed that performing particular cooperative robot movements will improve the localization accuracy. Further investigation about cooperative robot movements to increase the mutual information are carried out presently. Additionally, we plan to integrate the introduced approach into an absolute localization system using further sensors to improve the estimate of the robot poses. The idea is to equip one robot with a sensor for absolute localization. Thus, the other robots of the team are able to cooperatively estimate their absolute poses by combining the relative localization with the absolute localization. Moreover, we intend to apply small cameras for a close range tracking to enable an accurate docking of our mobile robots.

References

1. J. R. Spletzer and C. J. Taylor, A bounded uncertainty approach to multi-robot localization, in *IEEE/RSJ International Conference on Intelligent Robots and Systems (IROS '03)*, (Las Vegas, USA, 2003).
2. S. Thrun, W. Burgard and D. Fox, *Probabilistic Robotics (Intelligent Robotics and Autonomous Agents)* (The MIT Press, 2005).
3. S. Thrun, D. Fox, W. Burgard and F. Dellaert, Robust monte carlo localization for mobile robots, *Artificial Intelligence* (2000).
4. L. Zhang, R. Zapata and P. Lepinay, Self-adaptive monte carlo for single-robot and multi-robot localization, in *IEEE International Conference on Automation and Logistics (ICAL '09)*, (Shenyang, China, 2009).
5. B. Charrow, N. Michael and V. Kumar, Cooperative multi-robot estimation and control for radio source localization, in *Proc. of the Intl. Sym. on Experimental Robotics*, (Québec, Canada, 2012).
6. G. Hoffmann and C. Tomlin, Mobile sensor network control using mutual information methods and particle filters, *IEEE Transactions on Automatic Control* (January 2010).
7. L. Pfotzer, S. Bohn, C. Birkenhofer and R. Dillmann, Biologically-inspired locomotion with the multi-segmented inspection robot kairo-ii, in *Proceedings of CLAWAR2011*, (Paris, France, 2011).
8. J. Gutiérrez, D. Apostolopoulos and J. L. Gordillo, Numerical comparison of steering geometries for robotic vehicles by modeling positioning error, *Autonomous Robots* (2007).
9. A. Eliazar and R. Parr, Learning probabilistic motion models for mobile robots, in *In Proc. of the International Conference on Machine Learning (ICML '04)*, (ACM Press, 2004).

EFFICIENTLY USING RGB-D DATA TO SELF-LOCALIZE A SMALL WALKING ROBOT IN MAN-MADE ENVIRONMENTS

Piotr SKRZYPCZYŃSKI

Institute of Control and Information Engineering, Poznan University of Technology, ul. Piotrowo 3A, PL-60-965 Poznań
E-mail: piotr.skrzypczynski@put.poznan.pl

This paper presents a concept and preliminary experimental results concerning reliable incremental self-localization of a small walking robot in man-made environments from RGB-D data. The main idea of the presented approach is to reduce the 3D point clouds produced by the Kinect-like sensor to 2D virtual scans, which can be processed much faster considering the limited computing resources of the robot. Matching of the virtual scans is combined with initial estimation of the sensor orientation change from matching of highly descriptive photometric features extracted from the Kinect RGB images.

Keywords: Self-localization, 3D sensor, Photometric features, Walking robot

1. Introduction

Self-localization with regard to 6-DOF is particularly important in the case of walking robots, because the discrete nature of their motion causes sudden changes in the attitude, while leg slippages make any form of odometry almost useless. In a walking robot the 6-DOF pose contains the position in three dimensions, orientation, and pitch and roll angles: $\mathbf{x}_R = [x_r \; y_r \; z_r \; \theta_r \; \phi_r \; \psi_r]^T$. Moreover, if self-localization of a walking robot is considered, the common "2D assumption",[11] which states that the robot moves on a flat floor and a range sensor takes measurements in a plane parallel to that floor does not hold anymore.

Often point clouds yielded by 3D laser scanners are matched against each other in order to obtain the displacement between two poses of a mobile robot. This approach, known as scan matching, is characterized by high reliability and relative independence from the characteristics of the environment. However, implementation of 6-DOF scan matching with data from mechanically scanning 3D laser range finders is impossible in most of the legged robots, because of the tight size, mass, energy and computing resources limits in the walking machines. A solution to this problem is to use a compact 3D range sensor, which has no moving parts, such as Asus Xtion or Microsoft Kinect. These sensors are compact and affordable, and as shown in Ref. 10, the range measurement accuracy of Kinect up to the distance of about 3.5 m is comparable to that achieved by a 3D laser scanner. The Kinect sensor allows to obtain RGB images and range images of 640 $\times$ 480 resolution, which yields 307200 points in 3D space. By combining information from the range image and the RGB image the RGB-D image can be obtained. However, the range data stream obtained from the Kinect sensor is very intensive due to the high number of 3D points. Therefore, the commonly used Iterative Closed Points algorithm, which tries to establish correspondences between all the available points cannot be applied for real-time 3D data registration considering the limited processing power available on-board of a walking robot.

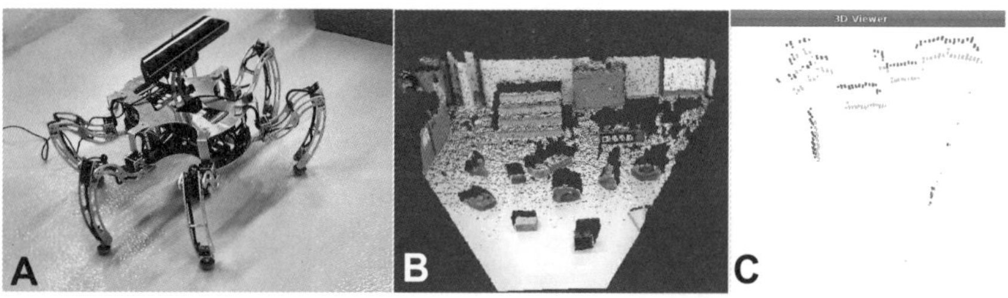

Fig. 1. Messor robot with the Kinect sensor (A), RGB-D image of the scene (B), 2D virtual scan of a cluttered area (C)

As it was shown in our previous work[9] the 6-DOF pose of a walking robot can be successfully estimated by matching scans from a Hokuyo URG-04LX 2D laser scanner, and by using data from an inertial measurements unit (IMU) to correct the scans for the varying attitude of the robot. This system works well in man-made environments, where flat walls are available, but has several practical shortcomings. Because the terrain mapping system of our Messor walking robot uses a tilted down URG-04LX scanner, another URG-04LX mounted horizontally on the robot is necessary to obtain environment profiles (2D scans) that are suitable for scan matching. The cost of a Hokuyo sensor (about 2000 USD) is quite high for the small-size and inexpensive Messor robot. Moreover, 2D scanning may be insufficient in some cluttered environments.

In this paper we show that a similar incremental self-localization method for man-made environments can be implemented using data from a 3D

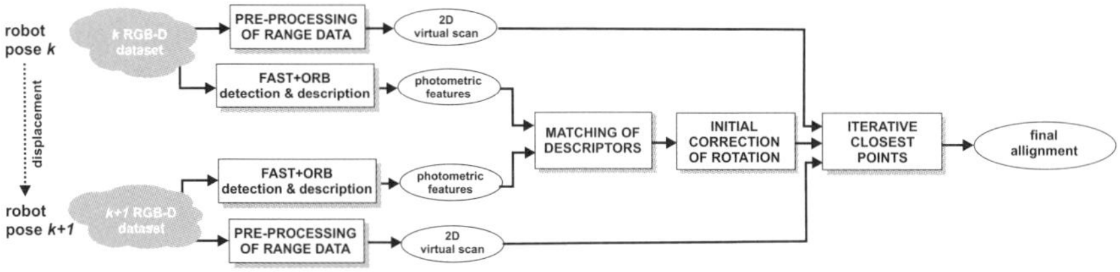

Fig. 2. Block scheme of the RGB-D data registration procedure

RGB-D sensor, namely, the Kinect (Fig. 1A). Our approach substantially reduces the amount of data obtained from Kinect by projecting the 3D point clouds (Fig. 1B) into virtual 2D scans (Fig. 1C). The gains due to using the RGB-D sensor instead of a 2D laser scanner on a walking robot are twofold: (i) processing of 3D point clouds provides the ability to observe walls partially occluded by clutter, which helps the scan matching algorithm to establish proper correspondences between the consecutive scans; (ii) the RGB images from Kinect can be used to establish correspondences between salient point features, and then these correspondences can be exploited to obtain an initial transformation that aligns the two RGB-D datasets. The RGB-D data registration procedure is summarized in Fig. 2.

2. Related work

Many Kinect-based self-localization systems use both the range and the RGB data. Typically, algorithms presented in the literature rely on the photometric data for salient feature detection, and then use the depth image to locate the keypoints in 3D space.[4] Alternatively, the photometric features and their associated range values can be used to obtain an initial alignment (which is similar to our approach, but we use only the RGB data), and then the final transformation is computed by jointly optimizing over the sparse feature matches and dense 3D point correspondences.[6] Yet another approach is presented in Ref. 3, where the RGB-D data are used to generate new point features, which are then tracked on RGB images for a number of frames to avoid frequent processing of the large amount of 3D data. Unfortunately, all the above-mentioned solutions require at some stage processing of the whole 3D datasets, in order to extract feature positions in 3D and/or to establish correspondences between dense point clouds. Such an approach is prohibitive to the on-board computer of our small walking robot. Therefore, we prefer to reduce the amount of 3D data significantly before we attempt to establish any correspondences between the datasets.

3. Matching of virtual scans

The range data reduction process is accomplished by using a simple heuristic, similar to the one described in Ref. 11, which extracts virtual 2D scans from distance images. At first, we choose from each column of the depth (D) image the point that has the largest depth (range) value. In a man-made environment, which is limited by walls, the most distant points most likely belong to these walls. Points located on the floor and obstacles partially occluding the walls are filtered out. Windows are not a problem, as in most cases the Kinect does not provide valid distance data for these areas, however, the virtual scans can be disturbed by open doors. Then, $\mathbf{p} = [x\ y\ z]^T$ coordinates of the points are calculated from the depth image using procedures from the Point Cloud Library and the calibration data of the Kinect sensor.[7] In a man-made environment we can assume that the surrounding walls have the same shape for all heights. Using this assumption we can correct the range measurements accounting for the changing attitude. The attitude is obtained from the IMU, which is used merely as a two-axis inclinometer, because the yaw angle estimate is unreliable. Finally, points are projected onto the plane perpendicular to the gravity vector, dropping the z value. The virtual 2D scans are then feed to the scan matching algorithm.

The 2D scan matching algorithm aims at finding the displacement $\mathbf{q}_i$ that minimizes the distances between the points of a given scan, rotated and translated by $\mathbf{q}_i$ and the points of another scan, treated as a reference. As in Ref. 9 we use the Point-to-Line ICP (PLICP) modification of this approach,[1] which minimizes the Euclidean distance provided by the normal to the locally defined line between scan points. The PLICP is more robust to outliers and converges quickly. Because PLICP is prone to errors due to poor initial estimate of the displacement between scans, in Ref. 9 this method was paired with the Polar Scan Matching algorithm,[2] which has larger average basin of convergence. However, using another algorithm to bootstrap the PLICP slows down the processing of scans, while this approach still does not solve the initial displacement estimate problems that arise due to environment symmetries or lack of distinguishable geometric features. Therefore, we exploit the photometric (RGB) data obtained from the Kinect sensor to compute the initial estimate of rotation with regard to the roll, pitch, and yaw angles of the robot. This initial rotation estimate usually is enough to transform the virtual scans to the convergence basin of the PLICP algorithm. Thus, the first guess of the robot displacement between the two poses at which the virtual scans $s(k)$ and $s(k+1)$ were taken is accomplished by using legged odometry and data from the robot's

IMU. Then the roll, pitch, and yaw angles are corrected by using data from matching RGB images. The final rotation and translation estimate $\hat{\mathbf{q}}_{k+1}$ at the $k+1$-th time instant is computed by the PLICP procedure, starting from the RGB-data corrected initial guess. The 2D pose estimate $\hat{\mathbf{x}}_{R(i)}$ with uncertainty $\mathbf{C}_R$ is computed compounding the local transformations head-to-tail:

$$\hat{\mathbf{x}}_{R(k+1)} = \hat{\mathbf{x}}_{R(k)} \oplus \hat{\mathbf{q}}_{k+1}, \quad \mathbf{C}_{R(k+1)} = \mathbf{J}_{1\oplus}\mathbf{C}_{R(k)}\mathbf{J}_{1\oplus}^T + \mathbf{J}_{2\oplus}\mathbf{C}_{q_{k+1}}\mathbf{J}_{2\oplus}^T, \quad (1)$$

where $\mathbf{C}_{q_{k+1}}$ is the uncertainty of the transformation, computed as in Ref. 9, $\oplus$ is the compounding operator, and $\mathbf{J}_{1\oplus}$, $\mathbf{J}_{2\oplus}$ are the Jacobians of this operation.

4. Estimating the initial rotations

We use the RGB images obtained from the Kinect sensor to compute the first guess of rotation between the two datasets used to generate virtual scans. The translation, which can be also computed from the image matching is not used because in the monocular case the translation is computed only up to the unknown scale.

To estimate the relative rotations between the two positions of the Kinect sensor we calculate the fundamental matrix $\mathbf{F}$ and the essential matrix $\mathbf{E}$ that relate the corresponding point pairs across two views of the same scene. The essential matrix takes into account also the camera calibration parameters. To calculate the fundamental matrix the 8-point algorithm[5] is used. The input data for this algorithm are pairs of point features obtained by matching characteristic keypoints in the two Kinect RGB images, which are two different views of the same scene.

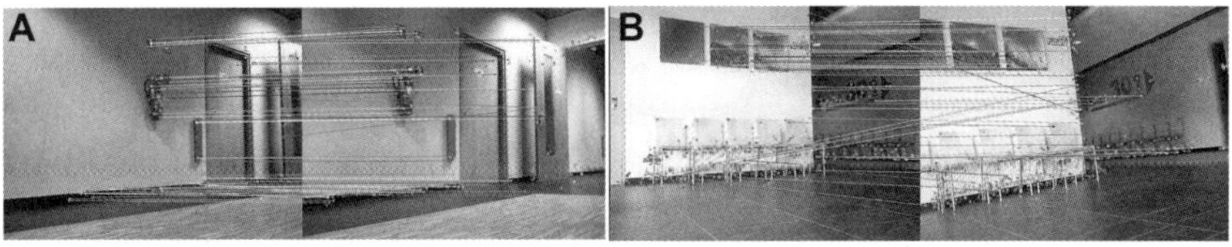

Fig. 3. Matching of point features: translation and rotation on flat floor (A), rotation in 3D (B)

Point features are detected on images already corrected for distortions (Fig. 3). Taking into account recent results concerning the speed and robustness of known point feature detectors and descriptors in mobile robotics[8] we decided to use the FAST (Features from Accelerated Segment Test) keypoint detector paired with the ORB (Oriented FAST and Rotated BRIEF) descriptor.

Because the set of matched points can contain some outliers, the 8-point algorithm is applied within the robust estimation framework based on the RANSAC scheme. Having the estimate of the fundamental matrix and the Kinect camera calibration matrix $\mathbf{K}$, the essential matrix can be calculated: $\mathbf{E} = \mathbf{K}^{\mathbf{T}}\mathbf{F}\mathbf{K}$. The computed essential matrix allows to calculate the homogeneous matrix of rotation between the two camera positions. To solve for rotations the method based on the Singular Value Decomposition (SVD) proposed in Ref. 5 is used. This way we can obtain rotations in 3D between the two consecutive RGB-D data sets on the robot's path (Fig. 3B). Points belonging to the first scan are then rotated to the new coordinates by the formula:

$$\mathbf{p}'_{s(k)}[i] = \mathbf{R}_{\text{vis}}\ \mathbf{p}_{s(k)}[i] \quad \text{for} \quad i = 1, 2, ..., n_{s(k)}, \tag{2}$$

where $\mathbf{R}_{\text{vis}}$ is the rotation matrix between the two camera positions, while $\mathbf{p}_{s(k)}$ are points of the first virtual scan. The newly computed scan $s'(k)$ is then used by the ICP algorithm.

5. Experimental results

The incremental self-localization system proposed in this paper has been evaluated in several indoor environments to test its robustness against situations that often cause problems with typical 2D scan matching.

We show results of two experiments demonstrating robustness to problems resulting from lack of well-defined geometric features (corridor experiment), and the ability of our system to self-localize in the presence of robot's attitude changes. Figure 4 presents results of matching virtual 2D scans from the Kinect sensor in a corridor. Registration with the walking robot's odometry resulted in a useless map (Fig. 4A). The PLICP algorithm was able to correct this situation only to some degree, as the corridor does not contain enough distinguishable geometric features (Fig. 4B). But using the image matching solved this problem (Fig. 4C), as the corridor contained some visually distinct objects (cf. Fig. 3A).

The experiment illustrated in Fig. 5 was conducted in a lab room, where the teleoperated Messor robot changed frequently its posture in order to introduce varying attitude (Fig. 5A). If the PLICP algorithm without attitude correction was used to match the virtual scans, the basic geometry of the room was extracted from the range data, but the precision of the map

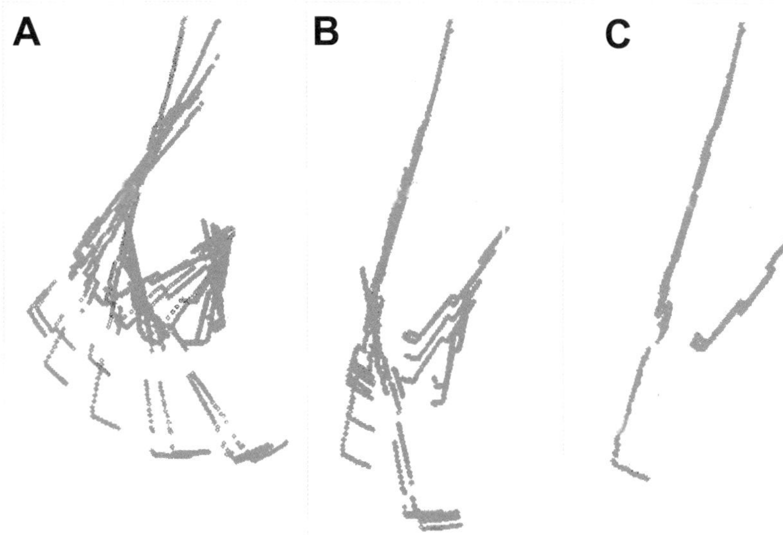

Fig. 4. Self-localization in a corridor with few geometric features: 2D virtual scans registered by dead reckoning (A), scans registered by using the PLICP (B), scans registered by using the PLICP with rotation corrected from images (C)

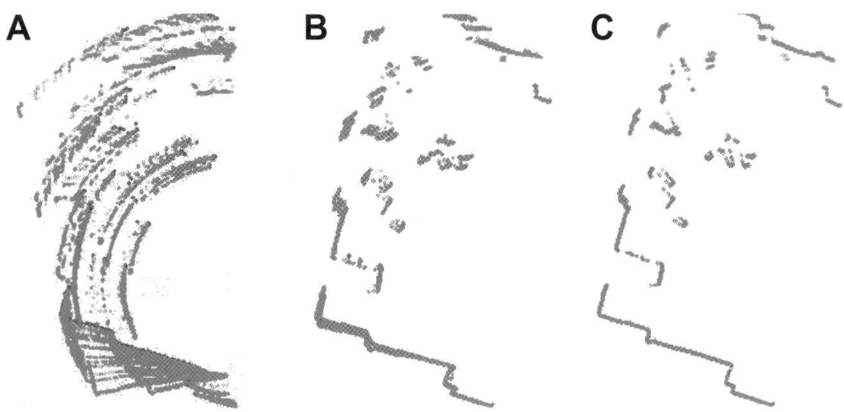

Fig. 5. Self-localization results in a lab: scans registered by dead reckoning (A), scans registered by using the PLICP (B), scans registered by using the proposed method (C)

is rather low (Fig. 5B). The map is much more precise, as indicated by the better overlapping consecutive scans, if the attitude correction is used (Fig. 5C).

6. Conclusions

This paper presents a practical implementation of the 2D scan matching system with the data obtained from a 3D RGB-D sensor. This system is tailored specifically for a small walking robot, addressing such issues as cost effectiveness, small mass of the sensory system, and the ability to use both geometric and photometric information. The main difficulty observed so far with this system is the small horizontal field of view of the Kinect sensor (only 58°), which causes problems when trying to self-localize in open spaces, like a large room. Therefore, in the self-localization system of our new walking robot we plan to combine the Xtion sensor with a low-cost 2D laser scanner.

References

1. Censi A. (2008) An ICP Variant Using a Point-to-Line Metric, Proc. IEEE Int. Conf. on Robotics and Automation, Pasadena, 19–25.
2. Diosi A., Kleeman L. (2007) Fast Laser Scan Matching Using Polar Coordinates, Journal of Intelligent and Robotic Systems, 26(10), 1125–1153.
3. Domínguez S., Zalama E., García-Bermejo J., Worst R., Behnke S. (2013) Fast 6D Odometry Based on Visual Features and Depth, Frontiers of Intelligent Autonomous Systems, SCI Vol. 466, Springer, 5–16.
4. Endres F., Hess J., Engelhard N., Sturm J., Cremers D., Burgard W. (2012) An Evaluation of the RGB-D SLAM System, Proc. IEEE Int. Conf. on Robot. and Automat., St. Paul, 1691–1696.
5. Hartley R., Zisserman A. (2004) Multiple View Geometry in Computer Vision, Cambridge Univ. Press.
6. Henry P., Krainin M., Herbst E., Ren X., Fox D. (2012) RGB-D Mapping: Using Kinect-Style Depth Cameras for Dense 3D Modeling of Indoor Environments, Int. Journal of Robotics Research, 31(5), 647–663.
7. Rusu R., Cousins S. (2011) 3D is Here: Point Cloud Library (PCL), Proc. IEEE Int. Conf. on Robot. and Automat., Shanghai, 1–4.
8. Schmidt A., Kraft M., Fularz M., Domagala Z. (2012) The Comparison of Point Feature Detectors and Descriptors in the Context of Robot Navigation, CLAWAR Workshop on Perception for Mobile Robot Autonomy (PEMRA'12), Poznan, (CD-ROM).
9. Skrzypczyński P. (2012) Laser Scan Matching for Self-Localization of a Walking Robot in Man-Made Environments, Industrial Robot: An International Journal, 39(3), 242–250.
10. Stoyanov T., Louloudi A., Andreasson H., Lilienthal A. (2011) Comparative Evaluation of Range Sensor Accuracy in Indoor Environments, Proc. 5th European Conf. on Mobile Robots (ECMR), Örebro, 19–24.
11. Wulf O., Arras K., Christensen H., Wagner B. (2004) 2D Mapping of Cluttered Indoor Environments by Means of 3D Perception, Proc. IEEE Int. Conf. on Robotics and Automation, New Orleans, 4204–4209.

RECOGNITION OF 3D OBJECTS FOR WALKING ROBOT EQUIPPED WITH MULTISENSE-SL SENSOR HEAD*

JANUSZ BĘDKOWSKI† AND KAROL MAJEK AND ANDRZEJ MASŁOWSKI
Institute of Mathematical Machines, ul. Ludwika Krzywickiego 34
Warsaw, 02-078, Poland

PIOTR KACZMAREK
Institute of Automatic Control and Robotics, ul. Św. A. Boboli 8
Warsaw, 05-525, Poland

In this work the system of the 3D object recognition developed for the walking robot equipped with relatively new sensor (Multisense-SL Sensor Head) is discussed. The novelty of this approach is new 3D descriptor (CSH - Complex Shape Histogram) that characterizes 3D cloud of points using semantic classification of local shapes. The system is composed of several algorithms (noise removal, sub-sampling, normal vector estimation, shape classification) implemented using General Purpose Graphic Processing Unit (GPGPU) that improves the computation time. The system is able to learn new 3D objects using Support Vector Machine as a classification method.

1. Introduction

One of the applications of mobile robots in today's world is reducing devastation caused by the natural or man-made disasters. In some situations, further damage could be prevented if there was a way of intervening directly at the disaster site. However it is often too dangerous or impossible to send human due to nuclear or chemical contamination, intense heat and air pressure or structural instability. Robots that are currently available are useful in such situations, but they are not always able to navigate in all environments and to perform required tasks. US government Defense Advanced Research Projects Agency (DARPA) seeks solution to that problem and organizes the Disaster Robotics Challenge. The goal of the DARPA DRC is to develop a robot, which is capable of operating in human engineered environment freely, using tools and vehicles designed for

* This work is supported by Polish National Centre of Science project „Methodology of semantic models building based on mobile robot's observations " number DEC-2011/03/D/ST6/03175 (2012-2015)

† Work partially supported by the European Community's Seventh Framework Programme (FP7/2007-2013) under grant agreement n°285417 "Integrated Components for Assisted Rescue and Unmanned Search operations"

human. While there is no specific requirement that the designed robot should be humanoid, it seems that walking robot is the way to achieve the goal. Tracked or wheeled platforms lack capability of easily climbing steep stairs and ladders and it would be hard for wheeled robots to steer a common vehicle without any modifications to it. Therefore, many researches are focused on the problem of the mobility in such difficult environment. The first stage of the competition, the Virtual Robotics Challenge, will take place entirely in simulator, therefore many of researchers can join the competition because of the decreased budget needed to obtain satisfactory results. Teams, participating in track B and C of the DRC challenge, will have opportunity to prove their skills and abilities by completing challenge tasks with simulated Atlas robot made by Boston Dynamics. The best teams will have an opportunity to work with real Atlas robot provided by the vendor, they will compete with other teams robots in physical challenge. In this paper we demonstrate the result concerns robot's perception based on 3D laser data. We were focused on the design and the implementation of the system capable of recognizing 3D objects on-line.

2. Related Work

Recently processing of cloud of points is popular because of the improvement of available sensors (Stereo Cameras, spinning 2D lasers[14], Prime Sense sensor – known also as KINECT). A limitation is usually the computational time, therefore there is a need to improve algorithms such as Nearest Neighborhood Search, Normal Vector Estimation or 3D data registration. Many researchers are using GPGPU(General Purpose Graphic Processing Unit) to achieve better performance. For example detailed analysis of GPGPU computation for data registration is shown in [15]. There are many techniques for 3D cloud of points characterization and there are many available 3D descriptors (some of them were evaluated in[1]) such as NARF Normal Aligned Radial Feature [2], Regional Point Descriptors (3D shape contexts and harmonic shape contexts) [3], Point Feature Histogram (PFH) [4,5], Fast Point Feature Histogram (FPFH) [6], Viewpoint Feature Histogram (VFH) [7], Concentric Ring Signature (CORS) [8], The Rotation Invariant Feature Transform (RIFT) [9], Spin image descriptor [10], Compact Covariance Descriptor [11]. Barcode shape descriptor [12] and Point Pair Feature [13]. Most of these 3D descriptors are efficient to characterize 3D cloud of point. PFH and FPFH descriptors are dedicated to classify local shapes, therefore they could be sufficient for our application but the computation time is large to obtain on-line capability. Other mentioned descriptors are suitable for the pattern recognition unfortunately they were difficult to integrate with our software framework. Most prominent solution was to use Point Pair Feature(PPF) descriptor, therefore we decided to improve it by the addition of semantic classification of local shapes. Proposed in this paper Complex Shape Histogram is an extension of PPF computes distances and angles between normal vectors for all pairs in cloud of point and store the result in the model descriptor. CSH additionally categorizes the shape around each query points similarly to PFH (Point Feature Histogram). Because PFH is time consumptive we proposed in this work simplified version of CSH that categorizes query points into two classes {flat, not flat}. Query point is classified as "flat shape" when surrounding (nearest neighbors) can be approximated by local plane (local planes are defined by the radius 20cm), in the other case this query point is classified as "not flat shape". This basic semantic classification distinguishes flat regions such as walls, floor, ceiling etc. from another shapes such as corner, cone, sphere etc. Semantic classification is an advantage of the CSH over another 3D descriptors, moreover CSH is able to describe this semantic information qualitatively. In this paper we introduced first semantic 3D descriptor and its practical application. In near future we will investigate the advantages and disadvantages of the approach. At current stage of the work we claim that the approach can efficiently distinguish different 3D objects, even when the information is not complete, or data is noisy.

3. Hardware Description

Boston Dynamics Atlas is a high mobility, humanoid robot. It includes 28 hydraulically-actuated degrees of freedom, two hands, arms, legs, feet and torso(figure 1). Robot is equipped with MultiSense-SL sensor head which consists of two stereo cameras and a spinning Hokuyo laser range finder. This is relatively new sensor available in Darpa Robotics Challenge. Proper utilization of sensor data will play essential role in the DARPA DRC competition. Sensor deliver 3D cloud of points and depth images.

Figure 1. Left – Humanoid robot Atlas being developed by Boston Dynamics, Inc. for DARPA Robotics Challenge. Right - Multisense-SL Sensor Head (combined sensor: Hokyo laser range finder mounted on the rotated head, stereocamera).

4. Complex Shape Histogram

Complex Shape Histogram takes into account all pairs of points from query cloud of points. Three features taken into account: a)Euclidean distance between points, b) combination of shapes (in simplified implementation of CSH:{flat, not flat}, {flat, flat}, {not flat, not flat}), c) angle between normal vectors computed for pair of points. The advantage of CSH over PPF is new feature - combination of shapes. This feature efficiently improves the unique characterization of 3D cloud of points composed of different shapes. A disadvantage of CSH is high computation complexity, therefore we implemented GPGPU based algorithm for fast CSH computation. In average for the cloud of 15000 data points CSH is computed within 4 seconds on GPGPU GF680GTX and the approximated method is computed within 1.7 second. An approximated method is implemented fully on GPGPU and provides some minor numerical artifacts. For the experiments we are not using an approximated method because the time of 4seconds is acceptable from the task point of view. The measured performance guarantee the real application of proposed approach. It is not necessary faster than another approaches but it is using semantics for object characterization, therefore it can be considered as further step in the research on 3D descriptors.

5. System of 3D object's recognition

Figure 2 shows the system structure for 3D objects recognition.

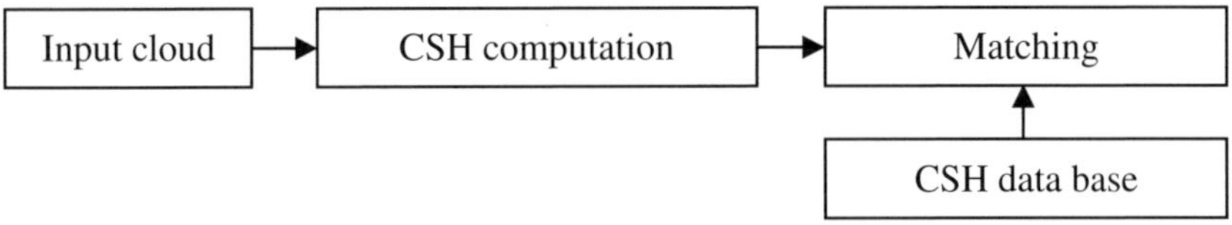

Figure 2. Simplified scheme of the system of 3D object's recognition.

Input cloud is obtained via simulator of Multisense-SL Sensor Head. This cloud of 3D points has to be processed (segmentation) to obtain points of the object by removing ground plane and surrounding obstacles. CSH computation is computed on GPGPU in following steps:

1. Normal Vector Computation using regular grid decomposition[15][16] (currently it is the fastest published method that computes normal vector estimation PCA/SVD from unorganized 3D cloud of points).
2. Classification of points into two classes: {flat, not flat}.
3. Computation of CSH features.

CSH database is composed of training sets. Each training set is composed of positive examples (CSH of query 3D object) and negative example (CSH of different 3D objects). Classification-matching is done via Support Vector Machine (SVM) with linear kernel function. Before matching SVM has to be trained to obtain a model describing query 3D object. Matching is performed by SVM respond for new query 3D object (described by CSH) that responds binary – {positive, negative}.

6. Experiments

Virtual Robotics Challenge is divided into three competition tasks. Task One is to drive a utility vehicle. Atlas robot controlled by competitors has to enter the Polaris Ranger EV, drive it along a roadway avoiding obstacles and then get out of it. While driving robot controls the vehicle by physically turning the steering wheel and pushing pedals. There are no any additional sensors attached to the car, only those mounted on Atlas head are available. The Task involves several perception challenges including road following, obstacle recognition, finding the vehicle and localizing its elements for ingression and control. In this paper we are focused on the problem of the vehicle recognition. For this purpose we are using rotated hokuyo laser because data delivered by this sensor are more reliable than data from stereo camera looking from our experience point of view. Real and simulated model of the vehicle are shown on figure 3.

Figure 3. Left – Polaris Ranger EV planned to be used as utility vehicle in Task One of DARPA VRC. Middle - model of Atlas robot located in front of virtual model of Polaris Ranger EV in simulated environment of DRCsim. Right – example of 3D cloud of points obtained via simulator of spinning hokuyo laser.

To train the system we prepared a training set composed of 112 observations of Polaris Ranger EV and 50 observations of different objects (10 observations for each object). For each observation we computed CSH – Complex Shape Histogram (examples are shown on figures 4,5,6). The system was trained using prepared different training sets. The most important observation is that in all cases the system had 100% of successful matching on each training set, therefore we claim that the CSH is able to transform 3D cloud of points into models that are linear separable in n-dimensional spaces (in our case model has n=300 values, this number was chosen experimentally, further research is needed to investigate proper amount of dimensions to obtain more satisfactory result). The matching decreases when SVM is being trained with wrong strategy. For example if we would like to recognize a vehicle based on 3D observation taken in front of vehicle (figure 5) we cannot train the system using 3D observations taken from the side of the vehicle (figure 6). At current state of the work we claim that it is not so obvious what kind of strategy of training we should take into the consideration. We observed that CSH is able to build efficient matching system, capable to recognize 3D object from database observed from the same location as during preparing training set.

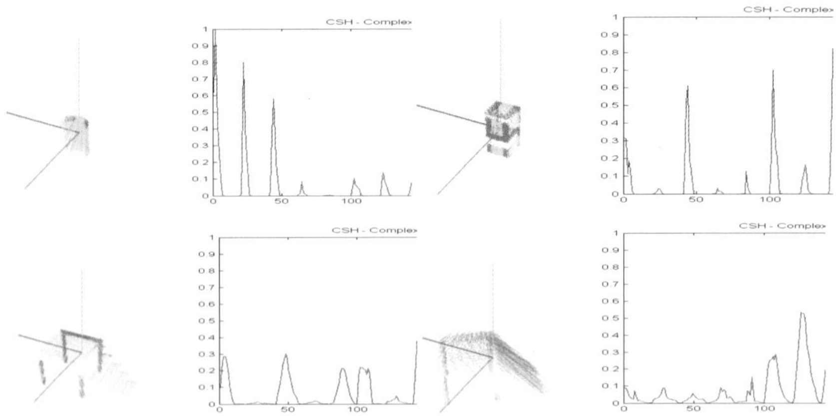

Figure 4. Complex Shape Histograms for four different objects. Red color corresponds to class "not flat", green color corresponds to class "flat". All CSH plots are different.

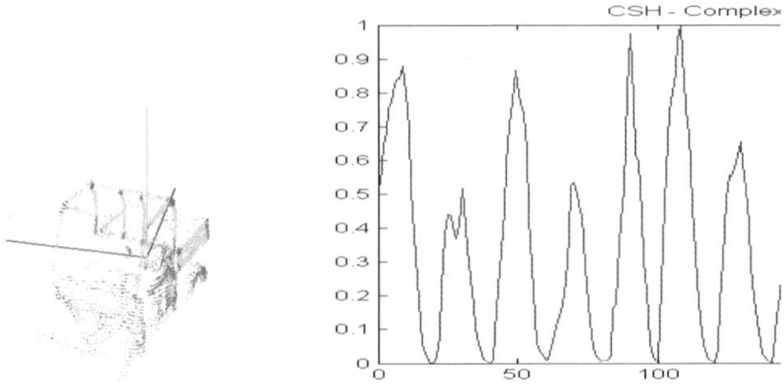

Figure 5. Complex Shape Histograms for Polaris Ranger EV observed from its front. Red color corresponds to class "not flat", green color corresponds to class "flat". All CSH plots are similar.

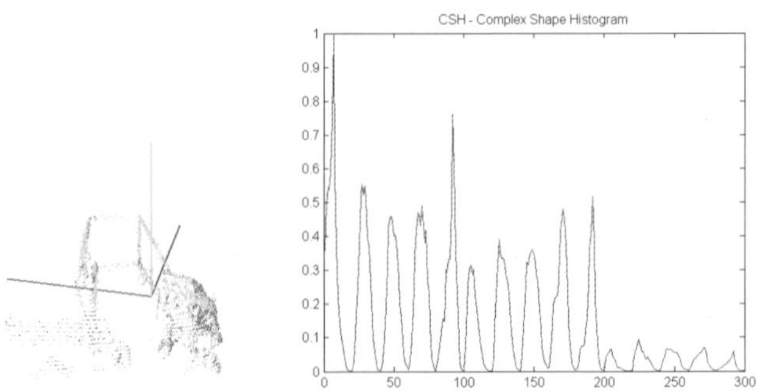

Figure 6. Complex Shape Histograms for Polaris Ranger EV observed from its side. Red color corresponds to class "not flat", green color corresponds to class "flat". All CSH plots are similar.

7. Conclusions and Future Work

In this paper new descriptor for 3D object represented by 3D cloud of points is shown. New descriptor (Complex Shape Histogram) extends PPF(Point Pair Feature) by new feature – combination of basic shapes related to the query points (it is inspired by PFH- Point Feature Histogram). In this paper simplified implementation of CSH is shown – the set of available shapes is decreased to {flat, not flat}. An efficient classification system for recognizing 3D objects is shown. The system is able to recognize Polaris Ranger EV observed via simulator of spinning hokuyo laser (Multisense-SL Sensor Head of simulated Atlas robot). Future work will be related with the full implementation of CSH including GPGPU based PFH for obtaining more classes of basic shapes.

Acknowledgments

This work is funded by Polish National Center of Science: project "Methodology of semantic models building based on mobile robot's observations" number DEC-2011/03/D/ST6/03175 (2012-2015).
Work partially supported by the European Community's Seventh Framework Programme (FP7/2007-2013) under grant agreement n°285417 "Integrated Components for Assisted Rescue and Unmanned Search operations"

References

[1] L. A. Alexandre, 3D descriptors for object and category recognition: a comparative evaluation, in: Workshop on Color-Depth Camera Fusion in Robotics at the IEEE/RSJ International Conference on Intelligent Robots and Systems (IROS), Vilamoura, Portugal, 2012.
[2] B. Steder, R. B. Rusu, K. Konolige, W. Burgard, NARF: 3D range image features for object recognition, in: Workshop on Defining and Solving Realistic Perception Problems in Personal Robotics at the IEEE/RSJ Int. Conf. on Intelligent Robots and Systems (IROS), Taipei, Taiwan, 2010.
[3] A. Frome, D. Huber, R. Kolluri, T. Bülow, J. Malik, Recognizing objects in range data using regional point descriptors, in: European Conference on Computer Vision, 2004, pp. 224–237.
[4] R. B. Rusu, N. Blodow, Z. C. Marton, M. Beetz, Aligning point cloud views using persistent feature histograms, in: Proceedings of the 21st IEEE/RSJ International Conference on Intelligent Robots and Systems (IROS), Nice, France, 2008.
[5] R. B. Rusu, Z. C. Marton, N. Blodow, M. Beetz, Learning Informative Point Classes for the Acquisition of Object Model Maps, in: Proceedings of the 10th International Conference on Control, Automation, Robotics and Vision (ICARCV), Hanoi, Vietnam, 2008.
URL http://files.rbrusu.com/publications/Rusu08ICARCV.pdf

[6] R. B. Rusu, N. Blodow, M. Beetz, Fast Point Feature Histograms (FPFH) for 3D Registration, in: Proceedings of the IEEE International Conference on Robotics and Automation (ICRA), Kobe, Japan, 2009.
URL http://files.rbrusu.com/publications/Rusu09ICRA.pdf
[7] R. B. Rusu, G. Bradski, R. Thibaux, J. Hsu, Fast 3D Recognition and Pose Using the Viewpoint Feature Histogram, in: Proceedings of the 23rd IEEE/RSJ International Conference on Intelligent Robots and Systems (IROS), Taipei, Taiwan, 2010.
[8] H. V. Nguyen, F. Porikli, Concentric ring signature descriptor for 3d objects, in: 2011 18th IEEE International Conference on Image Processing (ICIP), 2011, pp. 2893–2896. doi:10.1109/ICIP.2011.6116153.
[9] L. J. Skelly, S. Sclaroff, Improved feature descriptors for 3D surface matching, in: Proc. SPIE 6762, Two- and Three-Dimensional Methods for Inspection and Metrology V, 67620A, 2007, pp. 67620A–67620A–12.
doi:10.1117/12.753263. URL http://dx.doi.org/10.1117/12.753263
[10] A. E. Johnson, M. Hebert, Using spin images for efficient object recognition in cluttered 3d scenes, IEEE Transactions on Pattern Analysis and Machine Intelligence 21 (5) (1999) 433–449.
[11] D. Fehr, A. Cherian, R. Sivalingam, S. Nickolay, V. Morellas, N. Papanikolopoulos, Compact covariance descriptors in 3D point clouds for object recognition., in: ICRA, IEEE, 2012, pp. 1793–1798.
[12] A. Collins, A. Zomorodian, G. Carlsson, L. J. Guibas, A barcode shape descriptor for curve point cloud data, Comput. Graph. 28 (6) (2004)881–894. doi:10.1016/j.cag.2004.08.015.
URL http://dx.doi.org/10.1016/j.cag.2004.08.015
[13] B. Drost, M. Ulrich, N. Navab, S. Ilic, Model globally, match locally: Efficient and robust 3d object recognition, in: IEEE Conference on Computer Vision and Pattern Recognition (CVPR), 2010, pp. 998–1005.
doi:10.1109/CVPR.2010.5540108.
[14] A. Typiak, Use of Laser Rangefinder to detecting in surroundings of mobile robot the obstacles (2008) ISARC 2008 - Proceedings from the 25th International Symposium on Automation and Robotics in Construction, pp. 246-251
[15] J. Bedkowski, A. Maslowski, G. de Cubber, Real time 3D localization and mapping for USAR robotic application, Industrial Robot 39 (5) (2012) 464–474
[16] J. Bedkowski, Parallel Computing In Mobile Robotics For RISE, GPU Technology Conference 2012, May 14-17, San Jose, California, URL:
http://developer.download.nvidia.com/GTC/PDF/GTC2012/PresentationPDF/S0081-GTC2012-Mobile-Robotics-RISE.pdf

SECTION–12

SENSING AND SENSOR FUSION

HUMAN MOTION LOCALIZATION AND INTERACTION BASED ON WEARABLE INERTIAL AND CONTACT SENSORS

QILONG YUAN, I-MING CHEN

School of Mechanical and Aerospace Engineering, Nanyang Technological University, 639798, Singapore

The human body kinematics and kinetics with the environment are the crucial features to correctly represent the human- environment interaction. This paper introduces the framework of the system and method to track the spatial human motion interaction with the environment. In this work, the wearable sensors systems are used to measure the kinematic parameters of the body segment and the contact with the environment. Though monitoring the contacts, and capturing the body joint movement, the spatial location of the person can be updated during contacts. Methods on tracking the spatial location, velocity of a person are also discussed. In general, the system and method is able to fully track the spatial behavior, location and velocity of the person during various motions without volume and terrain limitations. The self-contained solution has litter restriction on the subject's movement. It can be applied to many daily activity related applications such as sports and exercises, personal navigation and entertainment etc.

1. INTRODUCTION

Tracking the human- environment motion interaction with the environment is very useful in many applications like sports/exercises [1, 2, 3, 4, 5, 6], monitoring the activity of kids or patients [7, 8, 9], personal navigation and entertainments [10, 11, 12, 13, 14], etc. To track the movement of human and the interaction with the environment in these applications which take places in outdoor environment or around the house, the capture devices should be able to cover a large volume in various surroundings, and there should be as few effects on the persons' movement. In this situation, the lab-based capture systems (optical, ultrasound, magnetic etc) with fixed small capture volume are not applicable. Therefore, it would be very useful if there are methods to fully track the person's spatial behavior and the interaction with the environment purely based on easily wearable body sensors.

In the past decades, with the development of the MEMS fabrication technology, the IMU sensors are packaged with small size and can provide accurate orientation, angular rate and the acceleration measurements. Thus, this type of sensor is quite suitable for daily living applications.

I-Ming Chen, phone: 67906203, e-mail: michen@ntu.edu.sg;
Qilong Yuan. phone: 65-98965107, e-mail: yu0017ng@ntu.edu.sg

Many companies and research groups have developed their own motion tracking systems using these inertial sensors [10, 15, 16]. However, these systems either don't have the localization function or use external positioning devices to track the persons location. This limits the applications of them for daily practical applicatons. The Xsens company developed the commercial IMU full body Mocap "MVN" and "MVN BIOMECH" [17, 18, 19]. Based on [20], the localization accuracy is moderate. They also haven't show the localization results during motions with flight phases, such as jumping or running.

In outdoor applications, especially sports and personal navigations, the localization accuracy is quite important. The system should be able to continuously track the spatial location and movement. Besides, for walking, jogging, the velocity is a key parameter to analyze the energy consumption and the performance level of the exerciser. The discussion on real-time velocity tracking is currently lacking. Thus, in order to resolve the human spatial motion and interaction tracking problem, we work on the tracking algorithm and the system development to build a self-contained system that can be more applicable in daily applications.

While using the wearable sensors to track the spatial motion of human, the internal joint movement can be easily available based on the orientation measurement from the sensors. The difficulties arise when we want to track the spatial location and the velocity of the subject purely based on the inertial sensor data. If we can monitor the person's contacts with environment, and monitor the body movement of the person, the kinematics of the body and these contacts can be implemented to figure out the location of the person.

In this work, we developed an inertial and contact sensor based motion capture system and the tracking algorithms to track the spatial motion, location and velocity of human during various daily activities. After calibration of the system, through monitoring the contacts with the environment and capture the body movement of the person, the 3D movement of the person in the environment is updated based on the localization and velocity tracking method.

2. Framework

While human acts with interaction with the environment, the kinematics of the human body is very crucial and important. Thus, the geometric parameters of the person: the hierarchy of the skeleton model, the limb dimensions and the angular displacements of the joints etc are key parameters to describe the person's behavior. Also, the features of the environments such as the ground conditions (floor, stairs, uneven etc) and the maps of the environment etc are also important useful. Initially, the system needs to be calibrated to initialize the skeleton model, the initial posture and the terrain constraint. Then the spatial motion tracking can be conducted based on the tracking algorithms.

The overall diagram of the spatial human motion tracking and interaction using pure self-contained wearable system is presented below:

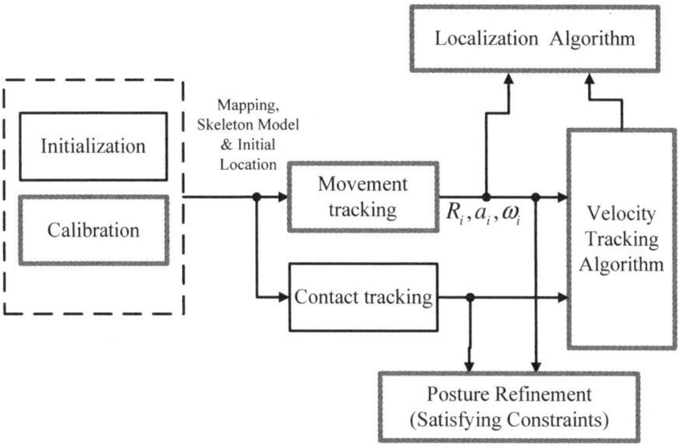

Figure 1: Tracking Process Diagram

2.1. *Initialization*

In order to accurately track the spatial motion of a person and its interaction with the environment using self-contained wearable device (see Figure 4), the initialization of the system and environment is needed. Thus, in the initialization:

(1) Firstly, the skeleton dimensions of the person need to be calibrated. This is completed based on a template based localization method as shown [21]. The 3D skeleton dimensions can be optimized based on several footprint matching process. Basically, this method determines the 3D limb length of the subject to best fit the sensor output with the pre-designed footprint locations. The dimensions of limbs can be solved based on the kinematic equations in these strides.

(2) Secondly, the initial location and pose of the person in the environment need to be defined.

(3) Thirdly, to map the sensor measurements to the body frames, the coordinate mapping between body and sensor frames need to be calibrated [1].

If the floor is regular (even or stairs), the terrains are added as constraints.

2.2. *Constraint Monitoring*

The constraints (mainly the contacts) define the relative pose between the body parts and the environment. To define the contact, the location of the contact point in the environment, P_e and on the body, P_b is needed. Thus, a contact point is described by $\{P_e, P_b\}$. Provided the precise posture of the person, the location of the person can be updated based on the known contact location. On the other hand, if the location of the person is known, the contact location can be estimated based on the kinematics from the root to the contact point.

2.3. *Localization*

After the initialization, the motion tracking and localization can be conducted in real time. For motions with continuous contacts like walking and climbing, as the person start to walk, the capture system can capture the limb posture and measure other kinematic date. With the well calibrated skeleton model and the coordinate mapping between body and sensor frame. The location as well as the velocity of the person can be updated based on the contact points and body kinematics [1, 4, 22]. This is illustrated in Figure 2. The localization result of the person can help us to figure out the feature of environments (the terrain condition, the maps etc.).

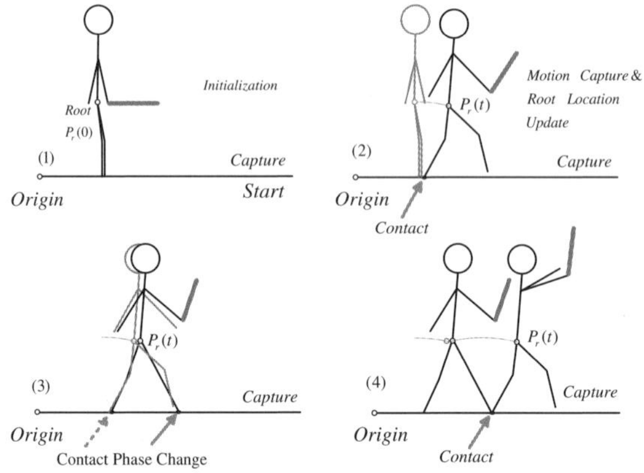

Figure 2: Localization with contacts

However, this is not sufficient for localization during movements with flight phases. Thus, additional information is needed in these situations.

2.4. *Velocity and Dynamic Localization*

During the motion where contacts are not available, for example, while jumping and jogging, there are flight phases without contacts. Therefore, the body motion switch between contact and non-contact. In this case, the localization during the flight phase is very critical.

As there is no absolute position reference available, the localization should be dependent on the kinematic data measured by the wearable system. As shown in Figure 3, if we can accurately monitor the velocity before jumping and accurately measure the root acceleration, then the velocity and location of the root can be updated based on the integrations of the acceleration during the short flight phase within a few seconds. Therefore, accurately tracking the velocity of

Figure 3: Localization without contacts

the root becomes a key issue in this flight localization. This method is discussed in [4].

2.5. *Constraints and Interaction*

During many activities, the key features of the movements and the constraints are also very important during the interaction. For example, when a person stands on the ground and uses his hand to open the door (the hand contacts with the door handle), the standing location as while as the touching point of the person are crucial to correctly represent the actual behavior of the person. The locations of the standing point and the handle defines the relative posture of the foot and the hand. This is a constraint to be satisfied. However, since the model and the sensor measurement errors are inevitable, and the slight errors in the skeleton model and the sensor estimation exist in the model. Thus, purely tracking the body movements will not satisfy the interaction conditions. In these scenarios, the captured posture needs to be refined based on the constraints in order to correctly represent the behavior.

3. DEVICES

The IMU sensors are used as the orientation measurement devices for posture measurement. As shown in Figure 4, a set of IMUs (maximum of 15 for full-body capture) are used to measure the posture of the body. Sensors are tightly attached on the limbs (head, chest, pelvis, upper arms, forearms, hands, shanks, thighs, and feet). This IMU sensors [23] measures the orientation, the angular rate and the acceleration of the body segments.

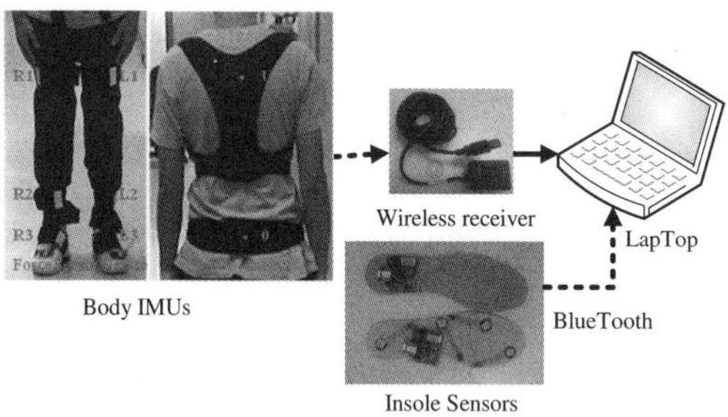

Figure 4: Attachment of sensors

Force Sensing Resistor (FSR) is use to detect the contact event of the limbs. The output is wirelessly sent to the PC through Bluetooth module.

All sensors communicate wirelessly with the PC through Therefore, there is little to the person's movement. It can be conveniently set up and applied for applications in various environments.

4. INTERACTIVE TRACKING

Localization and interactive tracking experiments have been conducted for various movements on different terrain conditions. Walking on the even ground, the stairs, irregular terrains have been tested. The spatial behavior and location of the person as well as the contacts are properly captured.

4.1. *Lab Environment*

The localization in the indoor environment is tested inside the lab. The subject walks inside the lab around three tables and go back to the initial point for several times. The tracked spatial behavior is shown in Figure 6. For each traveled path, the total distance is around 54m. The close loop error is (0.13±0.09m, 0.16±0.11m).

As the location and the behavior of the person are correctly tracked, the person's behavior and the interaction with the environment are correctly represented.

Figure 5: Localization in lab environment

4.2. *Irregular Terrain*

A tracking experiment is conducted around a roundabout. The terrain is not even since the height of the ground varies around the outer profile. The subject walks around the area and go back to the initial place for two times. The spatial behavior of the subject is shown in Figure 6.

It is good to know that after a localization experiment is conducted around the areas of interest, the terrain shape can be recovered based on the results. In Figure 6, the upper edge of the pink surfaces provides the detected ground surfaces during the experiment.

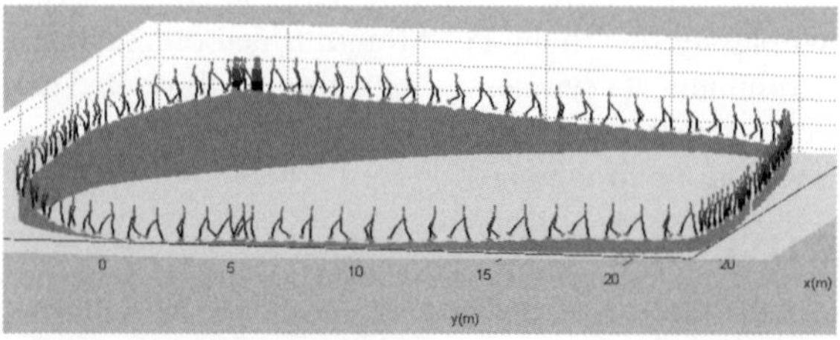

Figure 6: Result of irregular terrain experiment

The total distance for each round is 98.6m, and the close loop error for the each round is less than one meter along all direction.

With the proposed velocity tracking and localization algorithm, the localization and capture of dynamic behaviors such as running and jumping are also tracked [4, 6].

5. CONCLUSION

This paper introduces a system and method to track the human motion interaction with the environment based on inertial and contact sensors. Based on monitoring the interaction contacts, and measuring the kinematic quantities of the person, the 3D location and the posture of the body can be updated in real time.

Experimental results are conducted on different terrain conditions. The person's spatial behavior is correctly tracked. Localization results show that the accuracy is within 1% to 2% of the total traveled distance. The system is able to track the velocity and the location of the person in both slow behaviors and dynamic behaviors with flight phases. Motions like: walking, jumping, running can be accurately captured based on this method.

With the contact information, the correct interaction with the environment is achieved after refining the body posture based the constraints. Thus, the movement, the location and the velocity of the person as well as the correct interaction with the environments are tracked based on this system and method.

As no external assistive device is required for this system, it is easy to set up and calibrate. These characteristics show advantages in implementing this system into our daily activities or sports which take place in large and open terrains. With further improvement of the MEMS technology and more robust tracking algorithm of the inertial sensors, the accuracy of this type of system can be improved such that further improvement of the system and method can be expected to go for highly dynamic sports related applications.

ACKNOWLEDGMENT

This work was supported in part by the Agency for Science, Technology and Research, Singapore, under SERC Grant 092 149 0082 and Singapore Millenium Foundation Grant. The authors would like to thank the technical support received from Dr. Albert, Mr. WS Ang, Ms. LL Liu, Mr. B. Li etc.

REFERENCES

1. Qilong Yuan, I-Ming Chen, and S. P. Lee, "SLAC: 3D Localization of Human Based on Kinetic Human Movement Capture," in *Robotics and Automation (ICRA), 2011 IEEE International Conference on*, Shanghai, China, 2011.
2. O. Bebek, M. Suster, S. Rajgopal, M. Fu, X. Huang, M. Cavusoglu, D. Young, M. Mehregany, A. van den Bogert, and C. Mastrangelo, "Personal navigation via shoe mounted inertial measurement units," in *Intelligent Robots and Systems, IROS. IEEE/RSJ International Conference on*, 2010, pp. 1052-1058.
3. K. King, S. Yoon, N. Perkins, and K. Najafi, "Wireless MEMS inertial sensor system for golf swing dynamics," *Sensors and Actuators A: Physical,* vol. 141, pp. 619-630, 2008.

4. Qilong Yuan and I.-M. Chen, "Human Velocity and Dynamic Behavior Tracking Method for Inertial Capture System," *Sensors and Actuators A: Physical,* 2012.
5. Qilong Yuan and I.-M. Chen, "Human-like walking of humanoid robot based on biped kinematics and captured motion of human," presented at the Proceedings of the 13th World Congress in Mechanism and Machine Science, Guanajuato, MÉXICO, 2011.
6. Qilong Yuan and I.-M. Chen, "Simultaneous Localization and Capture with Velocity Information," in *Intelligent Robots and Systems. IEEE/RSJ International Conference on,* San Francisco, California, 2011.
7. K. Nguyen, I. Chen, Z. Luo, S. Yeo, and H. Duh, "A wearable sensing system for tracking and monitoring of functional arm movement," *Mechatronics, IEEE/ASME Transactions on,* vol. 16, pp. 213-220, 2009.
8. Vicon. (2010). *http://www.vicon.com.*
9. S. Scapellato, F. Cavallo, C. Martelloni, and A. M. Sabatini, "In-use calibration of body-mounted gyroscopes for applications in gait analysis," *Sensors and Actuators A: Physical,* vol. 123, pp. 418-422, 2005.
10. D. Vlasic, R. Adelsberger, G. Vannucci, J. Barnwell, M. Gross, W. Matusik, and J. Popovi "Practical motion capture in everyday surroundings," *ACM Transactions on Graphics (TOG),* vol. 26, pp. 35-44, 2007.
11. E. Bachmann, "Inertial and magnetic tracking of limb segment orientation for inserting humans into synthetic environments.," PhD Thesis, Naval Postgraduate School Monterey CA., 2000.
12. H. Luinge, "Inertial sensing of human movement," PhD Thesis, Universiteit Twente, 2002.
13. D. Roetenberg, H. Luinge, and P. Slycke, "Xsens MVN: full 6DOF human motion tracking using miniature inertial sensors," Xsens Technologies,2009.
14. Y. Xiaoping and E. R. Bachmann, "Design, Implementation, and Experimental Results of a Quaternion-Based Kalman Filter for Human Body Motion Tracking," *Robotics, IEEE Transactions on,* vol. 22, pp. 1216-1227, 2006.
15. M. A. D. Brodie, "Development of fusion motion capture for optimization of performance in alpine ski racing," PhD Thesis, Massey University, 2009.
16. MetaMotion. (2013). *http://www.metamotion.com.*
17. T. Cloete and C. Scheffer, "Benchmarking of a full-body inertial motion capture system for clinical gait analysis," in *30th Annual International IEEE EMBS Conference,* Canada, 2008, pp. 4579-4582.
18. R. Damgrave and D. Lutters, "The drift of the xsens moven motion capturing suit during common movements in a working environment," in *Proceedings of the 19th CIRP Design Conference,* 2009, pp. 338-342.
19. X. Technologies. (2010). *http://www.xsens.com.*
20. S. A. Skogstad, K. Nymoen, Y. de Quay, and A. Jensenius, "OSC Implementation and Evaluation of the Xsens MVN suit," in *Proceedings of New Interfaces for Music Expression, Oslo, Norway,* 2011.
21. Qilong Yuan, I-Ming Chen, and A. W. Sin, "Method to Calibrate the Skeleton Model Using Orientation Sensors," in *Robotics and Automation (ICRA), 2013 IEEE International Conference on,* Karlsruhe, Germany, 2013.
22. Qilong Yuan and I.-M. Chen, "3D Localization of Human Based on Inertial Capture System," *Robotics, IEEE Transactions on,* vol. Volume 29, 2013.
23. InterSense. (2011). *www.intersense.com.*

REAL-TIME DETECTION OF THE ACTIVITY OF A DOG

GERMAIN LEMASSON
UMR 6285 – Lab-STICC, Université de Bretagne Sud, France

PHILIPPE LUCIDARME
LISA – Université d'Angers, France

DOMINIQUE DUHAUT
UMR 6285 – Lab-STICC, Université de Bretagne Sud, France

This paper introduces our preliminary work with assistance dogs. Even when dogs are very well trained some problems may occur in practice, typical examples are dog escaping or running after a cat. Our long term objective is to take advantage of the technology to increase the safety of the dog and its owner. Our first work focuses on the activity classification of the dog. This paper presents preliminary results for recognizing four types of activity: walk, run, lay and sit down. Experiments and results on real data collected with low-cost gyroscopes and accelerometers are presented and discussed.

1. Introduction

Dog is the companion of human since 14000 years [1]. The man quickly recognized his ability and gave him some task. The first working dogs were guard, hunting dogs or sheepdogs. But gradually as the human society has evolved dogs have evolved within it, leading to the creation of several breed. New tasks were given to the dogs like police dogs, urban search and rescue dogs or detection dogs. It is only relatively recently that appeared a new type of dog, the service dog. The first ones appeared after WWI for blind soldiers; it is the guide dog that we know today [2]. And more recently dogs are trained for assisting person with reduced mobility. These dogs can help people in their daily life, they can do a large number of actions such like pulling a wheelchair, bringing something, picking up or holding stuff, switching light and much more. They are also an extraordinary psychological support and there owners become more autonomous. Another type of service dog appeared in 1997, dogs for mentally disabled person like autistic children [3]. It allows the child to socialize better and prevent him from endangering himself, for example on a ballad he prevents the child from running on the road. The demand for these two types of working dog is growing. Handi'Chien a French association, has trained more than 1,000 dogs in 20 years and gets more and more demand every year [4]. These dogs are trained from birth, and the training lasts 2 years. The first 6 months of their life, they spend it in a foster family where they socialize and learn basic orders. Then during 18 months they are trained in specialized centers where they learn about fifty orders. At the end of their training, they are given free of charge to their new owner. The persons asking for a dog are evaluated to see if they could actually take care of the dog, they were also given a short training on how to act with their dogs. Despite these precautions, unfortunately there still are some failures, the master does not necessarily have enough the authority or the dog cannot adapt [5]. It sometimes becomes complicated; the dog passes from one master, their coach, who is dynamic and active, to a disabled person who has less mobility. It is from this observation that is born the project on which our team is currently working. The fundamental questionof this project is: is it possible to increase the bond between dog and master device using robotics, mechatronics? Vernay et al. [4] discuss some scenario of interaction between a disabled person, a service dog and a robot. The first step of our project is to propose tools that might be interesting to use. We first thought of the remote detection of the dog's activity. This may be useful in two situations, when the dog is out of view during free time or if he escapes of control. Or if it keeps a mentally disabled person, combined with other tools, the dog could alert a third party if his master is endangered.

This paper describes the preliminary study we conducted and shows these results. In the first part we discuss previous works. In a second part, our work and preliminary results are presented. Before ending with a general conclusion and discussion about future works.

2. Related works

The use of electronic equipment on dog is quite novel; the best known is certainly the bark control collar. However at first we don't want work with negative stimuli or punishment. They are also some GPS collar that enables the owner of a dog to track it. We will certainly use GPS on our device but we chose to start with the detection of the activity of the dog. Ribeiro et al. [6] proposed a method to detect the pose of USAR dogs. It uses two 1- axis accelerometer to find the poses of USAR dogs. There method detects the transition between the poses. Results on static poses are really encouraging.

Inspired by this previous work we'll present our method which detects the static poses of the dogs but also his activity like walking and running. For this preliminary study we chose to detect only four basic activities: walking, running, laying down, sitting down.

3. Detection of the activity

The activities detection is the first tool of a set that we want to experiment. Some of the other tool will be GPS tracking, video streaming, and remotely giving orders. We rapidly decided to use a smartphone on the dog. It has numerous advantages: it is embed a small and powerful processor that already have many built-in devices, like GPS, camera, speaker, sensors, and communication capacities GSM, Bluetooth, Wi-Fi... So unlike Ribeiro et al. [6], which use two 1- axis accelerometer, we decided to take advantage of the sensor of the phone, in our case: 3-axis accelerometer, 3-axis gyroscope, 3-axis magnetometer.

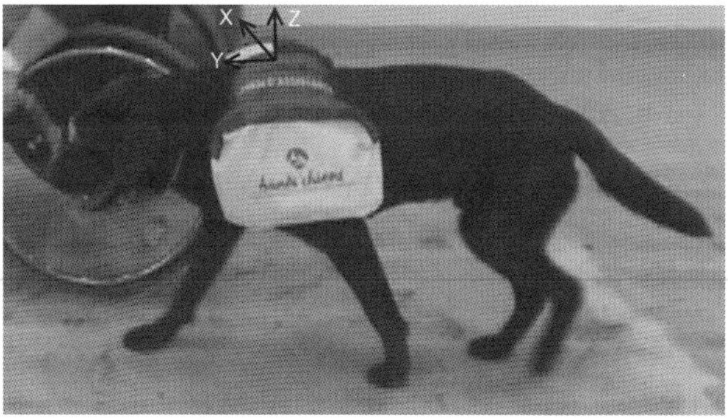

Figure 1 Dog with a smartphone on is back.

The smartphone is placed on the back of the dog. The orientation of the sensors is shown in Figure 1. Y to the dog's head, X to his right side and Z perpendicular to his back.

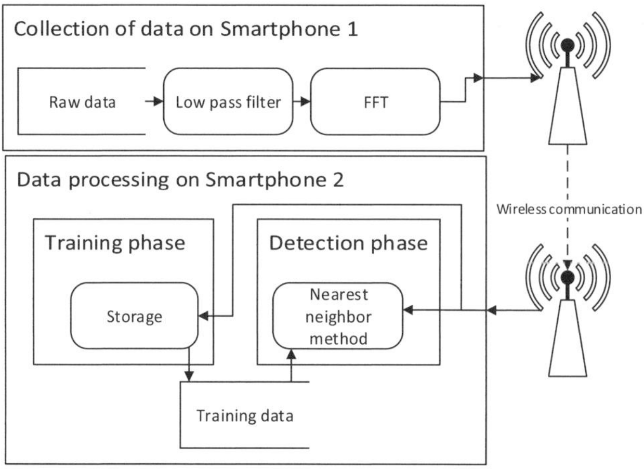

Figure 2 Data flowchart

Figure 2 describe the data flowchart of the presented method. Data are collected by the smartphone 1 on the back of the dog then sent to the smartphone 2 in the hand of the owner of the dog which performs the detection. The data collection and the data processing are explained in the following sections.

3.1. *Collection of data*

Among the nine available data (three axes for each of the three sensors), we chose to use only 2 of them: the accelerometer z-axis and the gyroscope z-axis. A first experiment on one of the dog of Handi'chien allows us to determine the significant data. We put the smartphone on the back of the dog and the trainer asks him to walk, run, sit down and lay down. After we analyzed the collected data and conclude that the accelerometer z-axis and the gyro z-axis are sufficient to determine our four basic activities. The single z-axis accelerometer can even be sufficient but we chose to add the z-axis gyroscope for more reliability.

Table 1 shows the collected data. The x-axis of the curves is the times in second (3 seconds) and the y-axis is the values returned by the z-axis of the sensors.

We clearly see the periodicity for the walk and run. And the mean of the accelerometer z-axis is higher when the dog is lying down than when it is sitting down due to the tilt of the smartphone.

In order to extract the periodicity of the movement, data are filtered with a low pass filter and a FFT (Fast Fourier Transform) [7] is performed on the two chosen axis. Then the maximum amplitude and its corresponding frequency are extracted. This provides four values. A fifth value is calculated: the average of the accelerometer z-axis. The window used for the processing is 2 second. It contains twice the pattern of the walk.

This processing is done locally in real time by the smartphone on the back of the dog. Every 2 seconds it sends wirelessly the fifth data to a second smartphone which perform the detection.

3.2. *Data processing*

In order to detect the activities we use the method of the nearest neighbor. It is a supervised learning method so a calibration phase is done in order to collect the training data. The owner of the dog makes him walk for 20 seconds, during this time all recovered data are stored. The operation is repeated for running, sitting position and lying position. Then we have 10 sets of data for each activity. The calibration last no more than 2-3 minutes. At the beginning of the detection phase, the calibration data are scaled and placed in a 5-dimensional space. Then each received data is scaled and we use the Minkowski distance (generalization of the Euclidean distance) to find its nearest neighbor. The activity is determined by the activity of its nearest neighbor.

After the collection of data on a second dog we saw that the curves were quite different.

Table 1 Data collected

	Walk	Run
z-axis accelerometer	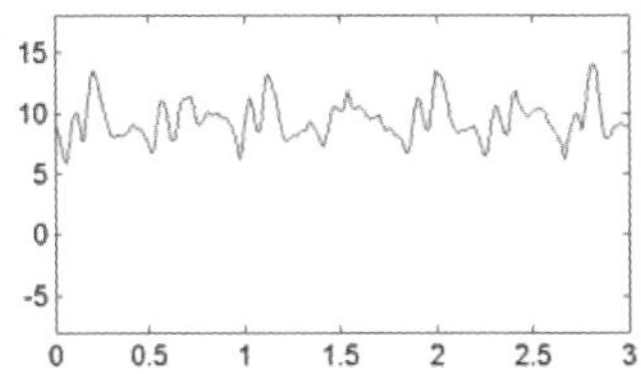	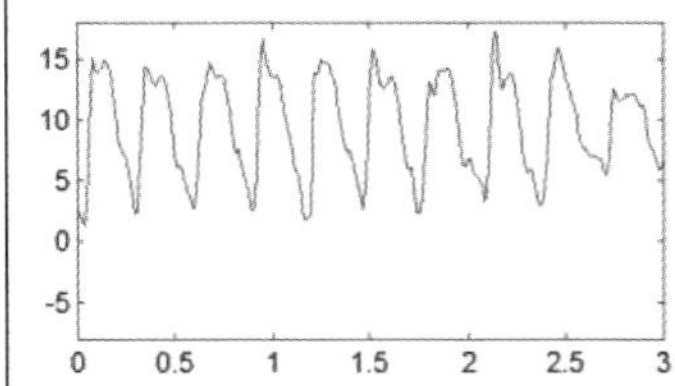
z-axis gyroscope	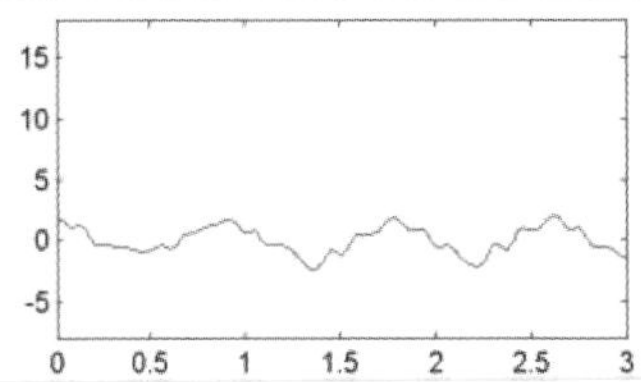	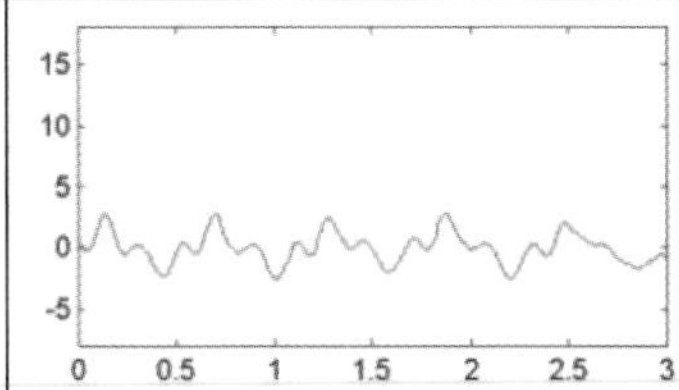

	Lay down	Sit down
z-axis accelerometer	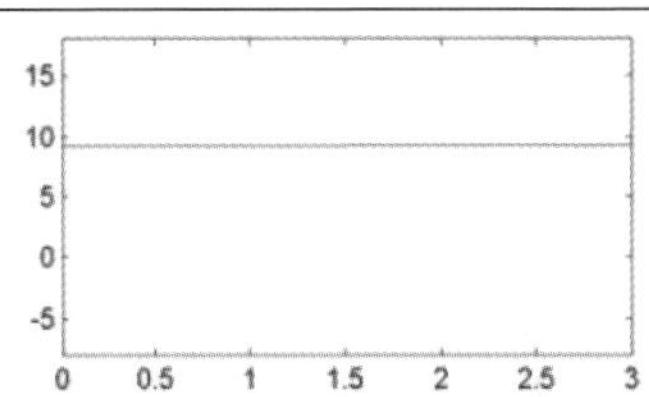	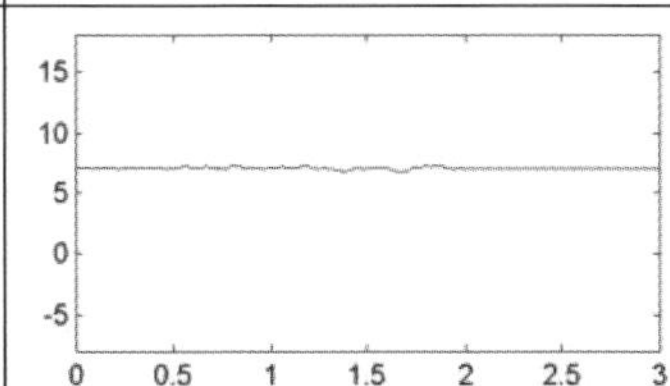
z-axis gyroscope	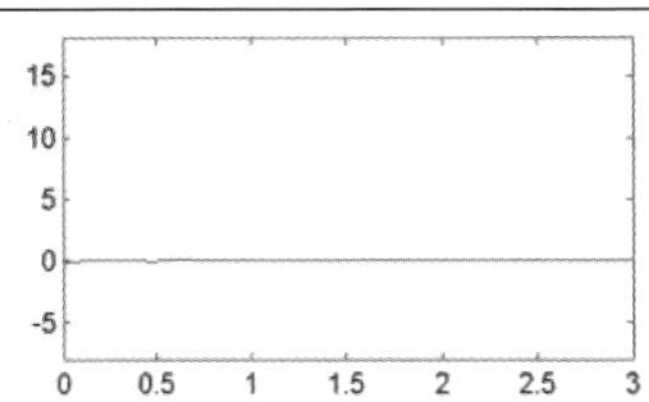	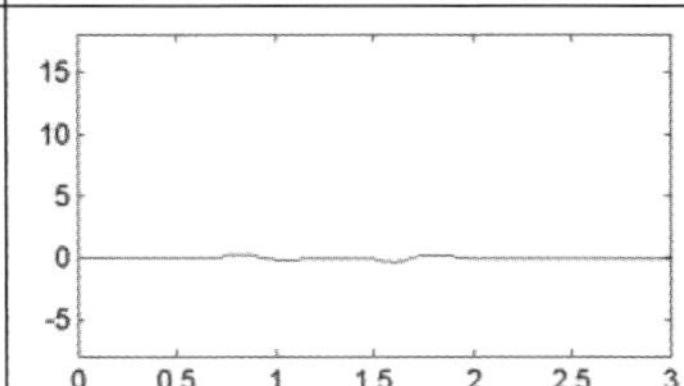

Table 2 shows the mean value of the data collected on the two dogs for the run. We can clearly see differences on frequency on the gyroscope. It confirms the Ribeiro et al. [6] hypothesis: result are (highly) dependent on the placement of the sensor. It is difficult to place the sensor at the same position on dogs that have different size and morphology. So we chose to do the calibration phase on each dog. We believe it may be relevant in practice, configuring the equipment of a dog last several minute (in regard to the training that last two years).

Table 2 Mean frequency and mean amplitude of the run

	Dog 1	Dog 2
Mean max frequency accelerometer	2.8035	2.8754
Mean max amplitude accelerometer	0.4983	0.6654
Mean max frequency gyroscope	1.5109	2.4624
Mean max amplitude gyroscope	0.2585	0.2611

4. Sample results

We tested our method for 8 minutes during which we ask the dog to walk, run, sit down and lie down. The activity changed every 30 seconds.

Table shows the result our experimentation. Errors in the lying position are mainly due to the position of the smartphone on the harness. Indeed, when the dog was lying but lifts high enough his head the harness and the smartphone fell slightly like if the dog was sitting down. The overall percentage of detection on our experiment is 83 %.

Table 3 Experimentation results

Activity	Correct	Incorrect	
Walk	55	5	91%
Race	50	10	83%
Sitting position	52	8	86%
Lying position	45	15	75%

5. Conclusion and future work

In this paper we presented a preliminary survey on dog's activities detection. The method uses 2 smartphones with a short calibration phase. The preliminary results, based on real set of data, are quite encouraging even if it's still early for a generalization.

In our future works we will try other types of classification such as SVM [8] or Artificial Neural Network [9]. We will also consider other data of the sensors, other axis, with the aim of expanding the number of detected activities. We also plan to experiment on more dogs to see how results can be generalized. Our middle term objective is to combine it with other tools such as GPS tracking, bark detection, video streaming… any ideas are welcome!

6. Acknowledgments

This work is supported by French National Research Agency (ANR-012-BLAN).

References

[1] A. Prestrude, "Dogs in service to humans," *Comparative psychology: A handbook*, pp. 386–392, 1998.

[2] E. Weiss and G. Greenberg, "Service dog selection tests: Effectiveness for dogs from animal shelters," *Applied Animal Behaviour Science*, vol. 53, no. 4, pp. 297–308, Jul. 1997.

[3] K. E. Burrows, C. L. Adams, and S. T. Millman, "Factors Affecting Behavior and Welfare of Service Dogs for Children With Autism Spectrum Disorder," *Journal of Applied Animal Welfare Science*, vol. 11, no. 1, pp. 42–62, 2008.

[4] D. Vernay, P. Rybarczyk, M.-C. Lebret, and Y. Rybarczyk, "Collaboration hommes, chiens et robots: quels scenarios?," 2011.

[5] R. Coppinger, L. Coppinger, and E. Skillings, "Observations on Assistance Dog Training and Use," *Journal of Applied Animal Welfare Science*, vol. 1, no. 2, pp. 133–144, Apr. 1998.

[6] C. Ribeiro, A. Ferworn, M. Denko, and J. Tran, "Canine Pose Estimation: A Computing for Public Safety Solution," in *Canadian Conference on Computer and Robot Vision, 2009. CRV '09*, 2009, pp. 37 –44.

[7] R. N. Bracewell, *The Fourier transform and its applications*, vol. 31999.

[8] M. A. Hearst, S. T. Dumais, E. Osman, J. Platt, and B. Scholkopf, "Support vector machines," *IEEE Intelligent Systems and their Applications*, vol. 13, no. 4, pp. 18–28, 1998.

[9] K. Mehrotra, C. K. Mohan, and S. Ranka, *Elements of artificial neural networks*. MIT Press, 1997.

ESTIMATION OF THE TRUNK ATTITUDE OF A HUMANOID BY DATA FUSION OF INERTIAL SENSORS AND JOINT ENCODERS

SIDDHARTHA KHANDELWAL

Intelligent Systems Laboratory, Halmstad University,
Halmstad, Box 823, S-301 18, Sweden
E-mail: siddhartha.khandelwal@hh.se

CHRISTINE CHEVALLEREAU

IRCCYN, CNRS, Ecole Centrale de Nantes,
Nantes, Cedex 3 - 44321, France
E-mail: christine.chevallereau@irccyn.ec-nantes.fr

The major problem associated with the walking of humanoid robots is to maintain its dynamic equilibrium while walking. To achieve this one must detect gait instability during walking to apply proper fall avoidance schemes and bring back the robot into stable equilibrium. A good approach to detect gait instability is to study the evolution of the attitude of the humanoid's trunk. Most attitude estimation techniques involve using the information from inertial sensors positioned at the trunk. However, inertial sensors like accelerometer and gyro are highly prone to noise which lead to poor attitude estimates that can cause false fall detections and falsely trigger fall avoidance schemes. In this paper we present a novel way to access the information from joint encoders present in the legs and fuse it with the information from inertial sensors to provide a highly improved attitude estimate during humanoid walk. Also if the joint encoders' attitude measure is compared separately with the IMU's attitude estimate, then it is observed that they are different when there is a change of contact between the stance leg and the ground. This may be used to detect a loss of contact and can be verified by the information from force sensors present at the feet of the robot. The propositions are validated by experiments performed on humanoid robot NAO.

Keywords: Humanoid walking; Dynamic equilibrium; Attitude estimation; Instability detetection; Sensor fusion; Inertial sensors; Joint encoders.

1. Introduction

There has been increasing academic and commercial interest in humanoid robots for various applications since the recent significant advancements in robot control technology. An effective trunk attitude control is one of the most important factors for a humanoid to continue dynamically stable walking which is necessary to accomplish any given task. Most indoor robots use inertial sensors and vision, or other external signals and cues, for determining their position and orientation. In most cases, humanoid robots like NAO use MEMS based inertial sensors to calculate its attitude due to cost, payload and other limitations. Most inertial measurement units (IMUs) consist of 3-axis gyros and accelerometers which measure the angular rate and specific forces of the vehicle respectively. Hence these IMUs do not need external sources for attitude measurement. Conventionally, attitude is computed by integrating the angular rate obtained from gyro readings. However, this method is inappropriate as the gyro bias error makes the attitude error to diverge.[1] On the other hand the attitude calculated by accelerometers is heavily influenced by mechanical disturbances caused by the movement of the robot. Accordingly, a number of fusion algorithms have been developed like Complementary filter method,[2] direction cosine matrix method,[3] Euler angle update method[4] and modified versions of linear and extended Kalman Filter (EKF)[5 6 7] in particular. In all these methods, different filtering or fusion techniques are applied to minimize the error in IMU measurements. However, accuracy of the attitude thus calculated is still poor considering the high level of noise present in the inertial sensor readings. There is a need to gather more information from other sensors to improve the accuracy of the estimated attitude for a better trunk control capability. We propose that the information from joint encoders present in the legs can be used to measure the trunk's attitude during different phases of humanoid walk. This extra sensor measurement, which is readily avaliable, is combined with the IMU data using an EKF to provide a highly improved estimate of the trunk's attitude. Another outcome is that if the joint encoders' attitude measure is compared separately with the IMU's attitude estimate, then it is observed that they are different when there is a change of contact between the stance leg and the ground. This may be used to detect a loss of contact and can be compared and validated by the information from force sensors located at the feet of the humanoid.

In this paper, the process model of the EKF is developed using the gyro data. Then, indivudial measurement models are developed for accelerometer and joint encoders for different phases of the humanoid walk. These individual measurement models are then combined to construct a general Measurement Model for the EKF. Different experiments are done on humanoid robot NAO and the results are discussed to show the relevance of using joint encoder data for obtaining an improved trunk attitude estimate.

2. EKF for Estimating the Trunk Attitude

A Kalman filter provides a simple yet effective way to fuse all sensor information that is provided to it.[8] We use an EKF as the described process is non-linear. The notations of Welch and Bishop[5] are used for constructing the EKF.

The orientation of the non-inertial frame n, attached to the trunk of the humanoid, relative to the inertial reference frame b, attached to the ground, can be described in terms of three consecutive rotations through three body-referenced Euler angles.[9] These are commonly written as ψ, θ and ϕ which are the yaw, pitch and roll angles respectively. In this paper, the states variables of the EKF provide a measure of only θ and ϕ. They are enough to detect the instability at the trunk. The yaw angle diverges as there is no other measurement to compensate for the gyro bias erros.[4] The overall transformation ${}^{n}R_b$ from frame b to n is calculated[49] and the elements of last row of ${}^{b}R_n$ are chosen as the state variables (x_1, x_2, x_3):

$$^{n}R_b = {}^{b}R_n^T = \begin{bmatrix} 1 & 0 & 0 \\ 0 & cos\phi & sin\phi \\ 0 & -sin\phi & cos\phi \end{bmatrix} \begin{bmatrix} cos\theta & 0 & -sin\theta \\ 0 & 1 & 0 \\ sin\theta & 0 & cos\theta \end{bmatrix} \begin{bmatrix} cos\psi & sin\psi & 0 \\ -sin\psi & cos\psi & 0 \\ 0 & 0 & 1 \end{bmatrix} \quad (1)$$

$$\therefore \quad x_1 = -sin\theta; \quad x_2 = sin\phi cos\theta; \quad x_3 = cos\phi cos\theta \quad (2)$$

Angles θ and ϕ are calculated from the state variables as $\theta = -sin^{-1}x_1$ and $\phi = tan^{-1}x_2/x_3$ respectively.

2.1. *Process Model of the EKF*

The process model is based on gyro measurements. There exists a relationship between the angular rates in the x, y, z axis, i.e ω_x, ω_y, ω_z and the time rate-of-change of the Euler angles:[9]

$$\begin{bmatrix} \dot{\phi} \\ \dot{\theta} \\ \dot{\psi} \end{bmatrix} = \begin{bmatrix} 1 & sin\phi tan\theta & cos\phi tan\theta \\ 0 & cos\phi & -sin\phi \\ 0 & sin\phi sec\theta & cos\phi sec\theta \end{bmatrix} \begin{bmatrix} \omega_x \\ \omega_y \\ \omega_z \end{bmatrix} \quad (3)$$

Differentiating the state variables in eq. (2) and using eq. (3), the state derivatives can be written in matrix form (refer eq. (4)) where u is skew-symmetric matrix of the angular rates and w represents process noise. This can then be expressed in discrete time as shown in eq. (5)(for $i = 1, 2, 3$).

$$\begin{aligned} \left[\dot{x_1}\ \dot{x_2}\ \dot{x_3}\right]^T &= \left[u_{\text{skew-symmetric}(\omega)}\right]\left[x_1\ x_2\ x_3\right]^T + w(t) && (4) \\ \Rightarrow x_{i,k} &= [I + (u_{k-1} + w_{k-1})\Delta t]x_{i,k-1} \\ &= f_i(x_{i,k-1}, u_{k-1}, w_{k-1}) && (5) \end{aligned}$$

The state variables so chosen meet a constraint, i.e. $x_1^2 + x_2^2 + x_3^2 = 1$. Imposing this constraint we obtain the process model as:

$$x_{i,k} = \frac{f_i(x_{i,k-1}, u_{k-1}, w_{k-1})}{\|f_i(x_{i,k-1}, u_{k-1}, w_{k-1})\|} \tag{6}$$

2.2. *Measurement Model of the EKF*

Individual measurement models are constructed for accelerometer and joint encoders for single and double support and then combined based on the type of support the humanoid has with the ground to form a general measurement model. (All variables of the measurement model are expressed in discrete time and have a subscript k.)

2.2.1. *Accelerometer*

The accelerometer is not used to measure the real acceleration of the robot but to measure the trunk attitude assuming that the IMU (or trunk) is in steady state. Here it measures only the gravity vector expressed in the frame n, as this frame is attached to the IMU. The effect of trunk velocity and acceleration are included in accelerometer noise v_{acc}. Thus the accelerometer measurements can be expressed as a function of the state variables (using eq. (2)) which gives the accelerometer measurement model as:

$$\begin{aligned} \left[acc_x \; acc_y \; acc_z\right]^T &= -g\left[-sin\theta \; sin\phi cos\theta \; cos\phi cos\theta\right]^T + v_{\text{acc}} \\ &= -g\left[x_1 \; x_2 \; x_3\right]^T + v_{\text{acc}} \\ \Rightarrow g_{\text{Acc}}(acc_x, acc_y, acc_z) &= h_{\text{Acc}}(x_1, x_2, x_3, v_{\text{acc}}) \end{aligned} \tag{7}$$

2.2.2. *Joint Encoders - Single Support (SS)*

The joint values are provided by the joint encoders present at the hip and legs. Modelling of the humanoid is done using the DH method[10] and the notation of Khalil[11] is used. (For eg. refer Fig. 1a.) In SS, the humanoid is considered as a serial structure from the stance foot to the trunk. By computing the DH table for each leg, the transformation matrix ${}^bT_n$ is calculated using the Direct Geometric Model (DGM). It must be noted that the rotation matrices ${}^bR_0$ and ${}^{lf}R_n$ must be known (frame lf is attached to the last link of the leg). The rotation matrix ${}^bR_n$ is simply the first 3 rows and columns of ${}^bT_n$:

$${}^bR_{n,\text{Leg}} = {}^bR_0 \;\; {}^0R_{lf}(q_i, v_{\text{joint}}) \;\; {}^{lf}R_n \tag{8}$$

where q_i are the joint angles used to calculate the DGM and v_{joint} is the noise at each joint. This rotation can also be represented using Euler angles. Hence comparing the last row of ${}^bR_{n,\text{Leg}}$ (R_{31}, R_{32}, R_{33}) with last row of ${}^bR_n$ in eq. (1) and substituting state variables using eq. (2); the joint encoder model for SS is obtained as:

$$\begin{aligned} \left[R_{31}\ R_{32}\ R_{33}\right]^T &= \left[x_1\ x_2\ x_3\right]^T \\ \Rightarrow g_{\text{Leg}}(q_i, v_{\text{joint}}) &= h_{\text{Leg}}(x_1, x_2, x_3) \end{aligned} \tag{9}$$

2.2.3. *Joint Encoders - Double Support (DS)*

In DS since both feet are on the ground, the trunk along with two legs acts as a closed-loop structure. The joint encoder model for DS is a combination of the joint encoders present on both the legs:

$$\begin{aligned} \left[{}_lR_{31}\ {}_lR_{32}\ {}_lR_{33}\ {}_rR_{31}\ {}_rR_{32}\ {}_rR_{33}\right]^T &= \left[x_1\ x_2\ x_3\ x_1\ x_2\ x_3\right]^T \\ \Rightarrow g_{\text{BothLegs}}({}_lq_i, {}_r q_i, v_{\text{joint}}) &= h_{\text{BothLegs}}(x_1, x_2, x_3) \end{aligned} \tag{10}$$

where the left subscripts l and r stand for the left and right leg respectively. But this closed-loop structure leads to kinematic coupling between the joints of both legs. Hence Jacobian V_{BothLegs} (used to compute the Kalman Gain) in closed loop is given as:

$$V_{\text{BothLegs}} = \begin{bmatrix} {}_lV_{\text{Leg}} & {}_{lr}V_{\text{Leg}} \\ {}_{rl}V_{\text{Leg}} & {}_rV_{\text{Leg}} \end{bmatrix} \tag{11}$$

The terms ${}_lV_{\text{Leg}}$ and ${}_rV_{\text{Leg}}$ are calculated from h_{Leg} as in eq. (9) for the given leg but the terms ${}_{lr}V_{\text{Leg}}$ and ${}_{rl}V_{\text{Leg}}$ are calculated using the kinematic model of closed-loop robots.[12] Inertial frame b is taken as the root of the closed-loop and trunk frame n is attached to the cut joint k (at the hip) such that the two legs act as the two branches of the loop. Applying the kinematic constraint at cut joint k:

$$\left[v\ \omega\right]^T = {}_lJ\ {}_l\partial q_i = {}_rJ\ {}_r\partial q_i \tag{12}$$

Using eq. (12), the terms ${}_{lr}V_{\text{Leg}}$ and ${}_{rl}V_{\text{Leg}}$ are derived and substituted back in eq. (11) to get V_{BothLegs}:

$${}_{lr}V_{\text{Leg}} = {}_lV_{\text{Leg}} \frac{{}_l\partial q_i}{{}_r\partial q_i} = {}_lV_{\text{Leg}}\ {}_lJ^{-1}\ {}_rJ \tag{13}$$

$$\therefore\ V_{\text{BothLegs}} = \begin{bmatrix} {}_lV_{\text{Leg}} & {}_lV_{\text{Leg}}\ {}_lJ^{-1}\ {}_rJ \\ {}_rV_{\text{Leg}}\ {}_rJ^{-1}\ {}_lJ & {}_rV_{\text{Leg}} \end{bmatrix} \tag{14}$$

2.2.4. *Generalized Measurement Model of the EKF*

Unlike accelerometer measurements which can be always considered, joint encoder measurements can be used only in certain phases of the humanoid walk. This is because of the fact that the rotation matrix ${}^bR_0$ is unknown when the foot is not in complete support (see section 2.2.2). Hence the type of contact a foot has with the ground determines the possibility to calculate the DGM of that leg. The FSRs present at the feet help to determine three types of contact, namely; SS where one foot is in full contact with the ground while the other foot is completely in air; DS where both feet are in full contact with the ground and Intermediate Support (IS) which is any other type of foot contact other than SS and DS. For IS, ${}^bR_0$ is unknown but for SS and DS, ${}^bR_{0,\mathrm{Leg}}$ and ${}^bR_{0,\mathrm{BothLegs}}$ are known respectively. Hence depending on the type of contact, different measurements models are constructed by simply combining the individual sensor measurement models that have been derived earlier. It is assumed that there is no slipping or sliding of the legs.

For IS, the general measurement model is simply the accelerometer model as in eq. (7). For SS, the general measurement model (refer eq. (15)) is a combination of accelerometer and joint encoder model of the supporting leg derived using eqs. (7) and (9). For DS, the general measurement model (refer eq. (16)) is a combination of accelerometer and joint encoder model for both legs derived using eqs. (7) and (10).

$$\left[g_{\mathrm{Acc}}\ g_{\mathrm{Leg}}\right]^T = h_{\mathrm{SS}}(x_1, x_2, x_3, v_{\mathrm{noise}}) \tag{15}$$

$$\left[g_{\mathrm{Acc}}\ g_{\mathrm{BothLegs}}\right]^T = h_{\mathrm{DS}}(x_1, x_2, x_3, v_{\mathrm{noise}}) \tag{16}$$

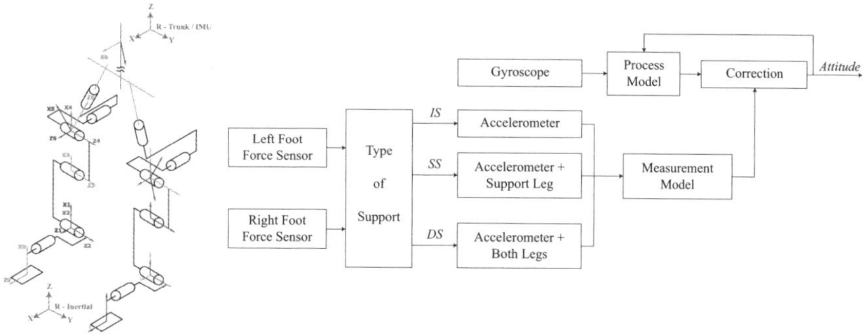

Fig. 1. (a) Modeling the legs and hip of NAO. (b) An overview of the proposed algorithm to estimate the trunk attitude by EKF.

3. Results and Discussion

The developed EKF is applied and tested on the humanoid robot NAO. Three different types of experiments are conducted, namely; 'Bending' where the humanoid bends its trunk with both feet in support by 15 deg, stays in this posture and then stands vertically again; 'Walking' where the humanoid walks slowly for a few seconds and then stops and 'Disturbed Walk' where the humanoid is slightly pushed after it starts walking but the push in not strong enough to make it fall. After a few sec. of disturbed walking, it stops and stands still. In all experiments the humanoid starts and stops in a vertical standing position.

3.1. *Comparing Joint Encoder and IMU Attitude Estimates*

3.1.1. *Joint Encoder Data Accuracy (Exp - Bending)*

As seen in Fig. 2, the attitude measurement by either the left or right leg joint encoders is quite accurate. In the entire movement since there is no change in roll angle, ϕ should be zero. The spikes seen in the roll angle calculated by the joint encoders is due to the sliding of legs. The left leg slides a lot more than the right. However, due to the noise in IMU data the estimated angle ϕ diverges.

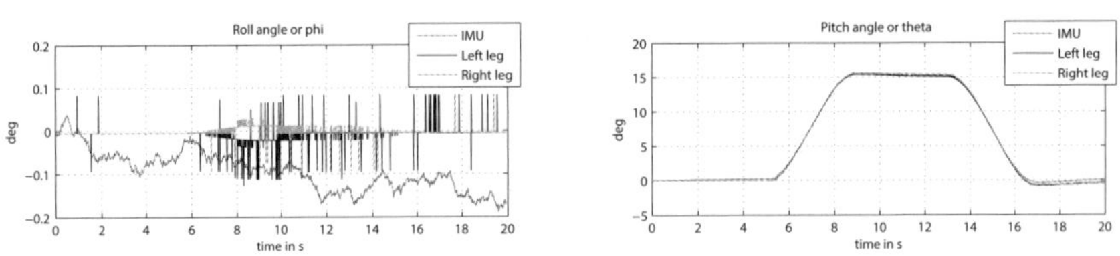

Fig. 2. Trunk attitude estimation by IMU and joint encoders for Bending.

3.1.2. *Detecting Type of Contact (Exp - Walking)*

It is observed that the attitude estimate given by the IMU and the joint encoders' attitude measure are different when there is a change of contact between the stance leg and the ground. As seen in Fig. 3, the black line (left leg joint encoders' attitude measure) approximately follows the red line (IMU's attitude estimate) when the left leg is in support and stops following the red when left foot goes from stance to swing. At this instant the green line (right leg joint encoders' attitude measure) starts following the red line indicating right leg support. The support phases can be verified by the FSR information from the feet. We believe that the presence of 'rare' left foot SS as shown by FSR is either due to the left foot FSR not working properly or too much sliding of the left foot or both reasons.

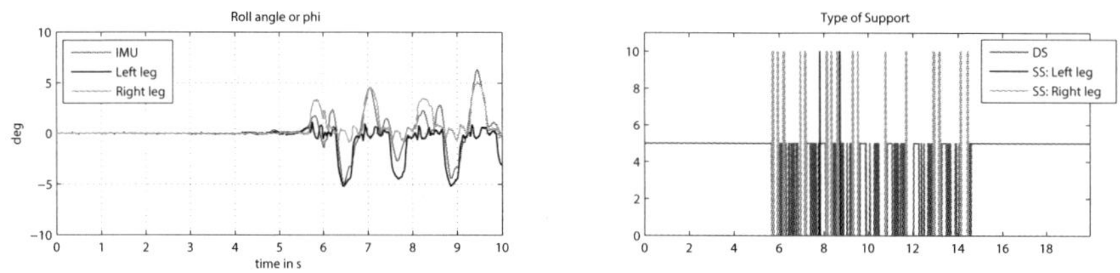

Fig. 3. Change of support by comparing IMU attitude estimate with joint encoders' attitude measure. ('Types of Support' is given by FSRs at feet: SS = 10; DS = 5.)

3.2. *Attitude Estimation by Data Fusion using EKF*

3.2.1. *Walking*

Even during slow walking, the inertial sensors are prone to noise whose effect can be clearly seen in the pitch angle θ (refer Fig. 4) which overshoots sometimes and does not come back to zero even when motion stops. Data fusion helps to improve the attitude estimate. Another observation seen in the roll angle is the difference for support on right and left leg (clear after 10 sec) which is probably connected to sliding of one leg which we believe to be the left leg as indicated by the spikes in roll angle in Fig. 2.

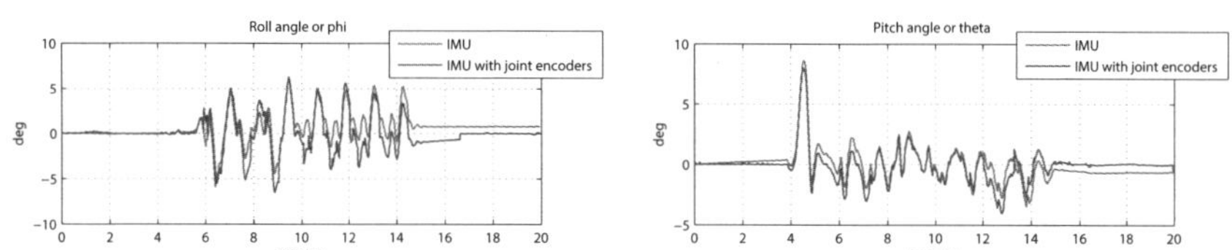

Fig. 4. Attitude estimation of trunk by EKF for Walking.

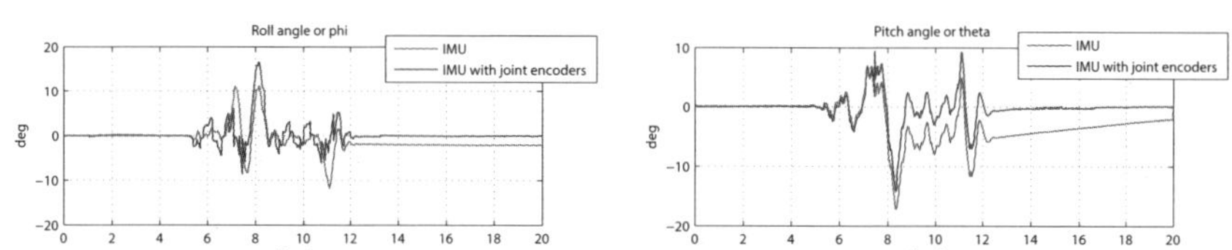

Fig. 5. Attitude estimation of trunk by EKF for Disturbed Walk.

3.2.2. *Disturbed Walk*

As shown in Fig. 5 the effect of noise that is introduced by the external push (at approx. 7s) is clearly visible in the estimation of roll and pitch angles by using only IMU data. Even after the motion stops, it takes a while for the IMU's attitude estimate to come back to zero. This can lead to a false fall detection. However, data fusion with joint encoders help to correct the effect of inertial noise to give an improved attitude estimate of the humanoid trunk.

4. Conclusion

Attiude estimates from inertial sensors that form a part of the IMU cannot be completely trusted as they suffer heavily from various kinds of noise. It is understood that there shall be many external disturbances which the humanoid shall encounter while walking on even or uneven terrains. Poor attitude estimates would risk false fall detections. An effective trunk attitude control capability requires good attitude measures. This can be provided by the joint encoders in the legs which are readily available on a humanoid platform. By fusing the noisy IMU data with joint encoders' data from the legs, an improved trunk attitude estimation can be achieved.

References

1. O. J. Woodman, *An Introduction to Inertial Navigation*, Tech. Rep. UCAM-CL-TR-696, Comp. Lab., Univ. of Cambridge, (2007).
2. J. R. Solar, J. Moya and I. P. Tsunekawa, Fall detection and management in biped humanoid robots, in *ICRA*, 2010.
3. L. Lou, X. Xu, J. Cao, Z. Chen and Y. Xu, Sensor fusion-based attitude estimation using low-cost MEMS-IMU for robot navigation, in *ITAIC*, 2011.
4. C. W. Kang and C. G. Park, Attitude estimation with accelerometers and gyros using fuzzy tuned Kalman filter, in *ECC*, 2009.
5. G. Welch and G. Bishop, *An Introduction to the Kalman Filter*, Tech. Rep. TR 95-041, Dept. of Comp. Science, Univ. of N. Carolina (2006).
6. F. Caron, E. Duflos, D. Pomorski and P. Vanheeghe, *Information Fusion* **7**, 221 (2006).
7. J. Crassidis, F. Markley and Y. Cheng, *Journal of Guidance, Control and Dynamics* **30**, 12 (2007).
8. P. S. Maybeck, *Stoc. Models, Estimation and Control* (Academic Press, 1979).
9. W. Phillips and C. Hailey, *Journal Of Aircraft* **38**, 718 (2001).
10. J. Denavit and R. Hartenberg, *Journal of Applied Mechanics* **22**, 215 (1955).
11. W. Khalil and J. Kleinfinger, A new geometric notation for open and closed-loop robots, in *ICRA*, 1986.
12. W. Khalil and E. Dombre, *M., Idn. and Cont. of Robots* (H. Penton, 2002).

SECTION–13

SERVICE ROBOT STANDARDS AND STANDARDIZATION

CHALLENGES OF THE CHANGING ROBOT MARKETS

GURVINDER SINGH VIRK[1]
University of Gävle, Sweden and CLAWAR Association Ltd, UK

CAROL HERMAN
Association for the Advancement of Medical Instrumentation, USA

ROGER BOSTELMAN
National Institute of Standards and Technology, USA

TAMÁS HAIDEGGER
Óbuda University, Hungary

Service robots are becoming an integral part of daily life, entering even the most complex scenarios, yet at a slower pace than previously anticipated. This paper presents an overview of the changing area of robotics and the new challenges being faced. The case is made for all stakeholders to focus on the bottlenecks preventing the realization of the mass markets in robotics expected since the late 1990s. Some of the key issues are discussed and possible strategies analysed to allow the emerging service robotics sector to reach its full potential. The goal is to create a unified vision and good working relationships with the key players throughout the world, active in both robotics R&D and robotics standardization. This paper presents an overview of these activities and provides suggestions for future plans. Along these lines, example cases of industrial, personal service, and medical robots are presented to highlight the desired development directions.

1. Introduction

There have been significant advances in industrial, service, and medical robotic technologies over recent years and many impressive systems have been developed. Examples include[2]: Rethink's Baxter[3] , Honda's Asimo[4], iRobot's

---

[1] Work partially supported by UK DTI and CLAWAR Association Ltd.

[2] Certain trade names and company products are mentioned as illustration examples. In no case does such an identification imply recommendation or endorsement, nor does it imply that the products are necessarily the best available for any purpose.

[3] www.rethinkrobotics.com/index.php/products/baxter/

[4] http://world.honda.com/ASIMO/

Roomba[5], and Intuitive Surgical's da Vinci Surgical System[6]. The current robot market is estimated at $45B/year made up of $25B for industrial robots and $20B for service robots and expected to increase to $79B by 2017 ($32B and $46B, respectively)[7]. Despite the many successful installations at industrial sites, robotics has not yet matured sufficiently to expand from its traditional roots to meet the global challenges of the rising service domain. Reasons include:

- the lack of ISO safety requirements for close robot–human interaction;
- the fact that although ISO 13482 [1] is the first ISO robot safety standard allowing robot-human contact, it will not be widely clear to manufacturers how the safety requirements should be implemented in practise;
- the difficulty of integration into existing systems for new technologies leading to wasted efforts due to reinventing-the-wheel scenarios;
- the lack of effective coordination of results that leads to many labs doing the same work because *normative knowledge* for robotics is missing.

This lack of knowledge and experience must be addressed collectively by stakeholders worldwide and steps to encourage open collaboration between robot research and robot standardization communities made so that globally acceptable guidelines for robot safety, benchmarking, and performance testing are reached to spur rapid technical and commercialisation developments. This can only be achieved through harmonisation between researchers, and standardisation and regulatory bodies as discussed in this paper.

2. Robot sectors

Although R&D covers a wide area of robot applications, the traditional robotics market has been manufacturing applications in industrial environments. An industrial robot is defined as an "*automatically controlled, reprogrammable multipurpose manipulator, programmable in three or more axes, which can be either fixed in place or mobile for use in industrial automation applications*" (see clause 3.10, ISO 10218-1 [2]). Safety has driven manufacturers to keep robots and humans apart by virtual or real cages for many years but this limits human/robot collaboration. However, moving industrial robotics sector from this central ethos appears to be technically challenging. Appropriate safety rules for designing, operating, and regulating the new service robots to allow close robot-human interaction are however essential if the new sector is to grow successfully. Examples are emerging, such as the recent legalisation of driverless cars in the USA (e.g., Nevada, California and Florida), the designation of cities in Japan as special zones for robot R&D (e.g., Fukuoka), and the use of the Dustcart robot for public garbage collection in Peccioli, Italy. Such adoption

---

[5] www.irobot.com/us/learn/home/roomba.aspx
[6] www.intuitivesurgical.com/
[7] www.marketandmarkets.com

of service robots in public areas is also increasing the likelihood of accidents, potential injuries and damage. As a consequence, litigation fears are escalating for companies developing new types of robots and there is growing urgency to publish international safety regulations to allow service robots to operate in complex real-world, human-occupied scenarios. Naturally, the specific requirements differ in the various domains, and key issues are cited here for the industrial, personal care, and medical robot sectors.

3. Robot standards development

Standards provide crucial communication, alignment, and compatibility at an international, national, industry, and individual organization level. Standards help make technologies accessible to all by harmonizing language, state-of-the-art knowledge, management, and industry best practices, which are all requirements in a highly regulated environment.

Although there are many organisations involved in international standardization activities (DICOM,[8] IEEE SA,[9] OMG,[10] ACM[11]) as well as national/regional efforts (ASME,[12] NIST,[13] and CEN[14]), the two main bodies are the International Organization for Standardization (ISO, www.iso.org), and International Electrotechnical Commission (IEC, www.iec.ch) which have been responsible for the majority of the international robot standards that are in current use.

3.1. *Industrial robots*

The main international standardization efforts for industrial robots are in ISO TC184/SC2/WG3 Industrial robot safety. WG3 has recently updated the safety requirements for industrial robots [2] by allowing limited collaborative modes within ISO 10218-1, -2. The intention is to develop a Technical Specification to enhance the collaborative operation (ISO/PDTS 15066) by allowing closer human–robot interaction [3], [4].

Other planned activities include addressing industrial robot and automated guided vehicle (AGV) standards. The National Institute of Standards and Technology (NIST)'s Next Generation Robotics and Automation Program (NGRA)[15] includes several projects to support these efforts with the objective

---

[8] DICOM: Digital Imaging and Communications in Medicine, www.dicom.com
[9] IEEE Standards Association,standards.ieee.org
[10] OMG: Object Management group, www.omg.org
[11] ACM. Advancing Computing as a Science & Profession, www.acm.org
[12] ASME: American Society of Mechanical Engineers Standards, www.asme.org/kb/standards
[13] NIST: NIST standards, www.nist.gov/srm
[14] CEN: European standards, www.cen.eu
[15] http://www.nist.gov/el/isd/ps/nextgenrobauto.cfm

"to develop and deploy advances in measurement science to safely increase the versatility, autonomy and rapid re-tasking of intelligent robots and automation technologies for smart manufacturing and cyber–physical systems applications."

Industry is interested in leveraging the dexterity and versatility of people and the precision and repeatability of robots by enabling collaboration in dynamic and reconfigurable manufacturing environments. Such collaborations, however, are not possible today. According to current standards, industrial robots are still not capable of safely interacting with their human co-workers in highly variable task scenarios. Today, low power, low capacity robots are being developed for human–robot collaborative activities in manufacturing. Generic example activities might include tasks that are repetitive, such as assembly line support; parts pick-and-placement; insertion of small, lightweight items in boxes for shipment and assembly of simple, easy to grasp parts without force-controlled connections. These robots may not currently have the same accuracy and repeatability as traditional robots used for automobile and aircraft manufacturing. Current ISO standards allow some human–robot collaboration (e.g., speed and separation monitoring) for industrial robots [5].

Mobile equipment is extensively used in manufacturing and there is a growing acceptance of either partially or fully autonomous equipment in the field. A major problem, however, is that (especially small) manufacturing facilities frequently operate with people and mobile equipment moving through the same cluttered, constantly-changing environment. Safety is of paramount concern, and standards are essential to reduce the risk of potential injury. The NIST Mobile Autonomous Vehicle Obstacle Detection/Avoidance Project develops standard test methods and performance measures for semi-autonomous and autonomous industrial vehicles that use advanced sensor and control systems and operator alerts to help improve standards. For example, this project performs measurements using AGV safety sensors and advanced three dimensional, non-contact sensors on standard-sized test pieces—similar to human leg and body profiles—for the ANSI/ITSDF[16] B56.5 AGV safety standard [6].

3.2. *Personal care robots*

ISO TC184/SC2/WG7 Personal care robot safety has been developing the ISO 13482 safety standard for personal care robots comprising mobile servant, physical assistant, and person carrier robots. The standard, to be published in August 2013, defines the safety requirements for close human–robot interaction including contact when the robot is operational. WG7 is intending to next explore the development of the following standards:

---

[16] Industrial Truck Standards Development Foundation website, www.itsdf.org.

- an informative guidance document on the usage of ISO 13482;
- a validation and verification document for ISO 13482;
- normative human injury classification for robots for different types of users (adults, children, elderly persons, pregnant women, etc).

3.3. *Medical robots*

Emerging technologies present challenges to the regulatory processes for all areas but medical robotics is especially difficult. The benefits and risks must be managed properly and leveraging of medical standards is essential. For example, the US Food and Drug Administration (FDA) struggles to keep its regulatory processes and procedures aligned with the rapid pace of new medical technologies. There was a time when simpler individual devices were developed and submitted for regulatory clearance. The devices still provided necessary tools for medical procedures that sustain, repair, or improve the patient's condition. As technology has advanced, so has the complexity of the review process as it has to be much more robust to meet the growing risks and challenges of maintaining patient safety.

Risks are part of doing business, and while there is no *acceptable level of risk* particularly in healthcare, managing risk is a daunting responsibility. Risk management is the key and utilizing medical device risk management standards throughout the process can make the difference. What makes a medical robot a robot is the level of autonomy which takes risk up to an even higher level than with other medical devices that only rely on human operation.

There is no harmonised definition of a medical robot as such, but experts working on this topic in IEC TC62/SC62A & ISO TC184/SC2 JWG9 *Medical electrical equipment and systems using robotic technology (Medical robots)*, have come up with the following possible suggestion: ROBOT or ROBOTIC DEVICE intended to be used as MEDICAL ELECTRICAL EQUIPMENT (MEE) or as MEDICAL ELECTRICAL SYSTEM (MES)[17].

The industrial robot definition includes "automatically controlled," although this robot type does not perform surgical procedures or autonomous functions on behalf of the human which can have life or death implications. The key issue is autonomy – the ability to perform intended tasks based on current state and sensing without human intervention.

Fortunately, the different standards groups are learning from each other. In July 2012, the Association for the Advancement of Medical Instrumentation (AAMI)[18], a US developer of medical devices standards, and the nuclear power

---

[17] The words in small capitals are formally defined IEC terms.

[18] AAMI: www.aami.org

industry held a workshop to exchange experience on managing risks in two of the most highly regulated industries in the world and determine the best practices applicable for each domain. In both industries, the risks are high, but the benefits are perceived to outweigh them. Also, the risk to operators/users and the public/patients must be considered and standards are instrumental in assisting manufacturers, operators and regulators alike. International standards not only drive efficiencies, they also drive safety and efficacy and are great leveraging tools for the medical device industry to ensure that better and safer devices are on the market.

3.4. *Boundary issues*

As new robots are developed, it is important to ensure boundaries between the various robot classes are clear. For example, the boundary between medical and non-medical robots is important because different regulatory frameworks are enforced. It must be noted that personal care robots may have both potential medical and non-medical applications. For example, assistive exoskeleton robots can be used for rehabilitation of injured people as a medical application, as well as physically helping, a healthy person to carry heavy loads in a non-medical application. Other applications are not so clear, namely an assistive exoskeleton for improving the degraded mobility of healthy elderly people. For this, detailed work on the *EC Machinery Directive* and the EC *Medical Device Directive* is needed to identify key issues and provide appropriate guidelines.[19]

4. Moral and social projections

4.1. *Performance metrics*

In industrial robotics, the usability of a robot system is linked to accuracy, repeatability, quality of service, and further well-defined, quantifiable metrics. In the case of service robotics, the overall system design might be application-oriented, which may mean lower precision and reliability. Service robots, especially personal service robots, must be more intuitive and require minimal maintenance and engineering skills to operate. The assessment of the core values of service robot systems will be via user acceptance, but this is hard to bind to any absolute scale. Nevertheless, performance metrics are getting defined for specific application domains, such as robot vacuum cleaners. Even in the case of the precision-oriented field of medical robots, accuracy is typically mis-measured and wrong/confusing numbers are reported [7].

---

[19] IEC has initiated recently setting up a new working group to identify these boundaries.

4.2. *Financial interests and applicability of safety standards*

Service robots are strongly application-oriented, and so their entire architecture may be defined by the target domain. The emergence of companies focusing on the full spectrum of design, development, manufacturing, and sales has created a practice of profit-oriented design, which raises ethical questions when assessment is undertaken, e.g., a surgical system applied as a life saving device.

It is commonly quoted that a medical product needs 10–15 years to grow from concept to full product. This long time to market must to be managed carefully to ensure adequate resources. Intuitive Surgical Inc. leads the market with their surgical system having been commercialised using the "razor and razor blade" model; they profit from robot sales, service contracts and also from selling many laparoscopic tools, since those are sterilisable only 8-10 times. This means that hospitals buying the daVinci system need to perform more surgeries to pay for it, while it generates further purchases of its supporting tools [8]. This has also induced the morally questionable phenomenon that buying a da Vinci robot increases the number of prostatectomy procedures performed locally [9].

Safety is a key issue in all human–robot interactions, yet it is most critical in surgery where many kinds of errors may lead to critical conditions in the operating room. According to Satava [10], errors in interventional medicine can be categorised as: 1) commission: doing the wrong thing, 2) omission: not doing the right thing, and 3) execution: doing the right thing incorrectly.

These errors (either systematic or specific) can be traced back to human decision making, therefore—in classical surgery—the error is always the human surgeon's responsibility. Extending this concept to robotic surgery leaves some ambiguity since it is generally not accepted to use statistical calculations to evaluate individual risk.

4.3. *Liability*

The concepts of acceptable risk and risk-benefit analysis (well established in industrial robotics) might be immoral in personal care or medical robot sectors. The concept of "acceptable risk" is also extremely hard to be introduced into an emotional context (e.g., medical applications), where relatives and friends would always assess and deal with hazards fundamentally differently. The relevant new standards must incorporate these human factors to ensure wide acceptance. In the meantime, manufacturers and governments should look into the statistics, and adjust local policies for different service robots in use. This is important, since a large number of deployed robots also means that there is an exponential rise in the number of hazardous incidents.

5. Conclusions

The paper has presented key issues that need to be addressed holistically by the robot stakeholder community now if more advanced service robots and assistive infrastructure are to be developed in the near future. This implies creating *normative data* and methods for designing and testing the new robot systems to ensure acceptable levels of safety and performance. It is needed to work together to realise this, since international collaboration is vital to maximize the likelihood of success of any adopted strategy. With each information-exchange program and publication, and each meeting of standards experts with close links to researchers, progress is being made. Both industry and regulatory bodies need better normative guidelines for risk management in a dynamic, innovative, and global environment. Newly developing service robot standards along with well-grounded quality and risk management standards are essential in this context.

References

[1] ISO/FDIS 13482:2013 Robots and robotic devices – Safety requirements for personal care robots.
[2] ISO 10218-1, -2:2011, Robots and robotic devices – Safety requirements for industrial robots – Part 1: Robots; Part 2: Robot systems and integration.
[3] HSE Research Report RR906, Collision and injury criteria when working with collaborative robots, UK, 2012.
[4] BG/BGIA U 001/1009e Report, Risk assessment according to machinery directive: design of workplaces with collaborative robots, Germany, 2009.
[5] Marvel J, "Performance metrics of speed and separation monitoring in shared workspaces". IEEE Trans. on Automation Science and Engineering, 2013. (10.1109/TASE.2013.2237904).
[6] Bostelman R, "Int. standards efforts towards safe accessibility technology for persons with disabilities: Cross-industry activities". Standards Eng J., pp. 1-20, 2010.
[7] Fitzpatrick, JM, "Fiducial Registration Error and Target Registration Error are Uncorrelated.," SPIE Med Imag, 7261-02, Orlando, FL, pp 1–12, 2009.
[8] Haidegger T, "Surgical Robot Prototyping—System Development, Assessment and Clearance," ch. 10. T. Sobh & X. Xiong, Eds. Prototyping of Robotic Systems: Applications of Design and Implementation, IGI Book, Bridgeport, CT, pp. 288–326, 2012.
[9] Makarov DV, Yu JB, Desai RA, Penson DF, Gross CP, "The association between diffusion of the surgical robot and radical prostatectomy rates". Med Care, 49(4), pp. 333-339, 2011.
[10] Satava RM, The nature of surgical error: a cautionary tale and a call to reason, Surgical Endoscopy, 19(8), pp. 1014–1016, 2005.

OFEX 2.0 : DATABASE FOR PERFORMANCE EVALUATION OF OBJECT FEATURE EXTRACTION ALGORITHMS

KI-YEOP SUNG AND SEUNGBIN MOON

Dept. of Computer Engineering, Sejong University
98 Gunja-dong Gwangjin-gu, Seoul, 143-747, Korea

In this paper, we present the revised database for the performance evaluation of object feature extraction algorithms. The aim is to establish evaluation procedure and performance measures for these algorithms utilizing image database. The image database, OFEX(Obejct Feature EXtraction) 1.0, is updated to OFEX 2.0, adding 3 more objects with various conditions. Experimental results are presented when SIFT (Scale Invariant Feature Transform) is employed.

1. Introduction

Object recognition and localization are important for the navigation of the mobile robots. In image processing, the meaningful information, called features, may be extracted from the raw image. The features indicate transformed data after unnecessary data has been removed. Transforming the image data into the set of features is called feature extraction. SIFT(Scale Invariant Feature Transform)[1] and SURF(Speeded-Up Robust Features)[2] are examples of best known feature extraction algorithms. Vision based object feature extraction algorithms are widely used for object recognition[3, 4] and localization[5, 6, 7, 8] for robots.

The purpose of this paper is to present performance evaluation procedure for the vision based feature extraction algorithms. We introduced OFEX 1.0 database which consists of three object images in 2009[9, 10, 11]. We decided to expand this database to six objects with various conditions, including composite conditions[12]. This work has been adopted as KOROS (Korea Robot Standard) standard.

This paper is structured as follows. In Sec. 2, the database, called OFEX(Object Feature EXtraction) 2.0 database, is described. The proposed evaluation procedure and performance measures are described in Sec. 3. Experimental results, employing SIFT algorithm, are shown in Sec. 4. The conclusion is given in the last section.

2. OFEX(Object Feature EXtraction) 2.0 DB

It is difficult to measure the performance of feature extraction algorithms. Most of the existing object databases[13, 14] contain images that do not reflect the environmental variations such as scale, pose and illumination. However, the robotic application usually has to take these variations into consideration, as the camera attached on the mobile robots experience these variations frequently. The first database, OFEX 1.0, contains only three basic objects with variations in scale, pose, occlusion, and illumination. It has been expanded to OFEX 2.0 with six basic objects with 29 images with various conditions. It also has 50 false positive images for each basic image to test the false positive rate. Table 1 shows three sets of OFEX 2.0, namely, training set, test set, and false positive set. Training set is used to register object images, and test set is used to obtain the recognition rates under various conditions. False positive set is used to test false acceptance rate (FAR) when objects in training and test set are similar but different.

Table 1. Categories of OFEX 2.0 database

Sets	Number of images
Training set	6
Test set	174
False positive set	300

Six basic objects used in OFEX 2.0 are shown in Figure 1. These objects were chosen considering the typical objects the robot will encounter in the real indoor environments.

Figure 1. Six basic objects in OFEX 2.0 database

Table 2 shows image acquisition conditions considering environmental variations including distance, illumination, pose, occlusion, cluttered, and composite condition.

Table 2. Image acquisition conditions

Environmental variations		Conditions	
		Training set	Test set
Distance		1.0m	0.5m, 1.5m
Illumination		200lx	60lx, 100lx, 400lx
Pose	Roll (X-axis)	0°	±20°
	Pitch (Y-axis)	0°	±45°
	Yaw (Z-axis)	0°	±45°
Occlusion		None	1/4, 2/4, 3/4
Cluttered		No background	Cluttered background
Composite		No condition	Composite condition

Three filters are used for occlusion conditions, as shown in Figure 2.

Figure 2. Image filters in occlusion conditions

We have five cluttered images for each basic object and ten images of composite condition where factors of distance, pose, occlusion, illumination, and cluttered can occur simultaneously. Finally, we built false positive sets, which are gathered from the internet web sites. Each object has 50 false positive images. Figure 3 shows the sample images for an object from OFEX 2.0 database.

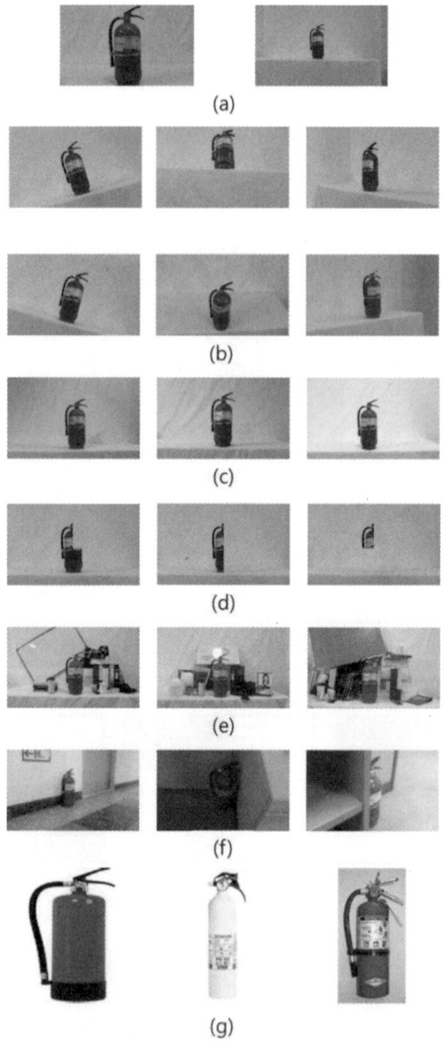

Figure 3. Sample images from OFEX 2.0 database
(a) distnace (b) pose (c) illumination (d) occlusion
(e) cluttered (f) composite condition (g) false positives

3. Performance measures

In order to compare the performance of the object feature extraction algorithms, we need to develop the performance measures. The following four measures are used to evaluate algorithms in this paper.

Recognition time: It is one of the most important performance measures since the real-time performance is needed in many robotic applications. It is evaluated as in Eq. (1).

$$\begin{aligned}&\text{Recognition Time(ms)} =\\&(\text{feature extraction time for test image}\\&+ \text{matching time between test and training images})\end{aligned} \quad (1)$$

As the number of features increases, the recognition time is generally expected to increase.

Feature matching rate: Since the goal of feature extraction algorithm is to find the robust features and match them, feature matching rate is an essential measure for evaluation. It is calculated as in Eq. (2).

$$\text{FMR}(\%) = \frac{\text{number of matched features}}{\text{number of training image features}} \quad (2)$$

The number of matched features is evaluated by matched features between test and training images. This number is used to decide the correct matching if it is larger than the predetermined value, called threshold.

Recognition rate: It is defined as the ratio of the number of correct test images to the number of total test images, as in Eq. (3).

$$\text{RR}(\%) = \frac{\text{number of correct test images}}{\text{number of total test images}} \quad (3)$$

False acceptance rate: False acceptance means that unregistered object is recognized as the training image. False acceptance rate is evaluated as in Eq. (4).

$$\text{FAR}(\%) = \frac{\text{number of false accepted images}}{\text{number of total test images}} \quad (4)$$

If the recognition rate is high while false acceptance rate remains low, then the algorithm may be considered better than others. Recognition rate and false acceptance rate generally shows trade-off relationship. In another words, if we maintain the lower false acceptance rate, recognition rate tends to be lower. The recognition time is also important if your application requires real-time performance.

4. Experimental results

In the experiment, we used SIFT algorithm to show the effectiveness of the performance measures introduced in this paper. SIFT algorithm is known to be one of the widely used feature extraction methods. We set the threshold, which decides the minimum percentage of features to be accepted as matching, as 7%.

We employed OFEX 2.0 database in the experiment. In the first experiment, we evaluated the performance measures for each object image set, as shown in Table 3.

Table 3. Experimental results for individual object sets

Performance Measures	Object image sets (Number of accepted images / Number of test images)		
	Fire extinguisher	Desktop PC	File drawer
Average number of features	338	193	662
Recognition Time(ms)	66.90	66.31	82.79
Feature Matching Rate	27.86	27.06	14.51
Recognition Rate	79.31	89.66	86.21
False Acceptance Rate	10	6	6

Performance Measures	Object image sets (Number of accepted images / Number of test images)		
	chair	wastebasket	electric heater
Average number of features	482	72	1946
Recognition Time(ms)	79.76	52.10	131.14
Feature Matching Rate	21.94	37.04	13.69
Recognition Rate	65.52	41.38	82.76
False Acceptance Rate	42	0	0

The electric heater has most number of features and it shows the slowest recognition time. On the other hand, the wastebasket has least number of features and it shows the fastest recognition time. Thus, one may conclude that the recognition time depends on the number of features. False acceptance rate for the chair is rather high compared to the other objects, as these chairs have many similar designs.

Table 4 shows experimental results for subsets with different conditions using OFEX 2.0 database. As expected, cluttered image and composite condition images have the highest number of features, while pose variation and occlusion images have lower number. Recognition rates are from 86.67% to

100% for all cases, except 38.33% for composite condition case. We may conclude that this feature extraction algorithm is weak for composite conditions.

Table 4. Experimental results for subsets with different conditions

Condition	Recognition time(ms)	Feature matching rate(%)	Recognition rate(%)
Distance	71.17	22.68	91.67
Illumination	60.67	24.02	100
Pose	58.78	59.90	91.66
Occlusion	50.00	43.53	100
Cluttered	107.37	16.06	86.67
Composite condition	95.13	6.07	38.33

Table 5 shows the performance of the algorithm when all test images in OFEX 2.0 database are employed. We found that false acceptance rate of 10.66% might be considered rather high. If we increase the threshold value, FAR could decrease, but the recognition rate is also expected to be decreased.

Table 5. Experimental results for overall object sets

Performance measure	values
Recognition Time	79.83ms
Feature Matching Rate	23.68%
Recognition Rate	74.14%
False Acceptance Rate	10.66%

5. Conclusion

We introduced the performance evaluation procedure for vision based object feature extraction algorithms employing OFEX 2.0 database. It is a useful means to compare any feature extraction algorithms. The experimental results showed that it could be used to analyze the performance of the given algorithm.

In the future, we are planning to expand OFEX 2.0 database to include more than 30 basic objects, so that the possibility of being optimized could be lowered.

Acknowledgments

This work was supported by National Strategic R&D Program for Industrial Technology from Ministry of Knowledge Economy, Korea.

References

1. D. Lowe, Distinctive Image Features from Scale-Invariant Keypoints, International Journal of Computer Vision, 60, 2, 91-110 (2004).
2. H. Bay, A. Ess, T. Tuytelaars and L. Van Gool, Surf: Speeded up robust features, In Proceedings of Computer Vision and Image Understanding, vol. 110 no. 3, 346-359 (2008).
3. D. Lowe, Object recognition from local scale-invariant features, In Proceedings of International Conference on Computer Vision, vol. 2, 1150 (1999).
4. H. Bay, B. Fase and L. Van Gool, Interactive museum guide: Fast and robust recognition of museum objects, In Proceedings of First international workshop on mobile vision (2006).
5. H. Tamimi, H. Andreasson, A. Treptow, T. Duckett and A. Zell, Localization of mobile robots with omnidirectional vision using particle filter and iterative SIFT, In Proceedings of 2nd European Conf. on Mobile Robots – ECMR'05, 758-765 (2005).
6. A.C. Murillo, J.J. Guerrero and C. Sagüés, SURF Features for Efficient Robot Localization with Omnidirectional Images, In Proceedings of 2007 IEEE Int'l Conf. Robotics and Automation, 3901–3907.7 (2007).
7. S. Se, D. Lowe and J. Little, Vision-based mobile robot localization and mapping using scale-invariant features. In Proceedings of the International Conference on Robotics & Automation, 2051-2058.9 (2001).
8. M. Cummins and P. Newman, FAB-MAP: Probabilistic Localization and Mapping in the Space of Appearance, International Journal of Robotics Research, Vol. 27, No. 6, 647-665 (2008).
9. M. Kang, W. Choo, S. Moon, Performance evaluation procedure for vision based object feature extraction algorithms, In Proceedings of Performance Metrics for Intelligent Systems (PerMIS'10) Workshop (2010).
10. M. Kang, S. Moon, S. Back and Y. Ryuh, Performance evaluation of object feature extraction algorithm for localization of robots, In Proceedings of Korea Automatic Control Conference, 832-835 (2009).
11. M. Kang and S. Moon, Database-based performance evaluation of SIFT algorithm, In Proceedings of annual fall Conf. Institute of Electronics Engineers of Korea (2009).
12. K. Sung and S. Moon, Image database for performance evaluation of object recognition algorithm for indoor service robots, In Proceedings of Korea information processing society conference, 662-664 (2012).
13. G. Griffin, A. D. Holub and P. Perona, The Caltech-256 Object Category Dataset. Technical Report. California Institute of Technology (2006).
14. B.C. Russell, A. Torralba, K.P. Murphy and W.T. Freeman, LabelMe: A Database and Web-Based Tool for Image Annotation. International Journal of Computer Vision, 157-173 (2008).

ETHICAL ASSESSMENT OF ROBOTS

M. O. TOKHI
Department of Automatic Control and Systems Engineering
The University of Sheffield, UK

Ethical assessment of a robot comprises a process of ethical analysis and estimation of ethical disturbance that may be caused by the robot. While misuse of a robot forms an element of ethical consideration and this may be addressed by the robot developer emphasising its proper use, it relates to the behaviour and responsibility of the user, whose actions at that level are regulated by laws and regulations. There are far more fundamental issues of ethical nature that need to be considered at the design and development stage of a robot, and for that a proper ethical assessment of the robot needs to be carried out. This will require ethical assessment of the robot from perspectives of society, application, economy/finance and environment. This paper discusses such issues and proposes a methodology for ethical assessment of robotics technology at the design stage in light of these issues.

1. Introduction

The development of a technology, including robotics, takes account of a number of factors. These include social, economical, application and environmental implications and associated issues. In this context the design of the technology is constrained and thus dictated by associated national and international standards and regulations. From the standards perspective the developer of the technology is required to carry out proper risk assessment so that the technology complies with safety requirements as emphasised in the standard. In this sense limits to, e.g. forces, torques, etc., are imposed through design and development of the technology so that safety limits are not violated.

Robot ethics, and in general technology ethics, is a further dimension of consideration for development of a technology. Ethical assessment of a robot comprises a process of analysis and estimation of ethical disturbance that may be caused by the robot. Such an assessment will need to be carried out as standard procedure at the design and development stage of the robot. Concerns have been raised in the literature on social and psychological implications of robots (Allenby, 2012; Sharkey and Sharkey, 2011; Stephen et al., 2012), and a proper ethical assessment of robots will help resolve such concerns.

Misuse of a robot is an element of the technology ethics for consideration by the developer at the design stage and further by emphasising proper use of the robot. However, the responsibility in respect of use of a robot after deployment is upon moral obligation of the user, who can be taken accountable for any misuse through laws and regulations of the land. From a moral agency perspective, considering robots as moral agents depending on their level of intelligence and autonomy that specifies their position on a moral agency scale, it has been highlighted in the literature that their accountability can accordingly be varied for which ethical norms, laws and regulations may be imposed (Arkin et al., 2012; Coeckelbergh, 2010).

This paper discusses fundamental issues of ethical nature for consideration at the design and development stage of a robot through a systematic process of ethical assessment. It is argued that such a process is essential from viewpoint of development of robotic standards, and thus a synergistic approach, incorporating ethical assessment into the standardisation work, is essential.

2. Ethical issues

The UK Robot Ethics (UKRE) group (www.robotethics.org.uk) formed as part of the BSI AMT/00-/02 (Robots and Robotic Devices) committee has identified various ethical issues of robotics, and has classified these into four main categories, namely societal, application, commercial/financial and environment. These are briefly described in this section.

2.1 Societal issues

Some of the fundamental societal robot ethics issues include the following

2.1.1 Privacy and confidentiality

It is expected that robots will be taking and storing personal information. This may be beneficial in the context of connectivity and linkages, e.g. health services, social care, family members, and robot manufacturers. However, the concern in this respect is about the collection and use of information concerning robot users, control of data storage and use by third party. Existing privacy laws on data collection, e.g. those applicable to CCTV, may be applied. However, on information storage issues that need resolving include the type of information the robot is allowed to record, who to have access to the information, who will be using the information, and how long to keep the data.

2.1.2 Respect for human dignity and human rights

Possible violation of human dignity and human rights through the design, manufacture and operation of robots will be a matter of concern, while the intention of the designer will be for the robots to promote human dignity. It has been highlighted that there are regulations and laws relevant to human dignity, and these may be traced in various national and international laws and human right conventions, such as international humanitarian law, law of armed conflict, etc. that may be helpful to develop suitable approaches for robot designs so that human dignity is not violated.

2.1.3 Respect for cultural diversity and pluralism

People are living in diverse cultural environments. Accordingly, appropriate account of different cultural norms, including respect of language, religion, gender, and formal interaction will be an important issue. Determining such norms will require public engagement via questionnaires, web resources, and meetings.

2.1.4 Dehumanisation of humans in the relationship with robot

Inappropriate control over human choice of a robot, for example, tedious repetitive task on an assembly line, has been highlighted as a potential ethical disturbance issue. Such a situation may be avoided by resting the ultimate authority with the human.

2.1.5 Responsibility and legal issues

It is envisaged that legal responsibility will rest with the designer, builder, operator, maintainer, and owner of the robot. In this sense the roles, responsibilities and legal liabilities of each need to be clearly identified for all stages of the robot's life-cycle. Moreover, it is required to define levels of responsibilities with individual user, organisation deploying the robot and organisation manufacturing the robot. Furthermore, it should always be possible to discover the person/organisation that is legally responsible for the robot.

2.1.6 Benefit and risk balance

Judgement of acceptable residual (economical, social, legal, psychological, physical, etc) ethical risks to help the design and deployment of robots is a challenging issue, as these will be varied across various sectors. To avoid potential conflicts it is envisaged in certain applications such as elder care, for the robot not to fully substitute a human carer.

2.1.7 Informed consent

It is standard practice in healthcare to inform the patient of risks associated with a surgical operation and/or with taking a drug. The extent of consent obtained depends on the extent of the risk involved. Such practice can be adopted for a robot with autonomy used in e.g. medical and domestic applications, to inform the user or person/guardian of the risks, benefits, constraints/limitations of use of the robot, and informed consent obtained.

2.1.8 *Anthropomorphisation of the robots*

It is necessary to exercise appropriate degree of humanisation in developing robots, particularly with children and other vulnerable people. Concerns have been raised over ethical issues concerning use of robots, for example, in interviewing children (Kyriakidou *et al.*, 2013). Moreover, inappropriate anthropomorphisation of a robot can lead to false expectation of users of capabilities of the robot.

2.2 *Application issues*

2.2.1 *Design principles and requirements*

In light of ethical considerations, design principles and requirements will largely be qualitative and hence not easy to quantify. In this respect a scale can be developed for a given application scenario against which to compare robot design, or generic ethical benchmarks that can be used for assessing the robot design.

Similarly, requirements for certain performance levels may be created, and large majority of these may not be quantified; define a level of autonomy and associated ethical requirement, i.e. levels of autonomy which are acceptable, for example an exoskeleton may want to take the person across the road and the person may not want so.

2.2.2 *Rehabilitation and medical applications*

There are a range of rehabilitation and medical applications of robots that require ethical consideration. Examples include physiotherapy augmentation of human, where control by human must be a consideration, replacement of a medical professional for treatment/diagnosis (although certain treatments may be done better with robot), which may lead to loss of certain skills with introduction of robots.

2.2.3 *Care applications*

There is a growing trend in development and deployment of child-minding and elder-care robots. Research has shown close binding and attachment to child-minding robots by children (Sharkey, 2008). Although exposure in the short-term may be enjoyable and entertaining, the lack of humanly care and attention that is needed and not provided by the robot will have detrimental impact on character development of the child in the long term. Similarly, although elder-care robots can help the elderly to maintain independence in their own home, their presence and over-reliance on them could lead to the risk of leaving the elderly in the exclusive care of the robot, and in social isolation.

2.2.4 Military applications

The development of robots for military applications, especially for combat is a matter of serious discussion and debate. Concerns are shown over the use of robots in combat scenarios, and these include how combatants and innocents may be discriminated from one another in close-contact encounter. Although, many of the ethical questions in this respect have been contained into the military command and control framework (by keeping human in the loop), i.e. where commanders are responsible for issuing orders and the soldiers are for carrying out those orders, the system becomes very complex as the number of robots deployed in combat scenarios increases.

2.3 Commercial/financial issues

Product manufacturing from an industrial perspective is largely based on cost-benefit models, and this respect there may be an oversight of ethical considerations. Thus, engagement with the industry sector is needed to ensure that only ethically accepted products are manufactured. This will further allow developing suitable ethical business models that are needed. At the standardisation level, on the other hand, embodying the standards development process with ethical consideration will ensure ethical issues to be taken into account in developing the standards.

2.4 Environmental issues

Previous studies have considered sustainable development in technology design, but quantification has been a problem. The use of scarce materials in development of robots, or development of bio-degradable robots (graceful and harmless degradation) could be helpful for the environment. Furthermore, a sense of responsibility towards biosphere will require life-cycle considerations of the robot.

3. Public engagement – questionnaire

A questionnaire survey was conducted among a group of researchers and academics from different institutions within the UK. In this section results of the survey on perception and acceptance of robots are presented and discussed. A total of 21 responses were received, and Table 1 shows results of the responses to the set of questions on perception and acceptance on percentage count. It is noted that robots were largely perceived by the group as machines that can provide help in the home environment and in that sense the group was un-decided whether the robots are friendly or not. On acceptance the range of questions were designed so as to determine the level of trust placed on the use of robots in performing certain tasks. It is noted that the group was largely agreed on the use of robots for tasks around the house. However, on questions related to intervention with human body, where the risk of injury was perceived to be relatively high, such as styling the hair, placing drops in the eye, giving injection, or filling/removing teeth the group was cautious and could not trust the robot as much. However, where the risk if injury was perceived low or none, such as taking blood pressure or pulse there was high trust placed on the use of robot. It is also interesting to note that the group could largely trust a robot for surgical operations in the hospital.

Table 1: Responses to questionnaire from a group of 21 researchers and academics

Question	Strongly disagree	Partially disagree	Do not know	Partially agree	Strongly agree
Perception:					
Do you consider robots are friendly		10%	67%	19%	5%
Do you agree a robot could help you at home				19%	81%
Acceptance:					
Would you be prepared to have a ROBOT visit you each day to confirm your well-being	10%	24%	14%	24%	29%
Would you trust a ROBOT to:					
▪ cut the grass	5%			19%	76%
▪ clean the windows or vacuum clean the house	5%			38%	57%
▪ exercise your pet	24%	14%	14%	29%	19%
▪ prepare you a meal	19%	5%	14%	33%	29%
▪ help you put your shoes and clothes on	14%	10%	10%	48%	19%
▪ clean your dishes	5%	10%		24%	62%
▪ cut and style your hair	29%	14%	19%	24%	14%
▪ measure your blood pressure and take your pulse	5%	10%	5%	29%	52%
▪ put drops in your eyes	14%	24%	10%	33%	19%
▪ give you an injection	14%	43%	10%	19%	14%
▪ cut your toe and finger nails	38%	14%		33%	14%
▪ clean your car		5%		19%	76%
▪ fill or remove your teeth	38%	14%	5%	33%	10%
If you were going into hospital for an operation would you be happy to learn that part or all of the operation was to be performed by a ROBOT	5%		19%	52%	24%

The questionnaire also contained the question: "If you would obtain a robot to perform just one task for you what would it be?" 67% indicated tasks of house-work nature, e.g. ironing, house cleaning, and general housework. Other tasks included help with exercise, transportation, shopping, and physical assistance, each scoring 7%. It appears from the above results that robots are perceived as technological beings like utility machines helping with tasks around the house that do not involve human body intervention. This implies that technological advances are still needed for robots to acquire certain levels of autonomy and sophistication to be trusted in terms of safety and to gain public ethical approval.

4. Ethical assessment – A formal approach

It is clear from the discussion and findings in the preceding sections that a standardised approach to ethical assessment of robots is needed. Essential components of the assessment will include

- *Ethical issues*: Engagement with industry, designers, developers and the public through questionnaires, web resources, meetings, seminars and presentations will help identifying ethical issues related to a planned robotic development. This is to cover a wide range of sectors including societal impact, application issues, financial implications and the environment. The issues related to these sectors will inevitably vary across communities, and prioritisation of the identified issues may need to be made in light of cost-benefit and the user's acceptance and perception.
- *Ethical norms*: Metrics and norms will allow containing the impact of identified ethical issues of a robot to acceptable levels. Ethical norms and metrics are largely expected to be qualitative. Moreover, varied requirements leading to such metrics can be expected across different communities, based on their educational/professional background, social background, and living environment. Accordingly, development of ethical norms to cover the full spectrum of stakeholders will be a challenge.
- *Ethical disturbance estimation and analysis*: This will comprise assessment of the identified ethical issues against associated set of ethical metrics and norms. Such an assessment will allow classifying identified issues from severe to low risk ethical disturbances, and hence help the design and development process of the robot.
- *Ethically compliant design, development, and deployment of robot*: The outcome of the ethical disturbance estimation and analysis process can feed into the design, development and deployment process of the robot, to help in development of ethically compliant robots for the specific community/group of people in their specific environment.

5. Conclusion

A process of ethical assessment for robot design, development and deployment has been formulated. The proposed process closely resembles the risk assessment approach adopted in the development of standards. Thus, the approach can be easily incorporated into standards development process in a smooth and natural way. In this manner, standards can be developed to help the industry to develop ethically compliant robots.

References

Allenby, B. (2012). Robot Ethics: The ethical and social implications of robotics, *Nature*, 481, (7379), pp. 26 – 27.

Arkin, R. C., Ulam, P. and Wagner, A. R. (2012). Moral decision making in autonomous systems: Enforcement, moral emotions, dignity, trust, and deception, *Proceedings of the IEEE*, **100**, (3), pp. 571–589.

Coeckelbergh, M. (2010). Robot rights? Towards a social-relational justification of moral consideration, *Ethics and Information Technology*, **12**, (3), pp. 209 – 221.

Kyriakidou, M. Sharkey, A. and Blades, M. (2013). Ethical dilemmas for children's robot crime interviewers. In Tokhi, M. O. (editor), *Proceedings of UKRE Workshop on Robot Ethics*, Sheffield, UK, 25 March 2013, pp. 14 – 19.

Sharkey, N. (2008). The ethical frontiers of robotics, *Science*, **322**, pp. 1800 – 1801.

Sharkey, A. and Sharkey, N. (2011). Children, the elderly, and interactive robots, *IEEE Robotics & Automation Magazine*, **18**, (1), pp. 32 – 38.

Stephan, K. D., Michael, K., Michael, M. G., Jacob, L. and Anesta, E. P. (2012). Social implications of technology: The past, the present, and the future, *Proceedings of the IEEE*, **100**, pp. 1752 – 1781.

AUTHOR INDEX

Ady, R. 68
Agouri, S. A. 179, 189
Agrawal, P. 171
Ahmad, S. 141
Akmeliawati, R. 141
Alamdari, A. 585, 593
Al-Homsy, A. 215
Almeshal, A. M. 179, 189
Altalmas, T. M. 141
Al-Yahmadi, A. 133
Amano, H. 51, 359
Araki, S. 359
Arevalo, J. C. 59, 223, 451
Aula, A. 141
Annunziata, S. 249, 257
Awad, M. 43
Ayaz, Y. 643

Bachmann, R. J. 199
Baptista, R. D. S. 153
Barasuol, V. 443, 511
Barbosa, R. S. 709
Bartsch, S. 495
Baudoin, Y. 3
Bastos, T. 85
Bhattacharya, B. 171
Bedkowski, J. 797
Berns, K. 153, 309
Bidaud, P. 519
Bidaud, Ph. 68, 101
Billingsley, J. 459
Bó, A. P. L. 153
Bohn, S. 781
Bolotnik, N. N. 19
Boonyaprapasorn, A. 578
Bonaccorso, F. 701
Bostelman, R. 833
Bouyarmane, K. 740
Buchli, J. 443, 555

Caldwell, D. 511
Caldwell, D. G. 443, 487
Cao, Q. 317
Carrillo, X. 59
Castillo-Castaneda, E. 399
Castro, T. S. 709
Cestari, M. 59, 223, 451
Chen, I.-M. 807
Chen, Y.-C. 593
Chernyak, V. 351
Chevallereau, C. 822
Chew, C.-M. 383, 761
Chou, Y.-C. 479
Chu, J. 729
Chugo, D. 109
Clady, X. 101
Claretti, E. 351
Contreras-Vidal, J. L. 291
Curaj, A. 651

Daffon, K. 547
Dehghani, A. 43
Dettmann, A. 495
Díaz-Hernández, C. A. 291
Dillmann, R. 563, 781
Dobashi, H. 392
Doroftei, I. 609
Drumwright, E. 555
Duhaut, D. 815
D'Urso, S. 701

Ermolov, I. L. 19, 27

Fang, B. 317
Fernandes, V. H. 85
Fernandes, V. H. B. 125
Filho, A. B. 85, 125
Focchi, M. 443
Frizera, A. 85
Fujimoto, H. 617
Fujiwara, K. 617

Garcia, E. 59, 85, 125, 223, 451
Garcia-Egea, T. 291
Gracia-Murillo, M. A. 399
Ghafoor, A. 643
Goher, K. M. 133, 179, 189
Goldman, D. I. 547
Golubev, Y. F. 427
Goto, Y. 241
Gradetsky, V. G. 19, 27
Granata, C. 68
Grigore, S. 651
Grinke, E. 419
Guan, Y. 685

Haidegger, T. 833
Haji, T. 51, 359
Han, C. S. 35
Han, J. S. 35
Hartmann, J. 215
Hashimoto, H. 109
Hashimoto, K. 487
Hassan, M. 375
Hassan, M. A. H. 659
Havoutis, I. 443
Hayashi, R. 359
Havoutis, I. 555
Hengst, B. 411
Heppner, G. 563, 781
Hérin, R. 585
Hérin, R. S. 593
Herman, C. 833
Hesse, F. 419
Hiep, D. X. 163
Hirai, K. 343
Hirata, Y. 435
Hirayama, T. 51
Huang, K.-J. 479
Huang, Z. 317

Ibanez, A. 519
Iinuma, N. 343
Ikeuchi, M. 301
Indrawibawa, I. N. 117
Ion, I. 609, 651
Iribe, M. 359
Ishigure, Y. 343
Ito, S. 529
Ito, T. 359
Iulia, D. 651

Jacobs, D. A. 463
Jatsun, S. F. 677
Jung, M. 309

Kaczmarek, P. 797
Kagawa, T. 11
Karpenko, A. 163
Kato, T. 343
Kawasaki, H. 343
Kazerooni, H. 8
Kern, N. I. 199
Kesper, P. 419
Khan, H. 511
Khandelwal, S. 822
Khurshid, A. 643
Kikuchi, K. 617
Kim, W. S. 35
Kinugasa, T. 51, 359
Kirchner, F. 495
Kishi, T. 301
Knyazkov, M. M. 19, 27
Koçer, B. 693
Koeda, M. 571

Korianov, V. V. 427
Kormushev, P. 487
Krovi, V. N. 585, 593
Kryczka, P. 487
Kryukova, A. A. 27
Kuniyoshi, K. 273

Laksanacharoen, P. 625
Langosz, M. 495
Laquini Jr, S. 125
Lee, H. D. 35
Lemasson, G. 815
Leng, C. 317
Leon, H. 77
Leon-Rodriguez, H. 367
Lim, D. H. 35
Lim, H. O. 487
Lin, P.-C. 479
Liu, D. 375
Longo, D. 701
Lopez-Coronado, J. 291
Lucidarme, P. 815
Lynch, K. M. 633

Maehle, E. 215
Maia, A. H. 85, 125
Majek, K. 797
Makino, R. 617
Malik, M. A. 643
Maneewarn, T. 578
Manoonpong, P. 281, 419, 625
Mao, Y. 729
Masłowski, A. 797
Matsuhira, K. 503
Matsui, Y. 571
Michols, R. J. 199
Mimura, N. 599
Mizoguchi, M. 471
Mizota, Y. 241
Moon, S. 841
Morimoto, J. 740
Morishita, Y. 265
Morisita, Y. 207
Moser, D. 43
Muscato, G. 701

Nagase, J.-Y. 327, 392
Nagata, K. 599
Nakamura, T. 207, 241, 265, 301, 435
Nakamura, Y. 6
Nakazawa, A. 529
Nasir, A. N. K. 179, 667
Naumov, G. S. 677
Nestinger, S. S. 351
Nishigami, N. 617
Nishikawa, S. 273
Nishimura, K. 471
Nishio, S. 529
Nozaki, H. 392

Oberlaender, J. 781
Oetomo, D. 749
Okada, T. 599
Onozawa, T. 529
Osawa, T. 265
Osumi, H. 435
Öztürk, M. 693

Padois, V. 519
Panchenko, A. V. 427
Pandy, M. 7
Park, L. J. 463
Pavlovsky, V. E. 427
Paskarbeit, J. 249, 257
Perez-Gutierrez, B. 77
Pfotzer, L. 563, 781
Poo, A. N. 383, 761

Qian, F. 547
Quack, L. 495
Quinn, R. D. 199

Radkhah, K. 537, 719

Roennau, A. 563, 781
Rosa Jr, N. 633

Saga, N. 327, 392
Sakio, S. 471
Salini, J. 68
Sandoval-Castro, X. Y. 399
Sannohe, D. 207, 265
Sanz-Merodio, D. 59, 223, 451
Sasaki, M. 529
Sayapin, S. 163
Schmidt, D. 309
Schneider, A. 249, 257
Semini, C. 443, 511, 555
Semyonov, E. A. 19, 27
Serizawa, T. 571
Shafiq, M. 133
Shibuya, K. 471, 503
Shimizu, T. 599
Shvartsman, R. 749
Silapunt, R. 578
Sidek, S. N. 141
Silva, M. F. 709
Skrzypczynski, P. 335, 789
Sukhanov, A. N. 19, 27
Sung, C. 11
Sung, K.-Y. 841
Suzuki, M. 773
Suzumori, K. 327

Takahashi, K. 241, 773
Takanishi, A. 487
Takesue, N. 617
Tan, B.-H. 383
Tan, Y. 749
Tanaka, K. 273
Tanaka, M. 207
Tanaka, T. 207, 265
Tesen, S. 392
Thung-Od, K. 578
TIRAMISU Consortium 3
Tokhi, M. O. 141, 179, 189, 659, 667, 849
Tokuda, K. 51, 359
Tomori, H. 435
Torres-Jara, E. 351
Triolo, R. J. 199
Tsagarakis, N. G. 487
Tsoi, Y. H. 93
Tsutsumi, K. 471

Uchida, H. 773
Ueki, S. 343
Uno, Y. 11
Uribe-Quevedo, A. 77
Uygun, R. 693

Vaillant, J. 740
Vasile, A. 651
Veneva, I. 232
Virk, G. S. 117, 833
Volkova, L. Y. 677
von Stryk, O. 537, 719

Wada, H. 599
Waldron, K. 375
Waldron, K. J. 463
Ward, P. 375
Wasielica, M. 335
Wasik, M. 335
Wörgötter, F. 281, 419, 625
Wu, N. 383, 761
Wu, W. 685

Xie, S. Q. 93
Xiong, R. 729
Xiong, X. 281

Yamada, T. 109
Yang, Y. 317
Yatsun, A. S. 677
Yokota, S. 109

Yoshida, K. 359
Yu, W.-S. 479
Yuan, Q. 807

Zahedi, S. 43
Zapolsky, S. 555
Zavala-De Paz, J. P. 399
Zhang, H. 685
Zhang, T. 547
Zhao, J. 153
Zhou, X. 685
Zhu, H. 685
Zhu, Q. 729
Zong, C. 101